THE BRAINSTEM

HANDBOOK OF CLINICAL NEUROLOGY

Series Editors

MICHAEL J. AMINOFF, FRANÇOIS BOLLER, AND DICK F. SWAAB

VOLUME 216

ELSEVIER

THE BRAINSTEM

Series Editors

MICHAEL J. AMINOFF, FRANÇOIS BOLLER, AND DICK F. SWAAB

Volume Editors

LEA TENENHOLZ GRINBERG AND AMOS D. KORCZYN

VOLUME 216

3rd Series

ELSEVIER

Elsevier
Radarweg 29, PO Box 211, 1000 AE Amsterdam, Netherlands
125 London Wall, London EC2Y 5AS, United Kingdom
50 Hampshire Street, 5th Floor, Cambridge, MA 02139, United States

ISBN: 978-0-443-15736-3

For Information on all Elsevier publications
visit our website at https://www.elsevier.com/books-and-journals

Publisher: Megan Ball
Acquisitions Editor: Megan McManus
Editorial Project Manager: Himanshi Yadav
Production Project Manager: Punithavathy Govindaradjane
Cover Designer: Gopalakrishnan Venkatraman

Typeset by MPS Limited, Chennai, India

Printed in Great Britain
Last digit is the print number: 10 9 8 7 6 5 4 3 2 1

Handbook of Clinical Neurology

Available titles
Vol. 79, The human hypothalamus: Basic and clinical aspects, Part I, D.F. Swaab, ed. ISBN 9780444513571
Vol. 80, The human hypothalamus: Basic and clinical aspects, Part II, D.F. Swaab, ed. ISBN 9780444514905
Vol. 81, Pain, F. Cervero and T.S. Jensen, eds. ISBN 9780444519016
Vol. 82, Motor neurone disorders and related diseases, A.A. Eisen and P.J. Shaw, eds. ISBN 9780444518941
Vol. 83, Parkinson's disease and related disorders, Part I, W.C. Koller and E. Melamed, eds. ISBN 9780444519009
Vol. 84, Parkinson's disease and related disorders, Part II, W.C. Koller and E. Melamed, eds. ISBN 9780444528933
Vol. 85, HIV/AIDS and the nervous system, P. Portegies and J. Berger, eds. ISBN 9780444520104
Vol. 86, Myopathies, F.L. Mastaglia and D. Hilton Jones, eds. ISBN 9780444518996
Vol. 87, Malformations of the nervous system, H.B. Sarnat and P. Curatolo, eds. ISBN 9780444518965
Vol. 88, Neuropsychology and behavioural neurology, G. Goldenberg and B.C. Miller, eds. ISBN 9780444518972
Vol. 89, Dementias, C. Duyckaerts and I. Litvan, eds. ISBN 9780444518989
Vol. 90, Disorders of consciousness, G.B. Young and E.F.M. Wijdicks, eds. ISBN 9780444518958
Vol. 91, Neuromuscular junction disorders, A.G. Engel, ed. ISBN 9780444520081
Vol. 92, Stroke – Part I: Basic and epidemiological aspects, M. Fisher, ed. ISBN 9780444520036
Vol. 93, Stroke – Part II: Clinical manifestations and pathogenesis, M. Fisher, ed. ISBN 9780444520043
Vol. 94, Stroke – Part III: Investigations and management, M. Fisher, ed. ISBN 9780444520050
Vol. 95, History of neurology, S. Finger, F. Boller and K.L. Tyler, eds. ISBN 9780444520081
Vol. 96, Bacterial infections of the central nervous system, K.L. Roos and A.R. Tunkel, eds. ISBN 9780444520159
Vol. 97, Headache, G. Nappi and M.A. Moskowitz, eds. ISBN 9780444521392
Vol. 98, Sleep disorders Part I, P. Montagna and S. Chokroverty, eds. ISBN 9780444520067
Vol. 99, Sleep disorders Part II, P. Montagna and S. Chokroverty, eds. ISBN 9780444520074
Vol. 100, Hyperkinetic movement disorders, W.J. Weiner and E. Tolosa, eds. ISBN 9780444520142
Vol. 101, Muscular dystrophies, A. Amato and R.C. Griggs, eds. ISBN 9780080450315
Vol. 102, Neuro-ophthalmology, C. Kennard and R.J. Leigh, eds. ISBN 9780444529039
Vol. 103, Ataxic disorders, S.H. Subramony and A. Durr, eds. ISBN 9780444518927
Vol. 104, Neuro-oncology Part I, W. Grisold and R. Sofietti, eds. ISBN 9780444521385
Vol. 105, Neuro-oncology Part II, W. Grisold and R. Sofietti, eds. ISBN 9780444535023
Vol. 106, Neurobiology of psychiatric disorders, T. Schlaepfer and C.B. Nemeroff, eds. ISBN 9780444520029
Vol. 107, Epilepsy Part I, H. Stefan and W.H. Theodore, eds. ISBN 9780444528988
Vol. 108, Epilepsy Part II, H. Stefan and W.H. Theodore, eds. ISBN 9780444528995
Vol. 109, Spinal cord injury, J. Verhaagen and J.W. McDonald III, eds. ISBN 9780444521378
Vol. 110, Neurological rehabilitation, M. Barnes and D.C. Good, eds. ISBN 9780444529015
Vol. 111, Pediatric neurology Part I, O. Dulac, M. Lassonde and H.B. Sarnat, eds. ISBN 9780444528919
Vol. 112, Pediatric neurology Part II, O. Dulac, M. Lassonde and H.B. Sarnat, eds. ISBN 9780444529107
Vol. 113, Pediatric neurology Part III, O. Dulac, M. Lassonde and H.B. Sarnat, eds. ISBN 9780444595652
Vol. 114, Neuroparasitology and tropical neurology, H.H. Garcia, H.B. Tanowitz and O.H. Del Brutto, eds. ISBN 9780444534903
Vol. 115, Peripheral nerve disorders, G. Said and C. Krarup, eds. ISBN 9780444529022
Vol. 116, Brain stimulation, A.M. Lozano and M. Hallett, eds. ISBN 9780444534972
Vol. 117, Autonomic nervous system, R.M. Buijs and D.F. Swaab, eds. ISBN 9780444534910
Vol. 118, Ethical and legal issues in neurology, J.L. Bernat and H.R. Beresford, eds. ISBN 9780444535016
Vol. 119, Neurologic aspects of systemic disease Part I, J. Biller and J.M. Ferro, eds. ISBN 9780702040863
Vol. 120, Neurologic aspects of systemic disease Part II, J. Biller and J.M. Ferro, eds. ISBN 9780702040870
Vol. 121, Neurologic aspects of systemic disease Part III, J. Biller and J.M. Ferro, eds. ISBN 9780702040887
Vol. 122, Multiple sclerosis and related disorders, D.S. Goodin, ed. ISBN 9780444520012
Vol. 123, Neurovirology, A.C. Tselis and J. Booss, eds. ISBN 9780444534880
Vol. 124, Clinical neuroendocrinology, E. Fliers, M. Korbonits and J.A. Romijn, eds. ISBN 9780444596024
Vol. 125, Alcohol and the nervous system, E.V. Sullivan and A. Pfefferbaum, eds. ISBN 9780444626196
Vol. 126, Diabetes and the nervous system, D.W. Zochodne and R.A. Malik, eds. ISBN 9780444534804
Vol. 127, Traumatic brain injury Part I, J.H. Grafman and A.M. Salazar, eds. ISBN 9780444528926
Vol. 128, Traumatic brain injury Part II, J.H. Grafman and A.M. Salazar, eds. ISBN 9780444635211
Vol. 129, The human auditory system: Fundamental organization and clinical disorders, G.G. Celesia and G. Hickok, eds. ISBN 9780444626301
Vol. 130, Neurology of sexual and bladder disorders, D.B. Vodušek and F. Boller, eds. ISBN 9780444632470
Vol. 131, Occupational neurology, M. Lotti and M.L. Bleecker, eds. ISBN 9780444626271

Vol. 132, Neurocutaneous syndromes, M.P. Islam and E.S. Roach, eds. ISBN 9780444627025
Vol. 133, Autoimmune neurology, S.J. Pittock and A. Vincent, eds. ISBN 9780444634320
Vol. 134, Gliomas, M.S. Berger and M. Weller, eds. ISBN 9780128029978
Vol. 135, Neuroimaging Part I, J.C. Masdeu and R.G. González, eds. ISBN 9780444534859
Vol. 136, Neuroimaging Part II, J.C. Masdeu and R.G. González, eds. ISBN 9780444534866
Vol. 137, Neuro-otology, J.M. Furman and T. Lempert, eds. ISBN 9780444634375
Vol. 138, Neuroepidemiology, C. Rosano, M.A. Ikram and M. Ganguli, eds. ISBN 9780128029732
Vol. 139, Functional neurologic disorders, M. Hallett, J. Stone and A. Carson, eds. ISBN 9780128017722
Vol. 140, Critical care neurology Part I, E.F.M. Wijdicks and A.H. Kramer, eds. ISBN 9780444636003
Vol. 141, Critical care neurology Part II, E.F.M. Wijdicks and A.H. Kramer, eds. ISBN 9780444635990
Vol. 142, Wilson disease, A. Członkowska and M.L. Schilsky, eds. ISBN 9780444636003
Vol. 143, Arteriovenous and cavernous malformations, R.F. Spetzler, K. Moon and R.O. Almefty, eds. ISBN 9780444636409
Vol. 144, Huntington disease, A.S. Feigin and K.E. Anderson, eds. ISBN 9780128018934
Vol. 145, Neuropathology, G.G. Kovacs and I. Alafuzoff, eds. ISBN 9780128023952
Vol. 146, Cerebrospinal fluid in neurologic disorders, F. Deisenhammer, C.E. Teunissen and H. Tumani, eds. ISBN 9780128042793
Vol. 147, Neurogenetics Part I, D.H. Geschwind, H.L. Paulson and C. Klein, eds. ISBN 9780444632333
Vol. 148, Neurogenetics Part II, D.H. Geschwind, H.L. Paulson and C. Klein, eds. ISBN 9780444640765
Vol. 149, Metastatic diseases of the nervous system, D. Schiff and M.J. van den Bent, eds. ISBN 9780128111611
Vol. 150, Brain banking in neurologic and psychiatric diseases, I. Huitinga and M.J. Webster, eds. ISBN 9780444636393
Vol. 151, The parietal lobe, G. Vallar and H.B. Coslett, eds. ISBN 9780444636225
Vol. 152, The neurology of HIV infection, B.J. Brew, ed. ISBN 9780444638496
Vol. 153, Human prion diseases, M. Pocchiari and J.C. Manson, eds. ISBN 9780444639455
Vol. 154, The cerebellum: From embryology to diagnostic investigations, M. Manto and T.A.G.M. Huisman, eds. ISBN 9780444639561
Vol. 155, The cerebellum: Disorders and treatment, M. Manto and T.A.G.M. Huisman, eds. ISBN 9780444641892
Vol. 156, Thermoregulation: From basic neuroscience to clinical neurology Part I, A.A. Romanovsky, ed. ISBN 9780444639127
Vol. 157, Thermoregulation: From basic neuroscience to clinical neurology Part II, A.A. Romanovsky, ed. ISBN 9780444640741
Vol. 158, Sports neurology, B. Hainline and R.A. Stern, eds. ISBN 9780444639547
Vol. 159, Balance, gait, and falls, B.L. Day and S.R. Lord, eds. ISBN 9780444639165
Vol. 160, Clinical neurophysiology: Basis and technical aspects, K.H. Levin and P. Chauvel, eds. ISBN 9780444640321
Vol. 161, Clinical neurophysiology: Diseases and disorders, K.H. Levin and P. Chauvel, eds. ISBN 9780444641427
Vol. 162, Neonatal neurology, L.S. De Vries and H.C. Glass, eds. ISBN 9780444640291
Vol. 163, The frontal lobes, M. D'Esposito and J.H. Grafman, eds. ISBN 9780128042816
Vol. 164, Smell and taste, Richard L. Doty, ed. ISBN 9780444638557
Vol. 165, Psychopharmacology of neurologic disease, V.I. Reus and D. Lindqvist, eds. ISBN 9780444640123
Vol. 166, Cingulate cortex, B.A. Vogt, ed. ISBN 9780444641960
Vol. 167, Geriatric neurology, S.T. DeKosky and S. Asthana, eds. ISBN 9780128047668
Vol. 168, Brain-computer interfaces, N.F. Ramsey and J. del R. Millán, eds. ISBN 9780444639349
Vol. 169, Meningiomas, Part I, M.W. McDermott, ed. ISBN 9780128042809
Vol. 170, Meningiomas, Part II, M.W. McDermott, ed. ISBN 9780128221983
Vol. 171, Neurology and pregnancy: Pathophysiology and patient care, E.A.P. Steegers, M.J. Cipolla and E.C. Miller, eds. ISBN 9780444642394
Vol. 172, Neurology and pregnancy: Neuro-obstetric disorders, E.A.P. Steegers, M.J. Cipolla and E.C. Miller, eds. ISBN 9780444642400
Vol. 173, Neurocognitive development: Normative development, A. Gallagher, C. Bulteau, D. Cohen and J.L. Michaud, eds. ISBN 9780444641502
Vol. 174, Neurocognitive development: Disorders and disabilities, A. Gallagher, C. Bulteau, D. Cohen and J.L. Michaud, eds. ISBN 9780444641489
Vol. 175, Sex differences in neurology and psychiatry, R. Lanzenberger, G.S. Kranz, and I. Savic, eds. ISBN 9780444641236
Vol. 176, Interventional neuroradiology, S.W. Hetts and D.L. Cooke, eds. ISBN 9780444640345
Vol. 177, Heart and neurologic disease, J. Biller, ed. ISBN 9780128198148
Vol. 178, Neurology of vision and visual disorders, J.J.S. Barton and A. Leff, eds. ISBN 9780128213773
Vol. 179, The human hypothalamus: Anterior region, D.F. Swaab, F. Kreier, P.J. Lucassen, A. Salehi and R.M. Buijs, eds. ISBN 9780128199756
Vol. 180, The human hypothalamus: Middle and posterior region, D.F. Swaab, F. Kreier, P.J. Lucassen, A. Salehi and R.M. Buijs, eds. ISBN 9780128201077
Vol. 181, The human hypothalamus: Neuroendocrine disorders, D.F. Swaab, R.M. Buijs, P.J. Lucassen, A. Salehi and F. Kreier, eds. ISBN 9780128206836
Vol. 182, The human hypothalamus: Neuropsychiatric disorders, D.F. Swaab, R.M. Buijs, F. Kreier, P.J. Lucassen, and A. Salehi, eds. ISBN 9780128199732
Vol. 183, Disorders of emotion in neurologic disease, K.M. Heilman and S.E. Nadeau, eds. ISBN 9780128222904
Vol. 184, Neuroplasticity: From bench to bedside, A. Quartarone, M.F. Ghilardi, and F. Boller, eds. ISBN 9780128194102
Vol. 185, Aphasia, A.E. Hillis and J. Fridriksson, eds. ISBN 9780128233849
Vol. 186, Intraoperative neuromonitoring, M.R. Nuwer and D.B. MacDonald, eds. ISBN 9780128198261
Vol. 187, The temporal lobe, G. Miceli, P. Bartolomeo, and V. Navarro, eds. ISBN 9780128234938

Vol. 188, Respiratory neurobiology: Physiology and clinical disorders, Part I, R. Chen and P.G. Guyenet, eds. ISBN 9780323915342
Vol. 189, Respiratory neurobiology: Physiology and clinical disorders, Part II, R. Chen and P.G. Guyenet, eds. ISBN 9780323915328
Vol. 190, Neuropalliative care: Part I, J.M. Miyasaki and B.M. Kluger, eds. ISBN 9780323850292
Vol. 191, Neuropalliative care: Part II, J.M. Miyasaki and B.M. Kluger, eds. ISBN 9780128245354
Vol. 192, Precision medicine in neurodegenerative disorders: Part I, A.J. Espay, ed. ISBN 9780323855389
Vol. 193, Precision medicine in neurodegenerative disorders: Part II, A.J. Espay, ed. ISBN 9780323855556
Vol. 194, Mitochondrial diseases, R. Horvath, M. Hirano, and P.F. Chinnery, eds. ISBN 9780128217511
Vol. 195, Motor system disorders, Part I: Normal physiology and function and neuromuscular disorders, D.S. Younger, ed. ISBN 9780323988186
Vol. 196, Motor system disorders, Part II: Spinal cord, neurodegenerative, and cerebral disorders and treatment, D.S. Younger, ed. ISBN 9780323988179
Vol. 197, Brain and crime, H. Swaab and G. Meynen, eds. ISBN 9780128213759
Vol. 198, Migraine biology, diagnosis, and co-morbidities, J.W. Swanson and M. Matharu, eds. ISBN 9780128233566
Vol. 199, Migraine management, J.W. Swanson and M. Matharu, eds. ISBN 9780128233573
Vol. 200, Paraneoplastic neurological disorders, B. Giometto and S.J. Pittock, eds. ISBN 9780128239124
Vol. 201, Hematopoietic stem cell transplantation for neurological diseases, M. Inglese and G.L. Mancardi, eds. ISBN 9780323902427
Vol. 202, Focal neuropathies, C. Chalk, ed. ISBN 9780323901086
Vol. 203, Neurological channelopathies, M.G. Hanna, ed. ISBN 9780323908207
Vol. 204, Inherited white matter disorders and their mimics, D.S. Lynch and H. Houlden, eds. ISBN 9780323992091
Vol. 205, Cell and gene therapies for neurologic diseases, H.J. Federoff and R.A. Barker, eds. ISBN 9780323901208
Vol. 206, Circadian rhythms, sleep, and sleep disorders, L. Ferini-Strambi and C. Cajochen, eds. ISBN 9780323909181
Vol. 207, Coma, stupor, and related disorders of consciousness, P.W. Kaplan and R.G. Geocadin, eds. ISBN 9780443134081
Vol. 208, Cerebral asymmetries, C. Papagno and P. Corballis, eds. ISBN 9780443156465
Vol. 209, Neuroglia in the healthy nervous system, Part I, A. Verkhratsky, L.D. de Witte, E. Aronica, and E.M. Hol, eds. ISBN 9780443191046
Vol. 210, Neuroglia in neurologic and psychiatric disorders, Part II, A. Verkhratsky, L.D. de Witte, E. Aronica, and E.M. Hol, eds. ISBN 9780443191022
Vol. 211, Cholinergic involvement in neurodegenerative diseases, A.C. Cuello, ed. ISBN 9780443190889
Vol. 212, Vestibular schwannomas, S.K.W. Lloyd, O. Pathmanaban, and D.G. Evans, eds. ISBN 9780128245347
Vol. 213, Placebo effects in neurology, F. Benedetti and A. Shaibani, eds. ISBN 9780443298844
Vol. 214, Immunotherapies for neurologic diseases, M.C. Dalakas, ed. ISBN 9780323908870
Vol. 215, Tic disorders and tourette syndrome, C. Ganos and Y. Worbe, eds. ISBN 9780443135545

All volumes in the 3rd Series of the **Handbook of Clinical Neurology** are published electronically, on Science Direct: http://www.sciencedirect.com/science/handbooks/00729752.

Foreword

Students, physicians, and scientists alike are generally apprehensive when thinking about the brainstem—its complexity is forbidding, its clinical importance is unparalleled, and its death is equated with the death of the brain and the person. For these reasons, we felt a particular need to develop a volume on the brainstem in the present series of the *Handbook of Clinical Neurology*.

We were delighted when Profs. Lea Tenenholz Grinberg and Amos D. Korczyn accepted our invitation and agreed to develop and edit such a volume. Dr. Grinberg, a board-certified neuropathologist and neuroscientist with a particular interest in neurodegenerative diseases, was professor of neurology and pathology at the University of California San Francisco, while this volume was in preparation, but subsequently moved to Florida and is now a senior associate consultant in the Departments of Laboratory Medicine and Pathology and Neurosciences at the Mayo Clinic in Jacksonville. Dr. Korczyn is emeritus professor in the Departments of Neurology, Physiology and Pharmacology at Tel Aviv University, Tel Aviv, Israel. His interests have also focused on neurodegenerative disease. Together, they have crafted an excellent volume that will be helpful for neurologists, neurosurgeons, and neuroscientists as an important compendium of the accumulating knowledge concerning this important anatomic structure.

The volume contains 22 chapters that are arranged in three sections. Section 1, with 5 chapters on physiology and semiology, contains chapters on history, on the control of pain and the autonomic nervous system, and the role of the brainstem in sleep, wakefulness, and consciousness. The 11 chapters in Section 2 on pathology relate the symptoms and signs of brainstem disease, as well as detailing various disorders affecting the brainstem. The 6 chapters in Section 3 focus on various techniques to evaluate brainstem function, such as evoked potentials, ultrasonography, imaging, functional and molecular imaging, and angiography, as well as a final chapter on brain death. As series editors, we read and commented on each of the chapters with great interest. All of the chapters are authored by acknowledged experts in the field, and we thank them and the volume editors for creating such a tremendous resource. We are confident that clinicians and researchers will find much in the volume to appeal to them.

The electronic availability of the volume on Elsevier's ScienceDirect site, as well as in print format, should ensure its ready accessibility and facilitate searches for specific information.

We are grateful to Kristi Anderson in San Diego, Megan McManus in Salt Lake City, Himanshi Yadav in Gurgaon, India, and Punithavathy Govindaradjane at Elsevier Global Book Production in Chennai, India, for their assistance in the development and production of the *Handbook of Clinical Neurology*.

Michael J. Aminoff
François Boller
Dick F. Swaab

Preface

The brainstem is small in size yet vast in functions. It sustains breathing, cardiovascular control, arousal, and balance—the elemental rhythms that make everything else possible—while concentrating pathways whose disruption can transform life. For clinicians, its compact, high-stakes architecture is both an inspiration and a challenge: a lesion measured in millimeters can redraw a patient's world. Precisely because of this complexity and centrality, the brainstem offers a privileged window where anatomy, pathophysiology, and bedside neurology meet.

This volume of the *Handbook of Clinical Neurology* was conceived as a clinically grounded, modern reference for the full spectrum of brainstem anatomy, function, and disorders. The past decades have transformed our field: high-resolution imaging and tractography, advanced electrophysiology and sleep studies, molecular and genetic insights, and innovative neuromodulatory approaches have all deepened what we can see and how we can intervene.

We structured this book with three commitments. First, to anchor clinical reasoning in functional neuroanatomy, linking nuclei, tracts, and neurotransmitter systems to recognizable clinical syndromes. Second, to weave mechanism and management tightly together, so that vascular, inflammatory, infectious, developmental, neoplastic, traumatic, paraneoplastic, and degenerative disorders are presented with both diagnostic pitfalls and practical therapeutic principles. Third, to translate modern methods such as high-resolution imaging and tractography, electrophysiology, vestibular and oculomotor testing, sleep studies, and emerging biomarkers into tools clinicians can actually use tomorrow morning. Throughout, we allow readers to move from nucleus to sign, from tract to symptom, and from pathophysiology to therapy.

This volume is the product of different experiences and one convergence. Lea Tenenholz Grinberg arrived here as a neuropathologist studying brain aging, "stumbling" upon the brainstem at autopsy and in clinicopathologic correlations as an underappreciated locus where neurodegenerative diseases may first take hold. Amos D. Korczyn arrived by way of Parkinson disease, drawn to nonmotor symptoms that patients often feel first, and more recently to the neuronal basis of consciousness. Despite our different paths, we share the conviction that careful attention to brainstem systems can sharpen diagnosis, reveal mechanisms, and open therapeutic possibilities earlier in the course of disease.

Despite our enthusiasm, this volume exists only because of a remarkable community. We thank our contributors, colleagues from around the world, who brought rigor, balance, and generosity to their chapters. They distilled complex subjects with clarity, highlighted what truly changes practice, and matched scholarship with a commitment to patient care. We are deeply grateful for the knowledge, time, precision, and judgment they invested.

We are also indebted to the series editors of the *Handbook of Clinical Neurology* for their trust, high standards, and for giving us this opportunity. We are grateful as well to the broader Elsevier editorial and production teams for the many invisible steps, permissions, figures, copyediting, and indexing, that turn manuscripts into a finished book.

On a personal note, editing this volume together has been both a pleasure and an honor. It goes beyond a professional collaboration and has truly been a family effort.

If the pages ahead help clinicians localize more confidently, recognize early and nonmotor manifestations sooner, and connect symptoms to systems more precisely, then we will have met our purpose. The pace of discovery in brainstem science is accelerating. We hope this volume serves as a bridge between innovation and bedside need and keeps the brainstem exactly where it belongs: at the center of clinical neurology.

Lea Tenenholz Grinberg
Amos D. Korczyn

List of contributors

M. Amans
Department of Neuroendovascular Surgery, University of California San Francisco, San Francisco, CA, United States

I. Asad
Department of Neurology, Beth Israel Deaconess Medical Center, Boston, MA, United States

S. Batista
Medical School, Federal University of Rio de Janeiro, Rio de Janeiro, Brazil

T. Beltraminelli
Neuro-Ophthalmology Unit, Department of Ophthalmology, Jules Gonin Eye Hospital, School of Biology and Medicine, University of Lausanne, Lausanne, Switzerland

R. Bertani
Division of Neurosurgery, Hospital das Clínicas, University of São Paulo Medical School, São Paulo, Brazil

J. Bogousslavsky
Genolier Swiss Medical Network Neurocenter, Clinique Valmont, Glion/Montreux, Switzerland

S. Bohnert
Institute of Forensic Medicine, University of Würzburg, Würzburg, Germany

G. Byczynski
Trinity College Institute of Neuroscience, School of Psychology, Trinity College Dublin, Dublin, Ireland

G. Calandra-Buonaura
Department of Biomedical and Neuromotor Sciences, University of Bologna; IRCCS Institute of Neurological Sciences of Bologna, Bologna, Italy

L.R. Caplan
Department of Neurology, Beth Israel Deaconess Medical Center, Boston, MA, United States

B. Çelik
Department of Neurology, Ankara University School of Medicine, Ankara, Turkey

D.G. Corrêa
Department of Radiology, Clínica de Diagnóstico por Imagem (CDPI)/DASA; Department of Diagnostic Imaging, Rio de Janeiro State University, Rio de Janeiro, Brazil

P. Cortelli
Department of Biomedical and Neuromotor Sciences, University of Bologna; IRCCS Institute of Neurological Sciences of Bologna, Bologna, Italy

L.S. Crawford
School of Medical Sciences (Neuroscience), Brain and Mind Centre, University of Sydney, Sydney, NSW, Australia

M. Duvvuri
Department of Neuroimaging and Neurointervention, Stanford University, Stanford, CA, United States

P.M. Enderby
School of Health and Related Research, University of Sheffield, Sheffield, United Kingdom

T. Freddi
Department of Radiology, Hospital do Coração, São Paulo, Brazil

D. Gomez-Andres
Pediatric Neurology Department, University Hospital Vall Hebron, Barcelona, Spain

L.T. Grinberg
Departments of Laboratory Medicine, Pathology and Neurosciences, Mayo Clinic, Jacksonville, FL, United States

H. Heinsen
Institute of Forensic Medicine, University of Würzburg, Würzburg, Germany

L.A. Henderson
School of Medical Sciences (Neuroscience), Brain and Mind Centre, University of Sydney, Sydney, NSW, Australia

H.I.L. Jacobs
Athinoula A. Martinos Center for Biomedical Imaging, Department of Radiology, Massachusetts General Hospital, Harvard Medical School, Boston, MA, United States

D. Kaski
SENSE Research Unit, Department of Clinical and Movement Neurosciences, UCL Institute of Neurology, London, United Kingdom

A. Kawasaki
Neuro-Ophthalmology Unit, Department of Ophthalmology, Jules Gonin Eye Hospital, School of Biology and Medicine, University of Lausanne, Lausanne, Switzerland

K.A. Keay
School of Medical Sciences (Neuroscience), Brain and Mind Centre, University of Sydney, Sydney, NSW, Australia

A.D. Korczyn
Departments of Neurology, Physiology and Pharmacology, Tel Aviv University, Tel Aviv, Israel

C. Krebs
Department of Cellular and Physiological Sciences, Faculty of Medicine, University of British Columbia, Vancouver, BC, Canada

S. Kusunoki
Department of Neurology, Faculty of Medicine, Kindai University, Osaka, Japan

E. Lavon
Hasharon Hospital, Rabin Medical Center, Clalit Health Services, Petah-Tikva; Gray Faculty of Medical & Health Sciences, Tel Aviv University, Tel Aviv, Israel

C. Machado
Department of Clinical Neurophysiology, Institute of Neurology and Neurosurgery, Havana, Cuba

K. Miyamoto
Department of Neurology, Wakayama Medical University, Wakayama, Japan

J. Moncayo-Gaete
School of Medicine. Universidad de las Américas, Quito, Ecuador

R. Moon
Department of Neurology, University of California San Francisco, San Francisco, CA, United States

A. Morgan
Murdoch Children's Research Institute, Melbourne, VIC, Australia

S. Mueller
Department of Neurology, Neurosurgery and Pediatrics, University of California San Francisco, San Francisco, CA, United States

F. Munell
Pediatric Neurology Department, University Hospital Vall Hebron, Barcelona, Spain

A. Ohry
Department of Rehabilitation Medicine, Gray Faculty of Medical & Health Sciences, Tel Aviv University, Tel Aviv, Israel

W.S. Paiva
Division of Neurosurgery, Hospital das Clínicas, University of São Paulo Medical School, São Paulo, Brazil

P.C. Prokopiou
Athinoula A. Martinos Center for Biomedical Imaging, Department of Radiology, Massachusetts General Hospital, Harvard Medical School, Boston, MA, United States

E. Raz
Department of Radiology and Neurosurgery, New York University Grossman School of Medicine, New York, NY, United States

M. Roig-Quils
Pediatric Neurology Department, University Hospital Vall Hebron, Barcelona, Spain

A. Silvani
Department of Biomedical and Neuromotor Sciences, University of Bologna, Bologna, Italy

T. Singhal
Ann Romney Center for Neurologic Diseases, Department of Neurology, Brigham and Women's Hospital, Harvard Medical School, Boston, MA, United States

S. Vanneste
Trinity College Institute of Neuroscience; Global Brain Health Institute, School of Psychology, Trinity College Dublin, Dublin, Ireland

P. Varlet
Department of Neuropathology, GHU Paris, Psychiatry and Neurosciences, Sainte-Anne Hospital, Paris, France

R. Yilmaz
Department of Neurology, Ankara University School of Medicine; Ankara University Brain Research Center, Ankara, Turkey

Contents

Foreword ix
Preface xi
List of contributors xiii

SECTION I Physiology and semiology

1. **History of the brainstem—A developing concept throughout the history of medicine** 1
E. Lavon and A. Ohry (Petah-Tikva and Tel Aviv, Israel)

2. **Structural and functional anatomy of the brainstem** 17
H. Heinsen, L.T. Grinberg and S. Bohnert (Würzburg, Germany and Jacksonville, United States)

3. **Autonomic blood pressure control** 59
P. Cortelli, G. Calandra-Buonaura and A. Silvani (Bologna, Italy)

4. **Brainstem in pain control** 75
L.A. Henderson, L.S. Crawford and K.A. Keay (Sydney, Australia)

5. **Brainstem control of sleep, wakefulness, alertness, and consciousness** 91
A.D. Korczyn and L.T. Grinberg (Tel Aviv, Israel and Jacksonville, United States)

SECTION II Pathology

6. **Locked-in syndrome** 97
L.R. Caplan and I. Asad (Boston, United States)

7. **The vestibular system: Unsteadiness, dizziness, and vertigo** 105
D. Kaski (London, United Kingdom)

8. **Oculomotor disorders and nystagmus** 123
C. Krebs (Vancouver, Canada)

9. **Pupillary disorders** 131
T. Beltraminelli and A. Kawasaki (Lausanne, Switzerland)

10. **Neurodevelopmental brainstem syndromes** 157
D. Gomez-Andres, M. Roig-Quils and F. Munell (Barcelona, Spain)

11. **Intrinsic pontine tumors: Pathology of brainstem tumors and exploration of diffuse intrinsic pontine glioma** 171
R. Moon, S. Mueller and P. Varlet (San Francisco, United States and Paris, France)

12. **Autoimmune maladies of the brainstem** 183
K. Miyamoto and S. Kusunoki (Wakayama and Osaka, Japan)

13. **Dysarthria** 193
P.M. Enderby and A. Morgan (Sheffield, United Kingdom and Melbourne, Australia)

14. **The brainstem in neurodegenerative diseases** 201
L.T. Grinberg (Jacksonville, United States)

15. Brainstem vascular syndromes **213**
J. Moncayo-Gaete and J. Bogousslavsky (Quito, Ecuador and Glion/Montreux, Switzerland)

16. Traumatic brainstem injury **237**
W.S. Paiva, R. Bertani and S. Batista (São Paulo and Rio de Janeiro, Brazil)

SECTION III Monitoring and intervention

17. Computed tomography and magnetic resonance imaging of the brainstem **251**
D.G. Corrêa and T. Freddi (Rio de Janeiro and São Paulo, Brazil)

18. Ultrasound studies of brainstem **285**
B. Çelik and R. Yilmaz (Ankara, Turkey)

19. Anatomic imaging: Angiography **303**
M. Duvvuri, E. Raz and M. Amans (Stanford, New York, and San Francisco, United States)

20. Brainstem evoked responses **317**
G. Byczynski and S. Vanneste (Dublin, Ireland)

21. Functional and molecular brainstem imaging **329**
H.I.L. Jacobs, P.C. Prokopiou and T. Singhal (Boston, United States)

22. Brain death and the brainstem **345**
C. Machado (Havana, Cuba)

Index **361**

Section I

Physiology and semiology

Handbook of Clinical Neurology, Vol. 216 (3rd series)
The Brainstem
L.T. Grinberg and A.D. Korczyn
https://doi.org/10.1016/B978-0-443-15736-3.00004-4

Chapter 1

History of the brainstem—A developing concept throughout the history of medicine

EITAN LAVON[1,2,*] AND AVI OHRY[3]

[1]*Hasharon Hospital, Rabin Medical Center, Clalit Health Services, Petah-Tikva, Israel*

[2]*Gray Faculty of Medical & Health Sciences, Tel Aviv University, Tel Aviv, Israel*

[3]*Department of Rehabilitation Medicine, Gray Faculty of Medical & Health Sciences, Tel Aviv University, Tel Aviv, Israel*

Abstract

The chapter explores the historical journey of the discovery and understanding of the brainstem, including its anatomy, physiology, and vital functions. It highlights key milestones, advancements in medicine and philosophy of science, from ancient Greece with figures like Galen, to the scientific leap starting at the Renaissance period with Vesalius, Mondino de Luzzi, Willis, and others providing detailed descriptions of the brainstem and cranial nerves. The 18th century marked a significant development with the invention of the microscope, leading to a deeper understanding of the brainstem's infrastructure and its role in regulating vital functions. Contributions from notable figures in neuroscience continued in the 19th and 20th centuries, shaping our modern concept of the brainstem.

INTRODUCTION

The history of the nervous system has been extensively studied and described in numerous publications in the past. The vast knowledge accumulated over thousands of years is based on the fundamental understanding of the anatomy, physiology, and pathology of various organs, including the brainstem, an organ whose unique functions greatly influence the human body. In this chapter, we will review the development of knowledge about the brainstem, from ancient times to the present day, while mentioning significant figures, important discoveries, symptoms, and syndromes that have contributed to the understanding of the anatomy and function of this organ.

Embryologists and anatomists described the brainstem as "two blossoms, budding from its lateral and dorsal aspects, flower into the cerebral hemispheres and cerebellum" (Encyclopaedia Britannica, 1960). The brainstem is interposed and interconnected with the other parts of the brain. The various sensory data, and the complicated motor nervous flow, pass through the brainstem. Brainstem nerves supply the head's skin, the eyeballs' muscles, and the sensory input from oral, and across also the pharyngeal cavities, ear, eye, and olfactory receptors. Ascending pathways from the spinal cord continue upward through the brainstem to the cerebellum and cerebral hemispheres. Descending pathways from these structures also cross the brainstem. The brainstem has a dominating and integrating role over the lower structures, to preserve the body's internal environment (milieu intérieur), a term coined by Claude Bernard (Haldane, 1929). It is beyond the scope of our intention to mention all the historical evidence concerning the development of the concept of the brainstem, but rather the most important facts.

THE BRAINSTEM – FIRST DISCOVERIES AND DESCRIPTIVE ANATOMY

The anatomy of the neurologic system was the basis of understanding the function of one, if not the most mysterious system of the human body. The documented history of human anatomy began thousands of years ago, with drawings and sculptures found in caves, some of which are 30,000 years old. Over the years, anatomic research of the nervous system has mainly focused on the cerebral hemispheres, cerebellum, and spinal cord. The brainstem, the functions of which were less clear, was relatively neglected. The description of the brainstem as a defined part of the nervous system came later, as will be described later.

In fact, for hundreds of years and more, macroscopic anatomy has been the only basis on which medical science

*Correspondence to: Eitan Lavon, MD, MHA, Hasharon Hospital, Rabin Medical Center, Clalit Health Services, 7 Kakal Street, Petah-Tikva 49372, Israel. E-mail: eitan@clalit.org.il.

relied. The development of anatomy as a science began in ancient Greece. The Greeks were the first to attempt to study the living body and the first to systematically understand and describe the various systems of the body. Some of the basic assumptions relied on beliefs, some on observations, and some on dissections of animals and later on humans as well. The scientific work that was continued by the scholars of Rome significantly declined after the fall of the Roman Empire. The political, religious, cultural, and philosophical changes that occurred during 1000 years of the Middle Ages led to almost a complete halt in the development of medical science. During the Renaissance period, the study of anatomy was revived and gradually progressed to where we are today.

As for the brainstem, development of knowledge occurred in the context of the history and anatomy of the cranial nerves. It can be divided into several periods and layers, starting from the discovery of macroscopic anatomy, through microscopic anatomy and neurophysiologic discoveries, understanding ontogenesis and genetic architecture, and up to the knowledge accumulated in modern surgery (Cucu, 2021).

As previously mentioned, the journey of discovering the anatomy of the brain began in ancient Greece. The existence of cranial nerves, even if their function was not clear, was known to Hippocratic physicians. It was the Greek philosopher Alcmaeon of Croton, who lived in the 5th century BC, who speculated that "thought and feelings" are located in the brain. Plato also believed that the soul is located in the brain. The philosophical understanding that the brain is the organ responsible for thinking and emotions, developed further with detailed anatomic descriptions. Among the first to contribute to the understanding of the anatomy of the brainstem was Herophilus of Chalcedon (c.330–c.260 BC), who described the origin of seven pairs of cranial nerves as well as the calamus scriptorius, a groove located in the posterior part of the brainstem. Rufus of Ephesus (110–180 AD), one of the great physicians of the ancient pre-Galenic period, described the lateral, third, and fourth ventricles as well as the mesencephalic aqueduct. Aretaeus of Cappadocia, his contemporary, distinguished between paralysis resulting from head injury and paralysis resulting from spinal cord injury, an observation that led him to postulate that some kind of crossing must take place above the craniovertebral junction. The first to differentiate between sensory and motor nerves was the Greek anatomist Erasistratus (c.304–c.250 BC). Some believe that Herophilus was the first to differentiate between the sensory and motor systems (Pearce, 2013). Both Herophilus and Erasistratus were Alexandrian-Hellenistic physicians who performed human dissections. Galen of Pergamon (c.129–c.216 AD) furthered the understanding of these two types of nerves and the anatomy of the cranial nerves.

During the Middle Ages, particularly from the 9th century onward, scientific and medical research in the Western world was primarily based on ancient Greek medicine. The influences of the Church and cultural restrictions greatly limited any rational inquiry and thought, and in fact, there was no opportunity to question the doctrines of Galen or Aristotle. Anatomic dissections were illegal, and therefore, the most detailed description of the human body during this period relied on Galen's work.

The foundation of universities in the 12th century, such as in Bologna, Oxford, and Montpellier, led to a renewed acceleration of medical and anatomic research. This was also aided by a special permission granted to universities in 1231 by Holy Roman Emperor Frederick II, allowing them to perform at least one dissection every five years. This permission led to the integration of human dissections as part of the anatomy curriculum for medical students in the new emerging faculties of medicine.

One of the significant figures in the development of neuroanatomic knowledge in the late medieval period was Mondino de Luzzi (c.1270–c.1336), also known as Mondino. Mondino, who was a physician and a professor of surgery at the University of Bologna and is regarded by many as the 'restorer of anatomy', performed the first known public human dissection in 1315, on an executed criminal. In 1316, Mondino published his book *Anathomia*. The book became a primary textbook for anatomy in most European universities for more than two centuries, describing not only the anatomic structures but also some aspects of physiology and pathology (Crivellato and Ribatti, 2006; Mavrodi and Paraskevas, 2014).

Dissections during this period aimed, among other things, to verify the observations of early anatomists such as Hippocrates, Galen, and Aristotle. Findings that did not align with traditional descriptions were referred to as anatomic variations. Additional errors made by Mondino were the misinterpretations of ancient texts and attempts to reconcile the observations of Galen and Aristotle. Despite these biases and the widely accepted assumption that Mondino himself conducted only a few dissections, the book was innovative in many ways and even served as a practical manual for dissections.

The process of tissue decay transformed dissections into a race against time, determining the order in which they were performed. Initially, the abdominal organs were dissected, followed by the thorax, head, and finally the extremities. Opening the skull itself was considered a sin, and hence, the chapter on the head was likely shorter. Nonetheless, *Anathomia* made a significant contribution to the development of neuroanatomy.

Mondino described the dura and the pia mater and falsely attributed cognitive functions to the cerebral ventricles and placed the ability to think at the choroid plexus. Mondino described the cranial nerves and, in the tradition of his predecessors, in particular Nemesius of Emesa (c.390 AD), proposed that the brain is divided into three sections or cells. The anterior cell served as the convergence point for the senses (sensus communis) as well as imagination, the middle cell housed the cognition or intellect, and the posterior cell contained the memory (Voogd and De Zeeuw, 2020; Schijns and Koehler, 2021). He referenced the incorrect Galenic belief regarding the existence of a "rete mirabile," a "marvelous network" of blood vessels at the base of the human brain, which he described as related to sleep. Despite inaccuracies in his observations and descriptions, Mondino made a significant contribution to the revival of anatomic science during the medieval period. A detailed commentary of Mondino's work was published in 1521 by the Italian surgeon and anatomist Berengario da Carpi (1460–1530). In this book, Berengario challenged some of the common assumptions of Galen's physiology, such as the existence of the rete mirabilis, and the vascular plexus in the base of the brain where the vital spirits evolve before released peripherally through hollow nerves, stating that he was unable to detect it in humans. He also noted that opposite to Galen's theory, nerves linked to the brain were solid. Berengario based his observations on a systematic dissection method, which he called "anatomia sensibilis." His special interest in brain anatomy led to significant contributions, including description of the meninges, cranial nerves, the ventricular system, choroid plexuses, pituitary stalk, and gland. Berengario was also known for his clinical skills, after successfully treating Lorenzo II de' Medici (1492–1519), Duke of Urbino and ruler of Florence, who had been wounded in a battle in 1517. The Duke suffered from occipital fracture and shock. Following that year, Berengario published in 1518 his *De fractura cranei*, a monumental work on cranio-cerebral surgery, describing fractures, hematomas, symptoms of brain concussion, and craniotomies (Parent, 2019).

The next significant leap occurred during the Renaissance. The renewed focus on visual arts, such as painting and sculpture, also encouraged anatomic research as a tool to help artists understand the structure of the human body. Evidence of artists such as Leonardo's and Michelangelo's involvement in medical research can be found in both artworks and in artistic representation of anatomic descriptions in medical and anatomic texts. One of the prominent figures of this era was the Brussels-born physician and anatomist Andreas Vesalius (1514–1564), who founded the branch of comparative anatomy and is often referred to as the father of modern anatomy. Among Vesalius's works was his masterpiece published in 1542, *De Humani Corporis Fabrica Libri Septem*. This anatomy textbook was based not only on Galenic anatomic and physiologic knowledge but also on numerous dissections conducted by Vesalius himself, most of which took place in Padua, Italy. In his works, Vesalius demonstrated that Galen had never dissected a human body himself.

Vesalius included illustrations of skulls and brains in his book, including various cross-sections at different depths, as well as detailed descriptions of specific structures such as the cerebellum, dura mater, ventricles, blood vessels, and the brainstem which he considered as the upper part of the spinal cord. Vesalius's works set a new standard and served as the foundation for the ongoing research of physicians, starting from his time and for at least two centuries thereafter (Finger, 2005; Scatliff and Johnston, 2014).

The knowledge of the anatomy of the nervous system, including the brain and its structures such as the brainstem, advanced during the Renaissance and became clearer with the emergence of discoveries and descriptions in medical literature. Johannes Eichmann (Dryander) (1500–1560) from Marburg, Germany, a "progenitor of rational scientific thought," published his *Anatomiae pars prior* in 1537. The book, which contained the earliest realistic images of posterior fossa anatomy, demonstrated the transition from the Galenic tradition and Mondino's works to the Vesalian systematic and precise description (Hanigan et al., 1990). The Italian anatomist Bartolomeo Eustachio (1520–1574), known best for his description of the Eustachian tube, added illustrations of the brainstem in his 1552 *Tabulae Anatomicae*. Other anatomists, some of whom are more familiar than others, contributed new descriptions of structures and nerves originating from the brainstem. As anatomic discoveries progressed, clinical correlations also began to emerge and a comprehensive and consolidated brainstem anatomy started to appear.

The trochlear nerve was described by the Italian philosopher and physician, Alessandro Achillini (1463–1512) (Davis, 2014). Gabriele Fallopius (1523–1562), described the abducens nerve and the semicircular canals of the inner ear. A less-known figure, Costanzo Varolio (1543–1575), an Italian anatomist and a papal physician to Gregory XIII, described the brainstem's various parts. The pons (*pons varolii*) was called after him. He discovered and named the structure as he began using a different method of brain dissection, notably starting from its base and up (Wijdicks, 2020).

The accessory nerve or in Latin *nervus accessorius* was described and named by the English physician, Thomas Willis (1621–1675), best known for the "circle of Willis." Willis is credited with the naming of nine cranial nerves (originally claimed to identify 10 pairs). Although his observations were partially challenged by other anatomists, Willis's nine cranial nerve classification was accepted in British anatomy text until the late 19th century (Storey, 2022). Willis also described "a liquid" in the aqueduct of Sylvius that connects the ventricles, and presumed that the consistency of the "liquid" is altered in "epidemic fever," i.e., meningitis (Hajdu, 2003). He played an important part in the history of anatomy, neurology, and psychiatry, and was one of the founders of The Royal Society, today the United Kingdom's National Academy of Sciences (Molnár, 2021).

MOVING ON TO NEW REVELATIONS AND INSIGHTS

The 18th century marked the beginning of a new era of discoveries and innovations in various fields, such as natural sciences, philosophy, geography, history, and archeology. It was often called the age of "enlightenment," challenging traditional and religious beliefs and even monarchies. It was an age of revolutions. The new perceptions also led to advancements in medical research, including neuroanatomy and neurophysiology, following the growth of medical schools in leading universities. A significant contribution to the ability to explore the intricacies of the nervous system was the development of the microscope in the second half of the 17th century by the English Robert Hooke (1635–1703) and the Dutch Antoni van Leeuwenhoek (1632–1723). The ability to demonstrate the microstructure of elements of the nervous system cast doubt on early theories that suggested that animal spirits or fluids were secreted from the brain via hollow nerves. This was further studied with systematic anatomic and physiologic investigations, observations, and experiments, using human and animal models. The electric properties of organisms were a focal point of interest for many researchers in the 18th century, among the most famous of whom was the Italian Luigi Galvani (1737–1798), one of the founders of electrophysiology. Despite significant progress, the brain was still considered by many as a sort of a gland, an enigmatic organ with an unclear function, a puzzle that would be explored in the coming years (Smith, 2010).

In the first decade of the 18th century, the Italian Domenico Mistichelli (1675–1715) and the Frenchman François Pourfour du Petit (1664–1741) were the first to describe the pyramidal decussation and the crossed association between brain injury and contralateral motor signs (Šimon, 2018). Raymond Vieussens (1635–1715) was a French anatomist, who described the brain's centrum semiovale (the white matter of the centrum ovale, also known as "Vieussens center"), and was the first to demonstrate the continuity of the white matter fibers from the centrum ovale to the brainstem. He also described the dentate nuclei, the pyramids, and the olivary nuclei in his *Neurographia universalis*, a work on neuroanatomy that is known for its excellent copperplate illustrations. The centrum semiovale is also known as "Vicq d'Azyr's center," named after the French physician Félix Vicq d'Azyr (1746–1794), who provided a more detailed description and is considered as one of the prominent neuroanatomists of the 18th century (Tubbs Loukas, et al., 2011).

The pyramidal crossing and its clinical-anatomic significance were still debated by the second half of the 19th century, by Franz Joseph Gall (1758–1828), John Cheyne (1777–1836), Félix Vicq d'Azyr (1746–1794), and Charles Bell (1774–1842). In 1853, Moritz Heinrich Romberg (1795–1873), a German neurologist, described the insecure gait (the Romberg's sign) in patients with tabes dorsalis (apparently in all proprioceptive disorders of the legs), commenting that "if he is ordered to close his eyes while in the erect posture, he at once commences to totter and swing from side to side" (Koehler et al., 2000). Romberg wrote and lectured on the "law of crossing conduction" (Rezende-Cunha and de Oliveira-Souza, 2011). Another interesting clinical observation was described by the French military surgeon, Pourfour du Petit, who noted soldiers with paralysis of the side of the body contralateral to that of an injury to the head.

In 1784, Félix Vicq d'Azyr discovered the locus coeruleus (LC). Vicq d'Azyr was one of the founders of the French Royal Society of Medicine (Société Royale de Médecine) in 1778. In 1789, Vicq d'Azyr was appointed physician to Queen Marie-Antoinette and in 1790, he presented to the Constituent Assembly a decisive plan to reform the teaching of medicine in France, paving the way to modern medical teaching (Parent, 2007). Later, the LC was "redescribed" by the German Johann Christian Reil (1759–1813) in 1809 (who coined the term psychiatry in 1808) and was named locus coeruleus (LC) by the German brothers Joseph and Karl Wenzel in 1812 (*De penitiori structura cerebri hominis et brutorum*) (Kutzer, 1988; Chandler et al., 2019). An important advance in understanding the anatomic organization of the LC was the application of the Falck–Hillarp technique, published in 1962. This method allowed visualization of monoamine neurons using formaldehyde fluorescence (Hökfelt, 2016).

The vestibular, glossopharyngeal, accessory, and hypoglossal nerves and nuclei, were described by Gall, a German neuroanatomist, who pioneered localization attempts of the cerebral functions (Rawlings and Rossitch, 1994). Gall is also known as the founder of organology, the pseudoscience that claimed to predict human skills, propensities, and personality traits from an examination of skull shape reflecting the underlying brain centers. His coworker Johann Gaspar Spurzheim (1776–1832) was the German physician who became a "phrenologist" and later went abroad to spread Gall's concepts, actually coined the term, while Gall himself used the term "cranioscopy." In his neurology book "The Physiognomical System" published in 1815 (Spurzheim, 1815), Spurzheim wrote that "The brain was pretty generally considered as the source of the medulla oblongata, of the spinal marrow, and of all the nerves… Till the present time, the difference between the nerves necessary to voluntary motion and to the sense of feeling has not been demonstrated; but, by several anatomic, physiologic, and pathologic proofs, I am convinced of its existence. This difference has been sought for from ancient times. Herophilus was the first who spoke of it… but entire nerves consisting of both these parts are distributed to the muscles or to the skin… Others still suppose that even the medulla oblongata and spinal marrow are prolongations of the brain and cerebellum, and give to all…" Gall's and Spurzheim's investigations established the division between gray and white matter and contributed to understanding that the cortex and its different regions have a role in various intellectual functions. In fact, Gall was an excellent neuroanatomist who did serious scientific research (Eling and Finger, 2019). His assistant Spurzheim later popularized the ideas and used the term phrenology. Gall and Spurzheim's most important book is *Anatomie et Physiologie du Système Nerveux en Général et du Cerveau en Particulier* (Anatomy and Physiology of the Nervous System in General and of the Brain in Particular), published in Paris, consisting of four volumes, 100 plates of brains and skulls of humans and lower animals (Gall and Spurzheim, 1810). Later on, Gall published alone a less expensive set of books (1822–1825): *Sur les Fonctions de Cerveau et sur Celles de Chacune de Ses Parties*, translated to English in 1835 (On the Functions of the Brain and Each of Its Parts) (Eling and Finger, 2019). Over the years, Gall criticized Spurzheim's work and eventually, they stopped collaborating (Whitaker and Jarema, 2017). The *Anatomica de Basi Encephali et*

Originibus Nervorum Cranio Egredientium Libri Quinque (1778) was compiled by Samuel Thomas von Soemmerring (1755–1830), a German physician, anatomist, anthropologist, paleontologist, and inventor. Soemmerring described and illustrated the brainstem and the origin of the 12 cranial nerves, establishing their routes (Bast, 1924; Stoyanov and Matev, 2018). His classification of cranial nerves (1778) is valid and accepted to date. Soemmerring earned a place of honor in the history of neuroanatomy for demonstrating the crossing of the optic nerve fibers (1788) and of the macula lutea in the retina (1791) (Hildebrand, 2005).

DISCOVERING THE VITAL CENTERS

Research in the 18th and 19th centuries led to additional discoveries about brainstem anatomy and its vital functions. Until the beginning of the 18th century, most anatomists relied on the findings of Willis, who postulated that the vital center is located in the cerebellum. He assumed the vagus nerve had its origin in the cerebellum and just passed through the medulla. The assumption of the cerebellum being of vital importance was based on experiments in which dissections were performed in the cerebellum, inadvertently affecting structures in the brainstem as well, or causing pressure effects by hemorrhages. Although several 17th century experiments reported fatal effect of removal of the cerebellum, other experiments conducted in the 18th century by François Pourfour du Petit, Antonio Maria Valsalva, and Giovanni Battista Morgagni, demonstrated that animals did not always die after this procedure. The recognition of the brainstem as a separate organ was not clear, and it was often regarded as a peduncle or appendix of the cerebellum. The Dutch physician Abraham Kaau-Boerhaave (1715–1758) was one of the first to acknowledge the vital role of the medulla oblongata. He observed that cardiac and respiratory function persisted after traumatizing the cerebellum. The Swiss physician Albrecht Haller (1708–1777) also demonstrated, contrary to the accepted assumption, that when other parts of the brainstem are spared from injury, there is no change in vital functions (Voogd and Koehler, 2018; Schijns and Koehler, 2021).

During the second half of the 18th century, the French physician and physiologist Anne Charles Lorry (1726–1783), known for his contributions to modern dermatology, concluded from animal studies that respiration and normal heart pulse persist after removing the cerebrum and cerebellum, hence suggesting that the area of the brain responsible for critical vital functions was located in the upper part of the spinal cord (Koehler, 2020). Following Lorry's and other 18th century researchers' findings, an additional significant contribution to the understanding of the function of the medulla oblongata was made by another French physician and physiologist, Julien Jean César Legallois (1770–1814). About half a century after Lorry, Legallois conducted a series of studies that included gradual removing slices of the brainstem of small animals such as rabbits, demonstrating that only a small area in the medulla oblongata is responsible for breathing (Wijdicks, 2019b). Published in 1806, Legallois's hypothesis, which posited that the medulla oblongata includes a respiratory center, was further reinforced by additional works that demonstrated the role of the medulla oblongata as an area responsible for vital functions. Some of these works were conducted by the famous Scottish physician, anatomist, and artist Sir Charles Bell (1774–1844) who induced respiratory arrest by lesioning the lateral medulla oblongata in donkeys, and by the English surgeon Sir Benjamin Brodie (1783–1862) who observed that the respiratory and cardiac centers are separated one from the other and noted that the heart stops only after compromising respiration. Many of those deductions remain largely accurate to this day, such as Cheyne–Stokes respiration. This phenomenon may be caused by damage to respiratory centers or by physiologic abnormalities in congestive heart failure. It is an abnormal pattern of breathing characterized by progressively deeper, and sometimes faster, breathing followed by a gradual decrease that results in a temporary stop in breathing called an apnea. Cheyne–Stokes respiration was named after John Cheyne (1777–1836) and William Stokes (1804–1878), both Irish physicians. Cheyne studied medicine at Edinburgh and later on moved to Dublin, where he eventually was appointed physician to the Meath Hospital and professor of medicine at the College of Surgeons. In 1820, he became Physician General of the Irish army. Cheyne founded the Dublin Hospital Reports where he first published a description of the Cheyne–Stokes respiration (1818). Stokes, also known for his contribution to the Stokes–Adams syndrome, was Regius Professor of Physic at Dublin University, succeeding his father Whitley Stokes, MD, who held this position. Thirty-six years after Cheyne, he mentioned Cheyne's description of abnormal respiration, caused by a diseased heart and aorta (Koehler et al., 2000; Pearce, 2002).

The area postrema, a caudal circumventricular organ, located on the dorsal surface of the medulla oblongata, was first described in 1896, by Magnus Gustaf Retzius (1842–1919), a Swedish physician-anatomist, anthropologist, and histologist. Retzius focused his main attention on the histology of the sense organs and nervous system, described it anatomically but did not elaborate on its functions (Sarikcioglu and Yildirim, 2008). Since 1951, when the area postrema was identified in the cat as a chemoreceptor trigger zone in vomiting responses, its role and significance were further studied. The area postrema detects chemical messengers originating from baroreceptors, osmoreceptors, stretch receptors, and others, incorporating them as neural inputs to the brainstem. These inputs have a role in regulating important functions such as blood pressure, appetite, feeding, salivation, metabolism, respiratory drive and rate, and others (Mirza and Das, 2023).

Marie-Jean-Pierre Flourens (1794–1867), a French physician and neurophysiologist, paved the way to determine the nervous system hierarchy. Flourens contributed to the understanding of cerebral localization and brain functions, using experimental methods in neuroanatomy. He continued Legallois' work on respiratory control functions of the medulla oblongata and reported that lesioning the medulla resulted in the death of the animal. By making high cervical transections in rabbits, Flourens identified a small area of 1 mm in the medulla, which he called a vital knot or node (noeud vital). Further studies of the respiratory centers in the 1920s and 1930s led to the identification of chemoreceptors in the medulla oblongata and their role in ventilatory response to blood oxygen levels. In 1954, I.R. Leusen reported that respiratory drive was also influenced directly by PCO_2 and pH of the cerebrospinal fluid (Yildirim and Sarikcioglu, 2007; Wijdicks, 2019b).

Eighteenth- and nineteenth-century clinicians, neuroanatomists, embryologists, and comparative anatomists contributed to enriching our understanding of the brainstem structure and functions. They "knew" (or rather believed) that the brainstem has an important vital role, but only during the 20th century, the ascending reticular formation, the respiratory center, and pressor centers were finally found and were recognized as having automatic and vital functions. The discovery of the Ascending Reticular Activating System was attributed to the works of the American Horace W. Magoun (1907–1991) with his Italian visiting Professor Giuseppe Moruzzi (1910–1986) and others, investigating and mapping the regions the brainstem necessary for regulating consciousness, arousal, and normal sleep (Wijdicks, 2019a; Taran et al., 2023). While it still may be controversial in some aspects (religious, cultural, and others), death is declared when these brainstem centers stop functioning (Saran and Padubidri, 2019).

MORE REVELATIONS AND THEORIES OF THE MECHANISMS OF THE NERVOUS SYSTEM

The 19th century was characterized by a rapid progression of scientific thought, theories, and precise observations. Additional detailed descriptions of the anatomic structure of the brainstem have continued to be published over the years, such as Hubert von Luschka's (1820–1875) observations. Luschka was a German anatomist who described several structures, including the foramina of Luschka, Luschka's crypts, Luschka's joints, and Ducts of Luschka. In 1855, he described three paired and three unpaired enlargements of the subarachnoid space, naming them as the subarachnoid sinuses, the same name used later by Ludwig Bruns (1858–1916) (Tubbs Vahedi, et al., 2011).

In 1858, the British physician Henry Gray, FRS (1827–1861), demonstrator of Anatomy at St. George's Hospital, and Surgeon to the St. George's and St. James's Dispensary, London, described in his *Anatomy, Descriptive and Surgical*, the pons, medulla oblongata, and several cranial nerve nuclei (Gray, 1858). The book, which became one of the most widely used resources for medical education, consisted of 750 pages and 363 images, created by Gray's colleague Henry Vandyke Carter (1831–1897). Gray died of smallpox three years later, at the age of 34 years (Coşkun et al., 2022). Gray's *Anatomy* was a comprehensive and detailed work, a culmination of hundreds of years of anatomic and surgical studies, starting with Vesalius's *De Humani Corporis Fabrica* (Ghosh and Kumar, 2019).

The Swiss-born Adolf Meyer (1866–1950), one of the most influential psychiatrists in the first half of the 20th century, who was appointed the first professor of psychiatry at Johns-Hopkins University, wrote that "higher and more complicated [frontal lobe] integrations or sublimation of the instinctive processes originating in the brainstem" (Garrison, 1929). Ludwig Edinger (1855–1918), a German neuropsychiatrist and anatomist, cofounder of the University of Frankfurt, was the first to describe the ventral and dorsal spinocerebellar tracts and to distinguish the paleocerebellum from the neocerebellum. "Edinger's tract" is the conductive pathway for basic perceptions of pain, temperature, pressure, and touch. The Edinger–Westphal nucleus or the accessory nucleus of the third oculomotor nerve is named after Edinger and the German neuropsychiatrist Karl Friedrich Otto Westphal (1833–1890). The structure and function of the ciliary ganglion have been studied and investigated since the time of Fallopius in the 16th century, who described a postganglionic plexiform junction between the third and the fifth nerves in the orbit, and through subsequent discoveries such as those made by Willis (1664) and others in the 18th and 19th centuries. The discovery of the origin of the third nerve parasympathetic outflow is attributed to the works of Edinger (1885) and Westphal (1887). This nucleus, which is one of the two nuclei of the oculomotor nerve, is located in the midbrain. It contributes the autonomic parasympathetic (i.e., visceral) component to the oculomotor nerve, providing innervation to the iris sphincter muscle and ciliary muscle to mediate the pupillary light reflex and accommodation, respectively. It is composed of parasympathetic preganglionic cell bodies that synapse in the ciliary ganglion. Studies in the last decades showed its important role in modulating metabolic, heart rate, and blood pressure responses to stress (Kozicz et al., 2011; Cano et al., 2021).

The German ophthalmologist Richard Perlia (1860–1931) found the convergence nucleus of Perlia and Spitzka, located between the somatic cell columns of the oculomotor. Edward Anthony Spitzka (1876–1922) was an American anatomist-pathologist who added to our knowledge of the midbrain centers. The Westphal–Pilcz syndrome or phenomenon, a neurotonic pupillary reaction consisting of constriction of the pupil in the case of voluntary contraction of the orbicularis oculi muscle, is named after Westphal's son, Alexander Karl Otto Westphal (1863–1941) and the neurologist Jan Pilcz (1870–1931). Pilcz was born in Russian-occupied Poland, graduated medicine in Zurich, and moved to Paris where he practiced under Jules Joseph Dejerine (1849–1917) and Joseph Babinski (1857–1932) at the L'hôpital Salpêtrière (Haynes, 1957; Magowska, 2015).

Otto Deiters (1834–1863) was a German physician and neuroscientist, who died aged 29 of typhus. Most of his scientific work was published posthumously by Max Schultze in the book *Untersuchungen über Gehirn und Rückenmark des Menschen und der Säugethiere* (Studies of Brain and Spinal Cord of Humans and Mammals, Braunschweig, 1865). Deiters' work included notes and drawings of dissected nerve cells, showing dendrites and axons, as well as drawings of microscopic sections of the spinal cord and the brainstem, including the lateral vestibular nucleus, named after him (Deiters' nucleus). Like others in the 19th century, Deiters' descriptions were consistent with the reticular theory, claiming that nerve cells were connected as a network, resulting in the collective action of the cerebrospinal axis. This theory was supported by the Italian Camillo Golgi's (1843–1926) finding, using his black reaction silver staining technique (1873), demonstrating neurons and their extensions. Toward the end of 19th century, the Spanish Santiago Ramon y Cajal (1852–1934), who used Golgi's method, suggested an alternative theory, stating that nerve cells develop and function as independent units, hence the neuron or neural theory. Both Golgi and Cajal won the Nobel Prize in 1906 for their work on the structure of the nervous system (Cimino, 1999; De Carlos et al., 2007; Deiters and Guillery, 2013; Voogd, 2016).

Another key figure of the late 19th and early 20th centuries was the Russian-Swiss neuroscientist Constantin von Monakow (1853–1930). The idea that damage to one part of the nervous system can have an effect at a distance, was not new to the researchers of the 19th century. It was described in Galen's treatise *De Locis Affectis*, written in the 2nd century, and mentioned further on in various works. The secondary brain damage was commonly explained by the "shock" effect due to edema, pressure, and decreased blood flow to the surviving brain tissue. Monakow blended these observations with the newly formulated neuron theory of the early 20th century, trying to explain ipsilateral paralyses and recovery of function. The connectivity of the sensory and motor pathways was explored in Monakow's works, in which he coined the term "diaschisis" ("splitting off," from Greek schizen "to split"). The concept of diaschisis was based on the idea that neural depression is caused by loss of inputs to structures connected to damaged territories in the brain, not only by local physical effects. He described how lesions in the motor cortex interrupt neural transmission to areas in the same hemisphere, to contralateral cortex (through commissural fibers), and downward to brainstem nuclei and spinal cord. Other observations revealed how lesions in the midbrain affected fibers between pontine neurons and the pyramidal tract. This was an important contribution to the effort made to understand brain dynamics and elasticity (Finger et al., 2004; Wiesendanger, 2006). Monakow's Brain Institute in Zurich attracted scholars from Europe, the United States, and Japan. One of them was Gennosuke Fuse (1880–1946), a Japanese anatomist, who worked with Monakow in his laboratory at the University of Zurich. Fuse, known for his famous atlas of the lower brainstem, is also memorized in the "Kölliker-Fuse nucleus," a nucleus in the rostral dorsolateral pons, part of the pontine respiratory nuclei that modulates respiratory pattern, especially the transition from inspiration to expiration. Rudolph Albert von Kölliker (1817–1905) was a Swiss anatomist, physiologist, and histologist, a prominent scientist of the 19th century, who coined the term Axon in 1896. He is also considered as one founders of microscopic anatomy (Jenkin et al., 2017; Bentivoglio et al., 2019).

Barrington nucleus (the pontine micturition center) was discovered by the English urologist and physiologist Frederick James Fitzmaurice Barrington (1884–1956). In 1925, when Barrington published this celebrated paper, his academic position was as a surgeon for genito-urinary diseases at University College Hospital (London). Barrington's experiments were made possible because of the development of the Horsley–Clarke stereotaxic instrument (in the first decade of the 20th century), which was developed by Victor Alexander Haden Horsley (1857–1916), professor of surgery, together with the physiologist Robert Henry Clarke (1850–1926) (Pereira et al., 2008). Barrington borrowed the first of these instruments for these experiments. Barrington collated the individual case histories of lesioned animals; he discovered a small region of the upper brainstem, within the lateral pontine tegmentum, controlling the lower urinary tract in cats. It is not clear how long the total series of experiments took, but since the study of individual cases sometimes took many months, the duration of the whole project must have lasted several years. Barrington concluded that various lesions in the brainstem were followed by inability to empty the bladder, or by loss of consciousness of wanting to micturate or defecate, without impairment of the performance of these functions, correlating with the location of the injury. Further studies of this nucleus, located between the LC and laterodorsal tegmental nucleus, showed it contains a small number of neurons that increase their activity in response to bladder distention. Discoveries of these sensory and motor neural pathways helped to shed light on the mechanism of spinal injuries and their effect on bladder emptying (Morrison, 2008; Verstegen, 2017). Other examples of structures revealed in the 19th century were the nucleus of Roller, named after Christian Friedrich Wilhelm Roller (1802–1878), a German neuropsychiatrist, who described in 1844 the lateral nucleus of the accessory nerve (Meling et al., 2020), and the nucleus of Darkschewitsch, named after Livery Osipovich Darkschewitsch (sometimes spelled Darkshevich), a Russian neurologist and scientist, (1858–1925), who in 1889 described a lateral interstitial midbrain. This nucleus is one of the accessory oculomotor (preoculomotor) nuclei, which are composed of interneurons concerned with eye movements and reflex gaze. Darkschewitsch was one of the leading scientists who placed the foundation for Russian neurology in the 19th century along with Korsakov, Kozhevnikov and Bekhterev (Bogdanov et al., 2009).

It is impossible to discuss the development of neurology in the 19th century without mentioning the significant contribution of French neurologist, pathologist, and artist Jean-Martin Charcot (1825–1893). Charcot who is considered by many as one of the prominent founders of modern neurology, spent most of his professional career at Salpêtrière Hospital in Paris, a hospital founded in the 17th century as an asylum for the poor and the insane, developing into a hospital at the time of the French revolution. In 1882, he was appointed to a new chair, created for him, Professor of diseases of the nervous system, the first neurologic professorship in Europe. Charcot's studies were focused on general medicine, diseases of the brain, spinal cord, and peripheral nervous system, as well as hysteria and hypnotism. His "anatomo-clinical" method was aimed to determine the correlation between symptoms in life and anatomic lesions in autopsies. Charcot and his students (some of them will be mentioned later in this chapter) were able to confirm that lesions in the brainstem were associated with weakness of the muscles controlling the face, mouth, and tongue. He also concluded that gray matter lesions cause weakness and muscle atrophy, whereas weakness caused by white matter lesions correlates with contractures and spasticity. In 1868, Charcot delivered three lectures describing for the first time the new entity of multiple sclerosis, including the histologic hallmarks of plaques, demyelination, and axonal loss, distinguishing it from paralysis agitans (Parkinson disease). In 1874, he gave two lectures synthesizing his discoveries and naming a new diagnosis of "amyotrophic lateral sclerosis" (Goetz, 2000; Lubetzki, 2018). As mentioned earlier, Charcot had a special interest in investigating hysteria, paving the way for his famous students Alfred Binet (1859–1911), Pierre Janet (1859–1947), and Sigmund Freud (1856–1939).

BRAINSTEM SYNDROMES

Toward the end of the 18th century and the beginning of the 19th century, the concept of the brain being divided into centers responsible for various functions began to solidify. The brainstem was discovered as a significant organ for regulating vital functions. Anatomy research also advanced, thanks to the development of new methods for tissue dissection, slicing using a microtome, and examination of frozen sections. The ability to comprehend microanatomy greatly contributed to the advancement of neuroscience. However, the pathologic descriptions alone were not sufficient to shed light on the connection between the symptoms and the damage to the nervous tissue. This section will review some of the known clinical syndromes originating from damage (i.e., stroke or other lesions) to different regions in the brainstem, according to the accepted classification, based on the anatomic division of the brainstem into the medulla oblongata, pons, and the midbrain. Although the literature elaborates on the notable and familiar syndromes (Marx and Thömke, 2009; Hurley, 2010), it is important to note that some of the patients present with unnamed combinations of crossed symptoms (Beucler et al., 2021). The neurologists and scientists whose names are enshrined in the names of syndromes were pivotal figures woven into the story of modern neurology. Many of them made significant contributions to the understanding of the structure and physiology of the nervous system, laying the foundation for modern knowledge. The low incidence of these syndromes today may suggest the possibility that some of the cases presented symptoms of neurosyphilis. Syphilis, known as "the Great Imitator," a common infection in the pre-antibiotic era, has a wide range of neurologic manifestation, from headaches to strokes, dementia, and psychiatric symptoms, caused by inflammatory processes in small to large blood vessels (Abkur et al., 2015; Ropper 2019).

MIDBRAIN

Weber syndrome (peduncular syndrome) – In 1863 the German-born physician Sir Hermann Weber (1823–1918) published a case of a patient with left oculomotor nerve palsy and right-sided crossed hemiplegia, caused by an acute hemorrhage in the left cerebral peduncle. Weber who was introduced to England by Queen Victoria and later on naturalized, was not the first one to describe third cranial nerve paralysis and contralateral hemiplegia, but in his article, he was the first to accurately describe the anatomic, clinical, and pathologic characteristics of a peduncular lesion. Although his main contributions were on tuberculosis and climatology, Weber will be remembered in the history of neurology due to the syndrome named after him by Charcot in 1891. The syndrome is commonly caused by an infarct of the paramedian mesencephalic or the peduncular perforating arteries, branches of the posterior cerebral artery, affecting the oculomotor nucleus and cerebral peduncle, but can also be caused by an aneurysm, tumor, or infections (Brown; Hassin, 1917).

Benedikt syndrome – Also known as paramedian midbrain syndrome, it is a condition caused most frequently by vascular injury of the tegmentum of midbrain, involving the fascicles of the oculomotor nerve and red nucleus. Also referred to in the past as modified Weber syndrome, consists of oculomotor palsy, contralateral ataxia, tremor, and chorea. This constellation of symptoms was first described in a lecture, delivered in 1889 by the Austro-Hungarian physician Moritz Benedikt (1835–1920), a professor of neurology at Vienna University, who served as a surgeon in the Austrian army during the Second Italian War of Independence (1859) and the Austro-Prussian war (1866). Benedikt described three similar cases of paralysis, hemiplegia, and tremors. In one of them, an autopsy revealed scattered tuberculomas in the brain and cerebellum. In 1893, Charcot was the first to coin the term "Benedikt's syndrome" when describing a similar patient (Triarhou, 2022).

Claude syndrome – Henri Charles Jules Claude (1869–1945) was a French neuropsychiatrist, a prominent figure in the development of psychoanalysis in France at the beginning of the 20th century. In 1912, Claude described a patient with an oculomotor lesion including a fixed and dilated pupil in addition to contralateral ataxia. An autopsy revealed an infarction of the right paramedian midbrain, involving the red nucleus, oculomotor fascicle, and the superior cerebellar peduncle. Considered a rare syndrome, usually caused by vascular occlusion, it was also associated with other etiologies such as malignancies, trauma, and neurosarcoidosis (Sciacca et al., 2019).

Nothnagel syndrome – This syndrome is caused mainly by a tumor in the tectum of the midbrain, affecting the superior cerebellar peduncle and unilateral or bilateral oculomotor, resulting in ocular palsies, paralysis of gaze, nystagmus, and ipsilateral cerebellar ataxia. Between 1879 and 1889, Hermann Nothnagel (1841–1905) reported four cases of tectal tumors, resulting clinically in ophthalmoplegia and unilateral gait ataxia. Nothnagel, an internist and neurologist who was born in Alt-Lietzegöricke (today Stare Łysogórki in Poland), ultimately became one of the prominent clinicians and researchers in Vienna at the end of the 19th century. He had a special interest in the physiology and pathology of the cerebellum, and conducted experiments using animal models, exploring various areas including the cerebral cortex, basal ganglia, postinfectious and alcohol neuritis, and more (Triarhou and Manto, 2023).

Parinaud syndrome – Henri Parinaud (1844–1905) was a French ophthalmologist, appointed by Charcot as an ophthalmologist in the Neurological Service, at the Salpêtrière Hospital in Paris. In 1883, Parinaud published a series of 10 cases, depicting disorders of binocular movement, such as gaze and convergence paralysis, attributed to lesions of the central nervous system rather than nuclei or peripheral nerves. Today's eponym "Parinaud syndrome" used since the 1930s, refers to a classical triad of upward gaze palsy, pupillary light-near dissociation, and convergence-retraction nystagmus, caused by pinealoma and other lesions of the dorsal midbrain. Parinaud was a humanist, who maintained a free clinic in Paris and composed music under a pseudonym (Ortiz et al., 2021).

PONS

Millard–Gubler syndrome – This is one of the classical crossed syndromes, caused by an infarct or a tumor in the caudal pons, involving the corticospinal tract, the abducens, and facial nerves. It is characterized by facial and abducens palsy, contralateral hemiplegia, and occasionally gaze palsy to the side of the lesion. The eponym was coined after the French physicians, Auguste Louis Jules Millard (1830–1915) and Adolphe-Marie Gubler (1821–1879) who almost simultaneously published a similar description of the syndrome in 1855 and 1856, respectively. The historical significance of these two leading physicians and the hospitals where they worked in Paris was eventually overshadowed by their contemporary, Charcot (Silverman et al., 1995).

Foville Syndrome – Achille Louis Francois Foville (1831–1887) was a French psychiatrist and neuroanatomist, who in 1858 was the first to describe a rare inferior pontine syndrome. In his paper "Notes on a Little-known Paralysis of Eye Muscles, and Its Relation to the Anatomy and Physiology of the Pons," described a patient with unilateral conjugate horizontal gaze paralysis, ipsilateral facial nerve paralysis, and crossed hemiparesis. The syndrome is due to a stroke of the paramedian caudal pons, involving cranial nerve nuclei, abducens, and facial. It is interesting to mention Foville's father, Achille Louis Foville (1792–1870) who published in 1844 a comprehensive, artistic, and detailed atlas of brain anatomy (Brogna et al., 2012; Mathis et al., 2020).

Raymond–Céstan syndrome – This pontine syndrome was reported by the French neurologists Fulgence Raymond (1844–1910) and Étienne Jacques Marie Raymond Céstan (1872–1934). Raymond succeeded Charcot as head of the Department of Neurology at the Salpêtrière Hospital, and in 1899 appointed Céstan, a student of Joseph Babinski, as the hospital's clinic director, both collaborating in the following years. Between 1901 and 1903, they described three patients with ipsilateral internuclear ophthalmoplegia, contralateral hemiparesis, and hemihypesthesia, as well as contralateral tremor, athetosis of fingers, ataxia, and gait disturbance characteristics of cerebellar involvement. The lesions causing the syndrome in these cases were tuberculomas of the pontine tegmentum and base, found in autopsy. Raymond was attributed with a syndrome named after him (Raymond syndrome), following his grand rounds description in 1895 of a woman who had abducens impairment and contralateral hemiparesis resulting from syphilis (Krasnianski, 2007; Marx and Thömke, 2009). Classic Raymond syndrome consists of ipsilateral abducens impairment, contralateral central facial paresis, and contralateral hemiparesis.

Locked-in syndrome – It is one of the recent pontine syndromes that have been described in 1966 by the American neurologists Fred Plum (1924–2010) and Jerome B. Posner (b. 1932), most commonly caused by infarction or hemorrhage in the pontine tegmentum and base. The classic form is characterized by quadriplegia, bulbar palsy, and total body sensory loss. Patients remain conscious, and cognition, hearing, blinking, and vertical eye movement are preserved. Despite Plum and Posner establishing the concept in the 20th century, it is worth mentioning that the earliest descriptions of the syndrome appeared already in the 19th century. Among them, in Alexander Dumas's (1802–1870) book *The Count of Monte Cristo* (1844), is Noirtier De Villefort, a completely paralyzed figure, capable of blinking only. Later, Emile Zola (1840–1902) described a woman who was paralyzed and "buried alive in a dead body" but could communicate via eye movements in his book *Thérèse Raquin* (1868). Additionally, a case report published in 1875 by Camille Darolle (1847–1915), a Parisian physician. He described a female patient suffering from rather acute headache and quadriplegia, being able to communicate solely by moving her eyes. After her passing a few days later, an autopsy revealed cerebral hemorrhage but also an occlusion of the basilar artery, resulting in pontine injury (Teive et al., 2022).

Lateral pontine syndrome (Marie–Foix syndrome) – This syndrome affects the nuclei of CN (Cranial Nerve) VII, and VIII, corticospinal, spinothalamic, and cerebellar tracts. It results in contralateral hemiparesis, contralateral loss of proprioception and vibration, ipsilateral limb ataxia, ipsilateral facial palsy, lateral hearing loss, vertigo, and nystagmus. The blood supply affected is the perforating branches of the basilar artery and the anterior inferior cerebellar artery (Balami et al., 2013). The syndrome was named after the French neurologist Charles Foix (1882–1927), a student of Pierre Marie (1853–1940) at the Hôpital de la Salpêtrière in Paris. Foix, who adopted Marie's scientific methods, wrote on various neurologic abnormalities and syndromes, but his best-renowned contributions are his studies on cerebrovascular diseases, detailing clinical presentation, anatomy, and pathology of blood vessels and brain lesions (Caplan, 1990).

Gasperini syndrome – Affects the nuclei of CN V, VI, VII, VIII, and the spinothalamic tract. It results in ipsilateral facial sensory loss, ipsilateral impaired eye abduction, ipsilateral facial palsy, ipsilateral nystagmus, vertigo, and contralateral hemisensory impairment. The blood supply involved derives from the pontine branches of the basilar artery and long circumferential artery of the anterior inferior cerebellar artery. It was first described in 1912 by the Italian Ubaldo Gasperini (1880–1918), who was an assistant professor of pathology at the University of Bologna. He reported a case of 49-year-old woman with a history of alcohol and nicotine abuse who was admitted to the hospital in a coma and presented a set of symptoms when she regained consciousness. Gasperini concluded that the lesion was situated in the right caudal posterior and medial pontine tegmentum, and named it "inferior and posterior pontine syndrome." Today, the syndrome is considered to be a variant of Foville's or Millard–Gubler's syndrome. Gasperini died in 1918 from Spanish flu, while serving as head of the Department of Infectious Diseases at the military hospital in Verona (Tacik, 2012).

Brissaud–Sicard syndrome is a very rare pontine stroke syndrome that involves the anterolateral and inferior pons. It affects the CN VII nucleus and corticospinal tract. It results in ipsilateral facial cramps and contralateral upper and lower limb hemiparesis. The blood supply affected is the posterior circulation. Rarely, the syndrome can arise due to brainstem glioma. Edouard Brissaud (1852–1909) was one of Charcot's favorite students. After Charcot died unexpectedly from myocardial infarction during a trip with some of his students and assistants, Brissaud was temporarily appointed as his successor at the Salpêtrière. He eventually was appointed as the chair of medical history and medical pathology. One of his important contributions was the attribution of Parkinson disease pathology to the substantia nigra, suggested in a lecture he gave at the Salpêtrière in 1894. This assumption was confirmed years later by the Russian neuropathologist Konstantin Tretiakoff (1892–1958) (Marques et al., 2018) who in 1919 published his doctoral dissertation *Contribution a l'Etude de L'Anatomie pathologique du Locus Niger de Soemmering avec quelques déductions relatives à la pathogénie des troubles du tonus musculaire et De La Maladie de Parkinson* (A Study of the Pathological Anatomy of the Locus Niger of Soemmering and Its Relevance to the Pathogenesis of Changes in Muscular Tone in Parkinson Disease), earning him a silver medal for this work by the University of Paris (Lees et al., 2008).

MEDULLA OBLONGATA

Wallenberg syndrome – Also known as lateral medullary syndrome, this syndrome was comprehensively described for the first time by Adolf Wallenberg (1862–1949), born in Danzig, Germany (now Gdansk, Poland). In 1895, Wallenberg published a paper on an "acute bulbar lesion" and later on in 1901, and described further the symptoms of the same patient suffering from a cluster of symptoms such as slow pulse, anesthesia of face, mouth, and half of the body, dizziness, and more. Wallenberg, who was a professor of internal medicine and became a self-taught renowned neurologist, concluded by localization of the neuropathologic signs that the lesion was in the lateral medulla. The syndrome is mostly caused by an infarction of the posterior inferior cerebellar artery, affecting the inferior cerebellar peduncle, vestibular nucleus, spinal trigeminal nucleus, and nucleus ambiguus. Prominent clinical features are ipsilateral trigeminal, glossopharyngeal, vagal, and accessory palsy; Horner syndrome; cerebellar ataxia; and contralateral loss of pain and temperature sense. In young persons, it is usually a dissection, and in elderly an occlusion of the vertebral artery leading to the occlusion of the last branch before the unifying to the basilar. Interestingly, Wallenberg himself suffered a basilar skull fracture in 1891, leaving him with chronic anosmia and possibly a behavior change, although he was portrayed as a modest, warm, and helpful person. Starting from the rise of the Nazis to power in 1933 and further intensified after the enactment of the Nuremberg Laws in 1935, the Jewish population of Danzig faced increasing violent attacks and discrimination. Wallenberg, grandson of a rabbi, was forced to leave his apartment and eventually was prohibited from conducting research and publishing his works. In August 1939, just days before the outbreak of World War II, he fled with his family to England. While in Oxford he was sponsored by and worked with the eminent neuroanatomist Wilfrid Le Gros Clark (1895–1971). In 1943, he was permitted to enter the United States and began to work at a state mental hospital near Chicago, where he became an honorary member of the Chicago Neurologic Society. Wallenberg died from coronary heart disease in 1948, aged 86 (Koehler et al., 2000; Zeidman and Mohan, 2014).

Babinski–Nageotte syndrome, also known as the hemimedullary syndrome, or "Wallenberg with hemiparesis," was named after Joseph Babinski (1857–1932) and Jean Nageotte (1866–1948) who described in 1902 a case of infarction in the hemimedulla. The syndrome, most frequently caused by intracranial portion of the vertebral artery and branches from the posterior inferior cerebellar artery, consists of a combination of medial and lateral medullary symptoms, with ipsilateral cerebellar ataxia, sensory deficits of the face, Horner syndrome, and contralateral hemiplegia and hemianesthesia. Babinski was one of Charcot's most prominent and famous students, along with Pierre Marie and George Gilles de la Tourette (1857–1904). Born in Poland, Babinski grew up in Paris, studied and worked under Charcot in La Salpêtrière. Best known for the eponym "Babinski sign," he recognized the importance of the clinical sign of extension of the great toe in lesions of the pyramidal tract, presumably a half-century after Félix Alfred Vulpian (1826–1887), neuropathologist at the Salpêtrière, observed it (Youl, 1989).

The French neuroanatomist Jean Nageotte, who also worked at the Salpêtrière in 1912, was a keen researcher of the anatomy of the nervous system as well as active clinician. He investigated nerve fibers, focusing on microscopic anatomy and the structure of the myelin sheath (Mathis et al., 2020). Eventually, Nageotte became the chair of comparative histology at the Collège de France.

Avellis syndrome – Georg Avellis (1864–1916) was a German laryngologist, who published in 1891 a series of 10 cases describing a clinical syndrome, later named after him, manifested with paralysis of soft palate and vocal cord and contralateral hemianesthesia and hemiparesis. This rare condition occurs usually because of occlusion of vertebral artery and infarction of tegmentum of the medulla. Other etiologies were associated with Avellis syndrome, such as trauma, infection, and some immune-inflammatory diseases.

Although named after Avellis, the first description of the set of symptoms consisting of this syndrome was published earlier by the Polish neurologist and anatomist Ludwik Maurycy Hirschfeld (1814–1876), who was one of the pioneering surgeons who diagnosed and localized brain lesions based on neurologic symptoms. As an anatomist, interested in neuroanatomy, he invented the "medullotom," a tool designated for preparing the medulla. Hirschfeld started his career as an apprentice in an anatomy studio, attending lectures in anatomy, and later on, at the age of 34, graduated from Paris Medical Faculty. In 1853, Hirschfeld published an atlas of the nervous system entitled *Neurology or Description and Iconography of the Nervous System and Human Sense Organs with Their Mode of Preparation*, a work that brought him fame, recognition from his peers, and an award from the French Academy of Sciences. Returning to Poland, Hirschfeld was appointed in 1859 as head of the Department of Anatomy at the Caesarian-Royal Medical-Surgical Academy in Warsaw and became a reformer of anatomy studies in Poland as well as a renowned clinician (Gonkowski et al., 2021).

Jackson syndrome – Similar to Avellis syndrome, this syndrome consists of ipsilateral paralysis of soft palate, pharynx, larynx, and ipsilateral paralysis and hemiatrophy of the tongue contralateral hemianesthesia, caused by infarction or tumor of the medulla oblongata. It was first described by the English neurologist John Hughlings Jackson (1835–1911), who published in 1884 a report "On a case of paralysis of the tongue from haemorrhage in the medulla oblongata," and in 1886 a case of "Paralysis of tongue, palate, and vocal cord." Jackson is most known for Jacksonian epilepsy and the Jacksonian March. Jackson's observation contributed to the theory of cerebral localization of functions, and to the idea that the brain is organized somatotopically. His works, among others, led the way to the iconic "homunculus," a map of cerebral organization published in 1937 by the American-Canadian neurosurgeon Wilder Penfield (1891–1976) and his research fellow and colleague the American Edwin Boldrey (1906–1988) (Berkowitz, 2018; Leblanc and Preul, 2020).

Dejerine syndrome – Also known as medial medullary syndrome, it is caused by an infarct of the medial medulla affecting the hypoglossal nerve nucleus. The syndrome consists of contralateral weakness and hemisensory deficit with ipsilateral hypoglossal palsy. It was described by Jules Joseph Dejerine (1849–1917), a Swiss-born French neurologist and pathologist who worked under Vulpian and eventually succeeded Charcot and Raymond as chair of neurology at La Salpêtrière. In 1879, Dejerine was the first to report 10 cases of acute inflammatory polyradiculoneuropathy, resembling acute myelitis without medullary lesions in autopsies. This condition was described more accurately in 1916 by Georges Charles Guillain (1876–1961), Jean Alexandre Barré (1880–1967), and André Strohl (1887–1977), referred later to as Guillain–Barré syndrome (Mathis and Vallat, 2017).

Other notable medullary syndromes were described and named after French physicians, such as *Vernet syndrome* (or jugular foramen syndrome) that involves the IX, X, and XI cranial nerves and is most often attributable to malignancy (paraganglioma), inflammation, aneurysm, Giant cell arteritis, thrombosis, or skull base fracture, first described by Maurice Vernet (1887–1974). *Collet–Sicard syndrome*, a dysfunction of cranial nerves IX, X, XI, and XII, with sparing of the sympathetic plexus, was independently described by Frédéric Justin Collet (1870–1966), a French pathologist and otolaryngologist and Jean-Athanase Sicard (1872–1929), a French neurologist and radiologist. *Villaret's syndrome*, a dysfunction of cranial nerves IX, X, XI, and XII, with an ipsilateral Horner syndrome, was first reported in 1917 by the neurologist Maurice Villaret (1877–1946), describing injured soldiers. An interesting syndrome named after Antonio Garcia Tapia (1875–1950), a Spanish otolaryngologist, who described it in 1904, is a rare complication of airway manipulation and head positioning during surgical anesthesia.

Tapia syndrome, a dysfunction of X and XII nerves, includes the extracranial involvement of the recurrent laryngeal nerve and the hypoglossal nerve and results in ipsilateral paralysis of the vocal cord and the tongue (Gutierrez, 2021).

SOME FINAL THOUGHTS

The journey of uncovering the anatomy, function, and pathology of the brainstem has spanned over many years, parallel to the development of science as a whole and specifically life and medical sciences. New discoveries, inventions, and evolving religious, social, cultural, and political changes have facilitated the deepening of research, scientific collaboration, and the expansion of knowledge. It is also important to note that the experiments conducted in previous centuries, did not always adhere to the ethical principles and norms accepted in our time, regarding experiments on animals and humans. However, they constitute a significant part of the creation of the extensive knowledge base that formed the basis for the development of neuroscience. This chapter attempts to briefly outline the main milestones and prominent figures that led to the recognition of the brainstem as an organ of significance in its own right. Due to the brevity of this section, it is impossible to describe the complete history, including that of the 20th century and beyond, with the development of posterior fossa and brainstem surgery starting from the late 19th century, through advances in medical imaging (CT and MRI etc.), and up to physiologic discoveries based on innovative and biologic methods; some of these will be detailed further in this book.

ACKNOWLEDGMENT

The authors are grateful to P.J. Koehler, MD, PhD, FAAN, Neurologist, Faculty of Health, Medicine and Life Sciences, Maastricht University, The Netherlands, for his wise and thoughtful remarks.

References

Abkur TM, Ahmed GS, Alfaki NO et al. (2015). Neurosyphilis presenting with a stroke-like syndrome. BMJ Case Rep 2015: bcr2014206988.

Balami JS, Chen RL, Buchan AM (2013). Stroke syndromes and clinical management. QJM: Monthly Journal of the Association of Physicians 106: 607–615.

Bast TH (1924). The life and work of Samuel Thomas von Sömmerring. Ann Med History 6: 369–386.

Bentivoglio M, Cotrufo T, Ferrari S et al. (2019). The original histological slides of Camillo Golgi and his discoveries on neuronal structure. Front Neuroanat 13: 3.

Berkowitz A (2018). You can observe a lot by watching: Hughlings Jackson's underappreciated and prescient ideas about brain control of movement. Neuroscientist 24: 448–455.

Beucler N, Boissonneau S, Ruf A (2021). Crossed brainstem syndrome revealing bleeding brainstem cavernous malformation: an illustrative case. BMC Neurol 21: 204.

Bogdanov EI, Mukhamedzyanov RZ, Sozinov AS et al. (2009). L. O. Darkshevich (1858–1925) (150th anniversary). J Neurol 256: 1028–1029.

Brogna C, Fiengo L, Türe U (2012). Achille Louis Foville's atlas of brain anatomy and the Defoville syndrome. Neurosurgery 70: 1265–1273.

Brown, G.H. Sir Hermann Weber. Royal College of Physicians. Available at: https://history.rcplondon.ac.uk/inspiring-physicians/sir-hermann-weber [Lancet, 1918; B.M.J., 1918; Times, 14 Nov. 1918].

Cano G, Hernan SL, Sved AF (2021). Centrally projecting Edinger-Westphal nucleus in the control of sympathetic outflow and energy homeostasis. Brain Sci 11: 1005.

Caplan LR (1990). Charles Foix – the first modern stroke neurologist. Stroke 21: 348–356.

Chandler DJ, Jensen P, McCall JG et al. (2019). Redefining noradrenergic neuromodulation of behavior: impacts of a modular locus coeruleus architecture. J Neurosci 39: 8239–8249.

Cimino G (1999). Reticular theory versus neuron theory in the work of Camillo Golgi. Physis Riv Int Stor Sci 36: 431–472.

Coşkun Ö, Chatzioglou GK, Öztürk A (2022). Henry Gray (1827-1861): the great author of the most widely used resource in medical education. Child's Nervous System 38: 1421–1423.

Crivellato E, Ribatti D (2006). Mondino de' Liuzzi and his Anothomia: a milestone in the development of modern anatomy. Clinical Anatomy 19: 581–587.

Cucu AI, Turliuc S, Costea CF et al. (2021). The brainstem and its neurosurgical history. Neurosurgical review 44: 3001–3022.

Davis MC, Griessenauer CJ, Bosmia AN et al. (2014). The naming of the cranial nerves: a historical review. Clinical Anatomy 27: 14–19.

De Carlos, Borrell J (2007). A historical reflection of the contributions of Cajal and Golgi to the foundations of neuroscience. Brain Research Reviews 55: 8–16.

Deiters VS, Guillery RW (2013). Otto Friedrich Karl Deiters (1834-1863). The Journal of Comparative Neurology 521: 1929–1953.

Eling P, Finger S (2019). Franz Joseph Gall on the cerebellum as the organ for the reproductive drive. Frontiers in Neuroanatomy 13: 40.

Finger S (2005). Chapter 5, Andreas Vesalius: The New "Human" Neuroanatomy'. Minds Behind the Brain: A history of the pioneers and their discoveries. Oxford University Press, pp. 53–68.

Finger S, Koehler PJ, Jagella C (2004). The Monakow concept of diaschisis: origins and perspectives. Archives of Neurology 61: 283–288.

Gall FJ, Spurzheim G (1810). Anatomie et physiologie du système nerveux en général et du cerveau en particulier (4 vol.). Paris: F. Schoell.

Garrison FH (1929). An introduction to the history of medicine, 4th ed. Philadelphia and London: Saunders.

Ghosh SK, Kumar A (2019). The rich heritage of anatomical texts during Renaissance and thereafter: a lead up to Henry Gray's masterpiece. Anatomy & cell biology 52: 357–368.

Goetz, C.G. (2000) Amyotrophic lateral sclerosis: early contributions of Jean-Martin Charcot. Muscle & nerve, 23: 336–343.

Gonkowski I, Gonkowski S, Makowska K (2021). Ludwik Maurycy Hirschfeld (1814-1876). Journal of Neurology 268: 2304–2306.

Gray Henry (1858). Anatomy: Descriptive and Surgical. London, England: John W. Parker and Son, West Strand.

Gutierrez S, Warner T, McCormack E (2021). Lower cranial nerve syndromes: a review. Neurosurgical review 44: 1345–1355.

Hajdu SI (2003). A note from history: discovery of the cerebrospinal fluid. Annals of Clinical and Laboratory Science 33: 334–336.

Haldane JS (1929). Claude bernard's conception of the internal environment. Science 69: 453–454.

Hanigan WC, Ragen W, Foster R (1990). Dryander of Marburg and the first textbook of neuroanatomy. Neurosurgery 26: 489–498.

Hassin GB (1917). Weber's Syndrome: With report of a case. JAMA LXIX, 2169–2170.

Haynes PR (1957). Accommodative responses associated with the Westphal-Piltz pupillary phenomenon. Am J Optom Arch Am Acad Optom 34: 405–415.

Hildebrand R (2005). Soemmerring's work on the nervous system: a view on brain structure and function from the late eighteenth century. Anatomy and embryology 210: 337–342.

Hökfelt T (2016). Early attempts to visualize cortical monoamine nerve terminals. Brain Res 1645: 8–11.

Hurley RA, Flashman LA, Chow TW et al. (2010). The brainstem: anatomy, assessment, and clinical syndromes. J Neuropsychiatry Clin Neurosci 22:iv-7.

Jenkin SEM, Milsom WK, Zoccal DB (2017). The Kölliker-Fuse nucleus acts as a timekeeper for late-expiratory abdominal activity. Neuroscience 348: 63–72.

Koehler PJ, Bruyn GW, Pearce J (2000). Neurological eponyms. Oxford: Oxford University Press, pp. 386 Illustrated.

Koehler PJ (2020). Early ideas on the localization of vital functions. World Neurology. Available at: https://worldneurologyonline.com/article/early-ideas-on-the-localization-of-vital-functions/

Kozicz T, Bittencourt JC, May PJ et al. (2011). The Edinger-Westphal nucleus: a historical, structural, and functional perspective on a dichotomous terminology. J Comp Neurol 519: 1413–1434.

Krasnianski M, Wendt M, Müller T et al. (2007). Raymond–Cestan syndrome in pontine ischemia. Clinical Neurology and Neurosurgery 109: 834–835.

Kutzer M (1988). Hirnforschung in der Nachfolge Samuel Thomas Soemmerrings: das neuroanatomische Werk von Joseph (1768-1808) und Karl (1769-1827) Wenzel. Soemmerring Forsch 3: 331–355.

Leblanc R, Preul MC (2020). Perspective: Edwin Boldrey and Penfield's homunculus. World neurosurgery 134: 454–456.

Lees AJ, Selikhova M, Andrade LA et al. (2008). The black stuff and Konstantin Nikolaevich Tretiakoff. Mov Disord 23: 777–783.

Lubetzki C (2018). 150 years since Charcot's lectures on multiple sclerosis. Lancet Neurol 17: 1041.

Magowska A (2015). Jan Piltz (1870-1930). J Neurol 262: 1099–1100.

Marques PT, Germiniani FMB, Camargo CHF et al. (2018). Édouard Brissaud: distinguished neurologist and Charcot's pupil. Arq Neuropsiquiatr 76: 490–493.

Marx JJ, Thömke F (2009). Classical crossed brain stem syndromes: myth or reality?. J Neurol 256: 898–903.

Mathis S, Vallat J-M (2017). Jules Dejerine and the peripheral nervous system. Neurology 89: 611–615.

Mathis S, Vallat J-M, Weis J (2020). When botany inspired pathology of the peripheral nervous system. Neurology 95: 532–536.

Mavrodi A, Paraskevas G (2014). Mondino de Luzzi: a luminous figure in the darkness of the Middle Ages. Croat Med J 55: 50–53.

Meling TR, Nouri A, May A et al. (2020). Upbeat vertical nystagmus after brain stem cavernoma resection: a rare case of nucleus intercalatus/nucleus of roller injury. J Neurol 267: 2865–2870.

Mirza M, Das JM (2023). Neuroanatomy, area postrema. StatPearls. Treasure Island, FL: StatPearls Publishing.

Molnár Z (2021). On the 400th anniversary of the birth of Thomas Willis. Brain 144: 1033–1037.

Morrison JFB (2008). The discovery of the pontine micturition centre. Experimental Physiology 93: 742–745.

Ortiz JF, Eissa-Garces A, Ruxmohan S et al. (2021). Understanding Parinaud's syndrome. Brain Sci 11: 1469.

Parent A (2007). Felix Vicq d'Azyr: anatomy, medicine and revolution. Can J Neurol Sci 34: 30–37.

Parent A (2019). Berengario da Carpi and the renaissance of brain anatomy. Frontiers in Neuroanatomy 13: 11.

Pearce JMS (2002). Cheyne-Stokes respiration. Journal of Neurology, Neurosurgery, and Psychiatry 72: 595.

Pearce JMS (2013). The neuroanatomy of Herophilus. Eur Neurol 69: 292–295.

Pereira EAC, Green AL, Nandi D et al. (2008). Stereotactic neurosurgery in the United Kingdom: the hundred years from Horsley to Hariz. Neurosurgery 63: 594–606discussion 606.

Rawlings CE, Rossitch E (1994). Franz Josef Gall and his contribution to neuroanatomy with emphasis on the brain stem. Surg Neurol 42: 272–275.

Rezende-Cunha F, de Oliveira-Souza R (2011). The pyramidal syndrome and the pyramidal tract: a brief historical note. Arquivos de Neuro-Psiquiatria 69: 836–837.

Ropper AH (2019). Neurosyphilis. N Engl J Med 381: 1358–1363.

Saran J, Padubidri JR (2019). Brain stem death - an overview. The Medico-legal journal 87: 18–20.

Sarikcioglu L, Yildirim FB (2008). Area postrema: one of the terms described by Magnus Gustaf Retzius. J Hist Neurosci 17: 109–110.

Scatliff JH, Johnston S (2014). Andreas Vesalius and Thomas Willis: their anatomic brain illustrations and illustrators. American Journal of Neuroradiology 35: 19–22.

Schijns OEMG, Koehler PJ (2021). On the threshold of scientific medicine: Gerard van Swieten and his perception of the pathophysiology in traumatic brain injury. European Neurology 84: 393–398.

Sciacca S, Lynch J, Davagnanam I et al. (2019). Midbrain, pons, and medulla: anatomy and syndromes. Radiographics 39: 1110–1125.

Silverman IE, Liu GT, Volpe NJ et al. (1995). The crossed paralyses. The original brain-stem syndromes of Millard-Gubler, Foville, Weber, and Raymond-Cestan. Arch Neurol 52: 635–638.

Šimon F (2018). On the origin of the term decussatio pyramidum. J Hist Neurosci 27: 101–105.

Smith CUM (2010). Chapter 9: understanding the nervous system in the 18th century. Handbook of Clinical Neurology 95: 107–114.

Spurzheim JG (1815). The physiognomical system of Gall and Spurzheim; founded on an anatomical and physiological examination of the nervous system in general, and of the brain in particular; and indicating the dispositions and manifestations of the mind. London: Baldwind Cradock and Joy.

Storey CE (2022). Then there were 12: The illustrated cranial nerves from Vesalius to Soemmerring. Journal of the history of the neurosciences 31: 262–278.

Stoyanov GS, Matev BK (2018). Samuel Thomas von Sömmerring's contributions on the cranial nerves and vomeronasal organ. Cureus 10: e2859.

Tacik P, Alfieri A, Kornhuber M et al. (2012). Gasperini's syndrome: its neuroanatomical basis now and then. Journal of the history of the neurosciences 21: 17–30.

Taran S, Gros P, Gofton T et al. (2023). The reticular activating system: a narrative review of discovery, evolving understanding, and relevance to current formulations of brain death. *Can J Anaesth* 70: 788—795.

Teive HAG, Ferreira MG, Coutinho L et al. (2022). The locked-in syndrome: the early French descriptions. Rev Neurol 178: 996–998.

Triarhou LC (2022). The brain masters of Vienna: psychology and neuroscience pioneers around the secession up to the Anschluss. Cham: Springer International Publishing.

Triarhou LC, Manto M (2023). Nothnagel syndrome. Cerebellum 22: 487–505.

Tubbs RS, Loukas M, Shoja MM et al. (2011). Félix Vicq d'Azyr (1746-1794): early founder of neuroanatomy and royal French physician. Child's Nerv Syst 27: 1031–1034.

Tubbs RS, Vahedi P, Loukas M et al. (2011). Hubert von Luschka (1820-1875): his life, discoveries, and contributions to our understanding of the nervous system. J Neurosurg 114: 268–272.

Verstegen AMJ, Vanderhorst V, Gray PA et al. (2017). Barrington's nucleus: Neuroanatomic landscape of the mouse 'pontine micturition center. J Comp Neurol 525: 2287–2309.

Voogd J (2016). Deiters' nucleus. its role in cerebellar ideogenesis : the Ferdinando Rossi memorial lecture. Cerebellum 15: 54–66.

Voogd J, De Zeeuw CI (2020). Cerebellum: what is in a name? historical origins and first use of this anatomical term. Cerebellum 19: 550–561.

Voogd J, Koehler PJ (2018). Historic notes on anatomic, physiologic, and clinical research on the cerebellum. Handbook of Clinical Neurology 154: 3–26.

Whitaker H, Jarema G (2017). The split between Gall and Spurzheim (1813-1818). Journal of the history of the neurosciences 26: 216–223.

Wiesendanger M (2006). Constantin von Monakow (1853-1930): a pioneer in interdisciplinary brain research and a humanist. C R Biol 329: 406–418.

Wijdicks EFM (2019a). The ascending reticular activating system. Neurocrit Care 31: 419–422.

Wijdicks EFM (2019b). Noeud vital and the respiratory centers. Neurocritical Care 31: 211–215.

Wijdicks EFM (2020). Historical awareness of the brainstem: From a subsidiary structure to a vital center. Neurology 95: 484–488.

Yildirim FB, Sarikcioglu L (2007). Marie Jean Pierre Flourens (1794 1867): an extraordinary scientist of his time. Journal of Neurology, Neurosurgery, and Psychiatry 78: 852.

Youl BD (1989). A neurological double entente. Archives of Neurology 46: 1167.

Zeidman LA, Mohan L (2014). Adolf Wallenberg: giant in neurology and refugee from Nazi Europe. Journal of the history of the neurosciences 23: 31–44.

Recommended further reading - Medical eponyms

Whonamedit? – A dictionary of medical eponyms. <https://www.whonamedit.com/>.

Handbook of Clinical Neurology, Vol. 216 (3rd series)
The Brainstem
L.T. Grinberg and A.D. Korczyn
https://doi.org/10.1016/B978-0-443-15736-3.00015-9

Chapter 2

Structural and functional anatomy of the brainstem

HELMUT HEINSEN[1,*], LEA TENENHOLZ GRINBERG[2], AND SIMONE BOHNERT[1]

[1]*Institute of Forensic Medicine, University of Würzburg, Würzburg, Germany*

[2]*Departments of Laboratory Medicine, Pathology and Neurosciences, Mayo Clinic, Jacksonville, FL, United States*

Abstract

The brainstem connects the forebrain to the spinal cord. It consists of three layers, the ventral basis, the intermediate tegmentum, and the dorsal tectum. Caudo-rostrally, the brainstem is divided into the medulla oblongata or myelencephalon, the metencephalon, and the mesencephalon. The tegmentum is the most conserved part of the brainstem and contains the motor and sensory nuclei of cranial nerves (CN) III–XII. They are arranged in a mediolateral direction, and, unlike the spinal cord, their afferent and efferent fibers are not macroscopically separated. The tegmental reticular formation, the brainstem core, has a lattice structure with manifold longitudinal and transverse fibers and stacked neurons. Its architecture provides maximum convergence and divergence of intrinsic and ascending/descending information from the spinal cord and forebrain. It coordinates brainstem-based vital functions, and its integrity is essential for arousal, consciousness, and goal-directed behavior. Its small aminergic and cholinergic telencephalic projection nuclei have a profound impact on mood, motivation, and cognition. Human telencephalic evolution manifests in volume increases of the cerebral crura, pons, cerebellar hemispheres, olives, pyramidal tract, and medial lemniscus. However, the spatial constraints of the posterior fossa limited the proportional expansion of the brainstem during human telencephalic evolution, rendering this evolutionarily conserved structure increasingly vulnerable to malfunction and disease.

MACROSCOPIC ANATOMY

Brainstem and spinal cord – similarities and differences

The brainstem connects the telencephalon with the spinal cord. At the level of the foramen magnum, the spinal cord leaves the spinal canal, enters the cranial cavity, and continues as the brainstem, changing its shape and form (Fig. 2.1A and B). Together with the cerebellum, the brainstem fills the lower cranial fossa of the skull. Unlike the spinal root fibers, the motor and sensory root fibers of the cranial nerves (CN) III–XII are macroscopically indistinguishable, and individual cranial nerves typically differ in brainstem entry/exit location, diameter, and number of rootlets (Fig. 2.1B).

Components and topographic anatomy of the brainstem

The brainstem consists of three parts (Fig. 2.2A and B): The *caudal myelencephalon or medulla oblongata, the intermediate metencephalon* with its prominent ventral pons and dorsal cerebellum, which is particularly prominent in humans, and the *rostral mesencephalon* with its impressive cerebral crura (Fig. 2.2A) and a tectum consisting of the characteristic quadrigeminal lamina (Fig. 2.2C – 8).

The mesencephalon is connected to the *diencephalon.* The diencephalon is in the medial cranial cavity. The brainstem, along with the cerebellum, rests in the inferior cranial cavity. Most of the inferior cavity is covered by the *cerebellar tentorium*, and only a relatively small slit, the tentorial incisure (Corsellis, 1958), opens a passage for the junction between the mesencephalon and the diencephalon. In cases of increased intracranial pressure due to space-occupying intracranial pathology, including tumors, hemorrhage, or edema, medial parts of the temporal lobe may herniate through the tentorial incisure, causing life-threatening emergencies and leaving an uncal notch on the medial parahippocampal gyrus (Rai et al., 2018).

A mediosagittal slice through the rostral brainstem reveals *topographic relationships and internal structures.* The splenium of the corpus callosum (Fig. 2.2C – 4), pineal body (5), posterior

*Correspondence to: Helmut Heinsen, Institute of Forensic Medicine, University of Würzburg, Versbacher Straße 3, 97078 Würzburg, Germany. Tel: +491744007031; E-mail: helmut.heinsen@uni-wuerzburg.de.

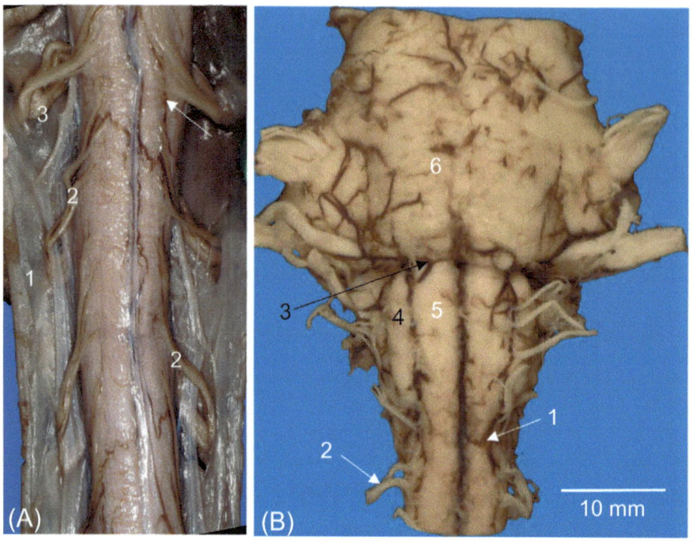

Fig. 2.1. (**A**) Spinal cord with (1) dura mater opened, (2) anterior root, and (3) posterior root with the spinal ganglion. Both roots unite to the segmental spinal nerve. The rootlets of the anterior root (arrow) exit the ventral surface of the spinal cord in the anterolateral sulcus. (**B**) Ventral aspect of a brainstem with (1) the pyramidal decussation, (2) the first right cranial nerve, (3) the bulbopontine sulcus, (4) right olive, (5) pyramis, and (6) pons.

commissure (6), and mammillary body (1,12) are already *telencephalic and diencephalic components*, respectively. The diencephalic third ventricle is continuous with the thin tubular *mesencephalic aqueduct* (7). The entrance of the aqueduct is marked by a line connecting the posterior commissure with the mammillary body. The aqueduct provides the passage for the *cerebrospinal fluid* (CSF), which flows in this thin tubular channel between the floor of the tectum (8) and the underlying mesencephalic tegmentum into the *fourth ventricle* between the base of the cerebellum and the ventral pontine and myelencephalic tegmentum.

Profiles of the brainstem in horizontal sections are divided into three layers. These are called tectum (dorsal part), tegmentum (intermediate part), and basis (ventral part) in horizontal/axial slices through the mesencephalon (Fig. 2.2D). In horizontal slices through the central parts of the pons, the term tectum is replaced by tegmentum of the fourth ventricle. Thus, the cerebellum forms the metencephalic tegmen. The myelencephalon caudal to the cerebellum does not have a tectum of nervous tissue. It is covered by a neuron-free tela choroidea.

Myelencephalon – medulla oblongata

The caudal part of the medulla oblongata and the rostral part of the spinal cord gradually merge into one another. The medulla oblongata gradually changes its shape from round to oval with a simultaneous increase in its diameter. The border between rostral spinal cord and caudal medulla oblongata is defined by a plane horizontal to the roots of the first cervical segment (Fig. 2.1B, arrow 2). Alternatively, the pyramidal decussation (Fig. 2.1B – 1) or a virtual horizontal plane parallel to the level of the foramen magnum serve to delineate spinal cord and medulla oblongata. The bulbopontine sulcus (Fig. 2.1B – 3) determinates the rostral part of the myelencephalon.

Ventral aspect

The anterior median fissure (Fig. 2.3 – 9) of the spinal cord terminates at the bulbopontine sulcus (Fig. 2.3 – 6). Fibers of the pyramidal decussation cross the median fissure at the caudal tip of the pyramids (Fig. 2.3 – 7). The pyramidal fiber bundles are part of the projection fibers that connect the isocortex to the spinal cord. The anterolateral sulcus (Fig. 2.1A, arrow) of the spinal cord is a shallow longitudinal depression on the anterolateral aspect of the spinal cord. The anterior roots of the spinal nerves (Fig. 2.1A – 1) leave the white matter of the spinal cord to meet the corresponding segmental dorsal roots (Fig. 2.1A – 3). Both roots unite to the spinal nerve. The rostral parts of the anterolateral sulcus on the brainstem separate the pyramids from the olives (Fig. 2.4 – 7,8). The numerous rootlets of the hypoglossal nerve (Fig. 2.3 – 20) leave the myelencephalon through this sulcus.

Dorsal aspect

The dorsal or posterior median sulcus divides the bilateral posterior funiculi into two symmetric parts (fascicles) (Fig. 2.5A – 31). The funiculi are subdivided by the intermediate posterior sulcus (Fig. 2.5A – 30) into the smaller medially located gracile fascicle (Fig. 2.5A – 13) and the more conspicuous cuneate fascicle (Fig. 2.5A – 14). The medial gracile fascicle is composed of sensory fibers from the lower regions of the body, while the lateral cuneate fascicle is composed of sensory fibers from the neck and upper parts of the body. Both fascicles terminate in tubercles, with the gracile tubercule (Fig. 2.5A – 11) marking the medial border of the caudal rhomboid fossa. The spinal nerves' dorsal roots enter the spinal cord's white matter through the posterolateral sulcus (Fig. 2.5A – 15) and terminate on neurons in the dorsal horn (Fig. 2.7A – 2). The trigeminal tubercle (Fig. 2.5A – 12) marks the

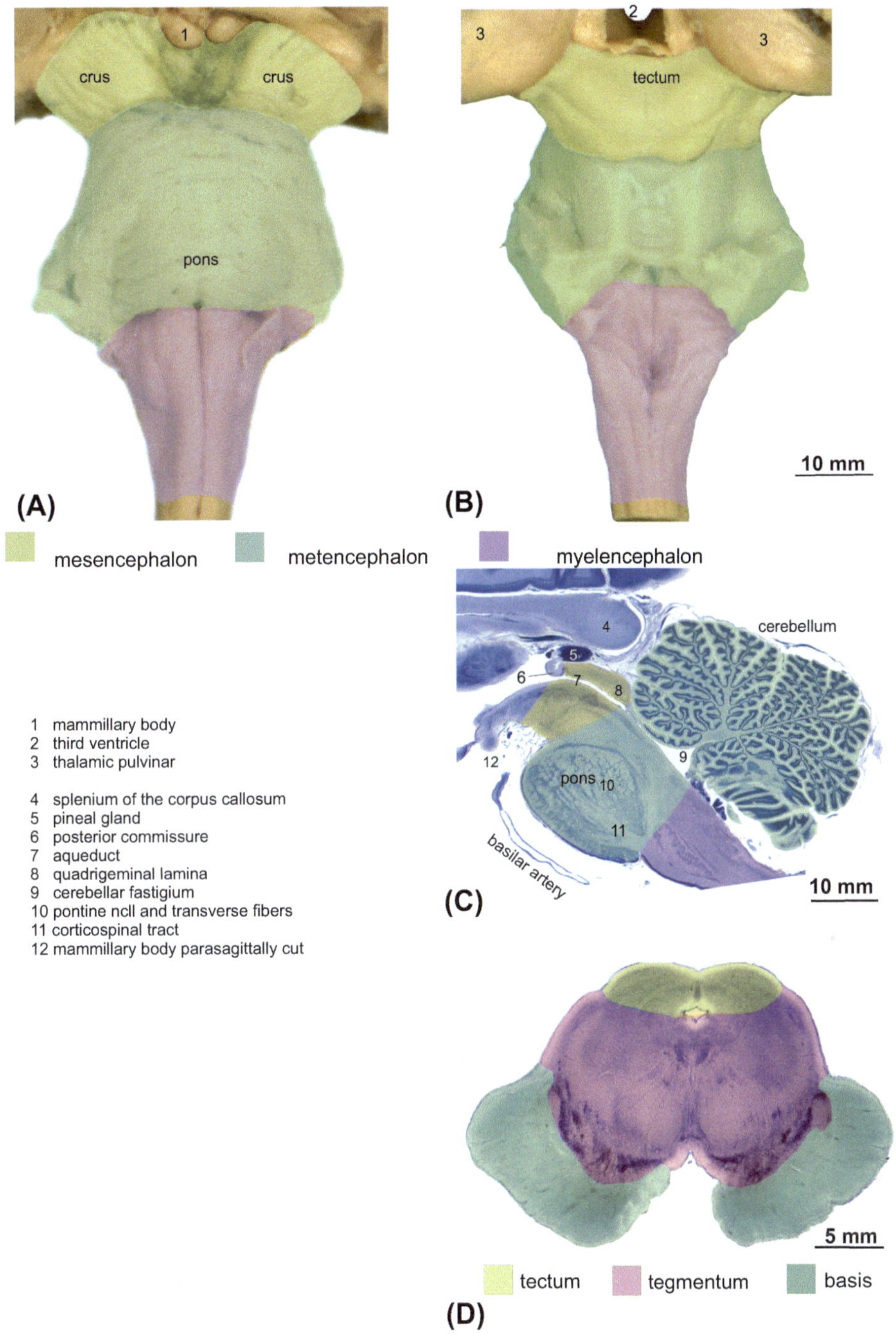

Fig. 2.2. (**A**) ventral aspect; (**B**) dorsal aspect of a human spinal cord with the (1) mammillary bodies in **A** and parts of the (2) third ventricle and bilateral (3) thalamic pulvinar in **B**; (**C**) mediosagittal gallocyanin-stained section through a human brainstem and adjacent telencephalon: (1) splenium of corpus callosum, (2) pineal body, (3) posterior commissure, (4) mammillary body, (5) aqueduct, (6) quadrigeminal lamina, (7) cerebellar fastigium, (8) pontine nuclei, and (9) corticospinal tract. (**D**) Mesencephalic tectum tegmentum and basis in axial (horizontal) plan of section; gallocyanin stain.

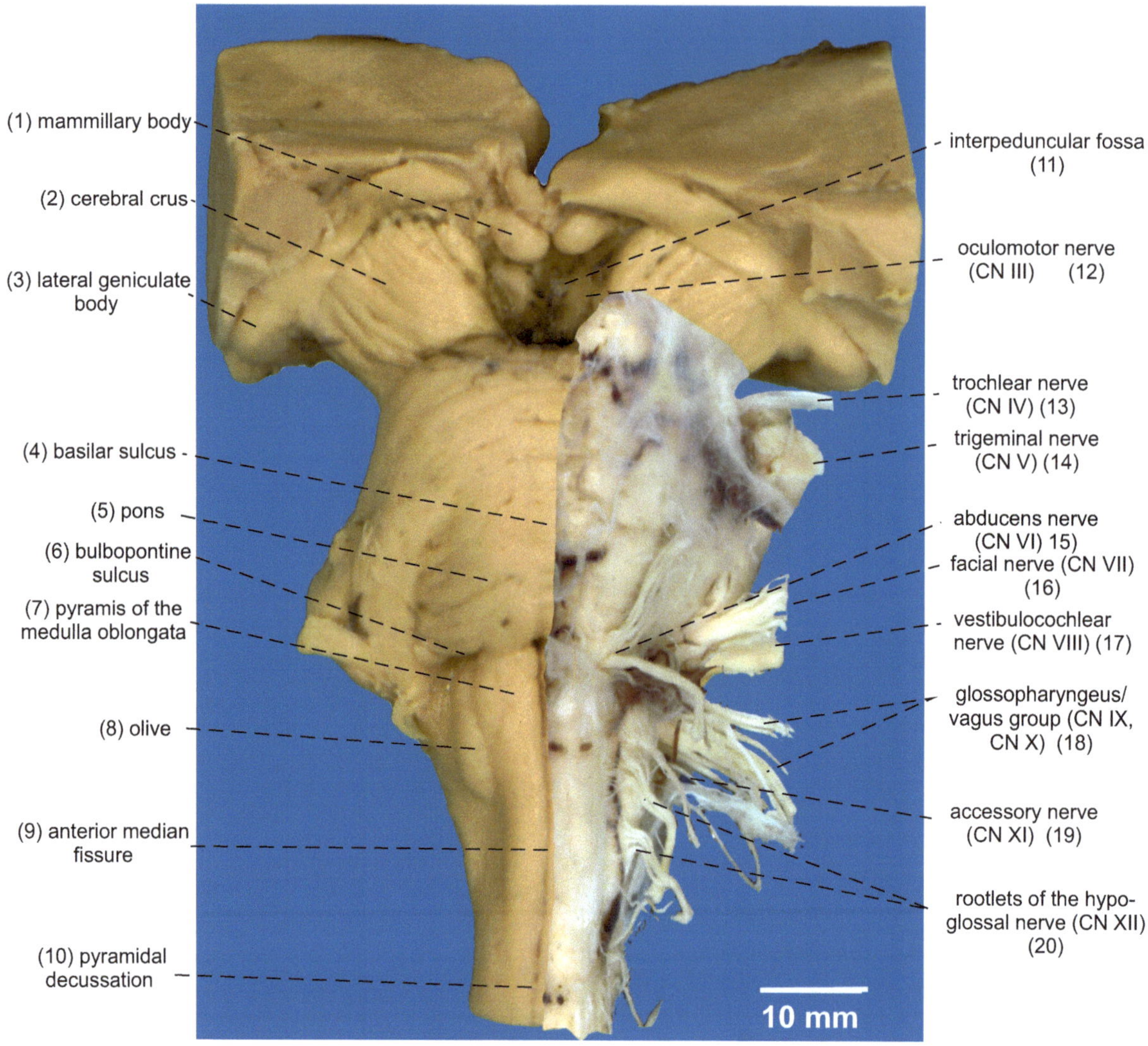

Fig. 2.3. Ventral aspect of the brainstem with arachnoid and cranial nerves partially left in situ.

end of the posterolateral sulcus, which is slightly elevated due to the underlying spinal nucleus of the trigeminal nerve. The shallow retro-olivary sulcus (Fig. 2.6 – 8) separates the olive (Fig. 2.6 – 7) from the cuneate fascicle's lateral margin (Fig. 2.6 – 18). The cranial nerves CN IX, CN X, and CN XI emerge/enter from/in this sulcus (Fig. 2.3 – 18,19).

Metencephalon

The metencephalon joins the myelencephalon rostrally. It consists of the ventrally located pons (Figs. 2.2C and 2.3 – 5) and the dorsal cerebellum (Fig. 2.4A). The pons (Fig. 2.2C) represents the base of the metencephalon, and the cerebellum the roof or tegmen. The tegmentum is hidden by the former components, which are especially conspicuous in primates.

Pons

The human pons is an impressive transverse protrusion consisting of transversely running myelinated fibers with closely attached neurons of the pontine nuclei (Fig. 2.2C – 10). The underlying longitudinal fibers of the pyramidal tract (Fig. 2.2C – 11) form two longitudinal protrusions that line the pontine basilar sulcus (Fig. 2.3 – 4). The transverse fibers cross the midline of the pons and pass in a laterodorsal course into the medial cerebellar peduncle (Fig. 2.6 – 15). The latter splices into medullary fibers, as it enters the cerebellum to supply predominantly the cerebellar hemispheres.

The equine pons is readily distinguishable from the human pons based on its size and shape (Fig. 2.4C – 4). In contrast to the human pons, the equine pons represents a flat plate that covers the corticospinal tract (Fig. 2.4C – 5). In Tupaia, the pons is composed of a thin layer of myelinated fibers oriented transversely (Fig. 2.4E – 4), which cover the myelinated fibers of the corticospinal tract (Fig. 2.4E – 5).

Cerebellum

The cerebellum of mammals is tripartite, comprising the central vermis (1), bilateral intermediate parts (2), and lateral hemispheres (3) (Fig. 2.4A, B and D). This tripartite organization is immediately evident in the cerebellum of Tupaia, which are

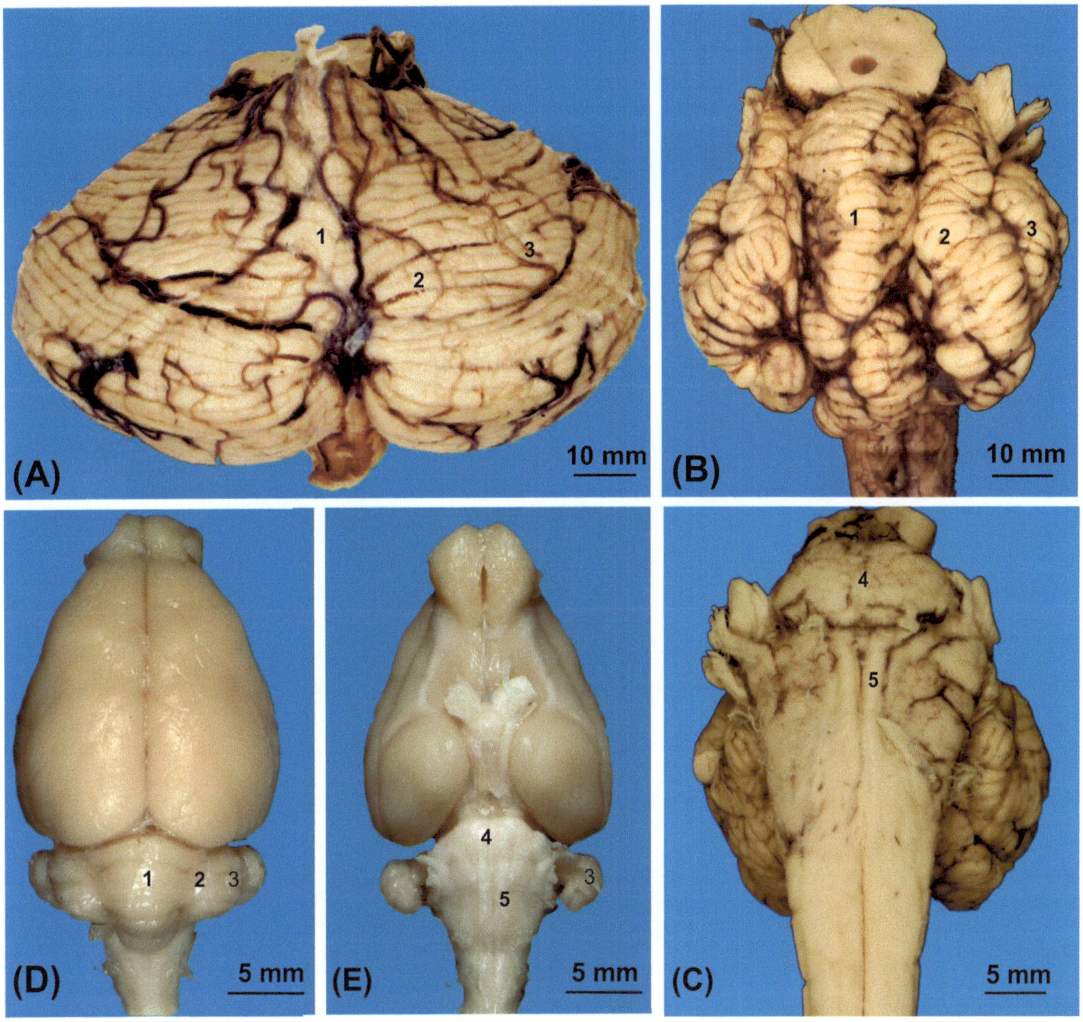

Fig. 2.4. (**A**) Formalin-fixed human cerebellum dorsal view; (**B**) formalin-fixed equine cerebellum, dorsal view; (**C**) equine brainstem, ventral view; (**D**) dorsal aspect of perfusion-fixed brain of Tupaia Belangeri, age 2 years; (**E**) ventral aspect of perfusion-fixed brain of Tupaia Belangeri, (1) vermis, (2) intermediate part, (3) hemisphere (4) pons, (5) pyramid, and (6) flocculus.

considered to have a common euarchontan ancestor with primates and dermoptera (Kundu et al., 2022). The human cerebellar surface is characterized by numerous delicate, predominantly transverse lobules (Fig. 2.4A). In dorsal view, the cerebellum, with its wide, sweeping hemispheres, covers almost the entire brainstem. The vermal cortex processes afferent spinal cord information. The vermis is prominent in carnivores and ungulates, and its size and curved course are typical for each species. The intermediate part in the equine cerebellum is clearly separated from the vermis and the hemispheres. In the dorsal aspect, the human vermian lobules are slightly elevated and continuous with those of the intermediate part and dominant hemispheres.

The cerebellum is connected to the brainstem by the superior, medial, and inferior cerebellar peduncles. The superior cerebellar peduncle (also called brachium conjunctivum) comprises the afferent ventral spinocerebellar tract and the efferent nucleo-olivary fibers from the dentate and interposed nuclei. The medial cerebellar peduncle or brachium pontis carries afferent reticulo- and cerebropontine fibers; the inferior cerebellar peduncle conveys posterior spinocerebellar, vestibulocerebellar, olivocerebellar, and reticulocerebellar fibers and the efferent cerebellovestibular tract.

Rhomboid fossa

To remove the cerebellum from the brainstem, the anterior, the middle, and the posterior cerebellar peduncles must be carefully dissected in different planes of section (Fig. 2.5). This is difficult without visual control. With the removal of the cerebellum, the rhomboid fossa with its characteristic relief is exposed (Fig. 2.5A and B). The superior cerebellar peduncles (Fig. 2.5A,B – 25) outline the rostral portions of the rhomboid fossa, while the cuneate (10) and gracile tubercles (11) demarcate the caudal portions. Superior cerebellar peduncles, gracile, cuneate tubercles and their corresponding fascicles converge in their rostro-caudal course, respectively (Fig. 2.5B). The rostral tip of the rhomboid fossa terminates at the junction of the mesencephalic aqueduct with the fourth ventricle. The superior

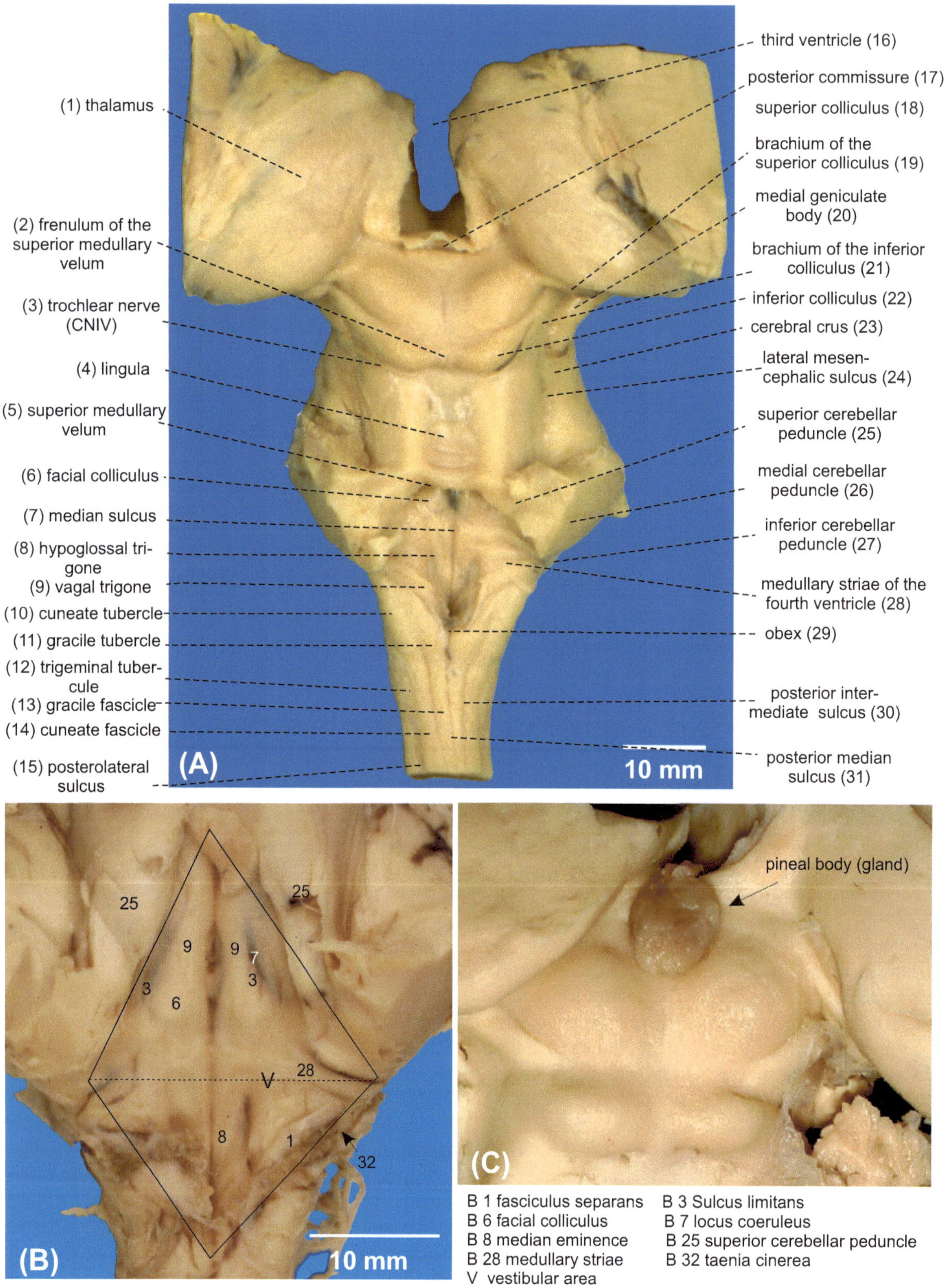

Fig. 2.5. Dorsal aspect of the brainstem. (**A**) Complete brainstem, (**B**) rhomboid fossa, and (**C**) tectum with pineal gland.

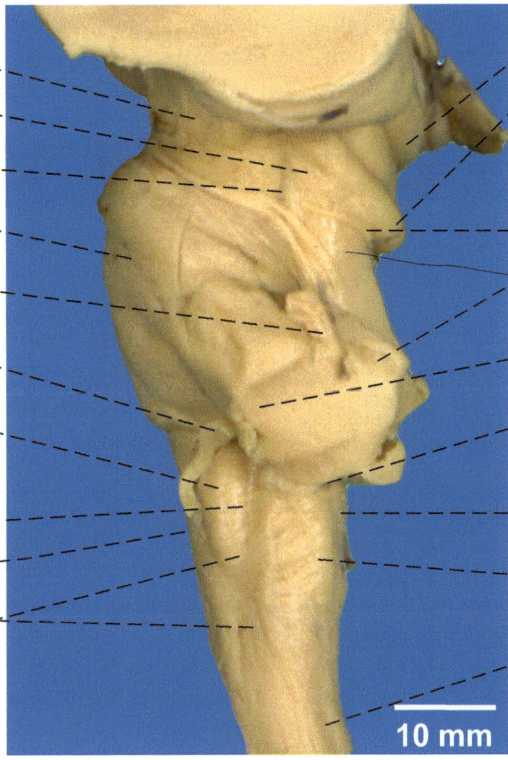

Fig. 2.6. Lateral aspect of the brainstem.

medullary velum (Fig. 2.5A – 5) with attached parts of the lingula cerebelli (Fig. 2.5A – 4) cover the entrance of the aqueduct. The caudal tip of the rhomboid fossa ends with the obex of the myelencephalon (Fig. 2.5A – 29). The obex is a small tissue bridge that marks the endpoint of the median sulcus of the spinal cord (Fig. 2.5A – 31). At this point, the fourth ventricle continues into the central canal of the spinal cord.

Halfway between caudal aqueduct and obex myelinated fibers, medullary striae of the fourth ventricle (Fig. 2.5B – 28) cover the floor of the rhomboid fossa. They run lateromedially and serve as a landmark for the division of the rhomboid fossa into a triangular superior and inferior rhomboid fossa. The median sulcus (Fig. 2.5A – 7) divides the fourth ventricle into two symmetric halves. In the upper rhombic fossa, two parallel ridges, medial eminences (Fig. 2.5B – 9), flank the median sulcus. A few millimeters rostral to the medullary striae, two hemispheric elevations bulge into the fourth ventricle on the right and left side, the facial colliculi (Fig. 2.5A andB – 6). They are caused by the axons leaving the facial nucleus. They arch as an inner knee over the abducens nucleus. The bilateral shallow limitans sulci (Fig. 2.5B – 3) extend caudo-rostrally lateral to both medial eminences in the superior rhomboid fossa. The numerous superficial neuromelanin-containing nerve cells of the locus coeruleus give the floor of the sulcus limitans a dark color (Fig. 2.5B – 7).

In the inferior rhomboid fossa, two triangular structures, the hypoglossal (Fig. 2.5A and B – 8) and the vagal trigone (Fig. 2.5A – 9), can be distinguished near the midline. The tips of these structures point caudally. The base of the larger hypoglossal trigone borders the fibers of the medullary striae (Fig. 2.5A and B – 28). The area postrema, a triangular field caudal to the vagal trigone (Fig. 2.5B, arrow), is separated from the latter by the funiculus separans (Fig. 2.5B – 1), which consists of ependyma.

The roof of the caudal rhomboid fossa does not contain neurons. It comprises the choroid plexus, and pia mater forming the tela choroidea. The pia mater attaches to the caudal parts of the cerebellum and the caudal edge of the rhomboid fossa. Removing the cerebellum tears the pia and creates the artificial taenia cinerea (Fig. 2.5B – 32). The vestibular area (Fig. 2.5B – V) extends between the taenia cinerea, sulcus limitans, and the inferior border of the medullary stria of the fourth ventricle.

The cerebellum covers the rhomboid fossa like a tent or roof, which creates the fourth ventricle that is filled with CSF. Caudal to the middle cerebellar peduncle, a bulge on both sides of the tela choroidea creates the lateral recesses of the fourth ventricle. CSF from the ventricle can flow into the subarachnoid space via the lateral apertures. Another wider opening in the caudal roof of the fourth ventricle (median aperture) allows for additional outflow of CSF.

Mesencephalon

Ventral aspect

The mesencephalon extends from the upper margin of the pons to an arbitrary coronal plane at the level of the posterior margin of the mammillary bodies (Fig. 2.2C – 12). An estimated 20 million predominantly myelinated axons on each side unite to form the crura cerebri (Fig. 2.3 – 2), which connect the telencephalic cortical areas to the pons, cerebellum, and spinal cord. Crura converge rostrocaudally. The medial parts of the crura border the interpeduncular fossa (Fig. 2.3 – 11). The medial mesencephalic sulcus (synonym: sulcus of the oculomotor nerve (Fig. 2.3 – 12)) runs along the medial border of both crura. The floor of the interpeduncular fossa, posterior perforating substance, appears to be perforated like a sieve by the entrance of numerous small perforating arteries emerging from the posterior cerebral artery.

Lateral aspect

The lateral mesencephalic sulcus (Fig. 2.6 – 3) forms a visible boundary between the ventral crura cerebri, while the adjacent lateral lemniscal trigone consists of a triangular field between the tectal plate (Fig. 2.6 – 11 and 12), the superior cerebellar peduncle (Fig. 2.6 – 14), and the cerebral crus (Fig. 2.6 – 1). Pathways encoding pain run just below the lateral lemniscal trigone (Fig. 2.6 – 2). It serves as an important landmark for neurosurgeons performing surgical severing of pathways for the therapy of unbearable pain, such as in tumor patients.

Dorsal aspect

The mesencephalic roof is represented by the tectal or quadrigeminal plate (Fig. 2.5A – 18 and 22). The superior medullary velum of the cerebellum (Fig. 2.5A – 2) extends at its rostral tip to the superior medullary frenulum (Fig. 2.5A – 2). It is covered by the cerebellar lingula (Fig. 2.5A – 2). On both sides next to the frenulum of the superior medullary velum, the two trochlear nerves emerge (Fig. 2.5A – 3). The trochlear nerve differs from the other 11 cranial nerves in that it exits on the dorsal side of the brainstem. Rostrally, the superior colliculi are covered by the pineal gland (Fig. 2.5C, arrow). The pineal gland is already part of the diencephalon.

The quadrigeminal plate of the midbrain consists of the paired superior and inferior colliculi (Fig. 2.5A – 18 and 22). The colliculi have a cortical-like structure (Fig. 2.18) and serve as important integration centers in lower vertebrates. The superior and inferior colliculi are connected to the ipsilateral lateral and medial geniculate bodies (Fig. 2.5A – 20) of the thalamus via the brachia of the superior and inferior colliculi (Fig. 2.5A – 19 and 21).

HISTO- AND ONTOGENESIS OF SPINAL CORD GRAY MATTER COMPARED WITH THE BRAINSTEM NUCLEI

Spinal cord

In vertebrates, the central nervous system (CNS) is derived from the embryonic ectoderm, which forms the groove and later the neural tube, which is lined by a monolayer of columnar cells.

The monolayer of the neural tube transforms into a matrix layer and gives birth to neuroblasts which migrate radially and accumulate in the mantle zone of the neural tube in form of a dorsal alar and a ventral basal plate separated by a sulcus limitans. The neuroblast of the basal plate emit axons to future muscle cells in somites, and with progressing differentiation, the descendants of the basal plate arrange themselves into the nuclei of the spinal cord anterior (motor) horn (Fig. 2.7A – 1). The neuroblast of the alar plate are contacted by axons of the neural crest-derived spinal ganglia (Fig. 2.1A – 3). They differentia, arrange themselves into nuclei of the sensory dorsal horn (Fig. 2.7A – 2).

Brainstem

Like the spinal cord, the wall of the early rhombencephalic groove is covered by a monolayer of ependymal cells. The monolayer transforms into a matrix layer. Neuroblasts from the matrix layer migrate into the mantle layer to form basal and alar plates separated by the sulcus limitans. As in the spinal cord, the sulcus limitans separates future sensory from future motor neurons. At variance with the spinal cord, the neuroblasts in the mantle layer are arranged mediolaterally.

In this process, groups of neuroblasts develop in the mantle zone and arrange themselves into four longitudinally extending neuroblast columns. They are named from medial to lateral as general somatic efferent (GSE; Fig. 2.7B – 1, hypoglossal nucleus), general visceroefferent (GVE; Fig. 2.7B – 2, dorsal vagus nucleus), general visceroafferent (GVA; Fig. 2.7B – 3, solitary tract nucleus), and general somatic afferent (GSA; Fig. 2.7B – 4, gracile and cuneate nuclei) columns.

In addition to these four columns, final cranial nerve nuclei differentiate including the special visceroefferent (SVE; Fig. 2.7B – 7, ambiguus nucleus) between the GSE and the GVE column, the special visceroafferent (Fig. 2.7B – 5, spinal nucleus of the trigeminal nerve) between the GVA and the GSA column, and the special somatic afferent (SSA; Fig. 2.7B – 6, vestibular nuclei, cochlear nuclei) dorsolateral to the GSA column.

This classification system links the origin's topography to functional aspects. The medial gracile and lateral cuneate nuclei emerge from the caudal part of the rhombencephalic alar plates. The fibers of the dorsal funiculus end in a somatotopic arrangement on the neuronal clusters found inside these nuclei (Strata et al., 2003; Qi and Kaas, 2006). The GSE nuclei of CN XII, VI, and IV stay in their ventral position, and their fibers leave the brainstem close to the midline. The migration of neuroblasts from matrix cells and their positioning near the ventricular wall remain largely consistent during the development of both the visceroafferent solitary and SSA cochlear and vestibular nuclei.

However, certain brainstem nuclei do not remain in the vicinity of their periventricular birthplace, where mature brainstem nuclei are anticipated to be located. Radial and tangential migration may result in shorter or longer distances from the periventricular birthplace to the destination. The branchiomotor (special visceral efferent) nuclei of CN VII, IX, X (ambiguus nucleus), and XI migrate radially.

The facial nucleus undergoes a caudal movement. During development the neurites of the CN VII arch around the abducens nucleus, indicating the path of the SVE facial nucleus and its complicated developmental movements. The nerve fibers' course is reminiscent of Ariadne's thread for Theseus' orientation in the labyrinth of the Minotaur in Greek mythology.

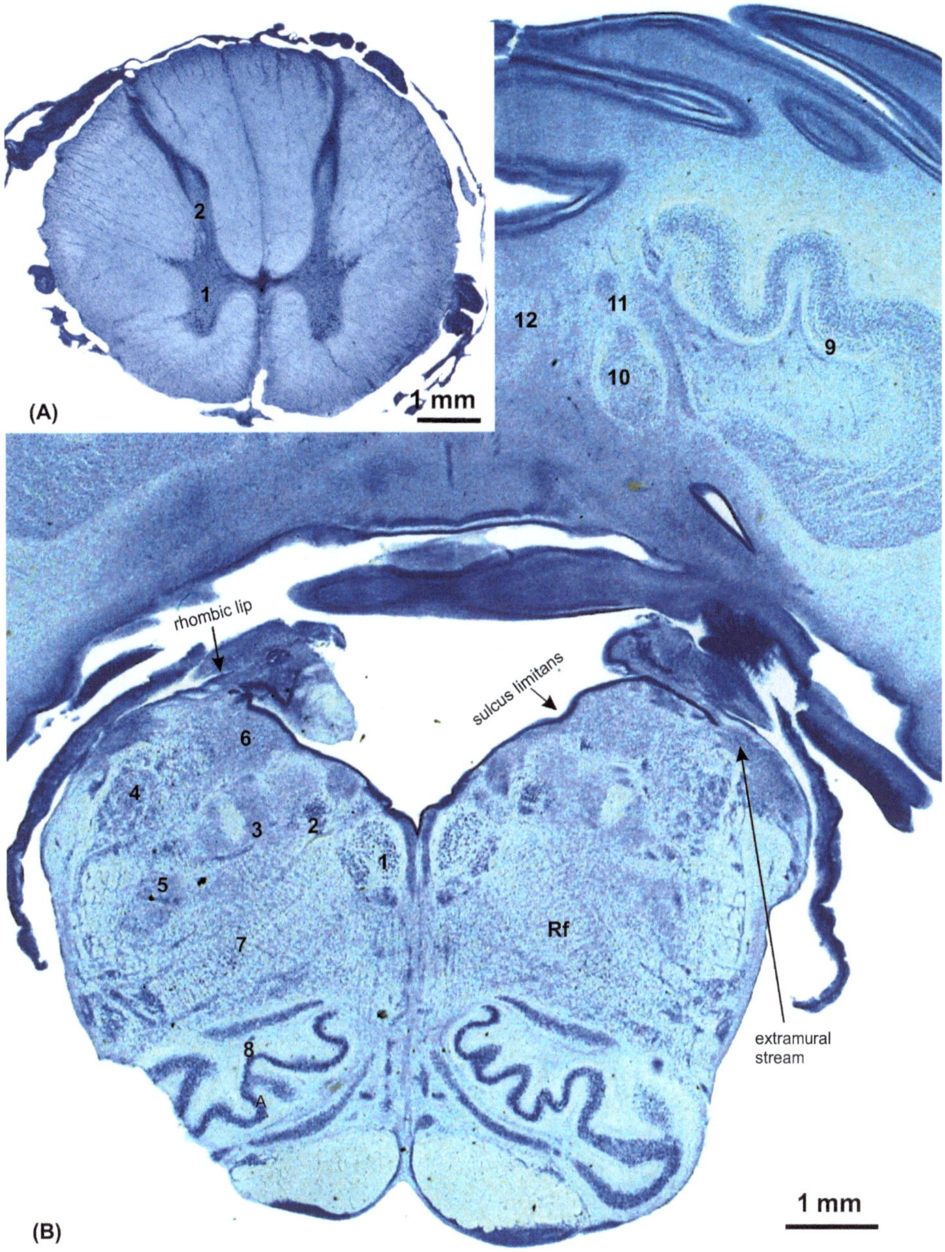

Fig. 2.7. (**A**) Horizontal section through the spinal cord of a 55-year-old male, (1) ventral horn, (2) dorsal horn; (**B**) horizontal section through the brainstem and cerebellum of a 22-week-old human fetus. The medial, intermediate, and lateral zones of the reticular formation (Rf) are well delineated. (1) Hypoglossal nucleus, (2) dorsal vagus nucleus, (3) solitary tract nucleus, (4) gracile and cuneate nuclei, (5) spinal nucleus of the trigeminal nerve, (6) vestibular and cochlear nuclei, (7) ambiguus nucleus. Note the similar configuration of inferior olivary nucleus (8) and expanding cerebellar dentate ncl (9). The cerebellar dentate ncl is medially flanked by the much less prominent globose (10), emboliform (11), and fastigial (12) ncll. Left arrow points to rhombic lip.

The roof the fourth ventricle is attached to the lateral wall of the caudal rhomboid fossa. The caudal rhombic lip (Fig. 2.7B, left arrow), located laterally to this junction, represents a region where neuroblasts proliferate and then migrate to their destination through shorter and longer streams. The progenitors in the posterior extramural migratory stream cross the midline of the myelencephalon and generate the contralateral external cuneate nucleus and the nucleus of the lateral funiculus. Progenitor cells within the intramural migratory stream migrate ventromedially to the ipsilateral mantle layer, contributing to the formation of the inferior olivary complex (Fig. 2.7B – 8). The extramural migratory stream located in the anterior section directs progenitor cells toward the basal plate region that lies dorsal to the cephalic flexure. Following migration, the cells mature into reticular tegmental and basal pontine nuclei, all of which have connections to the cerebellum. Therefore, the external cuneate nucleus, nucleus of the lateral funiculus, complex of the inferior olive, and basal pontine nuclei are grouped together as the precerebellar nuclei.

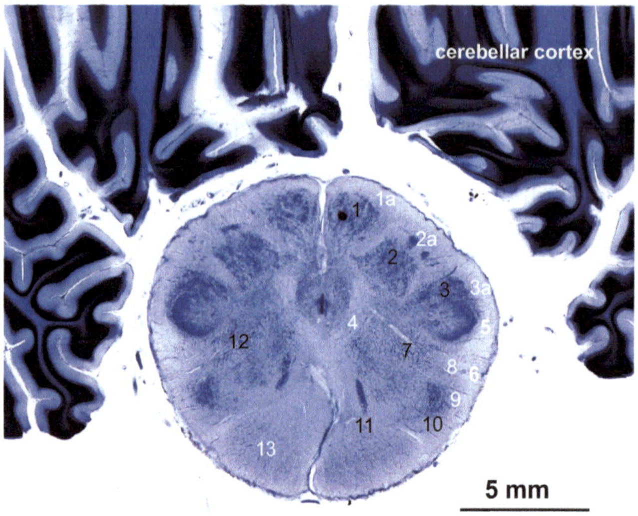

Fig. 2.8. Plane of section pyramidal decussation.

For more detailed information on the mitotic activity and cycles of the neural tube ependyma and migratory pathways, the reader is referred to the textbook by Nieuwenhuys et al. (2008). Theories of the embryologic origin of the various head components, including the muscles associated with the branchial arches, are still under revision (Blessing, 1997). Knowledge of the principle of developmental cell column formation and migration is a great help in recognizing and comparing brainstem nuclei in different species (Fig. 2.10A–C). However, it is necessary to reconcile recent studies on gene expression with the phenomenology (Puelles, 2007; Nieuwenhuys and Puelles, 2016; Puelles and Hidalgo-Sanchez, 2023).

It is of the utmost importance to have a comprehensive understanding of the origin, migration, and termination of cranial nerves to make an accurate diagnosis and localize the origin of neurologic deficits. It is, however, crucial to acknowledge that the intracranial and peripheral pathways of afferent or efferent cranial nerves frequently exhibit intricate complexities. Furthermore, embolic infarcts in the vertebrobasilar arterial system may manifest as complex symptoms. For further details regarding clinical symptoms, differential diagnosis, and pathoanatomic correlation, the interested reader is directed to Chapter 15 of this volume and Blumenfeld, 2018.

MICROSCOPIC ANATOMY AND FUNCTIONAL ASPECTS

Cyto- and myeloarchitecture in horizontal sections through selected regions of the brainstem

Unless otherwise stated, the horizontal gallocyanin-stained serial sections originate from a female aged 66 who succumbed to an acute myocardial infarction. The complete brain, including the brainstem, was excised from the cranium 11 hours after death and fixed in 4% formalin for a period of three months.

Pyramidal decussation

At the rostral level of the pyramidal decussation, the profile of the caudal myelencephalon is still round (Fig. 2.8). The dorsal column of the caudal spinal cord is replaced by the prominent special viscerosensory spinal trigeminal nucleus of the CN V (Fig. 2.8 – 3) and by descending sensory fibers forming the spinal trigeminal tract (3a). The fibers of the gracile (Fig. 2.8 – 1a) and cuneate (Fig. 2.8 – 2a) fascicles terminate on the second-order neurons of the corresponding (Fig. 2.8 – 1) gracile and (Fig. 2.8 – 2) cuneate nuclei. The axons of these second-order nuclei cross as internal arcuate fibers (Fig. 2.8 – 4) and continue their course as the ascending medial lemniscus through the brainstem to terminate on the diencephalic third-order ventral posterolateral thalamic nucleus.

The dorsal (5) and ventral (6) spinocerebellar tracts, the rubrospinal tract (8), and the spinothalamic tract (9) connect the spinal cord to the cerebellum, red nucleus, and thalamus. An isolated cluster of motor neurons is called lamina 9 of the spinal gray or supraspinal ncl (Fig. 2.8 – 11). These special visceromotor neurons send their axons to the accessory trunk of the CN XI, which innervates the sternocleidomastoid and trapezius muscles.

The ventral part of caudal myelencephalon is dominated by pyramidal tract profiles (Fig. 2.8 – 13). Their profiles appear asymmetric because at this level, fibers of the pyramidal tract cross.

In the sector spanning the spinal trigeminal nucleus and the pyramids, irregular clusters of bigger chromophile neurons and regions with smaller neurons are encountered. This region encompasses the medullary reticular formation (Fig. 2.8 – 12). The special visceromotor retroambiguus ncl (Fig. 2.8 – 7) is

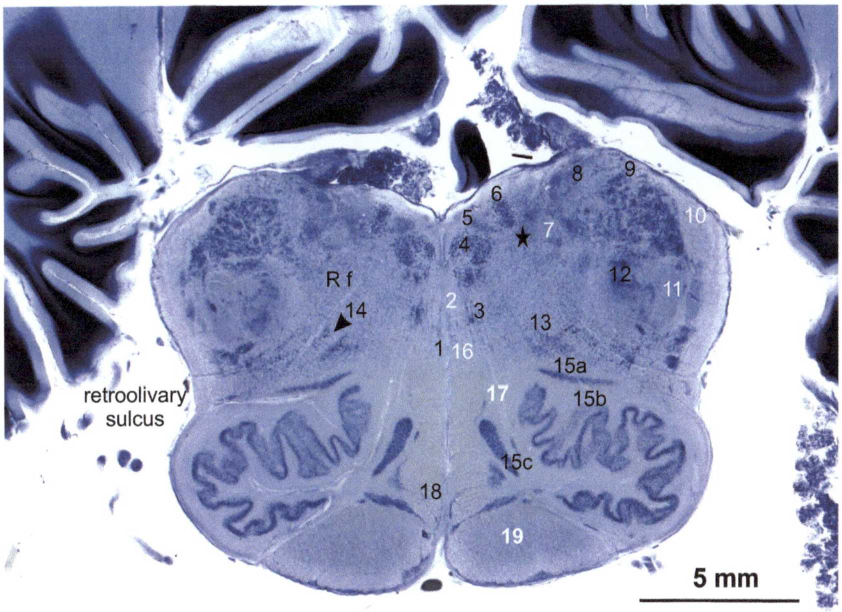

1 raphe obscurus ncl
2 tectospinal tract
3 ncl of Roller
4 hypoglossus ncl
5 intercalated ncl
6 dorsal motor ncl of vagus
7 solitary tract
asterisk solitary ncll
8 medial vestibular ncl
9 external cuneate ncl
10 inferior cerebellar peduncle
11 spinal trigeminal tract
12 spinal trigeminal ncl, interpolar part
13 lateral reticular ncl rostral part
14 ambiguus ncl, semicompact part
15a inferior olive, dorsal ncl
15b inferior olive, principal ncl
15c inferior olive, medial ncl
16 medial lemniscus
17 hypoglossal nerve
18 raphe pallidus ncl
19 pyramidal tract

Fig. 2.9. Plane of section central parts of the olivary complex.

embedded in the ventrolateral parts of the reticular formation. The lateral reticular ncl (Fig. 2.8 – 10), a precerebellar nucleus, is also considered part of the reticular formation.

Caudal end of the inferior rhomboid fossa

At the caudal end of the inferior rhomboid fossa, the profile of the myelencephalon is oval. The nuclear profiles of the hypoglossal (Fig. 2.9 – 4) and dorsal vagal (6) nerve bulge the floor of the rhomboid fossa to form the corresponding hypoglossus and vagus trigones (Fig. 2.5A – 8,9). The shallow sulcus limitans cannot be identified. Fixation artifacts, deformation, and tissue shrinkage during embedding of the brainstem are the most likely reasons. Nevertheless, the mediolateral arrangement of the somatomotor hypoglossus ncl (Fig. 2.9 – 4), the special visceromotor ambiguus ncl (14, arrowhead), the general visceromotor dorsal vagus ncl (Fig. 2.9 – 6), the general viscerosensory nuclear complex (asterisk) of the solitary tract (Fig. 2.9 – 7), the special viscerosensory ncl of the interpolar part of the spinal trigeminal ncl (Fig. 2.9 – 12), the spinal tract of CN V (Fig. 2.9 – 11), and the caudal somatosensory medial vestibular ncl (Fig. 2.9 – 8) is comprehensible.

The perihypoglossal nuclei consist of Roller's nucleus (Fig. 2.9 – 3) and the intercalated nucleus (Fig. 2.9 – 5). The latter is a caudal continuation of the precerebellar prepositus hypoglossus nucleus.

The plane of section passes the caudal part of the bilateral raphe obscurus nucleus (Fig. 2.9 – 1), which is part of the serotonergic raphe nuclei. These nuclei are in the reticular formation (Rf), which extends bilaterally from the midline to the spinal nuclei of CN V. Subnuclei of the reticular formation are visible in thick sections. The reticular formation may be difficult to distinguish from surrounding and embedded cranial nerve nuclei, including the solitary tract with associated nuclei (Fig. 2.9 – 7), and fragmented portions of the caudal ambiguus nucleus (Fig. 2.9 – 14). The ambiguus nucleus is embedded in the reticular formation of the ventromedial medulla oblongata, which is part of an extended network that controls respiration and blood circulation. Neurons that are part of the ventromedial medulla oblongata come close to the surface of the retro-olivary sulcus.

The rostral parts of the lateral reticular nucleus (Fig. 2.9 – 13), a precerebellar nucleus, are visible. This nucleus sends numerous mossy fibers to the cerebellar cortex.

The intermediate myelencephalon's basal region is dominated by the olivary complex, which comprises the dorsal, principal, and medial (Fig. 2.9 – 15a, b, and c, respectively) subnuclei. The main nucleus is typically folded and has a profile more like that of a cortex rather than a subcortical nuclear (generally round to oval) element. The arrangement of the olivary neurons in the brainstem provides a large surface area, even though they are confined to a limited volume in the ventral inferior cranial fossa.

The myelinated fibers are arranged into four distinct groups: the medial longitudinal fascicle (MLF) (2), the descending fibers of the tectospinal tract (Fig. 2.9 – 16), ascending fibers of the medial lemniscus (16), dorsal to the raphe pallidus ncl (Fig. 2.9 – 19), the descending fibers of the pyramidal corticospinal tract (20), and the axons from the hypoglossal nucleus (Fig. 2.10 – 17) that pass between the medial and principal olive before exiting the brainstem at the preolivary sulcus.

Entrance of the vestibulo-cochlear nerve and comparative anatomic aspects of the reticular formation

Horizontal sections through the rostral myelencephalon show few of the previously identified cranial nerve nuclei in the brainstem

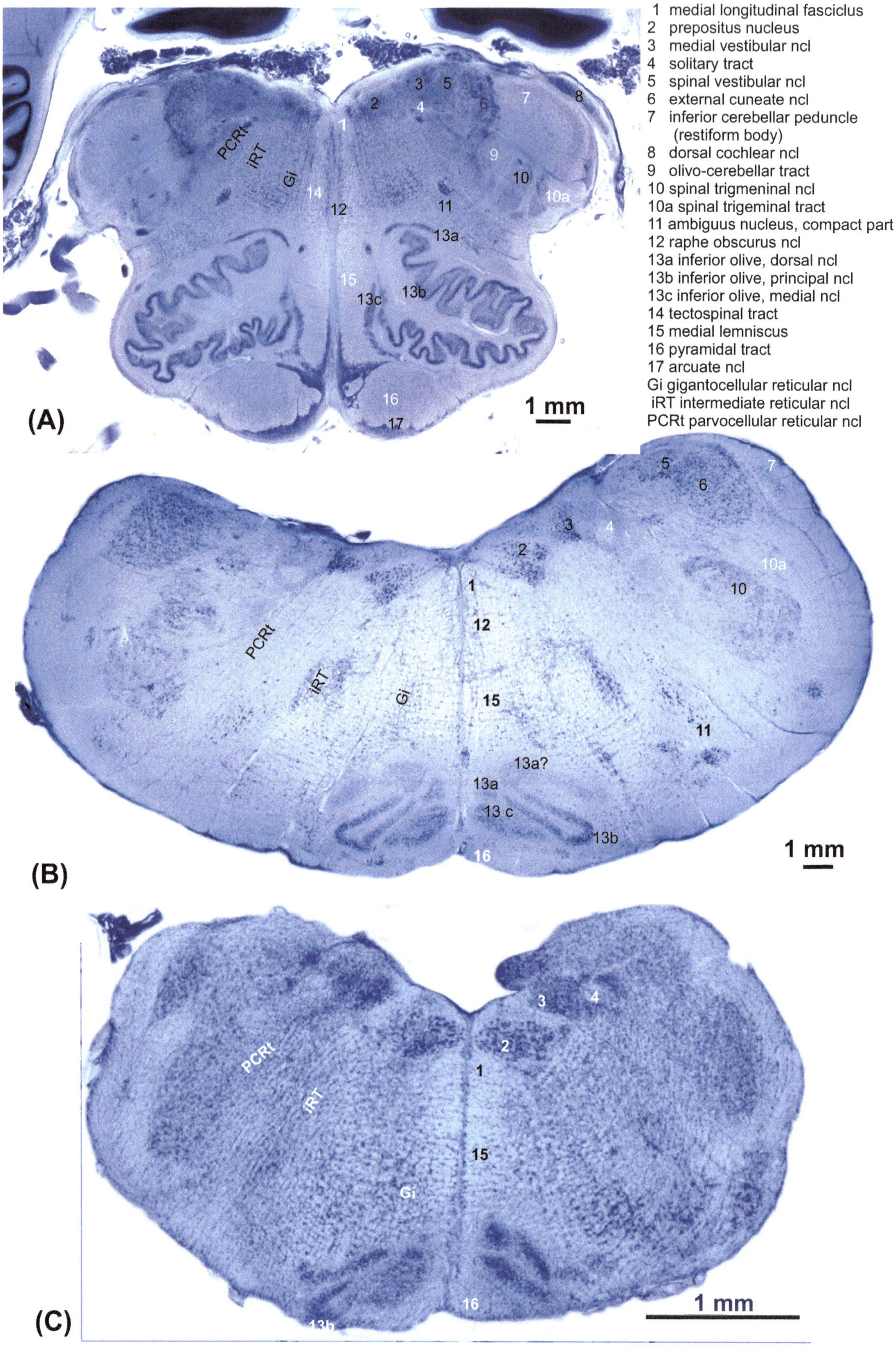

Fig. 2.10. Comparative anatomy of the human brainstem at the rostral hypoglossus-caudal prepositus hypoglossi level. (**A**) human, (**B**) horse, and (**C**) tree shrew (Tupaia belangeri). The tree shrew brainstem is characterized by high neuronal density, the horse brainstem by fiber tracts, and neuropil with low neuronal density and widely spaced nuclei. The human brainstem is intermediate in neuron and neuropil/fiber density. Neurons in the reticular formation of the tree shrew brainstem are not randomly distributed. They are arranged parallel to fiber tracts that run in a semicircular, longitudinal, or radial direction through the brainstem of Tupaia. In all three species, a lateral parvocellular, a ventrointermediate, and a medial gigantocellular nucleus can be identified.

tegmentum. The special visceromotor ambiguus nucleus (Fig. 2.10A – 11) and the general viscerosensory solitary tract with attached relay nuclei can be identified (Fig. 2.10A – 4). The vestibular area of the rhomboid fossa harbors the special sensory medial (Fig. 2.10A – 3) and spinal vestibular nuclei (Fig. 2.10A – 5). Lateral to the vestibular nuclei, the dorsal cochlear nucleus is closely associated with the vestibulo-cochlear nerve (CN VIII; Fig. 2.10A – 8). The spinal trigeminal tract and closely associated special viscerosensory nuclei of the CN V (Fig. 2.10A – 10) are located ventrolaterally to the vestibular nuclei.

Bilaterally on either side of the tegmental midline, vertically stacked neurons form the long thin columns of raphe obscurus (Fig. 2.10A – 12).

The precerebellar prepositus hypoglossal ncl (Fig. 2.10A – 2) and the external cuneate ncl (Fig. 2.10A – 6) send mossy fibers to the cerebellum. The dorsal, principal, and medial olivary ncll are also connected to the cerebellum by sending climbing fibers to the cerebellar Purkinje cells.

The olivocerebellar tract (Fig. 2.10A – 9) is a prominent fiber bundle that connects to the inferior cerebellar peduncle (Fig. 2.10A – 7), also known as the restiform body, before reaching the cerebellar Purkinje cells.

The tectospinal tract (Fig. 2.10A – 14), based on the medial lemniscus (Fig. 2.10A – 15), is crowned by the myelinated fibers of the MLF.

The arcuate nucleus of the brainstem (Fig. 2.10A – 17) sends myelinated axons to the cerebellum via external arcuate fibers, the medullary stria of the fourth ventricle, and the inferior cerebellar peduncle.

At this level of section, the reticular formation is a notable structure of the tegmentum. The reticular formation includes the raphe obscurus, gigantocellular (Gi), intermediate (iRT), and parvocellular reticular nuclei (PcRT). Upon closer examination, it is evident that the gigantocellular neurons are organized into vertical columns and horizontal rows. This is a defining characteristic of the bulbopontine reticular formation, reflecting the interaction between cyto- and myeloarchitectonic constraints. The borders between the bulbopontine reticular formation and other brainstem and precerebellar nuclei are either blurred or highly irregular, exhibiting species-specific differences. The equine reticular formation is distinguished by thick fibers and a low cell density. In contrast, the reticular formation of tree shrews exhibits a high cell density and numerous fibers coursing sagittally, obliquely, and horizontally. In all three species, the arrangement of fibers and neurons resembles a lattice, supporting maximal convergence and divergence of interneuronal connections rather than a reticular formation.

Cerebellum and comparative aspects

The macroscopic anatomy of the mammalian cerebellum is diverse and species-specific (Fig. 2.6). This is reflected in the metaphoric terminology of the cerebellar parts that is still used today (Fig. 2.11A and B). Larsell's studies of the ontogeny and fissuration of the cerebellum in different species have unraveled the bauplan of the vertebrate cerebellum (Jansen and Brodal, 1958; Larsell, 1970).

In mammalian ontogeny, the primitive cerebellar wulst, which forms the primordial tegmen of the metencephalon, is first divided by the shallow posterolateral fissure (Fig. 2.11A,B – f.p.l) into a caudal nodulofloccular part and a rostral corpus cerebelli. Next, the deep and prominent primary fissure (f.pr. a misnomer, it is the second fissure in ontogeny) divides the cerebellar body into anterior and posterior lobes. These lobes are further subdivided by subsequent fissures into a total of 10 lobules. Larsell used the abbreviation L and Roman numerals to designate the vermian lobules (Fig. 2.11, LI–LX), which form

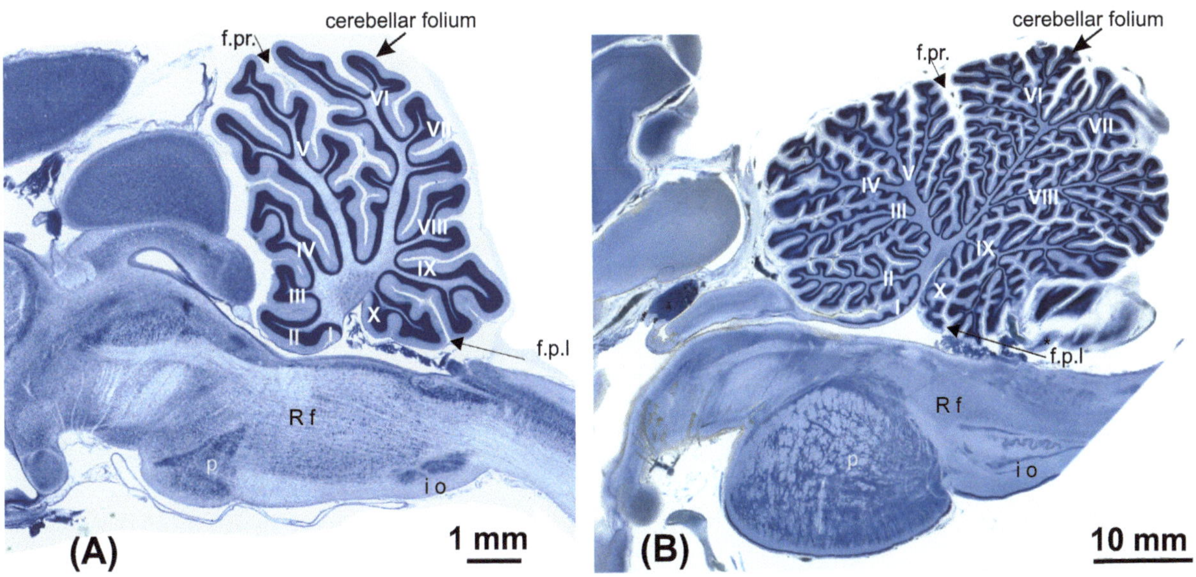

Fig. 2.11. Images of mediosagittal plane of section through the brainstem and cerebellar vermis of an (**A**) 2-year-old male Tupaia and (**B**) 55-year-old human male. *f.pl.*, posterolateral fissure, *f.pr.*, primary fissure.

the arbor vitae (tree of life) of the ancient anatomists. The anterior lobe processes predominantly spinal cord afferents (spinocerebellum), the posterior lobe isocortical afferents (neocerebellum), and the nodulofloccular part afferents from the vestibular nuclei (vestibulocerebellum).

The lobules in the hemispheres are designated HI–HX. LI is synonymous with the lingula in the ancient terminology still in use, LX with the nodulus, and HX with the paraflocculus. In contrast to the human cerebellum, the transitions between the medial and lateral (hemispheric) parts of the cerebellum are more distinct in other mammals (Fig. 2.4B and D). The cerebellum of mammals and birds consist of 10 lobules.The evolution of the cerebellum manifests in a general linear logarithmic increase in volume. The same is true for surface area. In primates, the hemispheric part experiences a major increase (MacLeod et al., 2003), in ungulates and particularly in whales (Korneliussen, 1967) it is the intermediate part.

The main branches, or lobules, may branch further into sublobules. Interestingly, the arbor vitae of mammals and birds generally consists of 10 lobules, numbered LI–LX. A vastly increased number of sublobules and innumerable folia characterize the human cerebellum with its extended cerebellar surface and billions of small granule cells (Oliveira-Pinto et al., 2016) populating the granule cell layer (stained dark blue in Fig. 2.11 of the cerebellar cortex).

Two fiber systems provide afferent information to the intracerebellar nuclei and the cerebellar cortex, the climbing fiber and the mossy fiber system. The former originates mainly from the inferior olive and together they carry information of somatosensory, vestibular, acoustic, and visual origin, as well as from the cerebral cortex and other brainstem and spinal motor centers.

The cerebellar cortex consists of three layers, and the cerebellar Purkinje cells are highly characteristic and diagnostic by their topographic location in the cortex, their size, and their dendritic arbor. Purkinje cells are the major projection neurons of the cerebellum.

Most Purkinje cells project topographically to the cerebellar nuclei. The subcortical cerebellar nuclei are the fastigial, emboliform, globose, and dentate nuclei. The nuclei are in the white matter of the vermal, intermediate, and hemispheric parts (Fig. 2.7B – 9–12). The neurons in the deep cerebellar nuclei are the gate for the information from the Purkinje cells to brainstem and thalamic neurons.

The majority of cerebellar dentate nuclear efferents target the red nucleus, the thalamus, and finally the cerebral cortex. Descending fibers from the red nucleus use the central tegmental tract to terminate on neurons in the principal ncl of the inferior olive. These in turn send climbing fibers back to the cerebellum. In this way, two closed neuronal circuits are formed that started with corticopontine fibers and ended with pontine neurons.

Genu (knee) of facial nerve

The characteristic elevation of the floor of the rhomboid fossa is caused by the facial nerve (Fig. 2.12 – 1,10), which winds like a bent knee around the abducens nucleus (Fig. 2.12 – 1,4).

The sulcus limitans separates the special somatosensory parvo- (Fig. 2.12 – 5) and magnocellular medial vestibular nucleus (Fig. 2.12 – 6), the superior vestibular nucleus (Fig. 2.12 – 7), and the special viscerosensory oral spinal trigeminal nucleus (Fig. 2.12 – 8) with its spinal trigeminal tract (9) from the somatomotor abducens nucleus (Fig. 2.12 – 4), and the special visceromotor facial nucleus (Fig. 2.12 – 11). The nuclear complex of the superior olive (Fig. 2.12 – 12) is located within the transverse fiber bundle of the trapezoid body medial to the fibers of the facial nerve (Fig. 2.12 – 10). This bundle originates from the cochlear nuclei. It comprises the medial and

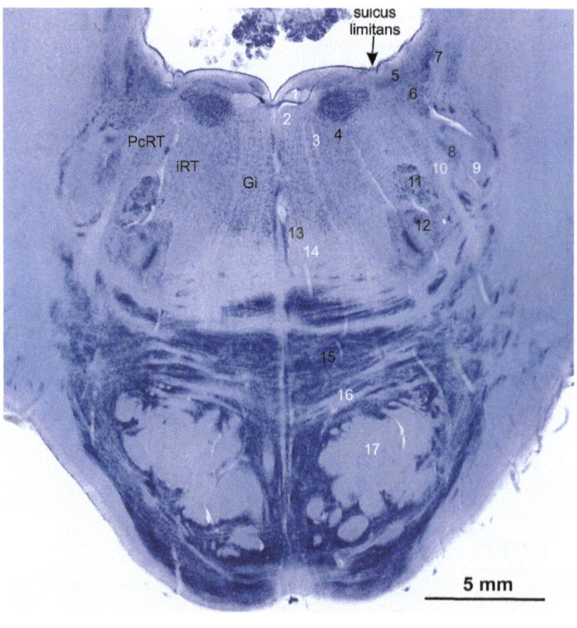

Fig. 2.12. Plane of section facial knee.

lateral superior olive, the periolivary nuclei, and the nucleus of the trapezoid body. The fibers cross the medial lemniscus (Fig. 2.12 – 14) in the ventral part of the tegmental area, then continue rostrally as the lateral lemniscus and terminate in the inferior colliculus.

The pontine neurons (Fig. 2.12 – 15) are oriented transversely. Topographically oriented corticopontine fibers (Fig. 2.12 – 16) are relayed in the pontine nuclei, cross to the contralateral side, and terminate according to their telencephalic cortical origin in the cerebellar cortex (Schmahmann et al., 2004).

Fibers directed to the spinal cord (corticospinal tract) and those terminating in brainstem motor nuclei (corticobulbar tract) assume a ventral position (Fig. 2.12 – 17).

The vestibular nuclei process information concerning the position and movement of the head derived from the labyrinthine semicircular canals. The lateral vestibulospinal tract, originating from the lateral vestibular nucleus, facilitates spinal reflex activity and inhibits flexor muscles (Nieuwenhuys et al., 2008). They contribute to the vestibulo-ocular and vestibulocollic reflex pathways.

Central parts of the pons

In this plane of section, the crossed and laterally ascending fibers of the middle cerebellar peduncle are the dominant feature of the image. The fibers of the superior cerebellar peduncle (Fig. 2.13 – 4) exhibit a different course and color staining than the those of the pontocerebellar peduncle.

The dorsal paramedian ncll are located near the midline close to the surface of the bilateral median eminences and are characterized by numerous small and hyperchromic cells (Fig. 2.13 – 1). They are considered to be the rostral continuation of the caudal dorsal paramedian nucleus, which is cholinergic and involved in receiving vestibular input and participating in the control of eye movements.

The oral dorsal paramedian nucleus overlies the MLF (Fig. 2.13 – 2) and the lateral part of the reticulotegmental nucleus of Bechterew (Fig. 2.13 – 5). The oral part of the special somatosensory superior vestibular nucleus (3) lies dorsolateral to the sulcus limitans.

The main sensory nucleus of the trigeminal nerve (Fig. 2.13 – 8) is visible ventrolateral to the superior vestibular nucleus. It is flanked medially by the motor trigeminal nucleus (Fig. 2.13 – 6) and the motor root of the trigeminal nerve (Fig. 2.13 – 7).

The gigantocellular reticular nucleus is no longer identifiable in the pontine tegmentum. The intermediate (iRT) and parvocellular reticular nuclei (PcRT) are displaced laterally by the central tegmental tract (9) and, to a lesser extent, by the rounded medial lemniscus (Fig. 2.13 – 12).

The MLF (10) closely contacts the reticulotegmental nucleus (Fig. 2.13 – 11).

Oral part of the superior rhomboid fossa

The left and right superior cerebellar peduncles (Fig. 2.14 – 10) flank the oral portion of the superior rhomboid fossa. Compared to the area of the entire brainstem, the tegmentum accounts for a rather small proportion in quantitative terms. The pons with its pontine nuclei (Fig. 2.14 – 13) crossing pontocerebellar fibers (Fig. 2.14 – 13a), corticobulbar and -spinal as well as the trigeminothalamic tracts (Fig. 2.14 – 17) and medial lemniscus

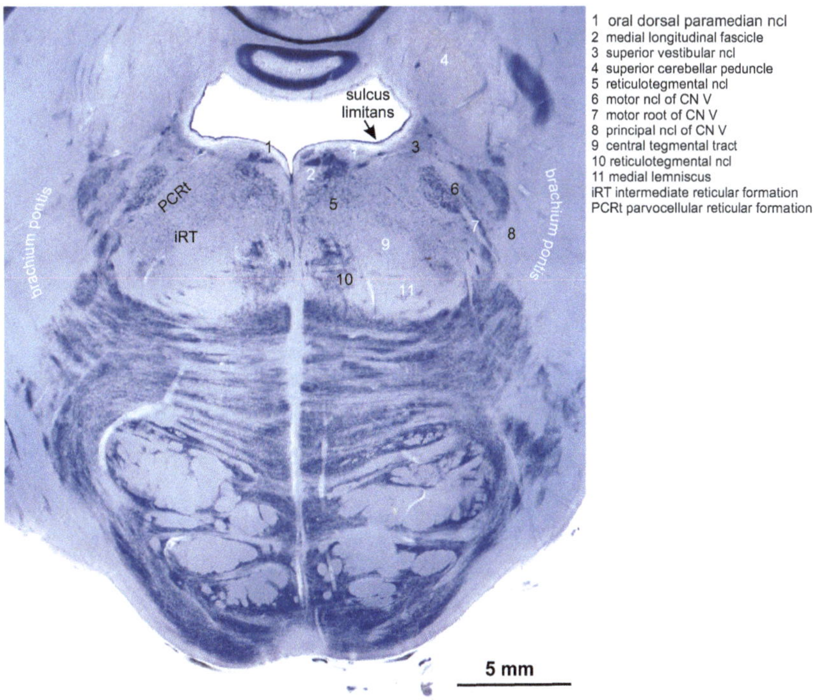

Fig. 2.13. Plane of section central parts of the pons with motor nucleus of CN V.

with their ascending fibers (Fig. 2.14 – 19) are predominant at this mesodiencephalic junction. The MLF (Fig. 2.14 – 1) can be identified due to its subventricular position near the midline. The profiles of the caudal part of the dorsal raphe (Fig. 2.14 – 2, arrow) are squeezed between the midline of the brainstem and the fibers of the MLF. Neurons of the lateral part of the reticulotegmental nucleus (Fig. 2.14 – 3) either cover or integrate into it. The ventral parts of the reticulotegmental nucleus branch out and extend more laterally (Fig. 2.14 – 5). The lateral and ventral parts of the reticulotegmental nucleus are not in direct contact with each other. They are separated from each other by the profiles of the paramedian raphe nucleus (Fig. 2.14 – 4, arrow).

The neurons in the locus coeruleus are distinguished by their size and dark pigmentation (Fig. 2.14 – 6). They mainly use norepinephrine as a neurotransmitter. Neuromelanin pigment, which increases with age, is generated and deposited intracellularly from catecholamine metabolites. The subcoeruleus nucleus (Fig. 2.14 – 12) mainly contains chromophilic unpigmented neurons.

The mesencephalic tract of CN V is near the locus coeruleus (Fig. 2.7, arrow). Its neurons parallel (Fig. 2.8, arrows) the tract, as it passes through the pontine tegmentum and rostral mesencephalon. The neurons of the trigeminal mesencephalic tract are considered immigrated intracerebral bipolar cranial ganglion neurons that transmit a proprioceptive information from the jaw-closing muscles and the periodontium.

The superior cerebellar peduncle (Fig. 2.14 – 10) has lateral convex and medial concave surfaces that are covered by the lateral (Fig. 2.14 – 9) and medial parabrachial nuclei (Fig. 2.14 – 11). Both nuclei converge ventrally, and the nucleus of Kölliker–Fuse (Fig. 2.14 – 15) is in this region. The latter is considered a generator of the respiratory cycle (Krohn et al., 2023).

Lateral to the Kölliker–Fuse nucleus, there are two identifiable structures: the central nucleus of the acoustic tract consisting of numerous small cells (Fig. 2.14 – 14) and the chromophilic pontine nuclei (Fig. 2.14 – 13). These structures are separated from each other by the fibers of the ascending spinothalamic tract (Fig. 2.14 – 18).

The trigeminothalamic tract (trigeminal lemniscus, Fig. 2.14 – 17) and the medial lemniscus (Fig. 2.14 – 19) are two other ascending myelinated fiber tracts located more medially.

The central tegmental tract (Fig. 2.14 – 16) originates from the parvocellular red nucleus and terminates in the principal inferior olive (Nieuwenhuys et al., 2008). The diameter of the central tegmental tract increases, as the distance from the red nucleus decreases. There is limited space available for the neurons of the oral pontine reticular formation (PnO).

The oral part of the pons (Fig. 2.14 – 20) is rich in fibers. The fibers of the CN V (Fig. 2.14 – 21) traverse this pontine region.

Caudal pole of the trochlear nucleus

The mesencephalic tectum at this plane of section is composed of the dorsomedial, lateral, and ventrolateral periaqueductal gray (PAG, Fig. 2.15 – 1, 3, and 4, respectively) surrounding the mesencephalic aqueduct as well as the dorsomedial (Fig. 2.15 – 2) and external (Fig. 2.15 – 6) cortex together with the central nucleus (Fig. 2.15 – 5) of the inferior colliculus. The ventral parts of the PAG are part of the mesencephalic tegmentum, along with the dorsal and median raphe nuclei Fig. 2.15 – 10, 12, respectively), and the trochlear nucleus (Fig. 2.15 – 11). The latter is embedded in the MLF (Fig. 2.15 – 13). The fibers of the superior cerebellar peduncles (Fig. 2.15 – 19) cross the midline in the eponymous decussation (Fig. 2.15 – 20) on their way to the red nucleus. Neurons forming the caudal linear raphe nucleus (Fig. 2.15 – 21) can be seen between the decussation of the cerebellar peduncles and the interpeduncular ncl (Fig. 2.15 – 22)

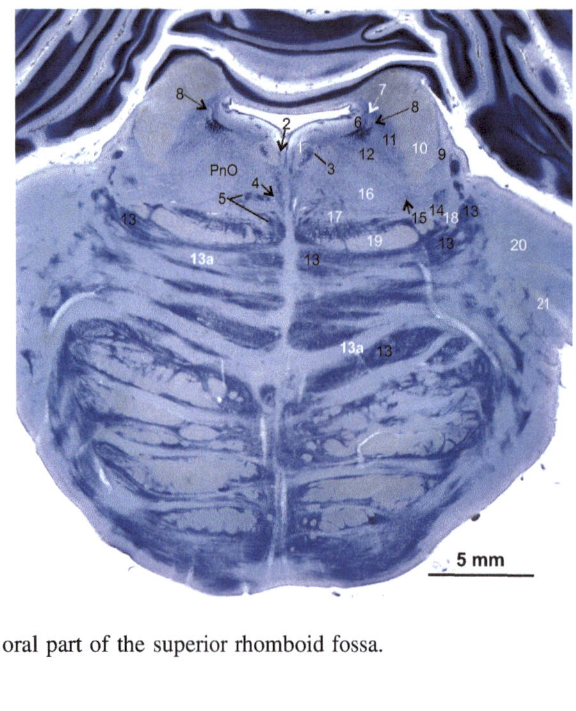

1 medial longitudinal fascicle
2 dorsal raphe ncl caudal part
3 reticulotegmental ncl lat. pt.
4 paramedian raphe ncl
5 reticulotegmental ncl
6 locus coeruleus
7 mesencephalic tract of N V
8 mesencephalic trigeminal nucleus
9 lateral parabrachial ncll
10 superior cerebellar peduncle#
11 medial parabchrial ncll
12 subcoeruleus ncl
13 pontine ncll
13a pontocerebellar fibers
14 central ncl of the acoustic tract
15 Koelliker-Fuse ncl
16 central tegmental tract
17 tirgeminothalamic tract
18 spinothalamic tract
19 medial lemniscus
20 medial cerebral peduncle
21 motor root of the N V

Fig. 2.14. Plane of section oral part of the superior rhomboid fossa.

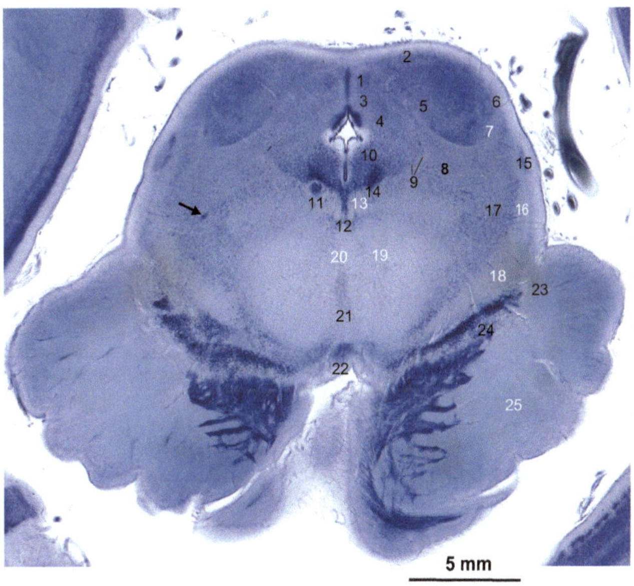

Fig. 2.15. Plane of section mesodiencephalic transition with oral pontine nuclei and caudal substantia nigra.

just below the floor of the interpeduncular fossa. The small interpeduncular ncl is part of the dorsal diencephalic conduction system, a highly conserved pathway linking the basal forebrain and the monoaminergic brainstem (Roman et al., 2020). The chromophilic neurons of the laterodorsal tegmental nucleus or Ch6 (Fig. 2.15 – 14) clearly stand out dorsal to the trochlear ncl and the MLF.

Laterally attached to the crossing fibers of the superior cerebellar peduncle is the pedunculopontine nucleus (Fig. 2.15 – 17). The tegmental pedunculopontine nucleus merges dorsomedially with the cuneiform nucleus (Fig. 2.15 – 8) without a clear boundary. A small cluster of chromophilic neurons is embedded in the dorsomedial parvocellular part of the left pedunculopontine nucleus, forming the most caudal part of the cholinergic nucleus Ch5n (arrow). The full extent of Ch5 is visible in Fig. 2.26E.

The caudal pars compacta of the substantia nigra (Fig. 2.15 – 23) is separated from the pedunculotegmental nucleus by the fibers of the medial lemniscus (Fig. 2.15 – 18). The fibers of the spinothalamic tract (Fig. 2.15 – 16) lie dorsal to the medial lemniscus, just below the surface of the lateral mesencephalic trigone (Fig. 2.6 – 2) and are accessible to neurosurgical intervention in patients with intractable pain. The course of the lateral lemniscus (Fig. 2.15 – 7) is marked by the central nucleus of the acoustic tract (Fig. 2.15 – 15). The interpeduncular fossa is lined by the cerebral peduncles (Fig. 2.15 – 25), which, together with the myriads of pontine neurons (Fig. 2.15 – 24), comprise the mesencephalic base.

The oculomotor complex

The seven-layered cortex of the superior colliculi (Fig. 2.16 – 1) represents the mesencephalic tectum in this plane of horizontal section. The mesencephalic aqueduct is surrounded by a broad rim of PAG matter. Several subnuclei can be distinguished in this rim (Fig. 2.16 –2). The deep white layer of the superior colliculus (Fig. 2.16 – 4) separates the lower layers of the superior colliculus from the PAG. Basolaterally, the deep white layer of the superior colliculus borders the mesencephalic tract of CN V (Fig. 2.16 – 5).

The individual nuclei of the oculomotor complex (Fig. 2.16 – 3) are located ventromedial to the PAG. Their profiles differ in size, shape, and staining characteristics.

Brownish staining, fine striations, and variable density and size of neurons characterize the parvo- (Fig. 2.16 – 13) and magnocellular (Fig. 2.16 – 15) portions of the red nuclei. Two motor systems meet in the red nucleus; the cerebellar fastigial nucleus projects to the magnocellular part, and the cerebral cortex to the parvocellular one (Cacciola et al., 2019). Projections of the red nucleus to the inferior olivary nucleus relay cerebral cortical information to the cerebellum via climbing fibers. This system is subject to considerable changes in evolution (ten Donkelaar, 1988; Stacho et al., 2024).

The precuneiform area (Fig. 2.16 – 6) and the mesencephalic reticular formation (Fig. 2.16 – 7) are in the dorsolateral quadrant of the mesencephalic tegmentum. The spinothalamic tract (Fig. 2.16 – 10) and the medial lemniscus (Fig. 2.16 – 11) separate mesencephalic reticular formation and precuneiform area from the nucleus of the brachium of the inferior colliculus (Fig. 2.16 – 9) and overlying brachium of the inferior colliculus (Fig. 2.16 – 8).

The ventral tegmentum is characterized by the presence of big pigmented and chromophilic substantia nigra neurons (Fig. 2.16 – 14). They are arranged either in single clusters or in clusters interconnected by thin cellular bridges which are stacked in three layers or tiers: dorsal (Fig. 2.16 – 14a), medial (Fig. 2.16 – 14b), and ventral (Fig. 2.16 – 14c). Together, they constitute the compact part of the substantia nigra. Most of the neurons in the tiers contain intracytoplasmic neuromelanin granules. As a result, the nigral pars compacta is visible to the naked eye as a narrow, curved band on fresh, unstained

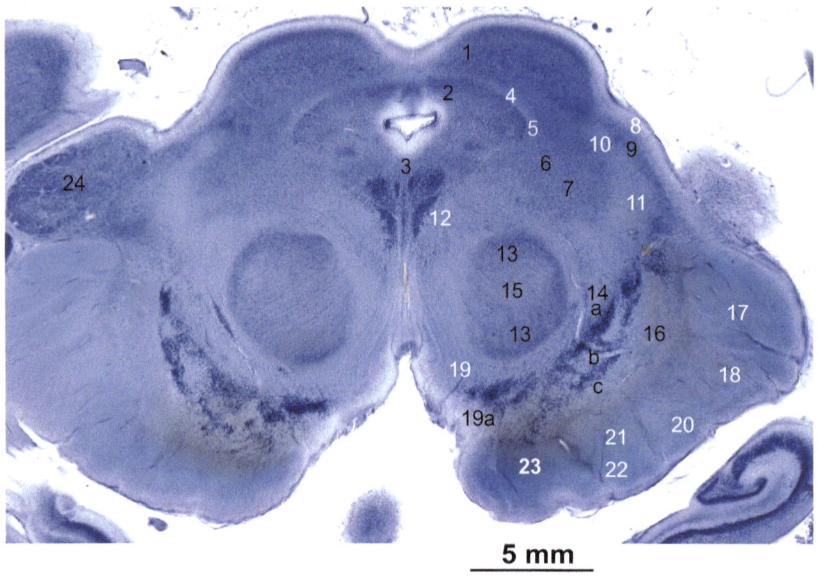

1 superior colliculus
2 periaqueductal gray (PAG)
3 oculomotor complex
4 deep white layer of superior coll.
5 mesencephalic tract of CN V
6 precuneiforme area
7 mesencephalic reticular formation
8 brachium of the inferior colliculus
9 ncl of the brachium of the inferior colliculus
10 spinothalamic tract
11 medial lemniscus
12 medial longitudinal fascicle
13 parvocellular red ncl
14 substantia nigra
14a dorsal tier of substantia nigra
14b medial tier
14c ventral tier
15 magnocellular red ncl
16 pars reticulata of the subst. nigra
17 occipitopontine tract
18 parietopontine tract
19 intracerebral fibers of N III
19a paranigral ncl
20 temporopontine tract
21 corticospinal tract
22 corticobulbar tract
23 fontopontine tract
24 medial geniculate body

Fig. 2.16. Plane of section caudal pole of the trochlear ncl.

sections through the midbrain (Fig. 2.20B). As in locus coeruleus neurons, the main transmitter of pigmented substantia nigra neurons, the monoamine dopamine, is catabolized and stored as neuromelanin.

The ventral parts of the pars compacta protrude into a cell-poor region of the substantia nigra, which is brownish in color in older individuals and is continuous with the internal pallidal globe. This is the substantia nigra, pars reticulata (Fig. 2.16 – 16).

The medial parts of the substantia nigra pars compacta are located just beneath the surface of the crura. Two or three compacta clusters are regularly pierced by the rootlets of the CN III (Fig. 2.16 – 19). The overall size of its neurons is smaller, and the density of these pigmented cells is higher. This is the paranigral nucleus of the ventral tegmental area (VTA; Fig. 2.16 – 19a).

The cerebral crura are made up of many axons from layer V pyramidal cells that project to targets in the brainstem and spinal cord. They are arranged according to their origin and destination, from lateral to medial: parietopontine (Fig. 2.16 – 17), occipitopontine (Fig. 2.16 – 18), temporopontine (Fig. 2.16 – 20), corticospinal (Fig. 2.16 – 21), corticobulbar (Fig. 2.16 – 22), and frontopontine tract (Fig. 2.16 – 23).

Oculomotor complex and premotor nuclei

The nuclear complex of the third cranial nerve in our case extended 7.2 mm in the midline of the mesencephalon between the inferior and superior colliculus. In vivo, this would be approximately 14 mm after correction for shrinkage factors associated with histologic procedures. The oculomotor subnuclei are closely attached to the MLF (Fig. 2.17 – 6). The central caudal nucleus (Fig. 2.17 – 3, CCN) is embedded in the fascicle, which is pierced by the rootlets of the CN III (Fig. 2.17 – 7). In serial sections from the caudal to the rostral pole, seven nuclear profiles could be identified in a combined cyto/myeloarchitectonic and immunohistochemical study (Che Ngwa et al., 2014). They were named central caudal nucleus (Fig. 2.17 – 3, CCN), dorsolateral (Fig. 2.17 – 1, DL), dorsomedial (Fig. 2.17 2, DM), lateral (Fig. 2.17 – 4, LAT) group embedded in the MLF (Fig. 2.17 – 6), central (Fig. 2.17 – 5, CEN) and ventral (Fig. 2.17 – 8, VEN) nuclear group, and the Edinger–Westphal nucleus (Fig. 2.17B – 9, EW). The Edinger–Westphal nucleus has been subdivided into (1) the cholinergic, preganglionic neurons supplying the ciliary ganglion (Edinger–Westphal preganglionic population, EWpg) and (2) the centrally projecting, peptidergic neurons (Edinger–-Westphal centrally projecting (EWcp) population; Kozicz et al., 2011). The nucleus of Perlia (Fig. 2.17B – 10, NP) and the nonpreganglionic CEN (Fig. 2.17 – 5) are considered as superior rectus and inferior oblique motoneurons, DL (Fig. 2.17 – 1) and VEN (8) as B- and A-group of medial rectus motoneurons, respectively, and DM (Fig. 2.17 – 2) as inferior rectus motoneurons.

On tangential sections through the rostral pole of the oculomotor nucleus complex (Fig. 2.17C), the small, elongated profiles visible are difficult to assign to the subnuclei listed in the previous paragraph. Embedded in the MLF (Fig. 2.17C – 6) are the interstitial nucleus of Cajal (Fig. 2.17C – 12) and the medial nucleus of Darkschewitsch (Fig. 2.17CI – 11). They represent the higher-level nuclei in the oculomotor system.

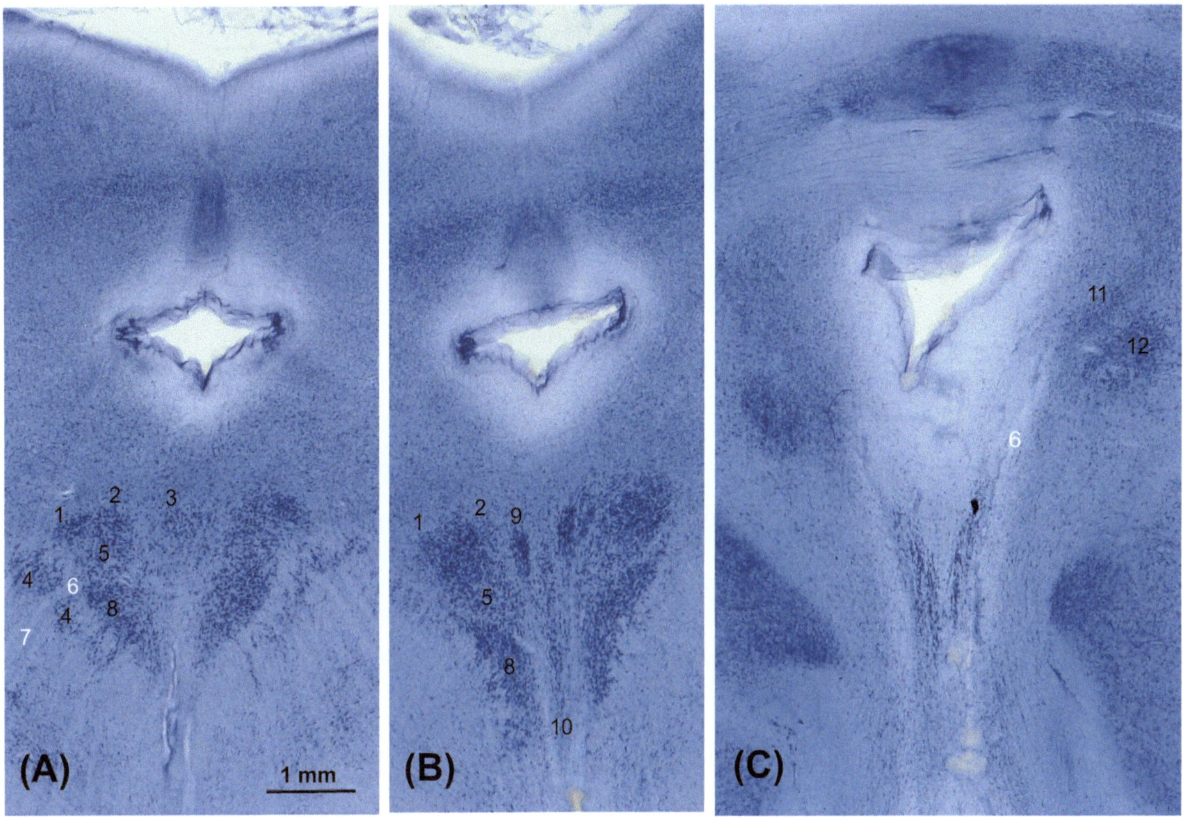

Fig. 2.17. (**A**) caudal, (**B**) central, and (**C**) rostral parts of the oculomotor complex.

Parasagittal aspect of the inferior and superior colliculi

Due to the geometry and curved shape of the superior and inferior colliculi, the plane of section is not strictly perpendicular, resulting in a blurred layer-wise arrangement and delineation of neurons.

Seven layers were described in the *superior colliculus* (Kaas and Huerta, 1988) and designated as follows: (Fig. 2.18, I) Stratum Zonale or SZ, (Fig. 2.18, II) Stratum Griseum Superficiale or SGS, (Fig. 2.18, III) Stratum Opticum or SO, (Fig. 2.18, IV) Stratum Griseum Intermediale or SGI, (Fig. 2.18, V) Stratum Album Intermediale or SAI, (Fig. 2.18, VI) Stratum Griseum Profundum or SGP, and (Fig. 2.18, VII) Stratum Album Profundum or SAP.

In addition, the layers may be classified as either fibrous or cellular. The fibrous layers are I, III, V, and VII, while the cellular layers are II, IV, and VI. The layers may also be grouped into three categories: superficial (Fig. 2.18, I–III), intermediate (Fig. 2.18, IV and V), and deep (VI and VII).

The superficial zone receives primarily visual retinal and corticotectal afferents and gives rise to ascending fibers. The deep zone receives heterogeneous multimodal auditory and somatosensory inputs and sends ascending and descending fibers to centers in the brainstem, thalamus, and spinal cord. Dendritic arbors of neurons in superficial layers are generally restricted to these layers. Large multipolar neurons in deeper layers extend their dendrites into the superficial layers. The connections between superficial and deep layers are formed by axon collaterals of small superficial L-neurons (Nieuwenhuys et al., 2008). The superior colliculus initiates and/or participates in eye movements, eye and head movements, eye-head-body movements, grasping movements, and sonar vocalization (Gandhi and Katnani, 2011).

The principal output tracts are the crossed predorsal bundle and the uncrossed tectospinal tract (Nieuwenhuys et al., 2008; Gandhi and Katnani, 2011).

The terminology of the *inferior collicular components* is equivocal (Huffman and Henson, 1990). We have adopted the designation of (Geniec and Morest, 1971; Morest and Oliver, 1984) who illustrated their findings in cats and humans. The central nucleus (Fig. 2.18 CN) is separated from the surface of the collicular convexity by the cortical fields dorsally (Fig. 2.18 – *DC*) and caudally (Fig. 2.18 – *CC*). The pericollicular tegmentum (Fig. 2.18 – *IT*, C) respectively, separate the central nucleus from the superior colliculus, from the lateral surface of the tectum, and from the underlying regions of the midbrain tegmentum. The feline dorsal (DC) and caudal (CC) cortical structures have been parcellated into four layers (I ic–IV ic; "ic" is here used to distinguish the inferior collicular layers from the superior ones). MeV, mesencephalic trigeminal tract, LL lateral lemniscus.

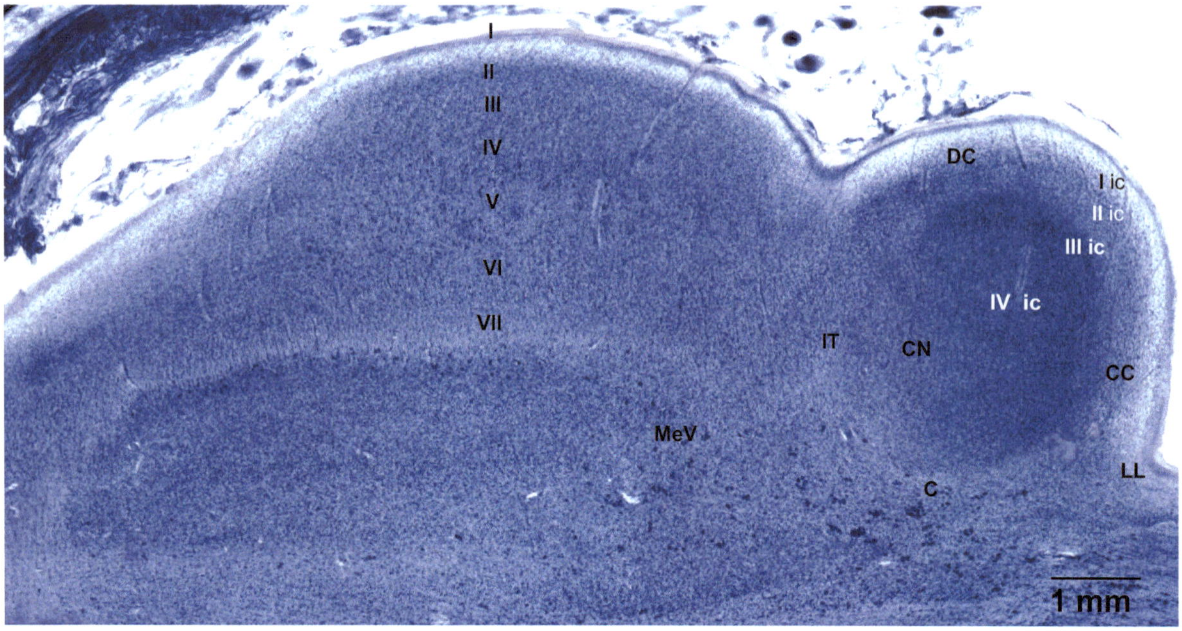

Fig. 2.18. Parasagittal section through the right tectum passing the inferior and superior colliculi.

The inferior colliculus is a major hub for ascending and descending fibers from the cochlear nuclei. Ascending mainly crossed and tonotopically arranged fibers from the cochlear nucleus are relayed in the superior olive, the nucleus of the lateral lemniscus, and in different components of the inferior colliculus.

Periaqueductal gray

Cytoarchitecturally, the PAG is not homogeneous (Mantyh, 1982). Classical Golgi, Nissl, histochemical, and cytochemical studies have revealed the arrangement of neurons in longitudinal columns: the dorsal PAG, the dorsolateral PAG, the lateral PAG, and the ventrolateral PAG. Electrophysiologic and chemical stimulation of columns located in the dorsal and lateral regions evoked the fight–flight response in experimental animals (Carrive and Morgan, 2012). Recently, the somatotopic representation of the body could be demonstrated in 7T functional magnetic resonance imaging (Tinoco Mendoza et al., 2023). Topographically, the PAG is located at the intersection of the descending limbic and ascending sensory systems, which include the entire CNS from the dorsomedial, orbitofrontal, and anterior cingulate cortices to laminae V, VII, and VIII of the spinal cord. Many of these connections are reciprocal. Subcortical and brainstem nuclei contribute additional information to this extended system. These include amygdala, hypothalamic, midbrain, pontine, and medullary afferents, as well as ascending neuronal circuits are relayed by thalamic midline and intralaminar nuclei (Fig. 2.19C). The pathways emanating from the thalamic intralaminar nuclei are widely distributed within the cerebral cortex and tend to favor layer I. Lesions of these nuclei can cause widespread and confused state and spatial neglect, bilateral influence on cortical activity, and modulate signal-to-noise ratios during attentional focusing and sensory discrimination (Raos et al., 1995).

Chemoarchitecture

Histochemical, immunohistochemical, and molecular biologic methods complement and expand our knowledge of the structure and function of the nervous system. Neuronal groups and nuclei of monoaminergic and cholinergic neurons have been identified in the brainstem. Neuromelanin pigment in neurons is considered a marker for noradrenergic (locus coeruleus, Fig. 2.20A) and dopaminergic (substantia nigra pars compacta, Fig. 2.20B) neurons (Saper and Petito, 1982).

Immunohistochemical methods are used to directly or indirectly identify serotonin transmission. The same is true for the neurotransmitter acetylcholine. These methods are expensive and time-consuming. They require nervous tissue that has not been subjected to prolonged agony, a short postmortem time, perfusion fixation, and human labor, expertise, and experience (Swaab and Uylings, 1988).

Monoaminergic and cholinergic nuclei share some common characteristics. The neurons are large, they have an extended dendritic arbor, and their axons are thin, long, and extensively arborized (Ramón-Moliner and Nauta, 1966; Mesulam et al., 1989; Arendt et al., 2015; Gilvesy et al., 2022). Extensive convergence and divergence of information is based on these characteristics (Kositzyn, 1964).

In contrast to rapid conduction and transmission at pre- and postsynaptic structures of peripheral and central synapses, monoamines and acetylcholine can be released (paracrine mechanism) from neurons in the brainstem and basal forebrain into the intercellular space of the target region. They then

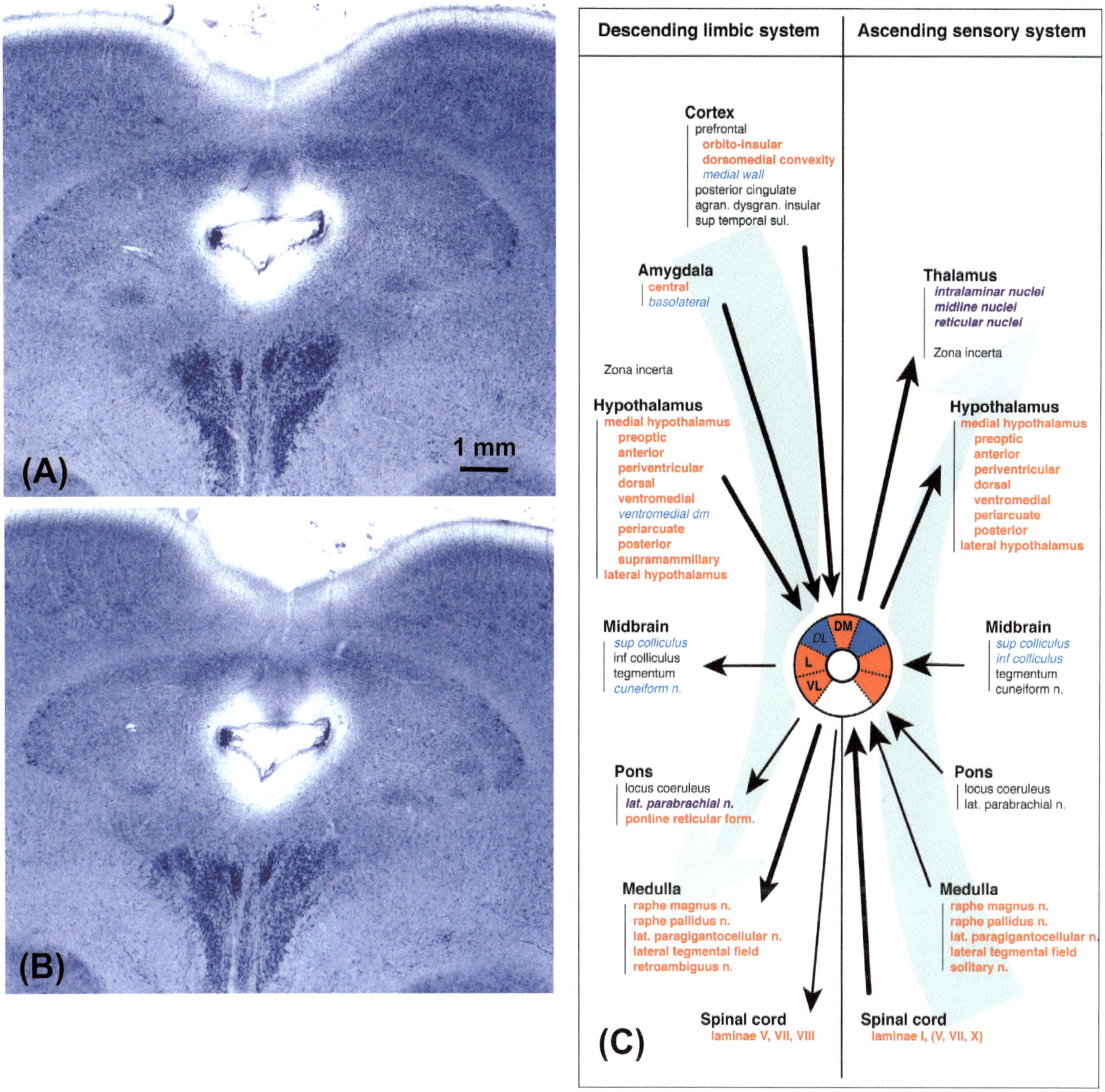

Fig. 2.19. (**A** and **B**) Two consecutive 400-µm serial sections through human periaqueductal gray. Note the change in size and shape of the nuclear profiles in the Nissl-stained sections. These profiles reflect mediolateral differences in connectivity (**C**) as well as rostrocaudal changes in distribution of receptors and chemoarchitecture within a short longitudinal distance. (C) Reproduced with permission from Carrive P, Morgan MM **2012** Periaqueductal gray. In: Mai JK, Paxinos G (eds**.**) The human nervous system. Amsterdam/Boston/Heidelberg: Elsevier, **pp**. 367–400.

diffuse and spread over long distances to the somas and processes of nerve cells, glial cells, or vessel walls. Depending on the receptor type, the transmitters can induce various effects, including excitation, inhibition, changes in metabolism and blood flow (Fuxe et al., 2007).

Monoaminergic neurons display autoinhibition by collaterals (Langer, 2015). The evaluation of effects following pharmacologic or electrical stimulation of monoaminergic neurons can be complicated by their colocalization (intraneuronal presence of different transmitters and neuroactive peptides) and their corelease (Vadodaria et al., 2018).

Thin, long axons, many of which lack a myelin sheath, have a relatively large surface area in a small volume. The surface area of a sphere or cylinder decreases with the square, but the volume decreases with the cube. This means that thin, including unmyelinated, fibers have a relatively large surface area relative to their volume, which is necessary for axonal flow and transport. Neural transmission requires lifelong integrity of

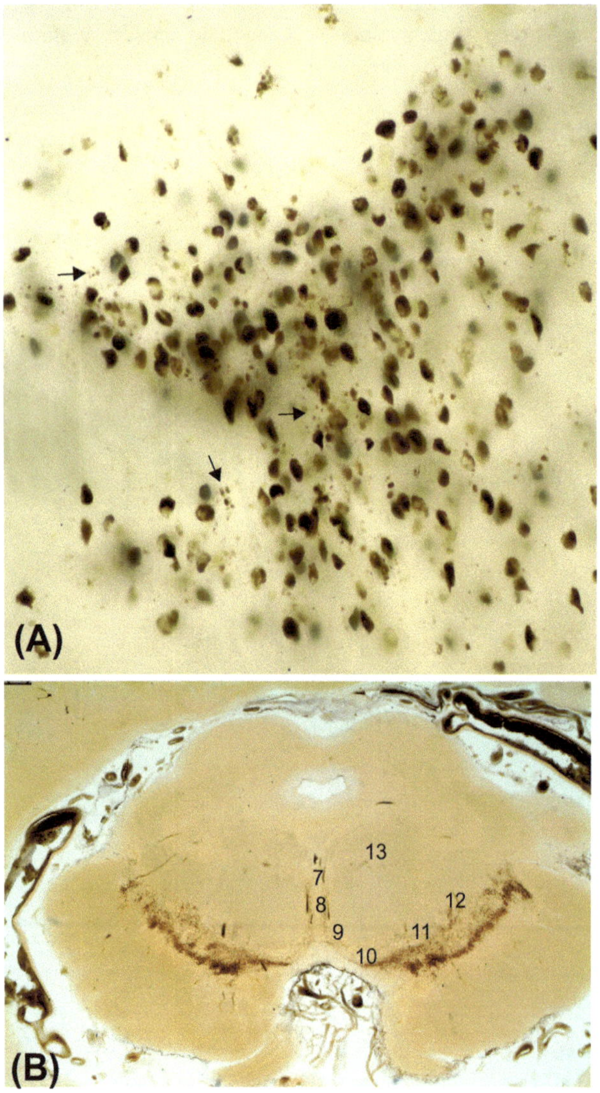

Fig. 2.20. (**A** and **B**) Unstained xylene-cleared 400 μm thick sections through the rostral parts of the locus coeruleus in (**A**) and the substantia nigra in (**B**). Accumulated neuromelanin in the perikaryon of both types of neurons marks the silhouettes of the neurons. In (**A**), scattered lipofuscin/neuromelanin granules varying in size and shape can be seen with higher magnification in the unstained neuropil between neuromelanin-containing neurons (arrows). Dying neuromelanin-rich neurons release their neuromelanin from the perikaryon. Images (A) and (B) are from a 79-year-old female with Alzheimer disease. The numbers in (**B**) correspond to the numbers in Fig. 2.22: (7) caudal linear ncl, (8) rostral linear ncl, (9) interfascicular ncl, (10) paranigral ncl, (11) ventral tegmental area, (12) parabrachial pigmented ncl, and (13) retrorubral field.

neuronal membranes and a continuous, finely tuned balance of ions and transmitters. In addition, humans have an exceptionally large brain, which correlates with body length (Stephan et al., 1988). This means that a limited number of 1–2 million monoaminergic and cholinergic neurons can/must modulate the activity of billions of cortical neurons (Fig. 2.21). Early selective vulnerability of monoaminergic and cholinergic neurons in human-specific neurodegenerative and psychiatric diseases may be the price of human evolution and lifestyle (Rapoport, 1990).

Monoaminergic neuronal groups and nuclei

The abbreviations A1–A7 are used to code and classify noradrenergic cell groups and nuclei, while A8–A15 are used for dopaminergic ones. Additionally, B1–B9 are used for serotoninergic cell groups and nuclei, and C1–C3 are used for adrenergic ones. Agostinelli et al. (2023) provide a combined imaging and immunohistochemical study of a human brainstem and their results can be easily compared with our illustrations (Fig. 2.22).

A1–A7 (noradrenergic cell groups and nuclei)

A1 extends in the myelencephalon from the pyramidal decussation to the obex. A2 is part of the dorsal vagal complex, and A3 could not be identified in humans (Fig. 2.22). A4 neurons lie close to the superior cerebellar peduncle, and A5 around the facial nucleus and between the latter and the superior olive. A6 is the nucleus (locus) coeruleus or caeruleus, and A7 includes neurons in the ventrolateral nucleus subcoeruleus and ventrolateral to the parabrachial nuclei.

A6 locus coeruleus

The majority of locus coeruleus neurons are spindle-shaped and pigmented (Fig. 2.20A), while those in the subcoeruleus are multipolar (Fig. 2.14 – 6,12). However, the neurons exhibit local differences in receptors, intracellular peptides, afferent and efferent connectivity.

Ascending somato- and viscerosensory information is relayed to locus coeruleus neurons via nuclei in the medulla oblongata, including the gigantocellular, paragigantocellular, solitary tract, and prepositus hypoglossal nuclei. Descending signals originate in the hypothalamus, central nucleus of the amygdala, and the prefrontal cortex (Felten and Sladek, 1983; Delbono et al., 2022).

The long, thin, and ramified ascending axons from locus coeruleus neurons innervate most parts of the cerebral cortex (Fig. 2.21 – 6 and 7) with focus on hippocampus (Fig. 2.21 – 3) and the telencephalic basal forebrain nuclei (Fig. 2.21 – 5), extended amygdala (Fig. 2.21 – 4) and septum (Fig. 2.21 – 5), diencephalon with focus on its paraventricular nucleus (Fig. 2.21 – 2) and thalamus (Fig. 2.21 – 1), and upper brainstem, including the cerebellum. Descending fibers that originate in A5 and A7 supply brainstem nuclei and the spinal cord.

The locus coeruleus plays an important role in the modulation of wakefulness, attention, cortical perception, and consciousness. Depending on the interplay between environmental sensory information and processed prefrontal information, signals from locus coeruleus neurons activate cortical and subcortical regions tonically (sleep–wake cycle) or phasically. Novel sensory input causes arousal, focused attention, and activation of working memory. On the other hand, stress

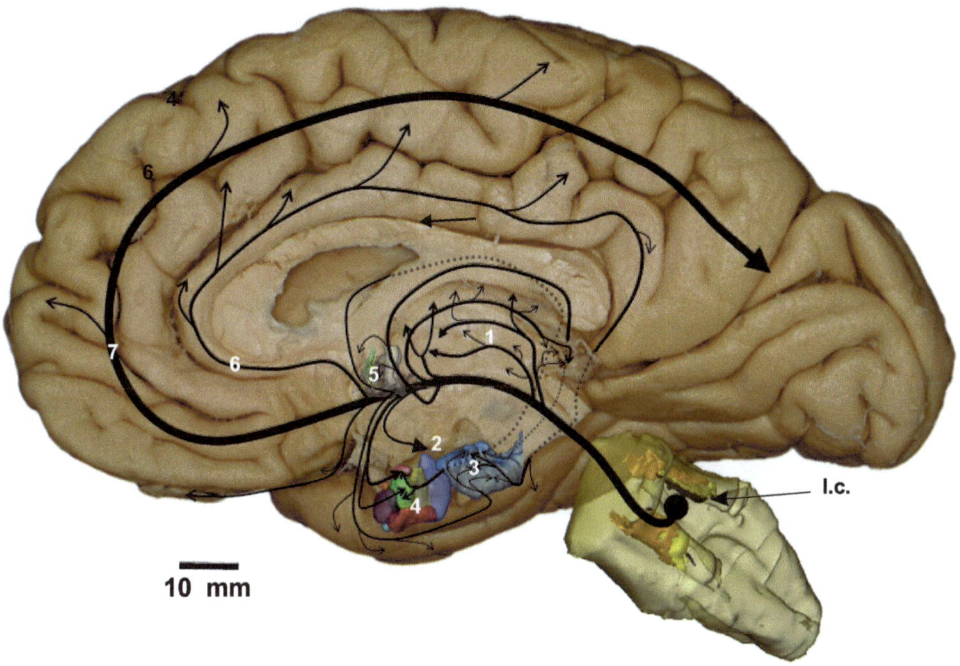

Fig. 2.21. A scaled 3D reconstruction of the right ncl locus coeruleus (arrow, l.c., anteroposterior length 12–16 mm) and its noradrenergic projections to (1) thalamic nuclei, (2) hypothalamus, (3) hippocampus, (4) amygdala complex, (5) basal forebrain complex, (6) mesocortical cingulate gyrus, and (7) isocortical fields.

suppresses working memory, enhances attention, and locus coeruleus activation of the hypothalamic paraventricular nucleus initiates stress reactions in the central autonomic nervous.

Pain perception is modulated by ascending locus coeruleus fibers to the anterior cingulate cortex (Fig. 2.21 – 6) and descending fibers to the spinal trigeminal nucleus and dorsal horn of the spinal cord, which are involved in processing nociceptive signals.

Memory consolidation is modulated by efferents to the hippocampus and hypothalamus, which facilitate memory formation by integrating emotional and autonomic with somatosensory experiences.

Recently, distinct functional subpopulations of locus coeruleus neuronal circuits based on the afferent and efferent connections were postulated to explain the diverse functions of this brainstem nucleus (Poe et al., 2020).

Dopaminergic neuronal groups and nuclei (A8–A15)

Dopamine-positive neurons are distributed in the mesencephalon (Fig. 2.23), di- and telencephalon. A8 (Fig. 2.20B – 3, Fig. 2.22C – 13) is in the retrorubral field of the caudal mesencephalon lateral to the superior cerebellar peduncle. A9 corresponds to the nigral pars compacta (Fig. 2.16 – 14a-c), and A10 is in the VTA (Fig. 2.23B – 11). A11 is associated with the mesencephalic PAG, A12 with the arcuate nucleus, A13 with the zona incerta, A14 with the paraventricular and supraoptic nuclei, and A15 with the periglomerular cells of the olfactory bulb. A11 through A15 are connected to neighboring structures. The functional significance of A9 and A10 is more complex.

Substantia nigra pars compacta dopaminergic neurons (A9)

Cortical regions with similar functions are reciprocally connected via cortico-striatal-globus pallidus-thalamus-cortical loops in a parallel and segregated manner (Alexander et al., 1986).

Motor, oculomotor, dorsolateral prefrontal, lateral orbitofrontal, and anterior cingulate loops were identified by electrophysiology. Each loop consists of nonoverlapping circumscribed cortical fields, striatal (putamen, caudate, ventral striatum sectors), segregated internal pallidal sectors, and thalamic nuclei (ventral lateral, ventral anterior, and mediodorsal nucleus). Neurons in the pallidal sectors also transmit impulses from the various cortical areas to segregated substantia nigra pars reticulata neurons, which in turn project to segregated regions of pars compacta neurons. Dopaminergic efferent pars compacta neurons target stellate cells in the caudate nucleus and putamen. Depending on the presence of D1 or D2 dopamine receptors, GABAergic stellate cells are either inhibited (D1 receptors) or disinhibited (D2 receptors). Knowing the pathways, transmitters, and receptors can explain the pathophysiology of clinical deficits in Huntington (HD) and Parkinson (PD) diseases.

The concept of parallel and segregated cortico-striato-pallido-thalamo-cortical loops cannot explain emotional and other influences on complex motor actions. Haber (2003) suggests that dopaminergic nigrostriatal together with mesolimbic and mesocortical projections could orchestrate parallel and segregated neuronal circuits.

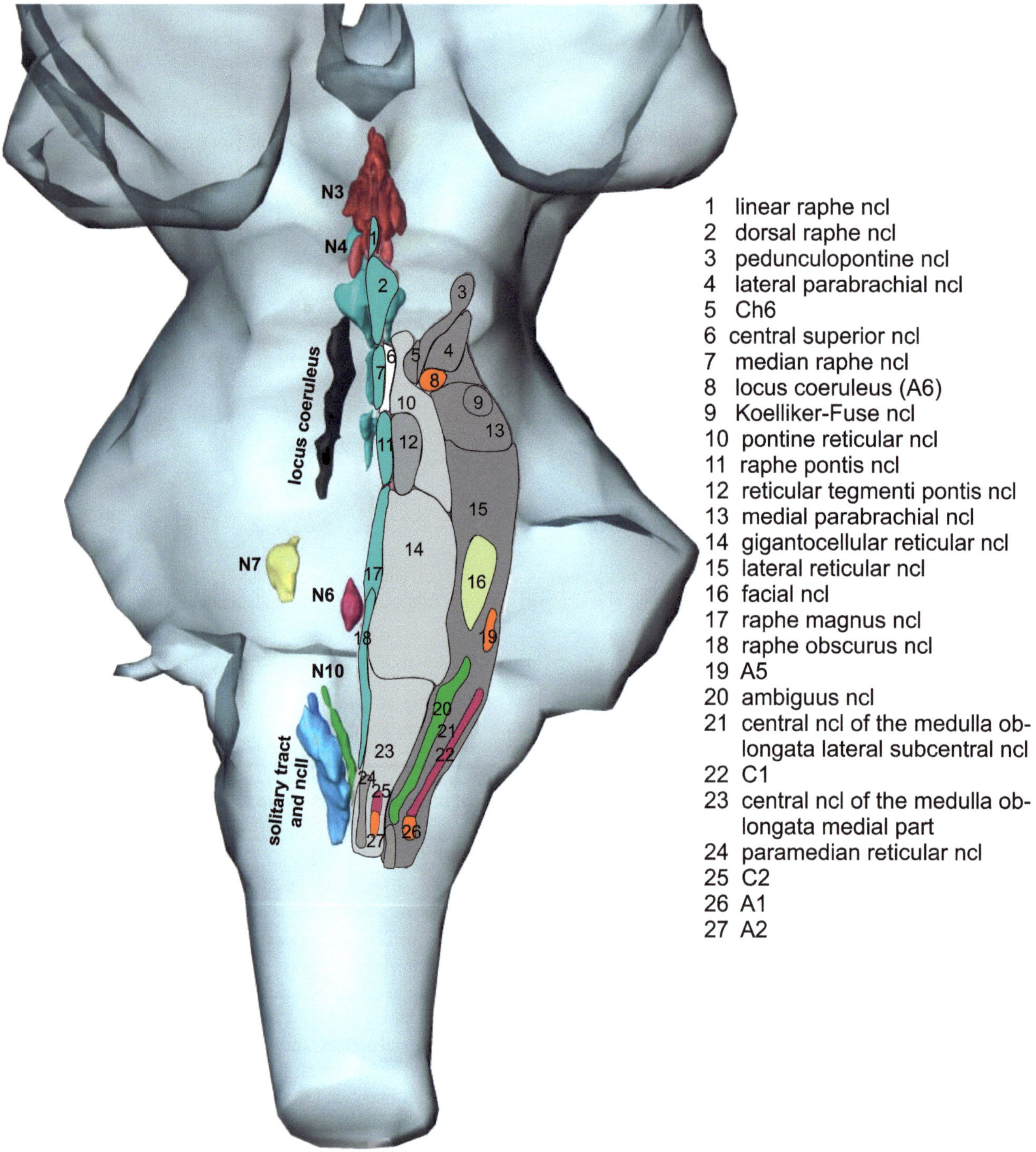

Fig. 2.22. The reticular formation of the brainstem: *light grey*, medial tegmental field; *dark gray*, lateral tegmental field; light *blue*, raphe nuclei; *orange*, noradrenergic cell groups A1–A6; *magenta*, adrenergic groups C1 and C2; *grey*, the cholinergic cell groups Ch5 and Ch6. Modified after Nieuwenhuys R, Voogd J, van Huijzen C (2008) The human central nervous system. Berlin: Springer.

Hassler (1937) divided the substantia nigra into dozens of subfields. Using 20 μm thick serial sections, he was unaware that his subnuclei were knotted in a neuronal network. Our 3D reconstructions showed individual, highly characteristic shapes with randomly distributed asymmetric neuron numbers in either the right or left substantia nigra (Di Lorenzo Alho et al., 2016). The 3D reconstructions show only the largest holes in these irregular contours. We had to make a compromise between detail and smoothing the 3D model.

Ventral tegmental area (A10)

The mesolimbic system is composed of dopaminergic neurons in the VTA (Fig. 2.23B – 10,11) that project to the nucleus accumbens, hippocampus, amygdala, and hypothalamus. The

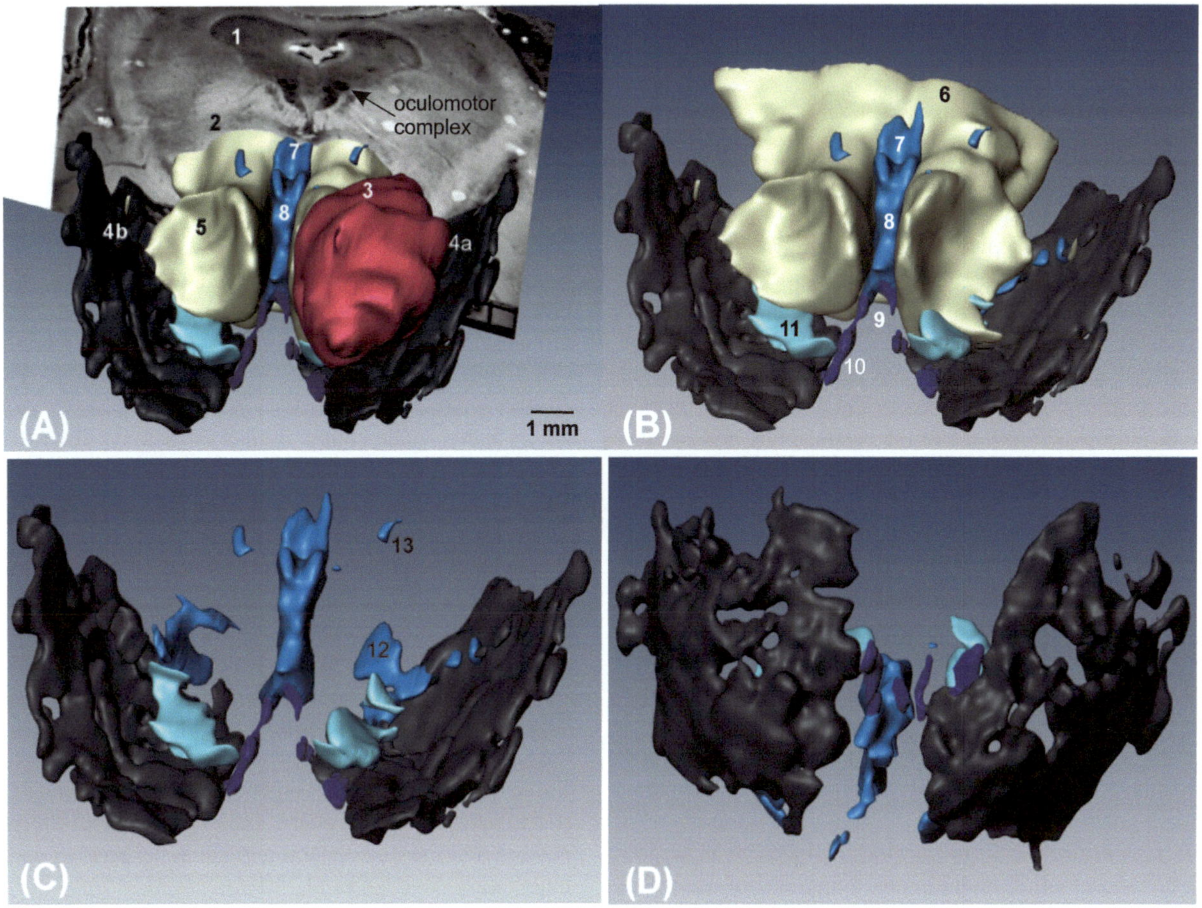

Fig. 2.23. (**A–D**) 3D reconstruction of selected nuclei in the ventral mesencephalic tegmentum and compact part of the substantia nigra; the aspect in figs. (**A–D**) is from rostro-dorsal to caudo-ventral, the axis of the brainstem is inclined rostro-ventrally and slightly rotated in a counterclockwise manner in (**A–C**) and by 180° in (**D**); the left red ncl (3) in (**D**), and both superior cerebellar peduncles (5) were faded out in (**B** and **C**), respectively. Black and white digital gallocyanin-stained images *(A – 1)* were used for manual segmentation of the nuclei reconstructed. *A – 1* is the most caudal section of a total of 25 serial 400 μm thick sections; *A – 2* right uncrossed superior cerebellar peduncle; *A – 3* left red ncl; *A – 4b* and *4a* right and left caudal pole of the compact part of the substantia nigra, respectively; *A – 5* left crossed cerebellar peduncle; *B – 6* left uncrossed superior cerebellar peduncle; *A – 7* caudal linear ncl; *A – 8* rostral linear ncl; *B – 9* interfascicular ncl; *10* paranigral ncl, *11* ventral tegmental area; *B*, *C – 12* parabrachial pigmented ncl; *C – 13* retrorubral field.

mesocortical system, on the other hand, refers to dopaminergic neurons with projections to the anterior cingulate, medial prefrontal, and orbitofrontal cortices (Morales and Margolis, 2017).

Electrophysiologic studies in primates have shown that dopamine encodes prediction errors (Schultz et al., 1997). Addictive drugs interfere with physiologic dopamine activity and induce widespread symptoms and behaviors associated with drug abuse and withdrawal (Schultz et al., 1997; Wolf, 2016; Paniccia et al., 2024).

Adrenergic neuronal groups (C1–C3)

C1 neurons are located in the rostral ventrolateral myelencephalon near the reticular paragigantocellular nucleus, C2 in the dorsal vagus nuclei, and single C3 neurons in the midline of the brainstem (Fig. 2.22).

Ascending projections terminate in several brain regions, including the parabrachial nuclei, hypothalamic paraventricular nucleus, thalamic midline nuclei, and amygdala. The adrenergic innervation of the hypothalamic paraventricular nucleus can directly regulate the hypothalamic-pituitary-adrenal axis. Additionally, it can indirectly regulate this axis through adrenergic projections to the noradrenergic myelencephalic and pontine nuclei, including the locus coeruleus and serotonergic raphe nuclei.

Descending fibers target preganglionic parasympathetic neurons in the dorsal vagus nucleus and presympathetic neurons in the spinal cord. The presympathetic C1-adrenergic neurons are important for circulatory control and stimulation of the respiratory center. C1 neurons are indispensable for the control of respiration and circulation.

Serotoninergic neuronal groups (B1–B9)

Serotonergic neuronal groups are populated by neuronal elements using transmitters other than serotonin, including GABA, glutamate, dopamine, noradrenaline and peptides such as TRH, CCK, somatostatin, enkephalin, galanin, substance P, and neurotensin (Fig. 2.22).

Neuronal groups B1–B4 are in the myelencephalon and caudal pons: B1 in the raphe pallidus (Fig. 2.9 – 18), B2 in the raphe obscurus (Figs. 2.10 – 12, B3 in the raphe magnus, and B4 between the floor of the fourth ventricle and the vestibular nuclei.

Neuronal groups B5–B9 are in the rostral pons and mesencephalon: B5, B6, and B8 are associated with the median raphe (Fig. 2.15 – 12) in the rostral pons and the linear raphe (Fig. 2.15 – 21) in the caudal midbrain, and group B7 with the dorsal raphe nucleus. The dorsal raphe nucleus is the largest nucleus and consists of several subnuclei (Fig. 2.15 – 10).

The efferent fibers of the serotonergic raphe nuclei innervate all parts of the CNS by means of a dense network of thin, branched fibers. The ascending fibers course in the central gray, midbrain tegmentum, and medial forebrain bundle (MFB) to innervate the mesencephalic colliculi, interpeduncular nucleus, hypothalamus, thalamic nuclei, subthalamic nucleus, striatum, septum, amygdala, olfactory cortex, and cerebral cortex. A strong efferent projection from the dorsal raphe nuclei targets the amygdala. Serotonin is an inhibitory antagonist of dopamine in the amygdala. It may play an important role in the regulation of emotion, anxiety, and depression. Efferents to the hippocampus may modulate memory formation.

Descending tracts originate from the caudal raphe nuclei and innervate the rhombencephalon and spinal cord. The raphe magnus innervates the dorsal horn and the zona intermedia. The raphe pallidus and obscurus innervate the anterior horn and the intermediolateral nucleus. Other raphe nuclei innervate the cerebellum, somatomotor, visceromotor, and sensory brainstem nuclei, and connect individual raphe nuclei.

Afferents to the raphe nuclei originate from the hypothalamus, habenular nuclei, tegmental gray, interpeduncular nucleus, and various dopaminergic and noradrenergic nuclei.

The number of drugs and substances with excitatory or inhibitory activity at human serotonin receptors reflects the widely distributed innervation of different regions and the physiologic role of serotonin in the central and peripheral nervous system (Kawashima, 2018). These include antidepressants, neuroleptics, hypnotics, tranquilizers, appetite suppressants, migraine treatments, antihypertensives, anticoagulants, antiemetics, psychedelics, and hallucinogens (Vadodaria et al., 2018).

Cholinergic neuronal groups (Ch1–8)

Acetylcholine (ACh) acts through the ionotropic nicotinic receptor as a rapid transmitter in somatomotor, branchiomotor, and visceroefferent neurons of the cranial and spinal nerves, and postganglionic parasympathetic neurons communicate with postsynaptic structures via acetylcholine Sarter and Parikh, 2005 (Fig. 2.22). The activity of the ACh-synthesizing enzyme cholineacetyltransferase (ChAT) can be detected histochemically in a nonspecific way and more specifically by immunohistochemistry using antibodies that bind to the vesicular acetylcholine transporter.

ACh-gated neuronal groups can be neurons in the telencephalon and brainstem. These groups have been labeled Ch1 through Ch8. Ch1–Ch4 are part of the basal forebrain complex (Mesulam et al., 1983; Mesulam, 1995; Heinsen et al., 2006; Grinberg and Heinsen, 2007). Ch4, the basal nucleus of Meynert, shows early vulnerability during Alzheimer disease (Kilimann et al., 2017). Abnormal acetylcholine transmission is also seen in Parkinson disease, Parkinson disease with dementia, and diffuse Lewy body disease (He et al., 2023).

The cholinergic neuronal groups in the brainstem include Ch5 in the pedunculopontine tegmental nucleus (2.15 arrow and 2.26E), Ch6 in the dorsolateral tegmental nucleus (Fig. 2.15 – 14), and Ch8 near the inferior colliculus. Ch5 and Ch6 project to the thalamic intralaminar nuclei, lateral geniculate body, lateral hypothalamus, and interpeduncular nucleus. Some fibers reach the basal forebrain and areas of the cortex (Mesulam et al., 1989). These projections are components of the so-called ascending (reticular) activation system. Interestingly, both the pedunculopontine and the dorsolateral tegmental ncl receive dopaminergic innervation (Ryczko et al., 2016).

Descending projections, especially from Ch5, reach the paramedian medulla oblongata.

The longitudinal organization of brainstem nuclei

The reticular formation, a lattice of longitudinally and transversally oriented fibers and stacked neurons providing maximal convergence and divergence of information between spinal cord, brainstem nuclei, and endbrain based on multiple transmitters and neuromodulators

Meynert used Flechsig's term Formatio reticularis (Forel, 1877) to describe the origin and course of sagittally oriented fibers in the tegmentum of the brainstem. Strands and groups of carmine-stained neurons are embedded in the reticular formation. The arrangement of neurons in the reticular formation appears diffuse, in contrast to most cranial nerve nuclei. Neurons in the reticular formation are classified as isodendritic. The dendritic trees are arranged in discrete units like mediolaterally flattened discs arising from neurons stacked anteroposteriorly in the longitudinal axis. The dendritic domain of a single reticular cell can be so large that it transcends cytoarchitectonically determined nuclear boundaries; long axons can descend and ascend over long distances that span the axis of the brainstem. This concept is mainly based on the Golgi impregnation of the brainstem of a 2-day-old mouse and the drawing presented Scheibel and Scheibel (1958). This may be true for young rats, but less than 2% of the reticular neurons project to both diencephalon and spinal cord (Waltzer and Martin, 1984; Jones and Yang, 1985).

Following high-frequency electrical stimulation of the brainstem in chloral-derivative anesthetized cats, the synchronized discharges of high-voltage slow waves in the electroencephalogram (EEG) are replaced by low-voltage fast activity, closely resembling the corresponding changes observed in the human EEG during the transition from a relaxed or drowsy state to attention or alertness. These changes are commonly referred to as desynchronization, activation, or the arousal response. They occur diffusely throughout the cortex and are thought to be mediated by the diffuse thalamic system. Similar effects can be observed with stimulation of the medial bulbar reticular formation, the pontine and midbrain tegmentum, and the dorsal hypothalamus and subthalamus. The same changes can be produced by natural or electrical stimulation of the spinal nerves, the trigeminal, splanchnic, vagus, acoustic, and some other cranial nerves (Brodal, 1981).

However, lesions in the central parts of the brainstem produce behavioral somnolence and electrocortical synchrony, and activation cannot be obtained with stimuli that are effective when the central part of the brainstem is intact (Brodal, 1981).

Cytoarchitecturally, the size, shape, and density of the constituent reticular neurons are neither uniform nor randomly distributed. Olszewski and Baxter (1982) parcellated the reticular formation into several subnuclei. We can confirm and extend the authors' parcellation by comparing our Nissl-stained thick sections (Fig. 2.10 A–C) with the corresponding figures and drawings in the Olszewski and Baxter atlas. A stereomicroscopic examination provides an idea of the three-dimensional arrangement and parcellation of neurons. However, it remains challenging to translate these 3D impressions and conclusions into objective scientific terms.

The introduction of new methods and interdisciplinary research has led to an accumulation of observations and data pertaining to the structure and function of the brainstem. Most of these findings were derived from animal studies. For more information on methodologic issues, readers are encouraged to consult Blessing's book on lower brainstem homeostasis (Blessing, 1997).

It is helpful to keep in mind Cajal's original observation that motor control systems descending from the cerebral hemispheres synapse principally on interneurons and premotor neurons rather than on the lower motor neurons themselves. Cranial lower motoneurons do not have axonal collateral projections to other cranial nerve nuclei. Integrated function is mediated via premotor neurons with complex projections to different cranial nerve nuclei. These premotor neurons receive inputs from interneurons in loops of increasing complexity. Swallowing, for example, involves coordinated contraction of muscles of the face and lips, jaw, pharynx, tongue and neck muscles, as well as the upper esophagus. Swallowing thus depends on motoneurons distributed through trigeminal, facial, glossopharyngeal, vagal, hypoglossal, and accessory nuclei. The relevant interneurons and premotor neurons are still poorly anatomically defined even in experimental animals, so that "swallowing centre," "pontine gaze centre," and "micturition centre" still refer to functional concepts as well as neuroanatomic realities (Blessing and Benarroch, 2012).

The hierarchical control of motor actions in the brainstem differs from the voluntary motor system represented by the corticospinal tract. Axons of telencephalic area 4 Betz cells terminate directly on motoneurons innervating hand muscles in the lateral motor column of the cervical spinal cord. Brodal describes in detail the effects of a circumscribed ischemic lesion of the corticospinal tract in the internal capsule (Brodal, 1973).

The subdivision of the reticular formation

To integrate previous observations on stimulation and lesion effects with morphology, Nieuwenhuys et al. (2008) subdivided the reticular formation of the brainstem into a medial, tegmental field (light gray) and a lateral tegmental field (Fig. 2.22).

The diencephalic hypothalamic gray matter is contiguous with the mesencephalic PAG. The latter comprises also the pontine central gray as defined by Olszewski and Baxter (1982), the parabrachial nuclei, and the dorsal vagal complex. This continuum was also called core of the neuroaxis by Nieuwenhuys and colleagues.

At the level of the brainstem, this core has two adjuncts, the *median paracore* and the (bilateral) *lateral paracore*. The intermediate zone is part of the lateral paracore and extends from the caudal myelencephalon to the caudal pons. It includes viscerosensory and visceromotor brainstem nuclei as well as aminergic nuclei.

The MFB is a complex structure that unites fibers from various regions of the CNS including isocortical, allocortical, subcortical, and hypothalamic regions. It connects the telencephalon with the brainstem and spinal cord in a reciprocal manner. The descending MFB diverges at the mesodiencephalic transition into a smaller medial and a larger lateral stream. The medial fiber stream passes through the medial parts of the mesencephalic and rhombencephalic tegmental area. The lateral stream of fibers extending from the MFB to the brainstem, or lateral paracore bundle descends through the lateral tegmental field of the pons and medulla oblongata. This stream passes through a bandlike array of grisea which are all involved in the regulation of specific, motivated behaviors and of related autonomic responses.

The MFB comprises numerous unmyelinated ascending and descending fibers (Nieuwenhuys et al., 1982) and the bundle of Schütz. It contains a range of transmitters and peptides, facilitating multiple connections and synchronizing the diverse nuclear elements in the longitudinal zones. Both bundles connect the hypothalamus, prefrontal, cingulate, orbitofrontal cortex, and the amygdala complex with the caudal brainstem.

Medulla oblongata/myelencephalon – an assembly of vital systems

The high contrast and specificity of our gallocyanin-stained neurons facilitated the delineation of cranial nerve nuclei in serial sections. The profiles of the brainstem and nuclei could be digitized, and their contours visualized using a computer-based program for three-dimensional reconstruction of serial sections, which enabled the illustration of the longitudinal organization of the brainstem. We have not separated the tracts from the corresponding nuclei.

The available space of the medulla oblongata is dominated by the cell columns of the caudal cranial nerves. The nuclei of the afferent and efferent cranial nerves are arranged in a mediolateral sequence, including CN V through CN XII. They vary significantly in size and shape. The *spinal tract, together with the trigeminal nucleus* (Fig. 2.24 – 10) and the *solitary tract* (Fig. 2.24 – 9) are prominent. They are characterized by irregular contours. The afferent fibers to both nuclei are somatotopically arranged, and the irregular outlines of the nuclei mirror the high innervation density of the face and oral cavity of the former and the extended volume and area of thoracic and abdominal viscera innervated by the latter (Topolovec et al., 2004; Berthoud et al., 2025). The branchiomotor *ambiguus nucleus* does not form a continuous neuronal cell column. Its caudal part splits into fragments of different size and form (Fig. 2.24 – 7). The elongated cell columns of the dorsal vagus nucleus and hypoglossal nucleus provide widespread cholinergic innervation to the heart, lungs, and gastrointestinal organs, and high motor innervation density of the tongue, respectively. The human ambiguus nucleus is relatively large. Respiration and speech require a finely tuned interaction of the laryngeal muscles. The *dorsal column gracile* (Fig. 2.24 – 12) and *cuneate nuclei* (Fig. 2.24 – 11) relay proprioceptive, tactile, and vibratory information (position sense and kinesthesia) from the ipsilateral side of the body primarily to the thalamus and somatosensory parietal cortex. They are major contributors to the kinesthetic universe of the primate and particularly the human telencephalic cortex.

The solitary tract, along with the associated nuclei—the parasympathetic dorsal vagus nucleus, the ambiguus nucleus, among others—play a crucial role in regulating *cardiovascular, respiratory, and gastrointestinal functions*. They also interact with noradrenergic A1 and A5 neurons, as well as adrenergic C1 neurons, which influence pituitary secretion of vasopressin and ACTH. Additionally, the brainstem nuclei in the medulla oblongata are vital for maintaining the body's homeostasis and survival. Krohn et al. (2023) have recently published a comprehensive overview on respiration and its control by higher order centers, as well as the integrative role of the brainstem. For further diagrams illustrating the vital structures and circuits in the caudal brainstem and the role of the ventrolateral medulla oblongata, the reader is directed to Chapters 3 of this handbook and the brainstem chapters of the books by Nieuwenhuys, Voogd, and Huijzen (Nieuwenhuys, 2008) and Mai and Paxinos (2012). While Krohn et al. (2023) summarized much of the available detailed data from research on hierarchical control of respiration in a comprehensive and complex scheme, we coregistered the images of Blessing and Benarroch (2012) with our gallocyanin-stained thick sections (Fig. 2.25). We were able to identify neuronal assemblies consisting of smaller paler neurons together with larger chromophilic neurons. In some of these neuronal assemblies, the neuronal elements appeared stacked in the ventrolateral medulla oblongata (Fig. 2.25 – Bo, RvR, CVRG). It is feasible that stacked respiratory premotor neurons and interneurons closely associated with smaller or larger parallel fibers, mirrors a basic principle of the organization of the brainstem reticular formation. In the brainstem of Tupaia, this basic feature is best seen. In other species, depending on body size, habitat, or behavior, it is more elaborate or modified (Fig. 2.10A–C).

METENCEPHALON: SYSTEMS FOR SPATIAL AND MENTAL MOVEMENTS

The *metencephalic tegmentum* encloses the somatomotor abducens, the special visceromotor complex of the facial nucleus, parts of the rostral viscerosensory trigeminal nucleus, the special somatosensory complex of the vestibular nuclei, and the cochlear nuclei. Unlike the medulla oblongata, the metencephalic motor nuclei are round to ellipsoid and they do not form columns. The nuclei and pathways in this area are either intimately connected to or embedded in the reticular formation.

The superior rhomboid fossa is devoid of cranial nerve nuclei and the mesencephalic oculomotor complex together with the trochlear nuclei mark the rostral termination of the medial somatomotor nuclear array. However, the *MLF*, an intrinsic fiber bundle consisting of ascending, descending, and crossing fibers, extends from the pretectal regions to the cervical spinal cord. It coordinates the action of mesencephalic- and metencephalic-based oculomotor nuclei with those of the muscles of the neck and trunk. Its ascending and descending fibers connect the rostral intermediate nucleus of the MLF, the interstitial nucleus of Cajal, the cortex of the superior colliculus, the motoneurons of the abducens, trochlear, and oculomotor ncll with nuclei of the vestibular complex, and the prepositus hypoglossal ncl. The cerebellum and vestibular ncll share common connections. Cerebellar efferences can modify the vestibulo-ocular reflex.

The size of the *vestibular complex* and the *precerebellar ncll* (Fig. 2.24) reflect the paramount importance of *spatial* awareness and equilibrium for upright posture and movement. This information is supplemented by *reflexive and voluntary eye movements*, including vergence, smooth pursuit, and saccades. Saccade-related efferences (goal-directed, rapid, precise ballistic movements from the frontal eye field and basal ganglia) originate in the frontal eye field and are transmitted to the superior colliculi, the rostral interstitial nucleus of the MLF (horizontal saccades), the paramedian pontine reticular formation (vertical saccades), and finally to the ocular motor neurons. The pontine reticulotegmental (Bechterew's) nucleus is a precerebellar nucleus. It is closely associated with the MLF (Fig. 2.14 – 3 and 5). In this way, vestibular information is indirectly complemented by stable visual information, which plays a paramount role in primate orientation and behavior.

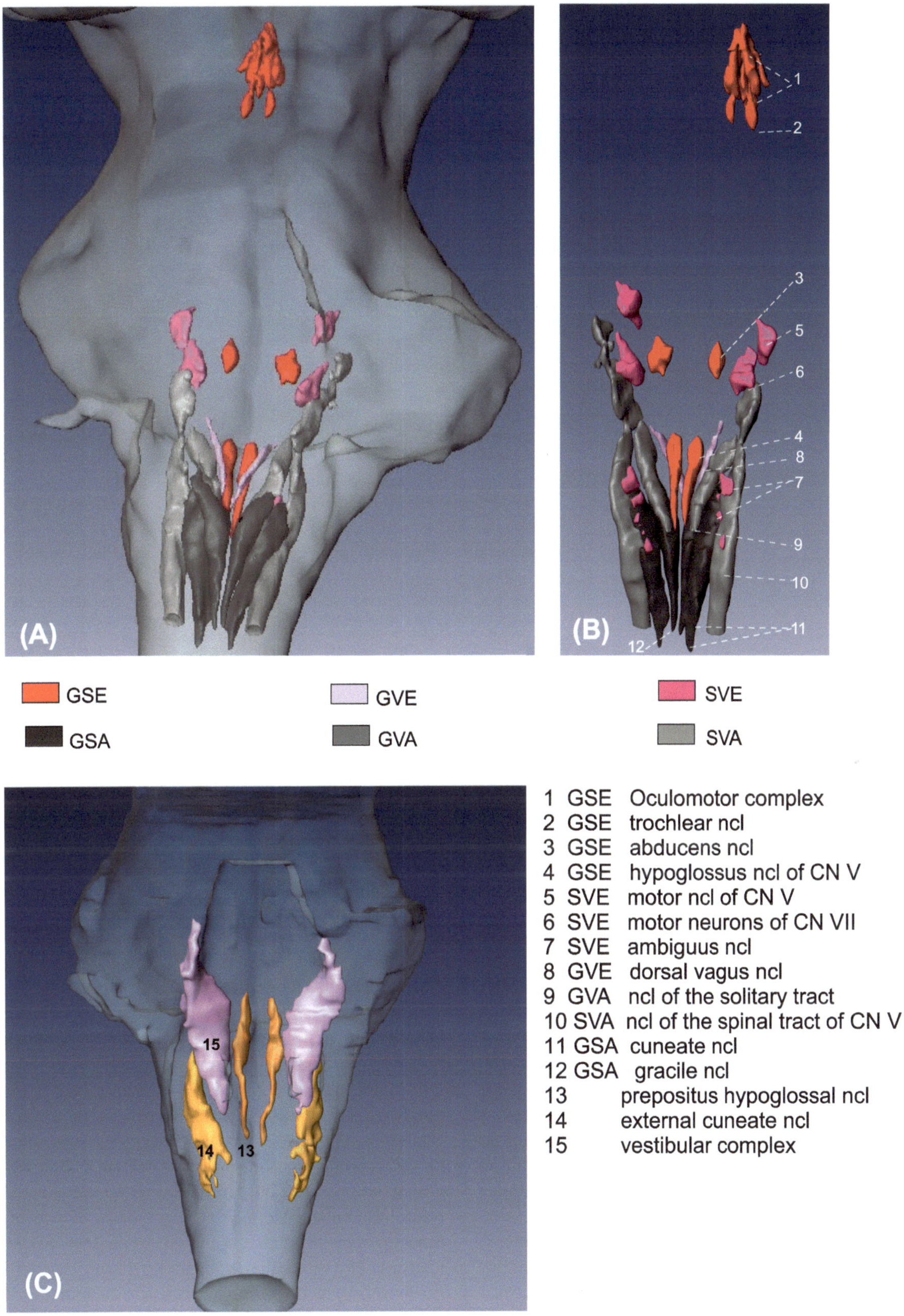

Fig. 2.24. Computer-assisted 3D reconstruction of brainstem nuclei in the three ascending sections of the human brainstem. (**A**) dorsal aspect of the brainstem; (**B**) ventral aspect, the contour of the brainstem has been faded out; (**C**) dorsal aspect of brainstem from an additional reconstruction with intact vestibular area.

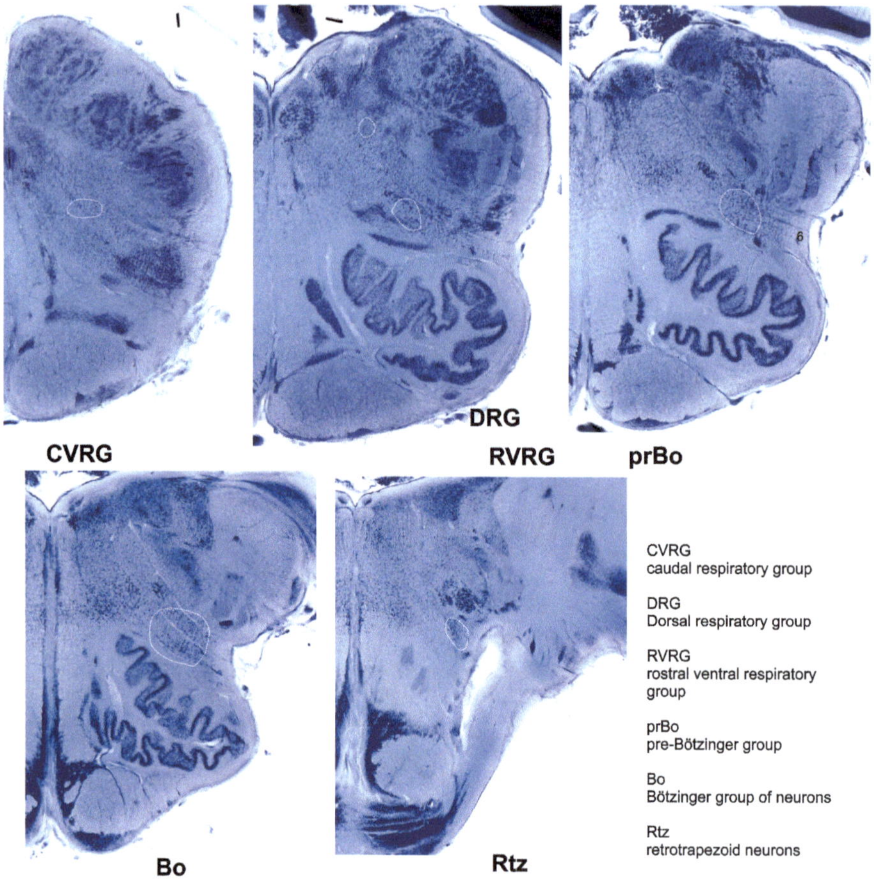

Fig. 2.25. The profiles and topographic coordinates of respiratory interneurons and premotor neurons coordinating inspiration and expiration (Fig. 29.4 in Blessing and Benarroch, 2012) were coregistered with corresponding sections from our case (circles with dotted outlines). Higher magnification shows chromophilic neurons of various sizes stacked parallel to obliquely running unstained fiber bundles. *Bo*, Bötzinger group of respiratory neurons; *CVRG*, caudal respiratory group of neurons; *DRG*, dorsal respiratory group; *prBo*, pre-Bötzinger group of respiratory neurons; *RVRG*, rostral ventral respiratory neurons; *Rtz*, retrotrapezoid nucleus. It is likely that the stacked arrangement of neurons in the marked region reflects the course of ascending/descending fibers in the medial forebrain bundle coursing in the lateral paracore.

The vestibular nuclei also have connections with vasomotor centers in the caudal and rostral ventrolateral medulla. These centers could coordinate rapid changes in body position with changes in blood pressure. Cohen et al. (2019) emphasize the role of the cerebellar nodulus in controlling symptoms of motion thickness.

Vergence, vestibulo-ocular reflex, and optokinetic responses are based on few brainstem elements except for their retinal input. Signals for smooth pursuit and saccades are generated in the cerebral cortex, and for saccades in the basal ganglia. Cortical information for smooth pursuit is transmitted to pontine nuclei, the vestibulocerebellum, and via vestibular nuclei to ocular motoneurons.

Diseases of the oculomotor system are described in Chapter 8 of this handbook. In cases of spinocerebellar ataxias (Rüb et al., 2013), loss of neurons has been observed. For observations on other lesions in various neurodegenerative diseases of the oculomotor system connected by the MLF, see Rüb et al. (2001, 2004, 2009).

Oculomotor and vestibular information are processed by the vermian spino- and vestibulocerebellum and forwarded by corresponding regions in the fastigial ncl. The *cerebellum* is subject of considerable evolutionary modification in shape, size, and function. In mormyroidae with their evolved novel active electrosensory system, the size of the cerebellum and hindbrain increased conspicuously, whereas the sizes of the telencephalon and optic tectum decreased (Sukhum et al., 2018). The weight of the cerebellum of an African elephant was 1172 g comprising 251 billion neurons (Herculano-Houzel et al., 2014). The total number of neurons in the elephant's brain were 257 billion. Therefore, the cerebellum contributed 97.5% to the total brain neuron number in the African elephant. The volume of the human brain attains only one-third of the volume of an elephant brain (Herculano-Houzel et al., 2014). However, the total number of neurons in the human brain amounts to 86 billion neurons. The cerebellum contributes 69 billion neurons, and the cerebral cortex 19 billion neurons or 20% of the total neuron number (Azevedo et al., 2009).

The morphology of the cerebellar cortex differs from that of other cortical regions by its almost uniform basic geometric structure, which has evolved in myxines (Larsell, 1947; Sugahara et al., 2021) and has been preserved and refined in mammals and birds. This greatly facilitated morphologic analysis down to the level of electron microscopy (Palay and Chan-Palay, 1974; Chan-Palay, 1977) and correlation with neurophysiology. As a result, the cerebellum was regarded as a neuronal machine (Eccles et al., 1967) or computer, as we would say today, 57 years after the publication of this book.

Nevertheless, the function of the cerebellum remains enigmatic. Cerebellar lesions result in a range of symptoms, including ataxia, dysmetria, intention tremor, dysarthria, hypotonia, loss of coordination, and hypotonia. The explanation of the cerebellum's role in motor coordination remains elusive (Nieuwenhuys et al., 2008). In his recent publication, Schmahmann (2019) hypothesizes that dysmetria is a main feature of cerebellar dysfunction and that the cerebellum plays a role in sensorimotor and vestibular control, as well as cognition, emotion, and autonomic function. The location and extent of a lesion are crucial in determining the three main areas of cerebellar clinical ataxiology: the nature of the cerebellar motor syndrome, the cerebellar vestibular syndrome, and the cerebellar cognitive-affective syndrome.

The mesencephalon, a region of higher-order life-support systems and the mediator of postnatal growth and maturation

Tegmentum

The *Edinger–Westphal ncl* of the oculomotor complex is a parasympathetic GVE nucleus. Through the parasympathetic ciliary ganglion, the Edinger–Westphal nucleus innervates the pupillary sphincter and is involved in the light reflex and convergence responses. The Edinger–Westphal ncl does not receive direct retinal input. It receives afferent fibers from the pretectal nuclei, which in turn are supplied by fibers from the optic tract. The absence of the pupillary reflex in comatose patients is indicative of a lesion in the pretectal region.

Immunohistochemical studies have shown that the Edinger–Westphal ncl consists of a mixed population of cholinergic and urocortin-positive neurons with ascending and descending projections from spinal cord to amygdala. The former are now called EWpg (preganglionic projecting) and the latter EWcp (centrally projecting) neurons (Kozicz et al., 2011; Lima et al., 2013). EWcp neurons are discussed to be involved in reward, stress, mood, and addiction (Xu et al., 2012; Xu, 2014; Spencer et al., 2015; Ryabinin and Giardino, 2017; Zuniga and Ryabinin, 2020; Cano et al., 2021). For more details, see Chapter 9 of this handbook.

The mesencephalic reticular formation

The mesencephalic reticular formation occupies a sector of the mesencephalic tegmentum dorsolaterally to the decussation of the cerebellar peduncles. Three nuclei are particularly noteworthy in this tegmental sector: the *pedunculopontine tegmental nucleus*, which occupies the ventrolateral part of the tegmental sector, the *laterodorsal tegmental nucleus*, close to the MLF, and the *cuneiform nucleus*, which is in the dorsomedial part.

Interestingly, both the pedunculopontine and the dorsolateral tegmental ncl receive dopaminergic innervation (Ryczko et al., 2016).

Both cuneiform and pedunculopontine ncll are considered part of the mesencephalic locomotor region, which complements voluntary movements with emotional features. Both are the target of electrostimulation after recovery from cerebral infarction and in the course of Parkinson disease (Noga et al., 2003; Goetz et al., 2016; Kiehn, 2016; Mazzone et al., 2016; Ryczko et al., 2016; Karachi and Francois, 2018; Di Giovanni et al., 2019; Tubert et al., 2019; Vitale et al., 2019; Hofer et al., 2022). Ch5 of the pedunculopontine ncl is small, and it is a challenge to localize the part of the nucleus in 3T imaging (Alho et al., 2017).

The *magnocellular part of the pedunculopontine tegmental nucleus* is distinguished by its spindle-shaped neurons, which have been shown to react with antibodies against ChAT (Mesulam et al., 1989). The authors hypothesize that ChAT-positive neurons in *the laterodorsal tegmental nucleus* or Ch6 in their nomenclature, as well as the pedunculopontine tegmental nucleus (Ch5), provide the principal cholinergic innervation of the human thalamus. Furthermore, they postulate that these cholinergic cells participate in the neural circuitry of the reticular activating, limbic, and potentially also extrapyramidal systems.

However, Fuller et al. (2011) conclude from their lesion experiments in rats, that the reticulo-thalamo-cortical pathway may play a very limited role in behavioral or electrocortical arousal, whereas the *projection from the parabrachial nucleus and precoeruleus region*, relayed by the basal forebrain to the cerebral cortex, may be critical for this process.

Snider et al. (2019) performed high-angle-resolution diffusion imaging in 16 patients with acute severe traumatic brain injury (TBI) who were comatose on admission and 16 matched controls. They used probabilistic tractography to measure the connectivity probability of the *ascending arousal network*, an axonal pathway connecting the brainstem tegmentum to the hypothalamus, thalamus, and basal forebrain. The authors provide preliminary evidence for reduced integrity of axonal pathways connecting the brainstem tegmentum to the hypothalamus and thalamus, but not to the basal forebrain, in patients with traumatic coma. These results confirm Jellinger's (2013) observation that only small peripheral lesions in the pontine tegmentum allow progressive remission of the unresponsive wakefulness syndrome. Kinney et al. (1994) systematically sampled and diagnosed the pathohistology of the entire brain and spinal cord of a case who died in a prolonged coma 10 years after a cardiopulmonary arrest. The authors concluded from the disproportionately severe damage in the thalamus compared to the cerebral cortex that the thalamus is critical for cognition and consciousness and may be less essential for arousal. It remains to be clarified whether the selective loss of neurons in the thalamic reticular nucleus (Ross et al., 1993) is due exclusively to topographic damage to cortical neurons (Necula et al., 2022) or to more diffuse afferents from the brainstem, basal forebrain, and other regions (Guillery and Harting, 2003).

The mesencephalic aminergic system and its importance for basal ganglia and cortical activity

The *raphe system* consists of several nuclei that supply serotonin to the spinal cord, brainstem, and forebrain. The *mesencephalic dorsal raphe serotoninergic ncl* is prominent and, like the noradrenergic locus coeruleus, supplies diencephalic and telencephalic gray matter with serotonin. Its actions depend on the receptor types in the target areas, including the cerebral cortex. Species differences in the primate cerebral cortex range from subtle to compelling (Gebhard et al., 1995).

Neurons in the *compact part of the substantia nigra (A9)* and the *VTA (A10)* or the mesolimbic system use dopamine as a transmitter. Nigrostriatal pathways supply dopamine to the striatum in a topographic manner. Dopamine inhibits D1 receptors and stimulates D2 receptors. In HD, GABA-ergic enkephalin-positive medium spiny cells degenerate, resulting in disinhibition of inhibitory pallidothalamic neurons and excessive excitation of thalamocortical neurons and involuntary excessive movements in HD patients. Parkinson disease is characterized by degeneration of dopaminergic substantia nigra neurons, which supply dopamine to the striatum, and symptoms such as rigidity and akinesia, which are mirror images of HD symptoms. These examples clearly demonstrate how the substantia nigra, a brainstem nucleus, affects telencephalic circuits.

VTA neurons project to the ventral striatum and also directly to allo- and isocortical regions. In humans, this cortical dopaminergic innervation is particularly elaborated in primates. It differs from rodents. In rats, the dopaminergic projections to the medial frontal cortex are distributed mainly to the deep cortical layers (5 and 6), while in primates these projections are preferentially distributed in the superficial layers (1–3). There are important regional differences in dopamine innervation as well: the primary motor area (M1) of rats is nearly devoid of dopaminergic fibers, while M1 is among the areas with most densest innervation by dopaminergic fibers in macaques and humans (Preuss, 2001).

The mesencephalic tectum

Superior and inferior colliculi feature cortical layers; in the former case up to seven layers were delineated and both colliculi were amply studied neurophysiologically. They are connected to di- and telencephalic regions as well as to the brainstem. They are involved in eye and head movements, body movements, vocalization, and even basic emotional expressions including pleasure and disgust.

The deliberate creation of lesions or transections in the lower and upper brainstem of experimental animals has been shown to result in deficits that are specific to the species in question. Quadrupedal animals are still able to execute automatic movements on treadmills, whereas humans are unable to do so (Magnus, 1924; Koehler and Lameris, 2016).

It is unethical to conduct such experiments on humans. Two studies were performed on two individuals with absent telencephalons due to brain maldevelopment during their three months of survival; Gamper, 1926) (Neilsen and Owman, 1949). The neuropathologic and clinical records of Gamper are notable for their thoroughness and documentation.

STRUCTURE AND FUNCTION OF THE HUMAN BRAINSTEM AFTER TRAUMATIC BRAIN INJURY, A COMMON EXAMPLE OF THE FATAL INTERACTION OF HIGH-TECH PRODUCTS WITH BIOLOGY

These clinical observations of Gamper and Neilsen resemble clinical cases after cerebral infarcts or TBI, with the diagnosis of *Minimally Conscious State* (MCS). After survival of traumatic brain injuries patients can recover through the following stages: (1) minimal levels of behavioral interaction that is characterized by the absence of nonreflex movements, (2) orientation of noxious stimuli, (3) pursuit eye movements that occur in direct response to moving or salient stimuli, and (4) movements or affective behaviors that occur appropriately in relation to relevant environmental stimuli (such as appropriate smiling or crying in response to the linguistic or visual content of emotional, but not to neutral, topics or stimuli, vocalizations or gestures that occur in direct response to the linguistic content of questions, reaching for objects that demonstrates a clear relationship between object location and direction of reach, touching or holding objects in a manner that accommodates the size and shape of the object) (Bruno et al., 2011). Stage 3 corresponds to the capabilities of the cases described in detail by Gamper and Neilsen and Owman.

The compulsive habit of car use is a major contributor to the annual death toll from road traffic accidents worldwide (Yalachkov et al., 2014). Jellinger selected 100 cases out of 600 with unresponsive wakefulness syndrome after blunt head injury (Jellinger, 2013). Seventy-seven percent were victims of traffic accidents, and 23% of falls. The author diagnosed paraffin-embedded tissue blocks, cut and stained with routine stains including H&E, cresyl violet, Klüver–Barrera, and GFAP. He concluded that only small peripheral lesions in the pontine tegmentum allow progressive remission after blunt TBI.

We could extend previous studies of the deleterious effects of TBI to a *complete serially sectioned brain*. This technique requires special microtomes, staining protocols, time, and manpower that are available in only a few institutions worldwide (Toga et al., 2006; Amunts et al., 2013, 2020).

A 59-year-old man was hit by a car at midnight. He was severely injured while crossing the street, and quickly transported to a nearby intensive care unit; fractures of the left leg and right pelvic ring were treated surgically. A tracheotomy was performed for artificial respiration; a gastric tube and a bladder catheter, as well as a subclavian catheter, were placed. The patient had to be permanently ventilated, infused and artificially fed. He died 21 days after admission to the intensive care unit from clinically diagnosed sepsis.

The forensic autopsy confirmed the radiologically diagnosed orbitofrontal fracture, bilateral circumscribed orbitofrontal cortical contusion foci, and bilateral extensive coup and contrecoup subarachnoid hemorrhages extending to the central region of the brain, fracture of the left femur, and left tibia with soft tissue bruises.

We selected four representative levels of brainstem pathology based on the hierarchy of brainstem structure and function.

Gallocyanin, a Nissl stain, stains nucleic acids in formalin-fixed tissue in a stoichiometric manner, as demonstrated by Sandritter (Einarson, 1935; Sandritter et al., 1954). The nuclei of neurons and glia contain a high concentration of DNA. The cytoplasm or perikaryon of neurons can be rich in RNA associated with the endoplasmic reticulum. In such instances, the Nissl bodies become discernible under the light microscope. Regions of the CNS with a transient, circumscribed lack of blood supply may appear unstained or pale when stained with Nissl stains (Fig. 2.26D – 10, and 2.27B – 5). Alternatively, they may appear extensively eosinophilic when stained with hematoxylin and eosin. Neurons whose axons are damaged or deliberately severed will exhibit chromatolysis of Nissl bodies, perikaryal swelling, and degeneration. This phenomenon has been employed in the past century to facilitate a comprehensive investigation of the interconnections between fiber tracts within the CNS and their underlying neuronal origins. Moreover, gallocyanin stains degenerating fibers (Fig. 2.27B – 7), autolytic tissue, and degenerating tissue rich in calcium salts (O'Sullivan et al., 2019).

An orienting view over different planes of sections shows chromophilic tissue interdigitating with pale regions. However, *nuclear profiles of the cranial nerve nuclei* could be identified in Fig. 2.26B: hypoglossus (1), dorsal vagus (2), and solitary tract ncll (3), and in Fig. 2.27B – 1, the oculomotor complex. The same is true for the locus coeruleus in Fig. 2.26D – 5, the compact part of the substantia nigra in Fig. 2.27B – 6, and the dorsally located paranigral nucleus of the tegmental area in Fig. 2.27B – 6a. however, with an obviously reduced neuron density. Median and dorsal raphe nuclei could be identified in Fig. 2.26F – 12 and 13, which exhibited a strongly reduced neuron density.

Neuronal density in the *reticular formation* was reduced in most parts of the brainstem, including the *central nucleus of the medulla oblongata* and its ventrolateral parts with few groups of chromophilic neurons spared in the latter (compare Fig. 2.26A – iRT, with the corresponding field in Fig. 2.26B), the *reticulo-tegmental nuclei of the pons* (Fig. 2.26D – 9), the *medial parabrachial* (Fig. 2.26D – 7) and Mesulam's Ch5 and Ch6 ncll (Fig. 2.26 F – 17). The *mesencephalic PAG* was almost devoid of neurons (Fig. 2.27B – 3).

Some of the observed neuronal degenerations are most likely a direct response to loss of neuronal input. This is true for the *somatotopic neuron loss* in the *gracile nuclei* (Fig. 2.26B – 4).

The *central parts of the pons* were devoid of pontine neurons. This region of pontine neuron degeneration was maximal in the central parts of the of the pontine base and decreased rostrally. The corticopontine fibers were strongly stained by gallocyanin (compare Fig. 2.27A – 7 with Fig. 2.27B – 7). The same was true for *the corticobulbar and -spinal fibers in the cerebral crura* (Fig. 2.27B – 7). The transverse fibers of the *corpus callosum* appeared spliced and disintegrated (Fig. 2.27B – 9). We also observed pallor of the central cortical white matter (here not documented photographically) with few spared glial cells along with extensive areas of *neuronal loss in all thalamic nuclei* including the pulvinar (Fig. 2.27B – 2). Intracerebral damage to corticopontine fibers apparently caused the striking *degeneration of pontine neurons*. The severe loss of neurons in the red nucleus (Fig. 2.27B – 4) could also be explained by impairment of the cortico-rubral and cerebello-rubral neuronal circuits. Interestingly, the cortex of the retrocommissural *hippocampus* (Fig. 2.27B – 8) appeared less affected.

Conclusions to be drawn from systemic studies of traumatic brain injury

These observations exemplify different pathogenetic factors associated with the course of TBI. *Traumatic rupture* or intentional axotomy in the CNS generally cause *retrograde degeneration of the parent neuron* (Barron, 2004) and *Wallerian degeneration* of the distal myelinated axonal segment. The degenerated distal axon stump can be visualized with special silver impregnations (Strich, 1961) or immunohistochemical stains (Johnson et al., 2013), while the Nissl-stained parent perikaryon shows chromatolysis.

In addition, *transneuronal degeneration* may evolve after degeneration of primarily traumatically lesioned neurons. Depending on the kind of innervation original targets of traumatically lesioned neurons degenerate because of missing input.

Finally, TBI can cause *rupture of vessels, inflammation, brain swelling*, and compression of intact vessels by *hematomas or edema*.

The greater loss of neurons in the left gracile nucleus is indicative of a transneuronal response resulting from deafferentation of proprioceptive and epicritic fibers from an extensive asymmetric traumatic skin and muscle lesion. The patient sustained an extensive skin and muscle lesion around the fractured left femur, the knee joint capsule was opened, the left tibia was fractured, and the lower leg muscles were bruised. In contrast, the traumatic skin and muscle damage over the right pelvic ring was less severe on the right side.

Extensive transneuronal degeneration of the pontine nuclei can be explained by direct TBI, as we observed in the telencephalic semi oval center. The unusual dark staining of the corticobulbar fibers in the cerebral crura could be an indicator of anterograde degeneration. On the other hand, dilated vessels in this region and in the mesencephalic–diencephalic transition zone could also be the result of cardiovascular failure exacerbated by several factors: (1) an almost complete loss of neurons in the PAG, which is a bottleneck structure for reciprocal communication between the cortex, amygdala, and brainstem engaged in the initiation of stress-induced cascades;

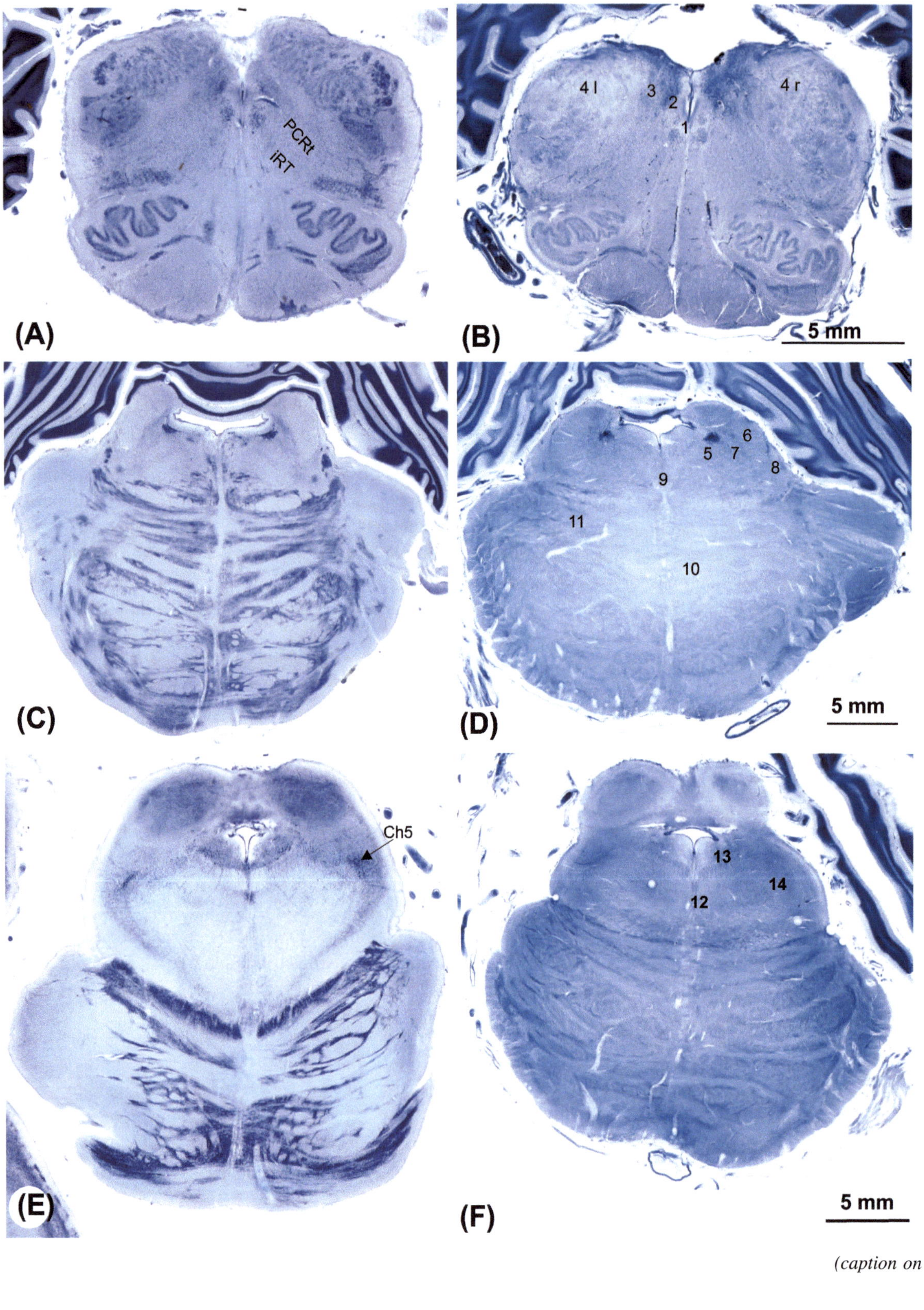

(caption on next page)

(2) probable lesions in the mesencephalic motor region, the parabrachial ncll; and (3) massive myocardial hypertrophy (heart weight 885 g) with extensive scar tissue in the anterior wall of the left ventricle; and (4) sepsis.

In contrast, brainstem nuclei including the locus coeruleus, the substantia nigra, and the reticular formation in the caudal myelencephalon appeared *less affected*. This raises the question of *early selective vulnerability* of other degenerated brainstem nuclei. Knowledge of this feature was and is of paramount importance for the elucidation of the initiation and progression of neuronal loss in common neurodegenerative diseases. In this point of view, a complete systemic neuropathologic examination of TBI victims should be given high priority to rationally intervene in the gradual and successive deterioration of life-essential neural circuits.

The question concerning the *morphologic basis of consciousness* remains unanswered (Brodal, 1964).

TELENCEPHALIC REALITY AND THE BIASED BRAINSTEM CENTERS OF REWARD AND HEDONIA, WITH A BRIEF EXCURSION INTO LITERATURE, PHILOSOPHY, AND SUSTAINABILITY

The defining characteristics that set humans apart from most other mammals include *bipedal terrestrial movement*, the *utilization of the hands*, and the *use and creation of tools*. If we compress the evolution of the primate species over the past billion years into a 24-hour timescale (de Schotten and Zilles, 2019), the earliest primates would have appeared less than a minute ago. It is estimated that the first primates to walk on two legs appeared less than 10 seconds ago, with the first drawing being created approximately 500 milliseconds ago and the first representation of a human face being carved approximately 250 milliseconds ago. The relatively brief span of our recorded history occurred in the last five milliseconds.

The *evolution of the human brain* is based on an increase in relative and absolute brain volume, a refinement of the primary sensory and motor fields, and the appearance of new poly- and transmodal association cortices. These innovations enabled humans to create a subjective internal or virtual view of their environment, to anticipate events, and to plan to prevent or escape from life-threatening situations (Mesulam, 1998).

Monoaminergic brainstem nuclei targeting telencephalic cortical and subcortical gray matter are thought to provide salience to abstract polymodal cortical content or recruit diverse transmodal information for decision-making (Mesulam, 1998). Noradrenaline from locus coeruleus neurons can focus attention, and serotonergic psychedelic drugs can induce visual hallucinations. In addition, Nguyen et al. (2021) distinguish between "liking" and "wanting". The former is associated with the *brainstem parabrachial nuclei*, a *hedonic hotspot* associated with medial prefrontal and orbitofrontal cortical fields (Berridge and Kringelbach, 2015); the latter is based on A9 and A10 cortically projecting dopaminergic neurons.

The brainstem and its components were and are essential to life and evolution. It effectively supports survival in different habitats with different modes of locomotion. These include swimming, crawling, walking, running, and flying. Unlike modern humans, animals are directly exposed to their natural environment. For example, some mammals can overcome a lack of food in the winter by hibernating. These examples show a successful economic and ecologic interaction between the telencephalon, the locomotor system, and the viscera of the trunk, mediated by the brainstem and its reticular formation.

Human evolution was accelerated by bipedalism, use of hands, instruments, and language. Instead of adapting to the environment man changed the latter, symbolized in religious narratives, by eating the forbidden fruit of knowledge from the tree in the paradise. In his story about the marionette theater, Heinrich von Kleist, a German poet who belonged to the Romantics, described three examples of how the natural grace of man is disturbed by reflection. In this way, Kleist addresses the tension between nature and reason or necessity and freedom. Readers interested in the philosophical position and dimensions of Kleist's story are referred to the freely accessible thesis of Mioc (2011).

Since Kleist's time, instruments have become innumerable, sophisticated, and gigantic. Kleist would wonder how present instruments for transportation and communication are commonly considered an indispensable biologic necessity of human

Fig. 2.26. Topographically matched gallocyanin-stained sections through the brainstem of a 66-year-old female control subject (**A, C, and E**) and a 59-year-old male with traumatic brain injury (**B, D, and F**), who died in an intensive care unit 20 days after collision with a motor vehicle. (**A and B**) Caudal medulla oblongata; the profiles of the hypoglossal (1), dorsal vagus (2), and solitary tract ncl (3) could be identified. The left gracile ncl (4 l) was almost completely devoid of neurons, while the right (4 r) shows segmental loss. Neurons of the parvocellular (PcRT) and intermediate reticular formation (iRT) were less well stained, but neuronal groups and clusters could still be delineated. (**C and D**) The pigmented locus coeruleus neurons (5) were clearly visible. The integrity of the lateral (7) and medial (8) parabrachial ncll (lateral andmedial to the superior cerebellar peduncle (6)) was difficult to assess as their outlines and staining characteristics are known to vary in a caudo-rostral direction. The reticulotegmental ncll (9) were obviously degenerated. The central pontine region (10) was pale, vascular profiles were dilated, and strands of pontine ncll (11) were absent in the central parts and pale with increasing staining toward the periphery. (**E and F**) The the neuron density in the median (12) und dorsal raphe ncl (13) was reduced. We could not unequivocally identify the Ch5, the compact part of the pedunculopontine ncl, and (Ch6), the laterodoral ncl close to the dorsal raphe ncl.

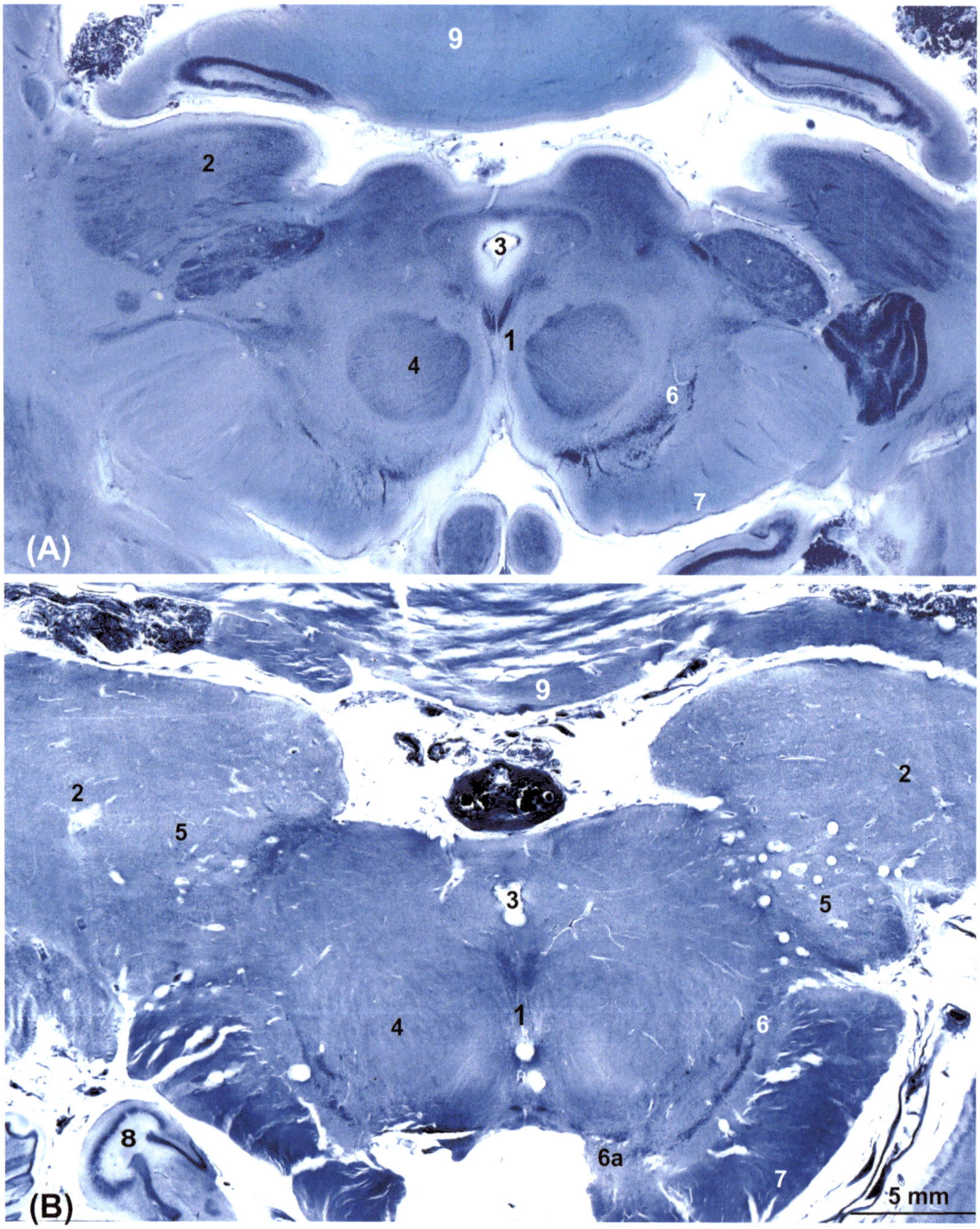

Fig. 2.27. Rostral mesencephalon of the female control case (**A**) and the male case with traumatic brain injury (**B**): (1) oculomotor complex, (2) thalamic pulvinar, (3) mesencephalic aqueduct with surrounding periaqueductal gray, (4) red nucleus, (5) penetrating vessels with perivascular edema, (6) substantia nigra, (7) right cerebral peduncle, (8) hippocampus, and (9) splenium of the corpus callosum. As in the caudal medulla oblongata, some nuclei appear to be preserved, including the nuclei of the oculomotor complex (1) and the substantia nigra (6). The same is true for the allocortical hippocampus (8) of the temporal lobe. On the other hand, most of the neurons of the periaqueductal gray matter (3) and the red nuclei (5) are degenerated. In contrast to the gray and white matter of the control case, dilated vascular-associated pale areas (2) predominated in this mesencephalic and adjacent thalamic pulvinar region. Another phenomenon is the disintegration and rupture of callosal association fibers in the splenium or in the deep telencephalic central myelin layer. This is a common finding after severe blunt traumatic brain injury.

existence. Unfortunately, life in general and modern human instruments in particular consume resources (Elhacham et al., 2020), energy, and inevitably generate *entropy* (Georgescu-Roegen, 1971, 1986).

Courtship and display behavior are basic natural elements for reproduction and survival. Combined with impressive, energy-consuming status symbols and the powerful products of the arms and military industries, they guarantee the existence of self-appointed industrial and economic experts, oligarchs, and potentates. They keep the population in check with overwhelming façades of illusions, gifts through exploitation of the weak and oppressed, or simply by force. Through the systemic production and use of instruments and the unrestrained consumption of carbon-based energy resources, humans have become a threat to their own existence (van Daalen, 2024).

To regain natural grace, Kleist encouraged his contemporaries to continue eating the forbidden fruit in order to finally attain divine abilities. This is, of course, a metaphoric suggestion. With today's knowledge, Kleist would most likely suggest finding out why closely related species prefer aggressive over cooperative and even altruistic behavior (Samuni et al., 2022), using natural abilities instead of relying on energy-consuming technological devices, and replacing consuming behavior with cultural activity. This could be a realistic strategy to harmonize brainstem-based reward systems with subjective *telencephalic virtual reality* and to recognize and accept the implications imposed by the four principal laws of thermodynamics (Table 2.1) (Georgescu-Roegen, 1971, 1986).

Table 2.1

Cranial nerves (CN) III–XII.

Cranial nerves (n(n) and nuclei (ncl(l)	Classification	Origin	Innervation m(m) muscle(s)
CN III oculomotor n	GSE	Somatoefferent ncll of the mesencephalic oculomotor ncl	Superior, inferior, medial rectus, inferior obliquus, levator palpebrae mm
	GVE	Westphal-Edinger ncll	sphincter pupillae and ciliar mm
CN IV trochlear n	GSE	Trochlear ncl	Superior oblique m
CN V trigeminal n	GSA	bipolar cells in semilunar ganglion	sensitivity in the facial region
	SVE	motor ncl of CN V	masticatory mm
CN VI abducens n	GSE	Abducens ncl	Lateral rectus m
CN VII facial n	SVE	Facial ncl	facial mm, platysma, stylohyoid m, digastric m
Intermediate n	GVE	superior salivatory ncl	nasal and lacrimal glands, sublingual and submandibular glands
	SVA	geniculate ganglion	taste anterior 2/3 of tongue,
	GSA	geniculate ganglion	skin of ear, parts of auditory duct, external surface of tympanic membrane
CN VIII vestibulo-	SSA	Vestibular ganglion	Equilibrium sense
cochnlear n	SSA	spiral ganglion	audition
CN IX glosso-	SVE	Ambiguus ncl	Stylopharyngeus m, muscles of the pharynx
pharyngeus n	GVE	inferior salivatory ncl	parotic gland, salivation
	SVA	inferior cervical ganglion	taste tongue caudal 1/3
	GSA	superior cervical ganglion	sensory innervation of caudal 1/3 of tongue
CN X vagus n	GSA	Superior cervical ganglion	Middle ear, tuba Eustachii
	SVE	ambiguus ncl	larynx and pharynx mm
	GVE	dorsal vagus ncl	parasympathic innervation of thoracic and abdominal viscera with chemo- and baroreceptors
	GVA	inferior cervical ganglion	of the aortic arch, cardiorespiratory system, and digestive tract sensibility from intra-abdominal region
	SVA		taste epiglottis
CN XI, accessorius n	GSA	Superior cervical ganglion	Sensibility external acoustic meatus, dura
	SVE	ambiguus ncl	pharyngeal and laryngeal mm
CN XII	GSE	Spinal cord anterior horn neurons	Sternocleidomastoid and trapezius mm
hypoglossus n	GSE	hypoglossus ncl	tongue mm

ABBREVIATIONS

CN cranial nerve
CNS central nervous system
Ncl nucleus
ncll nuclei

References

Agostinelli LJ, Seaman SC, Saper CB et al. (2023). Human brainstem and cerebellum atlas: chemoarchitecture and cytoarchitecture paired to MRI. J Neurosci 43: 221–239.

Alexander GE, DeLong MR, Strick PL (1986). Parallel organization of functionally segregated circuits linking basal ganglia and cortex. Annu Rev Neurosci 9: 357–381.

Alho A, Hamani C, Alho EJL et al. (2017). Magnetic resonance diffusion tensor imaging for the pedunculopontine nucleus: proof of concept and histological correlation. Brain Struct Funct 222: 2547–2558.

Amunts K, Lepage C, Borgeat L et al. (2013). BigBrain: an ultrahigh-resolution 3D human brain model. Science 340: 1472–1475.

Amunts K, Mohlberg H, Bludau S et al. (2020). Jülich-Brain: a 3D probabilistic atlas of the human brain's cytoarchitecture. Science 369: 988–992.

Arendt T, Brückner MK, Morawski M et al. (2015). Early neurone loss in Alzheimer's disease: cortical or subcortical? Acta Neuropathol Commun 3: 10.

Azevedo FA, Carvalho LR, Grinberg LT et al. (2009). Equal numbers of neuronal and nonneuronal cells make the human brain an isometrically scaled-up primate brain. J Comp Neurol 513: 532–541.

Barron KD (2004). The axotomy response. J Neurol Sci 220: 119–121.

Berridge KC, Kringelbach ML (2015). Pleasure systems in the brain. Neuron 86: 646–664.

Berthoud HR, Muenzberg H, Morrison CD et al. (2025). Gut-brain communication: Functional anatomy of vagal afferents. Curr Opin Neurobiol 93: 3058.

Blessing WW (1997). The lower brainstem and bodily homeostasis. New York/Oxford: Oxford University Press.

Blessing WW, Benarroch EE (2012). Lower brainstem regulation of visceral, cardiovascular, and respiratory function. In: Mai JK, Paxinos G (eds.) The human nervous system. Third edition. Amsterdam/Boston, MA: Elsevier, pp. 1058–1073.

Blumenfeld H (2018). Neuroanatomy through clinical cases. Massachusetts: Sinauer Associates.

Brodal A (1964). Anatomical points of view on the alleged morphological basis of consciousness. Acta Neurochir (Wien) 12: 166–186.

Brodal A (1973). Self-observations and neuro-anatomical considerations after a stroke. Brain 96: 675–694.

Brodal A (1981). Neurological anatomy in relation to clinical medicine. Third edition. New York/Oxford: Oxford University Press.

Bruno MA, Vanhaudenhuyse A, Thibaut A et al. (2011). From unresponsive wakefulness to minimally conscious PLUS and functional locked-in syndromes: recent advances in our understanding of disorders of consciousness. J Neurol 258: 1373–1384.

Cacciola A, Milardi D, Basile GA et al. (2019). The cortico-rubral and cerebello-rubral pathways are topographically organized within the human red nucleus. Sci Rep 9:12117.

Cano G, Hernan SL, Sved AF (2021). Centrally projecting Edinger-Westphal nucleus in the control of sympathetic outflow and energy homeostasis. Brain Sci 11.

Carrive P, Morgan MM (2012). Periaqueductal gray. In: Mai JK, Paxinos G (eds.) The human nervous system. Amsterdam/Boston/Heidelberg: Elsevier, pp. 367–400.

Chan-Palay V (1977). Cerebellar dentate nucleus: organization, cytology and transmitters. Berlin: Springer.

Che Ngwa E, Zeeh C, Messoudi A et al. (2014). Delineation of motoneuron subgroups supplying individual eye muscles in the human oculomotor nucleus. Front Neuroanat 8: 2.

Cohen B, Dai M, Yakushin SB et al. (2019). The neural basis of motion sickness. J Neurophysiol 121: 973–982.

Corsellis JA (1958). Individual variation in the size of the tentorial opening. J Neurol Neurosurg Psychiatry 21: 279–283.

de Schotten MT, Zilles K (2019). Evolution of the mind and the brain. Cortex 118: 1–3.

Delbono O, Wang ZM, Messi ML (2022). Brainstem noradrenergic neurons: identifying a hub at the intersection of cognition, motility, and skeletal muscle regulation. Acta Physiol (Oxf) 236: e13887.

Di Giovanni G, Chagraoui A, Puginier E et al. (2019). Reciprocal interaction between monoaminergic systems and the pedunculopontine nucleus: implication in the mechanism of L-DOPA. Neurobiol Dis 128: 9–18.

Di Lorenzo Alho AT, Suemoto CK, Polichiso L et al. (2016). Three-dimensional and stereological characterization of the human substantia nigra during aging. Brain Struct Funct 221: 3393–3403.

Eccles JC, Ito M, Szentágothai J (1967). The cerebellum as a neuronal machine. Berlin: Springer.

Einarson L (1935). Histological analysis of the Nissl-pattern and -substance of nerve cells. J Comp Neurol 61: 101–133.

Elhacham E, Ben-Uri L, Grozovski J et al. (2020). Global human-made mass exceeds all living biomass. Nature 588: 442–444.

Felten DL, Sladek JRJ (1983). Monoamine distribution in primate brain V. Monoaminergic nuclei: anatomy, pathways and local organization. Brain Res Bull 10: 171–284.

Forel A (1877). Untersuchungen über die Haubenregion und ihre oberen Verknüpfungen im Gehirne des Menschen und einiger Säugethiere, mit Beiträgen zu den Methoden der Gehirnuntersuchung. Arch Psychiatr Nervenkr 7: 393–495.

Fuller PM, Sherman D, Pedersen NP et al. (2011). Reassessment of the structural basis of the ascending arousal system. J Comp Neurol 519: 933–956.

Fuxe K, Dahlström A, Höistad M et al. (2007). From the Golgi-Cajal mapping to the transmitter-based characterization of the neuronal networks leading to two modes of brain communication: wiring and volume transmission. Brain Res Rev 55: 17–54.

Gamper E (1926). Bau und Leistungen eines menschlichen Mittelhirnwesens (Arhinencephalie mit Encephalocele) zugleich ein Beitrag zur Teratologie und Fasersystematik. Z Ges Neurol Psychiat 102: 154–235.

Gandhi NJ, Katnani HA (2011). Motor functions of the superior colliculus. Annu Rev Neurosci 34: 205–231.

Gebhard R, Zilles K, Schleicher A et al. (1995). Parcellation of the frontal cortex of the new world monkey Callithrix jacchus by eight neurotransmitter-binding sites. Anat Embryol 191: 509–517.

Geniec P, Morest DK (1971). The neuronal architecture of the human posterior colliculus. A study with the Golgi method. Acta Otolaryngol Suppl 295: 1–33.

Georgescu-Roegen N (1971). The entropy law and the economic process. Cambridge, Mass: Harvard University Press.

Georgescu-Roegen N (1986). The entropy law and the economic process in retrospect. East Econ J 12: 3–25.

Gilvesy A, Husen E, Magloczky Z et al. (2022). Spatiotemporal characterization of cellular tau pathology in the human locus coeruleus-pericoerulear complex by three-dimensional imaging. Acta Neuropathol 144: 651–676.

Goetz L, Piallat B, Bhattacharjee M et al. (2016). On the role of the pedunculopontine nucleus and mesencephalic reticular formation in locomotion in nonhuman primates. J Neurosci 36: 4917–4929.

Grinberg LT, Heinsen H (2007). Computer-assisted 3D reconstruction of the human basal forebrain. Dement Neuropsychol 2: 140–146.

Guillery RW, Harting JK (2003). Structure and connections of the thalamic reticular nucleus: advancing views over half a century. J Comp Neurol 463: 360–371.

Haber SN (2003). The primate basal ganglia: parallel and integrative networks. J Chem Neuroanat 26: 317–330.

Hassler R (1937). Zur Normalanatomie der Substantia nigra. Versuch einer architektonischen Gliederung. J Psychol Neurol (Lpz) 48: 1–55.

He G, Li Y, Deng H et al. (2023). Advances in the study of cholinergic circuits in the central nervous system. Ann Clin Transl Neurol 10: 2179–2191.

Heinsen H, Hampel H, Teipel SJ (2006). Nucleus subputaminalis: neglected part of the basal nucleus of Meynert–Response to Boban et al: computer-assisted 3D reconstruction of the nucleus basalis complex, including the nucleus subputaminalis (Ayala's nucleus). Brain 129: U1–U4.

Herculano-Houzel S, Avelino-de-Souza K, Neves K et al. (2014). The elephant brain in numbers. Front Neuroanat 8: 46.

Hofer AS, Scheuber MI, Sartori AM et al. (2022). Stimulation of the cuneiform nucleus enables training and boosts recovery after spinal cord injury. Brain 145: 3681–3697.

Huffman RF, Henson OW Jr (1990). The descending auditory pathway and acousticomotor systems: connections with the inferior colliculus. Brain Res Rev 15: 295–323.

Jansen J, Brodal A (1958). Kleinhhirn. Handbuch der Mikroskopischen Anatomie des Menschen14/8. Berlin Göttingen Heidelberg: Springer, pp. 1–323.

Jellinger KA (2013). Neuropathology of prolonged unresponsive wakefulness syndrome after blunt head injury: review of 100 post-mortem cases. Brain Inj 27: 917–923.

Johnson VE, Stewart W, Smith DH (2013). Axonal pathology in traumatic brain injury. Exp Neurol 246: 35–43.

Jones BE, Yang TZ (1985). The efferent projections from the reticular formation and the locus coeruleus studied by anterograde and retrograde axonal transport in the rat. J Comp Neurol 242: 56–92.

Kaas JH, Huerta MF (1988). The subcortical visual system of primates. Comp Primate Biol 4: 327–391.

Karachi C, Francois C (2018). Role of the pedunculopontine nucleus in controlling gait and sleep in normal and parkinsonian monkeys. J Neural Transm (Vienna) 125: 471–483.

Kawashima T (2018). The role of the serotonergic system in motor control. Neurosci Res 129: 32–39.

Kiehn O (2016). Decoding the organization of spinal circuits that control locomotion. Nat Rev Neurosci 17: 224–238.

Kilimann I, Hausner L, Fellgiebel A et al. (2017). Parallel atrophy of cortex and basal forebrain cholinergic system in mild cognitive impairment. Cereb Cortex 27: 1841–1848.

Kinney HC, Korein J, Panigrahy A et al. (1994). Neuropathological findings in the brain of Karen Ann Quinlan. The role of the thalamus in the persistent vegetative state. N Engl J Med 330: 1469–1475.

Koehler PJ, Lameris B (2016). The Magnus-Rademaker Scientific Film Collection: ethical issues on animal experimentation (1908-1940). J Hist Neurosci 25: 102–121.

Korneliussen HK (1967). Cerebellar corticogenesis in Cetaca, with special reference to regional variations. J Hirnforsch 9: 151–185.

Kositzyn NS (1964). Axo-dendritic relations in the brain stem reticular formation. J Comp Neurol 122: 9–17.

Kozicz T, Bittencourt JC, May PJ et al. (2011). The Edinger-Westphal nucleus: a historical, structural, and functional perspective on a dichotomous terminology. J Comp Neurol 519: 1413–1434.

Krohn F, Novello M, van der Giessen RS et al. (2023). The integrated brain network that controls respiration. eLife 12.

Kundu S, Pakrashi A, Kamalakannan M et al. (2022). Complete mitogenome of the endangered and endemic Nicobar treeshrew (*Tupaia nicobarica*) and comparison with other Scandentians. Sci Rep 12: 877.

Langer SZ (2015). α2-Adrenoceptors in the treatment of major neuropsychiatric disorders. Trends Pharmacol Sci 36: 196–202.

Larsell O (1947). The cerebellum of myxinoids and petromyzonts including developmental stages in the lampreys. J Comp Neurol 86: 395–445.

Larsell O (1970). The comparative anatomy and histology of the cerebellum: from monotremes through apes. Minneapolis: Minnesota University Press.

Lima FF, Sita LV, Oliveira AR et al. (2013). Hypothalamic melanin-concentrating hormone projections to the septo-hippocampal complex in the rat. J Chem Neuroanat 47: 1–14.

MacLeod CE, Zilles K, Schleicher A et al. (2003). Expansion of the neocerebellum in Hominoidea. J Hum Evol 44: 401–429.

Magnus R (1924). Körperstellung: experimentell-physiologische Untersuchungen über die einzelnen bei der Körperstellung in Tätigkeit tretenden Reflexe, über ihr Zusammenwirken und ihre Störungen. Berlin: Springer.

Mai JK, Paxinos G (2012). The human nervous system. Third edition. Elsevier Inc.

Mantyh PW (1982). The midbrain periaqueductal gray in the rat, cat, and monkey: a Nissl, Weil, and Golgi analysis. J Comp Neurol 204: 349–363.

Mazzone P, Vilela O, Viselli F et al. (2016). Our first decade of experience in deep brain stimulation of the brainstem: elucidating the mechanism of action of stimulation of the ventrolateral pontine tegmentum. J Neural Transm (Vienna) 123: 751–767.

Mesulam MM (1995). Cholinergic pathways and the ascending reticular activating system of the human brain. Ann N Y Acad Sci 757: 169–179.

Mesulam MM (1998). From sensation to cognition. Brain 121: 1013–1052.

Mesulam MM, Geula C, Bothwell MA et al. (1989). Human reticular formation: cholinergic neurons of pedunculopontine and laterodorsal tegmental nuclei and some cytochemical comparisons to forebrain cholinergic neurons. J Comp Neurol 281: 611–633.

Mesulam MM, Geula C, Bothwell MA et al. (1989). Human reticular formation: cholinergic neurons of pedunculopontine and laterodorsal tegmental nuclei and some cytochemical comparisons to forebrain cholinergic neurons. J Comp Neurol 281: 611–633.

Mesulam MM, Mufson EJ, Levey AI et al. (1983). Cholinergic innervation of cortex by the basal forebrain: cytochemistry and cortical connections of the septal area, diagonal band nuclei, nucleus basalis (substantia innominata), and hypothalamus in the rhesus monkey. J Comp Neurol 214: 170–197.

Mioc A (2011). Baroque expressions in romanticism: Heinrich von Kleist and Keats. London, ON: University of Western Ontario.

Morales M, Margolis EB (2017). Ventral tegmental area: cellular heterogeneity, connectivity and behaviour. Nat Rev Neurosci 18: 73–85.

Morest DK, Oliver DL (1984). The neuronal architecture of the inferior colliculus in the cat: defining the functional anatomy of the auditory midbrain. J Comp Neurol 222: 209–236.

Necula D, Cho FS, He A et al. (2022). Secondary thalamic neuroinflammation after focal cortical stroke and traumatic injury mirrors corticothalamic functional connectivity. J Comp Neurol 530: 998–1019.
Neilsen KC, Owman C (1949). Instincts and emotions in an anencephalic monster. J Nerv Ment Dis 110: 387–394.
Nguyen D, Naffziger EE, Berridge KC (2021). Positive affect: nature and brain bases of liking and wanting. Curr Opin Behav Sci 39: 72–78.
Nieuwenhuys R (2008). The human central nervous system. Fourth edition. New York: Springer.
Nieuwenhuys R, Geeraedts LM, Veening JG (1982). The medial forebrain bundle of the rat. I. General introduction. J Comp Neurol 206: 49–81.
Nieuwenhuys R, Puelles E (2016). Towards a new neuromorphology. Heidelberg: Springer Cham, pp. 1–329.
Nieuwenhuys R, Voogd J, van Huijzen C (2008). The human central nervous system. Berlin: Springer.
Noga BR, Kriellaars DJ, Brownstone RM et al. (2003). Mechanism for activation of locomotor centers in the spinal cord by stimulation of the mesencephalic locomotor region. J Neurophysiol 90: 1464–1478.
Oliveira-Pinto AV, Andrade-Moraes CH, Oliveira LM et al. (2016). Do age and sex impact on the absolute cell numbers of human brain regions? Brain Struct Funct 221: 3547–3559.
Olszewski J, Baxter D (1982). Cytoarchitecture of the human brain stem2nd ed. Basel: Karger.
O'Sullivan S, Heinsen H, Grinberg LT et al. (2019). The role of artificial intelligence and machine learning in harmonization of high-resolution post-mortem MRI (virtopsy) with respect to brain microstructure. Brain Inform 6: 3.
Palay SL, Chan-Palay V (1974). Cerebellar cortex. Berlin/Heidelberg/ New York: Springer Verlag.
Paniccia JE, Vollmer KM, Green LM et al. (2024). Restoration of a paraventricular thalamo-accumbal behavioral suppression circuit prevents reinstatement of heroin seeking. Neuron 112: 772–785. e779.
Poe GR, Foote S, Eschenko O et al. (2020). Locus coeruleus: a new look at the blue spot. Nat Rev Neurosci 21: 644–659.
Preuss TM (2001). The discovery of cerebral diversity: an unwelcome scientific revolution. In: Falk D, Gibson KR (eds.) Evolutionary anatomy of primate cerebral cortex. Cambridge: Cambridge University Press, pp. 138–164.
Puelles E (2007). Genetic control of basal midbrain development. J Neurosci Res 85: 3530–3534.
Puelles L, Hidalgo-Sanchez M (2023). The midbrain preisthmus: A poorly known effect of the isthmic organizer. Int J Mol Sci 24 11.
Qi HX, Kaas JH (2006). Organization of primary afferent projections to the gracile nucleus of the dorsal column system of primates. J Comp Neurol 499: 183–217.
Rai R, Iwanaga J, Shokouhi G et al. (2018). The tentorium cerebelli: a comprehensive review including its anatomy, embryology, and surgical techniques. Cureus 10: e3079.
Ramón-Moliner E, Nauta WJ (1966). The isodendritic core of the brain stem. J Comp Neurol 126: 311–335.
Raos VC, Dermon CR, Savaki HE (1995). Functional anatomy of the thalamic centrolateral nucleus as revealed with the [14C]deoxyglucose method following electrical stimulation and electrolytic lesion. Neuroscience 68: 299–313.
Rapoport SI (1990). Integrated phylogeny of the primate brain, with special reference to humans and their diseases. Brain Res Rev 15: 267–294.
Roman E, Weininger J, Lim B et al. (2020). Untangling the dorsal diencephalic conduction system: a review of structure and function of the stria medullaris, habenula and fasciculus retroflexus. Brain Struct Funct 225: 1437–1458.
Ross DT, Graham DI, Adams JH (1993). Selective loss of neurons from the thalamic reticular nucleus following severe human head injury. J Neurotrauma 10: 151–165.
Rüb U, Brunt ER, de Vos RAI et al. (2004). Degeneration of the central vestibular system in spinocerebellar ataxia type 3 (SCA3) patients and its possible clinical significance. Neuropathol Appl Neurobiol 30: 402–414.
Rüb U, Del Tredici K, Schultz C et al. (2001). The premotor region essential for rapid vertical eye movements shows early involvement in Alzheimer's disease-related cytoskeletal pathology. Vision Res 41: 2149–2156.
Rüb U, Heinsen H, Brunt ER et al. (2009). The human premotor oculomotor brainstem system – can it help to understand oculomotor symptoms in Huntington's disease? Neuropathol Appl Neurobiol 35: 4–15.
Rüb U, Schols L, Paulson H et al. (2013). Clinical features, neurogenetics and neuropathology of the polyglutamine spinocerebellar ataxias type 1, 2, 3, 6 and 7. Prog Neurobiol 104: 38–66.
Ryabinin AE, Giardino WJ (2017). Contribution of urocortin to the development of excessive drinking. Int Rev Neurobiol 136: 275–291.
Ryczko D, Cone JJ, Alpert MH et al. (2016). A descending dopamine pathway conserved from basal vertebrates to mammals. Proc Natl Acad Sci U S A 113: E2440–E2449.
Samuni L, Langergraber KE, Surbeck MH (2022). Characterization of Pan social systems reveals in-group/out-group distinction and out-group tolerance in bonobos. Proc Natl Acad Sci U S A 119: e2201122119.
Sandritter W, Diefenbach H, Krantz F (1954). Über die quantitative Bindung von Ribonukleinsäure mit Gallozyanin-chromalaun. Experientia 10: 210–216.
Saper CB, Petito CK (1982). Correspondence of melanin-pigmented neurons in human brain with A1-A14 catecholamine cell groups. Brain 105: 87–101.
Sarter M, Parikh V (2005). Choline transporters, cholinergic transmission and cognition. Nat Rev Neurosci 61: 48–56.
Scheibel ME, Scheibel A (1958). Structural substrate for integrative patterns in the brainstem reticular core. In: Jasper HH, Procter LD (eds.) Reticular formation of the brain (Henry Ford Hospital Symposium). Boston: Little Brown, pp. 31–55.
Schmahmann JD (2019). The cerebellum and cognition. Neurosci Lett 688: 62–75.
Schmahmann JD, Rosene DL, Pandya DN (2004). Motor projections to the basis pontis in rhesus monkey. J Comp Neurol 478: 248–268.
Schultz W, Dayan P, Montague PR (1997). A neural substrate of prediction and reward. Science 275: 1593–1599.
Snider SB, Bodien YG, Bianciardi M et al. (2019). Disruption of the ascending arousal network in acute traumatic disorders of consciousness. Neurology 93: e1281–e1287.
Spencer SJ, Emmerzaal TL, Kozicz T et al. (2015). Ghrelin's role in the hypothalamic-pituitary-adrenal axis stress response: implications for mood disorders. Biol Psychiatry 78: 19–27.
Stacho M, Häusler AN, Brandstetter A et al. (2024). Phylogenetic reduction of the magnocellular red nucleus in primates and inter-subject variability in humans. Front Neuroanat 18: 1331305.
Stephan H, Baron G, Frahm HD (1988). Comparative size of brains and brain components. Comparative primate biology, volume 4: neurosciences. New York: Alan R. Liss, pp. 1–38.
Strata F, Coq JO, Kaas JH (2003). The chemo- and somatotopic architecture of the Galago cuneate and gracile nuclei. Neuroscience 116: 831–850.

Strich S (1961). Shearing of nerve fibres as a cause of brain damage due to head injury: a pathological study of twenty cases. Lancet 278: 443–448.

Sugahara F, Pascual-Anaya J, Kuraku S et al. (2021). Genetic mechanism for the cyclostome cerebellar neurons reveals early evolution of the vertebrate cerebellum. Front Cell Dev Biol 9: 700860.

Sukhum KV, Shen J, Carlson BA (2018). Extreme enlargement of the cerebellum in a clade of teleost fishes that evolved a novel active sensory system. Curr Biol 28: 3857–3863. e3853.

Swaab DF, Uylings HBM (1988). Potentialities and pitfalls in the use of human brain material in molecular neuroanatomy. In: van Leeuwen F (ed.) Molecular Neuroanatomy. Amsterdam: Elsevier, pp. 403–416.

ten Donkelaar HJ (1988). Evolution of the red nucleus and rubrospinal tract. Behav Brain Res 28: 9–20.

Tinoco Mendoza FA, Hughes TES, Robertson RV et al. (2023). Detailed organisation of the human midbrain periaqueductal grey revealed using ultra-high field magnetic resonance imaging. Neuroimage 266: 119828.

Toga AW, Thompson PM, Mori S et al. (2006). Towards multimodal atlases of the human brain. Nat Rev Neurosci 7: 952–966.

Topolovec JC, Gati JS, Menon RS et al. (2004). Human cardiovascular and gustatory brainstem sites observed by functional magnetic resonance imaging. J Comp Neurol 471: 446–461.

Tubert C, Galtieri D, Surmeier DJ (2019). The pedunclopontine nucleus and Parkinson's disease. Neurobiol Dis 128: 3–8.

Vadodaria KC, Stern S, Marchetto MC et al. (2018). Serotonin in psychiatry: in vitro disease modeling using patient-derived neurons. Cell Tissue Res 371: 161–170.

van Daalen KR (2024). The 2024 Europe report of the Lancet Countdown on health and climate change: unprecedented warming demands unprecedented action. Lancet Public Health 97: e495–e522.

Vitale F, Capozzo A, Mazzone P et al. (2019). Neurophysiology of the pedunculopontine tegmental nucleus. Neurobiol Dis 128: 19–30.

Waltzer R, Martin GF (1984). Collateralization of reticulospinal axons from the nucleus reticularis gigantocellularis to the cerebellum and diencephalon. A double-labelling study in the rat. Brain Res 293: 153–158.

Wolf ME (2016). Synaptic mechanisms underlying persistent cocaine craving. Nat Rev Neurosci 17: 351–365.

Xu L (2014). Leptin action in the midbrain: from reward to stress. J Chem Neuroanat 61-62: 256–265.

Xu L, Scheenen WJ, Roubos EW et al. (2012). Peptidergic Edinger-Westphal neurons and the energy-dependent stress response. Gen Comp Endocrinol 177: 296–304.

Yalachkov Y, Naumer MJ, Plyushteva A (2014). The compulsive habit of cars. Trends Cogn Sci 185: 227–228.

Zuniga A, Ryabinin AE (2020). Involvement of centrally projecting Edinger-Westphal nucleus neuropeptides in actions of addictive drugs. Brain Sci 10.

Handbook of Clinical Neurology, Vol. 216 (3rd series)
The Brainstem
L.T. Grinberg and A.D. Korczyn
https://doi.org/10.1016/B978-0-443-15736-3.00014-7

Chapter 3

Autonomic blood pressure control

PIETRO CORTELLI[1,2], GIOVANNA CALANDRA-BUONAURA[1,2,*], AND ALESSANDRO SILVANI[1]

[1]*Department of Biomedical and Neuromotor Sciences, University of Bologna, Bologna, Italy*

[2]*IRCCS Institute of Neurological Sciences of Bologna, Bologna, Italy*

Abstract

The regulation of systemic arterial blood pressure is key to survival. The sympathetic and parasympathetic regulation of arterial blood pressure requires information related to the actual blood pressure values for feedback control as well as integration with the ongoing behavior, including physical exercise and sleep, and with the regulation of breathing and body temperature. The brainstem includes the core neural circuitry underlying control of autonomic preganglionic neurons; integration of the baroreceptor, chemoreceptor, and exercise pressor reflexes; and integration of behavioral responses with central autonomic commands associated with physical exercise, sleep, and arousal. As a result, alterations in the central autonomic network of the brainstem may cause severe impairment in the regulation of arterial blood pressure. These alterations lead to afferent baroreflex failure and to cardiovascular autonomic failure with neurogenic orthostatic hypotension and supine hypertension, or occur as a result of neurodegeneration in multiple system atrophy or of multiple sclerosis and brainstem lesions. A better understanding of the brainstem circuitry involved in the regulation of arterial blood pressure may also shed light on the mechanisms of arterial hypertension, which is a powerful risk factor for adverse cardiovascular events.

PHYSIOLOGY

The role of the brainstem in the integrative regulation of arterial blood pressure

The regulation of systemic arterial blood pressure is key to survival. Values of arterial blood pressure must be high enough to drive the microvascular blood flow that is required for systemic tissue metabolism and to drive glomerular filtration rate. However, inappropriately high values of arterial blood pressure increase cardiac O_2 consumption and promote ventricular hypertrophy, arterial remodeling, and atherosclerosis burden. Arterial hypertension is a powerful risk factor for adverse cardiovascular events (Lloyd-Jones et al., 2022).

Arterial blood pressure is regulated by both the sympathetic and parasympathetic branches of the autonomic nervous system. Nevertheless, it is the sympathetic nervous system that is fundamental to arterial blood pressure regulation (Joyner et al., 2008), and increased sympathetic activity presumably causes or exacerbates arterial hypertension (Guyenet et al., 2018). The sympathetic and parasympathetic regulation of arterial blood pressure requires information related to the actual arterial blood pressure values for feedback control as well as integration with the ongoing behavior, including physical exercise and sleep, and with the regulation of breathing and body temperature. The following sections will review long-established and more recent evidence on the brainstem neural mechanisms that underlie these functions (Fig. 3.1).

The role of the brainstem in the central autonomic network: an overview

THE CENTRAL AUTONOMIC NETWORK THAT CONTROLS CARDIOVASCULAR SYMPATHETIC OUTFLOW

The sympathetic ganglionic neurons that control the heart and resistance vessels lie in para- and prevertebral ganglia along the vertebral column. The adrenal medullary chromaffin cells, which mainly release adrenaline in the circulation, may be viewed as a neuroendocrine functional equivalent of sympathetic ganglionic neurons. The sympathetic preganglionic neurons involved in the control of the heart, systemic resistance vessels, and adrenal medulla are located in the intermediolateral column of the thoracic spinal cord (Silvani et al., 2016).

The neural structures that send direct projections to the sympathetic preganglionic neurons (presympathetic neurons) are mostly located in the brainstem, and particularly in the rostral ventromedial and ventrolateral medulla, the midline medulla, and the A5 noradrenergic cell group of the pons.

*Correspondence to: Giovanna Calandra-Buonaura, Professor of Neurology, DIBINEM, University of Bologna, Bellaria Hospital, Via Bellaria 3, 40139 Bologna, Italy. Tel: +393496367679; E-mail: giovanna.calandraunibo.it.

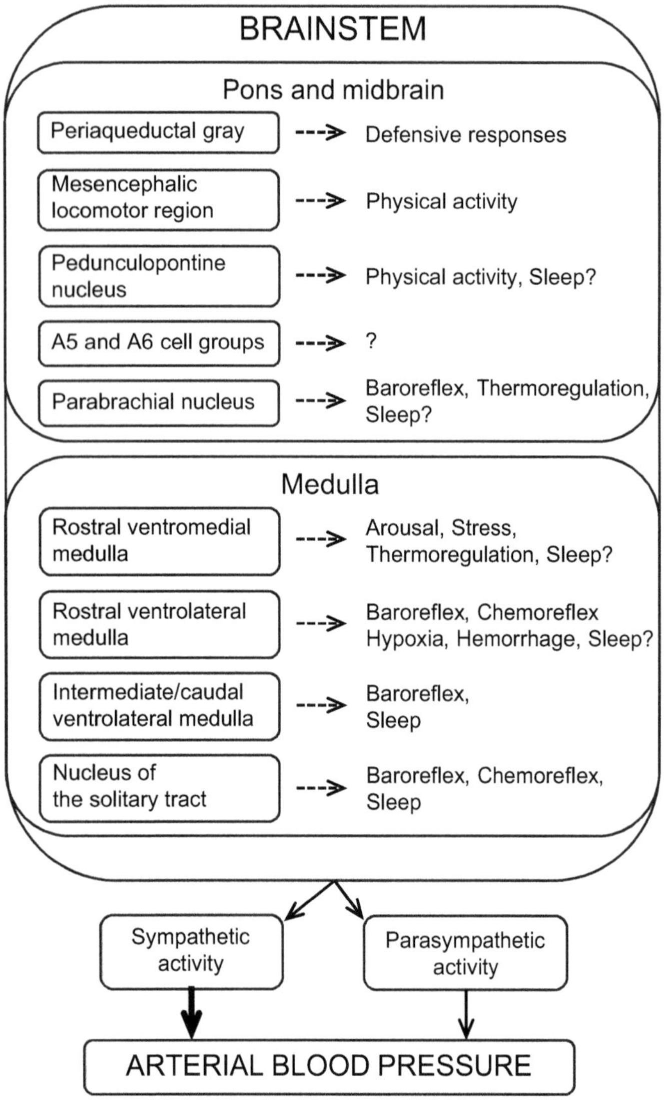

Fig. 3.1. Overview of the physiologic brainstem control of arterial blood pressure. The main functional domains of blood pressure control thought to be mediated by brainstem structures of the midbrain, pons, and medulla are indicated. Question marks indicate hypotheses and/or lack of robust functional evidence. Other brainstem structures potentially relevant for blood pressure control but with limited available functional evidence are discussed in the text and omitted from the figure for clarity.

Other brainstem structures exert indirect control of sympathetic preganglionic neurons. These structures include the nucleus of the solitary tract and the intermediate and caudal portions of the ventrolateral medulla, which are part of the medullary baroreflex circuit, as well as the midbrain periaqueductal gray, which contributes to regulate autonomic responses to physical and psychologic stressors. The nucleus of the solitary tract and the intermediate and caudal portions of the ventrolateral medulla modulate the activity of sympathetic preganglionic neurons through the presympathetic neurons in the rostral ventrolateral medulla, whereas the periaqueductal gray does so through the nucleus of the solitary tract and through the presympathetic neurons in the rostral ventromedial medulla (Benarroch, 1993; Dampney, 1994; Guyenet and Stornetta, 2022; Silvani et al., 2016).

Neural structures rostral to the brainstem contribute to direct and indirect control of sympathetic preganglionic neurons largely by modulating the activity of the brainstem central autonomic network. The presympathetic neurons in the hypothalamic paraventricular nucleus increase renal sympathetic nerve activity and arterial blood pressure at least in part via the rostral ventrolateral medulla, and may mediate sympathetic responses to changes in blood volume or plasma osmolality (Koba et al., 2018). Orexin/hypocretin neurons of the lateral hypothalamus include presympathetic neurons that project to sympathetic preganglionic neurons in the upper thoracic spinal cord, but the functional role of these projections for the control of arterial blood pressure is still uncertain (Bastianini and Silvani, 2018). On the other hand, the amygdala modulates sympathetic preganglionic neurons indirectly through the dorsomedial hypothalamus, which is critical for cardiovascular responses to stress, and through the periaqueductal gray, both pathways converging on presympathetic neurons in the rostral ventromedial medulla. The hypothalamic suprachiasmatic nucleus controls sympathetic preganglionic neurons indirectly via a multisynaptic pathway responsible for the circadian rhythm of sympathetic activity. Finally, the medial prefrontal and the insular cortices control sympathetic preganglionic neurons activity indirectly via multiple pathways that involve brainstem and subcortical forebrain structures (Silvani et al., 2016).

The central autonomic network that controls cardiac parasympathetic outflow

The parasympathetic regulation of arterial blood pressure is essentially exerted at the cardiac level. The parasympathetic ganglionic neurons that control the heart lie in interconnected epicardial ganglionated plexi. These parasympathetic ganglionic neurons receive monosynaptic projections from parasympathetic preganglionic neurons that lie in the medulla in, around, and between the nucleus ambiguus and the dorsal motor nucleus of the vagus (Silvani et al., 2016).

The neurons that send monosynaptic projections to parasympathetic preganglionic neurons are termed preparasympathetic neurons. The preparasympathetic neurons include neurons in the brainstem, and particularly in the nucleus of the solitary tract and the pre-Bötzinger complex of the medulla, which are involved in arterial baroreceptor reflex and respiratory control, respectively. Other brainstem pathways that may include preparasympathetic neurons relevant to cardiac control are the dorsal horn of the trigeminal nucleus in the medulla and the lateral and ventrolateral columns of the midbrain periaqueductal gray (Benarroch, 1993; Dampney, 1994; Silvani et al., 2016).

The central autonomic network that controls cardiovagal parasympathetic preganglionic neurons also includes structures rostral to the brainstem, which largely converge on the brainstem central autonomic network. The hypothalamic paraventricular nucleus integrates neuroendocrine, homeostatic, and stress responses and may control cardiovagal parasympathetic preganglionic neurons in the dorsal motor nucleus of the vagus both directly and by modulating the activity of the nucleus of the solitary tract. The hypothalamic suprachiasmatic nucleus is the master clock of the circadian rhythm and controls parasympathetic preganglionic neurons indirectly with multisynaptic projections that involve other hypothalamic structures (Dampney, 1994; Silvani et al., 2016). The hypothalamic neurons that release the orexin (hypocretin) neuropeptides, which play key roles in the control of sleep and of purposeful active wakefulness, may exert bidirectional effects on cardiovagal parasympathetic preganglionic neurons in the nucleus ambiguus and dorsal motor nucleus of the vagus. The effects of the orexin neurons on the nucleus ambiguus are indirect and depend on the modulation of preparasympathetic neurons projections, whereas those on the dorsal motor nucleus of the vagus are direct and mediated by orexin cotransmitters (Bastianini and Silvani, 2018). The amygdala is primarily involved in information processing related to negative emotions and may modulate cardiovagal parasympathetic preganglionic neurons indirectly, largely by projecting to the midbrain periaqueductal gray (Dampney, 1994; Silvani et al., 2016). The amygdala and the periaqueductal gray are also involved together with the nucleus of the solitary tract in mediating indirect effects of the medial prefrontal and insular cortex on cardiovagal parasympathetic preganglionic neurons. The medial prefrontal cortex is thought to promote cardiovagal activity, inhibiting fear or threat representations in the amygdala in contexts coded as "safe." The insular cortex may exert bidirectional effects on heart rate and arterial blood pressure, although discrimination of sympathetic and parasympathetic effects is not clear (Dampney, 1994; Silvani et al., 2016). These cortical structures may exert more specific autonomic control than subcortical forebrain and brainstem structures, but the magnitude of the responses tends to be greater from caudal than from rostral levels (Hoff et al., 1963).

The role of the medulla in the regulation of arterial blood pressure

The brainstem medulla includes several structures that are essential to the regulation of arterial blood pressure. As discussed in previous sections, the medulla includes the presympathetic neurons in the rostral ventromedial medulla and rostral ventrolateral medulla, the cardiovagal parasympathetic preganglionic neurons of the nucleus ambiguus and dorsal motor nucleus of the vagus, and the preparasympathetic neurons in the nucleus of the solitary tract and the pre-Bötzinger complex. The nucleus of the solitary tract is the main afferent relay for visceral information and is involved in mediating the arterial baroreceptor reflex (together with the intermediate and caudal portions of the ventrolateral medulla), the chemoreceptor reflex (together with medullary structures of the respiratory pattern generator), the diving reflex, and the exercise pressor reflex.

Functional neuroanatomy of the cardiovagal parasympathetic preganglionic neurons in the nucleus ambiguus and dorsal motor nucleus of the vagus

The cardiovagal parasympathetic preganglionic neurons in the nucleus ambiguus and the dorsal motor nucleus of the vagus are functionally heterogeneous (Silvani et al., 2016). The activity of separate groups of parasympathetic preganglionic neurons in the nucleus ambiguus selectively decreases heart rate, prolongs atrioventricular conduction, and decreases ventricular contractility (Gatti et al., 1996). These parasympathetic preganglionic neurons have relatively high basal activity with significant cardiac and respiratory modulation and receive inputs from baroreceptors, chemoreceptors, and nasopharyngeal receptors (McAllen and Spyer, 1978; Gilbey et al., 1984; McAllen et al., 2011). Respiratory modulation may be generated centrally via inputs from the respiratory pattern generator, including the medullary pre-Bötzinger complex, as well as peripherally, via the entrainment of baroreceptor and pulmonary stretch receptors to the lung inflation cycle (Eckberg, 2003). By contrast, the parasympathetic preganglionic neurons in the dorsal motor nucleus of the vagus have relatively low basal activity without cardiac or respiratory modulation and have little effect on heart rate, but may modulate ventricular contractility and excitability (Silvani et al., 2016).

The role of the ventrolateral medulla in the control of arterial blood pressure

The ventrolateral medulla extends from the facial motor nucleus to the spino-medullary junction and from the ventral medullary surface and lateral reticular nucleus to the nucleus ambiguus (Guyenet and Stornetta, 2022). The dorsal portion of the ventrolateral medulla consists of the ventral respiratory column and the rostral ventral respiratory group. The ventral respiratory column includes the Bötzinger region and the pre-Bötzinger complex, which may function as the breathing pacemaker, whereas the rostral ventral respiratory group includes inspiratory premotor neurons. Neurons involved in autonomic control of arterial blood pressure are mostly localized in the ventral portion of the ventrolateral medulla, with significant overlap with the respiratory-related dorsal portion (Guyenet and Stornetta, 2022). In humans, the rostral ventrolateral medulla can be identified in vivo by integrating functional magnetic resonance imaging with peroneal microneurography to measure muscle sympathetic nerve activity (Macefield and Henderson, 2019).

The presympathetic neurons in the rostral ventrolateral medulla are critical for the cardiovascular responses to baroreceptor and chemoreceptor inputs (Dampney, 1994; Silvani et al., 2016). The sympathetic responses to these inputs differ. For instance, arterial baroreceptor stimulation causes reflex vasodilation in skeletal muscles and, more modestly, in the skin, whereas chemoreceptor stimulation has a similar effect on skin blood vessels but evokes powerful reflex vasoconstriction in skeletal muscles (Dampney, 2016). A minority of presympathetic neurons in the rostral ventrolateral medulla send projections to spinal sympathetic preganglionic neurons that innervate functionally diverse tissues, such as heart and skeletal muscles, thus mediating global changes in sympathetic activity (Farmer et al., 2019).

Increased oxidative stress has been consistently reported in the rostral ventrolateral medulla of rodent models with different forms of hypertension, such as essential, salt-sensitive, or renovascular hypertension, or hypertension driven by angiotensin II, high-fat diet, low-level inflammation, or chronic intermittent hypoxia (Guyenet et al., 2018). Experimental downregulation of central antioxidant enzyme expression in the rostral ventrolateral medulla was reported to trigger sustained sympatho-excitation, baroreflex inhibition, and hypertension in mice (Gao et al., 2017). Nevertheless, pathophysiologic data on the contribution of the rostral ventrolateral medulla to hypertension in preclinical models should be interpreted with caution (Guyenet et al., 2018), as much evidence is still needed to translate these results to human subjects.

The rostral ventrolateral medulla includes the C1 group of adrenergic neurons (Dampney, 1994). The C1 neurons corelease adrenaline, glutamate, and neuropeptides, may be activated by danger signals including hypotension, hypoxia, hypoglycemia, pain, and inflammation, and are instrumental in triggering coordinated autonomic and neuroendocrine responses promoting survival (Guyenet and Stornetta, 2022; Stornetta and Guyenet, 2018). These neurons significantly contribute to regulate arterial blood pressure by controlling the sympathetic nervous system, particularly in conditions of anesthesia, hypoxia, or hemorrhage (Stornetta and Guyenet, 2018). In rats, C1 neurons mediate the compensatory increase in sympathetic outflow during hemorrhage, and a reduction in C1 neuron activity contributes to hypotension during decompensated hemorrhage (Souza et al., 2022).

The C1 neurons represent roughly half of the presympathetic neurons in the rostral ventrolateral medulla, but this figure may reflect the technical inability to detect weak expression of the adrenalin synthesis machinery in the non-C1 presympathetic neurons in the rostral ventrolateral medulla. These neurons, are also glutamatergic and may preferentially control the skeletal muscles vasculature (Guyenet and Stornetta, 2022). On the other hand, roughly half of the C1 neurons are not presympathetic neurons and generally project to structures rostral to the midbrain, including the orexin/hypocretin neurons of the hypothalamus, the hypothalamic paraventricular nucleus, and the paraventricular thalamic nucleus. The C1 cells also project to and activate the ventrolateral column of the midbrain periaqueductal gray and the pontomedullary noradrenergic cell groups (A1–A7), including the pontine locus coeruleus (Guyenet and Stornetta, 2022). The C1 neurons activate a cholinergic, predominantly sympathetic, anti-inflammatory pathway in response to acute stress in mice, attenuating experimental ischemia-reperfusion injury (Abe et al., 2017). Finally, the C1 cells project to parasympathetic preganglionic neurons in the dorsal motor nucleus of the vagus, although the relevance of this projection for the control of arterial blood pressure is uncertain (Guyenet and Stornetta, 2022).

Collectively, the diversity of C1 cell projections may underlie the variability in the reported effects of C1 cell modulation on arterial blood pressure. For instance, optogenetic stimulation of C1 cells at 5 Hz for 10 minutes was reported to decrease arterial blood pressure and heart rate in mice, a response usually associated with freezing behavior and possibly mediated by the ventrolateral periaqueductal gray (Abe et al., 2017). In a different experiment, C1 cell optostimulation at 20 Hz for 20 seconds in rats during quiet resting or sleep increased arterial blood pressure and elicited arousal from nonrapid-eye-movement (non-REM) sleep, possibly by means of C1 projections to the locus coeruleus (Burke et al., 2014). However, C1 neuron ablation had no effect on arousal induced by either hypoxia or hypercapnia (Souza et al., 2019), suggesting parallel and redundant pathways of chemoreceptor-driven arousal.

The majority of C1 cells, including presympathetic neurons and nonpresympathetic neurons, are under powerful inhibition by the arterial baroreflex, which is mostly exerted through neurons of the intermediate and caudal portions of the ventrolateral medulla that release gamma-aminobutyric acid (GABA) as neurotransmitter (Schreihofer and Guyenet, 2003; Parker et al., 2017). The tonic activity of C1 neurons under resting conditions permits both reflex decreases and increases in the activity of sympathetic preganglionic neurons in response to changes in baroreceptor input (Dampney, 2016).

The region between the rostral ventrolateral medulla and the caudal ventrolateral medulla, overlapping with the pre-Bötzinger complex, has been termed the intermediate ventrolateral medulla. The intermediate ventrolateral medulla harbors cardiovagal parasympathetic preganglionic neurons as well as GABAergic neurons involved in the arterial baroreflex control of presympathetic neurons, which receive direct input from barosensitive neurons in the nucleus of the solitary tract (Guyenet and Stornetta, 2022). GABAergic projections control the threshold excitability of rostral ventrolateral medulla neurons, whereas glycinergic projections, probably also from the caudal ventrolateral medulla, potentiate inhibition under conditions of increased synaptic activity within the rostral ventrolateral medulla (Gao et al., 2019).

The caudal ventrolateral medulla also includes the A1 group of noradrenergic neurons, which are critical for vasopressin release due to baroreceptor unloading. The A1 cells may also contribute to mediate hypertensive and natriuretic responses to acute hypernatremia (da Silva et al., 2013).

The role of the ventromedial and midline medulla in arterial blood pressure control

The rostral ventromedial medulla, which comprises the midline medullary raphe and the parapyramidal area, is a major site of presympathetic neurons, many of which express serotonin (Silvani et al., 2016). Stimulation of serotoninergic neurons of the rostral ventromedial medulla may increase arterial blood pressure (Pilowsky et al., 1986). The presympathetic neurons in the rostral ventromedial medulla are activated during arousal and stress, possibly through direct projections from the dorsomedial hypothalamus and from the lateral column of the periaqueductal gray (Silvani et al., 2016). On the other hand, the rostral ventromedial medulla does not play a significant role in baroreflex or chemoreflex control of arterial blood pressure (Dampney, 1994). Putative serotonergic neurons in the medullary raphe do not alter their firing depending on baroreflex input (McCall and Clement, 1989), and may rather provide a tonic excitatory input that increases the level of excitability of sympathetic preganglionic neurons (McCall, 1983).

The discharge of sympathetic preganglionic neurons that control cutaneous vasoconstriction in response to cold signals is governed primarily by glutamatergic and/or serotonergic presympathetic neurons in the raphe pallidus and parapyramidal area of the rostral ventromedial medulla, as well as by glutamatergic presympathetic neurons in the rostral ventrolateral medulla, which include C1 neurons (Morrison and Nakamura, 2019). This circuitry may be responsible for the skin, splanchnic, and renal vasoconstriction that occurs in humans with whole-body skin surface cooling together with an increase in arterial blood pressure, even in the absence of hypothermia and shivering (Wilson et al., 2007). The same circuitry may mediate skin vasodilation in the hands and feet during non-REM sleep in humans, which occurs together with a decrease in skin vasoconstrictor sympathetic nerve activity (Silvani and Dampney, 2013) and correlates with a decrease in arterial blood pressure (Kräuchi et al., 2012).

The midline medullary raphe includes both sympatho-excitatory and sympatho-inhibitory neurons that are inhibited and excited by baroreceptor input, respectively, but are not a necessary component of the arterial baroreflex pathway (McCall and Harris, 1987). In particular, nonserotonergic neurons in the raphe obscurus that are active in rapid-eye-movement (REM) sleep may mediate differential effects on sympathetic nerve activity: sympathetic nerve activity to the splanchnic and skeletal muscle circulation is decreased and increased, respectively, on passing from non-REM sleep to REM sleep together with an increase in arterial blood pressure (Futuro-Neto et al., 1982; Morrison and Gebber, 1984; Verret et al., 2005).

The role of the nucleus of the solitary tract in reflex control of arterial blood pressure

The nucleus of the solitary tract receives direct synaptic projections from the primary arterial baroreceptor neurons that innervate the carotid sinuses and the aortic arch. The barosensitive nucleus of the solitary tract neurons are critical for arterial baroreflex control (Silvani et al., 2016). These neurons send direct glutamatergic projections to the cardiovagal parasympathetic preganglionic neurons in the nucleus ambiguus and to the GABAergic neurons of the intermediate and caudal portions of the ventrolateral medulla, which, in turn, inhibit the presympathetic neurons in the rostral ventrolateral medulla (Dampney, 1994; Guyenet and Stornetta, 2022). As a result, increases in stretch of the baroreceptor afferent fibers increase cardiac parasympathetic outflow and decrease cardiac and vascular sympathetic outflow, whereas the opposite occurs in response to decreases in baroreceptor afferent stretch (Silvani et al., 2016). Astrocytes also contribute to glutamatergic regulation in the nucleus of the solitary tract to maintain baseline cardiovascular parameters and baroreflex and chemoreflex control (Martinez and Kline, 2021).

The arterial baroreflex function depends on the ongoing behavior. During exercise and psychologic stress, the maximum gain of the baroreflex sympathetic control increases and its operating range, that is, the range of arterial blood pressure values that produce reflex sympathetic changes, is reset toward higher values of arterial blood pressure (Dampney, 2016). The mechanisms that reset the baroreflex during psychologic stress are incompletely understood and may be mediated by the dorsomedial hypothalamus and the hypothalamic perifornical area (Dampney, 2017). The mechanisms that reset the baroreflex during exercise are thought to involve exercise-related central autonomic commands and the exercise pressor reflex (Raven et al., 2006). These mechanisms may result from the summation of increased GABAergic inhibitory inputs to the nucleus of the solitary tract, which resets the baroreflex to higher values of arterial blood pressure, from facilitation of inhibitory inputs to the intermediate and caudal portions of the ventrolateral medulla, which increases the gain of baroreflex sympathetic control, and from increased excitatory drive to presympathetic neurons in the rostral ventrolateral medulla and to sympathetic preganglionic neurons in the spinal cord, which resets the baroreflex toward higher sympathetic activity (Dampney, 2017).

The exercise pressor reflex increases sympathetic activity to the heart and blood vessels in response to activation of muscle metaboreceptor (group IV) and mechanoreceptor (group III) afferents. This counterbalances the effects on arterial blood pressure of the reduction in skeletal muscle vascular resistance, which results from flow–metabolism coupling during physical activity (Teixeira and Vianna, 2022). Activation of the muscle metaboreceptor afferents and sensitization of muscle mechanoreceptor afferents by metabolites such as thromboxane also inhibits cardiovagal parasympathetic preganglionic neurons activity, thereby decreasing the sensitivity of the baroreflex control of heart rate. Activation of muscle mechanoreceptor afferents causes renal vasoconstriction, which becomes exaggerated in patients with heart failure (Drew, 2017). The neural pathway of the exercise pressor reflex awaits further investigation, except for the occurrence of its first synapse in the medullary dorsal horn (Silvani et al., 2016; Teixeira and Vianna, 2022), and the responsiveness of a subset of neurons in the nucleus of the solitary tract to muscle mechanoreceptor afferents (Degtyarenko and Kaufman, 2006).

The modulation of the arterial baroreflex is also a key feature of cardiovascular control during sleep (Silvani and Dampney, 2013). In rats, the arterial baroreflex is reset to lower levels of renal sympathetic nerve activity during REM sleep and, to a lesser extent, during non-REM sleep, and is reset to lower values of heart rate during non-REM sleep (Nagura et al., 2004). This sleep-related baroreflex modulation occurs at least in part at the level of the nucleus of the solitary tract (Tang and Dworkin, 2010). Conversely, mild baroreflex stimulation inhibits cortical arousal and favors sleep, possibly through inhibition by the nucleus of the solitary tract of C1 neurons in the rostral ventrolateral medulla. Substantial increases or decreases in baroreflex activation cause arousal from sleep, possibly by activation by the nucleus of the solitary tract of neurons in the parabrachial nucleus and by disinhibition by the nucleus of the solitary tract of C1 neurons, respectively (Silvani et al., 2015). Accordingly, chemogenetic or optogenetic activation of barosensitive neurons in the nucleus of the solitary tract promoted non-REM sleep and decreased arterial blood pressure and heart rate in mice. Similar effects occurred with activation of GABAergic neurons in the caudal ventrolateral medulla, partly through the inhibition of C1 neurons in the rostral ventrolateral medulla (Yao et al., 2022).

The nucleus of the solitary tract receives direct axonal projections from neurons in the caudal vestibular nuclei, which may underlie effects of changes in static head position and head movements on arterial blood pressure and heart rate, as well as on respiration and gastrointestinal motility. Many of the cell bodies of the vestibulo-solitary neurons are surrounded by GABAergic processes, consistent with inhibitory control by Purkinje cells of the vestibulo-cerebellum (Gagliuso et al., 2019). Early evidence indicates that neurons in the medial and inferior vestibular nuclei increase firing during REM sleep in correspondence with bursts of rapid eye movements (Bizzi et al., 1964). These neurons are required for the occurrence of spontaneous phasic cardiovascular events in REM sleep (Morrison and Pompeiano, 1970), which include surges in arterial blood pressure and heart rate (Silvani and Dampney, 2013).

The nucleus of the solitary tract is also a main hub for the arterial chemoreceptor reflex, receiving direct projections from arterial chemoreceptors in the carotid and aortic glomi (Dampney, 2016). The chemosensitive neurons of the nucleus of the solitary tract send direct, presumably glutamatergic projections to the cardiovagal parasympathetic preganglionic neurons in the nucleus ambiguus and to presympathetic neurons in the rostral ventrolateral medulla (Silvani et al., 2016). As a result, decreases in the arterial partial pressure of O_2, increases in the arterial partial pressure of CO_2, and decreases in arterial pH sensed by arterial chemoreceptors increase both cardiac parasympathetic outflow and cardiac and vascular sympathetic outflow (Dampney, 2016). Simultaneous sympathetic and parasympathetic activation is thought to limit cardiac O_2 consumption in the face of limited decrease in cardiac output and to compensate the local vasodilatory effect of hypoxia, hypercapnia and acidosis, thereby preserving arterial blood pressure values (Kara et al., 2003; Paton et al., 2005). However, cardiac sympathetic and parasympathetic coactivation may be proarrhythmic (Paton et al., 2005). This effect may contribute to explain why the night, which is physiologically protective against sudden cardiac death, ceases being so in subjects with obstructive sleep apneas, which entail repeated transient episodes of hypoxia and hypercapnia (Gami et al., 2005). In rats, chronic sleep fragmentation caused by tracheal obstruction, mimicking upper airway collapse during obstructive sleep apneas, increases sympathetic activation (Ferreira et al., 2020). Nevertheless, carotid bodies are key in mediating persistent increases in arterial blood pressure in rats with repetitive hypoxia, which is a hallmark of obstructive sleep apneas in humans (Fletcher et al., 1992). Heightened carotid body activity may also be a contributing factor to essential and renovascular hypertension in animal models (Guyenet et al., 2018). The modulation of arterial chemoreceptor function also occurs physiologically, enhancing the chemoreflex ventilatory response to hypoxia during exercise (Dampney, 2016).

The potential proarrhythmic effect of the cardiac sympathetic and parasympathetic coactivation that is driven by the arterial chemoreflex (Paton et al., 2005) suggests an adaptive value for the inhibition of autonomic chemoreflex effects, which is exerted by pulmonary stretch receptors innervated by afferent vagal fibers (Kara et al., 2003). Autonomic modulation by stretch receptors also occurs downstream of the nucleus of the solitary tract in the baroreflex autonomic control circuit, including through baroreflex gating in the intermediate and caudal portions of the ventrolateral medulla by components of the central respiratory network (Guyenet et al., 2018). Conversely, there is evidence for a pauci-synaptic circuit that delays inspiration by activating postinspiratory neurons in response to baroreceptor input (Barnett et al., 2021). Alterations in cardiorespiratory coupling involving sympathetic activity have been repeatedly linked to hypertension in experimental models, but their contribution to increased sympathetic activity in conscious hypertensive animals and humans remains unclear (Guyenet et al., 2018).

Chemoreceptor autonomic control may also be mediated by central chemoreceptors, which are mainly located in the retrotrapezoid nucleus of the medulla and sense partial pressure of CO_2 and pH in the brain interstitium. The retrotrapezoid nucleus sends widespread projections to structures of the central respiratory network and is inhibited by pulmonary stretch receptors (Guyenet and Stornetta, 2022). However, selective activation of the retrotrapezoid nucleus during sleep increases ventilation without altering arterial blood pressure significantly (Burke et al., 2015). The C1 neurons in the rostral ventrolateral medulla are also marginally activated by hypercapnia, with limited effects on arterial blood pressure (Wenker et al., 2017).

Apneas, bronchoconstriction, and bradycardia may be elicited by the diving reflex, which is triggered by stimulation of nasopharyngeal receptors on trigeminal afferents (Panneton, 2013). Activation of these afferents modulates the activity of neurons in the nucleus of the solitary tract and of parasympathetic preganglionic neurons in the nucleus ambiguus, either directly or through a relay in the medullary dorsal horn of the facial nucleus (Silvani et al., 2016). Recent data identified a distinct subpopulation of nucleus ambiguus neurons, termed the cardiopulmonary neurons, that innervate cardiac and lung parasympathetic ganglionic neurons and underlie the diving reflex (Veerakumar et al., 2022).

The role of the pons and midbrain in the regulation of arterial blood pressure

As previously discussed, the upper brainstem includes different structures involved in the control of arterial blood pressure, such as the A5 noradrenergic cell group, parabrachial nucleus, Kölliker–Fuse nucleus, and locus coeruleus (A6 noradrenergic cell group) in the pons and the midbrain periaqueductal gray. Other rostral brainstem structures relevant to arterial blood pressure control include the pedunculopontine nucleus in the dorsal tegmentum of the rostral pons and midbrain (Mena-Segovia and Bolam, 2017) and its neighboring mesencephalic locomotor region (Garcia-Rill et al., 2019).

The A5 noradrenergic cell group

The pontine A5 cell group is a major source of noradrenergic innervation of sympathetic preganglionic neurons (Dampney, 1994). The A5 neurons send widespread projections to central autonomic network structures in and beyond the brainstem and receive afferent input from major cardiovascular regulatory centers including the nucleus of the solitary tract (Byrum and Guyenet, 1987; Dampney, 1994; Souza et al., 2022). These inputs plausibly mediate the baroreflex inhibition of most A5 neurons (Huangfu et al., 1991). Chemical lesions of A5 cells did not lead to significant alterations of arterial blood pressure under baseline conditions or in response to hypoxia or hypercapnia in unanesthetized rats. However, similar lesions attenuated the sympathoexcitation in response to stimulation of peripheral chemoreceptors or to hypercapnia in an in situ rat preparation (Taxini et al., 2017). On the other hand, chemogenetic stimulation of A5 neurons increased arterial blood pressure and cardiac baroreflex sensitivity in freely behaving rats without changing locomotor activity or respiration and preferentially increased visceral (splanchnic) over skeletal muscle (lumbar) sympathetic nerve activity in anesthetized rats (Souza et al., 2022). These data suggest that the A5 cell group may be important for patterned recruitment of the sympathetic efferents (Souza et al., 2022).

The parabrachial nucleus and Kölliker–Fuse nucleus

The parabrachial nucleus and the Kölliker–Fuse nucleus are interconnected with the nucleus of the solitary tract and represent a second-order relay of visceral afferent information to structures of the central autonomic network (Silvani et al., 2016).

The parabrachial nucleus may represent a hub for pain and aversion, including bad taste and thermoregulatory challenges (Chiang et al., 2019). In particular, the parabrachial nucleus receives inputs regarding taste and other viscerosensory modalities, respiratory control, fluid balance, and cardiovascular function primarily from the nucleus of the solitary tract. The parabrachial nucleus also receives inputs regarding respiratory control from brainstem chemosensory neurons and nociceptive, pruritic, and thermal information from the trigeminal and spinal dorsal horns (Chiang et al., 2019). In addition, relays in the external and dorsolateral subnuclei of the lateral parabrachial nucleus convey information from skin cold and warm receptors, respectively, to the median preoptic nucleus of the hypothalamus (Morrison and Nakamura, 2019).

Inhibition of the Kölliker–Fuse nucleus attenuates increases in arterial blood pressure, splanchnic sympathetic nerve activity, and phrenic nerve activity produced by hypercapnia or hypoxia in anesthetized rats, suggesting that this nucleus integrates and modulates breathing and sympathetic outflow during chemoreceptor activation (Damasceno et al., 2015; Mirzaei-Damabi et al., 2020).

In rats, the parabrachial nucleus and adjacent precoeruleus region are also necessary to maintain behavioral responsiveness by means of glutamatergic projections to the basal forebrain (Fuller et al., 2011). The central lateral parabrachial nucleus projects to the ventrolateral preoptic nucleus of the hypothalamus, which is a key sleep-promoting structure that sends a return projection to the whole parabrachial nucleus–precoeruleus nucleus complex (Fuller et al., 2011).

Neurons in the parabrachial nucleus and Kölliker–Fuse nucleus may contribute to modulate the arterial baroreflex during sleep. Reduced activity of these neurons during non-REM sleep may disinhibit the baroreflex at the level of the nucleus of the solitary tract, explaining why the arterial baroreflex is the least disrupted by central autonomic commands in this sleep state (Silvani and Dampney, 2013). On the other hand, projections from the nucleus of the solitary tract to the parabrachial nucleus may promote arousal from sleep with baroreflex activation, mediating a positive feedback loop that increases arterial blood pressure and arousal upon awakening (Silvani et al., 2015).

The locus coeruleus

Both increases and decreases in arterial blood pressure have been reported in response to stimulation of the locus coeruleus, making its direct role in the control of arterial blood pressure uncertain (Dampney, 1994). Nevertheless, the locus coeruleus has widespread projections to the brain and spinal cord and plays an important role in the control of sleep and arousal (Carter et al., 2010). The noradrenergic neurons of the locus coeruleus directly project to the ventrolateral preoptic nucleus of the hypothalamus, a key sleep-promoting region, and optogenetic activation of this pathway promotes rapid arousal from sleep (Liang et al., 2021). While contribution of parallel arousal pathways triggered by the locus coeruleus is also plausible (Liang et al., 2021), changes in the activity of neurons in the ventrolateral preoptic nucleus may drive the central autonomic network responsible for the increases in arterial blood pressure and heart rate upon arousal from sleep (Silvani and Dampney, 2013).

The pedunculopontine nucleus and the mesencephalic locomotor region

The pedunculopontine nucleus, long considered an interface between the basal ganglia and motor systems, is now thought to have a central role in updating the behavioral state as a result of changes in environmental contingencies (Mena-Segovia and Bolam, 2017). Single cholinergic neurons of the pedunculopontine nucleus are multisynaptically linked to central sympathetic circuits as well as to motor-related portions of the thalamus and the primary motor cortex (Krout et al., 2003). The pedunculopontine nucleus also includes neurons excited by static muscle contractions (Plowey et al., 2002). Chemical activation of the pedunculopontine nucleus increases arterial blood pressure together with muscle activity and splanchnic sympathetic nerve activity in rats, an effect attenuated by blockade of muscarinic cholinergic receptors in the rostral ventrolateral medulla (Padley et al., 2007).

The pedunculopontine nucleus also includes cholinergic neurons that fire much more during wakefulness and REM sleep than during non-REM sleep (el Mansari et al., 1989). Decreased activity of the circuitry involving cholinergic neurons in the pedunculopontine nucleus and presympathetic neurons in the rostral ventrolateral medulla may contribute to the decrease in arterial blood pressure during non-REM sleep (Silvani and Dampney, 2013), which is driven by the modulation of sympathetic vascular control (Lo Martire et al., 2018). The same circuitry may also contribute to increase arterial blood pressure during REM sleep (Silvani and Dampney, 2013). However, the cardiovascular effects of optogenetic activation of C1 neurons in the rostral ventrolateral medulla, which may be modulated by the pedunculopontine nucleus, are reduced in REM sleep (Abbott et al., 2013).

The mesencephalic locomotor region, originally identified as a region in which electrical stimulation elicited locomotion in decorticate cats (Shik and Orlovsky, 1976), is now considered to consist of glutamatergic bulbospinal neurons medial to the pedunculopontine nucleus (Garcia-Rill et al., 2019). In decerebrate and paralyzed rats, electrical stimulation of the mesencephalic locomotor region markedly increased arterial blood pressure, renal sympathetic nerve activity, and renal vascular resistance, with milder increases in lumbar sympathetic nerve activity and skeletal muscle vascular resistance (Koba et al., 2006). These results suggest that the mesencephalic locomotor region underlies the blood flow redistribution evoked by differential sympathetic outflow during exercise as a result of central autonomic commands (Koba et al., 2006). Recent data on rats indicate that optogenetic activation of a monosynaptic projection from the mesencephalic locomotor region to the rostral ventrolateral medulla elicits the locomotor and cardiovascular responses characteristic of exercise. During voluntary running in rats, the same pathway is spontaneously activated, and its optogenetic inhibition attenuates locomotor activity and the associated increase in arterial blood pressure (Koba et al., 2022). Despite some discrepancies that still await clarification, these data on the pedunculopontine nucleus and the mesencephalic locomotor region collectively support the view that projections from the upper brainstem to the rostral ventrolateral medulla play an important role in modulating arterial blood pressure during locomotion.

The midbrain periaqueductal gray

A strong connection between the periaqueductal gray and the rostral ventrolateral medulla was found with the analysis of functional magnetic resonance connectivity in humans (Macefield and Henderson, 2019). Data on animal models revealed that the midbrain periaqueductal gray is organized in longitudinal columns that mediate different types of defensive responses and associated cardiovascular adjustments. The lateral and dorsolateral columns of the periaqueductal gray generate active coping responses that require interaction with the environment in response to escapable threats (Bandler et al., 2000). These strategies are supported by increases in arterial blood pressure and heart rate, visceral vasoconstriction, skeletal muscle vasodilation, and increased ventilation (Silvani et al., 2016). Responses involving the lateral column of the periaqueductal gray are triggered primarily by physical stimuli such as skin nociceptor activation and mediated by direct descending projections to the medulla. Responses involving the dorsolateral column of the periaqueductal gray are elicited primarily by emotional or psychologic stressors and mediated by projections that involve the dorsomedial hypothalamic nucleus, the superior lateral parabrachial nucleus, and the mesencephalic cuneiform nucleus (Dampney, 2016). Conversely, the ventrolateral column of the periaqueductal gray is implicated in generating passive coping ("conservation-withdrawal") responses to stressors that are inescapable, such as severe traumatic injury or repeated defeat in social encounters (Bandler et al., 2000). These strategies involve behavioral quiescence and are associated sympatho-inhibition and decreases in arterial blood pressure and heart rate (Dampney, 2016).

More recent data indicate that the synaptic transmission in the ventrolateral periaqueductal gray is involved in cardiac baroreflex control, modulating the tachycardic response to arterial baroreceptor unloading (Lagatta et al., 2016).

CLINICAL IMPLICATIONS

Alterations in the central autonomic network of the brainstem may cause severe impairment in the regulation of arterial blood pressure. These alterations may lead to afferent baroreflex failure and to cardiovascular autonomic failure with neurogenic orthostatic hypotension and supine hypertension, or may occur as a result of neurodegeneration in multiple system atrophy or of multiple sclerosis and brainstem lesions (Fig. 3.2).

Afferent baroreflex failure

Afferent baroreflex failure is the clinical syndrome resulting from damage of the carotid baroreceptors, the afferent limb of the baroreflex, or its central connections at the level of the nucleus of the solitary tract. In absence of afferent inputs from the baroreceptors, the neurons of the nucleus of the solitary tract do not inhibit, through interneurons in the caudal ventrolateral medulla, the presympathetic neurons in the rostral ventrolateral

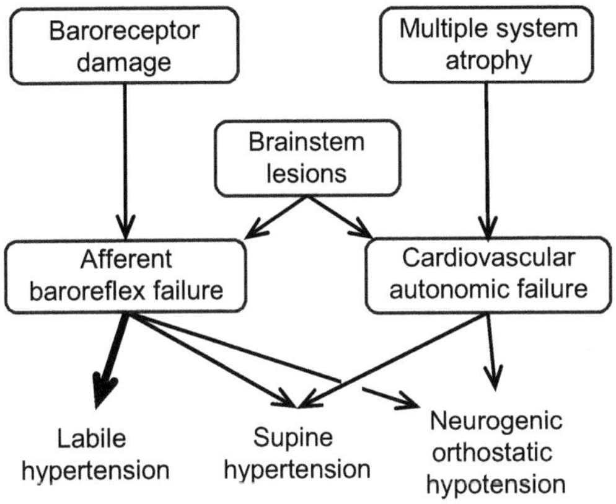

Fig. 3.2. Overview of impairments in the regulation of arterial blood pressure caused by alterations of the central autonomic network of the brainstem.

medulla that control basal and reflex sympathetic activity. As a result, the barosensitive presympathetic neurons in the rostral ventrolateral medulla become unrestrained, causing an uncontrolled release of norepinephrine that drives hypertensive crises and volatile hypertension, the clinical hallmark of the disease (Kaufmann et al., 2020).

Afferent baroreflex failure has been described as a complication of carotid endarterectomy or angioplasty causing damage to or stretch of the baroreceptors, or occurring after radiotherapy to the neck or radical neck surgery for cancer, or as the clinical manifestation of lesions of the nucleus of the solitary tract like stroke or tumors (Benarroch, 2008; Kaufmann et al., 2020; Norcliffe-Kaufmann et al., 2020; Lamotte et al., 2021).

Clinical manifestations and therapeutic options for afferent baroreflex failure

The acute phase of afferent baroreflex failure is characterized by severe hypertensive surges, with systolic arterial blood pressure values that may exceed 300 mm Hg and may be associated with tachycardia, headache, flushing, or diaphoresis. These hypertensive crises closely resemble those occurring in pheochromocytoma and may evolve over days and weeks into chronic volatile hypertension. The latter can also develop insidiously due to a gradual decline of baroreflex function. In chronic conditions, patients present normal to elevated baseline values of arterial blood pressure and sudden episodes of sympathetic activation, which are characterized by surges in arterial blood pressure lasting minutes to hours associated with tachycardia (Ketch et al., 2002). Other less frequent clinical manifestations of baroreflex failure include orthostatic tachycardia and severe bradycardia with hypotension (malignant vagotonia), orthostatic hypotension, and supine hypertension. Severe bradycardia and asystole attributable to increased parasympathetic tone in patients with intact efferent sympathetic and parasympathetic output to the heart and blood vessels are rare.

Diagnosis of afferent baroreflex failure is based on history (e.g., previous neck radiation or surgery and labile hypertension), and pharmacologic testing with continuous blood pressure and heart rate monitoring. Pharmacologic tests show exaggerated blood pressure increases associated with absent bradycardia in response to pressor agents such as phenylephrine or absent tachycardia in response to depressor agents such as nitroprusside.

Orthostatic hypotension may result from inefficient baroreflex unloading, hypovolemia due to pressor diuresis, or hypersensitivity to short-acting sympatholytic drugs for the therapy of hypertension (Biaggioni et al., 2019; Lamotte et al., 2021). Long-acting sympatholytic agents such as clonidine are the mainstay treatment to reduce the frequency and magnitude of life-threatening surges of arterial blood pressure in afferent baroreflex failure. However, the prognosis is usually poor, with high mortality and morbidity (Biaggioni et al., 2019; Lamotte et al., 2021).

Cardiovascular autonomic failure

Cardiovascular autonomic failure is the consequence of the impairment of the baroreflex-modulated sympathetic ganglionic neurons that release norepinephrine at the neurovascular junctions. At the brainstem level, cardiovascular autonomic failure is related to lesions or degeneration of the presympathetic neurons in the rostral ventrolateral medulla and frequently occurs in the context of a more generalized autonomic failure.

Neurogenic orthostatic hypotension and supine hypertension in cardiovascular autonomic failure

The clinical hallmark of cardiovascular autonomic failure is neurogenic orthostatic hypotension. In physiologic conditions, the transition from the supine to the orthostatic position causes a redistribution of 300–800 mL of blood volume to the lower limbs and the splanchnic circulation. This reduces venous return to the heart and the left ventricular filling pressure, with consequent decreases in stroke volume and cardiac output. In response to these changes, the baroreceptor reflex causes sympathetic activation and a reduction in the vagal tone to the heart, with the net effect of increasing vascular tone, heart rate and contractility, and to restore arterial blood pressure values, usually within 10 seconds. An impairment of the brainstem structures that control sympathetic outflow hinders the compensation for orthostatic stress, leading to the development of neurogenic orthostatic hypotension and tissue hypoperfusion.

The diagnosis of orthostatic hypotension requires a sustained decrease in systolic arterial blood pressure of at least 20 mmHg and/or in diastolic arterial blood pressure of at least 10 mmHg within 3 minutes of standing or head-up tilt to at least 60° on a tilt table, compared to the supine position. In mild or early cases, delayed orthostatic hypotension may be observed, occurring after 3 minutes and within 10 minutes of standing or head-up tilt test (Freeman et al., 2011).

Orthostatic hypotension is defined as neurogenic when it is associated with impaired efferent baroreflex sympathetic control. A blunted heart rate response to the assumption of the orthostatic position can be demonstrated by measuring the ratio between the increase in heart rate and the decrease in systolic arterial blood pressure at the third minute of the orthostatic test compared to the values in the supine position. If this ratio is less than 0.5, it can be hypothesized with sensitivity of 91% and specificity of 88% that the orthostatic hypotension is neurogenic (Norcliffe-Kaufmann et al., 2018; Guaraldi et al., 2020).

The direct method to assess sympathetic dysfunction in cardiovascular autonomic failure is the dosage of plasma catecholamines during supine rest and head-up tilt test. Under physiologic conditions, noradrenaline values double in the transition from supine to the upright position, while this does not occur in neurogenic orthostatic hypotension (Goldstein and Sharabi, 2009). Alternatively, abnormal cardiovascular responses to the Valsalva maneuver, showing the absence of arterial blood pressure recovery during phase II and of the typical arterial blood pressure increase ("overshoot") at the end of expiration, are also diagnostic for cardiovascular autonomic failure (Cheshire et al., 2021).

Orthostatic hypotension can be either symptomatic or asymptomatic. Symptoms of hypoperfusion may appear on transition from the supine position to standing or during prolonged standing and recover promptly upon acquisition of the sitting or lying position. These symptoms may be reported as "dizziness," "light-headedness," or "blurred vision." Symptoms may also include slowing of cognitive performance, a painful contracture of the neck defined "coat-hanger" pain due to hypoperfusion of the neck muscles, symptoms similar to angina pectoris caused by cardiac hypoperfusion, and syncope due to cerebral hypoperfusion. The reduced blood flow to the kidneys leads to oliguria during standing, often associated with nocturnal polyuria favored by the increased renal perfusion in the supine position (Mathias et al., 1999).

Neurogenic orthostatic hypotension may be associated with supine hypertension, defined as systolic arterial blood pressure ≥ 140 mmHg and/or diastolic arterial blood pressure ≥ 90 mmHg, measured after at least 5 minutes of rest in the supine position. A reduction in the physiologic drop of mean nocturnal arterial blood pressure compared to mean daytime arterial blood pressure (nondipper pattern, < 10% drop) or an increase in mean nocturnal arterial blood pressure compared to mean daytime arterial blood pressure (reverse dipper pattern) may also be observed (Fanciulli et al., 2018).

The association of neurogenic orthostatic hypotension and supine hypertension has serious long-term implications and has been associated with end-organ damage including renal failure, left ventricular hypertrophy, and cerebrovascular disease. Neurogenic orthostatic hypotension impacts on prognosis and survival in neurodegenerative synucleinopathies presenting cardiovascular autonomic failure like Parkinson disease, Lewy body dementia, and multiple system atrophy (Palma and Kaufmann, 2020). The co-occurrence of supine hypertension can contribute to worsen prognosis and survival (Palma et al., 2020).

Multiple system atrophy

Multiple system atrophy is a sporadic and fatal neurodegenerative disease, clinically characterized by autonomic dysfunction (cardiovascular autonomic failure and/or neurogenic bladder) associated with a parkinsonian syndrome that is poorly responsive to levodopa and/or a cerebellar syndrome (Wenning et al., 2022). The mean age at onset is 55–60 years, with an average survival time from the onset of motor symptoms of 8–9 years, although both young-onset cases before the age of 40 (Batla et al., 2018) and pathology-proven long-survival cases, with survival exceeding 15 years (Petrovic et al., 2012), have been described.

The neuropathologic hallmark of multiple system atrophy is the finding of widespread and abundant glial cytoplasmic inclusions in the central nervous system, which consist of misfolded α-synuclein aggregates, in association with neurodegenerative changes in striato-nigral or olivopontocerebellar structures (Trojanowski and Revesz, 2007). Several interconnected brainstem areas involved in vital functions, like cardiovascular and respiratory control, urinary function, and behavioral arousal, are also affected in multiple system atrophy (Benarroch, 2019), resulting in clinical manifestations that may predispose to sudden death (Giannini et al., 2016; Shimohata et al., 2016; Vichayanrat et al., 2022; McCarter et al., 2023).

Neurogenic orthostatic hypotension and supine hypertension in multiple system atrophy

Neurogenic orthostatic hypotension, the hallmark of the failure of brainstem sympathetic cardiovascular control, is a core clinical feature of multiple system atrophy, affecting 60%–80% of patients (Low et al., 2015; Wenning et al., 2013). Delayed neurogenic orthostatic hypotension is also observed, mainly in early disease stages (Pavy-Le Traon et al., 2016). Isolated neurogenic orthostatic hypotension may predate the motor onset of the disease, with 8%–28% of patients developing multiple system atrophy over 10 years (Kaufmann et al., 2017; Singer et al., 2017; Giannini et al., 2018).

In multiple system atrophy, neurogenic orthostatic hypotension reflects impaired baroreflex efferent control of arterial blood pressure in the upright position due to the degeneration of the presympathetic neurons in the rostral ventrolateral medulla, including the C1 group, that are also involved in baroreflex control (Benarroch, 2008). Sympathetic ganglionic neurons are affected only in a minority of patients (Treglia et al., 2011; Orimo et al., 2012).

Compared to conditions with prevalent postganglionic degeneration like Parkinson disease with neurogenic orthostatic hypotension, multiple system atrophy is characterized by near-normal plasma norepinephrine levels in the supine position and a significant pressor response to norepinephrine and also to atomoxetine, a norepinephrine reuptake inhibitor (Jordan et al., 2015). In patients with isolated autonomic failure, a supine norepinephrine level > 100 pg/mL has been reported as a sensitive and specific predictor of phenoconversion to multiple system atrophy, whereas abnormal findings at 123I-MIBG-scintigraphy are considered an exclusion criterion (Baschieri et al., 2017; Kaufmann et al., 2017; Singer et al., 2017; Wenning et al., 2022).

Multiple system atrophy is also characterized by the loss of noradrenergic A1 neurons in the caudal ventrolateral medulla, which mediate the vasopressin release that occurs during baroreflex withdrawal, such as in orthostatic stress. Impairment of reflex vasopressin secretion may contribute to orthostatic hypotension (Benarroch et al., 2006). Neurogenic orthostatic hypotension is symptomatic in approximately half of the patients with multiple system atrophy, and the frequency of syncope ranges between 19% and 65%. As multiple system atrophy is a progressive disorder, the development of mechanism of cerebrovascular regulation may contribute to tolerance to postural hypotension (Mathias et al., 1999).

Supine hypertension has been observed in 37%–46% of patients with multiple system atrophy, with severe hypertension occurring in 28% of patients and with up to 75% of patients showing a pathologic nocturnal nondipper or reverse-dipper profile (Fanciulli et al., 2014, 2016; Coon et al., 2015; Eschlböck et al., 2020). In these patients, supine hypertension is caused by residual peripheral sympathetic function due to postsynaptic vascular adrenoceptor hypersensitivity in a context of abolished baroreflex control and to the effects of pharmacologic treatment of neurogenic orthostatic hypotension.

Neurogenic orthostatic hypotension impacts on both prognosis and survival in multiple system atrophy, particularly when it occurs within 3 years from disease onset (Coon et al., 2015; Glasmacher et al., 2017; Vichayanrat et al., 2022). This may be due to both short-term consequences of neurogenic orthostatic hypotension, such as syncope and falls, and to long-term complications of hypotension and its associated supine hypertension, such as renal failure, left ventricular hypertrophy, and cerebral small vessel disease (Palma et al., 2020). In a single-center prospective study on patients with alpha-synucleinopathies and neurogenic orthostatic hypotension, including 35 patients with multiple system atrophy, the presence of supine hypertension was associated with an increased risk for target organ damage, cardiovascular events, and premature death (Palma et al., 2020). Patients with multiple system atrophy with prolonged survival typically develop autonomic failure later in the disease course (Calandra-Buonaura et al., 2013).

Sudden death in multiple system atrophy

The degeneration of brainstem autonomic structures devoted to cardiovascular and respiratory control may contribute to sudden death, which occurs in approximately 15% of patients with multiple system atrophy and most often during sleep (Shimohata et al., 2008; Zhang et al., 2018). Postulated causes of sudden death during sleep in patients with multiple system atrophy include airway obstruction from impaired vocal cord abduction in patients with nocturnal stridor, floppy epiglottis egress into the laryngeal inlet during inspiration, central hypoventilation, cardiac autonomic disturbances, or a combination of these factors (Shimohata et al., 2016). Brainstem structures known to integrate arterial blood pressure responses to hypoxia, such as the C1 neurons of the rostral ventrolateral medulla, or heart rate responses during sleep, such as the nucleus ambiguus, or hypercapnic arousal, such as the lateral parabrachial nucleus, undergo degeneration in multiple system atrophy. The loss of cholinergic parasympathetic preganglionic neurons in the nucleus ambiguus can also explain the impaired respiratory modulation of heart rate observed in these patients (Benarroch, 2019). Additional contributors to the risk of sudden death might include a blunted arousal threshold and an impaired cardiac response to sleep-disordered breathing during sleep. A study on a small sample of 10 patients with multiple system atrophy demonstrated blunted changes in heart rate and electroencephalographic spectral changes associated with periodic limb movements during sleep compared to patients with the restless legs syndrome, suggesting impaired cortical and autonomic arousal responses during sleep (Vetrugno et al., 2007). Abnormal cardiac responses to nocturnal hypoxia, not associated with daytime cardiovagal abnormalities, have also been observed in a study on 46 patients with multiple system atrophy undergoing overnight pulse oximetry. These patients presented fewer nocturnal surges in heart rate (i.e., pulse events) compared to age- and sex-matched controls despite similar respiratory event frequency (McCarter et al., 2023).

Brainstem lesions

Brainstem lesions may impair the autonomic control of the cardiovascular system, with clinical manifestations depending on the etiology and extent of the lesion. Orthostatic hypotension, sinus arrest requiring placement of a permanent cardiac pacemaker, syncope, labile hypertension, and impairment of parasympathetic functions have all been reported as rare presentations of lateral medullary infarction (Hong et al., 2013; Huynh et al., 2018).

Neurovascular and neuroinflammatory diseases involving lateral brainstem lesions, including multiple sclerosis, can lead to stress-induced cardiomyopathy, also known as neurogenic stunned myocardium, broken-heart syndrome, or takotsubo syndrome. This name derives from the Japanese word takotsubo (“octopus trap”), which refers to the characteristic transient apical ballooning of the left ventricle. It occurs predominantly in postmenopausal women and is often preceded by physical or emotional stress. The pathogenic mechanism is still unclear, but sympathetic hyperactivity with increased catecholamine release, coronary artery spasm, and cardiotoxicity have been hypothesized (Rogers et al., 2001; Androdias et al., 2017; Risseeuw et al., 2021).

References

Abbott SBG, Coates MB, Stornetta RL et al. (2013). Optogenetic stimulation of C1 and retrotrapezoid nucleus neurons causes sleep state-dependent cardiorespiratory stimulation and arousal in rats. Hypertension 61: 835–841.

Abe C, Inoue T, Inglis MA et al. (2017). C1 neurons mediate a stress-induced anti-inflammatory reflex in mice. Nat Neurosci 20: 700–707.

Androdias G, Bernard E, Biotti D et al. (2017). Multiple sclerosis broke my heart. Ann Neurol 81: 754–758.

Bandler R, Keay KA, Floyd N et al. (2000). Central circuits mediating patterned autonomic activity during active vs. passive emotional coping. Brain Res Bull 53: 95–104.

Barnett WH, Baekey DM, Paton JFR et al. (2021). Heartbeats entrain breathing via baroreceptor-mediated modulation of expiratory activity. Exp Physiol 106: 1181–1195.

Baschieri F, Calandra-Buonaura G, Cecere A et al. (2017). Iodine-123-meta-iodobenzylguanidine myocardial scintigraphy in isolated autonomic failure: potential red flag for future multiple system atrophy. Front Neurol 8: 225.

Bastianini S, Silvani A (2018). Clinical implications of basic research: the role of hypocretin/orexin neurons in the central autonomic network. Clin Transl Neurosci 2: 2514183X1878932.

Batla A, De Pablo-Fernandez E, Erro R et al. (2018). Young-onset multiple system atrophy: clinical and pathological features. Mov Disord 33: 1099–1107.

Benarroch EE (1993). The central autonomic network: functional organization, dysfunction, and perspective. Mayo Clin Proc 68: 988–1001.

Benarroch EE (2008). The arterial baroreflex: functional organization and involvement in neurologic disease. Neurology 71: 1733–1738.

Benarroch EE (2019). Multiple system atrophy: a disorder targeting the brainstem control of survival. Clin Auton Res 29: 549–551.

Benarroch EE, Schmeichel AM, Sandroni P et al. (2006). Involvement of hypocretin neurons in multiple system atrophy. Acta Neuropathol 113: 75–80.

Biaggioni I, Shibao CA, Diedrich A et al. (2019). Blood pressure management in afferent baroreflex failure: JACC Review Topic of the Week. J Am Coll Cardiol 74: 2939–2947.

Bizzi E, Pompeiano O, Somogyi I (1964). Vestibular nuclei: activity of single neurons during natural sleep and wakefulness. Science 145: 414–415.

Burke PGR, Abbott SBG, Coates MB et al. (2014). Optogenetic stimulation of adrenergic C1 neurons causes sleep state-dependent cardiorespiratory stimulation and arousal with sighs in rats. Am J Respir Crit Care Med 190: 1301–1310.

Burke PGR, Kanbar R, Viar KE et al. (2015). Selective optogenetic stimulation of the retrotrapezoid nucleus in sleeping rats activates breathing without changing blood pressure or causing arousal or sighs. J Appl Physiol 118: 1491–1501.

Byrum CE, Guyenet PG (1987). Afferent and efferent connections of the A5 noradrenergic cell group in the rat. J Comp Neurol 261: 529–542.

Calandra-Buonaura G, Guaraldi P, Sambati L et al. (2013). Multiple system atrophy with prolonged survival: is late onset of dysautonomia the clue? Neurol Sci 34: 1875–1878.

Carter ME, Yizhar O, Chikahisa S et al. (2010). Tuning arousal with optogenetic modulation of locus coeruleus neurons. Nat Neurosci 13: 1526–1533.

Cheshire WP, Freeman R, Gibbons CH et al. (2021). Electrodiagnostic assessment of the autonomic nervous system: a consensus statement endorsed by the American Autonomic Society, American Academy of Neurology, and the International Federation of Clinical Neurophysiology. Clin Neurophysiol 132: 666–682.

Chiang MC, Bowen A, Schier LA et al. (2019). Parabrachial complex: a hub for pain and aversion. J Neurosci 39: 8225–8230.

Coon EA, Sletten DM, Suarez MD et al. (2015). Clinical features and autonomic testing predict survival in multiple system atrophy. Brain 138: 3623–3631.

da Silva EF, Freiria-Oliveira AH, Custódio CHX et al. (2013). A1 noradrenergic neurons lesions reduce natriuresis and hypertensive responses to hypernatremia in rats. PLoS One 8: e73187.

Damasceno RS, Takakura AC, Moreira TS (2015). Respiratory and sympathetic chemoreflex regulation by Kölliker-Fuse neurons in rats. Pflug Arch 467: 231–239.

Dampney RAL (1994). Functional organization of central pathways regulating the cardiovascular system. Physiol Rev 74: 323–364.

Dampney RAL (2016). Central neural control of the cardiovascular system: current perspectives. Adv Physiol Educ 40: 283–296.

Dampney RAL (2017). Resetting of the baroreflex control of sympathetic vasomotor activity during natural behaviors: description and conceptual model of central mechanisms. Front Neurosci 11: 461.

Degtyarenko AM, Kaufman MP (2006). Barosensory cells in the nucleus tractus solitarius receive convergent input from group III muscle afferents and central command. Neuroscience 140: 1041–1050.

Drew RC (2017). Baroreflex and neurovascular responses to skeletal muscle mechanoreflex activation in humans: an exercise in integrative physiology. Am J Physiol Regul Integr Comp Physiol 313: R654–R659.

Eckberg DL (2003). The human respiratory gate. J Physiol 548: 339–352.

Eschlböck S, Delazer M, Krismer F et al. (2020). Cognition in multiple system atrophy: a single-center cohort study. Ann Clin Transl Neurol 7: 219–228.

Fanciulli A, Göbel G, Ndayisaba JP et al. (2016). Supine hypertension in Parkinson's disease and multiple system atrophy. Clin Auton Res 26: 97–105.

Fanciulli A, Jordan J, Biaggioni I et al. (2018). Consensus statement on the definition of neurogenic supine hypertension in cardiovascular autonomic failure by the American Autonomic Society (AAS) and the European Federation of Autonomic Societies (EFAS) endorsed by the European Academy of Neurology (EAN) and the European Society of Hypertension (ESH). Clin Auton Res 28: 355–362.

Fanciulli A, Strano S, Ndayisaba JP et al. (2014). Detecting nocturnal hypertension in Parkinson's disease and multiple system atrophy: proposal of a decision-support algorithm. J Neurol 261: 1291–1299.

Farmer DGS, Pracejus N, Dempsey B et al. (2019). On the presence and functional significance of sympathetic premotor neurons with collateralized spinal axons in the rat. J Physiol 597: 3407–3423.

Ferreira CB, Schoorlemmer GH, Rocha AA et al. (2020). Increased sympathetic responses induced by chronic obstructive sleep apnea are caused by sleep fragmentation. J Appl Physiol 129: 163–172.

Fletcher EC, Lesske J, Behm R et al. (1992). Carotid chemoreceptors, systemic blood pressure, and chronic episodic hypoxia mimicking sleep apnea. J Appl Physiol 72: 1978–1984.

Freeman R, Wieling W, Axelrod FB et al. (2011). Consensus statement on the definition of orthostatic hypotension, neurally mediated syncope and the postural tachycardia syndrome. Clin Auton Res 21: 69–72.

Fuller P, Sherman D, Pedersen NP et al. (2011). Reassessment of the structural basis of the ascending arousal system. J Comp Neurol 519: 933–956.

Futuro-Neto HA, Coote JH (1982). Desynchronized sleep-like pattern of sympathetic activity elicited by electrical stimulation of sites in the brainstem. Brain Res 252: 269–276.

Gagliuso AH, Chapman EK, Martinelli GP et al. (2019). Vestibular neurons with direct projections to the solitary nucleus in the rat. J Neurophysiol 122: 512–524.

Gami AS, Howard DE, Olson EJ et al. (2005). Day-night pattern of sudden death in obstructive sleep apnea. N Engl J Med 352: 1206–1214.

Gao H, Korim WS, Yao ST et al. (2019). Glycinergic neurotransmission in the rostral ventrolateral medulla controls the time course of baroreflex-mediated sympathoinhibition. J Physiol 597: 283–301.

Gao L, Zimmerman MC, Biswal S et al. (2017). Selective *Nrf2* gene deletion in the rostral ventrolateral medulla evokes hypertension and sympathoexcitation in mice. Hypertension 69: 1198–1206.

Garcia-Rill E, Saper CB, Rye DB et al. (2019). Focus on the pedunculopontine nucleus. Consensus review from the May 2018 brainstem society meeting in Washington, DC, USA. Clin Neurophysiol 130: 925–940.

Gatti PJ, Johnson TA, Massari JV (1996). Can neurons in the nucleus ambiguus selectively regulate cardiac rate and atrio-ventricular conduction? J Auton Nerv Syst 57: 123–127.

Giannini G, Calandra-Buonaura G, Asioli GM et al. (2018). The natural history of idiopathic autonomic failure: the IAF-BO cohort study. Neurology 91: e1245–e1254.

Giannini G, Calandra-Buonaura G, Mastrolilli F et al. (2016). Early stridor onset and stridor treatment predict survival in 136 patients with MSA. Neurology 87: 1375–1383.

Gilbey MP, Jordan D, Richter DW et al. (1984). Synaptic mechanisms involved in the inspiratory modulation of vagal cardio-inhibitory neurones in the cat. J Physiol 356: 65–78.

Glasmacher SA, Leigh PN, Saha RA (2017). Predictors of survival in progressive supranuclear palsy and multiple system atrophy: a systematic review and meta-analysis. J Neurol Neurosurg Psychiatry 88: 402–411.

Goldstein DS, Sharabi Y (2009). Neurogenic orthostatic hypotension: a pathophysiological approach. Circulation 119: 139–146.

Guaraldi P, Baschieri F, Barletta G et al. (2020). Validation of the new index of baroreflex function to identify neurogenic orthostatic hypotension. Auton Neurosci 229: 102744.

Guyenet PG, Stornetta RL (2022). Rostral ventrolateral medulla, retropontine region and autonomic regulations. Auton Neurosci 237: 102922.

Guyenet PG, Stornetta RL, Holloway BB et al. (2018). Rostral ventrolateral medulla and hypertension. Hypertension 72: 559–566.

Hoff EC, Kell JF, Carroll MN (1963). Effects of cortical stimulation and lesions on cardiovascular function. Physiol Rev 43: 68–114.

Hong JM, Kim TJ, Shin DH et al. (2013). Cardiovascular autonomic function in lateral medullary infarction. Neurol Sci 34: 1963–1969.

Huangfu DH, Koshiya N, Guyenet PG (1991). A5 noradrenergic unit activity and sympathetic nerve discharge in rats. Am J Physiol Regul Integr Comp Physiol 261: R393–R402.

Huynh TR, Decker B, Fries TJ et al. (2018). Lateral medullary infarction with cardiovascular autonomic dysfunction: an unusual presentation with review of the literature. Clin Auton Res 28: 569–576.

Jordan J, Shibao C, Biaggioni I (2015). Multiple system atrophy: using clinical pharmacology to reveal pathophysiology. Clin Auton Res 25: 53–59.

Joyner MJ, Charkoudian N, Wallin BG (2008). A sympathetic view of the sympathetic nervous system and human blood pressure regulation. Exp Physiol 93: 715–724.

Kara T, Narkiewicz K, Somers VK (2003). Chemoreflexes – physiology and clinical implications. Acta Physiol Scand 177: 377–384.

Kaufmann H, Norcliffe-Kaufmann L, Palma J-A et al. (2017). Natural history of pure autonomic failure: a United States prospective cohort. Ann Neurol 81: 287–297.

Kaufmann H, Norcliffe-Kaufmann L, Palma J-A (2020). Baroreflex dysfunction. N Engl J Med 382: 163–178.

Ketch T, Biaggioni I, Robertson R et al. (2002). Four faces of baroreflex failure: hypertensive crisis, volatile hypertension, orthostatic tachycardia, and malignant vagotonia. Circulation 105: 2518–2523.

Koba S, Hanai E, Kumada N et al. (2018). Sympathoexcitation by hypothalamic paraventricular nucleus neurons projecting to the rostral ventrolateral medulla. J Physiol 596: 4581–4595.

Koba S, Kumada N, Narai E et al. (2022). A brainstem monosynaptic excitatory pathway that drives locomotor activities and sympathetic cardiovascular responses. Nat Commun 13: 5079.

Koba S, Yoshida T, Hayashi N (2006). Differential sympathetic outflow and vasoconstriction responses at kidney and skeletal muscles during fictive locomotion. Am J Physiol Heart Circ Physiol 290: H861–H868.

Krout KE, Mettenleiter TC, Loewy AD (2003). Single CNS neurons link both central motor and cardiosympathetic systems: a double-virus tracing study. Neuroscience 118: 853–866.

Kräuchi K, Gompper B, Hauenstein D et al. (2012). Diurnal blood pressure variations are associated with changes in distal-proximal skin temperature gradient. Chronobiol Int 29: 1273–1283.

Lagatta DC, Ferreira-Junior NC, Deolindo M et al. (2016). Ventrolateral periaqueductal grey matter neurotransmission modulates cardiac baroreflex activity. Eur J Neurosci 44: 2877–2884.

Lamotte G, Coon EA, Suarez MD et al. (2021). Natural history of afferent baroreflex failure in adults. Neurology 97: e136–e144.

Liang Y, Shi W, Xiang A et al. (2021). The NAergic locus coeruleus-ventrolateral preoptic area neural circuit mediates rapid arousal from sleep. Curr Biol 31: 3729–3742e5.

Lloyd-Jones DM, Allen NB, Anderson CAM et al. (2022). Life's Essential 8: updating and enhancing the American Heart Association's Construct of Cardiovascular Health: a Presidential Advisory from the American Heart Association. Circulation 146: e18–e43.

Lo Martire V, Silvani A, Alvente S et al. (2018). Modulation of sympathetic vasoconstriction is critical for the effects of sleep on arterial pressure in mice. J Physiol 596: 591–608.

Low PA, Reich SG, Jankovic J et al. (2015). Natural history of multiple system atrophy in the USA: a prospective cohort study. Lancet Neurol 14: 710–719.

Macefield VG, Henderson LA (2019). Identification of the human sympathetic connectome involved in blood pressure regulation. NeuroImage 202: 116119.

el Mansari M, Sakai K, Jouvet M (1989). Unitary characteristics of presumptive cholinergic tegmental neurons during the sleep-waking cycle in freely moving cats. Exp Brain Res 76: 519–529.

Martinez D, Kline DD (2021). The role of astrocytes in the nucleus tractus solitarii in maintaining central control of autonomic function. Am J Physiol Regul Integr Comp Physiol 320: R418–R424.

Mathias CJ, Mallipeddi R, Bleasdale-Barr K (1999). Symptoms associated with orthostatic hypotension in pure autonomic failure and multiple system atrophy. J Neurol 246: 893–898.

McAllen RM, Salo LM, Paton JFR et al. (2011). Processing of central and reflex vagal drives by rat cardiac ganglion neurones: an intracellular analysis. J Physiol 589: 5801–5818.

McAllen RM, Spyer KM (1978). The baroreceptor input to cardiac vagal motoneurones. J Physiol 282: 365–374.

McCall RB (1983). Serotonergic excitation of sympathetic preganglionic neurons: a microiontophoretic study. Brain Res 289: 121–127.

McCall RB, Clement ME (1989). Identification of serotonergic and sympathetic neurons in medullary raphe nuclei. Brain Res 477: 172–182.

McCall RB, Harris LT (1987). Sympathetic alterations after midline medullary raphe lesions. Am J Physiol Regul Integr Comp Physiol 253: R91–R100.

McCarter SJ, Coon EA, Benarroch EE et al. (2023). Nocturnal pulse event frequency is reduced in multiple system atrophy. Ann Neurol 93: 205–212.

Mena-Segovia J, Bolam JP (2017). Rethinking the pedunculopontine nucleus: from cellular organization to function. Neuron 94: 7–18.

Mirzaei-Damabi N, Hatam M, Yeganeh F et al. (2020). Roles of glutamate and GABA of the Kölliker-Fuse nucleus in generating the cardiovascular chemoreflex. Pflug Arch 472: 1051–1063.

Morrison AR, Pompeiano O (1970). Vestibular influences during sleep. VI. Vestibular control of autonomic functions during the rapid eye movements of desynchronized sleep. Arch Ital Biol 108: 154–180.

Morrison SF, Gebber GL (1984). Raphe neurons with sympathetic-related activity: baroreceptor responses and spinal connections. Am J Physiol Regul Integr Comp Physiol 246: R338–R348.

Morrison SF, Nakamura K (2019). Central mechanisms for thermoregulation. Annu Rev Physiol 81: 285–308.

Nagura S, Sakagami T, Kakiichi A et al. (2004). Acute shifts in baroreflex control of renal sympathetic nerve activity induced by REM sleep and grooming in rats. J Physiol 558: 975–983.

Norcliffe-Kaufmann L, Millar Vernetti P, Palma J-A et al. (2020). Afferent baroreflex dysfunction: decreased or excessive signaling results in distinct phenotypes. Semin Neurol 40: 540–549.

Norcliffe-Kaufmann L, Kaufmann H, Palma J-A et al. (2018). Orthostatic heart rate changes in patients with autonomic failure caused by neurodegenerative synucleinopathies. Ann Neurol 83: 522–531.

Orimo S, Suzuki M, Inaba A et al. (2012). 123I-MIBG myocardial scintigraphy for differentiating Parkinson's disease from other neurodegenerative parkinsonism: a systematic review and meta-analysis. Parkinsonism Relat Disord 18: 494–500.

Padley JR, Kumar NN, Li Q et al. (2007). Central command regulation of circulatory function mediated by descending pontine cholinergic inputs to sympathoexcitatory rostral ventrolateral medulla neurons. Circ Res 100: 284–291.

Palma J-A, Kaufmann H (2020). Management of orthostatic hypotension. Continuum 26: 154–177.

Palma J-A, Redel-Traub G, Porciuncula A et al. (2020). The impact of supine hypertension on target organ damage and survival in patients with synucleinopathies and neurogenic orthostatic hypotension. Parkinsonism Relat Disord 75: 97–104.

Panneton MW (2013). The mammalian diving response: an enigmatic reflex to preserve life? Physiology 28: 284–297.

Parker LM, Le S, Wearne TA et al. (2017). Neurochemistry of neurons in the ventrolateral medulla activated by hypotension: are the same neurons activated by glucoprivation? J Comp Neurol 525: 2249–2264.

Paton JFR, Boscan P, Pickering AE et al. (2005). The yin and yang of cardiac autonomic control: vago-sympathetic interactions revisited. Brain Res Rev 49: 555–565.

Pavy-Le Traon A, Piedvache A, Perez-Lloret S et al. (2016). New insights into orthostatic hypotension in multiple system atrophy: a European multicentre cohort study. J Neurol Neurosurg Psychiatry 87: 554–561.

Petrovic IN, Ling H, Asi Y et al. (2012). Multiple system atrophy-parkinsonism with slow progression and prolonged survival: a diagnostic catch. Mov Disord 27: 1186–1190.

Pilowsky PM, Kapoor V, Minson JB et al. (1986). Spinal cord serotonin release and raised blood pressure after brainstem kainic acid injection. Brain Res 366: 354–357.

Plowey ED, Kramer JM, Beatty JA et al. (2002). In vivo electrophysiological responses of pedunculopontine neurons to static muscle contraction. Am J Physiol Regul Integr Comp Physiol 283: R1008–R1019.

Raven PB, Fadel PJ, Ogoh S (2006). Arterial baroreflex resetting during exercise: a current perspective. Exp Physiol 91: 37–49.

Risseeuw F, Masrori P, Baar I et al. (2021). The brain–heart link: a case report of a critically located multiple sclerosis lesion in the brainstem leading to recurrent Takotsubo syndrome. Front Cardiovasc Med 8: 674118.

Rogers ER, Phan TG, Wijdicks EFM (2001). Myocardial injury after hemorrhage into the lateral medulla oblongata. Neurology 56: 567–568.

Schreihofer AM, Guyenet PG (2003). Baro-activated neurons with pulse-modulated activity in the rat caudal ventrolateral medulla express GAD67 mRNA. J Neurophysiol 89: 1265–1277.

Shik ML, Orlovsky GN (1976). Neurophysiology of locomotor automatism. Physiol Rev 56: 465–501.

Shimohata T, Aizawa N, Nakayama H et al. (2016). Mechanisms and prevention of sudden death in multiple system atrophy. Parkinsonism Relat Disord 30: 1–6.

Shimohata T, Ozawa T, Nakayama H et al. (2008). Frequency of nocturnal sudden death in patients with multiple system atrophy. J Neurol 255: 1483–1485.

Silvani A, Calandra-Buonaura G, Benarroch EE et al. (2015). Bidirectional interactions between the baroreceptor reflex and arousal: an update. Sleep Med 16: 210–216.

Silvani A, Calandra-Buonaura G, Dampney RAL et al. (2016). Brain–heart interactions: physiology and clinical implications. Philos Trans R Soc Math Phys Eng Sci 374: 20150181.

Silvani A, Dampney RAL (2013). Central control of cardiovascular function during sleep. Am J Physiol Heart Circ Physiol 305: H1683–H1692.

Singer W, Berini SE, Sandroni P et al. (2017). Pure autonomic failure: predictors of conversion to clinical CNS involvement. Neurology 88: 1129–1136.

Souza GMPR, Stornetta RL, Stornetta DS et al. (2019). Contribution of the retrotrapezoid nucleus and carotid bodies to hypercapnia- and hypoxia-induced arousal from sleep. J Neurosci 39: 9725–9737.

Souza GMPR, Stornetta RL, Stornetta DS et al. (2022). Adrenergic C1 neurons monitor arterial blood pressure and determine the sympathetic response to hemorrhage. Cell Rep 38: 110480.

Stornetta RL, Guyenet PG (2018). C1 neurons: a nodal point for stress? Exp Physiol 103: 332–336.

Tang X, Dworkin BR (2010). Baroreflexes of the rat. VI. Sleep and responses to aortic nerve stimulation in the dmNTS. Am J Physiol Regul Integr Comp Physiol 298: R1428–R1434.

Taxini CL, Moreira TS, Takakura AC et al. (2017). Role of A5 noradrenergic neurons in the chemoreflex control of respiratory and sympathetic activities in unanesthetized conditions. Neuroscience 354: 146–157.

Teixeira AL, Vianna LC (2022). The exercise pressor reflex: an update. Clin Auton Res 32: 271–290.

Treglia G, Stefanelli A, Cason E et al. (2011). Diagnostic performance of iodine-123-metaiodobenzylguanidine scintigraphy in differential diagnosis between Parkinson's disease and multiple-system atrophy: a systematic review and a meta-analysis. Clin Neurol Neurosurg 113: 823–829.

Trojanowski JQ, Revesz T (2007). Proposed neuropathological criteria for the post mortem diagnosis of multiple system atrophy. Neuropathol Appl Neurobiol 33: 615–620.

Veerakumar A, Yung AR, Liu Y et al. (2022). Molecularly defined circuits for cardiovascular and cardiopulmonary control. Nature 606: 739–746.

Verret L, Léger L, Fort P et al. (2005). Cholinergic and noncholinergic brainstem neurons expressing Fos after paradoxical (REM) sleep deprivation and recovery. Eur J Neurosci 21: 2488–2504.

Vetrugno R, D'Angelo R, Cortelli P et al. (2007). Impaired cortical and autonomic arousal during sleep in multiple system atrophy. Clin Neurophysiol 118: 2512–2518.

Vichayanrat E, Valerio F, Koay S et al. (2022). Diagnosing premotor multiple system atrophy: natural history and autonomic testing in an autopsy-confirmed cohort. Neurology 99: e1168–e1177.

Wenker IC, Abe C, Viar KE et al. (2017). Blood pressure regulation by the rostral ventrolateral medulla in conscious rats: effects of hypoxia, hypercapnia, baroreceptor denervation, and anesthesia. J Neurosci 37: 4565–4583.

Wenning GK, Geser F, Krismer F et al. (2013). The natural history of multiple system atrophy: a prospective European cohort study. Lancet Neurol 12: 264–274.

Wenning GK, Stankovic I, Vignatelli L et al. (2022). The Movement Disorder Society criteria for the diagnosis of multiple system atrophy. Mov Disord 37: 1131–1148.

Wilson TE, Sauder CL, Kearney ML et al. (2007). Skin-surface cooling elicits peripheral and visceral vasoconstriction in humans. J Appl Physiol 103: 1257–1262.

Yao Y, Barger Z, Saffari Doost M et al. (2022). Cardiovascular baroreflex circuit moonlights in sleep control. Neuron 110: 1–14.

Zhang L, Cao B, Zou Y et al. (2018). Causes of death in Chinese patients with multiple system atrophy. Aging Dis 9: 102–108.

Handbook of Clinical Neurology, Vol. 216 (3rd series)
The Brainstem
L.T. Grinberg and A.D. Korczyn
https://doi.org/10.1016/B978-0-443-15736-3.00003-2

Chapter 4

Brainstem in pain control

LUKE A. HENDERSON*, LEWIS S. CRAWFORD, AND KEVIN A. KEAY

School of Medical Sciences (Neuroscience), Brain and Mind Centre, University of Sydney, Sydney, NSW, Australia

Abstract

One of the most powerful systems in the brain is that which alters the processing of noxious information. The processing of noxious stimuli into the experience of pain is important for survival, as it warns an individual of danger and motivates them to either escape the source of pain, or to confront and control it, that is, fight/flight. While pain is important for survival, it can be powerfully modulated in a range of situations, such as intense emotional state changes, stress, during shifts in attention and expectation, and after confrontational responses where it aids in rest and recuperation. Given its biologic importance, the circuits responsible for pain modulation are located in regions of the brainstem considered phylogenetically primitive. While in many situations, higher cortical regions can drive the modulation of pain, these higher regions are thought to contact the brainstem which in turn modulates incoming noxious information from the spinal cord and caudal brainstem. Understanding the brainstem circuits that control noxious information is vital for our understanding of how the brain modulates pain in various situations. The following is a review of the brainstem circuits that control noxious inputs and considers how higher brain regions can drive these brainstem sites to modulate overall intensity of pain.

INTRODUCTION

The capacity of the brain to modulate inputs is ubiquitous to all sensory systems and is usually linked to changes in an individual's behavior. One of the most powerful of these sensory modulatory systems is that which alters the processing of somatosensory information, and noxious information in particular. If sensory modulatory systems allow incoming noxious information to reach higher brain centers, it can then be interpretated as pain – an experience that warns an individual of danger and motivates them to either escape its source, or to confront it with the goal of exerting control over it, that is, the classically described active defensive behaviors (Fields, 1999). The proposed function of pain is to initiate behavioral responses, and it has been suggested that pain is more like the states of hunger and thirst, which drive fluid and food seeking behaviors, rather than vision and hearing, which are not so tightly coupled to specific behavioral repertoires (Wall, 1979).

While the perception of pain is critical for survival, the brain possesses robust circuits that can increase or decrease the intensity of perceived pain under certain conditions. Indeed, it was suggested in the late 1970s by Patrick Wall that the immediate phase in response to tissue damage is one where pain is commonly not reported. Wall's explanation was "that injury may be followed by circumstances in which treatment of the injury does not have the highest biologic priority. Pain does not occur in this phase. The three obvious high priority behaviors are fighting, escaping and obtaining aid" (Wall, 1979). It was subsequently discovered that such pain-related behaviors and the modulation of noxious information are mediated by neural circuits contained within the brainstem (Basbaum and Fields, 1984; Lovick, 1993) and can be expressed without the need for inputs originating above the brainstem level (Reynolds, 1969; Cheng et al., 1986). The presence of such critical circuitry in phylogenetically primitive parts of the brain is not surprising given the role of these behavioral and sensory responses for survival.

While pain can be modulated following injury, the perception of pain can also be strongly modulated in a number of situations, such as during intense changes in emotional state, anxiety, attention, stress, expectations, and even subtly by other sensory inputs such as olfaction and vision (Miron et al., 1989; Villemure et al., 2003). While this wide range of complex behavioral states arise from activity in multiple higher order brain regions, their common inclusion of pain modulation is driven via descending inputs to brainstem pain modulatory regions that in turn exert dynamic control over nociceptive transmission at the site at which noxious inputs first enter the central nervous system, that is, the dorsal horn for the body and the spinal trigeminal nucleus (SpV) for the head

*Correspondence to: Luke A. Henderson, University of Sydney, 88 Mallett Street, Camperdown 2050, NSW, Australia. Tel: +612-9351-7063; Fax: +612-9351-6556; Email: luke.henderson@sydney.edu.au.

(Villanueva et al., 1996; Ossipov et al., 2014). Understanding the brainstem circuits responsible for the modulation of noxious information is critical for our understanding of how the brain modulates pain in a variety of situations. The following is a review of the brainstem circuits that modulate noxious inputs during a number of experimental paradigms, as well as these circuits influence over chronic pain states.

PAIN INHIBITS PAIN: THE ROLE OF THE SUBNUCLEUS RETICULARIS DORSALIS

One of the first brainstem circuits described as being important in pain modulation is that responsible for conditioned pain modulation (CPM), that is, pain inhibits pain. Experimentally, CPM is produced when pain from one body location (test stimulus) reduces the intensity of pain in another location (conditioned stimulus). In animal experiments, this phenomenon is termed diffuse noxious inhibitory control (DNIC) and was first reported in the late 1970s by Le Bars and colleagues (1979a). In these studies, they recorded neural activity in the dorsal horn during painful stimulation and found that the initial increases in pain-related activity were blocked completely by introducing a second thermal, mechanical, or chemical noxious stimulus in another body region. Furthermore, they found that this inhibitory effect on dorsal horn neural activity (DNIC) did not occur when the initial noxious stimulus was then paired with a nonnoxious somatosensory stimulus (Le Bars et al., 1979a).

Following these initial findings, subsequent animal experiments began to define the neural circuits responsible for DNIC. It was found that the removal of supraspinal inputs eliminated the DNIC effect, showing that the circuits responsible for DNIC depend critically on supraspinal structures (Le Bars et al., 1979b; Cadden et al., 1983; Morton et al., 1987). To determine whether sites in the brainstem were critical for DNIC expression, a series of lesioning studies were performed in which brainstem sites previously implicated in pain modulation were targeted. Bouhassira and colleagues found that while neurotoxic lesions to the midbrain, encompassing the periaqueductal gray (PAG) matter, cuneiform nucleus, and the parabrachial nuclei (PB) reduced the DNIC effect, these reductions were not significantly different to those of a control group (Bouhassira et al., 1990). In contrast, an earlier investigation had found that microinjections of morphine in the PAG reduced DNIC ability in approximately half of the recorded dorsal horn neurons, suggesting that the PAG was in fact capable of modifying DNIC expression (Dickenson and Le Bars, 1987). Subsequently, it was found that unilateral lesions in the dorsal medulla, specifically encompassing the region of the subnucleus reticularis dorsalis (SRD), depressed DNIC responses by almost 50%, and it was suggested that if a bilateral SRD lesions did not leave the animal in a poor physiologic state, it would likely completely abolish DNIC expression (Bouhassira et al., 1992). It is now largely accepted that the SRD region is the key neural site responsible for mediating DNIC analgesic effects (Fig. 4.1).

Investigations in human subjects also support the involvement of the brainstem and in particular the SRD in CPM expression. In tetraplegic patients, dual noxious stimulation does not depress the nociceptive flexion reflex, showing that CPM requires supraspinal involvement (Roby-Brami et al., 1987). Further, it was found that the spinoreticular pathway which projects to brainstem areas including the SRD, but not the spinothalamic tract which has no brainstem projections,

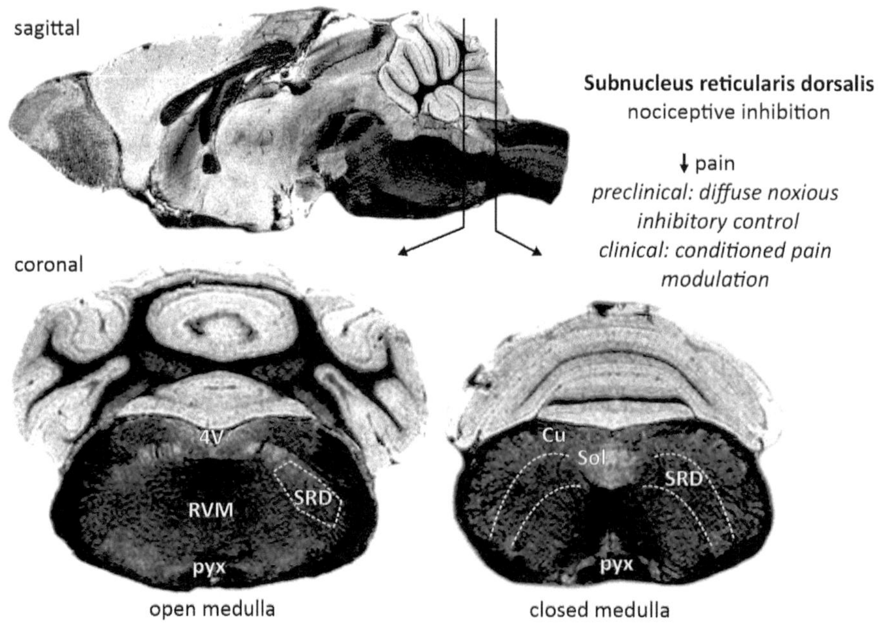

Fig. 4.1. Subnucleus reticularis dorsalis (SRD). Location of the SRD in the dorsolateral open and closed medulla. Preclinical and clinical studies have revealed that this medullary region is critical for the expression of analgesia during the application of two noxious stimuli at distant locations. The location is indicated on a high-resolution T2-weighted anatomic magnetic resonance imaging scan of a postmortem rat. *4V*, fourth ventricle; *Cu*, cuneate nucleus; *pyx*, pyramids; *RVM*, rostral ventromedial medulla; *Sol*, nucleus of the solitary tract.

primarily subserves the expression of CPM (De Broucker et al., 1990). Although these studies implied brainstem involvement in CPM, only relatively recently has brain imaging shown robust activity changes in the brainstem during CPM in awake humans. In the first study of this kind, using a paradigm in which repeated heat noxious stimuli were applied to the face (test stimuli), it was shown that the addition of a sustained leg noxious stimulus (conditioning stimulus) evoked a significant reduction in the perceived intensity of the noxious face stimuli in approximately 40%–50% of individuals (Youssef et al., 2016b) (Fig. 4.2A). CPM sensitivity was not related to the temperature of the delivered stimulus, the perceived intensity of the test stimuli, or the perceived intensity of the conditioning stimulus. Consistent with animal experiments, brainstem functional magnetic resonance imaging (fMRI) revealed that CPM analgesia was associated with a profound reduction in signal intensity in the primary afferent synapse, that is, the SpV where orofacial noxious afferents first terminate, only in those individuals that displayed a CPM analgesic response (Fig. 4.2B). Also consistent with animal experiments, the area encompassing the SRD also displayed significant signal change increases during the test noxious stimuli that were eliminated during the combined test and conditioning stimuli in only those individuals that displayed a CPM analgesic response.

Furthermore, similar changes were observed in the dorsolateral pons in the region encompassing the PB, that is, reductions during dual noxious stimuli in those displaying a CPM analgesia. Though animal experiments show that excitotoxic lesions of the PB do not alter CPM-related dorsal horn activity (Bouhassira et al., 1990), the PB is a major target of nociceptive dorsal horn neurons (Keay et al., 1997; Gauriau and Bernard, 2002). Although the human fMRI data appears inconsistent with the results of animal experiments, it has been suggested that the PB may play an indirect role in CPM and may potentially provide a modulatory influence on the response (Dickenson and Le Bars, 1987). Consistent with this idea, are data showing that although signal intensity changes in spinal trigeminal nucleus and SRD were both strongly correlated to the degree of CPM analgesia, signal changes in the PB were only mildly correlated with CPM analgesia. More recently, a role for the A5 cell group in the ventral pons in CPM analgesia has been described in rodents. Following activation of an inhibitory opsin expressed in A5 neurons, dorsal horn neuron firing rates were no longer inhibited when both test and conditioning noxious stimuli were applied simultaneously (Kucharczyk et al., 2023). This observation is consistent with the earlier human imaging study which described apparently identical changes in the area of the trigeminal nerve entry into the pons, that is, at approximately the same location as the A5 cell group.

Data from both animal experiments and human studies have revealed that arguably the most primitive pain modulatory circuitry, relies heavily on a loop from the primary afferent synapse to the SRD region of the medulla, and back. This loop is likely influenced by higher brainstem regions such as the PB and A5 cell group. In addition, there is brain imaging evidence that CPM analgesia is associated with activation in areas above the level of the brainstem including the dorsolateral prefrontal and cingulate cortices (Youssef et al., 2016a). While describing these influences is beyond of the scope of this chapter, it is important to note that an important function of such descending inputs is likely the setting of the sensitivity of brainstem pain modulatory regions, and they can therefore have a significant influence on the ability of an individual to express analgesic or hyperalgesic responses.

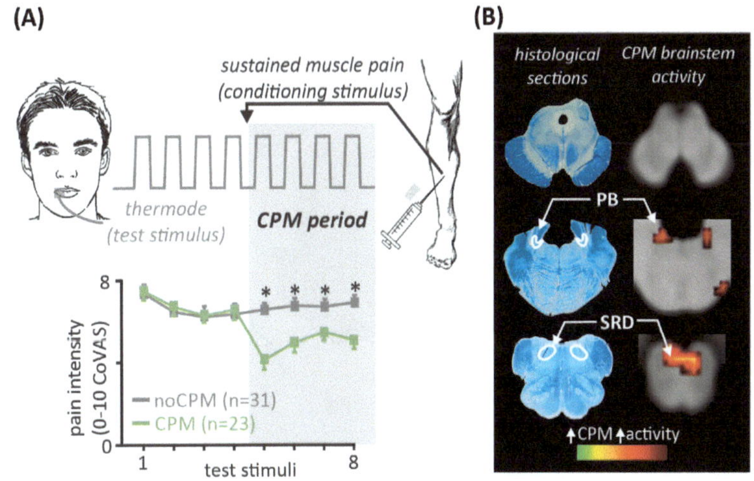

Fig. 4.2. Conditioned pain modulation (CPM), i.e., *pain inhibits pain*. (**A**) Eight brief noxious thermal test stimuli were applied to the face, and between test stimuli 4 and 5, a sustained noxious stimulus was applied to a distant body site (bolus of hypertonic saline injected into tibialis anterior muscle (CPM period: gray shading)). In 23 of the 54 participants, the conditioning stimulus resulted in a significant reduction in pain intensity of the test stimuli, i.e., CPM analgesia. (**B**) CPM analgesia was associated with brainstem functional magnetic resonance imaging signal changes in the region of the parabrachial nucleus (PB) and subnucleus reticular dorsalis (SRD). Modified with permission from Youssef, AM, Macefield VG, Henderson LA (2016b) Pain inhibits pain; human brainstem mechanisms. Neuroimage 124: 54–62.

PLACEBO ANALGESIA AND NOCEBO HYPERALGESIA – THE ROLE OF THE MIDBRAIN PERIAQUEDUCTAL GRAY, PARABRACHIAL NUCLEUS, AND ROSTRAL VENTROMEDIAL MEDULLA

Behavioral responses to noxious stimuli involve the integration of somatic motor changes, with changes in autonomic and sensory function. Profound differences in the composition of these integrated responses exist depending on the body tissue being stimulated. For example, cutaneous pain tends to evoke active defensive behaviors, that is, flight and fight, that are supported by increased heart rate and blood pressure and generalized arousal and hyper-reactivity, whereas pain originating in deeper tissues such as muscle and viscera evokes passive defensive behaviors, that is, quiescence which is accompanied by decreased blood pressure, falls in heart rate, withdrawal behaviors, and hyporeactivity (Depaulis et al., 1992; Bandler and Keay, 1996). Animal experiments, and more recently human imaging studies, have shown that different regions of the midbrain PAG matter code these differential responses to pain originating in different body tissues. That is, cutaneous pain activates the lateral column of the PAG (lPAG) and stimulation of this region evokes active defensive behaviors, whereas deep pain activates the ventrolateral PAG (vlPAG) and stimulation produces passive defensive behaviors (Fig. 4.3) (Keay and Bandler, 2001; Tinoco Mendoza et al., 2023). Importantly, superficial and deep pain are each associated with profound analgesia, although it has been shown that lPAG stimulation evokes a nonopiate-mediated analgesia, whereas vlPAG stimulation evokes an opiate-mediated analgesia (Heinricher et al., 2009; Sims-Williams et al., 2017).

In regard to the analgesic responses, the PAG has few projections to the spinal cord dorsal horn or the SpV, rather it exerts its effects indirectly via projections to an area of the medulla, the rostral ventromedial medulla (RVM) (Fields and Heinricher, 1985; Heinricher et al., 1989). The RVM encompasses both the midline nucleus raphe magnus and adjacent nucleus reticularis gigantocellularis, each of which have direct projections to the dorsal horn and SpV (Fields and Basbaum, 1978; Holstege and Kuypers, 1982; Mason and Fields, 1989; Chiang et al., 1994; Fields et al., 1995; Ren and Dubner, 2011; Aicher et al., 2012). Animal experiments have revealed that the RVM contains three distinct cell populations: "OFF" cells which inhibit noxious inputs upon stimulation, "ON" cells which facilitate noxious inputs upon stimulation, and "NEUTRAL" cells which adapt to act as either "ON" or "OFF" cell class depending on the nature of noxious stimuli (Fields et al., 1983; Heinricher et al., 1989; Khasabov et al., 2015). It was originally hypothesized that the PAG-RVM-dorsal horn/SpV system had an exclusively inhibitory effect on nociceptive transmission, stemming from electrophysiologic studies in

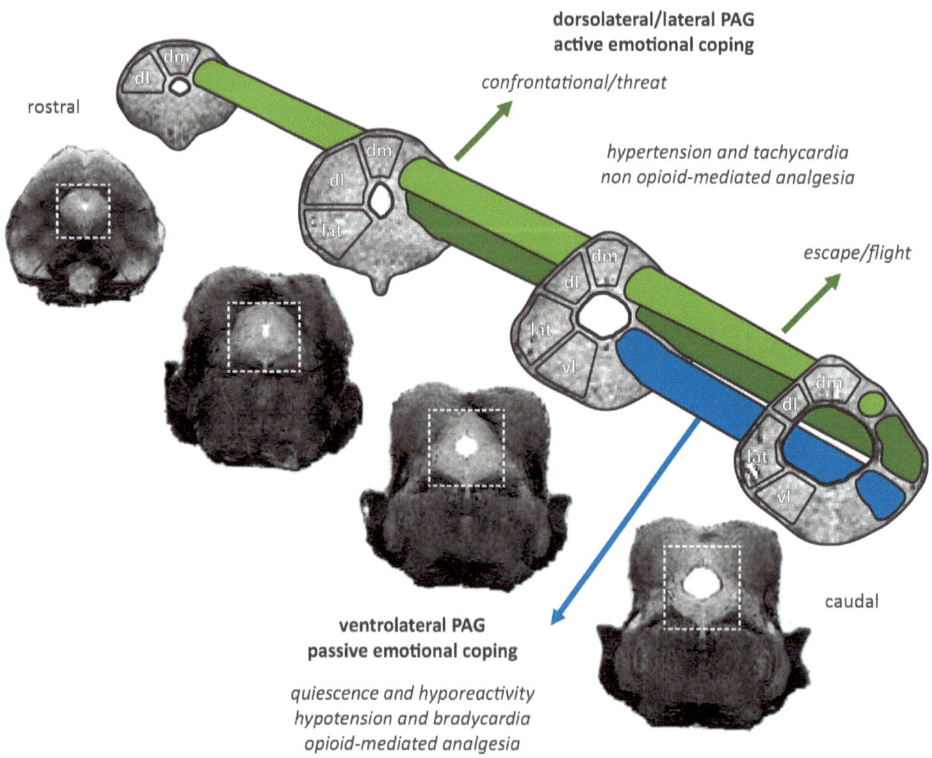

Fig. 4.3. Illustration of the dorsomedial (dm), dorsolateral (dl), lateral (lat), and ventrolateral (vl) neuronal columns within the periaqueductal gray (PAG). The location of the PAG is indicated on a high-resolution T2-weighted anatomic magnetic resonance imaging scan of a postmortem rat. Activation of the dorsolateral and lateral PAG evokes active emotional coping strategies including defensive reactions (confrontation, threat, escape, flight), hypertension, tachycardia, and a nonopioid-mediated analgesia. In contrast, activation of the vlPAG evokes passive emotional coping strategies including quiescence, hyporeactivity, hypotension, bradycardia, and an opioid-mediated analgesia.

animals which reported that electrical stimulation of either the PAG or the RVM strongly suppressed nociceptive activity in the dorsal horn (Sessle et al., 1976; Sessle and Hu, 1981; Waters and Lumb, 1997; Ren and Dubner, 2011). Consistent with these reports, direct electrical stimulation of the PAG in a human was reported to reduce the intensity of their chronic pain and that this analgesic effect was reversed by the administration of the opiate antagonist naloxone (Adams, 1976).

While it is clear that direct stimulation of the PAG can produce analgesia, it is now recognized that the PAG-RVM-dorsal horn/SpV system can also facilitate or enhance nociceptive transmission (Zhuo and Gebhart, 1997; Sugiyo et al., 2005; Guo et al., 2006). This shift in thinking stemmed from the finding that high-intensity electrical stimulation of the RVM inhibits, whereas lower intensity RVM stimulation facilitates activity in dorsal horn neurons and that these effects are accompanied by reduced or enhanced pain behaviors, respectively (Zhuo and Gebhart, 1992). Similar effects are also reported for the processing of orofacial noxious stimuli in the RVM-SpV circuit (Sugiyo et al., 2005). Given that the RVM can either inhibit or facilitate the processing of noxious information, some studies have explored whether these opposing effects are generated from anatomically distinct RVM regions. One study has suggested that inhibitory effects are driven specifically by the nucleus raphe magnus (Wei et al., 1999), whereas other studies report no inhibitory versus facilitatory arrangements (Zhuo and Gebhart, 1992; Zhuo and Gebhart, 1997). Furthermore, it appears that altered nociceptive processing may depend selectively on the balance between the activity in different RVM cell types. For example, the analgesic effects of morphine resulting from activation of OFF-cells and inhibition of ON-cell firing which ultimately reduces nociceptive transmission in the dorsal horn/SpV (Heinricher et al., 1994) and pain behaviors are greater when ON-cells synchronously enter a period of activity and OFF-cells a period of silence (Heinricher et al., 1989).

Placebo/nocebo responses

While some pain modulatory states are driven by brainstem circuit loops, which are modulated by higher centers, others are driven by high levels of the cerebral cortex, for example, placebo hypoalgesia and nocebo hyperalgesia. Placebo hypoalgesia refers to decreased pain sensitivity resulting from the application of an inert treatment such as sham procedures and inert substances, whereas its counterpart, nocebo hyperalgesia refers to increased pain sensitivity similarly due to an inert treatment. Perhaps the earliest recognized meta-analysis exploring the analgesic effects of placebo reactions was performed by Henry Beecher who asserted that pain in 35% of individuals could be satisfactorily relieved by placebo administration alone (Beecher, 1955). Despite ongoing dispute regarding the interpretation of his analysis, more recent investigations support the finding that significant placebo responses can be observed in between 30% and 50% of participants (Levine and Gordon, 1984; Benedetti, 1996; Vachon-Presseau et al., 2018; Crawford et al., 2021; Wang et al., 2022). What makes an individual respond to a placebo treatment is subject to vigorous debate, however, the effectiveness of conditioning procedures and expectations of a treatment's effectiveness are critical factors (Amanzio and Benedetti, 1999; Colloca and Benedetti, 2006). It is accepted that expectation is likely coded in higher brain centers, and it is becoming increasingly clear that it is the PAG-RVM-dorsal horn/SpV circuitry that is critical for the expression of placebo hypoalgesia or nocebo hyperalgesia responses.

Most studies exploring the underlying neural circuitry responsible for placebo and nocebo have been performed in humans, with only a handful of animal experiments (Keller et al., 2018). Early studies investigating the neurobiology of placebo responses reported that brain opioid systems were critical, since naloxone administration blocked placebo analgesic effects evoked in individuals with pain following dental procedures (Levine et al., 1978). A role for endogenous opioids in the generation of placebo analgesia is now well established, with positron emission tomography studies reporting increased opioid receptor binding in a number of pain processing regions including the cingulate, insula, and dorsolateral prefrontal cortices, and the nucleus accumbens (Zubieta et al., 2005; Wager et al., 2007; Scott et al., 2008). In addition, two of these studies also reported placebo-related increases in brainstem opioid binding in the region encompassing the PAG (Wager et al., 2007; Scott et al., 2008). The observation of PAG opioid binding implicated the vlPAG in placebo analgesia given that it evokes an opioid-mediated analgesic effect following direct stimulation (Bagley and Ingram, 2020).

Although the opioid system does appear critical for placebo analgesia, particularly when expectation-based designs are implemented, there is compelling evidence for the involvement of nonopioid systems in these responses. Some of the earliest studies reported that naloxone triggered only a partial antagonism of the placebo response (Grevert et al., 1983; Gracely, 1987), and there is now evidence that both the endogenous cannabinoid and opioid systems may be involved. In classical conditioning or subliminal cue designs, the cannabinoid system appears pivotal (Benedetti et al., 2011). Animal experiments have shown that the vlPAG contains dense concentrations of opioid receptors, whereas the lPAG contains dense concentrations of cannabinoid receptors; therefore, it is likely that different neurobiologic circuitries are recruited depending on the nature of analgesia elicited (Grevert et al., 1983; Benedetti et al., 2005). Nocebo hyperalgesia has been conceptualized as the functional mirror-image of placebo analgesia, and it has been reported that nocebo hyperalgesia responses can be abolished by administering cholecystokinin antagonists, which have antiopioid properties (Benedetti et al., 1997). While such studies provide evidence for the neurotransmitters involved in generating placebo and nocebo responses, the precise circuitry underpinning these pain modulatory responses have only just begun to be defined, and particularly those in the brainstem.

A number of human studies have used both positron emission tomography and fMRI to define the neural circuits underpinning placebo analgesia. Whilst some of these studies have reported signal changes in the general vicinity of the PAG during analgesic responses (Petrovic et al., 2002; Wager et al., 2004; Eippert et al., 2009; Grahl et al., 2018), none had the spatial acuity to accurately and robustly define the complete brainstem circuits responsible for either placebo analgesic or nocebo hyperalgesic responses. The recent advent of ultra-high-field human fMRI now allows for the exploration of activation patterns in small brain structures including within subregions of the PAG (Tinoco Mendoza et al., 2023). In a recent series of investigations, high-spatial resolution fMRI was used to determine the brainstem circuitry responsible for placebo and nocebo using a conditioning-based placebo and nocebo design (Crawford et al., 2021). Consistent with previous studies, it was found that approximately 40% of individuals displayed a robust placebo analgesia and 50% a robust nocebo hyperalgesia response. Whether an individual displayed an analgesic or hyperalgesia response was not associated with the stimulus temperature or the individual expectancy of a change in pain intensity. Analysis of signal changes in the brainstem revealed that placebo- and nocebo-evoked changes in perceived pain intensities were associated with opposing signal changes in the PAG and RVM (Fig. 4.4A). That is, in the PAG, signal intensity decreases were associated with greater placebo analgesia and in the RVM signal increases were associated with greater placebo responses. The opposite relationships were seen for nocebo hyperalgesia. Critically, rather than identifying the vlPAG as the site responsible for placebo analgesia, activity changes in the lateral and dorsolateral PAG column were most closely associated with both placebo analgesia and nocebo hyperalgesia (Fig. 4.4B).

In addition to the PAG-RVM circuit, placebo and nocebo responses were also associated with signal changes in the locus coeruleus and PB, and for placebo responses, only signal changes were seen in the region of the SRD, that is, two key regions involved in CPM expression. The PB is an important site for the integration and relay of ascending nociceptive information, and the spino-parabrachial tract is known to be

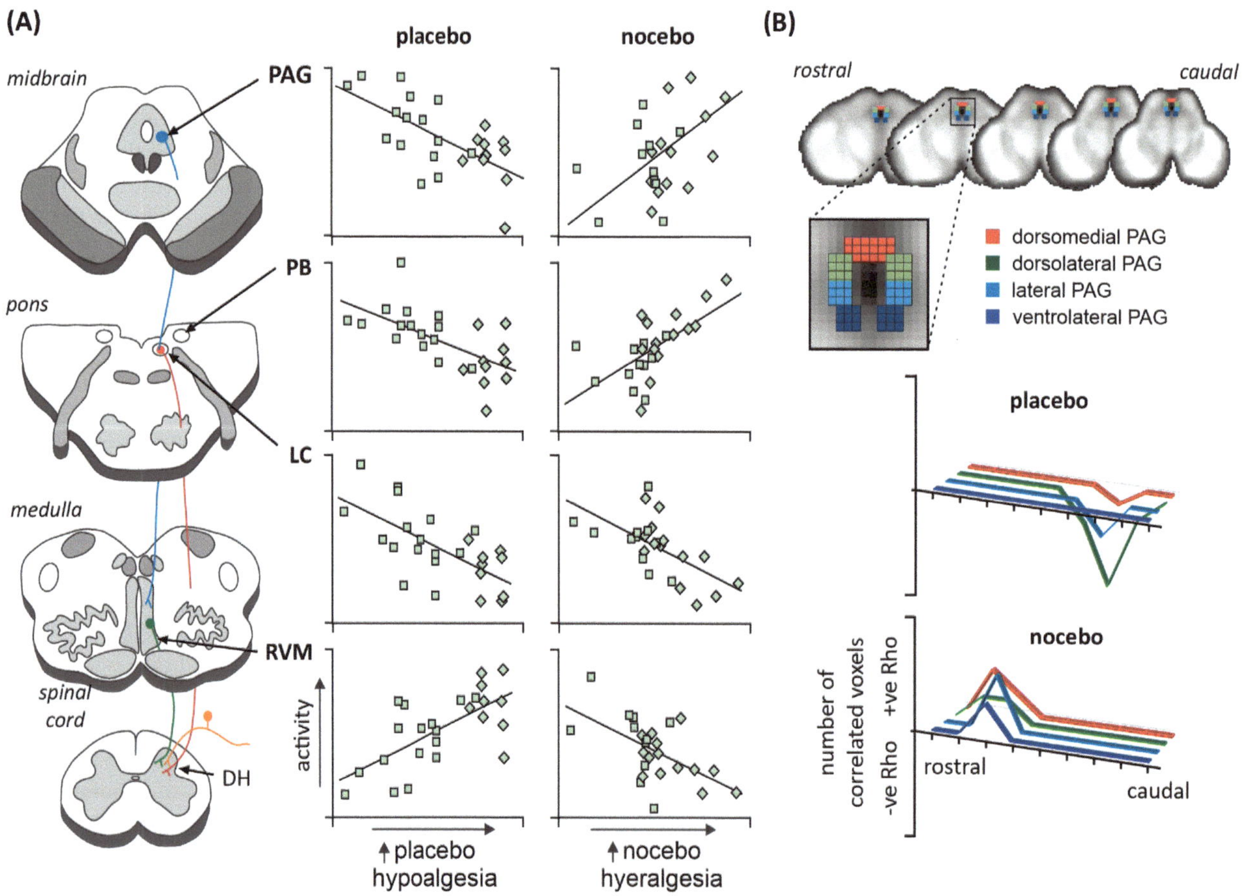

Fig. 4.4. Placebo and nocebo brainstem signal changes. (**A**) Functional magnetic resonance imaging signal changes during placebo analgesia and nocebo hyperalgesia. Opposing correlations between signal changes during hypoalgesia or hyperalgesia responses occurred in the midbrain periaqueductal gray (PAG) matter, parabrachial nucleus (PB), and rostral ventromedial medulla (RVM). In contrast, the locus coeruleus (LC) displayed the same direction correlations during both hypo- and hyperalgesia. (**B**) Placebo hypoalgesia and nocebo hyperalgesia were associated with signal changes primarily within the lateral and dorsolateral columns of the PAG. Modified with permission from Crawford LS, Mills EP, Hanson T, et al. (2021) Brainstem mechanisms of pain modulation: a within-subjects 7T fMRI study of placebo analgesic and nocebo hyperalgesic responses. J Neurosci 41: 9794–9806.

critical for the mediating affective-motivational aspects of pain (Rodriguez et al., 2017). The PB also provides modulatory influences over the PAG-RVM-dorsal horn/SpV circuit (Gauriau and Bernard, 2002; Petrovic et al., 2004; Roeder et al., 2016; Chen et al., 2017; Stroman et al., 2018). Parabrachial neurons that project to the lateral PAG are excitatory, and their activation produces a robust analgesia (Chiang et al., 2020), and inhibition of parabrachial neurons attenuates withdrawal responses to mechanical stimuli as well as RVM ON- and OFF-cell activity (Rodriguez et al., 2017; Chen and Heinricher, 2019). Overall, it is clear that placebo analgesia and nocebo hyperalgesia responses are driven through the PAG-RVM-dorsal horn/SpV pathway, and this pain modulatory axis can be strongly modulated by regions such as the locus coeruleus, parabrachial, and potentially the region of the SRD.

STRESS-INDUCED ANALGESIA – THE ROLE OF THE PERIAQUEDUCTAL GRAY, LOCUS COERULEUS, AND ROSTRAL VENTROMEDIAL MEDULLA REGIONS

In addition to driving placebo and nocebo responses, the cerebral cortex also drives other pain modulatory responses. For example, acutely stressful events can evoke a reduction in perceived pain intensity, that is, stress-induced analgesia. Similar to placebo analgesia, stress-induced analgesia is thought to involve the PAG-RVM-dorsal horn/SpV axis (Butler and Finn, 2009). Preclinical studies have shown that injection of the μ-opioid receptor antagonist Naltrexone into the vlPAG attenuates stress-induced analgesia (Helmstetter and Landeira-Fernandez, 1990; Wiedenmayer and Barr, 2000), stress-induced analgesia during social conflict is associated with decreases in PAG opioid immunoreactivity, and baseline PAG opioid immunoreactivity is correlated with the degree of stress-induced analgesia expressed (Kulling et al., 1988; Kulling et al., 1989). Injection of a cannabinoid receptor 1 antagonist into the dorsal PAG prevents stress-induced analgesia in rats, and inhibition of the enzymes involved in endogenous cannabinoid degradation enhances stress-induced analgesia (Hohmann et al., 2005). Blocking opioid receptor function in the RVM also attenuates stress-induced analgesia (Foo and Helmstetter, 2000).

In addition to the PAG-RVM-dorsal horn/SpV circuitry, there is also evidence that the locus coeruleus plays a critical role in stress-induced analgesia. This stems from findings in animal experiments that systemic administration of clonidine, an α2-adrenoceptor agonist, potentiates stress-induced analgesia (Bodnar et al., 1983) and noradrenaline antagonist injections into the lumbosacral spinal cord attenuates hindpaw, but not forepaw stress-induced analgesia (Watkins et al., 1984). Furthermore, in another preclinical study, it was found that stress-induced analgesia is impaired in mice treated with *N*-(2-chloroethyl)-*N*-ethyl-2-bromobenzylamine, a selective neurotoxin for noradrenergic neurons in the locus coeruleus (Uchiyama et al., 2022). At the level of the spinal cord, the locus coeruleus produces antinociception through a presynaptic action on α2-adrenoreceptors, an effect that can be attenuated via direct injection of α2-adrenoreceptor antagonists into either the locus coeruleus or the spinal cord (Guo et al., 1996).

It is not known whether the same circuits are responsible for stress-induced analgesia in humans, although recent human imaging studies reported that attention-related analgesia is associated with locus coeruleus signal increases and altered signal covariation between the locus coeruleus, PAG, and RVM (Brooks et al., 2017; Oliva et al., 2021; Oliva et al., 2022), and placebo analgesia, nocebo hyperalgesia, and acupuncture analgesia are all associated with signal changes within the area of the locus coeruleus (Napadow et al., 2009; Crawford et al., 2021). There is now growing and compelling evidence of a role for the locus coeruleus in pain processing in humans from studies using fMRI. For example, a recent investigation identified that locus coeruleus–RVM fMRI signal coupling prior to noxious stimulation correlates with subsequent perceived pain intensity during actual noxious stimulation (Stroman et al., 2018). Furthermore, during noxious stimulation, signal changes in the locus coeruleus and dorsal horn inversely correlate with interindividual pain intensity ratings (Khan and Stroman, 2015). Whilst these studies strongly suggest that the locus coeruleus and the PAG-RVM-dorsal horn/SpV circuit are involved in mediating stress-induced analgesia, further studies, particularly in humans, are needed to verify their roles.

Brainstem pain modulation and analgesic clinical trials

Defining how the brainstem modulates pain is not only important for an overall understanding of how the brain modulates sensory inputs, it also has significant clinical implications. For example, the placebo effect has been described as the "villain" of randomized control trials with many potential analgesics showing minimal benefit over a placebo arm in human clinical trials (Wampold et al., 2007). The active to placebo differential must exceed a "minimum clinically important difference" which essentially means that one must first dissect out the exact placebo effect size (Chuang-Stein et al., 2011). Of course, particularly with respect to analgesic treatments, the placebo effect is driven by changes in neural circuits that are also potentially recruited during any analgesic treatment itself. Thus, any positive effect induced by an analgesic treatment likely reflects an interplay between its active properties combined with any placebo effects. Given the significant time, effort, and costs associated with running clinical trials, studies have attempted to determine what makes an individual respond to any given placebo treatment and whether someone can be identified as a "placebo responder" based on one or more psychologic and/or biologic measures.

There is some evidence for psychologic or biologic determinants of the placebo response, and moreover, that these determinants are trait-based. For example, the personality trait of dispositional optimism is reported to be related directly to the magnitude of placebo analgesia elicited in response to a cold pressor task (Geers et al., 2010). Additionally, significant correlations between placebo responsiveness and the personality trait of acquiescence have also been reported (Mcnair and Barrett, 1979; Mcnair et al., 1979), although the general consensus is a reliable trait that underpins a placebo responder has not been identified (Brody, 2000). In addition, some studies have explored the consistency of placebo responses in the one individual over time. One might expect that if an individual could be identified as a "placebo responder" that they would respond similarly over time. In one study, in which two different placebo test protocols were administered to the same individual on two separate occasions, it was found that there was a strong correlation between analgesic responses to the same placebo paradigm between sessions, but not between different paradigms on two separate occasions (Whalley et al., 2008). An earlier study in which placebo administration was interleaved with morphine administration in individuals with postoperative pain, found that less than half of the participants demonstrated a consistent response, vacillating between being either a responder or nonresponder.

Although these studies concluded that placebo responses were not consistent over time and thus a "placebo responder" biomarker could not be identified, they both relied on a single pain intensity measurement at either baseline and/or during the placebo testing to determine the degree of analgesic response. We now know that in approximately 50% of individuals, administration of multiple identical intensity noxious heat stimuli evokes considerably varied pain intensity ratings (Crawford et al., 2023a), and that in approximately 50% of individuals with chronic pain, the intensity of ongoing pain varies dramatically even over the space of 12 minutes (Mills et al., 2020). Whilst it looks unlikely that it will be possible to identify placebo responders or nonresponders using relatively simple personality tests, it is becoming clear that some individuals respond to pain in a consistent manner, while others show significant fluctuations in their responses. By extension, future studies that employ a combination of brain imaging, genetic, endocrine, and psychologic markers may provide a more complete picture of the multiple interacting mechanisms likely responsible for an individual's consistency or inconsistency to placebo interventions. If one could identify who will be a placebo nonresponder, it could revolutionize clinical trial design, particularly in the identification of novel analgesics, and result in more timely and effective pain treatment development.

Brainstem pain modulation and chronic pain

There is growing evidence that people with chronic pain display significantly impaired endogenous analgesic ability. That is, if an individual is less able to modulate incoming nociceptor information, they are more likely to develop chronic pain following injury (Henderson and Di Pietro, 2016). Chronic pain occurs at surprisingly high rates following accidental physical trauma, planned surgical interventions, chemotherapy treatments, or as a consequence of disease processes such as diabetes and multiple sclerosis (Kehlet et al., 2006; Fischer and Waxman, 2010). Chronic pain that occurs following nervous system injury, that is, neuropathic pain, is notoriously difficult to treat and is considered a disease in its own right (Siddall and Cousins, 2004). It is well established in preclinical models and in humans that chronic pain is associated with significant anatomic, biochemical, and activity changes at the level of the primary afferent synapse, including excitotoxic cell death due to increased afferent drive (Azkue et al., 1998; Moore et al., 2002; Scholz et al., 2005; Wilcox et al., 2015a, 2015b). There is also evidence from animal experiments that chronic neuropathic pain is associated with neuroplasticity within the brainstem pain-modulation system (Porreca et al., 2002; Heinricher et al., 2009), leading to a shift in dorsal horn/SpV output which contributes to the initiation and maintenance of hypersensitivity (Pertovaara et al., 1996; Wang et al., 2013; Martins et al., 2015) and complex behavioral disturbances (Mor et al., 2010; Boorman et al., 2019). It is likely that individuals with an ability to robustly activate descending analgesic systems are able to limit the effects of peripheral injury on the dorsal horn/SpV, and thereby reduce the propensity for an individual to develop chronic pain.

Importantly, chronic pain is often associated with altered responses to acute pain and impaired CPM analgesia (Yarnitsky, 2010; Lewis et al., 2012; Crofford, 2015). Since CPM can be produced using a relatively simple and quick paradigm in clinical and experimental settings, numerous studies have used CPM analgesic ability as a proxy for an individual's endogenous analgesic ability. It has been shown that chronic pain conditions including headache/migraine, arthritis, fibromyalgia, temporomandibular disorder, irritable bowel syndrome, and neuralgias are all associated with impaired CPM ability (Yarnitsky, 2010; Lewis et al., 2012), and it has been suggested that reduced CPM analgesia ability is a risk factor for developing postsurgical chronic pain (Dursteler et al., 2021; Larsen et al., 2021). Furthermore, when the patients' ongoing chronic pain is considered the test stimulus, the application of a conditioning stimulus does not result in a reduction in ongoing pain intensity in patients with peripheral neuropathic pain (Witting et al., 2003), or central poststroke pain (Tuveson et al., 2009).

A few studies have also reported that CPM analgesic ability may also predict treatment efficacy with inefficient CPM predicting of favorable response to duloxetine treatment in individuals with painful diabetic neuropathy (Yarnitsky et al., 2012). Reduced baseline CPM ability is also associated with reduced pain three months following spinal cord stimulation (Campbell et al., 2015) and an overall reduction in spinal cord stimulation effectiveness (Ramaswamy et al., 2019). In contrast to these centrally acting treatments, topical application of an anti-inflammatory in individuals with osteoarthritis produced greater analgesic responses in those with enhanced CPM analgesic ability (Edwards et al., 2016). Furthermore, osteoarthritis treatment with total joint replacement; chronic neuropathic pain treatment with morphine, ketamine, and placebo; and fibromyalgia treatment with pregabalin are all associated with increased CPM ability and greater pain relief

(Kosek and Ordeberg, 2000a, 2000b; Niesters et al., 2013; Wodehouse et al., 2018).

It is becoming increasingly clear that the levels of activity in an individual's endogenous analgesic circuitry play a significant role in their propensity for developing chronic pain, as well as the potential success of analgesic treatments. Despite the importance of analgesic circuitry function in chronic pain, the precise role that the brainstem plays in chronic pain remains unknown and no human studies have specifically investigated the role of CPM-related brainstem changes in chronic pain. However, numerous preclinical and some recent human brain imaging studies have begun to explore the role of the PAG-RVM-dorsal horn/SpV pathway in chronic pain. Preclinical models of chronic neuropathic pain show vlPAG neuronal excitability changes as well as spontaneous and stimulus-evoked firing patterns associated with the presence of pain behaviors (Du et al., 2013; Ho et al., 2013; Samineni et al., 2017b). Altered RVM function has also been described with reports of ON- and OFF-cell altered firing patterns, reduced thresholds and altered responsivity to mechanical and thermal stimuli (Carlson et al., 2007), and reduced spontaneous discharge rates of OFF-cells and increased spontaneous firing rates in ON-cells, in animal models of chronic neuropathic pain (Goncalves et al., 2007; Silva et al., 2013). Furthermore, PAG or RVM deactivation or RVM-DH pathway disruption can reduce pain behaviors (Pertovaara et al., 1996; Wang et al., 2013), and there is even evidence that disrupting the RVM prior to injury can protect against the development of pain behaviors (Porreca et al., 2001).

Recent human brain imaging investigations have also begun to explore the role of the PAG-RVM-dorsal horn/SpV circuit in chronic pain. Anatomic MRI studies have reported volumetric and diffusion changes in the PAG, RVM, and SpV in individuals with chronic orofacial pain including trigeminal neuropathy, temporomandibular disorder, and in individuals with migraine (Wilcox et al., 2015a, 2015b; Marciszewski et al., 2018). A recent series of studies using functional MRI also showed changes in signal covariation (functional connectivity) within the PAG-RVM-dorsal horn/SpV circuitry in chronic pain. In individuals with chronic orofacial neuropathic pain, it was found that signal coupling between the RVM and both the PAG and SpV was significantly enhanced compared with pain-free controls (Mills et al., 2018). In contrast, in individuals with painful temporomandibular disorder, there was no change in PAG-RVM coupling, despite an increase in RVM coupling with the SpV, suggesting a difference in the function of this brainstem circuit in neuropathic compared with nonneuropathic chronic pain (Mills et al., 2021) (Fig. 4.5A). In addition to altered overall signal coupling in individuals with chronic orofacial neuropathic pain, a recent study also showed that moment-to-moment fluctuations in PAG-RVM-SpV signal coupling follows spontaneous changes in pain intensity (Mills et al., 2020). That is, when an individual's pain intensity was spontaneously high, so too were RVM connectivity strengths with the PAG and SpV, and vice versa (Fig. 4.5B).

In addition to altered PAG-RVM-dorsal horn/SpV pathway function, preclinical studies have reported altered functioning of the locus coeruleus in chronic pain models. Some investigations report that following nerve injury, the locus coeruleus acts to inhibit the dorsal horn (Jasmin et al., 2003; Hayashida et al., 2012), whereas others suggest a facilitatory role contributing to the maintenance of altered sensory and affective responses (Kaushal et al., 2016; Boorman et al., 2019). Furthermore, following nerve injury, noradrenergic cell ablation evokes a reduction in mechanical sensitivity for at least 4 weeks (Kaushal et al., 2016). Consistent with these findings, individuals with chronic orofacial neuropathic pain display enhanced signal coupling between the RVM and both the locus coeruleus and SRD (Mills et al., 2018). However, no change in locus coeruleus signal coupling with the SpV was found, suggesting that a direct descending noradrenergic projection from the locus coeruleus to the dorsal horn/SpV is unlikely to be involved in chronic neuropathic pain maintenance in humans, in spite of its likely role in preclinical models.

There is also preclinical evidence of altered SRD function in chronic neuropathic pain. In rats with neuropathic pain, blocking SRD activity significantly reduces noxious-evoked and spontaneous activity of dorsal horn neurons (Sotgiu et al., 2008). Furthermore, noxious stimulation increases noradrenaline release in the SRD, and blocking this release can significantly attenuate the allodynia and hyperalgesia associated with neuropathic pain (Martins et al., 2010; Martins et al., 2015). These effects are unlikely to be due to changes in tonic drive since basal activity in the locus coeruleus activity appears unchanged by nerve injury (Viisanen and Pertovaara, 2007; Boorman et al., 2019). Furthermore, preclinical investigations have identified reduced or abolished DNIC following nerve injury (Bannister et al., 2015; Patel and Dickenson, 2020).

These results provide evidence of altered brainstem pain modulatory signal coupling in chronic pain and likely underpin enhanced descending facilitation of information processing at the level of the primary afferent synapse which underpins increased nociceptive transmission to higher brain centers. Further human brain imaging investigations are required to determine the precise roles of each of the brainstem modulatory regions in different forms of chronic pain, and how such changes may underpin the reduced endogenous ability reported to occur in individuals with chronic pain.

CAN WE MODIFY BRAINSTEM ANALGESIC CIRCUITS FOR PAIN RELIEF?

Direct brainstem effects

It is clear that brainstem pain modulatory circuits are altered in individuals with chronic pain; the important question is whether it is possible to begin manipulating these circuits to provide pain relief? Several of the currently available pharmacologic interventions used for pain relief such as opioids, morphine, and fentanyl, are known to act directly on brainstem modulatory circuits and in particular in the PAG (Loyd et al., 2007; Lueptow et al., 2018). Whilst it is known that opiate receptors are distributed throughout all levels of the brain (Burns et al.,

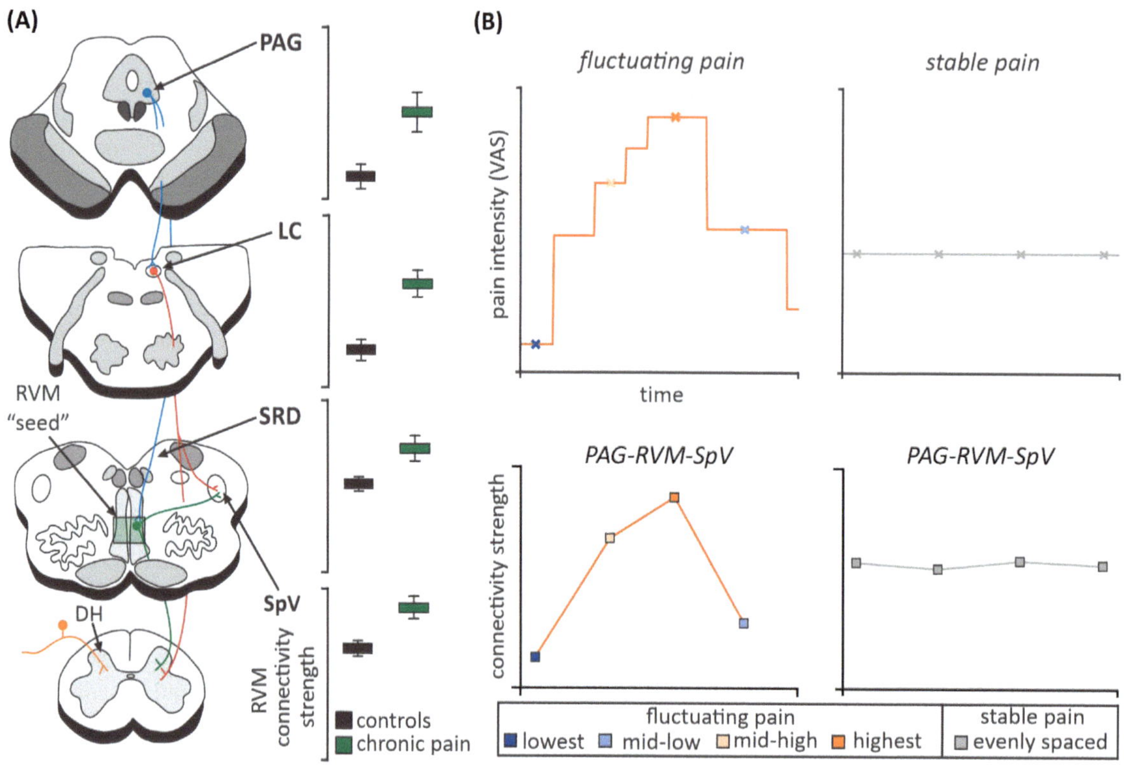

Fig. 4.5. Brainstem pain modulation circuitry changes in chronic pain. (**A**) Brainstem regions that show significantly greater signal covariation with the rostral ventromedial medulla (RVM) in individuals with chronic pain. Increased signal covariations occurred in the periaqueductal gray (PAG) matter, locus coeruleus (LC), subnucleus reticularis dorsalis (SRD), and the spinal trigeminal nucleus (SpV). (**B**) An example of two individuals with fluctuation and stable chronic pain intensities measured over a 12-minute period. Fluctuation pain was associated with graded changes in resting functional magnetic resonance signal covariations, i.e., connectivity strengths. That is, as pain increased, PAG-RVM-SpV connectivity increased. Those individuals with stable pain displayed no PAG-RVM-SpV connectivity changes over the 12-minute period.

2019), since electrical, pharmacologic, and chemogenetic manipulation of the vlPAG produces analgesia (Bennett and Mayer, 1979; Jones and Gebhart, 1988; Carstens et al., 1990; Samineni et al., 2017a) and infusion of opioids into the vlPAG relieves pain by suppressing drive onto the RVM (Vaughan et al., 1997), it is likely that this brainstem circuit is that which is responsible for the pain-relieving properties of opiates in humans. Although opiates are effective at relieving many forms of chronic pain (Ballantyne and Mao, 2003), they have significant deleterious side-effects and strong addictive potential (Kosten and George, 2002; Benyamin et al., 2008). Given these concerns, the drive to develop novel, effective, nonopiate treatments has increased.

A nonopiate pharmacologic treatment that has recently gained interest is the use of cannabinoids for pain relief. The analgesic effects of cannabinoids are well established in preclinical models of pain (Soliman et al., 2021), and these effects are thought to arise from interactions between lipid signaling molecules called endocannabinoids that bind to cannabinoid receptors or other endocannabinoid-sensitive receptors. It is well established that cannabinoid receptors are highly expressed in the dorsolateral and ventrolateral columns of the PAG (Wilson-Poe et al., 2012), and activation of cannabinoid receptors in the vlPAG produces antinociception and anti-hyperalgesia (Palazzo et al., 2012). Cannabinoid agonist injections into the vlPAG produce a mild analgesia that is about one-third the potency of morphine (Palazzo et al., 2001; Wilson-Poe et al., 2013). Although cannabinoids are seemingly less effective than opioids in reducing acute pain when delivered directly into the brainstem, there is preclinical evidence that they may have increased efficacy in chronic pain (Donvito et al., 2017). Indeed, there is growing evidence that *Δ^9-tetrahydrocannabinol* (THC: the main psychoactive component of cannabis) and/or *cannabidiol* (a nonpsychoactive component) might be effective in relieving pain in some chronic pain conditions (Henderson et al., 2021). Furthermore, cannabinoids may reduce the requirement for patients to use conventional analgesics including opiates (Boehnke et al., 2016). Whilst cannabinoids might be useful for activating brainstem pain modulatory circuits for pain relief, THC has well-documented side-effects including dizziness, appetite stimulation, drowsiness, altered mood, anxiety, and impaired cognition and attention, whereas cannabidiol has little side-effects even at very high doses (Chesney et al., 2020).

Indirect brainstem effects

Whilst both opiates and cannabinoids likely act directly on brainstem pain modulatory circuits, there are a number of emerging techniques that may "naturally" recruit cortical sites that in turn modulate brainstem pain modulatory circuits to produce pain relief. As mentioned above, the use of placebo analgesia has been of clinical interest for many years, and it is consistently reported that approximately 40%–50% of pain-free individuals express placebo analgesia (Crawford et al., 2021) and that chronic pain patients in placebo groups experience on average, approximately a 20%–35% reduction in pain (Enck et al., 2011; Tuttle et al., 2015). Since placebo responses require the integration of cognitive function, prior experience, and expectations, it is likely that brainstem circuits responsible for placebo analgesia are driven by descending inputs from higher-order cortical regions. Indeed, human brain imaging studies have linked activity changes in the dorsolateral prefrontal and anterior cingulate cortices to greater placebo analgesic responses (Wager et al., 2004; Eippert et al., 2009; Crawford et al., 2023b, 2023c), and transient disruption of the dlPFC through noninvasive stimulation can both block or amplify the generation of placebo analgesia (Krummenacher et al., 2010; Tu et al., 2021).

While many consider using placebo analgesic techniques to be unethical due to the inherent deception, reliable placebo effects can also be produced without deception. For example, clinicians can promote positive expectations combined with disclosure of expected treatment effects to simply enhance already existing placebo effects (Finniss et al., 2010). Furthermore, it has recently been reported that a proportion of individuals still express a placebo analgesia response if when they are aware they are being given an inert treatment, that is, open-label placebo analgesia (Klinger et al., 2018). This open-label technique may provide a way for placebo analgesia to be used in an open and ethical way. In addition, given that placebo likely driven by activity changes in higher brain centers, it is possible that techniques that directly target these higher brain regions, such as noninvasive stimulation or pharmacologic agents, could be used to control brainstem pain modulation circuits for pain relief.

Other less typical ways to recruit the brainstem pain modulatory systems are also being explored. Probably the oldest known method of altering pain intensity is through the use of music, relying on its ability to modulate an individual's cognitive and emotional state (Garza-Villarreal et al., 2014). Indeed, music can likely alter pain perception through various mechanisms including distraction (Fauerbach et al., 2002), pleasure (Roy et al., 2008), and a sense of control (Brown et al., 1989). Whilst the mechanism via which music ultimately alters the activity in brainstem pain modulatory circuits remains to be resolved, a human fMRI study reported that music-induced analgesia is associated with activity changes in the dorsolateral prefrontal cortex as well as in the PAG, RVM, and dorsal horn (Dobek et al., 2014). Music-induced analgesia has also been reported to occur in individuals with chronic pain and be associated with changes in fMRI signal covariation between the dorsolateral prefrontal cortex and amygdala (Pando-Naude et al., 2019).

In addition to auditory stimuli, olfactory stimuli have also been reported to evoke analgesia (Pehlivan and Karadakovan, 2019). For example, lavender essence aromatherapy can evoke an olfactory-analgesia after episiotomy (Vakilian et al., 2011), reduce postoperative analgesic demand (Kim et al., 2006), and has also been reported to reduce the severity of migraine attacks (Sasannejad et al., 2012). Whilst the circuits responsible for olfactory-analgesia are unknown, olfactory cortex neurons project directly and indirectly to the hypothalamus (Price et al., 1991), an area that is known to modulate the function of the PAG (Ellison and Flynn, 1968; Huang and Flynn, 1975). Indeed, it was recently shown that fMRI signal covariation between the hypothalamus and the lPAG regulates the expression of placebo-analgesia (Crawford et al., 2023b), and lavender administration results in activation of the prefrontal cortex and hypothalamus (Wang et al., 2005). It is likely that these less conventional ways to induce analgesia are using activity changes in higher brain regions that subsequently influence brainstem pain modulatory circuits to alter ascending noxious information at the dorsal horn/SpV.

CONCLUSIONS AND FUTURE DIRECTIONS

There is a considerable body of evidence identifying brainstem regions critical for the modulation of incoming noxious information. Whilst uncovering the location and function of these circuits in pain-free animal experiments and humans can inform us as to their roles in modulating acute pain, this has limited clinical value since by definition, acute pain resolves relatively quickly. However, it is becoming increasingly clear from preclinical and human studies that there is considerable plasticity in the structure and function of brainstem pain-modulation circuits in chronic pain, although there appears to be differences in the relative contributions of different brainstem regions to different chronic pain conditions. In neuropathic pain conditions, evidence suggests that altered brainstem pain modulatory circuit control over the dorsal horn/SpV results in a pronociceptive state, contributes to a reduced analgesic ability and therefore increasing sensitivity to incoming noxious information. It has been proposed that the reduced ability of an individual to limit the potentially damaging effects of increased afferent drive, results in a series of neural and nonneural changes that contribute to a brain state that maintains a constant perception of pain. This idea is also consistent with findings that the brainstem state prior to injury is critical for the subsequent development of chronic pain. Whilst evidence for such plasticity is growing, more studies are needed before we can fully understand the role of the brainstem in the initiation and maintenance of various chronic pain conditions.

Recent technologic advancements, particularly in human brain imaging, will allow for a more in-depth exploration of the brainstem circuitry responsible for pain modulation. For example, ultra-high-field fMRI now allows for submillimeter spatial resolution and thus will likely result in a more detailed view of the brainstem nuclei involved in pain modulation in awake humans. Furthermore, using ultra-high-field MRI to explore brainstem pain-modulatory circuitry structure and function in individuals with various neuropathic and nonneuropathic chronic pain conditions could result in personalized pain-modulation circuitry profiles. This could drive personalized treatments that could involve increasing an individual's pain modulatory ability to either reduce or eliminate ongoing chronic pain, or prevent the development of chronic pain following "planned" injuries such as surgery.

DISCLOSURE

The authors report no conflicts of interest in this work.

References

Adams JE (1976). Naloxone reversal of analgesia produced by brain stimulation in the human. Pain 2: 161–166.

Aicher SA, Hermes SM, Whittier KL et al. (2012). Descending projections from the rostral ventromedial medulla (RVM) to trigeminal and spinal dorsal horns are morphologically and neurochemically distinct. J Chem Neuroanat 43: 103–111.

Amanzio M, Benedetti F (1999). Neuropharmacological dissection of placebo analgesia: expectation-activated opioid systems versus conditioning-activated specific subsystems. J Neurosci 19: 484–494.

Azkue JJ, Zimmermann M, Hsieh TF et al. (1998). Peripheral nerve insult induces NMDA receptor-mediated, delayed degeneration in spinal neurons. Eur J Neurosci 10: 2204–2206.

Bagley EE, Ingram SL (2020). Endogenous opioid peptides in the descending pain modulatory circuit. Neuropharmacology 173: 108131.

Ballantyne JC, Mao J (2003). Opioid therapy for chronic pain. N Engl J Med 349: 1943–1953.

Bandler R, Keay KA (1996). Columnar organization in the midbrain periaqueductal gray and the integration of emotional expression. Prog Brain Res 107: 285–300.

Bannister K, Patel R, Goncalves L et al. (2015). Diffuse noxious inhibitory controls and nerve injury: restoring an imbalance between descending monoamine inhibitions and facilitations. Pain 156: 1803–1811.

Basbaum AI, Fields HL (1984). Endogenous pain control systems: brainstem spinal pathways and endorphin circuitry. Annu Rev Neurosci 7: 309–338.

Beecher HK (1955). The powerful placebo. J Am Med Assoc 159: 1602–1606.

Benedetti F (1996). The opposite effects of the opiate antagonist naloxone and the cholecystokinin antagonist proglumide on placebo analgesia. Pain 64: 535–543.

Benedetti F, Amanzio M, Casadio C et al. (1997). Blockade of nocebo hyperalgesia by the cholecystokinin antagonist proglumide. Pain 71: 135–140.

Benedetti F, Amanzio M, Rosato R et al. (2011). Nonopioid placebo analgesia is mediated by CB1 cannabinoid receptors. Nat Med 17: 1228–1230.

Benedetti F, Mayberg HS, Wager TD et al. (2005). Neurobiological mechanisms of the placebo effect. J Neurosci 25: 10390–10402.

Bennett GJ, Mayer DJ (1979). Inhibition of spinal cord interneurons by narcotic microinjection and focal electrical stimulation in the periaqueductal central gray matter. Brain Res 172: 243–257.

Benyamin R, Trescot AM, Datta S et al. (2008). Opioid complications and side effects. Pain Physician 11: S105–S120.

Bodnar RJ, Simone DA, Kordower JH et al. (1983). Capsaicin treatment and stress-induced analgesia. Pharmacol Biochem Behav 18: 65–71.

Boehnke KF, Litinas E, Clauw DJ (2016). Medical cannabis use is associated with decreased opiate medication use in a retrospective cross-sectional survey of patients with chronic pain. J Pain 17: 739–744.

Boorman DC, Kang JWM, Keay KA (2019). Peripheral nerve injury attenuates stress-induced Fos-family expression in the locus coeruleus of male Sprague-Dawley rats. Brain Res 1719: 253–262.

Bouhassira D, Bing Z, Le Bars D (1990). Studies of the brain structures involved in diffuse noxious inhibitory controls: the mesencephalon. J Neurophysiol 64: 1712–1723.

Bouhassira D, Villanueva L, Bing Z et al. (1992). Involvement of the subnucleus reticularis dorsalis in diffuse noxious inhibitory controls in the rat. Brain Res 595: 353–357.

Brody H (2000). The placebo response. Recent research and implications for family medicine. J Fam Pract 49: 649–654.

Brooks JC, Davies WE, Pickering AE (2017). Resolving the brainstem contributions to attentional analgesia. J Neurosci 37: 2279–2291.

Brown CJ, Chen ACN, Dworkin SF (1989). Music in the control of human pain. Music Ther 8: 47–60.

Burns JA, Kroll DS, Feldman DE et al. (2019). Molecular imaging of opioid and dopamine systems: insights into the pharmacogenetics of opioid use disorders. Front Psychiatry 10: 626.

Butler RK, Finn DP (2009). Stress-induced analgesia. Prog Neurobiol 88: 184–202.

Cadden SW, Villanueva L, Chitour D et al. (1983). Depression of activities of dorsal horn convergent neurones by propriospinal mechanisms triggered by noxious inputs; comparison with diffuse noxious inhibitory controls (DNIC). Brain Res 275: 1–11.

Campbell CM, Buenaver LF, Raja SN et al. (2015). Dynamic pain phenotypes are associated with spinal cord stimulation-induced reduction in pain: a repeated measures observational pilot study. Pain Med 16: 1349–1360.

Carlson JD, Maire JJ, Martenson ME et al. (2007). Sensitization of pain-modulating neurons in the rostral ventromedial medulla after peripheral nerve injury. J Neurosci 27: 13222–13231.

Carstens E, Hartung M, Stelzer B et al. (1990). Suppression of a hind limb flexion withdrawal reflex by microinjection of glutamate or morphine into the periaqueductal gray in the rat. Pain 43: 105–112.

Chen Q, Heinricher MM (2019). Plasticity in the link between pain-transmitting and pain-modulating systems in acute and persistent inflammation. J Neurosci 39: 2065–2079.

Chen Q, Roeder Z, Li MH et al. (2017). Optogenetic evidence for a direct circuit linking nociceptive transmission through the parabrachial complex with pain-modulating neurons of the rostral ventromedial medulla (RVM). eNeuro 4: ENEURO.0202-17.2017.

Cheng ZF, Fields HL, Heinricher MM (1986). Morphine microinjected into the periaqueductal gray has differential effects on 3 classes of medullary neurons. Brain Res 375: 57–65.

Chesney E, Oliver D, Green A et al. (2020). Adverse effects of cannabidiol: a systematic review and meta-analysis of randomized clinical trials. Neuropsychopharmacology 45: 1799–1806.

Chiang CY, Hu JW, Sessle BJ (1994). Parabrachial area and nucleus raphe magnus-induced modulation of nociceptive and nonnociceptive trigeminal subnucleus caudalis neurons activated by cutaneous or deep inputs. J Neurophysiol 71: 2430–2445.

Chiang MC, Nguyen EK, Canto-Bustos M et al. (2020). Divergent neural pathways emanating from the lateral parabrachial nucleus mediate distinct components of the pain response. Neuron 106: 927–939e5.

Chuang-Stein C, Kirby S, Hirsch I et al. (2011). The role of the minimum clinically important difference and its impact on designing a trial. Pharm Stat 10: 250–256.

Colloca L, Benedetti F (2006). How prior experience shapes placebo analgesia. Pain 124: 126–133.

Crawford L, Mills E, Meylakh N et al. (2023a). Brain activity changes associated with pain perception variability. Cereb Cortex 33: 4145–4155.

Crawford LS, Meylakh N, Macey PM et al. (2023b). Stimulus-independent and stimulus-dependent neural networks underpin placebo analgesia responsiveness in humans. Commun Biol 6: 569.

Crawford LS, Mills EP, Peek A et al. (2023c). Function and biochemistry of the dorsolateral prefrontal cortex during placebo analgesia: how the certainty of prior experiences shapes endogenous pain relief. Cereb Cortex 33: 9822–9834.

Crawford LS, Mills EP, Hanson T et al. (2021). Brainstem mechanisms of pain modulation: a within-subjects 7T fMRI study of placebo analgesic and nocebo hyperalgesic responses. J Neurosci 41: 9794–9806.

Crofford LJ (2015). Chronic pain: where the body meets the brain. Trans Am Clin Climatol Assoc 126: 167–183.

De Broucker T, Cesaro P, Willer JC et al. (1990). Diffuse noxious inhibitory controls in man. Involvement of the spinoreticular tract. Brain 113: 1223–1234.

Depaulis A, Keay KA, Bandler R (1992). Longitudinal neuronal organization of defensive reactions in the midbrain periaqueductal gray region of the rat. Exp Brain Res 90: 307–318.

Dickenson AH, Le Bars D (1987). Supraspinal morphine and descending inhibitions acting on the dorsal horn of the rat. J Physiol 384: 81–107.

Dobek CE, Beynon ME, Bosma RL et al. (2014). Music modulation of pain perception and pain-related activity in the brain, brain stem, and spinal cord: a functional magnetic resonance imaging study. J Pain 15: 1057–1068.

Donvito G, Bagdas D, Toma W et al. (2017). The interaction between alpha 7 nicotinic acetylcholine receptor and nuclear peroxisome proliferator-activated receptor-alpha represents a new antinociceptive signaling pathway in mice. Exp Neurol 295: 194–201.

Du L, Wang SJ, Cui J et al. (2013). Inhibition of HCN channels within the periaqueductal gray attenuates neuropathic pain in rats. Behav Neurosci 127: 325–329.

Dursteler C, Salazar Y, Rodriguez U et al. (2021). Conditioned pain modulation predicts persistent pain after knee replacement surgery. Pain Rep 6: e910.

Edwards RR, Dolman AJ, Martel MO et al. (2016). Variability in conditioned pain modulation predicts response to NSAID treatment in patients with knee osteoarthritis. BMC Musculoskelet Disord 17: 284.

Eippert F, Bingel U, Schoell ED et al. (2009). Activation of the opioidergic descending pain control system underlies placebo analgesia. Neuron 63: 533–543.

Ellison GD, Flynn JP (1968). Organized aggressive behavior in cats after surgical isolation of the hypothalamus. Arch Ital Biol 106: 1–20.

Enck P, Klosterhalfen S, Weimer K et al. (2011). The placebo response in clinical trials: more questions than answers. Philos Trans R Soc Lond B Biol Sci 366: 1889–1895.

Fauerbach JA, Lawrence JW, Haythornthwaite JA et al. (2002). Coping with the stress of a painful medical procedure. Behav Res Ther 40: 1003–1015.

Fields HL (1999). Pain: an unpleasant topic. Pain 82.

Fields HL, Basbaum AI (1978). Brainstem control of spinal pain-transmission neurons. Annu Rev Physiol 40: 217–248.

Fields HL, Bry J, Hentall I et al. (1983). The activity of neurons in the rostral medulla of the rat during withdrawal from noxious heat. J Neurosci 3: 2545–2552.

Fields HL, Heinricher MM (1985). Anatomy and physiology of a nociceptive modulatory system. Philos Trans R Soc Lond B Biol Sci 308: 361–374.

Fields HL, Malick A, Burstein R (1995). Dorsal horn projection targets of ON and OFF cells in the rostral ventromedial medulla. J Neurophysiol 74: 1742–1759.

Finniss DG, Kaptchuk TJ, Miller F et al. (2010). Biological, clinical, and ethical advances of placebo effects. Lancet 375: 686–695.

Fischer TZ, Waxman SG (2010). Neuropathic pain in diabetes – evidence for a central mechanism. Nat Rev Neurol 6: 462–466.

Foo H, Helmstetter FJ (2000). Activation of kappa opioid receptors in the rostral ventromedial medulla blocks stress-induced antinociception. Neuroreport 11: 3349–3352.

Garza-Villarreal EA, Wilson AD, Vase L et al. (2014). Music reduces pain and increases functional mobility in fibromyalgia. Front Psychol 5: 90.

Gauriau C, Bernard JF (2002). Pain pathways and parabrachial circuits in the rat. Exp Physiol 87: 251–258.

Geers AL, Wellman JA, Fowler SL et al. (2010). Dispositional optimism predicts placebo analgesia. J Pain 11: 1165–1171.

Goncalves L, Almeida A, Pertovaara A (2007). Pronociceptive changes in response properties of rostroventromedial medullary neurons in a rat model of peripheral neuropathy. Eur J Neurosci 26: 2188–2195.

Gracely RH (1987). Verbal descriptor measures of pain clarify mechanisms of analgesia due to narcotics, brain stimulation, and placebo. Anesth Prog 34: 113–116.

Grahl A, Onat S, Buchel C (2018). The periaqueductal gray and Bayesian integration in placebo analgesia. Elife 7: e32930.

Grevert P, Albert LH, Goldstein A (1983). Partial antagonism of placebo analgesia by naloxone. Pain 16: 129–143.

Guo TZ, Jiang JY, Buttermann AE et al. (1996). Dexmedetomidine injection into the locus ceruleus produces antinociception. Anesthesiology 84: 873–881.

Guo W, Robbins MT, Wei F et al. (2006). Supraspinal brain-derived neurotrophic factor signaling: a novel mechanism for descending pain facilitation. J Neurosci 26: 126–137.

Hayashida K, Peters CM, Gutierrez S et al. (2012). Depletion of endogenous noradrenaline does not prevent spinal cord plasticity following peripheral nerve injury. J Pain 13: 49–57.

Heinricher MM, Barbaro NM, Fields HL (1989). Putative nociceptive modulating neurons in the rostral ventromedial medulla of the rat: firing of on- and off-cells is related to nociceptive responsiveness. Somatosens Mot Res 6: 427–439.

Heinricher MM, Morgan MM, Tortorici V et al. (1994). Disinhibition of off-cells and antinociception produced by an opioid action within the rostral ventromedial medulla. Neuroscience 63: 279–288.

Heinricher MM, Tavares I, Leith JL et al. (2009). Descending control of nociception: specificity, recruitment and plasticity. Brain Res Rev 60: 214–225.

Helmstetter FJ, Landeira-Fernandez J (1990). Conditional hypoalgesia is attenuated by naltrexone applied to the periaqueductal gray. Brain Res 537: 88–92.

Henderson LA, Di Pietro F (2016). How do neuroanatomical changes in individuals with chronic pain result in the constant perception of pain? Pain Manag 6: 147–159.

Henderson LA, Kotsirilos V, Cairns EA et al. (2021). Medicinal cannabis in the treatment of chronic pain. Aust J Gen Pract 50: 724–732.

Ho YC, Cheng JK, Chiou LC (2013). Hypofunction of glutamatergic neurotransmission in the periaqueductal gray contributes to nerve-injury-induced neuropathic pain. J Neurosci 33: 7825–7836.

Hohmann AG, Suplita RL, Bolton NM et al. (2005). An endocannabinoid mechanism for stress-induced analgesia. Nature 435: 1108–1112.

Holstege G, Kuypers HG (1982). The anatomy of brain stem pathways to the spinal cord in cat. A labeled amino acid tracing study. Prog Brain Res 57: 145–175.

Huang YH, Flynn JP (1975). Unit activities in the hypothalamus and midbrain during stimulation of hypothalamic attack sites. Brain Res 93: 419–440.

Jasmin L, Boudah A, Ohara PT (2003). Long-term effects of decreased noradrenergic central nervous system innervation on pain behavior and opioid antinociception. J Comp Neurol 460: 38–55.

Jones SL, Gebhart GF (1988). Inhibition of spinal nociceptive transmission from the midbrain, pons and medulla in the rat: activation of descending inhibition by morphine, glutamate and electrical stimulation. Brain Res 460: 281–296.

Kaushal R, Taylor BK, Jamal AB et al. (2016). GABA-A receptor activity in the noradrenergic locus coeruleus drives trigeminal neuropathic pain in the rat; contribution of NAalpha1 receptors in the medial prefrontal cortex. Neuroscience 334: 148–159.

Keay KA, Bandler R (2001). Parallel circuits mediating distinct emotional coping reactions to different types of stress. Neurosci Biobehav Rev 25: 669–678.

Keay KA, Feil K, Gordon BD et al. (1997). Spinal afferents to functionally distinct periaqueductal gray columns in the rat: an anterograde and retrograde tracing study. J Comp Neurol 385: 207–229.

Kehlet H, Jensen TS, Woolf CJ (2006). Persistent postsurgical pain: risk factors and prevention. Lancet 367: 1618–1625.

Keller A, Akintola T, Colloca L (2018). Placebo analgesia in rodents: current and future research. Int Rev Neurobiol 138: 1–15.

Khan HS, Stroman PW (2015). Inter-individual differences in pain processing investigated by functional magnetic resonance imaging of the brainstem and spinal cord. Neuroscience 307: 231–241.

Khasabov SG, Malecha P, Noack J et al. (2015). Activation of rostral ventromedial medulla neurons by noxious stimulation of cutaneous and deep craniofacial tissues. J Neurophysiol 113: 14–22.

Kim JT, Wajda M, Cuff G et al. (2006). Evaluation of aromatherapy in treating postoperative pain: pilot study. Pain Pract 6: 273–277.

Klinger R, Stuhlreyer J, Schwartz M et al. (2018). Clinical use of placebo effects in patients with pain disorders. Int Rev Neurobiol 139: 107–128.

Kosek E, Ordeberg G (2000a). Abnormalities of somatosensory perception in patients with painful osteoarthritis normalize following successful treatment. Eur J Pain 4: 229–238.

Kosek E, Ordeberg G (2000b). Lack of pressure pain modulation by heterotopic noxious conditioning stimulation in patients with painful osteoarthritis before, but not following, surgical pain relief. Pain 88: 69–78.

Kosten TR, George TP (2002). The neurobiology of opioid dependence: implications for treatment. Sci Pract Perspect 1: 13–20.

Krummenacher P, Candia V, Folkers G et al. (2010). Prefrontal cortex modulates placebo analgesia. Pain 148: 368–374.

Kucharczyk MW, Di Domenico F, Bannister K (2023). A critical brainstem relay for mediation of diffuse noxious inhibitory controls. Brain 146: 2259–2267.

Kulling P, Frischknecht HR, Pasi A et al. (1988). Social conflict-induced changes in nociception and beta-endorphin-like immunoreactivity in pituitary and discrete brain areas of C57BL/6 and DBA/2 mice. Brain Res 450: 237–246.

Kulling P, Siegfried B, Frischknecht HR et al. (1989). Beta-endorphin-like immunoreactivity levels in the hypothalamus, the periaqueductal grey and the pituitary of the DBA mouse: determination by ELISA and relationship to nociception. Physiol Behav 46: 25–28.

Larsen DB, Laursen M, Edwards RR et al. (2021). The combination of preoperative pain, conditioned pain modulation, and pain catastrophizing predicts postoperative pain 12 months after total knee arthroplasty. Pain Med 22: 1583–1590.

Le Bars D, Dickenson AH, Besson J-M (1979a). Diffuse noxious inhibitory controls (DNIC). I. Effects on dorsal horn convergent neurones in the rat. Pain 6: 283–304.

Le Bars D, Dickenson AH, Besson JM (1979b). Diffuse noxious inhibitory controls (DNIC). II. Lack of effect on non-convergent neurones, supraspinal involvement and theoretical implications. Pain 6: 305–327.

Levine JD, Gordon NC (1984). Influence of the method of drug administration on analgesic response. Nature 312: 755–756.

Levine JD, Gordon NC, Fields HL (1978). The mechanism of placebo analgesia. Lancet 2: 654–657.

Lewis GN, Rice DA, Mcnair PJ (2012). Conditioned pain modulation in populations with chronic pain: a systematic review and meta-analysis. J Pain 13: 936–944.

Lovick TA (1993). The periaqueductal gray-rostral medulla connection in the defence reaction: efferent pathways and descending control mechanisms. Behav Brain Res 58: 19–25.

Loyd DR, Morgan MM, Murphy AZ (2007). Morphine preferentially activates the periaqueductal gray-rostral ventromedial medullary pathway in the male rat: a potential mechanism for sex differences in antinociception. Neuroscience 147: 456–468.

Lueptow LM, Fakira AK, Bobeck EN (2018). The contribution of the descending pain modulatory pathway in opioid tolerance. Front Neurosci 12: 886.

Marciszewski KK, Meylakh N, Di Pietro F et al. (2018). Altered brainstem anatomy in migraine. Cephalalgia 38: 476–486.

Martins I, Carvalho P, de Vries MG et al. (2015). Increased noradrenergic neurotransmission to a pain facilitatory area of the brain is implicated in facilitation of chronic pain. Anesthesiology 123: 642–653.

Martins I, Costa-Araujo S, Fadel J et al. (2010). Reversal of neuropathic pain by HSV-1-mediated decrease of noradrenaline in a pain facilitatory area of the brain. Pain 151: 137–145.

Mason P, Fields HL (1989). Axonal trajectories and terminations of on- and off-cells in the cat lower brainstem. J Comp Neurol 288: 185–207.

Mcnair DM, Barrett JE (1979). Two Bass Scale factors and response to placebo and anxiolytic drugs. Psychopharmacology (Berl) 65: 165–170.

Mcnair DM, Gardos G, Haskell DS et al. (1979). Placebo response, placebo effect, and two attributes. Psychopharmacology (Berl) 63: 245–250.

Mills EP, Akhter R, Di Pietro F et al. (2021). Altered brainstem pain modulating circuitry functional connectivity in chronic painful temporomandibular disorder. J Pain 22: 219–232.

Mills EP, Alshelh Z, Kosanovic D et al. (2020). Altered brainstem pain-modulation circuitry connectivity during spontaneous pain intensity fluctuations. J Pain Res 13: 2223–2235.

Mills EP, Di Pietro F, Alshelh Z et al. (2018). Brainstem pain-control circuitry connectivity in chronic neuropathic pain. J Neurosci 38: 465–473.

Miron D, Duncan GH, Bushnell CM (1989). Effects of attention on the intensity and unpleasantness of thermal pain. Pain 39: 345–352.

Moore KA, Kohno T, Karchewski LA et al. (2002). Partial peripheral nerve injury promotes a selective loss of GABAergic inhibition in the superficial dorsal horn of the spinal cord. J Neurosci 22: 6724–6731.

Mor D, Bembrick AL, Austin PJ et al. (2010). Anatomically specific patterns of glial activation in the periaqueductal gray of the sub-population of rats showing pain and disability following chronic constriction injury of the sciatic nerve. Neuroscience 166: 1167–1184.

Morton CR, Maisch B, Zimmermann M (1987). Diffuse noxious inhibitory controls of lumbar spinal neurons involve a supraspinal loop in the cat. Brain Res 410: 347–352.

Napadow V, Dhond RP, Kim J et al. (2009). Brain encoding of acupuncture sensation – coupling on-line rating with fMRI. Neuroimage 47: 1055–1065.

Niesters M, Aarts L, Sarton E et al. (2013). Influence of ketamine and morphine on descending pain modulation in chronic pain patients: a randomized placebo-controlled cross-over proof-of-concept study. Br J Anaesth 110: 1010–1016.

Oliva V, Gregory R, Davies W-E et al. (2021). Parallel cortical-brainstem pathways to attentional analgesia. Neuroimage 226: 117548.

Oliva V, Hartley-Davies R, Moran R et al. (2022). Simultaneous brain, brainstem, and spinal cord pharmacological-fMRI reveals involvement of an endogenous opioid network in attentional analgesia. Elife 11: e71877.

Ossipov MH, Morimura K, Porreca F (2014). Descending pain modulation and chronification of pain. Curr Opin Support Palliat Care 8: 143–151.

Palazzo E, Luongo L, Bellini G et al. (2012). Changes in cannabinoid receptor subtype 1 activity and interaction with metabotropic glutamate subtype 5 receptors in the periaqueductal gray-rostral ventromedial medulla pathway in a rodent neuropathic pain model. CNS Neurol Disord Drug Targets 11: 148–161.

Palazzo E, Marabese I, De Novellis V et al. (2001). Metabotropic and NMDA glutamate receptors participate in the cannabinoid-induced antinociception. Neuropharmacology 40: 319–326.

Pando-Naude V, Barrios FA, Alcauter S et al. (2019). Functional connectivity of music-induced analgesia in fibromyalgia. Sci Rep 9: 15486.

Patel R, Dickenson AH (2020). A study of cortical and brainstem mechanisms of diffuse noxious inhibitory controls in anaesthetised normal and neuropathic rats. Eur J Neurosci 51: 952–962.

Pehlivan S, Karadakovan A (2019). Effects of aromatherapy massage on pain, functional state, and quality of life in an elderly individual with knee osteoarthritis. Jpn J Nurs Sci 16: 450–458.

Pertovaara A, Wei H, Hamalainen MM (1996). Lidocaine in the rostroventromedial medulla and the periaqueductal gray attenuates allodynia in neuropathic rats. Neurosci Lett 218: 127–130.

Petrovic P, Kalso E, Petersson KM et al. (2002). Placebo and opioid analgesia – imaging a shared neuronal network. Science 295: 1737–1740.

Petrovic P, Petersson KM, Hansson P et al. (2004). Brainstem involvement in the initial response to pain. Neuroimage 22: 995–1005.

Porreca F, Burgess SE, Gardell LR et al. (2001). Inhibition of neuropathic pain by selective ablation of brainstem medullary cells expressing the mu-opioid receptor. J Neurosci 21: 5281–5288.

Porreca F, Ossipov MH, Gebhart GF (2002). Chronic pain and medullary descending facilitation. Trends Neurosci 25: 319–325.

Price JL, Slotnick BM, Revial MF (1991). Olfactory projections to the hypothalamus. J Comp Neurol 306: 447–461.

Ramaswamy S, Wodehouse T, Langford R et al. (2019). Characterizing the somatosensory profile of patients with failed back surgery syndrome with unilateral lumbar radiculopathy undergoing spinal cord stimulation: a single center prospective pilot study. Neuromodulation 22: 333–340.

Ren K, Dubner R (2011). The role of trigeminal interpolaris-caudalis transition zone in persistent orofacial pain. Int Rev Neurobiol 97: 207–225.

Reynolds DV (1969). Surgery in the rat during electrical analgesia induced by focal brain stimulation. Science 164: 444–445.

Roby-Brami A, Bussel B, Willer JC et al. (1987). An electrophysiological investigation into the pain-relieving effects of heterotopic nociceptive stimuli. Probable involvement of a supraspinal loop. Brain 110: 1497–1508.

Rodriguez E, Sakurai K, Xu J et al. (2017). A craniofacial-specific monosynaptic circuit enables heightened affective pain. Nat Neurosci 20: 1734–1743.

Roeder Z, Chen Q, Davis S et al. (2016). Parabrachial complex links pain transmission to descending pain modulation. Pain 157: 2697–2708.

Roy M, Peretz I, Rainville P (2008). Emotional valence contributes to music-induced analgesia. Pain 134: 140–147.

Samineni VK, Grajales-Reyes JG, Copits BA et al. (2017a). Divergent modulation of nociception by glutamatergic and GABAergic neuronal subpopulations in the periaqueductal gray. eNeuro 4: ENEURO.0129-16.2017.

Samineni VK, Premkumar LS, Faingold CL (2017b). Neuropathic pain-induced enhancement of spontaneous and pain-evoked neuronal activity in the periaqueductal gray that is attenuated by gabapentin. Pain 158: 1241–1253.

Sasannejad P, Saeedi M, Shoeibi A et al. (2012). Lavender essential oil in the treatment of migraine headache: a placebo-controlled clinical trial. Eur Neurol 67: 288–291.

Scholz J, Broom DC, Youn DH et al. (2005). Blocking caspase activity prevents transsynaptic neuronal apoptosis and the loss of inhibition in lamina II of the dorsal horn after peripheral nerve injury. J Neurosci 25: 7317–7323.

Scott DJ, Stohler CS, Egnatuk CM et al. (2008). Placebo and nocebo effects are defined by opposite opioid and dopaminergic responses. Arch Gen Psychiatry 65: 220–231.

Sessle BJ, Dubner R, Greenwood LF et al. (1976). Descending influences of periaqueductal gray matter and somatosensory cerebral cortex on neurones in trigeminal brain stem nuclei. Can J Physiol Pharmacol 54: 66–69.

Sessle BJ, Hu JW (1981). Raphe-induced suppression of the jaw-opening reflex and single neurons in trigeminal subnucleus oralis, and influence of naloxone and subnucleus caudalis. Pain 10: 19–36.

Siddall PJ, Cousins MJ (2004). Persistent pain as a disease entity: implications for clinical management. Anesth Analg 99: 510–520.

Silva M, Amorim D, Almeida A et al. (2013). Pronociceptive changes in the activity of rostroventromedial medulla (RVM) pain modulatory cells in the streptozotocin-diabetic rat. Brain Res Bull 96: 39–44.

Sims-Williams H, Matthews JC, Talbot PS et al. (2017). Deep brain stimulation of the periaqueductal gray releases endogenous opioids in humans. Neuroimage 146: 833–842.

Soliman N, Haroutounian S, Hohmann AG et al. (2021). Systematic review and meta-analysis of cannabinoids, cannabis-based medicines, and endocannabinoid system modulators tested for antinociceptive effects in animal models of injury-related or pathological persistent pain. Pain 162: S26–S44.

Sotgiu ML, Valente M, Storchi R et al. (2008). Contribution by DRt descending facilitatory pathways to maintenance of spinal neuron sensitization in rats. Brain Res 1188: 69–75.

Stroman PW, Ioachim G, Powers JM et al. (2018). Pain processing in the human brainstem and spinal cord before, during, and after the application of noxious heat stimuli. Pain 159: 2012–2020.

Sugiyo S, Takemura M, Dubner R et al. (2005). Trigeminal transition zone/rostral ventromedial medulla connections and facilitation of orofacial hyperalgesia after masseter inflammation in rats. J Comp Neurol 493: 510–523.

Tinoco Mendoza FA, Hughes TES, Robertson RV et al. (2023). Detailed organisation of the human midbrain periaqueductal grey revealed using ultra-high field magnetic resonance imaging. Neuroimage 266: 119828.

Tu Y, Wilson G, Camprodon J et al. (2021). Manipulating placebo analgesia and nocebo hyperalgesia by changing brain excitability. Proc Natl Acad Sci U S A 118: e2101273118.

Tuttle AH, Tohyama S, Ramsay T et al. (2015). Increasing placebo responses over time in U.S. clinical trials of neuropathic pain. Pain 156: 2616–2626.

Tuveson B, Leffler AS, Hansson P (2009). Influence of heterotopic noxious conditioning stimulation on spontaneous pain and dynamic mechanical allodynia in central post-stroke pain patients. Pain 143: 84–91.

Uchiyama S, Yoshihara K, Kawanabe R et al. (2022). Stress-induced antinociception to noxious heat requires alpha(1A)-adrenaline receptors of spinal inhibitory neurons in mice. Mol Brain 15: 6.

Vachon-Presseau E, Berger SE, Abdullah TB et al. (2018). Brain and psychological determinants of placebo pill response in chronic pain patients. Nat Commun 9: 3397.

Vakilian K, Atarha M, Bekhradi R et al. (2011). Healing advantages of lavender essential oil during episiotomy recovery: a clinical trial. Complement Ther Clin Pract 17: 50–53.

Vaughan CW, Ingram SL, Connor MA et al. (1997). How opioids inhibit GABA-mediated neurotransmission. Nature 390: 611–614.

Viisanen H, Pertovaara A (2007). Influence of peripheral nerve injury on response properties of locus coeruleus neurons and coeruleospinal antinociception in the rat. Neuroscience 146: 1785–1794.

Villanueva L, Bouhassira D, Le Bars D (1996). The medullary subnucleus reticularis dorsalis (SRD) as a key link in both the transmission and modulation of pain signals. Pain 67: 231–240.

Villemure C, Slotnick BM, Bushnell MC (2003). Effects of odors on pain perception: deciphering the roles of emotion and attention. Pain 106: 101–108.

Wager TD, Rilling JK, Smith EE et al. (2004). Placebo-induced changes in FMRI in the anticipation and experience of pain. Science 303: 1162–1167.

Wager TD, Scott DJ, Zubieta JK (2007). Placebo effects on human mu-opioid activity during pain. Proc Natl Acad Sci U S A 104: 11056–11061.

Wall PD (1979). On the relation of injury to pain. The John J. Bonica lecture. Pain 6: 253–264.

Wampold BE, Imel ZE, Minami T (2007). The story of placebo effects in medicine: evidence in context. J Clin Psychol 63: 379–390discussion 405-8.

Wang J, Eslinger PJ, Smith MB et al. (2005). Functional magnetic resonance imaging study of human olfaction and normal aging. J Gerontol A Biol Sci Med Sci 60: 510–514.

Wang R, King T, De Felice M et al. (2013). Descending facilitation maintains long-term spontaneous neuropathic pain. J Pain 14: 845–853.

Wang Y, Chan E, Dorsey SG et al. (2022). Who are the placebo responders? A cross-sectional cohort study for psychological determinants. Pain 163: 1078–1090.

Waters AJ, Lumb BM (1997). Inhibitory effects evoked from both the lateral and ventrolateral periaqueductal grey are selective for the nociceptive responses of rat dorsal horn neurones. Brain Res 752: 239–249.

Watkins LR, Johannessen JN, Kinscheck IB et al. (1984). The neurochemical basis of footshock analgesia: the role of spinal cord serotonin and norepinephrine. Brain Res 290: 107–117.

Wei F, Dubner R, Ren K (1999). Nucleus reticularis gigantocellularis and nucleus raphe magnus in the brain stem exert opposite effects on behavioral hyperalgesia and spinal Fos protein expression after peripheral inflammation. Pain 80: 127–141.

Whalley B, Hyland ME, Kirsch I (2008). Consistency of the placebo effect. J Psychosom Res 64: 537–541.

Wiedenmayer CP, Barr GA (2000). Mu opioid receptors in the ventrolateral periaqueductal gray mediate stress-induced analgesia but not immobility in rat pups. Behav Neurosci 114: 125–136.

Wilcox SL, Gustin SM, Macey PM et al. (2015a). Anatomical changes at the level of the primary synapse in neuropathic pain: evidence from the spinal trigeminal nucleus. J Neurosci 35: 2508–2515.

Wilcox SL, Gustin SM, Macey PM et al. (2015b). Anatomical changes within the medullary dorsal horn in chronic temporomandibular disorder pain. Neuroimage 117: 258–266.

Wilson-Poe AR, Morgan MM, Aicher SA et al. (2012). Distribution of CB1 cannabinoid receptors and their relationship with mu-opioid receptors in the rat periaqueductal gray. Neuroscience 213: 191–200.

Wilson-Poe AR, Pocius E, Herschbach M et al. (2013). The periaqueductal gray contributes to bidirectional enhancement of antinociception between morphine and cannabinoids. Pharmacol Biochem Behav 103: 444–449.

Witting N, Svensson P, Jensen TS (2003). Differential recruitment of endogenous pain inhibitory systems in neuropathic pain patients. Pain 103: 75–81.

Wodehouse T, Poply K, Ramaswamy S et al. (2018). A pilot study investigating whether quantitative sensory testing alters after treatment in patients with fibromyalgia. Br J Pain 12: 250–256.

Yarnitsky D (2010). Conditioned pain modulation (the diffuse noxious inhibitory control-like effect): its relevance for acute and chronic pain states. Curr Opin Anaesthesiol 23: 611–615.

Yarnitsky D, Granot M, Nahman-Averbuch H et al. (2012). Conditioned pain modulation predicts duloxetine efficacy in painful diabetic neuropathy. Pain 153: 1193–1198.

Youssef AM, Macefield VG, Henderson LA (2016a). Cortical influences on brainstem circuitry responsible for conditioned pain modulation in humans. Hum Brain Mapp 37: 2630–2644.

Youssef AM, Macefield VG, Henderson LA (2016b). Pain inhibits pain; human brainstem mechanisms. Neuroimage 124: 54–62.

Zhuo M, Gebhart GF (1992). Characterization of descending facilitation and inhibition of spinal nociceptive transmission from the nuclei reticularis gigantocellularis and gigantocellularis pars alpha in the rat. J Neurophysiol 67: 1599–1614.

Zhuo M, Gebhart GF (1997). Biphasic modulation of spinal nociceptive transmission from the medullary raphe nuclei in the rat. J Neurophysiol 78: 746–758.

Zubieta JK, Bueller JA, Jackson LR et al. (2005). Placebo effects mediated by endogenous opioid activity on mu-opioid receptors. J Neurosci 25: 7754–7762.

Handbook of Clinical Neurology, Vol. 216 (3rd series)
The Brainstem
L.T. Grinberg and A.D. Korczyn
https://doi.org/10.1016/B978-0-443-15736-3.00018-4

Chapter 5

Brainstem control of sleep, wakefulness, alertness, and consciousness

AMOS D. KORCZYN[1,*] AND LEA TENENHOLZ GRINBERG[2]

[1]*Departments of Neurology, Physiology and Pharmacology, Tel Aviv University, Tel Aviv, Israel*

[2]*Departments of Laboratory Medicine, Pathology and Neurosciences, Mayo Clinic, Jacksonville, FL, United States*

Abstract

The brainstem is crucial for maintaining wakefulness, sleep regulation, and consciousness. It orchestrates alertness through the ascending reticular activating system, integrating sensory information and projecting to the cortex. Consciousness emerges from intricate interactions involving the brainstem, though the brainstem alone does not generate consciousness. Clinical syndromes resulting from brainstem lesions highlight distinctions between wakefulness, alertness, and consciousness. Disorders such as sleep paralysis, parasomnias, vegetative state, minimally conscious state, coma, and locked-in syndrome illustrate how brainstem dysfunction affects consciousness and behavior. Unique states like lucid dreaming and hypnagogia further demonstrate brainstem contributions, revealing a distinct form of internally directed consciousness ("consciousness II"). Understanding these phenomena provides insights into the brainstem's essential role in the complex neural network underlying human consciousness and cognitive function.

The brainstem, strategically situated at the base of the brain, plays a vital role in regulating many essential bodily functions, including swallowing, breathing, and cardiovascular control. These functions are carried out in coordination with other parts of the central nervous system. This chapter focuses specifically on the brainstem's involvement in maintaining alertness, regulating the sleep–wake cycle, and modulating levels of consciousness. While the brainstem does not directly generate consciousness, which emerges from the integrated activity of the entire central nervous system, it is indispensable in supporting the conditions necessary for consciousness to arise. Damage to various brainstem regions, as well as to other brain areas, can impair these functions and alter conscious states. Due to the brainstem's compact and intricate anatomy, identifying the precise nuclei or pathways underlying higher functions remains a major challenge. While animal studies have provided valuable insights into the brainstem's role in sleep regulation, they fall short in capturing the complexity of human alertness and attention. As a result, much of our current understanding is derived from clinical observations in humans, though these, too, are limited. Brain lesions rarely affect a single structure in isolation, and the underlying pathology is often incompletely understood. In this review, we examine the brainstem's contributions to wakefulness and alertness by analyzing relevant clinical syndromes. Observations from patients with altered states of consciousness, often due to trauma or stroke, highlight that wakefulness, alertness, and consciousness, though frequently conflated, are in fact distinct processes that can become dissociated.

SLEEP AND WAKEFULNESS

Sleep is a vital physiologic state marked by reduced sensory responsiveness and disconnection from the environment. It supports essential processes such as memory consolidation and cellular repair. Sleep is actively regulated, cycling through non- rapid eye movement (NREM) and REM stages.

The brainstem is crucial for sleep regulation, coordinating inhibitory and excitatory pathways. Key sleep-promoting regions in the brainstem, such as the medullary and pontine reticular formation, interact with hypothalamic and thalamic structures. During sleep, activity in wake-promoting neurotransmitters decreases, while gamma-aminobutyric acidergic and galaninergic neurons increase, enabling sleep. Wakefulness is a state of sustained cortical activation and environmental engagement, necessary for perception, learning, and adaptive behavior. The brainstem, particularly the ascending reticular activating system (ARAS), is essential for maintaining

*Correspondence to: Amos D. Korczyn, MD, MSc, Departments of Neurology, Physiology and Pharmacology, Tel Aviv University Medical School, Tel Aviv University, P.O. Box 39040, Tel Aviv 6997801, Israel. Tel: +972524266726, E-mail: Amoskor@tauex.tau.ac.il.

wakefulness. The ARAS receives input from sensory pathways and projects to the thalamus and cortex, supporting vigilance and responsiveness. Key brainstem nuclei for wakefulness include the locus coeruleus (noradrenergic), dorsal and median raphe (serotonergic and cholinergic), pedunculopontine and laterodorsal tegmental nuclei (Brown et al., 2012; Lew et al., 2021).

ALERTNESS

Alertness is a fundamental aspect of consciousness, referring to a state of heightened sensory awareness and responsiveness to external stimuli. It reflects the brain's capacity to maintain wakefulness and sustain attention, serving as a prerequisite for higher-order cognitive processes such as perception, reasoning, and decision-making. The brainstem plays a pivotal role in maintaining alertness and overall wakefulness, both of which are essential components of consciousness. Central to this function is the arousal system, particularly the ARAS.

Anatomically, the ARAS integrates afferent sensory information from the body, receiving inputs from somatosensory pathways, auditory signals via cranial nerves, and visual stimuli. This broad input allows the system to respond dynamically to changes in the environment. Once activated, the RAS projects to the thalamus and cerebral cortex, promoting sustained alertness and awareness (Petersen and Posner, 2012). Several key brainstem nuclei contribute to the regulation of wakefulness and alert states including the locus coeruleus, the raphe nuclei, which produce serotonin, and the pedunculopontine nucleus, which is cholinergic and influences REM sleep and alertness.

AWARENESS AND CONSCIOUSNESS

Consciousness refers to the state of being aware of, and able to think about, our surroundings and ourselves (Zeman, 2002). In fact, "consciousness" and "awareness," as such, do not exist. Rather, these terms refer to the ability of the organism to be conscious of, or aware of, something, either in the environment or internally, or in other words, refer to direct attention to a target. At its core, consciousness involves the integration of sensory information, emotional responses, and cognitive processes, allowing an individual to navigate and interact with its environment. In humans, consciousness refers, in addition, to our ability to think of our own existence (Tononi and Koch, 2015). Human consciousness includes a range of mental states, from full to partial alertness and self-awareness, to other states such as sleep and dreaming, or coma. The existence of consciousness in another being at a given time is concluded from the reaction of this organism. However, if there is no reaction, we cannot tell whether consciousness exists in this organism. On the other hand, the existence of a reaction itself does not necessarily signify the existence of mental awareness. The human condition is unique because humans can report their feelings and internal reactions, and of course we are aware of our own subjective perception, feelings, and reactions. Thus, all discussion of consciousness in nonhumans is anthropomorphization, i.e., purely speculative and nonscientific and shall not be dealt with here. As humans, we are not only conscious but have meta-consciousness, i.e., we are able to think about the fact that we are conscious and discuss it. This unique ability, presumably cortical, is lost during somnolence and sleep. In essence, we assume the existence of meta-consciousness to be solely human (Seth and Bayne, 2022). The study of consciousness includes various disciplines: neuroscience, psychology, philosophy, and cognitive science, each contributing to a broader view of how consciousness arises from brain activity and how it manifests in behavior and thought. Sleep provides a unique opportunity to study the complex contribution of the brainstem to these functions.

CONSCIOUSNESS AND SLEEP

By definition, alertness is lost when falling asleep, but other brainstem functions, like motor activity and reactions, may be maintained. Consciousness is a dynamic and multifaceted phenomenon that changes throughout the different stages of sleep. It is a condition of the person being awake and alert, in voluntary control of hirself and hir activities, and able to direct one's attention to a target and later recall the thoughts and happenings. During sleep, consciousness of the environment is lost, and awareness of, and reaction to, environmental stimuli is minimal. During drowsiness and sleep, dreams appear, and attention is directed toward them without being aware of their nonreality. In other words, the person is conscious of the dream's content, although not able to judge their reality. Thus, a different aspect of consciousness appears, which we may call consciousness II. Its expressions are not subject to critical cortical analysis and judgment of reality, and usually not entered into the memory store. Consciousness II is a condition of the person being asleep or at least drowsy, i.e., not in full control of hir movements, yet with spontaneous thoughts (dreams) which are not under voluntary control, and which are mostly not recalled upon awakenings (Table 5.1). The cortical function of meta-consciousness is absent, as it is in all other stages of sleep. The brainstem plays a critical role in modulating physiologic states, particularly in controlling the delicate balance between wakefulness, sleep, and dreaming. Understanding how consciousness manifests during sleep

Table 5.1

Relation of consciousness to different clinical states.

	Consciousness	Consciousness II
Normal awake	Exists	Absent
Dreamless sleep	Absent	Absent
Dreams	Absent	Exists
Lucid dreams	Partly exists	Exists
Sleep terrors	Absent	Exists
Somnambulism	Minimal	Exists
RBD	Absent	Exists

provides deeper insight into the brain's functional architecture and reveals the complex interplay between sleep, cognition, and neural processing (Jellinger, 2009).

PARASOMNIAS AS A MODEL TO UNDERSTAND THE BRAINSTEM CONTRIBUTIONS TO CONSCIOUSNESS

During sleep, the brain responds primarily to internal stimuli, with limited responsiveness to the external environment. Simple actions, such as shifting position or waking to urinate, may be performed without conscious awareness or later recollection, suggesting that such acts are processed subcortically and are not encoded into cortical memory. In REM sleep, the brain generates vivid, often fantastical dreams that reflect a form of consciousness II. Dream content during this phase is typically an amalgam of emotions, memories, and imagination. While dreamers may experience a fleeting awareness of the dream's unreal nature, they generally lack meta-consciousness, the self-reflective capacity to recognize the dream as a product of their own mind or to exert control over it. This absence of meta-awareness is what prevents voluntary intervention in the dream narrative. In contrast, during wakefulness, we possess the ability to flexibly direct attention and consciously shift between internal and external targets, a capacity that is notably diminished during sleep (Siclari, 2025). Brainstem pathology underlies many sleep disorders in humans, including parasomnias. Parasomnias are sleep disorders characterized by abnormal behaviors, experiences, or movements during sleep, sleep–wake transitions, or arousals from sleep (Howell, 2012). These sleep disorders can provide insight into the neuronal basis of consciousness.

The hypnagogic state is a normal transitional phase between wakefulness and sleep, often referred to as the presleep or sleep onset period. It is particularly relevant in discussions of consciousness because it illustrates a unique hybrid state: although the person is not fully asleep, there is diminished awareness of external stimuli, replaced by vivid sensory hallucinations, including visual patterns, auditory snippets, or tactile sensations. Despite the presence of hallucinations, individuals in the hypnagogic state typically remain calm and undisturbed, neither startled by these internal events nor fully conscious of their surroundings. In this context, consciousness is preserved, but directed inwardly, much like during REM sleep, qualifying it as an instance of "consciousness II," where awareness is internally focused but detached from reality. Although the brainstem, through its control of arousal and alertness, likely plays a role in regulating this state, the exact anatomic underpinnings remain poorly defined in current literature. One area of active research involves the dorsolateral pontine tegmentum, particularly structures like the subcoeruleus nucleus and locus coeruleus, which are known to regulate REM transitions and dream-related atonia. These same structures may transiently activate or deactivate in the hypnagogic phase, contributing to the shift in conscious focus (Mahowald and Schenck, 2005).

Sleep paralysis is a striking example of neural dissociation, where components of REM sleep and wakefulness coexist abnormally. It occurs when a person regains conscious awareness while the body remains in a state of REM atonia, leading to a temporary inability to move or speak. This state often includes intense hallucinations, frequently involving a sense of presence, pressure on the chest, or even auditory or visual illusions. This condition offers a unique window into the functional architecture of consciousness, particularly how distinct brain systems involved in motor control, sensory processing, alertness, and emotion may activate out of sync. The brainstem's REM-on and REM-off nuclei, especially in the pons, fail to deactivate properly, allowing the motor suppression mechanisms (especially those involving glycinergic inhibition in the medulla) to persist despite the reactivation of cortical awareness. Sleep paralysis also highlights that the brainstem structures do not always operate in synchrony. The pedunculopontine nucleus, subcoeruleus area, and parabrachial complex are particularly important in this regulatory failure. Emotional responses during this state, including panic or fear, may be heightened due to amygdala activation in the absence of full cognitive control.

Sleep Terrors: These are episodes of intense fear, screaming, and physical thrashing that occur during deep non-REM sleep, and they are often accompanied by vivid dreams. These dreams likely reflect increased brain activity, though people typically have no conscious memory of them upon waking. The brainstem appears to play a key role by enabling just enough cortical activity for dream production, while limiting motor activity and reality testing. In this context, we once again see a form of "consciousness II." Notably, studies have shown that adrenergic activity can occur before motor arousal in sleep terror episodes, suggesting that changes in autonomic regulation, centered in the brainstem, may be crucial in triggering these events (Ledard et al., 2020).

Sleepwalking (somnambulism) is a phenomenon occurring during deep NREM sleep. The brainstem plays a significant role in the pathophysiology of somnambulism, primarily through its involvement in regulating arousal and motor control during NREM sleep. Sleepwalking occurs due to incomplete arousal from deep NREM sleep (typically stage 3), leading to a dissociation between wakefulness and sleep states. Sleepwalking involves performing complex behavior, thus being an example of the interaction between awareness and consciousness. During sleepwalking, the subjects are only minimally aware of their actions and of the environment (and thus can walk around without hitting objects) and have almost no memory of the event upon waking. This phenomenon illustrates that while consciousness is significantly reduced during deep sleep, the brain, presumably the cortex, can still perform complex motor tasks, in which the brainstem participates. We know nothing about the subjects' mental experience during the event; thus, this seems to be another example of consciousness–behavioral dissociation, caused by brainstem dysfunction. The brainstem hosts important NREM sleep centers like the parafacial zone (Anaclet and Fuller, 2017), and here is evidence of changes in respiratory and heart rate during sleepwalking suggesting autonomic dysregulation.

REM Sleep Behavior Disorder (RBD): RBD is a physiologic condition occurring during deep sleep in which the muscle atonia, that typically characterizes REM sleep, is absent, allowing individuals to physically act out their dreams (Dauvilliers et al., 2018). This functional disorder highlights the brainstem's crucial role in allowing consciousness and normal motor activity during sleep. Individuals with RBD may be partially aware of their actions within the dream, yet they are unaware that they are acting them out in reality. In other words, their sense of reality is absent. The experience thus belongs to consciousness II.

CHANGES IN CONSCIOUSNESS STATE AS A MODEL TO UNDERSTAND THE BRAINSTEM CONTRIBUTIONS TO CONSCIOUSNESS

One particularly intriguing phenomenon is that of lucid dreaming, a state in which the sleeper becomes aware that they are dreaming yet continues to remain in the dream without being able to awaken voluntarily. In this condition, the individual remains asleep and largely disconnected from their external surroundings, yet a form of secondary consciousness ("consciousness II") emerges, directing attention toward the dream's content. Unlike typical dreaming, the person is at least partially aware of being asleep and dreaming, even if control over the dream remains limited. Several brainstem structures have been implicated in maintaining this neurophysiologic balance. In particular, pontine regions, such as the subcoeruleus area and the sublaterodorsal tegmental nucleus, play essential roles in initiating and sustaining sleep stages. These nuclei are involved in producing the muscle atonia characteristic of REM sleep, effectively paralyzing the body to prevent the physical enactment of dream content.

Vegetative State (VS) is a condition which follows severe brainstem damage, in which a person is mostly immobile and unconscious but still maintains basic life functions such as breathing and circulation. Patients in a VS appear at times to be awake; they may open their eyes spontaneously, or in response to stimulation, show facial expressions, or have periods of apparent alertness during the day. However, these behaviors do not reflect conscious awareness or purposeful interaction with their surroundings. Thus, the concept of "alertness" is different from how it is usually described, i.e., the person is not "alert" in the sense of being aware of themselves or their environment and is not attentive. This situation is frequently referred to as wakefulness without awareness, again demonstrating separate effects of the brainstem on alertness, motor control, and awareness.

A Minimally Conscious State is a condition of severely altered consciousness in which a person demonstrates minimal, inconsistent but reproducible responses to stimuli, such as following simple commands, gesturing, and probably implying a degree of self-awareness that is not present in VS. However, it is not clear whether these movements are voluntary. Like in VS, this tragic situation results from traumatic brain injury, stroke, or other conditions that cause severe damage to the brainstem, but less severe than in VS. These behaviors suggest a level of alertness and cognitive function that goes beyond reflexive actions, but not of consciousness of any kind.

Coma. A coma is a state of deep unresponsiveness, usually due to very severe bilateral hemispheric or brainstem injury. The severity of the coma depends on the extent of the damage. Most cases of coma are due to mechanical or vascular lesions, but it may also be the consequence of metabolic impairments, such as hypoglycemia or toxic effects, for example, barbiturate or opiate overdose. In these latter cases, the prognosis may be good, with possible complete recovery, which is never the case in traumatic or vascular damage to the brain, which may improve to some degree but never to complete resolution.

Locked-in syndrome. This rare and tragic condition occurs when the pyramidal tracts, leading from the motor cortex to the brainstem motor neurons and the spinal cord, are injured, frequently in the brainstem. This cognitive-motor dissociation causes the patient to be awake and conscious yet completely paralyzed.

CONCLUSIONS

The brainstem has a critical role in maintaining wakefulness and alertness, which are essential for conscious awareness. The study of the involvement of the brainstem in regulating consciousness is difficult because of the limitation of observations in humans, who are the only beings that can report on their experiences. We depend on the subjective reports of the individuals, the reliability of which cannot be evaluated. This is emphasized by the fact that consciousness is a product of the activity of the whole brain, with different structures interacting with each other. Thus, the brainstem is only a part of the brain's very complex machinery that underlies consciousness. Its specific contribution is in its control of alertness and sleep, both of which are determinants of the ability of the organism to be able to receive information, store it, and react to it. However, the different clinical manifestations described above illustrate the effects of the brainstem on the motor system and alertness, thus influencing cognition and attention to inner or outer physical and mental phenomena. A peculiar situation occurs during dreams, where consciousness is limited and different in many aspects from the consciousness which we are experiencing during wakefulness (Table 5.2). During dream sleep, there is dissociation between alertness and motor activity, which are suppressed, and the mental experience, which is active but not subjected to voluntary manipulation. Thus, the consciousness then (consciousness II) is unlike the normal consciousness which is manifested during wakefulness. The various clinical manifestations described above demonstrate how the brainstem can affect wakefulness, alertness, motor activity, and consciousness in many ways.

Table 5.2

Clinical manifestations in different wakefulness states.

	Appearance	Voluntary movements	Involuntary movements	Ability to inhibit movements
Wakefulness	Alert	Exists	Exists	Exists
Normal deep sleep	Relaxed	None	Minimal	None
Dreaming	Variable	None	Sporadic	None
Sleep paralysis	Asleep	None	None	None

References

Anaclet C, Fuller PM (2017). Brainstem regulation of slow-wave-sleep. Curr Opin Neurobiol 44: 139–143.

Brown RE, Basheer R, McKenna JT et al. (2012). Control of sleep and wakefulness. Physiol Rev 92: 1087–1187.

Dauvilliers Y, Schenck CH, Postuma RB et al. (2018). REM sleep behaviour disorder. Nat Rev Dis Primers 4: 19.

Howell MJ (2012). Parasomnias: an updated review. Neurotherapeutics 9: 753–775.

Jellinger KA (2009). Functional pathophysiology of consciousness. Neuropsychiatrie 23: 115–133.

Ledard N, Artru E, Colmenarez Sayago P et al. (2020). Adrenergic reactions during N3 sleep arousals in sleepwalking and sleep terrors: the chicken or the egg? J Sleep Res 29: e12946.

Lew CH, Petersen C, Neylan TC et al. (2021). Tau-driven degeneration of sleep- and wake-regulating neurons in Alzheimer's disease. Sleep Med Rev 60: 101541.

Mahowald MW, Schenck CH (2005). Insights from studying human sleep disorders. Nature 437: 1279–1285.

Petersen SE, Posner MI (2012). The attention system of the human brain: 20 years after. Annu Rev Neurosci 35: 73–89.

Seth AK, Bayne T (2022). Theories of consciousness. Nat Rev Neurosci 23: 439–452.

Siclari F (2025). Consciousness in non-REM-parasomnia episodes. J Sleep Res 34: e14275.

Tononi G, Koch C (2015). Consciousness: here, there and everywhere? Philos Trans R Soc Lond B Biol Sci 370: 20140167.

Zeman A (2002). Consciousness: a user's guide. Yale University Press.

Section II

Pathology

Handbook of Clinical Neurology, Vol. 216 (3rd series)
The Brainstem
L.T. Grinberg and A.D. Korczyn
DOI: https://doi.org/10.1016/B978-0-443-15736-3.00012-3

Chapter 6

Locked-in syndrome

LOUIS R. CAPLAN* AND IKRAM ASAD

Department of Neurology, Beth Israel Deaconess Medical Center, Boston, MA, United States

Abstract

The prerequisites for the locked-in stste are the presence of two components: one negative-sufficient paralysis of structures that render usual communications (mouth and oral apparatus for speech, and limbs for writing and pointing accurately) and one positive- preservation of consciousness. Lesions of the pontine base cause the syndrome. The most common pontine lesions are infarction related to basilar artery occlusion and pontine hemorrhages. Occasionally midbrain vascular lesions can cause the syndrome.

INTRODUCTION

The *Locked-in syndrome* (LIS) was first called to medical attention By Plum and Posner in 1966. They defined the syndrome as consisting of quadriplegia, lower cranial nerve paralysis, and mutism with preservation of consciousness, vertical gaze, and upper eyelid movement (Plum and Posner, 1972). Other terms that sometimes have been applied are *deefferented* state (Laureys et al., 2005), *pseudocoma*, and *coma vigilante* (Chisholm and Gillett, 2005).

An early description of such an individual is contained in Alexandre Dumas' famous and very popular novel *The Count of Monte Cristo* (Dumas, 1998; Williams, 2003). Dumas describes a stroke that occurred to Monsieur Noirtier de Villefort, who lost the ability to speak or use his limbs. Dumas describes the condition eloquently as "when the soul is trapped in a body that no longer obeys its commands." Although his intellectual faculties are intact, he is physically powerless: *Sight and hearing were the only two senses which, like two sparks, still lit up this human matter, already three quarters moulded for the tomb. Moreover, only one of these two senses could reveal to the outside world the inner life, which animated this statue... He was a corpse with living eyes, and at times, nothing could be more terrifying than this marble face out of which anger burned, or joy shone* (Laureys et al., 2005). Noirtier de Villeforte was able to communicate through blinking, closing his eyes once for "yes" and several times for "no." His granddaughter Valentine used a dictionary to help him spell out letter by letter what he wanted to say (Dumas, 1998; Williams, 2003). A movie in 2007, *The diving bell and the butterfly* concerned an individual with the LIS and further brought it to the attention of the public.

The prerequisites for the LIS are the presence of two components: one negative-sufficient paralysis of structures that render usual communications (mouth and oral apparatus for speech, and limbs for writing and pointing accurately) and one positive- preservation of consciousness. These two prerequisites are most often found in central nervous system brain lesions that involve the brainstem. The oral speech producing structures are innervated by bulbar nuclei and supranuclear pathways in the brainstem. These structures are located within the rostral midbrain and pontine tegmentum. Descending motor pathways that control voluntary motion travel through the basal portions of the midbrain and pons. The commonest location of a lesion that produces paralysis of bulbar motor control and bilateral limb paralysis involves the ventral midbrain and pons on both sides. Consciousness requires preservation of structures in the tegmentum of the brainstem mostly the rostral pons that project to the two cerebral hemispheres, and preservation of at least one of the cerebral hemispheres.

Peripheral nervous system lesions could cause the locked-in state by paralyzing bulbar and limb muscles in patients that remain conscious and sensate. These are rare. Myasthenia

*Correspondence to: Louis R. Caplan, MD, Professor of Neurology, Harvard University, Neurologist, Beth Israel Deaconess Medical Center Boston, 330 Brookline Ave, Boston, MA 02215, United States. Tel: +1-617-680-7573; Fax: +1-617-632-8920; E-mail: lcaplan@bidmc.harvard.edu.

gravis often affects bulbar muscles but rarely produces very severe limb paralysis. Guillain–Barré syndrome often leads to limb paralysis but rarely involves bulbar muscles to prevent oral communication. Motor neuron disease could paralyze bulbar and limb muscles but progresses slowly and would not be a diagnostic problem. These entities will not be covered in this chapter which focuses on the brainstem.

LIS is a devastating condition. It is characterized by paralysis of all voluntary muscles in the body apart from eye movements and blinking. This condition is particularly challenging to diagnose. Patients are often misdiagnosed as being in a vegetative state. Unlike coma and the vegetative state, patients with LIS are fully conscious and aware of their surroundings, but unable to communicate or move. The LIS presents a unique set of challenges for neurologists, who must navigate the complex clinical, ethical, and emotional issues that arise in caring for these patients. In addition to providing medical care, neurologists must also work closely with patients and their families to ensure that they receive the support they need to cope with the profound physical and emotional effects of the condition.

RELEVANT BRAINSTEM ANATOMY

Although the anatomy of the nuclear and supranuclear motor pathways has been known for centuries, the brainstem structures that relate to consciousness is more recent. Moruzzi and Magoun are usually credited for describing the "reticular activating system" (RAS) in the brainstem (Moruzzi and Magoun, 1995). The RAS as it was described was a widespread region that traversed the brainstem that was composed of many nuclei and pathways. More recent experiments performed in rodents found that the RAS was subdivided into functionally discrete regions, and that injury to a region in the pontine tegmentum could disrupt arousal and cause coma (Fuller et al., 2011; Fischer et al., 2016). Parvizi and Damasio analyzed the magnetic resonance imagings (MRIs) of human comatose patients and found that the tegmental lesions were mostly bilateral and were located either in the pons alone or in the upper pons and the midbrain (Parvizi and Damasio, 2003). They posited that pontine tegmental lesions alone could cause coma but did not designate a single discrete area within the pontine tegmentum as critical (Parvizi and Damasio, 2003). A further study compared 12 coma-causing brainstem lesions to 24 control brainstem lesions to attempt to identify a site associated with coma (Fischer et al., 2016). These authors used resting-state functional connectivity to identify a network of regions functionally connected to this brainstem site. Injury to a small region in the left pontine tegmentum near the medial parabrachial nucleus was significantly associated with coma. This site is functionally connected to two cortical regions, the ventral anterior insula and the pregenual anterior cingulate cortex that become disconnected in disorders of consciousness. They posited that the left pontine tegmentum and the cerebral regions that it projects to are very important in maintaining the conscious state (Fischer et al., 2016). Clearly, bilateral pontine tegmental lesions and probably unilateral left pontine lesions can result in coma. Preservation of the left pontine tegmentum is likely to be essential in retaining consciousness in LIS patients who have lesions that cause bulbar and limb paralysis.

VASCULAR LESIONS ASSOCIATED WITH THE LOCKED-IN STATE

The two most common brainstem lesions that lead to a locked-in state are infarcts and hemorrhages.

Brainstem infarctions

Infarctions that localize to both sides of the pons are almost always due to occlusive lesions within the basilar artery or to embolism to the basilar artery (Kim, 2011; Caplan, 2014a). Occasionally, spontaneous or traumatic dissection of the basilar artery causes the infarction (Ruecker et al., 2010; Kim et al., 2011; Caplan, 2014a). Basilar artery occlusions that involve the proximal and/or middle portions of the artery are most likely to cause a large paramedian pontine infarct. When the basilar artery occludes, flow reduction is usually maximal in the paramedian branches of the artery that supply the paramedian basis pontis and the paramedian pontine tegmentum. Collateral circulation develops quickly in patients with basilar artery occlusive disease, especially when the stenosis develops gradually. Sudden occlusions that accompany embolism puts more stress on the vasculature to adapt. The collateral circulation supplies mostly the lateral regions of the brainstem. Collaterals come mainly from the intracranial vertebral arteries. Blood often goes from one or both intracranial vertebral arteries to their posterior inferior cerebellar artery branches; then blood courses around the brainstem through and around the cerebellum. Anastamoses develop with the anterior inferior and superior cerebellar arteries. Blood travels retrograde toward the basilar artery and fills the basilar artery above the region of thromboembolism. Another prominent collateral pathway is from the internal carotid artery to its posterior communicating artery branch into the posterior cerebral arteries and down into the rostral basilar artery above the occlusion.

Brainstem infarcts can be heralded by abnormal body movements often falsely attributed to convulsions. The movements are very variable and can include small twitches, fasciculation-like very small movements, and larger motions such as shivering, shuddering, jerking, or a tremulous, like intermittent shaking movement. At times these movements are intermittent. Sometimes movement of the limbs or painful stimuli can precipitate a flurry of abnormal movements. The limbs are paralyzed and may be hypotonic at onset. Plantar responses are extensor. Often pain and temperature sensation is spared because the cortico-spinal pain and temperature tracts course laterally in the brainstem. Position and vibratory sense can be affected when they are testable. Inability to voluntarily influence the depth and rate of respirations also develops in patients with basis pontis infarcts.

Weakness of bulbar muscles is also very common and a very important cause of morbidity and mortality. The face, pharynx, larynx, and tongue are most often involved. The pattern may be that of crossed motor loss, for example, one side of the face and the contralateral limbs but, more often the bulbar muscle weakness is bilateral. The bilateral involvement is usually due to involvement of corticobulbar fibers in the dorsal part of the basis pontis near the central tegmental tracts. Bulbar symptoms include facial weakness, dysphonia, dysarthria, dysphagia, and limited jaw movements. Some patients become totally unable to speak, open their mouth, protrude their tongue, swallow, or move parts of their face at will or on command. Secretions pool in the pharynx and aspiration is an important and serious complication. Occasionally the patient may "bulldog" firmly on a throat stick, making it difficult to open the mouth. This is explained by the jaw being held tightly shut by tonic masseter muscle contraction.

Basilar artery occlusions that involve the rostral portions of the basilar artery (most often embolic) cause infarcts mostly limited to the rostral pons, midbrain, and thalamus (Caplan, 2014a). When the basal regions of the midbrain are involved, including the cerebral peduncles, patients may have paralysis of their limbs. They are often hypersomnolent and may be difficult to arouse but invariable are not deeply comatose. These patients can be thought initially to be in a locked-in state, but further testing shows that they become awake and usually conversant. Some patients with midbrain infarcts have had a persistent LIS (Karp and Hurtig, 1974; Zakaria and Flaherty, 2006).

Brainstem hemorrhages

Hematomas that cause a locked-in state are characteristically centered at the tegmento-basal junction and usually extend dorsally into the tegmentum (Caplan, 2014b). These hematomas resulted from rupture of the large median arteries, which are much larger in diameter than the lateral and dorsal penetrating pontine artery branches. Hematomas are located predominantly in the midportion of the pons and commonly involve the tegmentum medially on one or both sides. The largest hematomas usually dissect dorsally and empty into the IVth ventricle. The tegmental lesions may be asymmetric. When both sides of the tegmentum are included in the bleeding, the patient is invariably comatose.

The most common clinical findings in patients with large paramedian pontine hemorrhages are decreased alertness often progressing to coma, loss of horizontal eye movements, small but reactive pupils, and quadriplegia. Among fatal cases, the incidence of stupor and coma is quite high. The medial pontine tegmentum must be injured bilaterally for coma to develop.

Cranial nerve abnormalities are invariably present. The corticobulbar pathways located near the tegmento-basal junction near the midline are destroyed by the hematomas whose epicenter lies in the region of these descending tracts. As a result, there is severe, usually bilateral, weakness of the face, palate, tongue, and pharynx. Testing of these muscles is often difficult, especially if the patient is stuporous or comatose. Manifestations of bulbar palsy are puffing out of the cheeks with expiration, pooling of secretions in the mouth and pharynx, and decreased eyelid tone. Occasional patients, whose early clinical signs are dominated by a hemiplegia, have a unilateral lower motor neuron-type facial palsy on the side opposite the hemiplegia. Other early symptoms and signs include decreased hearing and numbness of the face. Taste loss can often be found if tested (Goto et al., 1983). Respiratory abnormalities are also quite common

Motor signs are invariably present because of involvement of the basis pontis usually bilaterally. All patients with large paramedian pontine hematomas have bilaterally abnormal motor function. Quadriplegia is the most common abnormality. Some patients have a hemiplegia at onset but examination shows some weakness and hyperreflexia of the contralateral limbs. Most often the limbs are stiff and hyperreflexic and Babinski signs are present. Abnormal spontaneous motor movements are very common as they are in pontine infarcts. The abnormalities may be characterized as shivering, tremors, spontaneous stiffening of a limb or frank decorticate, or decerebrate posturing. Decerebrate movements of the arms and legs are quite common and may occur spontaneously or only after painful stimuli. Although the hematomas undoubtedly involve pontocerebellar fibers in the basis pontis, ataxia has not been described in patients with large paramedian pontine hematomas, probably because the severe weakness and stupor or coma preclude testing for ataxia.

Neuro-opthalmologic abnormalities are helpful in clinical localization. The pupils are usually small or pinpoint but retain reaction to light, in contrast to thalamic hematomas in which the pupils are also small but do not react normally to light. Anisocoria is common in patients with pontine hematomas. The eyes usually rest in the midline and are conjugate but one or both eyes can be deviated inward or outward at rest. Sometimes the eyes are skewed with one eye resting above the other in the vertical plane. The hematomas usually involve all of the major centers that relate to horizontal gaze including the VIth nerve nuclei, and the medial longitudinal fasciculus (MLF), and the paramedian pontine reticular formation (PPRF) bilaterally. Usually, horizontal gaze, both voluntary and reflex, is lost to both sides. In patients with asymmetric hematomas that involve the tegmentum on one side, a unilateral ipsilateral horizontal gaze paresis or a "one and a half" syndrome may be found. In a "one and a half" syndrome, there is an ipsilateral conjugate gaze paresis associated with adduction paresis to the contralateral side due to involvement of both the MLF and the PPRF on the same side. Spontaneous ocular bobbing also may be present. Bobbing is a term used to denote sudden brisk downward movement of the eyes followed by slower return to midposition (Fisher, 1964). Bobbing may be asymmetric in which case the downward excursion is usually in the eye with the more severe adduction paralysis. The downward eye movements are occasionally slow in which case the term "dipping" has been used instead of bobbing. These abnormal downward movements of the eyes can usually be elicited by ice-water caloric testing. Bobbing and dipping are explained by the preservation of vertical gaze control centers in the upper brainstem. Downward movements occur when horizontal motions are lost.

Hypertension is the most common cause of pontine hemorrhage. In patients with a bleeding diathesis, the hemorrhage occasionally involves the pons and can be large. Cavernomas can involve the pons but very rarely cause a locked-in state. Arteriovenous malformations are a very rare cause.

Secondary brainstem hemorrhages have also been reported to cause locked-in-like states. Zervas and Caplan reported the cases of two young women who had severe brain trauma causing subdural hematomas and marked increased in intracranial pressure (Caplan and Zervas, 1977). When the subdural hematomas were drained and measures were taken to reduce intracranial pressure, each patient had bilateral third nerve palsies and quadriparesis and were uncommunicative. When the arm and leg on one side began to move, the patients could accurately respond to queries and directions using their active limb to signal. Both patients survived. The cause of the temporary locked-in state was almost surely a secondary (Duret) hemorrhage in the midbrain. Henri Duret working in the lab of Charcot studied patients who died of head injuries and found that many had hematomas in the midbrain and pons. Later research by Duret (Duret, 1919) and by others (Attwater, 1911; Dill, 1939) showed that secondary hematomas in the pons and midbrain, now usually called Duret hemorrhages, were caused by sudden pressure elevations that distorted and compressed the brainstem.

The prevailing theory about the mechanism of Duret hemorrhages is that sudden increases of pressure in the supratentorial compartment cause compression of the brainstem often with elongation of the midbrain and tearing of midline vessels. Duret hemorrhages most often involve the midbrain. The hemorrhages are most often located in the midline usually below the aqueduct of Sylvius in the region of the IIIrd nerve nuclei and then spread in the midline toward the base. The midbrain is usually elongated and may be compressed from one side causing a Kernohan's notch in the region of the cerebral peduncle.

NONVASCULAR BRAINSTEM CAUSES OF LOCKED-IN SYNDROME

Infections can lead to abscess formation within the pons. Syphilitic tumors and tuberculomas can localize to the brainstem. Infection may cause pontine injury by infecting the basilar artery. This is most common in tuberculosis or fungal infection and very rare in bacterial infections.

Traumatic brain injury is listed as the second most common cause of LIS in some series after primary vascular causes (Leon-Carrion et al., 2002; Parvizi and Damasio, 2003; Laureys et al., 2005). Most of the traumatic brain injury-related instances of the LIS are due to trauma to vertebrobasilar arteries (Cabezudo et al., 1986; Schoeggl et al., 2001; Leon-Carrion et al., 2002) and only rarely due to direct compression, impingement, or contusion to the brainstem itself (Ahn and Aarabi, 2007).

Brain tumors that involve the brainstem can cause enough damage to the basis pontis and tegmentum to result in a locked-in state. These cases would develop gradually and would not be diagnostic puzzles. Pontine gliomas in children often involve the central portion of the pons. Other tumors rarely cause the syndrome, for example, fibrillary astrocytoma, pontine reticulum cell sarcoma, metastasis from an adenocarcinoma of the lung, and metastasis from a malignant melanoma have been reported in the literature as precipitating causes (Patterson and Grabois, 1986; Inci and Ozgen, 2003; Das, 2023).

Metabolic and demyelinating conditions can very rarely cause temporary or persistent locked-in states. The most likely condition is central pontine myelinolysis (Heckmann and Dinkel, 2013; Das, 2023). This condition invariably involves the central portion of the pons. It is most often found when hyponatremia is corrected aggressively. Multiple sclerosis (Forti et al., 1982; Keme-Ebi and Asindi, 2008; Das, 2023) and acute demyelinating encephalomyelitis may also involve the central pons but most often with other Central Nervous System (CNS) lesions.

DIFFERENT NOMENCLATURE OF LOCKED-IN SYNDROME

The classic form of LIS is described as complete body anesthesia and motor paralysis (quadriplegia and bulbar palsy) with preservation of blinking and vertical eye movement. This state is usually secondary to an insult to the brainstem with predominant damage to ventral pons or caudal ventral midbrain structures as described in the paragraphs above. Vertical eye movements, hearing, blinking, and cognition are often preserved.

Two other forms (Bauer et al., 1979; Lesenfants et al., 2018; Das, 2023) of LIS have been described: total LIS with complete body paralysis with loss of blinking and eye movements, with preservation of consciousness only seen on the electroencephalogram (Lesenfants et al., 2018). The last (third) form is similar to the classic form, except for preservation of a slight amount of motor function (Bauer et al., 1979).

DIFFERENTIAL DIAGNOSIS

The first step is to attempt to localize the lesion. Pupillary reflexes, eye movements, facial, jaw, pharyngeal and tongue motor function, limb postures and strength and response to stimuli, and deep tendon and plantar reflexes should be examined carefully. Brain imaging can usually confirm a brainstem lesion.

Other considerations are the *Persistent Vegetative State*: and the closely allied *Minimally Conscious State*. The pathology is usually bilateral cerebral hemisphere damage. The brainstem is rarely the site of additional lesions. Brainstem reflexes are preserved. In these patients, markedly reduced consciousness limits communication rather than paralysis of bulbar or limb muscles. These patients open their eyes and have preserved eye-opening but there is no awareness of their self or their surroundings. The limbs do not move well to commands but usually are not hyperreflexic. Respirations are intact.

In patients who are labeled *Akinetic Mutes*, there is preserved consciousness, with decreased motor response (without paralysis), and intact brainstem reflexes. The eye movements are intact; there is the slowness of speech, and greatly diminished ability to produce spontaneous speech. The lesions are usually in bilateral basal ganglia or thalamic regions. There is marked reduced spontaneity and delayed responses. The patients can move their limbs and speak but stimuli are needed to induce these responses. Brainstem functions are preserved.

Catatonic patients may initially be confused with those that are locked-in.

They often have had a previous history of psychiatric disease. There are no major lesions on brain imaging that would cause the decreased responsiveness. Consciousness and brainstem reflexes are preserved. Often the limbs are held stiffly and patients may be mute.

MANAGEMENT

Prognosis depends on the etiology of the lesion. At times ischemic regions clear allowing return of some motor function. Edema around hemorrhages may improve allowing better neurologic function. Initially, acute management of patients with the LIS is focused on securing a patent airway, maintaining adequate oxygenation, and stabilizing cardiac function.

The issues in caring for quadriparetic individuals are also pertinent to those who are locked-in. Attention to turning and skin care help prevent bedsores. Urinary and defecatory dysfunction demand attention. Passive limb movements are important in maintaining tone. Locked-in patients cannot swallow and require feeding usually through an inserted percutaneous endoscopic gastrostomy. Attention must also be directed at respiratory function and maintaining adequate oxygenation. Chest physiotherapy, breathing exercises, and positional changes are used to maintain adequate cardiovascular and respiratory function.

An effort should be made to establish some effective communication. Often this is by a retained movement of the face or limb, for example, blinking, moving the eyes up or down, a flicker of a finger, etc. Recent research by engineers has importantly advanced potential communication strategies. Newer technology includes infrared eye movement sensors and computer-modulated voice prosthetics to improve communication (Doble et al., 2003).

References

Ahn ES, Aarabi B (2007). Posttraumatic locked-in syndrome from a pontomedullary contusion: case report. J Trauma 63: 420–423.

Attwater H (1911). Pontine hemorrhages. Guy's Hosp Rep 65: 339–389.

Bauer G, Gerstenbrand F, Rumpl E (1979). Varieties of the locked-in syndrome. J Neurol 221: 77–91.

Cabezudo JM, Olabe J, Lopez-Anguera A et al. (1986). Recovery from locked-in syndrome after posttraumatic bilateral distal vertebral artery occlusion. Surg Neurol 25: 185–190.

Caplan LR (2014). Chapter 8: Basilar artery. Vertebrobasilar ischemia and hemorrhage: clinical findings, diagnosis, and management of posterior circulation disease. Cambridge: Cambridge University Press, pp. 263–319.

Caplan LR (2014). Chapter 14: Parenchymatous hemorrhage. Vertebrobasilar ischemia and hemorrhage: clinical findings, diagnosis, and management of posterior circulation disease. Cambridge: Cambridge University Press, pp. 517–522.

Caplan LR, Zervas NT (1977). Survival with permanent midbrain dysfunction after surgical treatment of traumatic subdural hematoma: the clinical picture of a duret hemorrhage? Ann Neurol 1: 587–589.

Chisholm N, Gillett G (2005). The patient's journey: living with locked-in syndrome. BMJ 331: 94–97.

Das JM (2023). Locked-in syndrome. StatPearls [Internet]. Treasure Island, FL: StatPearls Publishing.

Dill LV, Isenhour CE (1939). Etiologic factors in experimentally produced pontile hemorrhages. Arch Neurol Psychiatr: 1146–1152.

Doble JE, Haig AJ, Anderson C et al. (2003). Impairment, activity, participation, life satisfaction, and survival in persons with locked-in syndrome for over a decade: follow-up on a previously reported cohort. J Head Trauma Rehabil 18: 435–444.

Dumas A (1998). The count of Monte Cristo. Wordsworth Classics Ltd.

Duret H (1919). Traumatismes cranio-cerebraux. Paris: Librairie Felix Alcan.

Fischer DB, Boes AD, Demertzi A et al. (2016). A human brain network derived from coma-causing brainstem lesions. Neurology 87: 2427–2434.

Fisher CM (1964). Ocular bobbing. Arch Neurol 11: 543–546.

Forti A, Ambrosetto G, Amore M et al. (1982). Locked-in syndrome in multiple sclerosis with sparing of the ventral portion of the pons. Ann Neurol 12: 393–394.

Fuller PM, Sherman D, Pedersen NP et al. (2011). Reassessment of the structural basis of the ascending arousal system. J Comp Neurol 519: 933–956.

Goto N, Yamamoto T, Kaneko M et al. (1983). Primary pontine hemorrhage and gustatory disturbance: clinicoanatomic study. Stroke 14: 507–511.

Heckmann JG, Dinkel HP (2013). Recovery of locked-in syndrome in central pontine myelinolysis. Am J Case Rep 14: 219–220.

Inci S, Ozgen T (2003). Locked-in syndrome due to metastatic pontomedullary tumor – case report. Neurol Med Chir (Tokyo) 43: 497–500.

Karp JS, Hurtig HI (1974). "Locked-in" state with bilateral midbrain infarcts. Neurology 30: 176.

Keme-Ebi IK, Asindi AA (2008). Locked-in syndrome in a Nigerian male with multiple sclerosis: a case report and literature review. Pan Afr Med J 1: 4.

Kim JS (2011). Chapter 8: Brainstem infarction syndromes. Posterior circulation stroke: advances in understanding and management. Singapore: Springer, pp. 46–58.

Kim BM, Kim SH, Kim DI et al. (2011). Outcomes and prognostic factors of intracranial unruptured vertebrobasilar artery dissection. Neurology 76: 1735–1741.

Laureys S, Pellas F, Van Eeckhout P et al. (2005). The locked-in syndrome: what is it like to be conscious but paralyzed and voiceless? Prog Brain Res 150: 495–511.

Leon-Carrion J, van Eeckhout P, Del Rosario Dominguez-Morales M et al. (2002). The locked-in syndrome: a syndrome looking for a therapy. Brain Inj 16: 571–582.

Lesenfants D, Habbal D, Chatelle C et al. (2018). Toward an attention-based diagnostic tool for patients with locked-in syndrome. Clin EEG Neurosci 49: 122–135.

Moruzzi G, Magoun HW (1995). Brain stem reticular formation and activation of the EEG. 1949. J Neuropsychiatry Clin Neurosci 7: 251–267.

Parvizi J, Damasio AR (2003). Neuroanatomical correlates of brainstem coma. Brain 126: 1524–1536.

Patterson JR, Grabois M (1986). Locked-in syndrome: a review of 139 cases. Stroke 17: 758–764.

Plum F, Posner JB (1972). The diagnosis of stupor and coma. Contemp Neurol Ser 10: 1–286.

Ruecker M, Furtner M, Knoflach M et al. (2010). Basilar artery dissection: series of 12 consecutive cases and review of the literature. Cerebrovasc Dis 30: 267–276.

Schoeggl A, Reddy M, Bavinzski G (2001). A lateral mass fracture of C1 associated with left vertebral artery and mid-basilar artery occlusion. J Neurotrauma 18: 737–741.

Williams AN (2003). Cerebrovascular disease in Dumas' the count of Monte Cristo. J R Soc Med 96: 412–414.

Zakaria T, Flaherty ML (2006). Locked-in syndrome resulting from bilateral cerebral peduncle infarctions. Neurology 67: 1889.

Handbook of Clinical Neurology, Vol. 216 (3rd series)
The Brainstem
L.T. Grinberg and A.D. Korczyn
https://doi.org/10.1016/B978-0-443-15736-3.00016-0

Chapter 7

The vestibular system: Unsteadiness, dizziness, and vertigo

DIEGO KASKI*

SENSE Research Unit, Department of Clinical and Movement Neurosciences, UCL Institute of Neurology, London, United Kingdom

Abstract

This chapter provides a detailed overview of the vestibular system's anatomy, physiology, and clinical relevance, emphasizing how inner-ear sensory structures and brainstem pathways interact to maintain balance, gaze stability, and spatial orientation. It begins with the peripheral vestibular apparatus, describing the semicircular canals and otolith organs in detecting angular and linear accelerations. These signals are transmitted via the vestibulocochlear nerve to the vestibular nuclei at the pontomedullary junction, where they integrate with visual and somatosensory inputs to drive coordinated reflexes – most notably the vestibulo-ocular (VOR) and vestibulo-spinal reflexes – that stabilize vision and posture. We then examine central vestibular pathways, including reciprocal connections between the brainstem, cerebellum, and cortex. These circuits enable adaptive mechanisms like VOR gain recalibration and support functions such as self-motion perception. A range of common vestibular disorders is discussed – benign paroxysmal positional vertigo (BPPV), vestibular neuritis, Ménière's disease, vestibular migraine, and persistent postural-perceptual dizziness – covering their pathophysiology and clinical features. Diagnostic tools include bedside exams (e.g., head impulse test), imaging, and objective measures like caloric testing, video head impulse testing, and vestibular-evoked myogenic potentials. Treatment strategies are reviewed, highlighting the effectiveness of repositioning maneuvers for BPPV and vestibular rehabilitation for compensation after vestibular loss. The chapter concludes with emerging advances, including vestibular prosthetics and high-resolution imaging, which are redefining diagnosis and treatment. These innovations underscore the vestibular system's vital role in equilibrium and point to promising therapies for chronic dizziness and imbalance.

INTRODUCTION TO THE VESTIBULAR SYSTEM

The vestibular system is a sophisticated sensory organ fundamental to human vison, balance, spatial orientation, and coordination. Situated within the inner ear, it consists of a complex network of structures and neural pathways that provide critical information about head movements and position relative to gravity. This sensory input is indispensable for maintaining posture, stabilizing vision during motion, and enabling seamless navigation through our environment. Despite its often-overlooked role, the vestibular system underpins everyday functions, and its impairment can lead to profound physical and psychologic consequences.

Anatomically, the vestibular system resides within the bony labyrinth of the inner ear, alongside the cochlea. It comprises two main components: the semicircular canals and the otolith organs. The three semicircular canals are arranged orthogonally to detect angular motion in all planes. They are filled with endolymph, a fluid that shifts in response to head rotations. Hair cells within the canals, topped by stereocilia and embedded in a gelatinous structure called the cupula, transduce mechanical stimuli into neural signals. Complementing these are the otolith organs – the utricle and saccule – which detect linear accelerations and head position relative to gravity. These structures contain a dense layer of calcium carbonate crystals (otoconia) that exert shear forces on underlying hair cells, enabling the detection of tilts and translational motion.

The vestibular system communicates with the brain through the vestibulocochlear nerve (cranial nerve [CN] VIII), transmitting signals to multiple regions, including the brainstem, cerebellum, and cortex. These connections integrate vestibular input with visual and proprioceptive information, creating a coherent sense of balance and orientation. For instance, the vestibulo-ocular reflex (VOR) ensures stable vision during head movements by coordinating eye movements that counteract

*Correspondence to: Diego Kaski, MBBS, PhD, SENSE Research Unit, Department of Clinical and Movement Neurosciences, UCL Institute of Neurology, 33 Queen Square, London WC1N 3BG, United Kingdom. Tel: +44 7880 726240; E-mail: d.kaski@ucl.ac.uk.

head motion. This reflex exemplifies the vestibular system's remarkable speed and precision, essential for activities such as reading while walking or tracking moving objects.

The significance of the vestibular system becomes evident when it malfunctions. Disorders of the vestibular system such as vestibular neuritis, benign paroxysmal positional vertigo (BPPV), Meniere's disease, or bilateral vestibulopathy (BVP) can lead to dizziness, vertigo, imbalance, and nausea (Mahmud and Kaski, 2024). These symptoms often disrupt daily life, impair mobility, and increase the risk of falls. Beyond physical effects, chronic vestibular dysfunction has been linked to anxiety, depression, and cognitive impairments, emphasizing its impact on overall well-being. Accurate diagnosis and targeted rehabilitation strategies, such as vestibular physiotherapy or repositioning maneuvers for BPPV, underscore the need for clinical expertise in managing these disorders.

Recent research has expanded our understanding of the vestibular system's broader roles. Emerging evidence suggests its involvement in spatial memory, navigation, and even autonomic regulation (Britton and Arshad, 2019). Advances in imaging, neurophysiology, and computational modeling have illuminated its intricacies, opening new avenues for therapeutic interventions. Innovative diagnostic tools and treatments, including virtual reality-based rehabilitation and vestibular implants (Starkov et al., 2020), hold promise for restoring function in patients with intractable deficits.

In this chapter, we will explore basic vestibular neurophysiology and anatomy, discuss common disorders that impair this network, with a focus on brainstem pathology, and review current therapeutic strategies, looking also towards the future.

Balance and posture

The vestibular system is the cornerstone of balance, working in concert with the visual and somatosensory systems. When head movement occurs, sensory hair cells within the inner ear convert mechanical stimuli into electrical signals, which are relayed to the brain via the vestibulocochlear nerve. These signals are integrated in the brainstem and cerebellum to adjust muscle tone and generate reflexive movements that stabilize the body. For instance, the vestibulospinal reflex activates limb muscles to counterbalance shifts in the center of gravity, preventing falls. Impairments in the vestibular system can disrupt these mechanisms, leading to dizziness, vertigo, and unsteadiness. Chronic conditions, such as BVP, can severely affect balance, forcing reliance on compensatory strategies, including visual cues or proprioceptive feedback (Bronstein and Lempert, 2010).

Spatial orientation

Spatial orientation, or the ability to perceive and navigate through our environment, is another critical function of the vestibular system. By continuously monitoring head position relative to gravity and motion, the system provides a sense of where we are in space. This capability is crucial for activities ranging from walking and running to more complex tasks like driving or piloting.

The vestibular system achieves this by integrating its sensory input with visual and proprioceptive information in the brain. The otolith organs, in particular, play a key role by detecting gravitational forces, allowing us to differentiate between being upright, lying down, or tilting. This integration occurs primarily in the temporoparietal cortex but likely involves a widespread cortical network (Brandt and Dieterich, 1999). Emerging evidence also links vestibular function to cognitive processes, including spatial navigation and decision-making.

Deficits in the vestibular system can impair spatial orientation, causing disorientation, veering while walking, or an inability to judge distances and direction accurately. This can significantly impact mobility and quality of life.

Gaze stabilization

The vestibular system ensures clear and stable vision during head movement through the VOR. This reflex coordinates eye movements to compensate for head motion, enabling steady foveation. For example, when the head turns to the right, the VOR moves the eyes to the left at a similar velocity, ensuring that the visual target remains in focus, akin to a "steadycam."

Without this mechanism, even minor head movements would blur vision, making everyday activities like reading, walking, or driving impossible. Conditions such as vestibular neuritis or cerebellar dysfunction can compromise the VOR, leading to symptoms such as oscillopsia, where the visual field appears to "bounce."

Integration of the vestibular system within the brainstem

The brainstem serves as the central hub for processing vestibular signals, integrating sensory input from the inner ear with motor responses to maintain balance, gaze stabilization, and posture. The vestibular nuclei, located in the pontomedullary junction, are the primary relay centers where signals from the semicircular canals and otolith organs converge. These nuclei have extensive connections with other regions of the central nervous system (CNS), enabling precise coordination of reflexive and voluntary movements.

The vestibular nuclei are functionally divided into four major subnuclei: the superior, lateral, medial, and inferior (or descending) vestibular nuclei. Each plays a distinct role in processing vestibular information. For instance, the superior and medial nuclei are crucial for the VOR. The lateral vestibular nucleus is essential for the vestibulospinal reflex, which adjusts muscle tone in the limbs to maintain balance, while the inferior nucleus integrates vestibular, proprioceptive, and cerebellar input to fine-tune posture and head motion detection.

The brainstem also facilitates the integration of vestibular signals with inputs from the visual and somatosensory systems. This multimodal processing ensures coherent spatial perception and effective motor responses. Furthermore, projections from the brainstem to higher cortical centers, such as the thalamus and temporoparietal cortex, contribute to conscious awareness of spatial orientation (Brandt and Dieterich, 1999; Kaski et al., 2016; Dieterich and Brandt, 2018; Kirsch et al., 2019).

ANATOMY OF THE VESTIBULAR SYSTEM

Peripheral vestibular structures

The inner ear is located within the petrous temporal bone and comprises the bony and membranous labyrinths. The bony labyrinth in each ear is composed of the vestibule, three semicircular canals, and the cochlea filled with perilymph. The three semicircular canals are superior (anterior), lateral (horizontal), and posterior semicircular canals located posterior and superior to the vestibule, and these detect angular head acceleration. The membranous labyrinth lies within the bony labyrinth and is filled with endolymph. The membranous labyrinth comprises the utricle and saccule, three semicircular ducts, and the cochlear duct. The utricle and saccule are also known as the otolith organs and detect linear head acceleration.

The vestibulocochlear nerves carry vestibular information from the semicircular canals, utricle, and saccule, and pass through the internal acoustic meatus together with the facial nerve (Fig. 7.1). The nerve fibers converge in the vestibular ganglion (also known as Scarpa's ganglion), located in the internal auditory canal. This ganglion contains the cell bodies of bipolar sensory neurons. The vestibular nerve enters the brainstem at the pontomedullary junction and travels through the inferior cerebellar peduncle to reach the vestibular nuclei or the flocculonodular lobes of the cerebellum.

Anterior inferior cerebellar arteries (AICA) and posterior inferior cerebellar arteries (PICA) are the major arteries involved in the vestibular system vascularization. The labyrinthine artery most commonly branches from the AICA, while it occasionally branches from the basilar artery and supplies the CN VIII (vestibulocochlear) and CN VII (facial). As the

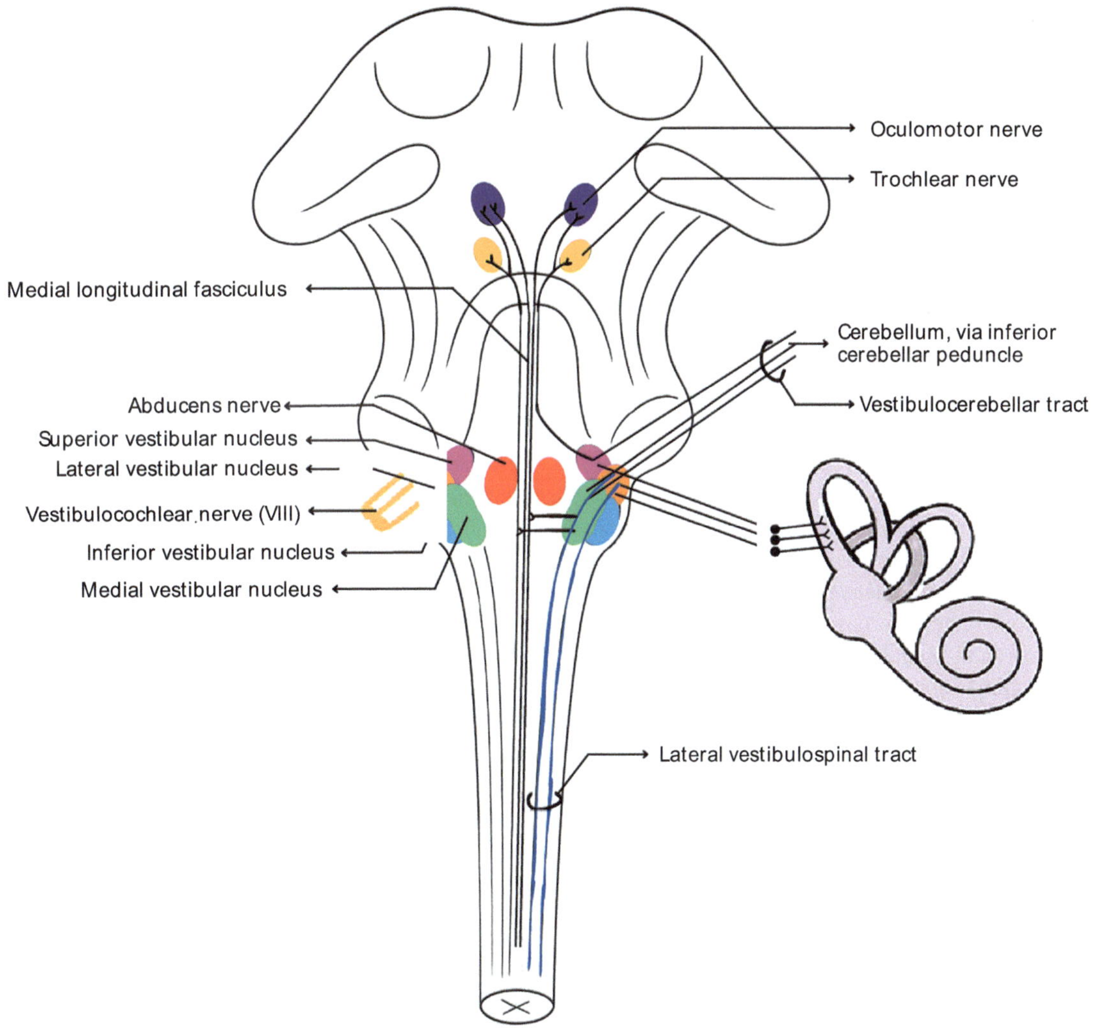

Fig. 7.1. Brainstem anatomy of the vestibular system.

labyrinthine artery travels with the CN VIII and CN VII through an opening called the internal auditory meatus to enter the inner ear, it divides into the anterior vestibular, cochlear, and vestibulocochlear arteries. The anterior vestibular artery supplies most of the utricle, the vestibular nerve, and the ampullae of the anterior and horizontal semicircular canals. The cochlear artery is the main vasculature of the cochlea.

The vestibulocochlear artery supplies the saccule and a portion of the cochlea and semicircular canals. The AICA supplies the flocculus of the cerebellum. An AICA infarction may result in multiple symptoms such as vertigo, hearing loss, facial weakness and sensory loss, Horner syndrome, and gait and limb ataxia. The PICA arises from the vertebral arteries. At the level of the upper part of the cerebellar tonsils, the PICA descends through the lower vermis, dividing into two branches; the medial branch supplies the inferior vermis (nodulus, uvula, pyramids), inner part of the inferior semilunar lobule, and cerebellar tonsils. The lateral branch supplies the inferior lateral portion of the cerebellar hemisphere.

Infarcts in the lateral territory of the PICA are rare and involve the anterolateral region of the caudal part of the cerebellar hemisphere. The most common symptoms include unsteadiness, gait ataxia, limb ataxia, and ipsilateral body pulsion. There is usually no nystagmus, nor is dysarthria observed unless the brainstem is also involved.

Central vestibular structures

The central vestibular pathways integrate sensory inputs from the inner ear to maintain balance, posture, and gaze stability. These pathways are centered around the vestibular nuclei in the brainstem, which consist of the superior, medial, lateral, and inferior nuclei. Each nucleus has specific roles and extensive connections to other regions of the CNS.

The superior vestibular nucleus is primarily involved in gaze stabilization. It plays a key role in the VOR by relaying signals to the ocular motor nuclei through the medial longitudinal fasciculus (MLF), enabling stable vision during head movements. The medial vestibular nucleus processes head movement information and contributes to postural control. It projects bilaterally to the cervical spinal cord via the medial vestibulospinal tract, regulating neck muscles to stabilize the head.

The lateral vestibular nucleus, or Deiters' nucleus, gives rise to the lateral vestibulospinal tract, which descends ipsilaterally to influence extensor muscles in the limbs. This pathway is essential for maintaining upright posture and balance by adjusting muscle tone. The inferior vestibular nucleus integrates sensory inputs from the vestibular organs and the cerebellum. Its projections to the spinal cord and brainstem modulate postural reflexes and support equilibrium.

Connections with the cerebellum are particularly important for fine-tuning these processes. The vestibular nuclei have reciprocal links with the vestibulocerebellum, including the flocculonodular lobe and the fastigial nucleus. This interaction refines balance and eye movement control. Furthermore, the vestibular nuclei communicate with the oculomotor, trochlear, and abducens nuclei through the MLF, ensuring coordinated eye movements during head motion.

Projections to the spinal cord via the medial and lateral vestibulospinal tracts facilitate the adjustment of muscle tone and reflexes necessary for balance and posture. Together, these pathways create an integrated system for sensory-motor coordination, enabling spatial orientation and stability.

VESTIBULAR PHYSIOLOGY

There is no single balance 'organ'. Balance is achieved and maintained via an integrated sensorimotor system. Orienting sensory information is derived from the paired vestibular labyrinths, visual input and somatosensory afferents (joint, tendon and muscle position sense, and superficial sensation). This information converges on the brainstem vestibular nuclei and is integrated with modulating influences from the reticular activating system and higher centers: cortex, cerebellum, and extrapyramidal system. Effector pathways from the vestibular nuclei project to the oculomotor nuclei generating compensatory eye movements (i.e., the efferent limb of the VOR, a three-neuron reflex arc that is under substantial cerebellar modulation; (Fig. 7.2). Other efferents pass to the neck, trunk, and limb muscles, the vestibulospinal tracts, which modulate spinal reflex arcs. Efferents also project to the vestibular cortex from the brainstem.

The vestibular system (1) stabilizes gaze in space during head movement (e.g., reading a sign while walking), (2) controls posture when the head and body are static (e.g., while standing and during passive motion), and (3) facilitates perception of orientation and motion.

Vestibulo-ocular reflexes

The vestibular and oculomotor systems have evolved in parallel to function in three similar planes, namely Yaw – head rotation about the vertical z-axis, Pitch – head flexion/extension about the horizontal y-axis, and roll – lateral head tilt about the horizontal x-axis.

This alignment, between spatial planes of the three semicircular canals and the planes of the three sets of extraocular muscles, is such that when the two paired labyrinths are excited, the two eyes move conjugately (i.e., they are yoked), producing an appropriately directed compensatory eye movement. This is the VOR. For example, the paired horizontal semicircular canals function as a gauge of rotational acceleration in the yaw plane and are connected via the brainstem nuclei to the set of extraocular muscles whose primary direction of pull is also in the horizontal plane – the lateral and medial recti. The action of these muscles is to produce a compensatory eye movement in the horizontal plane in an equal and opposite direction to that of head movement, thus maintaining stable gaze.

The neuronal circuitry is similar for the vertical planes of pitch and roll: the anterior and posterior semicircular canals, sited diagonally to the sagittal plane are connected to the paired extraocular muscles whose primary action is aligned to the

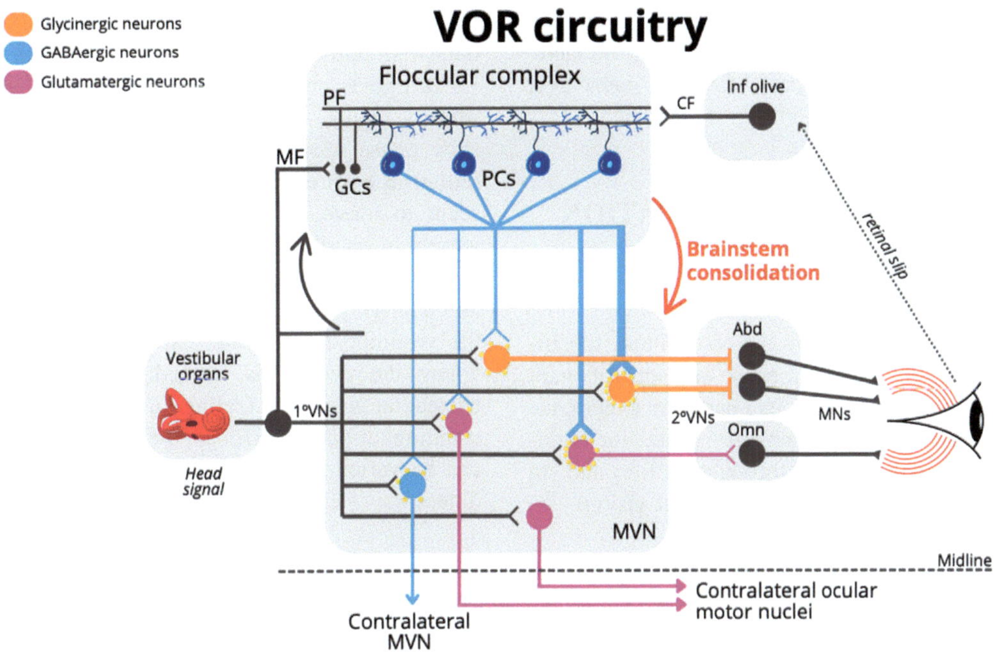

Fig. 7.2. Vestibular-ocular reflex (VOR) circuitry . Whilst it is often taught as a simple three-neuron arc, the VOR is in fact a more complex system under dynamic control of the cerebellum that must adjust the VOR response in relation to gaze position and target position and correct for retinal slip. *Abd*, abducens nucleus; *CF*, climbing fibers; *GCs*, Golgi cells; *Inf Olive*, inferior olive; *MF*, mossy fibers; *MNs*, motorneurons; *MVN*, medial vestibular nucleus; *Omn*, oculomotor nucleus; *PC*, Purkinje cells; *PF*, parallel fibers; *VNs*, vestibular neurons (Carcaud et al., 2017). Adapted from Carcaud, J., et al. (2017). Long-lasting visuo-vestibular mismatch in freely-behaving mice reduces the vestibulo-ocular reflex and leads to neural changes in the direct vestibular pathway. eNeuro 4: ENEURO.0290-16.2017.

spatial planes of each canal. Thus, the superior and inferior recti move the eye up or down when that eye is abducted; the inferior and superior oblique muscles move the eye up and down when it is adducted.

The system is driven by the relationship between endolymph and cristae. The relevant principles were described by Ewald in 1892: (1) Head and eye movements always take place in the plane of the canal that is stimulated and in the direction of endolymph flow; (2) in the horizontal canals, endolymph flow toward the ampulla (ampullopetal) causes a greater response than endolymph flow away from the ampulla (ampullofugal); and (3) in the vertical canals, endolymph flow away from the ampulla (ampullofugal) causes a greater response than endolymph flow toward the ampulla (ampullopetal).

Angular accelerations are sensed by the semicircular canals, which contain endolymph fluid and sensory hair cells embedded in the crista ampullaris within the ampulla. When the head rotates, the endolymph lags behind due to inertia, deflecting the gelatinous cupula and bending the stereocilia on the hair cells. This mechanical deflection opens ion channels, leading to depolarization or hyperpolarization of the hair cells, depending on the direction of movement. This process is critical for detecting angular head acceleration.

Linear accelerations and gravitational forces are detected by the otolith organs. These structures contain maculae, where hair cells are embedded in a gelatinous layer topped with otoconia (calcium carbonate crystals). When the head moves linearly or tilts, the weight of the otoconia creates shear forces on the gelatinous layer, bending the stereocilia of the hair cells. Similar to the semicircular canals, this bending modulates ion channel activity, generating a neural signal proportional to the acceleration. Hair cells play a central role in mechanotransduction, converting mechanical stimuli into electrical signals that are transmitted to the brain via the vestibular nerve. This process underpins the vestibular system's ability to sense motion and maintain balance.

Integration of vestibular inputs with visual and proprioceptive information

The vestibular system provides real-time data on head position and motion, while the visual system tracks environmental cues, and proprioception delivers feedback from muscles and joints. This multisensory integration occurs in the brainstem, cerebellum, and higher cortical centers, ensuring coherent perception and motor responses. For example, the VOR combines vestibular and visual inputs to stabilize gaze during head movements, maintaining a clear visual field. Similarly, the cerebellum integrates vestibular and proprioceptive signals to adjust ocular motor (VOR) function and fine-tune motor control. There is further sensory integration in the parietal cortex to generate self-motion perception and spatial awareness (Brandt and Dieterich, 1999; Kaski et al., 2016).

This multisensory processing is crucial for adapting to sensory conflicts, such as during rapid motion or unstable surfaces. Impairments in this integration can cause dizziness, imbalance, and disorientation, highlighting its importance in daily life.

BRAINSTEM–VESTIBULAR INTEGRATION

Vestibulo-ocular reflex and its neuroanatomic pathways

The VOR is subject to modulation by the cerebellum, specifically the flocculus and nodulus, which refine and adapt its gain (the ratio of eye velocity to head velocity) based on experience and environmental demands. This plasticity is vital for maintaining accurate gaze stabilization throughout life and in changing visual environments (e.g., near to distant vision).

Vestibulo-spinal reflexes and posture control

The vestibulo-spinal reflexes (VSR) are essential for maintaining posture and balance by mediating adjustments in muscle tone and body position in response to head movements and gravitational forces. These reflexes are primarily driven by the otolith organs, which as we have already seen, detect linear acceleration and head position relative to gravity.

The neuroanatomic pathways of the VSR involve signals from the vestibular organs transmitted to the vestibular nuclei. From there, descending pathways such as the lateral vestibulospinal tract and medial vestibulospinal tract project to the spinal cord. The lateral vestibulospinal tract primarily influences extensor muscles to maintain upright posture, while the medial vestibulospinal tract contributes to head stabilization and coordination of head and neck movements.

The integration of vestibular, proprioceptive, and visual inputs is critical for effective postural control. For example, during a sudden perturbation (e.g., being pushed), the VSR elicits compensatory adjustments in limb and trunk muscles to counteract instability, helping to prevent falls. Dysfunction in these pathways can result in postural instability, as seen in disorders like vestibular neuritis or any central vestibulopathy.

The brainstem plays a pivotal role in integrating multisensory information for spatial orientation. The vestibular nuclei, located in the medulla and pons, serve as the primary hub for processing signals from the vestibular system. These nuclei receive convergent input from the visual, proprioceptive, and somatosensory systems, allowing for a comprehensive representation of body position and motion in space.

CLINICAL CORRELATES OF VESTIBULAR DYSFUNCTION

Dizziness is best defined as the sensation of disturbed or impaired spatial orientation without a false or distorted sense of motion (Bisdorff et al., 2009). Vertigo, on the other hand, refers to the sensation of self-motion when no self-motion is occurring or the sensation of distorted self-motion during an otherwise normal head movement (Bisdorff et al., 2009).

Vertigo is either of internal origin – a mismatch between sensory inputs – or it may be linked to retinal slip during the slow phase of nystagmus (i.e. the image moves across the retina) in each case the surroundings (oscillopsia) or the head seem to move. Dizziness is diagnostically complex, with a variety of conditions, not always primarily vestibular in origin, giving rise to this sensation. Vertigo more specifically points to an underlying dysfunction along the vestibular network – from the vestibule in the inner ear, through to the cerebral cortex where the percept is formed. The Bárány Society defines unsteadiness as the feeling of being unstable while sitting, standing, or walking (Bisdorff et al., 2009). As such, the causes can be varied, and vestibulopathy is but one cause of unsteadiness.

Clinical assessment overview

Clinical examination

As in any other clinical examination in medical practice, the physician is guided by the symptoms reported by the patient. One would not spend too much time examining the VOR; for example, if a patient's dizzy spells are always preceded by chest pain and palpitations that would suggest a cardiovascular origin. Peripheral and central vestibular disorders can usually be diagnosed with a good history and examination.

Gait and posture

Examination of gait with the eyes open is very useful for detecting neurologic disorders that compromise balance and gait and can therefore predispose to falls. It is indeed an important part of the general neurologic examination, and there are many specific gait types (e.g., polyneuritic, tabetic, spastic, parkinsonian, cerebellar, apraxic, and functional; Mermelstein et al., 2024). The clinician should observe the general speed of gait, size of the base of support, regularity of step production and turning. Difficulties in initiating gait can be seen in gait apraxia, small vessel disease, and parkinsonism.

Increasing the base of support is an automatic reaction adopted when healthy people (walking on a boat, drunkenness) or patients perceive that their balance capabilities are reduced and is thus not pathognomonic of cerebellar disease. Broad-based gait is observed in any condition limiting balance, including acute vertigo (both peripheral and central), bilateral vestibular failure, polyneuropathies, cerebellar disease, and many pyramidal and extrapyramidal disorders. The relatively narrow gait present in patients with idiopathic Parkinson disease may indicate reduced adaptability (to their objective imbalance) in their motor system.

Unilateral vestibular disorders, except in the acute vertigo phase, do not cause significant abnormalities of gait. A full gait assessment will include tandem gait (most health individuals can manage 10 accurate heel–toe steps in three trials) and a test of Romberg stance (Mermelstein et al., 2024).

Nystagmus, eye movements, positional testing

Clinical examination of eye movements

The vestibular system provides a powerful input to the ocular motor system, so eye movements must be examined in detail. In a patient suspected to have a peripheral vestibular disorder (labyrinth and VIIIth nerve), one can consider that the examination has two broad aims:

1. To search for direct ocular signs indicative of a peripheral vestibular disorder, typically nystagmus, BPPV, or a positive head impulse test (a bedside assessment of the VOR; Bronstein and Lempert, 2017).
2. To exclude a CNS lesion by making sure that nonvestibular-mediated eye movements are normal.

The oculomotor examination should include

- cover test,
- search for spontaneous and gaze-evoked nystagmus,
- convergence,
- smooth pursuit,
- saccades,
- VOR, and
- positional maneuvers.

If the patient reports diplopia, or if at any point during the examination the two eyes appear dysconjugate, examination of the individual CN III, IV, and VI and a cover test is mandatory.

Spontaneous nystagmus

The search for spontaneous nystagmus begins in the primary straight ahead position of the eyes. In the presence of nystagmus, one must try to define (1) the waveform (e.g., saw-tooth or "jerk" nystagmus) or pendular nystagmus (in which there is no "fast-phase") and (2) the direction of the beat. As spontaneous nystagmus is often enhanced by convergence, it is convenient to examine convergence at the same time, by slowly moving an object in and out along the visual axis. Absence of convergence occurs in midbrain lesions, but it must be remembered that reduced/absent convergence is very common in normal people above the age of 60 years.

In the context of a symptomatic patient, with acute vertigo, severe unsteadiness, and nausea, the nystagmus can be either peripheral or central, but if the patient comes as a routine ambulatory patient and does not look acutely ill, the nystagmus is more likely to be of central origin. A clinically visible nystagmus of peripheral vestibular origin can only be seen in the acute phase of the disorder (e.g., vestibular neuritis, Ménière's disease). The nystagmus is essentially horizontal with a minor torsional (rotary) component. The fast phase beats contralesionally but in the acute irritative phase of Ménière's attack, it can beat to the affected side.

Soon after the acute stage of a peripheral vestibular lesion, the nystagmus is usually not visible in primary gaze but only on gaze deviation in the direction of the fast phase. A useful classification for the severity of nystagmus is Alexander's law:

First degree: only visible on gaze deviation in the direction of the fast phase.
Second degree: when present in primary gaze as well as the direction of the fast phase.
Third degree: when it is also visible when the eyes are deviated in the opposite direction to that of the fast phase.

Gaze-evoked nystagmus

Alexander's law applies to most peripheral vestibular nystagmus. Downbeat nystagmus (DBN) in primary gaze, which is always of central origin, often does not enhance on looking down but on looking sideways, or with convergence.

Nystagmus on gaze deviation can also indicate a central lesion. This is what is frequently called gaze paretic nystagmus, not to be confused with the broader term "gaze-evoked nystagmus." The term "gaze paretic" implies that the patient has difficulty in holding gaze in an eccentric position in the orbit. The eyes drift in centripetally with a slow phase, which progressively decreases in velocity, until a new saccade refixates the eccentric target. This nystagmus, usually of larger amplitude than peripheral vestibular nystagmus, is caused by damage to the gaze holding mechanisms mediated by ipsilateral brainstem and cerebellar structures (Zee, 2021). An example of vestibular and gaze paretic nystagmus present in the same patient is Brun's nystagmus caused by a large cerebellopontine angle lesion (e.g., stroke, inflammation, or indeed a vestibular schwannoma) (Mishra and Pandey, 2022).

Smooth pursuit eye movements

When tracking an object with our eyes, we use a combination of fast (saccadic) and slow-phase (pursuit) eye movements. Strictly speaking, only the latter should be considered smooth pursuit. When the target moves slowly (< 10–15 degrees/second), the eye is capable of following it almost exclusively with smooth pursuit movements (i.e., with a perfect gain of 1). When targets move at velocities greater than 40 degrees/second, the proportion of tracking carried out by saccades increases progressively and the gain decreases. Clinically, smooth pursuit looks abnormal, "broken up," or "choppy," when the observer detects too much saccadic tracking; this gives a cogwheel, jerky appearance to the pursuit movements.

The presence of normal pursuit essentially rules out a central vestibulo-cerebellar disorder. Similarly, a patient with balance symptoms and unequivocally broken pursuit movements almost certainly has a neurologic rather than a labyrinthine disorder. This, however, assumes a technically correct examination, ensuring the patient is able to see the target fully, and the target must be moved smoothly and slowly. A single cycle (to-and-fro) is insufficient because patients may not understand the task and jump their eyes ahead of target; the patient should be encouraged to "glue" their eyes to the target. Pursuit should be assessed in both the horizontal and vertical planes.

Saccadic eye movements

Saccades are fast movements of the eyes (200–500 degrees/second), which allow us to shift gaze from one object of interest to another.

These refixations occur incessantly so that gross saccadic defects are often detected when simply speaking to the patient. Merely stating that a patient has "abnormal saccades" is insufficient to make a topographic diagnosis. There are at least four independent properties of saccades that can all be assessed clinically:

1. Latency (normal or increased)
2. Velocity (normal, slow, or absent saccades, e.g., a gaze palsy)
3. Accuracy (normo-, hypo-, or hyper-metric)
4. Binocular conjugacy (conjugate or dysconjugate, e.g., internuclear ophthalmoplegia [INO]).

Saccadic latency refers to the delay between the onset of a visual stimuli or target and the onset of the saccadic eye movement. The initiation of a saccade relies on loss of inhibitory ("break") signals from frontal eye fields in the frontal lobe to the brainstem oculomotor ("omnipause") cells. Increased saccadic latency is therefore most commonly seen with frontal lobe disorders, such as corticobasal or Huntington disease (Sekar et al., 2024).

Saccadic velocity is tested by asking the patient to look fully to the right, left, up, and down and assessing whether these movements are rapid – when very slow, they appear similar to pursuit movements. The examiner's finger should be placed at 20–30 degrees to the right–left, up–down of primary gaze. The patient is instructed to look at the target when, for instance, the right or left finger flick, or when the hand opens up. It is customary to reinforce the visual stimulus with a simultaneous verbal command such as "right," "left" (from the patient's point of view!).

Saccadic accuracy can be assessed at the same time as testing velocity. The presence of two to three corrective saccades consistently is considered saccadic hypometria and severely hypometric saccades look fragmented, often called "multiple step" saccades. If the initial saccade is too large (saccadic hypermetria) and travels past the target, the patient makes corrective saccades in the opposite direction of the initial saccade. This is a reliable cerebellar sign (Sekar et al., 2024).

Saccadic conjugacy can only be assessed if there is no individual IIIrd, IVth, or VIth nerve palsy. A common source of dysconjugacy in the horizontal plane is the INO in which the abducting eye saccades are fast and large, whereas saccades in the adducting eye are slow and small. INOs are caused by lesions of the MLF, which connects abducens interneurons of the VI nucleus with medial rectus motor neurons in the contralateral III nucleus.

Optokinetic nystagmus

The optokinetic nystagmus (OKN) system in humans is not a truly separate oculomotor system but rather combines slow-phase movements from the smooth pursuit system and the resetting saccadic eye movements. You might see OKN when staring at the eyes of an individual staring out of a window as a train pulls away from the station. It can be tested using a drum or by moving a large written print (newspaper or a poster) in front of a patient's eyes but little additional information is gained if pursuit and saccades have been assessed already.

Vestibulo-ocular reflex

The VOR is assessed at the bedside. The clinical maneuvers rely either on a slow doll's head maneuver assessed either by direct observation of the eyes or ophthalmoscopy and measurement of visual acuity or a fast version of the doll's head, the "head impulse," or "head thrust" test (Halmagyi and Curthoys, 1988). The VOR is only irreplaceable during high accelerations of the head. Neither pursuit, optokinetic, nor cervical mechanisms can fully take over from the VOR during brisk head movements. To carry out a bedside assessment, the patient needs to be seated with the examiner at close range to observe the eyes carefully. The patient is instructed to fixate on a target such as the examiner's nose and the head is turned briskly in discrete steps of +10–15 degrees across the midline by the examiner, thus producing head velocities of several hundred degrees per second. A fast right head (nose) turn will make a patient with a right-sided vestibular loss introduce one or more catch-up saccades toward the target (i.e., toward the left), to refixate. These catch-up saccades can easily be identified by a trained observer. The test is clearly useful for identifying an acute, unilateral, peripheral vestibular deficit, for instance, in a patient with vestibular neuritis. In patients with chronic uncompensated, incomplete unilateral lesions, the test is often negative or inconclusive. A study comparing caloric testing with the clinical head impulse test in an outpatient setting shows that the overall sensitivity of the head impulse test is 34%. However, specificity is high (100%) and all patients with a canal paresis of greater than 87% had a positive head impulse test (Burston et al., 2018).

Vestibulo-ocular reflex suppression

There are times when activating the VOR is not desirable, such as when tracking a moving object with the eyes and head together (e.g., watching a tennis ball move from one side of the court to another as a spectator). Here, the human brain can suppress the VOR with mechanisms closely related to pursuit mechanisms and pathways. Abnormalities in VOR are this indicative of central dysfunction, as discussed already for smooth pursuit. Clinical examination of VOR suppression (VORS) involves oscillating the patient while they fixate a head-fixed target. Any obvious breakthrough nystagmus indicates poor suppression. In the clinic, this can be achieved by asking the patient to clasp the hands together in front and put the thumbs up as a target; the patient can then oscillate side to side "en bloc," by themselves if standing up, or the examiner can oscillate the patient's swivel chair if sitting down. As with pursuit, the oscillation should be slow, say two to five seconds from left to right and vice versa (0.1–0.25 Hz) and the examiner should have a good view of the patient's eyes.

Patients with peripheral vestibular lesions have a normal VORS (or even "supranormal," as they are trained to suppress their own nystagmus). In central lesions, the abnormality of pursuit and VORS go hand in hand (i.e., a patient with a right cerebellar lesion will have abnormal rightward pursuit and abnormal VORS, with visible nystagmus breaking through, when turning the head rightward). Remember that is there is complete loss of the VOR, however, there will be no VOR to suppress!

Positional maneuvers: Dix–Hallpike and Roll

BPPV is among the most frequent causes of vertigo, responsible for approximately one-quarter of cases involving dizziness and vertigo. Positional maneuvers are essential in assessing patients with balance disorders, especially given the availability of effective treatments for BPPV. While the condition is often suspected based on the patient's history, it requires confirmation through the Dix–Hallpike and supine roll maneuvers (Kaski and Bronstein, 2014). These maneuvers, when performed with the affected ear in the downward position, elicit characteristic nystagmus (refer to the clinical disorders section on BPPV for more details). The Dix–Hallpike maneuver is particularly useful for differentiating peripheral nystagmus in BPPV from central or atypical positional nystagmus (Fig. 7.3).

DIAGNOSTIC AND ASSESSMENT METHODS

Role of imaging in brainstem and vestibular system evaluation

The inner ear structures are commonly evaluated structurally using specialized imaging modalities such as magnetic resonance imaging (MRI) and computed tomography (CT). High-resolution MRI with contrast is the gold standard for assessing the vestibular nerve and inner ear structures. It is particularly valuable in detecting vestibular schwannomas, inflammatory conditions, or demyelinating lesions. Diffusion-weighted imaging helps in diagnosing acute vestibular disorders such as labyrinthine infarction/ischemia but can miss up to 20% of vascular vertigo syndromes if performed early (within 24 hours) (Saber Tehrani et al., 2014). Vascular imaging may also help in the diagnosis of vestibular paroxysmia, but this is a rare entity (Strupp et al., 2016). 3D constructive interference in steady state (CISS) or fast imaging employing steady-state acquisition (FIESTA) sequences provide detailed visualization of the CNs and membranous labyrinth (Nagururu et al., 2022).

High-resolution CT is ideal for detecting bony abnormalities of the inner ear, such as semicircular canal dehiscence or otosclerosis. CT is also used for evaluating trauma-related fractures or congenital anomalies affecting the vestibular system.

Emerging imaging technologies are enhancing diagnostic capabilities. Ultrahigh-field MRI (7 T) offers unprecedented resolution for visualizing brainstem microanatomy. Additionally, vestibular functional MRI (fMRI) is expanding understanding of central

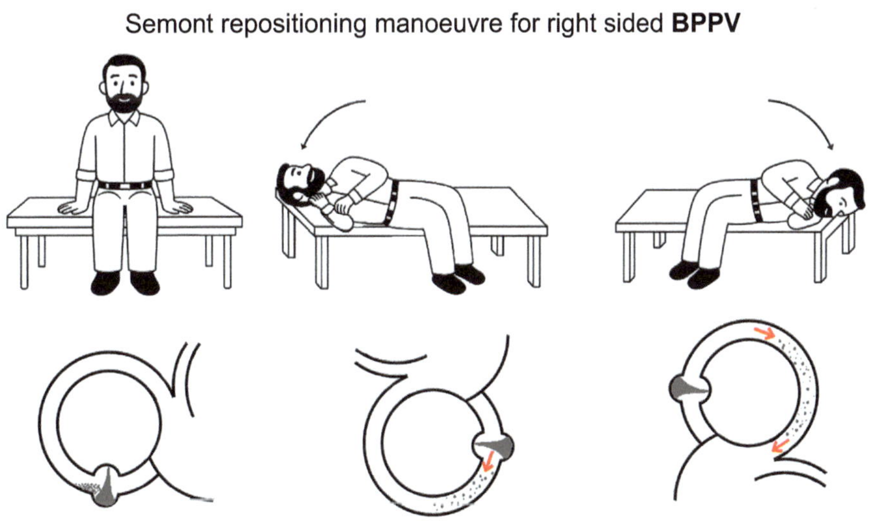

Fig. 7.3. Underlying pathophysiology of **benign paroxysmal positional vertigo** (BPPV), side-lying right-sided Dix–Hallpike maneuver, and Semont liberatory treatment maneuver for right posterior canal BPPV Start by seating the patient facing the clinician. To test the right posterior canal, turn the head 45 degrees to the left, and then lie the patient down quickly onto their right, keeping the head turned 45 degrees. This is the side-lying Dix–Hallpike maneuver. Observe for nystagmus. Note that in BPPV, there may be a short latency before the otolith particles move down the canal. Once the nystagmus (and vertigo) has settled, a sign that the otoliths have reached the lowest point in the canal, the clinician can perform a Semont treatment maneuver, by rapidly moving the patient, with the head still turned 45 degrees to the left, 180 degrees so that they are now lying on their left shoulder. Note the position of the head is toward the floor on this side. It is important to do this treatment maneuver quickly and smoothly to ensure inertial forces can be overcome, and the otoliths can migrate "up and over" within the canal.

vestibular processing, while AI-driven image analysis is improving lesion detection and diagnostic accuracy, but these modalities are not yet in routine clinical use (Thylur et al., 2017).

Objective measures of vestibular function

Quantitative analysis

Bithermal caloric testing remains the mainstay test of peripheral vestibular function in many countries (now being superseded by video head impulse test [vHIT]). Following a thermal water (or air) irrigation to the ear canal, nystagmus will appear (beating away from the irrigated side with cold water (30°C)), and toward the irrigated side with warm water (44°C). The nystagmus ceases 90–140 seconds after onset of irrigation and returns for up to a further 60 seconds after removal of optic fixation. Total or partial canal paresis is the complete or partial loss of labyrinthine function in one ear, seen when there is a total absence of nystagmus (or reduced in partial loss) following both 30°C and 44°C irrigations, even in the absence of optic fixation. Ideally, if total loss is observed, the test should be repeated using cold water at 20°C for 60 seconds to confirm the result. It may reflect an ipsilateral lesion of the labyrinth, VIIIth nerve, or brainstem vestibular nuclei, and confirms unilateral vestibular hypofunction.

Video head impulse test

The vHIT, using a lightweight video goggle with an embedded infrared camera to track eye movements and inertial accelerometer in the frame to track head movements, assesses the VOR during a rapid high-velocity low-amplitude (10°–20°) head rotation (Curthoys et al., 2023). The main advantage is the fast, simple, and accurate assessment of the functional status of each of the six semicircular canals individually. It is faster to carry out than caloric test, and more comfortable for patients, as it does not induce nystagmus or nausea. It provides a quantitative measurement of VOR gain by calculating the ratio of the eye velocity to the head velocity during the head impulse. Normal VOR gain is close to 1.0 but patients with peripheral vestibulopathy have a reduced VOR gain during the head turn (to their affected ear). The lack of slow phase compensation for head turn results in movement of the eyes with the head, and occurrence of corrective eye movements (corrective catch-up or refixation saccades) for head impulses toward the affected ear. Corrective catch-up saccades that occur during (covert saccades) or after the head impulse test (overt saccades) usually indicate peripheral vestibulopathy.

Note that in some neurologic patients, the presence of refixation saccades with a normal VOR gain may represent an oculomotor disorder rather than vestibular dysfunction (Koohi et al., 2022).

Posturography

Alterations in vestibular function significantly impact posture, a vital function maintained by a complex sensorimotor feedback system. Postural control involves coordinated reflexes to manage physiologic sway, where balance is maintained by shifting the body's center of gravity over the base of support. The Romberg test is a common clinical assessment for postural stability. Advanced tools like moving platform posturography address limitations of static platforms by controlling sensory inputs. The EQUITEST platform, which uses sway referencing to modulate visual and support surfaces, is valuable for rehabilitation through serial measurements but offers limited utility as a diagnostic tool (Di Fabio, 1995).

Vestibular evoked myogenic potentials

Cervical vestibular evoked myogenic potentials (cVEMP) are neural event-related potentials that serve as a short-latency (< 10 ms) ipsilateral vestibulocollic reflex, fine-tuning voluntary motor commands directed at the neck muscles (Curthoys et al., 2019). This reflex is initiated by a loud acoustic stimulus, triggering a saccular response through the inferior vestibular nerve, vestibular nuclei, vestibulospinal tracts, and motor neurons of the sternocleidomastoid muscle. cVEMP is a valuable tool for assessing brainstem function, as it evaluates the integrity of pathways involving the saccule and inferior vestibular nerve. Absent cVEMP responses are considered pathologic in individuals under 60 years when appropriate stimuli are applied. Abnormalities can be linked to conditions such as acoustic neuroma, vestibular neuritis, Tullio phenomenon, superior semicircular canal dehiscence (with a lowered cVEMP threshold), and Ménière's disease.

Ocular VEMP (oVEMP), reflecting the utricular–ocular reflex, also assesses brainstem function. Responses are recorded via electrodes beneath the eyes on the superior oblique muscle, with bone-conducted vibration eliciting a short-latency negative (n) excitatory component peaking at approximately 10 ms (n10). These tests are helpful in evaluating vestibular and brainstem function in certain clinical contexts.

THERAPEUTIC IMPLICATIONS

Specific vestibular disorders and treatment options

This section provides an overview of the most common vestibular disorders and briefly discusses evidence-based treatment options for these. It should be noted that recent evidence supports the clinical finding that traditional "vestibular sedatives," for example, antidopaminergics and antihistamines, have no effect on vestibulo-ocular function or the perception of self-motion and that their effect is primarily antiemetic.

BPPV

BPPV is still recognized as the most common cause of vertigo worldwide, its prevalence increasing with age, and an overall prevalence of 11–140 per 100,000 population (Bhattacharyya et al., 2017). The diagnosis of BPPV should be suspected in a patient with brief (< 1 min duration) attacks of severe, triggered

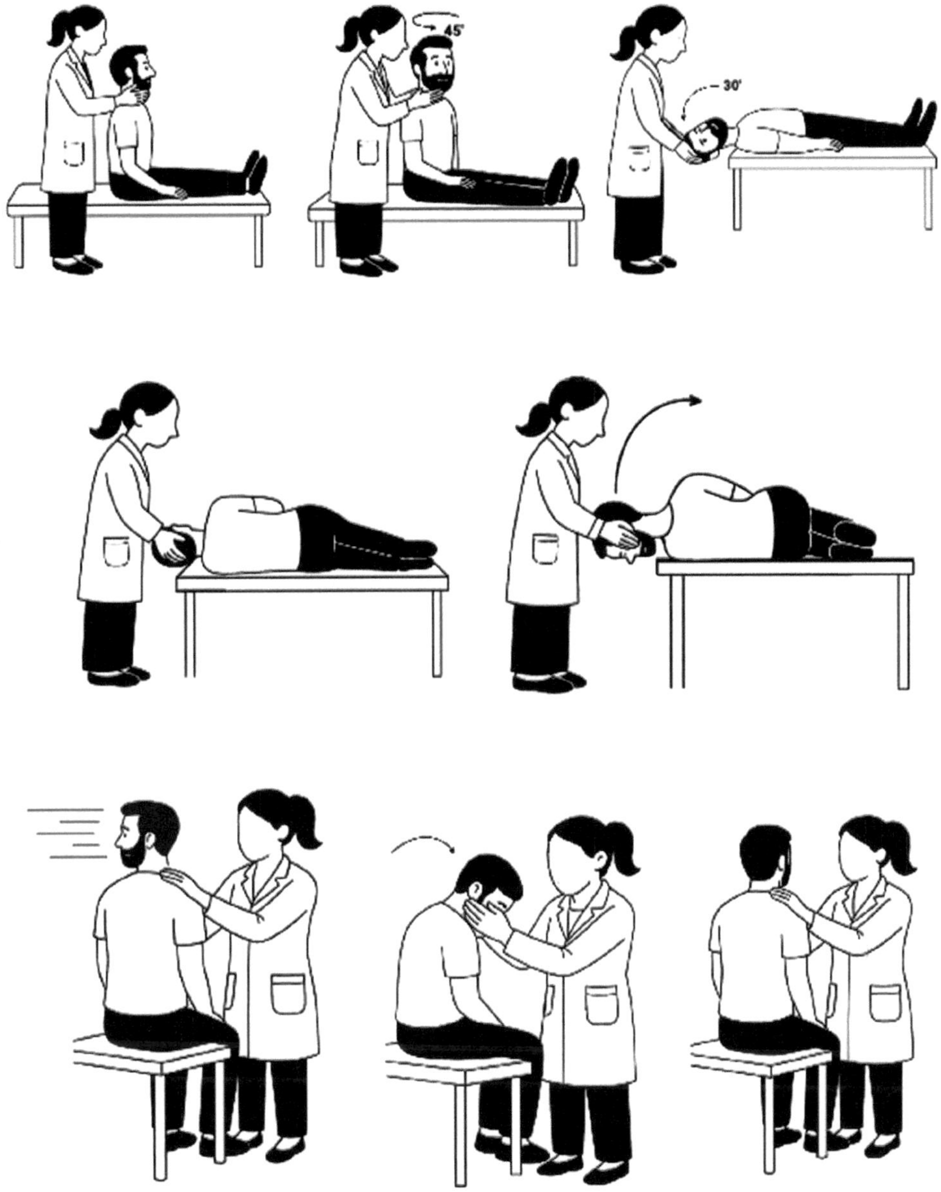

Fig. 7.4. Epley treatment maneuver Seat the patient on a couch so that, when supine, their head extends around 30 degrees over the end of the couch. The patient's head is turned 45 degrees toward the clinician and moved gently into the lying position with the head hanging some 30 degrees over the head of the couch. In this way, the posterior semicircular canal of the lowermost ear is moved directly through its plane of orientation. The patient's eyes are observed for nystagmus for up to 1 minute. In central positional nystagmus, sometimes seen in multiple sclerosis, posterior fossa lesions, or cerebellar vascular disease, typically there is no latency to the onset of nystagmus, frequently little vertigo, and no adaptation or fatigability. The direction of nystagmus may be toward the uppermost ear, torsional, or may be vertical. Note that positional nystagmus may be the only sign of posterior fossa pathology, but is also common in patients with vestibular migraine, in the acute phase (Rocha et al., 2023). See Kaski and Bronstein (2014) for a review of BPPV positional maneuvers.

by specific head movement (e.g., turning over in bed, bending down, or reaching up), but can only be confirmed with a positive Dix–Hallpike or side-lying maneuver (Fig. 7.3). In many patients, particularly those over 65 years, BPPV can present as a primary disorder of balance and gait, sometimes without associated vertigo. We therefore recommend exploring the possibility of BPPV in all patients with an unexplained gait disorder or balance complaint. This includes screening in the acute setting where examination findings are normal and there is no spontaneous nystagmus on examination (Li et al., 2023) ().

Multiple systematic reviews and meta-analyses have confirmed the safety and efficacy of repositioning maneuvers, such as the Epley Fig. 7.4 and Semont maneuvers Fig. 7.3, for treating posterior canal BPPV (Liu et al., 2016; Zhang et al.,

2017; Alashram, 2024). One meta-analysis of nine randomised controlled trials (RCTs), comparing repositioning maneuvers to sham procedures, found that the treatment arms were more likely to demonstrate symptom resolution (OR 4.6, 95% confidence interval [CI]: 2.8–7.6) and a negative Dix–Hallpike test (OR 5.2, 95% CI: 3.0–8.8) at the time of the first follow-up (Liu et al., 2016).

There are very few RCTs assessing the efficacy of medication for the symptoms of BPPV. One study randomizing 156 patients to either the Semont maneuver, flunarizine (10 mg/day), or no treatment reported symptom resolution of 57.7% with flunarizine, 94.2% with the Semont maneuver, and 34.6% with no treatment (Salvinelli et al., 2004). A further double-blind, placebo-controlled RCT in 72 patients with posterior canal BPPV suggested that combining 24 mg of betahistine twice daily for one week with the Epley maneuver improved quality of life above the Epley maneuver alone (Sayin et al., 2022). The underlying mechanism of this response is unclear, and is not routinely done. Replication of these data is required before recommending incorporation into clinical practice. Benzodiazepines have not shown any difference when compared to placebo for BPPV (Hunter et al., 2022).

Vestibular neuritis

Vestibular neuritis can be thought of as an acute unilateral vestibular "paresis" and has variously been known as vestibular neuronitis, vestibular neuro-labyrinthitis, or acute peripheral vestibulopathy. The syndrome is also often referred to as 'labyrinthitis' but this is incorrect because there is neither hearing loss nor tinnitus because vestibular neuritis spares the cochlear nerve. Vestibular neuritis is characterized by an acute onset of rotational vertigo, oscillopsia, postural instability, nausea, or vomiting, and the absence of additional hearing or neurologic signs. The presence of positive ocular motor signs enables a confident diagnosis to be made, namely unidirectional horizontal nystagmus (with a smaller torsional component) beating away from the lesion side, an abnormal head impulse test toward the lesion side, and gait ataxia. The latter is typically a moderate degree of unsteadiness rather than an inability to stand unaided, as seen in certain acute cerebellar syndromes such as stroke.

A randomised controlled trial (RCT) found that dimenhydrinate 50 mg was superior to lorazepam 2 mg for the relief of vertigo in patients with vestibular neuritis presenting to the emergency department. In a small prospective study of 39 patients, vestibular exercises were shown to improve central vestibulo-spinal compensation after the acute symptoms had resolved (Marill et al., 2000). Indeed, clinical experience suggests that patients with vestibular neuritis (or acute unilateral peripheral vestibulopathy as it is more appropriately referred to) recover faster with a good explanation of the cause of the symptoms, reassurance, and encouragement of early mobilization and head movements. This is backed by empiric data showing that increased autonomic arousal, heightened reliance on vision for balance (visual dependency), and neuroticism personality traits are poor prognostic factors (Cousins et al., 2017; Bronstein and Dieterich, 2019; Trinidade et al., 2023).

Corticosteroids, antivirals, and benzodiazepines are not routinely used, as there is no evidence that they improve long-term outcomes (Leong et al., 2021).

Ménière's disease

Ménière's disease is sometimes referred to as endolymphatic hydrops, reflecting the underlying pathologic process. It is an inner ear disorder characterized by prolonged attacks of vertigo of greater than 30 minutes duration, hearing loss, tinnitus, and aural fullness. The diagnostic criteria from the Classification Committee of the Bárány Society requires formal audiometric testing for a definitive diagnosis, typically revealing low-frequency losses that may fluctuate over time. As many as 30% of patients with Ménière's disease develop drop attacks (otolithic crises) that are characterized by a sudden fall without warning and without loss of consciousness, with no associated neurologic symptoms and normal neurologic examination (Joffily et al., 2024). There is increasing recogition of Meniere's subtypes, based on inflammatory profiles that in time may influence the therapuetic process.e

Medical treatment options are in fact very limited, not least because the underlying pathophysiology of this disorder is poorly understood. Options include the following:

1. Salt restriction (2–3 g) and diuretics have traditionally been considered the best medical therapy for Ménière's disease, reportedly controlling vertigo in 58% of patients and stabilizing hearing in 69%. The effect of diuretics has not, however, been systematically replicated (Webster et al., 2023a).
2. Betahistine has been used in the treatment of vertigo for over 50 years. A recent Cochrane review evaluated the benefits and harms of systemic pharmacologic interventions versus placebo or no treatment in people with Ménière's disease. Seven RCTs (548 participants) addressed this comparison but no meaningful conclusions could be drawn from the numerical results due to methodological differences. The evidence overall was low- or very low-certainty throughout (Webster et al., 2023b). Doses as high as 480 mg/day have been used successfully in individual patients. An RCT comparing placebo (n=74), low-dose betahistine (2×24 mg daily, n=73), or high-dose betahistine (3×48 mg daily, n=74) over nine months, found no significant difference in incidence of attacks between the groups (Adrion et al., 2016).
3. Intratympanic corticosteroid and gentamicin injections have limited evidence from small trials in the treatment of Ménière's disease, and many of these lack adequate placebo groups (Webster et al., 2023c, 2023d). The most recent Cochrane reviews included 10 studies with a total of 952 participants receiving corticosteroids. All studies used the corticosteroid dexamethasone, with doses ranging from approximately 2 to 12 mg. Intratympanic corticosteroids

appear to make little or no difference to the number of people who report an improvement in their vertigo at 6 to ≤12 months follow-up (intratympanic corticosteroids 96.8%, placebo 96.6%, risk ratio [RR] 1.00, 95% CI: 0.92–1.10; 2 studies; 60 participants; low-certainty evidence) or at more than 12 months follow-up (intratympanic corticosteroids 100%, placebo 96.3%; RR 1.03, 95% CI: 0.87–1.23; 2 studies; 58 participants; low-certainty evidence). Studies that included a placebo group also show high improvements.

Vestibular migraine

Vestibular migraine is perhaps the most common cause of recurrent episodic vertigo, particularly in younger individuals, with strong female predominance. vestbular migraine (VM) remains underdiagnosed, particularly in primary care. It is described as recurrent episodes of vestibular symptoms – most commonly an internal sensation of movement and unsteadiness – which are associated with migraine or migrainous features (particularly motion sensitivity), when other vestibular disorders have been ruled out. Vestibular migraine episodes can occur in the absence of headache, and each attack may last seconds to days, but minutes to hours is the norm. Clinical examination in between attacks is usually normal, as are laboratory investigations and brain imaging. During an attack, it is not uncommon to see nystagmus, which can be of any type (spontaneous, gaze-evoked, or positional) (von Brevern et al., 2005; Rocha et al., 2023).

A single RCT has compared venlafaxine, valproate, and flunarizine and found that a decrease in dizziness handicap inventory total score was shown following treatment with all three medications, with no obvious differences between the groups (Liu et al., 2017). A further RCT compared flunarizine combined with betahistine to betahistine alone, revealing an improvement in the frequency of vertigo in the flunarizine group, but no difference in duration and severity of symptoms (Lepcha et al., 2014). For a comprehensive review of treatment options, see Smyth et al. (2022).

Persistent postural-perceptual dizziness

Persistent postural-perceptual dizziness (PPPD) is a chronic functional vestibular disorder that affects balance and is the commonest cause of continuous dizziness in younger individuals. Patients with PPPD often describe a sense of internal motion (even without objective movement of the body), akin to a feeling of "rocking or swaying as if on a boat," a sense of unsteadiness, vibrations inside the body, and walking "as if drunk." Symptoms tend to be worse in busy or visually rich environments but can also persist even when the patient is lying in bed at night (Staab et al., 2017). There may be accompanying symptoms such as difficulty concentrating, short-term memory loss, and a sensation of derealization (feeling detached from the world) or depersonalization (feeling detached from oneself). PPPD is commonly precipitated by events that disrupt or threaten balance or cause vertigo, unsteadiness, or dizziness. The most common triggers we see include peripheral or central vestibular disorders, traumatic brain injuries, other medical illnesses and panic attacks, but spontaneous onset is only recognized.

Currently, no placebo-controlled randomized trials have been conducted to evaluate the efficacy of pharmacologic treatments, particularly selective serotonin reuptake inhibitors (SSRIs) and serotonin-norepinephrine reuptake inhibitors (SNRIs) for PPPD (Webster et al., 2023e). The lack of high-quality evidence leaves significant uncertainty about the role of these medications in managing this condition. Although some studies have suggested potential benefits based on open-label designs or small cohorts, these findings are limited by methodological weaknesses, including the absence of control groups and potential biases. As a result, it remains unclear whether SSRIs or SNRIs can reliably improve PPPD symptoms or whether their use may lead to adverse effects.

There is, however, good clinical evidence that psychologically informed rehabilitation therapy or "cognitive physical therapy" (Herdman et al., 2022; Kaski, 2024) is effective in managing patients with PPPD. Treatment must always start with a thorough and rational explanation of the disorder, in terms that patients can understand, and that aligns with their own beliefs of the underlying cause or trigger. In my experience, prognosis is good and medications are rarely required, or effective, in reducing symptom burden or improving quality of life if a coherent and thorough explanation of the nature of the symptoms is offered, and treatment addressing the underlying misperceptions are initiated.

Oscillopsia

Oscillopsia refers to the illusion of oscillation of the visual world and, in the clinical setting, denotes the presence of nystagmus. Patients with oscillopsia may describe blurring, "wobbly," or "jumpy" vision. Further in-depth information with regard to subtypes and etiology is beyond the scope of this chapter. For a review of the causes and therapy of oscillopsia, the reader is referred to the 2004 EFNS Task Force guidance (Straube et al., 2004). Off-license pharmacologic treatments of oscillopsia used by some specialists include the following, although no double-blind clinical trials are available (Tilikete and Vighetto, 2011):

1. Aminopyridines, for example, 3,4-diaminopyridine (DAP) (20 mg tds) and 4-aminopyridine (AP) (10 mg tds), may be used for downbeat nystagmus (DBN).
2. Modified amino acid acetyl-DL-leucine (5 g/day) may be used for cerebellar ataxia.
3. Gabapentin, valproate, clonazepam, memantine, and biperiden can be tried for continuous (rather than positional or paroxysmal) oscillopsia. There is variable response to these drugs, so each should be tried individually before being combined.

4. In one case series, treatment with baclofen (5–10 mg PO tds) resulted in an improvement in several patients with upbeat nystagmus (Dieterich et al., 1991).

Rehabilitation strategies for vestibular dysfunction

Patients with persistent chronic vestibular symptoms may have PPPD, uncompensated unilateral peripheral vestibular dysfunction, or less commonly, bilateral vestibular hypofunction. For both unilateral and bilateral dysfunction, intensive vestibular rehabilitation physiotherapy is the mainstay of treatment. Every effort should be made to ensure optimal sensory input for balance: correction of visual/ophthalmic disorders such as cataracts, glaucoma or refractive errors, appropriate management of joint/orthopedic problems (e.g., arthritis), and medical management of fluctuating vestibular conditions such as Ménière's disease or BPPV.

Compensation cannot be effective in the face of inadequate sensory input or fluctuating vestibular activity. Importantly, antiemetics and/or vestibular sedative drugs should not be used for chronic symptoms – there is evidence that they impair vestibular compensation. Notwithstanding this fact, there is widespread and inappropriate prescription of these drugs for chronic vestibular symptoms both in primary and tertiary care.

CNS plasticity underpins vestibular rehabilitation and symptomatic vestibular compensation. The Cawthorne–Cooksey exercises are systematic exercises aimed at stimulating visual, vestibular, and proprioceptive input on a repetitive basis to enhance compensation. They were initially introduced empirically in the 1940s (Cawthorne, 1946; Cooksey, 1946). In the 1970s and 1980s, animal experimentation provided a scientific basis for their value, providing evidence to support the rationale for specific physiotherapy for chronic vestibular symptoms, although it is not appropriate for all chronic dizziness (Whitney et al., 2016). Subsequently, a range of vestibular rehabilitation programs have been devised, containing a number of key elements that benefit from a tailored approach:

- A detailed explanation of the rationale of exercises and the aim of therapy to ensure patient motivation and compliance.
- A graded approach, with increasing sensory input and speed of task, to expedite compensation.
- Emphasis on exercises that are functionally relevant and that actually provoke dizziness in individual patients.
- Repeated short but frequent repetitions of individual exercises to promote gradual habituation.

Many studies have shown the efficacy of these approaches. Recent research has highlighted that customized exercises with a program devised specifically for each patient, based on individual vestibular limitations, are most effective (Whitney et al., 2016). Mechanical exercise programs including optokinetic stimulation, visual flow stimuli, and rotational stimuli have been shown to be especially helpful and to optimize vestibular compensation. Other strategies shown to enhance compensation include virtual reality t'ai chi and support surface translations (Sulway and Whitney, 2019).

Traditionally, vestibular rehabilitation physiotherapy has been used in the treatment of unilateral peripheral vestibular disorders that have failed to compensate. Many factors are associated with failure of compensation, and most can be targeted with physiotherapy, by specific evaluations including standardized health report questionnaires, scored balance assessment tests, and dynamic posturography. These measures also document recovery, support patient motivation and cooperation, and direct ongoing physiotherapy.

It cannot be overemphasized that a detailed explanation of the underlying pathologic processes, with a basic understanding of vestibular compensation, is crucial if the patient is to accept that physiotherapy rather than some alternative, technically sophisticated option is more appropriate. Early intervention may be beneficial, but neither long-standing symptoms nor old age are negative prognostic factors – enthusiastic treatment is always worth trying.

Recent work has also confirmed the value of vestibular rehabilitation physiotherapy in the management of migraine, bilateral vestibular failure, and central vestibular pathologies including brainstem and cerebellar disease (Harrell et al., 2023).

EMERGING CONCEPTS AND FUTURE DIRECTIONS

Advances in vestibular prosthetics and their impact on central adaptation

Vestibular prosthetics represent a ground-breaking development in the management of vestibular dysfunction, offering potential solutions for individuals with severe balance disorders, such as BVP (van de Berg et al., 2017, 2011; Starkov et al., 2020). These devices aim to replace or supplement the lost function of the vestibular system by delivering artificial sensory inputs to central vestibular pathways, thereby facilitating balance and spatial orientation. Advances in vestibular prosthetic technology and the understanding of their impact on central adaptation have been transformative, with significant implications for patient outcomes.

Vestibular prosthetics are typically implantable devices that interface with the vestibular nerve or its end organs, using motion sensors to detect head movements and deliver corresponding electrical stimulation. Early prototypes were adapted from cochlear implants, with electrodes targeting the semicircular canals. These devices demonstrated proof-of-concept efficacy in restoring rudimentary vestibular function. Recent advances, however, have focused on improving device precision, miniaturization, and biocompatibility, enabling more accurate and selective stimulation of vestibular pathways (van de Berg et al., 2011).

One of the most significant developments has been the refinement of electrode arrays and stimulation algorithms. Modern vestibular prosthetics can selectively stimulate individual semicircular canals or otolith organs, closely mimicking the natural encoding of head motion. This selectivity reduces unwanted side-effects, such as vertigo or oscillopsia, and improves the fidelity of sensory substitution. Advanced algorithms now allow for real-time adjustments to stimulation parameters based on patient-specific needs, optimizing the device's ability to integrate with the CNS.

The impact of vestibular prosthetics on central adaptation is profound. Central adaptation refers to the CNS's ability to reorganize and adjust its processing to accommodate changes in sensory input, a critical mechanism for restoring balance and spatial orientation in vestibular disorders. The use of prosthetics requires the brain to integrate artificial signals, which differ in temporal and spatial dynamics from natural vestibular inputs. Studies using animal models and early clinical trials have shown that the CNS can adapt to these novel inputs through neuroplasticity, particularly in the vestibular nuclei, cerebellum, and associated cortical areas.

However, the extent and speed of adaptation vary among individuals, influenced by factors such as age, the duration of vestibular loss, and the presence of other sensory deficits. Research has shown that prolonged exposure to consistent prosthetic stimulation can enhance central adaptation, with patients demonstrating improvements in postural control, gait stability, and visual–vestibular interaction. Furthermore, rehabilitation programs tailored to augment prosthetic use, including balance training and multisensory integration exercises, have been shown to accelerate adaptation and optimize outcomes.

Emerging technologies, such as closed-loop systems, have further enhanced the potential of vestibular prosthetics. These systems incorporate feedback from physiologic markers, such as eye movements or postural sway, to adjust stimulation parameters dynamically. This biofeedback-driven approach not only improves the accuracy of the artificial sensory inputs but also promotes more robust central adaptation by aligning stimulation with real-time functional needs.

Insights from functional imaging studies of brainstem–vestibular interactions

Functional imaging has provided invaluable insights into the complex interplay between the brainstem and the vestibular system, elucidating mechanisms underlying balance, spatial orientation, and multisensory integration. While the brainstem is a central hub for processing vestibular inputs, its deep anatomic location and small size historically limited detailed investigation. Advances in functional imaging modalities, including fMRI, positron emission tomography, and magnetoencephalography, have allowed researchers to study brainstem–vestibular interactions with increasing precision, offering a deeper understanding of their role in health and disease.

Functional imaging studies have demonstrated the critical role of the vestibular nuclei, located in the brainstem, as the primary relay for vestibular inputs from the semicircular canals and otolith organs. These nuclei integrate sensory signals with inputs from visual, somatosensory, and proprioceptive systems, coordinating outputs to the cerebellum, thalamus, and cortical regions. For instance, fMRI studies in healthy individuals undergoing caloric or galvanic vestibular stimulation have identified activation in the vestibular nuclei and adjacent regions, highlighting their central role in processing both static and dynamic vestibular inputs (Dieterich and Brandt, 2008; Smith et al., 2024).

Another important insight is the dynamic modulation of vestibular processing in the brainstem by higher cortical centers and external sensory stimuli. Functional imaging has revealed reciprocal interactions between the brainstem vestibular nuclei and cortical regions, such as the temporoparietal junction and the insular cortex. These findings suggest that the vestibular system is not a simple feedforward system but is dynamically influenced by attention, sensory context, and motor planning. For example, studies using task-based fMRI have shown that vestibular nuclei activity decreases during concurrent visual or proprioceptive dominance, supporting the concept of sensory reweighting.

Functional imaging has also shed light on the neural underpinnings of vestibular disorders, such as vestibular migraine and PPPD. Altered patterns of activation and connectivity between the brainstem and higher vestibular centers have been identified in these conditions, providing insights into pathophysiology. For instance, hyperactivation of the vestibular nuclei and disrupted brainstem–cerebellar connectivity observed in PPPD patients suggest maladaptive plasticity that perpetuates dizziness and spatial disorientation.

Emerging techniques, such as high-field 7 T fMRI (Theysohn et al., 2014), offer enhanced spatial resolution to study brainstem–vestibular interactions in even greater detail. Combined with advanced methods like dynamic causal modeling and connectivity analysis, these approaches are elucidating the bidirectional flow of information between the brainstem and cortical regions, advancing both basic science and clinical applications (Tak et al., 2018).

CONCLUSION

The vestibular system is integral to everyday human function, from being able to cross the road safely, getting to the bathroom at night, or even paying with your children or grandchildren. The tiny end organs within the labyrinth that detect head motion feed a deep and wide network that involves early brainstem processing of the head motion and gravity signals, before ultimate processing by in the cortex. When this system malfunctions, the sensations of vertigo, dizziness, unsteadiness, and spatial disorientation are as deep and wide as the networks that underpin these. Advances in our understanding of the mechanisms that underpin the common causes of vestibular dysfunction have led to novel treatment options for these debilitating symptoms; the future holds significant promise in this exciting field.

References

Adrion C, Fischer CS, Wagner J et al. (2016). Efficacy and safety of betahistine treatment in patients with Meniere's disease: primary results of a long term, multicentre, double blind, randomised, placebo controlled, dose defining trial (BEMED trial). BMJ 352: h6816.

Alashram AR (2024). Semont maneuver for posterior canal benign paroxysmal positional vertigo: a systematic review of randomized controlled trials. Eur Arch Otorhinolaryngol 281: 3985–3999.

Bhattacharyya N, Gubbels SP, Schwartz SR et al. (2017). Clinical practice guideline: benign paroxysmal positional vertigo (Update). Otolaryngol Head Neck Surg 156: S1–S47.

Bisdorff A, Von Brevern M, Lempert T et al. (2009). Classification of vestibular symptoms: towards an international classification of vestibular disorders. J Vestib Res 19: 1–13.

Brandt T, Dieterich M (1999). The vestibular cortex. Its locations, functions, and disorders. Ann N Y Acad Sci 871: 293–312.

Britton Z, Arshad Q (2019). Vestibular and multi-sensory influences upon self-motion perception and the consequences for human behavior. Front Neurol 10: 63.

Bronstein AM, Dieterich M (2019). Long-term clinical outcome in vestibular neuritis. Curr Opin Neurol 32: 174–180.

Bronstein AM, Lempert T (2010). Management of the patient with chronic dizziness. Restor Neurol Neurosci 28: 83–90.

Bronstein AM, Lempert T (2017). Dizziness: a practical approach to diagnosis and management, pp. 212. Cambridge University Press.

Burston A, Mossman S, Mossman B et al. (2018). Comparison of the video head impulse test with the caloric test in patients with sub-acute and chronic vestibular disorders. J Clin Neurosci 47: 294–298.

Carcaud J, França de Barros F, Idoux E et al. (2017). Long-lasting visuo-vestibular mismatch in freely-behaving mice reduces the vestibulo-ocular reflex and leads to neural changes in the direct vestibular pathway. eNeuro 4: ENEURO.0290-16.2017.

Cawthorne T (1946). Vestibular injuries. Proc R Soc Med 39: 270–273.

Cooksey FS (1946). Rehabilitation in vestibular injuries. Proc R Soc Med 39: 273–278.

Cousins S, Kaski D, Cutfield N et al. (2017). Predictors of clinical recovery from vestibular neuritis: a prospective study. Ann Clin Transl Neurol 4: 340–346.

Curthoys IS, Grant JW, Pastras CJ et al. (2019). A review of mechanical and synaptic processes in otolith transduction of sound and vibration for clinical VEMP testing. J Neurophysiol 122: 259–276.

Curthoys IS, McGarvie LA, MacDougall HG et al. (2023). A review of the geometrical basis and the principles underlying the use and interpretation of the video head impulse test (vHIT) in clinical vestibular testing. Front Neurol 14: 1147253.

Dieterich M, Brandt T (2008). Functional brain imaging of peripheral and central vestibular disorders. Brain 131: 2538–2552.

Dieterich M, Brandt T (2018). The parietal lobe and the vestibular system. Handb Clin Neurol 151: 119–140.

Dieterich M, Straube A, Brandt T et al. (1991). The effects of baclofen and cholinergic drugs on upbeat and downbeat nystagmus. J Neurol Neurosurg Psychiatry 54: 627–632.

Di Fabio RP (1995). Sensitivity and specificity of platform posturography for identifying patients with vestibular dysfunction. Phys Ther 75: 290–305.

Halmagyi GM, Curthoys IS (1988). A clinical sign of canal paresis. Arch Neurol 45: 737–739.

Harrell RG, Cassidy AR, Klatt BN et al. (2023). Vestibular rehabilitation in cerebellar ataxia with neuropathy and vestibular areflexia syndrome (CANVAS)—a case report. J Otol 18: 199–207.

Herdman D, Norton S, Murdin L et al. (2022). The INVEST trial: a randomised feasibility trial of psychologically informed vestibular rehabilitation versus current gold standard physiotherapy for people with Persistent Postural Perceptual Dizziness. J Neurol 269: 4753–4763.

Hunter BR, Wang AZ, Bucca AW et al. (2022). Efficacy of benzodiazepines or antihistamines for patients with acute vertigo: a systematic review and meta-analysis. JAMA Neurol 79: 846–855.

Joffily L, Simonyan S, Khalil S et al. (2024). Vestibular drop attacks diagnosed with valuable insights from CCTV. Lancet 404: 787–788.

Kaski D (2024). Introduction to the assessment and management of persistent postural-perceptual dizziness. J Laryngol Otol 138: S27–S31.

Kaski D, Quadir S, Nigmatullina Y et al. (2016). Temporoparietal encoding of space and time during vestibular-guided orientation. Brain 139: 392–403.

Kaski D, Bronstein AM (2014). Epley and beyond: an update on treating positional vertigo. Pract Neurol 14: 210–221.

Kirsch V, Boegle R, Keeser D et al. (2019). Beyond binary parcellation of the vestibular cortex – a dataset. Data Brief 23: 103666.

Koohi N, Mendis S, Lennox A et al. (2022). Video head impulse testing: pitfalls in neurological patients. J Neurol Sci 442: 120417.

Leong KJ, Lau T, Stewart V et al. (2021). Systematic review and meta-analysis: effectiveness of corticosteroids in treating adults with acute vestibular neuritis. Otolaryngol Head Neck Surg 165: 255–266.

Lepcha A, Amalanathan S, Augustine AM et al. (2014). Flunarizine in the prophylaxis of migrainous vertigo: a randomized controlled trial. Eur Arch Otorhinolaryngol 271: 2931–2936.

Li Y, Smith RM, Whitney SL et al. (2023). We should be screening for benign paroxysmal positional vertigo (BPPV) in all older adults at risk of falling: a commentary on the World Falls Guidelines. Age Ageing 52: afad206.

Liu F, Ma T, Che X et al. (2017). The efficacy of venlafaxine, flunarizine, and valproic acid in the prophylaxis of vestibular migraine. Front Neurol 8: 524.

Liu Y, Wang W, Zhang AB et al. (2016). Epley and Semont maneuvers for posterior canal benign paroxysmal positional vertigo: a network meta-analysis. Laryngoscope 126: 951–955.

Mahmud M, Kaski D (2024). Vestibular neurology for the generalist. J Neurol Neurosurg Psychiatry 95: 1196–1206.

Marill KA, Walsh MJ, Nelson BK (2000). Intravenous Lorazepam versus dimenhydrinate for treatment of vertigo in the emergency department: a randomized clinical trial. Ann Emerg Med 36: 310–319.

Mermelstein S, Barbosa P, Kaski D (2024). Neurological gait assessment. Pract Neurol 24: 11–21.

Mishra A, Pandey S (2022). Upbeat and Brun's nystagmus in cerebellopontine tuberculoma. Ann Indian Acad Neurol 25: 1194–1196.

Nagururu NV, Akbar A, Ward BK (2022). Using magnetic resonance imaging to improve diagnosis of peripheral vestibular disorders. J Neurol Sci 439: 120300.

Rocha MF, Sacks B, Al-Lamki A et al. (2023). Acute vestibular migraine: a ghost diagnosis in patients with acute vertigo. J Neurol 270: 6155–6158.

Saber Tehrani AS, Kattah JC, Mantokoudis G et al. (2014). Small strokes causing severe vertigo: frequency of false-negative MRIs and nonlacunar mechanisms. Neurology 83: 169–173.

Salvinelli F, Trivelli M, Casale M et al. (2004). Treatment of benign positional vertigo in the elderly: a randomized trial. Laryngoscope 114: 827–831.

Sayin I, Koç RH, Temirbekov D et al. (2022). Betahistine add-on therapy for treatment of subjects with posterior benign paroxysmal positional vertigo: a randomized controlled trial. Braz J Otorhinolaryngol 88: 421–426.

Sekar A, Panouilleres MTN, Kaski D (2024). Detecting abnormal eye movements in patients with neurodegenerative diseases – current insights. Eye Brain 16: 3–16.

Smith LJ, Wilkinson D, Bodani M et al. (2024). Cognition in vestibular disorders: state of the field, challenges, and priorities for the future. Front Neurol 15: 1159174.

Smyth D, Britton Z, Murdin L et al. (2022). Vestibular migraine treatment: a comprehensive practical review. Brain 145: 3741–3754.

Staab JP, Eckhardt-Henn A, Horii A et al. (2017). Diagnostic criteria for persistent postural-perceptual dizziness (PPPD): consensus document of the committee for the Classification of Vestibular Disorders of the Barany Society. J Vestib Res 27: 191–208.

Starkov D, Guinand N, Lucieer F et al. (2020). Restoring the high-frequency dynamic visual acuity with a vestibular implant prototype in humans. Audiol Neurootol 25: 91–95.

Straube A, Leigh RJ, Bronstein A et al. (2004). EFNS Task Force – therapy of nystagmus and oscillopsia. Eur J Neurol 11: 83–89.

Strupp M, Lopez-Escamez JA, Kim JS et al. (2016). Vestibular paroxysmia: diagnostic criteria. J Vestib Res 26: 409–415.

Sulway S, Whitney SL Advances in vestibular rehabilitation. Adv Otorhinolaryngol 82: 164–169.

Tak S, Noh J, Cheong C et al. (2018). A validation of dynamic causal modelling for 7T fMRI. J Neurosci Methods 305: 36–45.

Theysohn JM, Kraff O, Eilers K et al. (2014). Vestibular effects of a 7 Tesla MRI examination compared to 1.5 T and 0 T in healthy volunteers. PLoS One 9: e92104.

Thylur DS, Jacobs RE, Go JL et al. (2017). Ultra-high-field magnetic resonance imaging of the human inner ear at 11.7 Tesla. Otol Neurotol 38: 133–138.

Tilikete C, Vighetto A (2011). Oscillopsia: causes and management. Curr Opin Neurol 24: 38–43.

Trinidade A, Cabreira V, Goebel JA et al. (2023). Predictors of persistent postural-perceptual dizziness (PPPD) and similar forms of chronic dizziness precipitated by peripheral vestibular disorders: a systematic review. J Neurol Neurosurg Psychiatry 94: 904–915.

van de Berg R, Guinand N, Stokroos RJ et al. (2011). The vestibular implant: quo vadis? Front Neurol 2: 47.

van de Berg R, Guinand N, Ranieri M et al. (2017). The vestibular implant input interacts with residual natural function. Front Neurol 8: 644.

von Brevern M, Zeise D, Neuhauser H et al. (2005). Acute migrainous vertigo: clinical and oculographic findings. Brain 128: 365–374.

Webster KE, Lee A, Galbraith K et al. (2023d). Intratympanic corticosteroids for Ménière's disease. Cochrane Database Syst Rev 2: CD015245.

Webster KE, Galbraith K, Lee A et al. (2023c). Intratympanic gentamicin for Ménière's disease. Cochrane Database Syst Rev 2: CD015246.

Webster KE, George B, Lee A et al. (2023a). Lifestyle and dietary interventions for Ménière's disease. Cochrane Database Syst Rev 2: CD015244.

Webster KE, Harrington-Benton NA, Judd O et al. (2023e). Pharmacological interventions for persistent postural-perceptual dizziness (PPPD). Cochrane Database Syst Rev 3: CD015188.

Webster KE, Galbraith K, Harrington-Benton NA et al. (2023b). Systemic pharmacological interventions for Ménière's disease. Cochrane Database Syst Rev 2: CD015171.

Whitney SL, Alghwiri AA, Alghadir A (2016). An overview of vestibular rehabilitation. Handb Clin Neurol 137: 187–205.

Zee DS (2021). Eye movements in general neurology and its subspecialties: introduction to the topical collection. Neurol Sci 42: 387–388.

Zhang X, Qian X, Lu L et al. (2017). Effects of Semont maneuver on benign paroxysmal positional vertigo: a meta-analysis. Acta Otolaryngol 137: 63–70.

Handbook of Clinical Neurology, Vol. 216 (3rd series)
The Brainstem
L.T. Grinberg and A.D. Korczyn
https://doi.org/10.1016/B978-0-443-15736-3.00019-6

Chapter 8

Oculomotor disorders and nystagmus

CLAUDIA KREBS*

Department of Cellular and Physiological Sciences, Faculty of Medicine, University of British Columbia, Vancouver, BC, Canada

Abstract

The precision of visual perception depends on accurately directing the fovea to objects of interest, a process mediated by six supranuclear oculomotor systems that all converge in the brainstem: saccadic and smooth pursuit, vestibular optokinetic, fixation, and vergence. This chapter provides a clinically oriented overview of brainstem neuroanatomy relevant to eye movement control and links it to characterisitic patterns of gaze disturbance. After outlining the organization and functions of the oculomotor (III), trochlear (IV), and abducens (VI) nuclei and their connections via the medial longitudinal fasciculus, the brainstem substrates for horizontal and versitcal gaze are discussed. A review of the circuitry and physiology of saccades, smooth pursuit, the vestibulo-ocular and optokinetic reflexes, fixation micro-movements, and vergence, emphasizes how shared and parallel pathways support adaptive, stable gaze in dynamic environments. Finally, specific lesion in the midbrain, pons, medulla, and at the cerebella-medullary junction are discussed and how they give rise to distinct forms of nystagmus and gaze palsy. This provides a powerful localization approach in conditions such as demyelinating, vascular, degenerative, and vestibular disorders. By integrating systems-level physiology with detailed brainstem anatomy and clinical approaches, this chapter offers a framework for understanding oculomotor disturbances and for using eye movement examination as a high-yield tool for brainstem localization and diagnosis.

INTRODUCTION

The ability to perceive the world clearly and efficiently is intricately tied to the precision and coordination of eye movements. Normal visual perception depends on directing the fovea, the central region of the retina responsible for sharp vision, toward objects of interest with remarkable accuracy. The oculomotor system achieves this through a complex interplay of neural circuits and pathways, involving cortical and subcortical regions and six distinct supranuclear systems: the saccadic, smooth pursuit, vestibular, optokinetic, fixation, and vergence systems.

Gaze disorders, arising from disruptions at various levels of the central nervous system, offer unique insights into the functionality of these systems. From sudden jerky saccades that shift gaze to new targets to the smooth pursuit movements that track moving objects, these disorders manifest in patterns that aid clinicians in precisely localizing lesions or dysfunctions within the oculomotor pathway. Common causes of gaze impairments include cerebral infarcts, multiple sclerosis, neurodegenerative diseases like progressive supranuclear palsy, and metabolic disorders such as Wernicke's encephalopathy. Early and accurate diagnosis of these conditions can significantly influence treatment outcomes.

This chapter integrates detailed insights into the brainstem's neuroanatomy with the functional aspects of the oculomotor system. By exploring the cortical commands and brainstem execution pathways, clinicians and researchers can develop systematic diagnostic strategies and therapeutic approaches that enhance early recognition and management of gaze disorders.

DETAILED BRAINSTEM NEUROANATOMY IN OCULOMOTOR CONTROL

The brainstem's role in oculomotor control is anchored by its well-defined neuroanatomic structures. Three primary cranial nerve nuclei – oculomotor (CN III), trochlear (CN IV), and abducens (CN VI) – reside within the brainstem and are essential for eye movement. The oculomotor nucleus, located in the midbrain near the superior colliculus (SC), innervates most extraocular muscles and includes the visceral motor (parasympathetic) Edinger–Westphal nucleus, which mediates pupillary constriction and lens accommodation. The trochlear nucleus, found in the caudal midbrain, is unique because its

*Correspondence to: Claudia Krebs, MD, PhD, FAAA, Department of Cellular and Physiological Sciences, 2350 Health Sciences Mall, Vancouver, BCV6T 1Z3, Canada. Tel: +1-604-827-5694; E-mail: claudia.krebs@ubc.ca.

fibers cross the midline before exiting dorsally, allowing for contralateral control of the superior oblique muscle. The abducens nucleus, situated in the pontine tegmentum, controls the lateral rectus muscle and collaborates with the oculomotor nucleus via the medial longitudinal fasciculus (MLF) to coordinate horizontal gaze.

A detailed understanding of brainstem neuroanatomy and its role in oculomotor control has profound clinical implications. Assessments of saccadic dynamics, smooth pursuit, fixation stability, and vestibulo-ocular reflex (VOR) function – when interpreted in the context of brainstem integrity – allow clinicians to localize lesions with high precision.

BRAINSTEM AS THE INTEGRATIVE CENTER IN OCULOMOTOR CONTROL

Although supranuclear systems coordinate eye movements at multiple levels, the brainstem serves as the final common pathway that executes and integrates these signals. The brainstem is not merely a relay station; it actively participates in adaptive error correction, plasticity, and sensorimotor integration (Scudder et al., 2002; Gandhi and Katnani, 2011). This dynamic processing is crucial for maintaining accurate and stable vision in a constantly changing environment. The control of horizontal and vertical gaze relies on distinct yet interconnected neuroanatomic substrates within the brainstem.

HORIZONTAL GAZE

Horizontal gaze primarily involves structures in the pons and midbrain that coordinate the conjugate movement of both eyes to the same side. Central to this process is the abducens nucleus, which innervates the lateral rectus muscle, allowing the eye to move outward, or abduct. Additionally, the abducens nucleus sends signals through the MLF to the contralateral oculomotor nucleus, ensuring that the medial rectus muscle contracts in tandem for coordinated eye movement. This intricate interplay allows for smooth horizontal saccades and is critically supported by the paramedian pontine reticular formation (PPRF). The PPRF acts as a key generator of horizontal saccades, and damage to this area can selectively impair these rapid eye movements while leaving smooth pursuit and vergence relatively intact (Moschovakis et al., 1996; Scudder et al., 2002).

The MLF itself is a pivotal pathway, not only for horizontal gaze but also for the VOR, which stabilizes the eyes during head movements. Lesions in the MLF result in internuclear ophthalmoplegia (INO), a condition where the affected eye fails to adduct while the other eye exhibits nystagmus during abduction (Zee, 1992). Horizontal gaze stability, particularly in eccentric positions, is maintained by nuclei such as the nucleus prepositus hypoglossi and the medial vestibular nucleus (MVN; McFarland and Fuchs, 1992). These structures integrate eye position signals to prevent drift when the eyes are held away from the primary position.

VERTICAL GAZE

Vertical gaze is governed by a complex network of midbrain structures. The rostral interstitial nucleus of the medial longitudinal fasciculus (riMLF) plays a crucial role in generating vertical saccades and torsional quick phases. Damage to this nucleus impairs the ability to perform vertical saccades and often results in torsional nystagmus, disrupting the precise alignment needed for vertical gaze shifts. Complementing the riMLF, the interstitial nucleus of Cajal (INC) is essential for maintaining vertical gaze stability (Yu and Wang, 2025). Lesions here can lead to vertical gaze-evoked nystagmus and ocular tilt reactions, which include skew deviation and torsional nystagmus, reflecting the delicate balance the INC maintains in vertical positioning (Strupp et al., 2014, 2021).

Another critical structure in vertical gaze control is the posterior commissure. Lesions in this area often impair upward gaze, disrupt convergence, and cause convergence-retraction nystagmus, a hallmark of Parinaud's syndrome (Ortiz et al., 2021).

The oculomotor and trochlear nuclei are also deeply involved in vertical gaze mechanics. The oculomotor nucleus governs the superior and inferior rectus muscles, facilitating upward and downward movements, while the trochlear nucleus controls the superior oblique muscle, contributing to downward and torsional gaze.

The MLF extends its role into vertical gaze by facilitating the vertical VOR, ensuring that the eyes move appropriately in response to vertical head movements. The integration of these brainstem pathways allows for the precise control of both horizontal and vertical gaze, maintaining stable vision even as the head moves or when tracking objects across different visual fields.

SACCADES AND GAZE CONTROL

Saccadic eye movements are rapid, ballistic shifts in gaze that reorient the fovea toward targets of interest in the visual field. These movements are essential for efficient visual exploration, allowing humans to engage in tasks such as reading, scanning environments, and focusing on objects during dynamic activities like driving. Saccades occur at a high frequency, typically two or three times per second, underscoring their significance in active visual perception (Ibbotson and Krekelberg, 2011).

The neural circuitry underlying saccadic movements is distributed across cortical, subcortical, and brainstem structures. Key cortical regions include the frontal eye fields (FEFs), supplementary eye fields (SEFs), and parietal eye fields, which work in conjunction with subcortical structures such as the SC, cerebellum, and basal ganglia. These areas form a complex network responsible for initiating, guiding, and refining saccadic movements (Bruce and Goldberg, 1985; Munoz and Everling, 2004; Sommer and Wurtz, 2004).

Saccades are controlled by two primary error signals: position error, the difference between the current and desired gaze position, and velocity error, reflecting dynamic tracking adjustments. These signals converge into a tightly integrated system that adapts to both external and internal demands.

There are three main types of saccades: Reactive saccades are triggered reflexively in response to sudden peripheral stimuli. These saccades exhibit short latencies of about 200 ms and are essential for responding to unexpected changes in the visual environment. Voluntary saccades are guides by internal goals rather than external stimuli; they are guided by cortical areas such as the FEFs and the SEFs and have a longer latency (> 250 ms). The voluntary saccades include memory-guided saccades, predictive saccades, and antisaccades. Adaptive saccades comprise an adaptation that adjusts saccade amplitude and direction to maintain accuracy despite changing physiologic or environmental conditions. This is particularly important in conditions such as muscle weakness where the visual system recalibrates to avoid visual disturbances (Munoz and Everling, 2004; Leigh and Zee, 2015).

Saccades are executed through burst neurons in the brainstem that produce a precise command sequence, initiating eye movement and sustaining fixation at the new target location. During a saccade, omnipause neurons inhibit unnecessary movements, ensuring smooth transitions. This integration of cortical, subcortical, and cerebellar inputs ensures saccadic accuracy and adaptability (Wurtz, 2008).

Saccadic inhibition prevents eye movements toward irrelevant or distracting stimuli, critical during tasks requiring sustained fixation or suppression of reflexive saccades, such as in the antisaccade task.

Despite their speed, saccades do not disrupt visual perception due to mechanisms like saccadic suppression, which temporarily diminishes visual sensitivity during eye movement. This ensures perceptual continuity and prevents motion blur. Efference copies of motor commands also play a role by allowing the brain to predict and compensate for retinal shifts induced by saccades.

There are two critical error signals in saccadic control: position error and velocity error. These signals converge in the brainstem, where the interplay between burst and omnipause neurons allows for rapid correction and adaptation (Scudder et al., 2002). Abnormal saccadic dynamics, which are often seen in disorders such as Parkinson disease (White et al., 1983) and schizophrenia (Thakkar et al., 2017), have been linked to disruptions in these brainstem circuits.

SMOOTH PURSUIT EYE MOVEMENTS

Smooth pursuit eye movements are critical for maintaining the fovea's alignment with moving visual targets. Unlike saccades, which are rapid and discrete, smooth pursuit is characterized by its continuous and steady tracking of moving objects. Smooth pursuit movements are triggered when motion is detected in a target and serve to minimize retinal slip, which is the relative motion of an image on the retina (Sharpe, 2008; Lisberger, 2015). This ensures that the target remains within the foveal region, where visual acuity is highest. By reducing the image displacement across the retina, smooth pursuit enhances visual clarity and stability during dynamic interactions with the environment.

The dynamics of smooth pursuit movements are relatively slow compared to saccades, with velocities typically reaching up to 100 degrees per second. The process begins with an initiation phase, triggered by the detection of target motion. This phase is characterized by a latency of 100–150 ms and involves rapid acceleration of the eyes to match the target's velocity (Orban De Xivry and Lefèvre, 2007). Once the target's speed is achieved, the system enters the steady-state phase, where eye movement stabilizes to maintain alignment with the moving object. However, when a target moves unpredictably or accelerates suddenly, the pursuit mechanism incorporates corrective saccades. These small, rapid eye movements realign the fovea with the target, ensuring the gaze remains accurately fixed. This interplay between continuous tracking and intermittent corrections allows smooth pursuit to adapt to diverse motion patterns and maintain precise visual focus.

Smooth pursuit relies on a sophisticated network of sensory and motor areas within the brain (Krauzlis, 2004). The primary visual cortex, or V1, is responsible for processing initial motion signals and relaying this information to higher-order areas. The middle temporal area and medial superior temporal area are specialized for motion perception, encoding the direction and speed of moving targets. These signals are integrated within the FEFs, particularly the smooth pursuit region (FEFsem), which plays a pivotal role in initiating and controlling pursuit by coordinating sensory inputs with motor commands. The cerebellum also contributes significantly by fine-tuning these movements to ensure smooth and precise tracking. Through its continuous feedback and error correction capabilities, the cerebellum ensures that pursuit remains adaptable to varying conditions and target dynamics.

Smooth pursuit is governed by two principal error signals: position error and velocity error. Position error represents the difference between the target's location and the current gaze position, driving the initial phase of tracking (Badler et al., 2019). Velocity error, on the other hand, reflects the discrepancy between the target's speed and the eye's tracking velocity, becoming the dominant factor during the steady-state phase. The smooth pursuit system adaptively modulates its response, maintaining high levels of accuracy and stability.

Smooth pursuit enhances motion perception, improving the ability to judge the velocity and trajectory of moving targets (Krauzlis, 2004; Lisberger, 2015). This close relationship between pursuit and perception underscores the shared neural substrates that facilitate both action and sensory processing, highlighting the system's role in seamless interaction with dynamic visual stimuli.

Although smooth pursuit and saccades are distinct systems, they frequently interact to optimize gaze control. During pursuit, corrective saccades occur to bring the gaze back to the target when tracking accuracy diminishes. Additionally, shared neural pathways, including those in the FEFs and SC, govern both systems. This overlap allows for the integration of position and velocity signals across pursuit and saccadic networks, enhancing the overall efficiency of gaze stabilization and target tracking (Badler et al., 2019; Goettker and Gegenfurtner, 2021).

Abnormalities in smooth pursuit are linked to various neurologic and psychiatric conditions, including schizophrenia (Tregellas et al., 2004) and Parkinson disease (Li et al., 2024). Deficits in pursuit can serve as early diagnostic markers, providing critical insights into underlying pathologies. Clinical tests that evaluate smooth pursuit dynamics are integral to assessing oculomotor function, aiding in the diagnosis and management of disorders affecting gaze control.

THE VESTIBULO-OCULAR REFLEX

The vestibulo-ocular reflex, or VOR, is an essential mechanism that stabilizes gaze during head movements, ensuring that visual information remains clear and steady. This reflex achieves stabilization by generating compensatory eye movements in the direction opposite to head motion, thereby keeping the visual target fixed on the fovea. The VOR operates with remarkable speed and precision, often responding within milliseconds of head movement onset. It is indispensable for activities such as walking, running, and navigating dynamic environments, where constant adjustments to gaze are necessary to maintain a stable field of vision (Cullen, 2012; Branoner et al., 2016).

The VOR functions as a closed-loop control system that relies on sensory input from the vestibular organs within the inner ear. These inputs are processed to adjust eye position in response to head motion. The semicircular canals detect angular head movements and are particularly sensitive to rotational motion, while the otolith organs detect linear accelerations and changes in head orientation relative to gravity. These sensory signals are transmitted to the extraocular muscles, which execute compensatory eye movements to counteract head motion (Angelaki and Cullen, 2008). The VOR operates across multiple axes, including the horizontal VOR, which stabilizes gaze during side-to-side head movements; the vertical VOR, which compensates for up-and-down motions; and the torsional VOR, which maintains visual stability during head tilts. This multiaxis functionality ensures comprehensive stabilization of gaze relative to the position of the head.

Signals from the vestibular apparatus are first processed by the vestibular nuclei, which serve as the primary integration centers. From here, motor commands are relayed to the oculomotor and abducens nuclei, which control the activity of the extraocular muscles. The cerebellum, particularly the flocculus and nodulus, plays a critical role in refining these responses. Through its ability to adaptively modulate the reflex, the cerebellum ensures that the VOR remains accurate and effective despite changes in sensory input or motor demands (Cullen and Roy, 2004).

In cases of vestibular dysfunction, injury or environmental changes, the VOR can adjust to maintain stability, often by relying more heavily on visual or proprioceptive feedback (Brandt et al., 1997).

Although the VOR operates independently of direct visual input, it interacts with other oculomotor systems to achieve seamless gaze stabilization. The optokinetic reflex, for instance, complements the VOR by using visual cues to stabilize gaze during sustained motion or low-frequency head movements (Cullen and Roy, 2004). Together, these systems ensure that gaze remains stable across a broad range of movement frequencies. Additionally, the VOR works in tandem with saccades and smooth pursuit movements to redirect gaze or track moving objects during dynamic activities(Angelaki and Cullen, 2008).

Dysfunction in the VOR can lead to a range of symptoms, including vertigo, a sensation of spinning caused by abnormal vestibular input; oscillopsia, the perception of unstable visual surroundings due to inadequate compensatory eye movements; and gaze instability, characterized by difficulty maintaining fixation during head movements. Clinicians often assess VOR function through tests such as the head impulse test, which evaluates the reflex's response to rapid head rotations. Such assessments are invaluable in diagnosing vestibular disorders and other conditions that affect gaze stabilization.

In rehabilitation, vestibular therapy employs targeted exercises to recalibrate the VOR in patients with vestibular deficits, improving their balance and visual stability (Hillier and McDonnell, 2011). Advances in neuroprosthetics have led to the development of vestibular implants, which aim to restore VOR function in individuals with severe dysfunction (Della Santina et al., 2007; Wiboonsaksakul et al., 2024).

THE OPTOKINETIC SYSTEM

The optokinetic system is a vital mechanism for maintaining visual stability during sustained motion of the visual environment. Its primary function is to minimize retinal image slip, which it achieves by generating a combination of slow, continuous eye movements that track moving visual scenes and quick resetting saccades (Glasauer and Straka, 2022; Robinson, 2022a). These complementary movements allow the eyes to stabilize the visual field, even as the environment moves, ensuring a coherent visual experience. The optokinetic system works in close collaboration with other oculomotor mechanisms, particularly the VOR (Robinson, 2022a). While the VOR excels at stabilizing gaze during rapid or transient head movements, the optokinetic system assumes a more dominant role during prolonged or low-frequency motions, where it effectively sustains visual stability over time.

Optokinetic eye movements consist of two distinct phases that function in tandem to manage the dynamic visual scene. The slow-phase movement, driven by global motion cues across the visual field, allows the eyes to smoothly follow the motion of the scene. This phase closely resembles smooth pursuit movements but differs in that it is guided by broad environmental motion rather than a single object. The second phase, characterized by quick-phase saccades, resets the gaze after the eyes have reached their physical limits, enabling continuous tracking of the scene. This cyclical pattern of slow and quick phases creates a characteristic nystagmus, which is essential for stabilizing vision during large-scale movements, such as when observing a moving train or the drifting of clouds across the sky (Robinson, 2022b). Together, these movements ensure that the visual environment remains stable and intelligible, regardless of its complexity or scale.

The optokinetic system relies on a sophisticated network of cortical and subcortical pathways to process and respond to visual motion. Motion signals are first detected by the retina, where specialized ganglion cells encode directional information about the moving visual field. These signals are then transmitted to the accessory optic system (AOS) (Giolli et al., 2006), which includes the nucleus of the optic tract (Hoffmann, 1989; Yakushin et al., 2000b, 2000a) and the dorsal terminal nucleus (Schmidt et al., 1993), all located in the midbrain. The AOS plays a critical role in processing global motion cues, ensuring that the system responds to the overall movement of the environment rather than isolated elements. The cerebellum, particularly the flocculus and nodulus, further refines these responses by modulating the gain and accuracy of optokinetic movements, allowing for precise control and adaptation (Voogd and Barmack, 2006). Finally, brainstem oculomotor centers transform the processed sensory input into motor commands that direct the activity of the extraocular muscles, completing the feedback loop necessary for effective gaze stabilization.

The optokinetic system functions synergistically with the VOR to ensure seamless gaze stabilization across different conditions. During short-duration movements or rapid changes in head position, the VOR predominates, providing immediate compensation for head motion. However, during prolonged or low-frequency motion, such as when following a moving landscape, the optokinetic system becomes more prominent. It compensates for the limitations of the VOR by using visual cues to sustain gaze stability over extended periods. This complementary relationship between the two systems highlights their integrated roles in maintaining a stable visual field in dynamic environments, ensuring that the brain receives continuous and accurate sensory input.

THE FIXATION SYSTEM

The fixation system is a cornerstone of oculomotor control, enabling the stabilization of gaze on a target to maximize visual acuity and support detailed visual analysis. This system maintains the alignment of the fovea – the central retinal region with the highest visual resolution – on a stationary or slowly moving object, even in the presence of subtle head or body movements. Fixation is integral to various visual tasks, including reading, object recognition, and visual inspection. By ensuring that the gaze remains steady, the fixation system provides the foundation for clear and stable perception, allowing for efficient interaction with the visual environment.

Despite its apparent stillness, fixation is an active process that involves subtle micro-movements. These include tremors, drifts, and microsaccades, each contributing to the maintenance of steady gaze and the prevention of sensory adaptation (Martinez-Conde et al., 2004; Rolfs, 2009). Tremors are high-frequency, low-amplitude oscillations occurring continuously during fixation. Though their impact on visual perception is minimal, they play a role in maintaining retinal stimulation. Drifts are slow, irregular movements that cause the gaze to wander slightly from the fixation point. This wandering is beneficial, as it prevents sensory adaptation by continually refreshing the retinal image. Microsaccades are small, rapid, corrective movements that return the fovea to the fixation point when it drifts too far. These micro-movements enhance perception by ensuring that visual input remains dynamic and focused. Together, these mechanisms are essential for maintaining fixation while avoiding visual fading, a phenomenon that can occur when retinal stimulation becomes static over extended periods.

The SC plays a central role, particularly its rostral region, which contains fixation neurons that are tonically active during fixation. These neurons inhibit saccadic burst neurons, ensuring that the gaze remains stable (Sommer and Wurtz, 2000). The frontal eye fields also contribute by suppressing competing motor commands and engaging attentional mechanisms that prioritize the fixation target. Additionally, omnipause neurons located in the PPRF inhibit saccadic activity, further supporting gaze stability. This collaboration among neural structures ensures that fixation is precise and resilient, even in dynamic environments.

The interaction between fixation neurons in the SC, omnipause neurons in the PPRF, and cortical inputs ensures that the fovea remains on target. Disruptions in these brainstem circuits, as observed in conditions like nystagmus or fixation instability in Parkinson disease (Sekar et al., 2024), underscore the critical importance of brainstem control in maintaining visual stability.

THE VERGENCE SYSTEM AND BRAINSTEM CONTRIBUTIONS TO BINOCULAR ALIGNMENT

The vergence system is essential for binocular vision, aligning both eyes on a single visual target to ensure that its image falls on corresponding retinal points in each eye. This precise alignment enables depth perception, achieved through stereopsis, and maintains a single, fused image of the visual world. Unlike other types of eye movements, vergence is disconjugate, meaning the two eyes move in opposite directions. Convergence occurs as the eyes move inward to focus on closer objects, while divergence involves outward movement of the eyes to focus on distant objects. This unique property of vergence allows for seamless adaptation to objects at varying depths, supporting clear and unified visual perception across different spatial planes (Mays, 1984).

Vertical vergence adjusts eye alignment in the vertical plane, ensuring stability for targets at different heights (Sukkar et al., 2024). Torsional vergence compensates for rotational misalignments of the eyes, maintaining visual stability even during head tilts. These types of movements often operate in conjunction with other visual processes, such as accommodation – the adjustment of the lens for near vision – and pupillary constriction.

The brainstem processes binocular disparity and blur cues to guide vergence movements. These signals, derived from the visual cortex, are integrated with proprioceptive and vestibular information within the brainstem to ensure precise alignment. The supraoculomotor area in the midbrain near the oculomotor nucleus houses neurons dedicated to generating vergence motor commands. The cerebellum, particularly its vermis and fastigial nuclei, refines the accuracy and smoothness of vergence movements. Disruptions in these pathways can lead to strabismus or convergence insufficiency, emphasizing the importance of brainstem integrity for stable binocular vision.

NYSTAGMUS

An intricate network of brainstem and cerebellar structures maintains ocular stability. Specific lesions can lead to distinct nystagmus patterns, each reflecting dysfunction in a particular region of this complex system. By examining the neuroanatomic correlates of different types of nystagmus, we can trace the disruptions in the midbrain, pons, and medulla, which helps to localize lesions to the brainstem.

Midbrain lesions and nystagmus

At the level of the rostral midbrain, lesions affecting the INC, a crucial center for vertical and torsional gaze control, can produce torsional nystagmus (Rambold et al., 2000). The INC integrates signals from the vestibular system and helps stabilize gaze, particularly in the vertical and torsional planes. When this nucleus is damaged, patients experience involuntary rotations of the eyes around the visual axis, often accompanied by difficulty maintaining stable vertical gaze.

Further posteriorly in the midbrain, disruption of the MLF, which interconnects the oculomotor, trochlear, and abducens nuclei, results in INO (Zee, 1992). While INO itself does not produce nystagmus directly, it often leads to dissociated nystagmus in which the abducting eye displays a high-frequency movement due to impaired coordination between the medial and lateral rectus muscles.

Pontine lesions and nystagmus

Descending into the caudal pons, lesions affecting the PPRF can result in horizontal gaze palsy with associated gaze-evoked nystagmus (Klarendic et al., 2021). The PPRF is a key center for horizontal saccade generation, and damage to this area impairs the ability to maintain stable fixation when looking laterally. Patients with such lesions may exhibit difficulty in holding eccentric gaze, causing their eyes to drift back toward the primary position, followed by corrective saccadic movements that manifest as nystagmus (Kato et al., 1994).

Further laterally in the dorsolateral pons, lesions involving the vestibular nuclei, particularly the MVN, disrupt vestibular inputs to the ocular motor system and lead to Bruns nystagmus (Dabboucy et al., 2021; Mantokoudis et al., 2021). This distinctive pattern is seen in patients with pathology localizing to the cerebellopontine angle. It presents as a combination of low-frequency, high-amplitude nystagmus when looking toward the lesion (due to gaze-holding failure) and high-frequency, low-amplitude nystagmus when looking away from the lesion (due to vestibular imbalance).

Medullary lesions and nystagmus

At the level of the rostral medulla, lesions in the vestibulocerebellar pathways, particularly involving the flocculus, paraflocculus, and nodulus, frequently give rise to downbeat nystagmus (Strupp et al., 2011). These cerebellar structures provide critical modulation of the VOR and help stabilize gaze during head movements. Damage to these regions results in an imbalance in vertical gaze control, leading to downward drifting of the eyes with corrective upbeat movements.

Lesions in the caudal medulla, particularly those affecting the perihypoglossal nuclei and the paramedian tract cell groups, can lead to upbeat nystagmus. These structures integrate sensory and motor signals to regulate vertical eye movements. Damage here disrupts the neural integration required for smooth vertical gaze holding, causing a characteristic upward drift with downward corrective saccades.

Finally, periodic alternating nystagmus arises from dysfunction in the nodulus and uvula, which are located at the cerebellar–medullary junction (Dibartolomeo and Yee, 1988). These structures play a pivotal role in the velocity storage mechanism, a neural process that extends the duration of vestibular responses to head motion. When impaired, patients exhibit nystagmus that periodically changes direction, typically cycling every few minutes.

CONCLUSION

The six supranuclear systems – the saccadic, smooth pursuit, vestibular, optokinetic, fixation, and vergence systems – operate in dynamic concert to ensure precise gaze control and stable vision. The brainstem, as the integrative center, plays an indispensable role in executing these commands, adapting to errors, and mediating sensorimotor feedback. A deep understanding of brainstem neuroanatomy – from the nuclei of cranial nerves III, IV, and VI, to the essential white matter tracts and integrative centers such as the SC, MLF, and PPRF – provides critical insights into the pathophysiology of oculomotor disorders.

An understanding of oculomotor control and its brainstem pathways opens new avenues for clinical intervention and diagnostic accuracy.

References

Angelaki DE, Cullen KE (2008). Vestibular system: the many facets of a multimodal sense. Annu Rev Neurosci 31: 125–150.

Badler JB, Watamaniuk SNJ, Heinen SJ (2019). A common mechanism modulates saccade timing during pursuit and fixation. J Neurophysiol 122: 1981–1988.

Brandt T, Strupp M, Arbusow V et al. (1997). Plasticity of the vestibular system: central compensation and sensory substitution for vestibular deficits. Adv Neurol 73: 297–309.

Branoner F, Chagnaud BP, Straka H (2016). Ontogenetic development of vestibulo-ocular reflexes in amphibians. Front Neural Circuits 10: 91.

Bruce CJ, Goldberg ME (1985). Primate frontal eye fields. I. Single neurons discharging before saccades. J Neurophysiol 53: 603–635.

Cullen KE (2012). The vestibular system: multimodal integration and encoding of self-motion for motor control. Trends Neurosci 35: 185–196.

Cullen KE, Roy JE (2004). Signal processing in the vestibular system during active versus passive head movements. J Neurophysiol 91: 1919–1933.

Dabboucy B, Yazbeck M, Bassim M et al. (2021). Bruns nystagmus occurring after resection of a giant cerebellopontine angle acoustic neuroma. Neurochirurgie 67: 500–502.

Della Santina CC, Migliaccio AA, Patel AH (2007). A multichannel semicircular canal neural prosthesis using electrical stimulation to restore 3-D vestibular sensation. IEEE Trans Biomed Eng 54: 1016–1030.

Dibartolomeo JR, Yee RD (1988). Periodic alternating nystagmus. Otolaryngol Neck Surg 99: 552–557.

Gandhi NJ, Katnani HA (2011). Motor functions of the superior colliculus. Annu Rev Neurosci 34: 205–231.

Giolli RA, Blanks RHI, Lui F (2006). The accessory optic system: basic organization with an update on connectivity, neurochemistry, and function. Progress in brain research. Elsevier, pp. 407–440.

Glasauer S, Straka H (2022). Low gain values of the vestibulo-ocular reflex can optimize retinal image slip. Front Neurol 13: 897293.

Goettker A, Gegenfurtner KR (2021). A change in perspective: the interaction of saccadic and pursuit eye movements in oculomotor control and perception. Vision Res 188: 283–296.

Hillier SL, McDonnell M (2011). Vestibular rehabilitation for unilateral peripheral vestibular dysfunction. Cochrane Database Syst Rev CD005397.

Hoffmann K-P (1989). Chapter 15: Control of the optokinetic reflex by the nucleus of the optic tract in primates. Progress in brain research. Elsevier, pp. 173–182.

Ibbotson M, Krekelberg B (2011). Visual perception and saccadic eye movements. Curr Opin Neurobiol 21: 553–558.

Kato I, Nakamura T, Kanayama R et al. (1994). Slow saccades and quick phases of nystagmus after pontine lesions. Acta Otolaryngol (Stockh) 114: 95–98.

Klarendic M, Hribar M, Urbancic NB et al. (2021). Central nystagmus in progressive supranuclear palsy: a neglected clinical feature? Parkinsonism Relat Disord 84: 15–22.

Krauzlis RJ (2004). Recasting the smooth pursuit eye movement system. J Neurophysiol 91: 591–603.

Leigh RJ, Zee DS (2015). The neurology of eye movements. 5th edition. Oxford University Press.

Li H, Li C, Ma W et al. (2024). Eye movement disorders: a new approach to preliminary screening of Parkinson's disease. Neuroscience 563: 202–211.

Lisberger SG (2015). Visual guidance of smooth pursuit eye movements. Annu Rev Vis Sci 1: 447–468.

Mantokoudis G, Korda A, Zee DS et al. (2021). Bruns' nystagmus revisited: a sign of stroke in patients with the acute vestibular syndrome. Eur J Neurol 28: 2971–2979.

Martinez-Conde S, Macknik SL, Hubel DH (2004). The role of fixational eye movements in visual perception. Nat Rev Neurosci 5: 229–240.

Mays LE (1984). Neural control of vergence eye movements: convergence and divergence neurons in midbrain. J Neurophysiol 51: 1091–1108.

McFarland JL, Fuchs AF (1992). Discharge patterns in nucleus prepositus hypoglossi and adjacent medial vestibular nucleus during horizontal eye movement in behaving macaques. J Neurophysiol 68: 319–332.

Moschovakis A, Scudder C, Highstein S (1996). The microscopic anatomy and physiology of the mammalian saccadic system. Prog Neurobiol 50: 133–254.

Munoz DP, Everling S (2004). Look away: the anti-saccade task and the voluntary control of eye movement. Nat Rev Neurosci 5: 218–228.

Orban De Xivry J, Lefèvre P (2007). Saccades and pursuit: two outcomes of a single sensorimotor process. J Physiol 584: 11–23.

Ortiz JF, Eissa-Garces A, Ruxmohan S et al. (2021). Understanding Parinaud's syndrome. Brain Sci 11: 1469.

Rambold H, Helmchen C, Büttner U (2000). Vestibular influence on the binocular control of vertical-torsional nystagmus after lesions in the interstitial nucleus of Cajal. Neuroreport 11: 779–784.

Robinson DA (2022). The function and phylogeny of eye movements. Progress in brain research. Elsevier, pp. 1–14.

Robinson DA (2022). The behavior of the optokinetic system. Progress in brain research. Elsevier, pp. 215–230.

Rolfs M (2009). Microsaccades: small steps on a long way. Vision Res 49: 2415–2441.

Schmidt M, Zhang H-Y, Hoffmann K-P (1993). OKN-related neurons in the rat nucleus of the optic tract and dorsal terminal nucleus of the accessory optic system receive a direct cortical input. J Comp Neurol 330: 147–157.

Scudder C, Kaneko C, Fuchs A (2002). The brainstem burst generator for saccadic eye movements. Exp Brain Res 142: 439–462.

Sekar A, Panouillères MTN, Kaski D (2024). Detecting abnormal eye movements in patients with neurodegenerative diseases – current insights. Eye Brain 16: 3–16.

Sharpe JA (2008). Neurophysiology and neuroanatomy of smooth pursuit: lesion studies. Brain Cogn 68: 241–254.

Sommer MA, Wurtz RH (2000). Composition and topographic organization of signals sent from the frontal eye field to the superior colliculus. J Neurophysiol 83: 1979–2001.

Sommer MA, Wurtz RH (2004). What the brain stem tells the frontal cortex. I. Oculomotor signals sent from superior colliculus to frontal eye field via mediodorsal thalamus. J Neurophysiol 91: 1381–1402.

Strupp M, Hüfner K, Sandmann R et al. (2011). Central oculomotor disturbances and nystagmus: a window into the brainstem and cerebellum. Dtsch Arztebl Int 108: 197–204.

Strupp M, Kremmyda O, Adamczyk C et al. (2014). Central ocular motor disorders, including gaze palsy and nystagmus. J Neurol 261: S542–S558.

Strupp ML, Straumann D, Helmchen C (2021). Central ocular motor disorders: clinical and topographic anatomical diagnosis, syndromes and underlying diseases. Klin Monbl Augenheilkd 238: 1197–1211.

Sukkar M, Khatirnamani A, Wibble T (2024). Visually induced vertical vergence as a motion processing biomarker associated with postural instability. Neuroscience 555: 106–115.

Thakkar KN, Diwadkar VA, Rolfs M (2017). Oculomotor prediction: a window into the psychotic mind. Trends Cogn Sci 21: 344–356.

Tregellas JR, Tanabe JL, Miller DE et al. (2004). Neurobiology of smooth pursuit eye movement deficits in schizophrenia: an fMRI study. Am J Psychiatry 161: 315–321.

Voogd J, Barmack NH (2006). Oculomotor cerebellum. Progress in brain research. Elsevier, pp. 231–268.

White OB, Saint-Cyr JA, Sharpe JA (1983). Ocular motor deficits in Parkinson's disease. I. The horizontal vestibulo-ocular reflex and its regulation. Brain J Neurol 106: 555–570.

Wiboonsaksakul KP, Leavitt Brown OM, Cullen KE (2024). Restoring vestibular function during natural self-motion: progress and challenges. eLife 13: e99516.

Wurtz RH (2008). Neuronal mechanisms of visual stability. Vision Res 48: 2070–2089.

Yakushin SB, Gizzi M, Reisine H et al. (2000a). Functions of the nucleus of the optic tract (NOT). II. Control of ocular pursuit. Exp Brain Res 131: 433–447.

Yakushin SB, Reisine H, Büttner-Ennever J et al. (2000b). Functions of the nucleus of the optic tract (NOT). I. Adaptation of the gain of the horizontal vestibulo-ocular reflex. Exp Brain Res 131: 416–432.

Yu, M., Wang, S.-M., 2025. Neuroanatomy, interstitial nucleus of Cajal, in: StatPearls [Internet]. StatPearls Publishing, Treasure Island, FL.

Zee DS (1992). Internuclear ophthalmoplegia: pathophysiology and diagnosis. Baillieres Clin Neurol 1: 455–470.

Handbook of Clinical Neurology, Vol. 216 (3rd series)
The Brainstem
L.T. Grinberg and A.D. Korczyn
https://doi.org/10.1016/B978-0-443-15736-3.00001-9

Chapter 9

Pupillary disorders

TIM BELTRAMINELLI AND AKI KAWASAKI*

Neuro-Ophthalmology Unit, Department of Ophthalmology, Jules Gonin Eye Hospital, School of Biology and Medicine, University of Lausanne, Lausanne, Switzerland

Abstract

This chapter provides a comprehensive overview of various conditions affecting the pupil, encompassing anatomic, developmental, and pathologic aspects. The relevant anatomy and embryology of the iris is described, highlighting the sphincter and dilator pupillae muscles which control pupil size. A few examples of iris pathologies (congenital and acquired) are given for illustrating ocular disorders which affect pupil size and shape. The majority of this chapter discusses the autonomic pathways, parasympathetic and sympathetic, to control pupillary movement. Included are details of the anatomic pathways in the brainstem, in the orbit, and around to eye, as well as points of debate regarding the anatomy. Finally, the clinical manifestations of a sympathetic and parasympathetic deficit to the eye are presented with clinical examples.

INTRODUCTION

By simply noting the shape of each pupil, observing its size in dim and bright lights, and checking the light reflex, the examiner can elicit important clues to underlying ocular or neural or orbital pathology. Sometimes, a pupillary abnormality is the principal sign of a life-threatening condition such as cerebral aneurysm or carotid dissection. Other times, pupils can signal high anxiety or even pain. Thus, a working knowledge of the pupillary anatomy and physiology is useful when faced with abnormal pupils. This chapter considers the mechanic aspects (iris structure and muscles) of pupils and then delves into the autonomic innervation of the sphincter pupillae and the dilator pupillae and the related disorders. Finally, the significance and the detection of a relative afferent pupillary defect is described and illustrated.

THE IRIS

Pupillary disorders encompass a wide array of conditions, some of which can be ascribed to specific anatomic or developmental abnormalities of the iris. These conditions can affect the size, shape, or reactivity of the pupil and should always be considered in the differential diagnosis of pupillary disorders. Understanding the intricate anatomy of the iris is fundamental to understanding the complexities of pupillary disorders; therefore, in this first part of the chapter, we will review the clinically relevant anatomy of the iris as well as the embryology of its tissues.

Iris anatomy

The iris is the most anterior part of the uveal tract, which also includes the ciliary body and the choroid. The iris has a flat, circular shape with a central aperture, the pupil. If the body of the iris, which is an ensemble of stroma and epithelium, gives the eyes their distinctive color, the muscles of the iris have the capacity to dynamically adjust the pupil to control how much light reaches the retina (Forrester et al., 2021). These are the sphincter pupillae and the dilator pupillae.

The anterior aspect of the iris is composed of a discontinuous anterior layer that features fibroblasts and melanocytes. Underneath this first layer lies the iris stroma, a thick vascularized tissue that forms the bulk of the iris. The stroma is composed of collagen fibers, blood vessels, nerves, and pigment cells. Within this fibrous network are embedded the sphincter pupillae and the dilator pupillae, collectively called the internal muscles of the eye.

The sphincter pupillae muscle is a smooth muscle located near the pupillary margin. The muscular fibers are spindle-shaped and arranged in a continuous ring around the pupil forming a circular band (Lai, 1972b; Gabella, 1974). On the other hand, the dilator pupillae muscle consists of myoepithelial cells, which are specialized cells exhibiting both muscle and epithelial characteristics (Lai, 1972a; Tamura and Smelser, 1973). The basal aspect of these cells contains the contractile cellular elements actin and myosin, while the apical aspect exhibits epithelial features (Kivela et al., 1992). The muscular

*Correspondence to: Aki Kawasaki, MD, PhD, Department of Ophthalmology, School of Biology and Medicine, University of Lausanne, Avenue de France 15, 1001 Lausanne, Switzerland Tel: +41-21-626-86-60; E-mail: aki.kawasaki@fa2.ch.

fibers of the dilator pupillae are radially arranged extending from the iris root toward the sphincter pupillae.

The posterior aspect of the iris features the double-layered posterior pigmented epithelium, which provides a dark surface that absorbs excess light, preventing unwanted light scatter within the eye. The iris is stabilized by its peripheral attachment to the ciliary body at the iris root, visible only via gonioscopy techniques. This connection helps maintain the structural integrity of the anterior segment of the eye and defines the boundary between the anterior and posterior chambers.

The innervation of the internal muscles of the eye is provided by the autonomic nervous system, composed of the parasympathetic and sympathetic branches. The parasympathetic system mediates contraction of the sphincter pupillae by the release of acetylcholine at the neuromuscular junction and its binding to muscarinic receptors (Mitchelson, 2012). The parasympathetic pathway arises in the midbrain and reaches the anterior uvea through the short ciliary nerves. On the other hand, the contraction of the dilator pupillae is mediated by the sympathetic pathway and the release of noradrenaline which acts on α1-adrenergic receptors on the muscle cells (Schwinn and Afshari, 2006). The detailed anatomy of the autonomic pathways to the eye is discussed later in this chapter; see Anatomy of the parasympathetic pathway and Anatomy of the sympathetic pathway sections. The coordination between the sympathetic and parasympathetic systems allows for precise regulation of pupil size, ensuring that the eye adapts quickly to changing light conditions and visual demands. This coordination is essential for several reflexes that maintain visual acuity and protect the eye from damage due to excessive light exposure.

Iris embryology

The eye originates from three main embryonic structures, namely the surface ectoderm, the neural ectoderm, and the mesenchyme derived from neural crest cells. Herein, we give a broad overview of the major developmental steps that lead to the formation of the iris.

The iris begins to form from the rim of the optic cup and the surrounding mesenchyme. This process occurs in concert with the development of other anterior segment structures, such as the cornea and the ciliary body, and is contingent upon correct closure of the embryonic fissure throughout its length. For this reason, developmental anterior segment dysgenesis syndromes are often accompanied with iris abnormalities (Davis-Silberman and Ashery-Padan, 2008). Among the most important genetic drivers of the embryogenesis of the eye is the PAX6 gene, mutations of which may lead to a variety of ocular anomalies.

The anterior portion of the inner layer of the optic cup gives rise to the pigmented epithelium of the iris. The sphincter pupillae and dilator pupillae are both neuroectodermal in origin. However, while the former differentiates early from neuroectodermally derived anterior iris epithelial cell (around week 13–14), the latter arises later in embryonic development (around 6 months) as an extension of the anterior epithelial cells and remains, even in adulthood, tightly associated and never independent of the epithelium. Indeed, because of this embryologic origin the dilator pupillae is considered as a myoepithelial structure. Overall, the origin of smooth muscle cells from neuroectoderm represents a distinct aspect of ocular muscle development, as other smooth muscles in the body originate from mesoderm (Donadon and Santoro, 2021). Finally, the fibrous and vascular components of the iris stroma, as well as iris melanocytes, arise from migrating neural crest cells.

The amount of melanin produced by melanocytes, but not the number of melanocytes, in the anterior iris influences its pigmentation and ultimately determines the final color of the iris. The eyes of a newborn tend to be light blue or gray, because of the relative lack of melanin, which generally increases during the first years of life to reach their genetically determined density.

Iris anomalies

Abnormalities of pupil shape may arise from ocular causes, and in particular, from iris anomalies which could be either congenital or acquired.

The most common congenital iris anomaly affecting the pupil is iris coloboma. A coloboma arises from an incomplete closure of the embryonic fissure and typically manifests as an infero-nasal defect in the iris which is visible on close clinical examination of the iris and confirmed at the slit lamp (Fig. 9.1A). In presence of an iris coloboma, a thorough ophthalmologic examination is warranted to look for associated posterior (i.e., retinal, choroidal, or optic nerve) colobomata. An iris coloboma may also be found with other systemic defects, such as in the CHARGE syndrome (coloboma, heart defect, choanal atresia, retardation of growth, genital anomalies, and ear anomalies) (Kaeser and Kawasaki, 2010). Because of the close embryologic association between the iris and other anterior segment structures, iris anomalies may arise from broader anterior segment dysgenesis syndromes, such as the iridocorneal endothelial syndromes or Axenfeld–Rieger spectrum syndrome. These conditions generate abnormal irides, often with multiple atrophic holes resulting in a pseudo-polycoria (Idrees et al., 2006) (Fig. 9.1B).

Aniridia is a rare congenital disorder where the iris is either partially underdeveloped or completely missing. Patients appear to have pharmacologically dilated pupils, but careful ophthalmologic examination shows that only peripheral stumps of hypoplastic iris and its musculature are present (Hingorani et al., 2012). This condition is generally bilateral and inherited in an autosomal dominant manner in about two-thirds of cases. The genetic causes are found in the PAX6 gene. More worrisome are sporadic cases of aniridia because of its known association with Wilms' tumor (nephroblastoma), known as Wilm's tumor, aniridia, genitourinary anomalies, and mental retardation (WAGR) syndrome (Fischbach et al., 2005). This is due to the close proximity between the PAX6 gene and the gene responsible for Wilms' tumor, the WT1 gene. Therefore, patients having sporadic aniridia should be carefully monitored for development of malignancies, in particular a Wilms' tumor. Screening by renal or abdominal ultrasound is recommended every 3–4 months until the age of 5–7 years as the median age of presentation of Wilms'

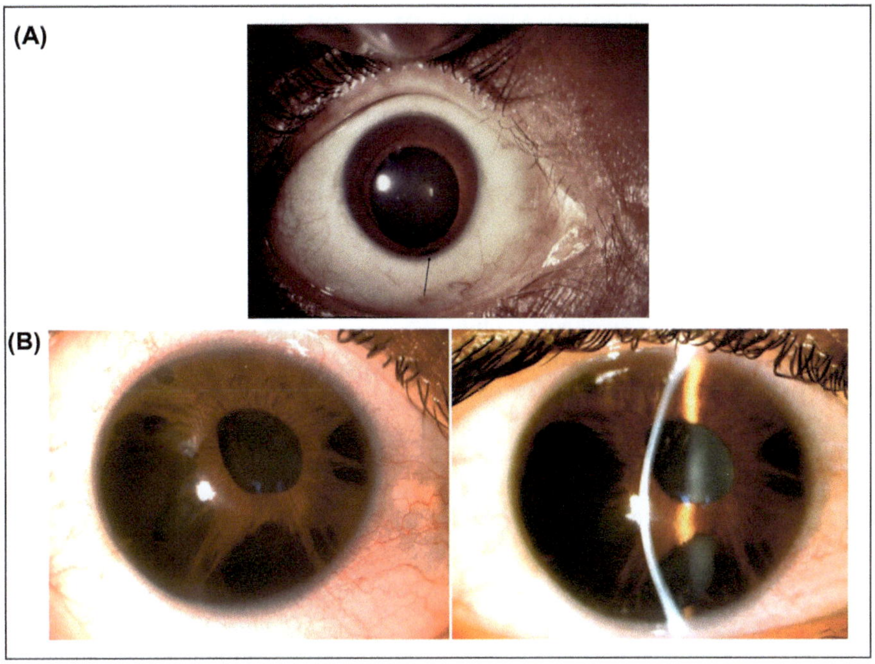

Fig. 9.1. Examples of iris anomalies affecting pupil shape. (**A**) Anterior segment photograph demonstrating incomplete iris coloboma. Observe the notch in inferior pupillary border and partial thickness iris defect near iris root (arrow) (Lingam et al., 2021). (**B**) Slit lamp examination. Slit lamp photograph shows clinical features of essential iris atrophy and keratoconus in the right eye (De Maria et al., 2019). (**A**) From Lingam G, Sen AC, Lingam V, et al. (2021) Ocular coloboma – a comprehensive review for the clinician. Eye (Lond) 35: 2086–2109. Available at: https://www.nature.com/articles/s41433-021-01501-5/figures/2. Open Access. Reprinted under the terms of the Creative Commons Attribution 4.0 International License (http://creativecommons.org/licenses/by/4.0/), which permits unrestricted use, distribution, and reproduction in any medium. (**B**) De Maria M, Iannetta D, Moramarco A., et al. (2019) Iridocorneal endothelial syndrome in a patient with keratoconus – a case report. BMC Ophthalmol 19: 221. Available at: https://bmcophthalmol.biomedcentral.com/articles/10.1186/s12886-019-1215-x/figures/1. Open Access. Reprinted under the terms of the Creative Commons Attribution 4.0 International License (http://creativecommons.org/licenses/by/4.0/), which permits unrestricted use, distribution, and reproduction in any medium.

tumor is 22 months and most form in children less than 8 years old (Kalish et al., 2017; Liu and Suson, 2020).

Common causes of acquired iris anomalies occur in four major categories: inflammatory, traumatic, ischemic, and postsurgic. Anterior segment inflammation is known as uveitis. In its acute stages, uveitis can result in an abnormally small pupil from the action of inflammatory mediators or through the formation of synechiae (called posterior synechiae) between the anterior surface of the crystalline lens and the posterior aspect of the iris. Longstanding or chronic repetitive uveitis can also lead to atrophic changes in the iris stroma, best seen in transillumination at the slit lamp. Traumatism is another major category of abnormal pupils. Indeed, after an ocular contusion, the pupil may be unresponsive to both light and near stimulation (see later for more details on the pupillary responses). Oftentimes, the causative lesion cannot be identified even by slit lamp examination, while in some cases a radial tear in the sphincter pupillae muscle clearly orients toward a traumatic origin. In such cases, the traumatized sphincter will not contract to direct pharmacologic stimulation with full-strength (1%–4%) pilocarpine. Ocular trauma can also be the cause of an irido- or cyclodialysis which may artificially create an additional "pupillary space" in the eye and deform the actual pupil.

Lack of perfusion in the anterior segment of the eye with consequent iris ischemia is another well-known cause of pupillary dysfunction. Vascularization of the iris is provided by the long posterior ciliary arteries and the anterior ciliary arteries. Both originate from the ophthalmic artery, but the former pierce the sclera and travel in the suprachoroidal space toward the iris, while the latter irrigate the extraocular muscles and anastomose with the long posterior ciliary arteries to provide the vascularization of the iridial tissues. A variety of local and systemic disorders can result in hypoperfusion or occlusion of these arteries, such as acute angle closure glaucoma, carotid occlusive disease, diabetes, giant cell arteritis, to name a few. The scenario of a patient with unilateral mydriasis and acute eye pain or localized headache commonly evokes consideration of an acute oculomotor nerve palsy from aneurysmal compression or other intracranial pathology but it is important to remember that ocular disorders causing acute iris ischemia, notably acute angle-closure glaucoma, may present with these findings. In the latter, the examination will reveal a red eye having a firm consistency upon digital palpation due to the abrupt rise in intraocular pressure.

Finally, unequal pupil size can be the result of ophthalmic surgical procedures (iatrogenic traumatism). For example, during cataract surgery, mechanic damage to the iris stroma or to the iris muscles may occur and postoperatively the pupil is slightly deformed in shape, slightly larger in size compared to the other pupil, and slightly reduced in its motility. A peculiar and still unexplained syndrome, called Urrets–Zavalia syndrome, is the presentation of fixed complete mydriasis after ocular surgery, typically after cataract surgery (Magalhaes et al., 2016). It is speculated that iris damage results from transiently increased intraocular pressure but no clear pathophysiologic mechanisms have been identified so far.

PUPIL LIGHT REFLEX

In this section, we will review the neural components of pupillary constriction to light. The pupillary light reflex is an oligosynaptic reflex in which the afferent pupillomotor signal from the eye activates the efferent motor limb at the midbrain to evoke contraction of the sphincter pupillae muscle (Fig. 9.2). Damage to either the afferent or efferent limb of the light reflex can result in subnormal pupillary constriction in response to a direct light stimulus but only deficits in the efferent limb affect both direct and consensual light responses of the affected pupil.

Afferent limb

The afferent limb of the pupil light reflex is also called the retino-tectal tract. This tract originates in the melanopsin-expressing ganglion cells (MGCs) which are intrinsically photosensitive neurons capable of phototransduction. While MGCs account for only 0.5% of all retinal ganglion cells, they synapse at many subcortic nuclei and connect extensively. In this way, MGCs mediate all nonvisual light-evoked physiologic responses such as the pupil light reflex, circadian photo-entrainment, melatonin secretion, and mood regulation, among others (Hattar et al., 2003). The optic nerve is composed of the axons of MGCs and of all other retinal ganglion cells and thus carries two separate fiber tracts: the retino-tectal tract for pupil and the retino-geniculate tract for vision.

MGCs can be activated by light by two different mechanisms: intrinsic direct activation via melanopsin photoexcitation and extrinsic trans-synaptic activation via light signals from rod and cone photoreceptors in the outer retina (Joyce et al., 2018) (Fig. 9.3). The hyper-rapid response kinetics of rods and cones activate a stimulus-synchronized firing from MGCs such that the first action potential appears immediately with light onset and cell firing stops immediately with light offset (Dacey et al., 2005). The pupil light reflex evoked in routine clinical examination is largely driven by extrinsic activation of MGCs, reflecting more the activity of rods and cones than melanopsin. The intrinsic melanopsin response to light displays

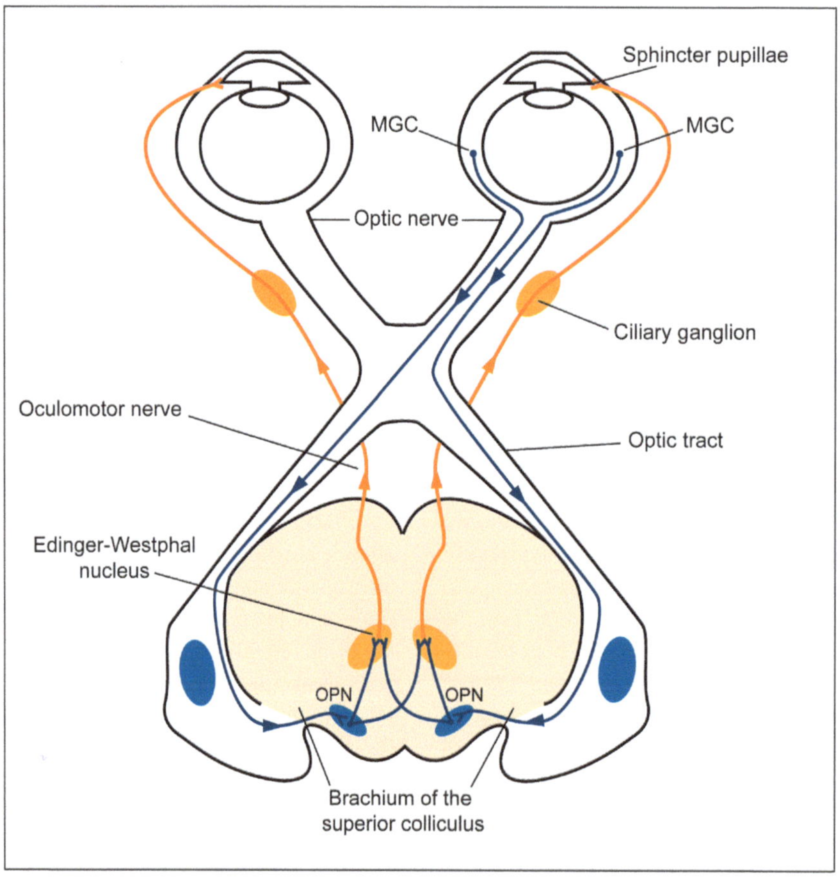

Fig. 9.2. Schema of the pupillary light reflex. The afferent limb is shown in blue, while the efferent limb (oculoparasympathetic pathway) in orange. The pupil light reflex is initiated when melanopsin ganglion cells of the retina are activated. The olivary pretectal nucleus acts as an integrating center and then transmits a luminance signal to both Edinger–Westphal nuclei (EW). The EW is the origin of the neural efferent limb of the pupil light reflex. The sphincter pupillae muscle is the effector organ whose contraction leads to pupillary constriction.

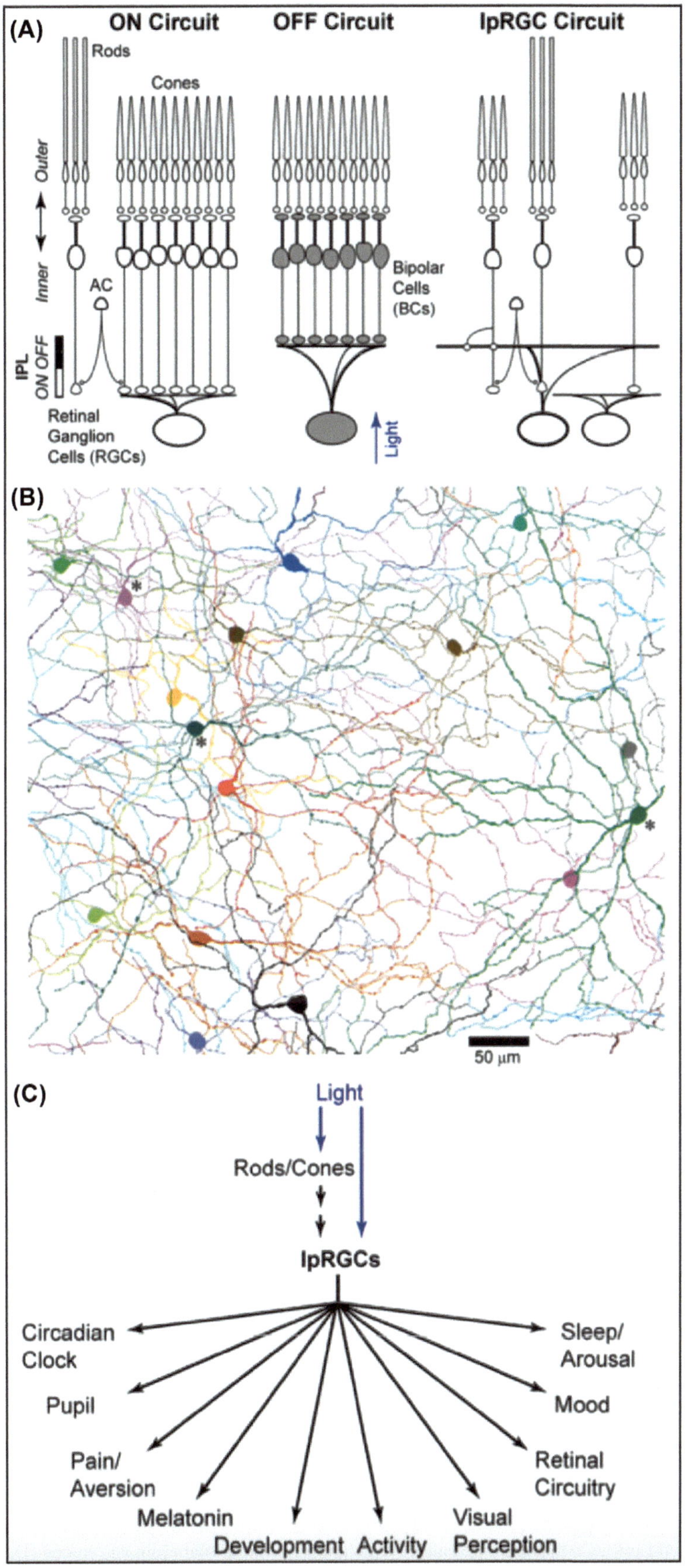

Fig. 9.3. Retinal connections of MGCs (referred in this figure as intrinsically photosensitive retina ganglion cells, or ipRGCs) and their widespread influence on physiologic systems. (**A**) A highly simplified schematic of the retina in cross-section, oriented with the inner aspect

much slower temporal kinetics, so that neuronal discharge is delayed in its onset and may continue long after light offset. In quantitative analysis of differential pupil response kinetics to different light stimuli, the melanopsin contribution to the pupil light reflex is usually evaluated from pupillary behavior after light offset (McDougal and Gamlin, 2010).

An anatomic aspect of the retina bears reminding as it has clinical implications. The nasal hemiretina has greater area with more ganglion cells per square millimeter compared to the temporal hemiretina (Curcio et al., 1987). This hemiretinal asymmetry is relayed to the chiasm in form of more decussating axons (nasal hemiretina) than nondecussating axons (temporal hemiretina) (Kardon et al., 2006). This nasal/temporal asymmetry explains the larger temporal visual field compared with the nasal visual field of each eye. It is also the origin of the relative afferent pupillary defect (RAPD) seen with optic tract lesions. At the distal part of the optic tract, the pupillomotor signal in the retino-tectal tract segregates from the visual signal to enter the brachium of the superior colliculus and synapses in the olivary pretectal nucleus (OPN) of the dorsal midbrain.

The pretectum of the dorsal midbrain contains highly interconnected cell groups. Several nuclei identified in the pretectum are the OPN, the nucleus of the optic tract, and the anterior, medial, and posterior pretectal nuclei. The OPN is the principal synaptic site of the retino-tectal tract and serves to integrate light information with other nonlight inputs to the pretectum. The OPN neurons encode a luminance signal, which is relayed to the ipsilateral and contralateral Edinger–Westphal nucleus (EW) to excite the parasympathetic pathway to the sphincter pupillae. Simultaneously, the luminance signal inhibits the sympathetic pathway to the dilator pupillae. The luminance signal displays both a linear relationship to the light stimulus and an intensity-dependent time delay. Thus, damage to the afferent limb results in a loss of amplitude and increased latency of the pupil light reflex.

Efferent limb

The efferent limb of the pupillary light reflex is mediated by the parasympathetic system, from the EW in the midbrain to the sphincter pupillae of the iris. The efferent limb is classically described as a two-neuron pathway with preganglionic parasympathetic signal originating in the EW and postganglionic parasympathetic fibers originating in the ciliary ganglion located in the orbit. The anatomic details are further described in Parasympathetic Innervation to the Pupil section.

Relative afferent pupillary defect

The RAPD is the interocular difference in retinal light signal, as detected through the pupil light reflex. The RAPD indicates interruption anywhere along the afferent limb, and in the clinical setting, it is most often associated with optic nerve disease. A few points about the physiologic basis and clinical testing of this pupil sign are discussed.

If a moderately bright light stimulus is continuously placed onto the eye, the outer retinal photoreceptors (rods and cones) will show adaptation to the light and attenuate their light signal to the MGCs. In response, the pupil will start to re-dilate during the light stimulation, a phenomenon known as "pupillary escape" (Cox, 1992). In clinical testing of pupils, when light stimulation evokes a minimal pupillary constriction and pupillary escape is the predominant action of the pupil, this is eponymously called a Marcus Gunn pupil, or simply an afferent pupil defect. As it is a uniocular clinical sign, it is possible to observe bilateral Marcus Gunn pupils in patients with bilateral optic nerve disease. However, pupillary escape shows high intraindividual variability from influences such as age, baseline pupil size, and mental arousal. For this reason, the Marcus Gunn pupil is reliably observed only when optic nerve dysfunction is advanced.

Detection of an afferent pupil defect is enhanced by comparing the pupil light response when the right eye is stimulated versus when the left eye is stimulated. Using the alternating light technique, the afferent pupil defect of one eye is always "relative" to the other eye; hence, a RAPD is always unilateral (Video 9.1). The alternating light test should be performed in a darkened room using a bright focal light source directed along the central visual axis. The patient should not look at the light source, as this induces an accommodative pupillary constriction. As the light is shined on one eye for 3 seconds, an immediate pupillary constriction (direct and consensual) followed by some pupillary escape is usually observed. Then the light is quickly alternated to the other eye for 3 seconds. An obvious RAPD is detected if one pupil dilates immediately when the light comes back to that eye. However, a small RAPD may show equal or nearly equal amplitude of constriction, and only an asymmetry in the timing of the pupillary escape signals its presence.

(nearer the center of the eye) down. The outer photoreceptors (i.e., rods/cones) drive bipolar cells (BCs). In the inner plexiform layer (IPL), BCs synapse with retinal ganglion cells (RGCs). Left: ON circuitry. Rods (top) drive rod BCs, whose signals pass through amacrine cells (ACs) to cone BCs. In the inner IPL, ON cone BCs convey signals to ON RGCs. ON RGCs show greater depolarization when light intensity increases. Center: OFF circuitry. In the outer IPL, OFF cone BCs provide synaptic input to OFF RGCs. OFF RGCs show greater depolarization when light intensity decreases. Right: A sample of circuits for outer- and inner-stratifying ipRGCs, which are both ON. ON cone BCs make ectopic synapses with the former and conventional synapses with the latter. Rod pathways also drive ipRGCs. IpRGCs make chemical and electrical synapses with ACs (not shown). (**B**) En face view of mouse M1 ipRGCs that were revealed by melanopsin immunolabeling and traced (Berson et al., 2010). Asterisks mark cells with somata displaced from the RGC layer to above the IPL. (**C**) A sample of ipRGC influences (Do, 2019). Reprinted with permission from Do, MTH (2019) Melanopsin and the intrinsically photosensitive retinal ganglion cells: biophysics to behavior. Neuron 104: 205–226.

Another technique to better observe a small RAPD is use of neutral density filters (NDFs) during alternating light test. NDFs decrease light transmission and create an "artificial" RAPD in the eye covered with the filter. In the patient with a small RAPD, adding a filter (0.2 or 0.3 log-unit) to the suspected eye with bring out the RAPD. Conversely, placing the same filter over the good eye will eliminate the afferent pupillomotor asymmetry. Larger RAPDs can be quantified by gradually increasing the amount of NDF over the good eye until pupil responses are equal during alternating light test (Video 9.2). Quantification of the RAPD provides one objective means to follow optic nerve disease, especially in patients who perform unreliably on psychophysic visual tests like visual acuity and visual field.

A potential confounder in the clinical alternating light test is a physiologic phenomenon called contraction anisocoria (Cox and Drewes, 1984). This is a transient anisocoria that occurs during the pupil light reflex because the amplitude of the direct pupil response amplitude is greater than the consensual one (Lowenstein, 1955). Unequal output from the Edinger–Westphal nucleus is purportedly at its origin. A large unilateral contraction anisocoria might be mistaken as a small RAPD in the opposite eye because during the clinical alternating light test, we observe and compare only direct pupil responses. Bilateral pupillography easily detects contraction anisocoria which may be found in 70%–80% of healthy persons.

The RAPD represents the interocular difference in retinal light signal transmitted by MGCs to the OPN. Thus, the RAPD is not a disorder of the pupil per se but rather a pupil biomarker of neuroretinal damage. As a general rule, an RAPD is not clinically present with retinal disease limited to rods and cones until visual loss is clinically significant. These include central retinal artery occlusion, retinal detachment involving the macula, and advanced unilateral retinitis pigmentosa. On the other hand, lesions of the optic nerve, even subtle, may cause a notable RAPD, while visual function remains quite preserved. This is common with compressive lesions like optic nerve sheath meningioma or recovered demyelinating optic neuritis. Conversely, the hereditary mitochondrial optic neuropathies like Leber hereditary optic neuropathy and dominant optic atrophy can cause significant visual loss without a deficit in pupillary function (Moura et al., 2013). This visual–pupillary dissociation is attributed to a differential susceptibility of visual retinal ganglion cells and MGCs to disease pathophysiology.

An RAPD can be observed in the eye opposite an optic tract lesion. In the patient with an optic tract lesion, light signals from the temporal hemiretina of the ipsilateral eye and from the nasal hemiretina of the contralateral eye are blocked at the tract lesion and an alternating light test is actually comparing the functioning nasal hemiretina of the ipsilateral eye against the functioning temporal hemiretina of the contralateral eye (Kardon et al., 2006) (Fig. 9.4). As the nasal hemiretina has greater area, it evokes a larger pupil response than the temporal hemiretina and thus the RAPD is detected in the eye opposite a tract lesion. In rare instances, it is possible to observe an RAPD from a lesion precisely located in the brachium of the superior colliculus. Each brachium contains the homonymous fibers of MGCs (retino-tectal tract) but does not contain visual fibers, as these had diverged to synapse in the lateral geniculate nucleus. Called a midbrain RAPD, it shares the same physiologic basis of nasal-temporal asymmetry but is distinguished from an optic tract RAPD by absence of a homonymous visual field defect.

Light-near dissociation

The amplitude of pupillary constriction to a sufficiently bright light stimulus is generally greater than that to an accommodative stimulus (near effort). Pupillary light-near dissociation (LND) is a condition in which the pupil light reflex is abnormally weak and the pupillary constriction to a near effort is better.

The most common cause of LND is damage to the afferent pupillomotor signal (retino-tectal tract) with sparing of the afferent limb of the pupil near reflex. Commonly, this is damage to the optic nerve. Patients with optic neuropathy are identified by visual loss and RAPD, thus checking for LND is not necessary. A second common site of afferent damage causing LND is the dorsal midbrain. A lesion at this site interrupts luminance signals from the two OPNs and essentially de-afferents the EW, thus a dorsal midbrain LND is usually bilateral (Fig. 9.5). Unlike LND associated with optic neuropathy, patients with LND from a dorsal midbrain lesion have normal vision but may demonstrate ocular motor deficits such as vertical gaze palsy, slow vertical saccades, lid retraction (Collier's sign), convergence-retraction nystagmus, and skew deviation. This constellation of clinical findings is known as Parinaud syndrome. In occasional patients with Parinaud syndrome the signs localizing to the dorsal midbrain are subtle and may comprise only bilateral LND and slow vertical saccades. For this reason, neuroimaging is recommended for any patient with normal vision and unexplained bilateral LND, even in absence of other neurologic deficits. Common causes of dorsal midbrain injury are ischemic stroke, brainstem hemorrhage, pineal region tumors, and obstructive hydrocephalus.

A rare cause of LND is the Argyll Robertson pupil, which is a sign of tertiary syphilis involving the central nervous system. The cause of LND in Argyll Robertson pupil is probably selective deafferentation of the EW from subependymal gliosis around the aqueduct and near the posterior commissure. Unlike dorsal midbrain syndrome in which the pupils are midsized, Argyll Robertson pupils are quite small (2 mm or less) and often irregular in shape. The pupils do not react to light but despite their small size, demonstrate a robust constriction upon near effort (Fig. 9.6). Pupillary dilation is also poor, both in response to darkness and after instillation of mydriatics (Thompson and Kardon, 2006).

Another mechanism leading to LND is damage to the efferent motor limb of the pupil light reflex which is followed

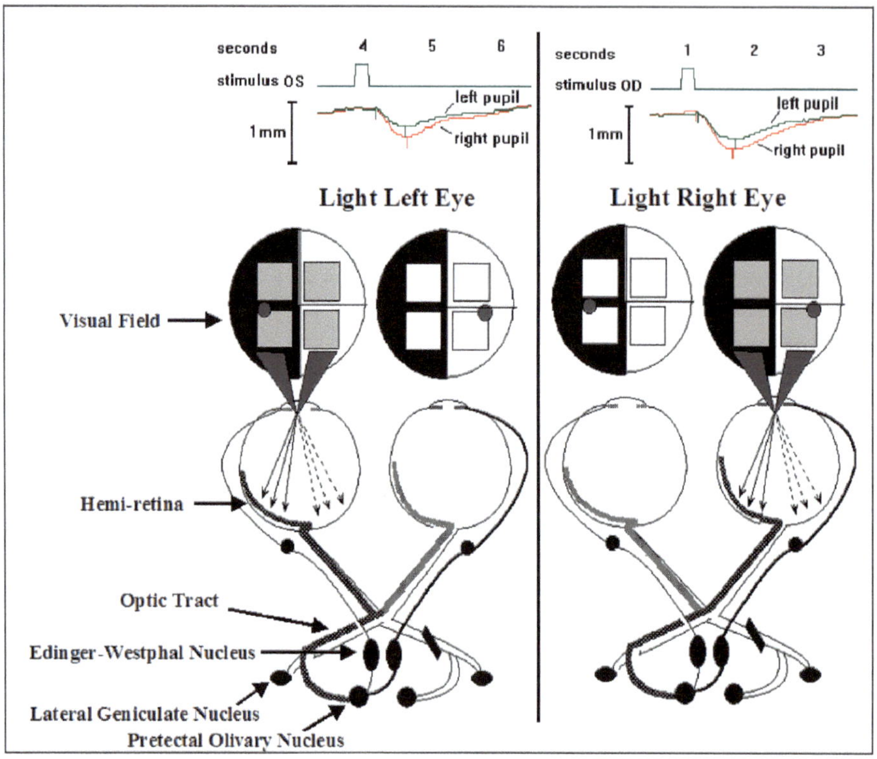

Fig. 9.4. Schema to explain a left relative afferent pupillary defect associated with a right optic tract lesion. Diagram (top) shows the pupil light reflex resulting from stimulating the left eye (OS) with a full-field stimulus (left) compared with right eye (OD) stimulation (right) in a patient with a right optic tract lesion (black bar). The resulting left homonymous hemianopic visual filed defect is shown as black shading above each eye. The functioning hemiretinas in each eye are shown as gray and black lines. Note that the pupil contraction is less for a full-field light stimulus given to the left eye (contralateral to the optic tract lesion). Because of the pattern of homonymous hemiretinal loss in a complete tract lesion, a full-field light stimulus alternated between the two eyes is really comparing the remaining functioning nasal hemiretinal of one eye (right eye, in this example) with the functioning temporal hemiretina of the other (left eye, in this example). Reprinted with permission from Kardon R, Kawasaki A, Miller NR (2006) Origin of the relative afferent pupillary defect in optic tract lesions. Ophthalmology 113: 1345–1353.

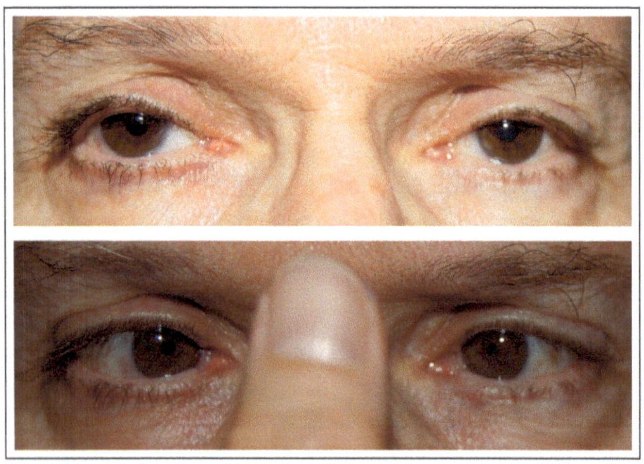

Fig. 9.5. Patient with bilateral light-near dissociation. Top: Pupils are small (about 3.5 mm) in response to a bright light shined on both eyes. Bottom: In response to a near effort (accommodation) with bright light removed, the pupils are smaller than under bright light stimulation.

by aberrant re-innervation of the iris sphincter by non-sphincter motor fibers. The most common cause of this form of LND is an idiopathic tonic (Adie) pupil. This disorder is fully discussed in the section "Clinical manifestations of a parasympathetic deficit to the eye." Autonomic neuropathies involving parasympathetic nerves may denervate the sphincter pupillae but aberrant re-innervation and thus LND, do not occur in these syndromes. When parasympathetic damage occurs in the setting of chronic compression or trauma of the oculomotor nerve, the denervated sphincter pupillae might become aberrantly re-innervated from somatic motor fibers. This leads to a simultaneous co-contraction of the sphincter pupillae with attempted eye movement. Specifically, if motor fibers destined to the medial rectus muscle aberrantly innervate the sphincter pupillae, the pupil will constrict upon any attempted adduction of the eye, for example, upon convergence and also upon gaze to the side opposite the nerve palsy.

Physiologic anisocoria

Unequal pupil size, termed anisocoria that is not associated with ocular or neurologic disease is referred to as physiologic. In dim light or darkness, nearly 20% of the normal population exhibits anisocoria of 0.4 mm or more. Under room light, this prevalence drops to around 10% (Loewenfeld, 1977). Typically,

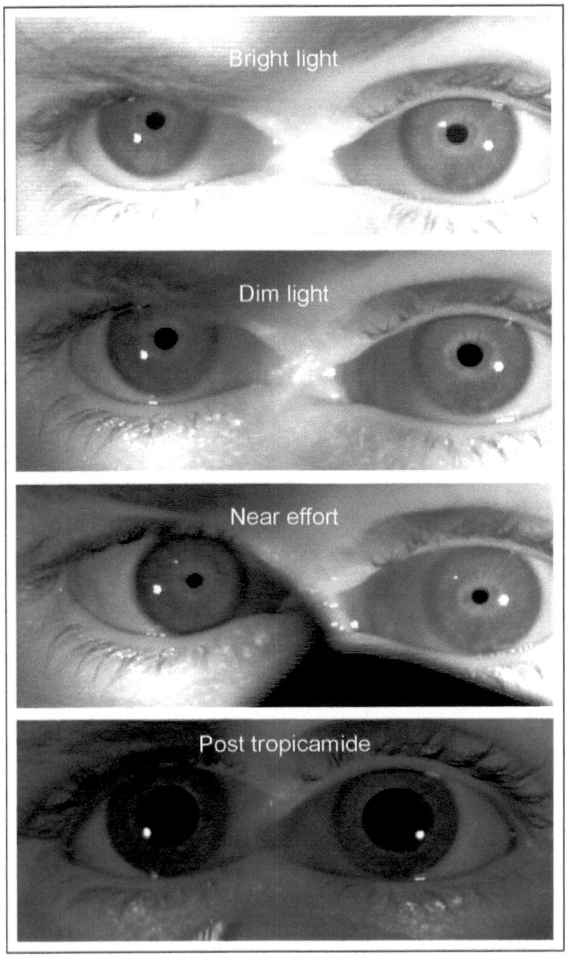

Fig. 9.6. Infrared photos of a patient with Argyll Robertson pupils demonstrating bilateral light-near dissociation. The eyes are artificially brought closer by use of prisms. Top: Pupils are small (almost 2 mm) under bright light stimulation. Second frame: Pupils remain smallish (2 mm) under dim light conditions. Third frame: Pupils are smaller (slightly more than 1 mm) in response to near effort as compared to their response to light stimulation (top frame). Bottom: Pupils dilate in response to topical tropicamide, an anticholinergic antimuscarinic agent which blocks parasympathetic stimulation of the sphincter pupillae. This indicates that secondary iris pathology, such as syphilitic uveitis, is not mechanically restricting the sphincter pupillae.

the anisocoria is 0.6 mm or less and rarely more than 1 mm (Lam et al., 1987). The degree of pupillary inequality in physiologic anisocoria can change from day to day or even hour to hour (Lam et al., 1996). Physiologic anisocoria usually diminishes slightly in bright light and is generally thought to result from asymmetry in the supranuclear inputs to the EW, and variations in stress levels, fatigue, sleepiness may affect the degree of anisocoria. In physiologic anisocoria, the pupils react normally to full-strength topical pharmacologic agents that act directly on the iris muscles and the posttest anisocoria is decreased or abolished. Physiologic anisocoria never demonstrates supersensitivity to dilute pharmacologic agents.

PARASYMPATHETIC INNERVATION OF THE PUPIL

Anatomy of the parasympathetic pathway

Parasympathetic innervation to the sphincter pupillae arguably originates in the Edinger Westphal nucleus (EW), a cluster of preganglionic parasympathetic neurons located in the midbrain at the level of the superior colliculus. In primates, EW lies dorsally and separated from the oculomotor subnuclei complex, which eventually gives rise to the oculomotor nerve (Wu et al., 2022). There is controversy on whether EW intrinsically possesses preganglionic cholinergic neurons expressing acetylcholine or if these neurons are part of the oculomotor complex. Animal studies have shown differences in the distribution of acetylcholine-expressing cells and the predicted anatomic location of the EW. In monkeys a co-localization between the predicted location of EW and distribution of cholinergic neurons exists. In several other mammals, including humans, a different noncholinergic cell population expressing the peptide urocortin-1 has been identified in the predicated anatomic location of EW, raising doubt that this nucleus is the parasympathetic origin of the pupil light reflex (Horn et al., 2009; Kozicz et al., 2011). In humans the parasympathetic preganglionic neurons expressing acetylcholine are located more dorsally and thus closer to the subnuclei of the oculomotor complex (Fig. 9.7) (Ryabinin et al., 2005; Yamaguchi, 2017; May et al., 2020). Acknowledging the interspecies anatomic variation of the EW and the absence of a consensus opinion about the location of acetylcholine-positive neurons in humans, for the rest of this section, we will maintain the traditional notion of the EW as the origin of the parasympathetic output to the pupil and ciliary muscle.

Another controversy on the EW exists. It has been debated whether the functional action of preganglionic parasympathetic neurons is selective or not. In an elegant study using anterior chamber and vitreous injection of tracers to study synaptic connections within the EW of macaque monkeys (May et al., 2020), May et al. confirmed previous observations that acetylcholine-expressing neurons are located into a single column situated dorsally to the oculomotor nucleus (Warwick, 1954) and additionally, they demonstrated at the ultrastructural level that neurons destined for pupillary constriction can be differentiated from neurons mediating accommodation. The former receives more supranuclear and infranuclear inhibitory inputs (May et al., 2020). This recent evidence strongly suggests that a distinction between accommodation neurons and pupillary neurons is already present at the preganglionic level in the midbrain, at least in nonhuman primates.

Several non-OPN structures project to the EW modulate pupil size. One is the superior colliculus, a key structure for coordinating eye and head movements, processing optical stimuli, and orienting attention-guided eye movements (May et al., 2021). A study from Wang and colleagues found that microstimulation of the superior colliculus in monkey causes bilateral transient pupillary dilation without inducing eye

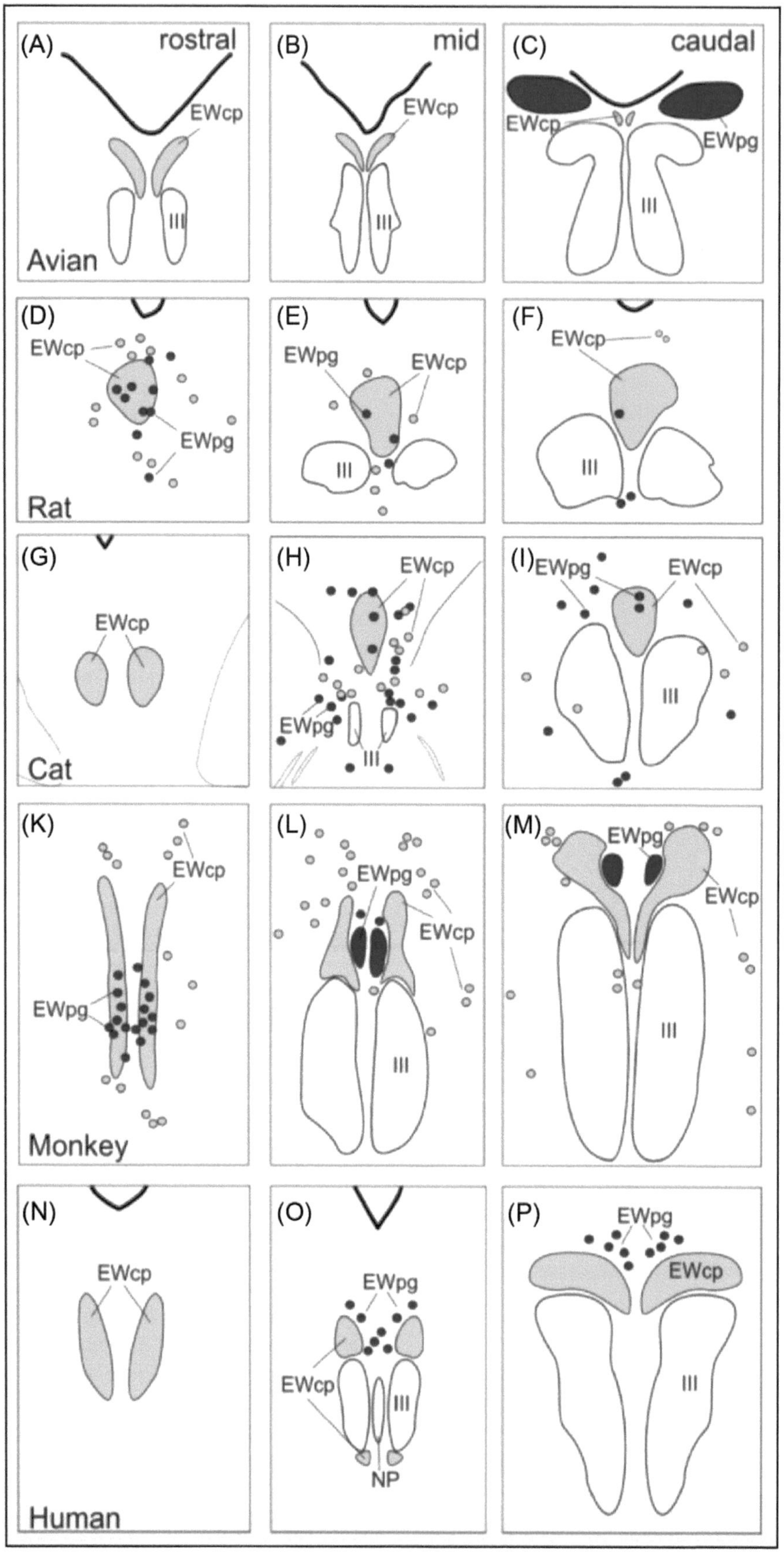

Fig. 9.7. Schema of the interspecies variation of the location of acetylcholine-expressing oculoparasympathetic neurons (EWpg, black circles) shown in rostral-caudal view at upper midbrain level. Note that the noncholinergic cell population expressing the peptide urocortin-1 (EWcp, light gray circles) does not project to the sphincter pupillae or ciliary muscle. Line drawings showing the organization of Edinger–Westphal preganglionic (EWpg) and Edinger–Westphal centrally projecting (EWcp) in several selected species: avian, rat, cat, monkey, and human. Representative rostral (left column), middle (middle column), and caudal (right column) sections are shown. The EWcp is indicated by light gray shading and EWpg is indicated by dark gray shading. Scattered cells located outside the nuclear boundaries are indicated by appropriately shaded circles. Reprinted with permission from Kozicz T, Bittencourt JC, May PJ, et al. (2011) The Edinger-Westphal nucleus: a historical, structural, and functional perspective on a dichotomous terminology. J Comp Neurol 519: 1413–1434.

movement (Wang et al., 2012a). Another study has shown lens shape changes in response to changes in fixation targets due to connections between the superior colliculus and EW (May et al., 2022). While the exact location of the synaptic connection between these structures might again vary depending on species, this tracer study using rabies virus showed that the superior colliculus provides input to, among others, a population of neurons involved in accommodation and lens shape changes that reside in the supraoculomotor area.

The periaqueductal gray (PAG) is another area that projects to the EW. The PAG surrounds the cerebral aqueduct in the midbrain and is crucially involved in autonomic functions (e.g., control of blood pressure, bladder control), fear and anxiety, and pain modulation (Silva and McNaughton, 2019). A study by Klooster and Vrensen showed a direct connection between EW and PAG, thereby providing an anatomic substrate for pupillary movements originating in the PAG that are related to fear and anxiety (Klooster and Vrensen, 1998).

In summary, in additional to excitatory luminance signals from the OPN, multiple other inputs to the EW permit continuous adjustment of the pupil size in response to ambient light, emotional state, and complex visual stimuli, aiding in maintaining focus and clarity of vision during rapid changes in fixation. However, the strength of signal from the different inputs at the EWN is not evenly proportional and an abrupt increase in light intensity relayed to the OPN is a signal that overpowers all other modulatory inputs to the EW to evoke a reliable and reproducible pupil light reflex.

The final parasympathetic motor signal from the EW travels with somatic motor fibers as the CN III, the oculomotor nerve. The fascicule from each subnucleus of the oculomotor complex travels in topographic manner through the ventral midbrain, with parasympathetic fascicle positioned medially relative to the somatic motor fascicles, passing through the red nucleus and the medial portions of the cerebral peduncles before exiting the midbrain into the interpeduncular fossa space. In the subarachnoid space, the motor fibers take a more ventral and lateral position while the parasympathetic fibers remain in a medial location. This anatomic configuration of the oculomotor nerve predisposes parasympathetic fibers to compressive damage from a posterior communicating artery aneurysm. The oculomotor nerve then passes along the lateral side of the posterior clinoid process, and reaching the cavernous sinus, it maintains a lateral position in the dural wall of the sinus as it continues forward to the superior orbital fissure (SOF) where it typically divides into superior and inferior divisions. Entering the orbital space, it is the inferior division that contains the parasympathetic fibers which synapse at the ciliary ganglion.

The orbital ciliary ganglion is located behind the ocular globe on the lateral side of the ophthalmic artery. Anatomically, it measures approximately 3 millimeters in diameter and is situated between the lateral rectus muscle and the optic nerve, nestled in the connective tissue close to the ocular globe (Haladaj, 2020). As previously discussed, there is differentiation between parasympathetic pupillary and accommodative fibers at the preganglionic level in the EW. At the level of the ciliary ganglion level, animal studies confirm a continued anatomic separation of pupillary and accommodative functions in the postganglionic parasympathetic neurons of the ciliary ganglion. Warwick's early study in monkeys reported only 3% of the parasympathetic neurons in the ciliary ganglion are destined for the innervation of the sphincter pupillae and the vast majority of the neurons innervate the ciliary muscle (Warwick, 1954). This functional division of neurons varies with species, as the percentage of neurons destined to the pupil is closer to 10% in felines. While there is clear bias towards accommodative neurons at the ciliary ganglion, it is not clear if there is a simple 1:1 relation between preganglionic acetylcholine-expressing neurons and postganglionic neurons in the ciliary ganglion.

Another interesting aspect is that, among different animal species, the size of neuronal cells of the ciliary ganglion varies according to the degree of visual organ development (Kuder et al., 2003). The main function of the ciliary ganglion is to relay the parasympathetic signal to the target muscular cells via acetylcholine signaling. However, studies in avian species and recently in macaques showed that this ganglion also receives different quantities of GAD-inputs and therefore it could represent a site of neuronal integration of preganglionic signaling rather than just a parasympathetic relay (Fig. 9.8) (Barnerssoi et al., 2017).

In addition to parasympathetic fibers that synapse there, the ciliary ganglion contains two other afferent roots: a sensory root from the nasociliary (V_1) nerve destined for the sensory innervation of the anterior segment of the eye and a postganglionic sympathetic root that joins the ganglion, either as a solitary branch or in association with the nasociliary branch. Noteworthy is that these two major afferent roots just pass through the ciliary ganglion without synapsing, and that together with the postganglionic parasympathetic fibers they form the short ciliary nerves. There are two different types of ciliary nerves, long and short. The former, usually two in number, directly arise from the nasociliary nerve, while the latter emerge from the ciliary ganglion and are grouped into two clusters totaling 6–10 short ciliary nerves. Along with the long ciliary nerves, they penetrate the sclera around the optic nerve. Both nerve types travel forward between the choroid and sclera, eventually forming a plexus in the ciliary muscle. This plexus supplies the cornea, ciliary body, and iris for the sensory aspect and targets the sphincter pupillae and the ciliary muscle, causing sphincter pupillae constriction and accommodation, respectively, for the motor component. Anatomically, the long ciliary nerves are located in the meridians at 3 and 9 o'clock in the globe while the short ciliary nerves form a 360° net around the sclera. Due to this anatomic peculiarity, it is possible to damage only sectors of short ciliary nerves while sparing others.

In summary, the parasympathetic innervation of the sphincter pupillae and ciliary muscle is generated in the midbrain and advances towards the orbit in close association with the oculomotor nerve. In the eye, it reaches the target muscles via the short ciliary nerves (Fig. 9.9).

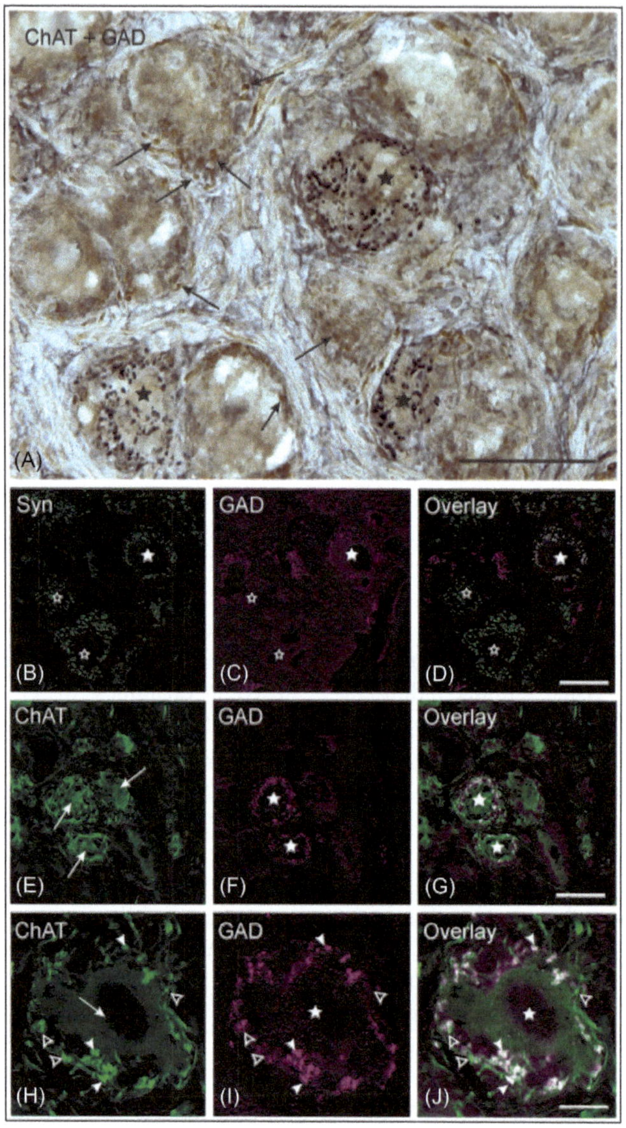

Fig. 9.8. Histologic section of the monkey ciliary ganglion showing different glutamic acid decarboxylase (GAD)-positive input to cholinergic postganglionic neurons. Immunostained cryosections demonstrating γ-aminobutyric acid-positive terminals. Stars mark a subpopulation of choline acetyltransferase (ChAT)-positive, postganglionic ciliary ganglion (CG) neurons that receive a dense supply of GAD-positive boutons. (**A**) Combined immunoperoxidase staining of a monkey CG section showing ChAT-positive neurons (brown) surrounded by ChAT-positive profiles (arrows) and/or GAD-positive bouton-like structures (black). Autofluorescent lipofuscin granules could be distinguished from immunostaining in all neurons by their color. (**B–D**) Confocal images of double-immunofluorescence labeling for GAD and synaptophysin (Syn). (**B**) Numerous green-fluorescing Syn-positive terminals supply every CG neuron. (**C**) A subpopulation of CG neurons receives dense input from GAD-positive bouton-like structures colored in magenta (filled star). The open star indicates CG neurons without GAD input. (**D**) The overlay of both images shows that both Syn-expression (green) and GAD-expression (magenta) colocalize in boutons (white structures), proving that GAD-positive boutons represent terminals. (**E–J**) Confocal images of double-immunofluorescence staining of monkey CG sections for ChAT (green) and GAD (magenta). Images **G** and **J** show the overlay of **E** and **F** and **H** and **I**, respectively. All CG neuron somata show ChAT-immunoreactivity (**E** and **H**, arrows). None of the CG neuron somata stain positively for GAD (**F** and **I**, arrows). A higher magnification of an example cell (**H–J**) shows that all GAD-positive bouton endings (**J**, solid arrowheads) colocalize ChAT (**G** and **J**, white, arrowheads), but few ChAT-positive endings lack GAD immunoreactivity (**H–J**, open arrowheads). Scale bars=50 μm in **A**; 40 μm in **D** (applies to **B–D**); 50 μm in **G** (applies to **E–G**); 10 μm in **J** (applies to **H–J**). Reprinted with permission from Barnerssoi M, May PJ, Horn AK (2017) GABAergic innervation of the ciliary ganglion in macaque monkeys – a light and electron microscopic study. J Comp Neurol 525: 1517–1531.

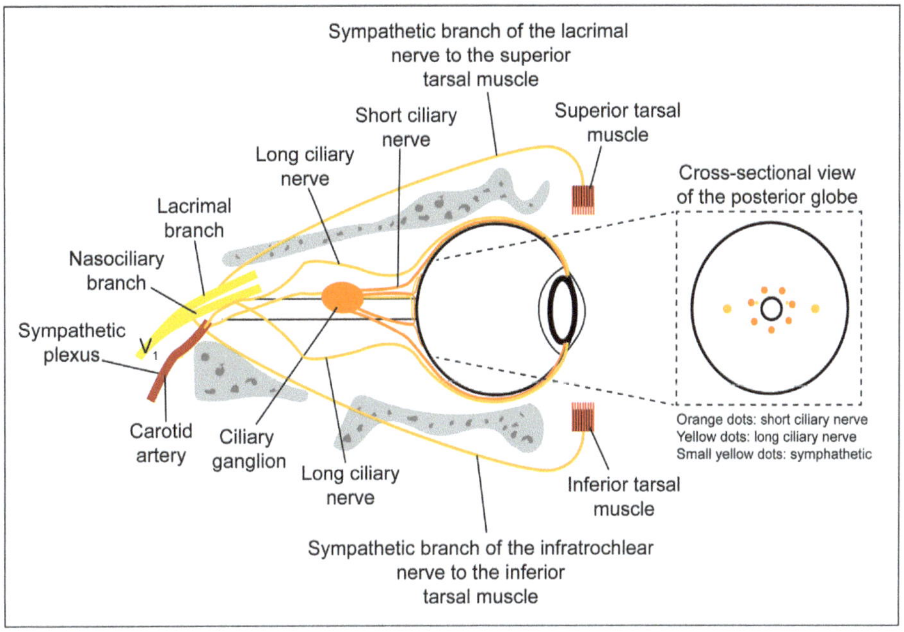

Fig. 9.9. Schema of the orbital and ocular parasympathetic and sympathetic pathways. Left side: Sagittal representation of the sympathetic and parasympathetic, as well as other relevant orbital structures, to the eye. Right side: View of the posterior pole of the eye, showing different penetration of short and long ciliary nerves into the sclera.

Clinical manifestations of a parasympathetic deficit to the eye

A large pupil that constricts poorly can arise from congenital or acquired disorders of the iris, as previously discussed, or from denervation of the iris. Owing to the widespread distribution of the short ciliary nerves in the sclera, lying just beneath the retina, surgical procedures like scleral buckling for retinal detachment repair, cryotherapy to the sclera, or laser treatments like panretinal photocoagulation can damage these parasympathetic nerves (Yilmaz et al., 2016). Weakness of the sphincter pupillae and ciliary muscle is sometimes referred to as internal ophthalmoplegia because the two internal smooth muscles of the eye are affected but not striated extra-ocular muscles. Confusion often arises in the literature as internal ophthalmoplegia is sometimes also called tonic pupil. However, for this chapter, we will refer to tonic pupil as a poorly reactive pupil due to postganglionic parasympathetic denervation followed by re-innervation.

Tonic pupil

Tonic pupil, or Adie's pupil when idiopathic, predominantly affects young women aged between 20 and 40 years. It is typically described as an abnormally large pupil having poor or absent reaction to light but an exaggerated response to near vision accommodation, a clinical sign called LND (see Light-near dissociation section for more details). Other characteristic clinical signs include sectorial sphincter palsy, loss of pupillary ruff, and pupillary vermiform movements (Kawasaki, 2014; Xu et al., 2022). However, these clinical signs of tonic pupil evolve and vary from the hyperacute phase through the chronic phase, as described below.

At the moment of injury to postganglionic oculoparasympathetic nerves, the sphincter pupillae and ciliary muscles are acutely denervated. The patient may notice a new anisocoria and may also complain of photophobia and blurred vision while reading. The examination reveals a large pupil that is unresponsive to both light and near stimulation as well as loss of accommodation amplitude. During this hyperacute stage (initial 48–72 hours), the appearance of the pupil resembles pharmacologically induced mydriasis. However, the presence of sectorial sphincter palsy orients toward a tonic pupil rather than a pharmacologic mydriasis. Sectorial sphincter palsy is observable under high magnification at the slit lamp, which shows paralyzed sectors of the iris adjacent to sectors that are constricting normally (Video 9.3). The pathophysiologic basis of sectorial sphincter palsy can be found in the anatomic distribution of short ciliary nerves. As these parasympathetic fibers are numerous and broadly traverse the sclera, any insult typically results in partial damage of nerves and sectorial denervation of the sphincter pupillae, and not complete sphincter paralysis.

The hyperacute phase is followed by a subacute phase of denervation (days to weeks after denervation) in which compensatory changes at the receptor level in the neuromuscular junction occur. In response to reduced neurotransmitter stimulation, cells of the receptor organ often increase the number of postsynaptic receptors, a process known as upregulation (Jones and Vrbova, 1974; Steward, 2014). This makes the receptor organ supersensitive to their neurotransmitter, starting as early as 4–5 days after denervation (Kardon and Bergamin, 2005). In the case of parasympathetic denervation the sphincter pupillae upregulates its muscarinic acetylcholine receptors, thereby increasing the sensitivity of the muscular fibers to acetylcholine stimulation.

Denervation supersensitivity is the basis for the dilute pilocarpine (0.125% or less) test, in which a parasympathetically denervated sphincter will exhibit an exaggerated contraction to dilute pilocarpine (Fig. 9.10). However, using a dilution to 0.125%, up to 40% of normal eyes can show some degree of sphincter contraction, hence increasing false positive results (Younge and Buski, 1976; Yoo et al., 2021). For this reason, use of 0.0625% dilution of pilocarpine as diagnostic test for tonic pupil is recommended (Cohen and Zakov, 1975; Leavitt et al., 2002). It should be remembered that the dilute pilocarpine test is not specific for tonic pupil, because even preganglionic parasympathetic injury, for example, oculomotor nerve injury, can lead to denervation supersensitivity. However, the clinician will use other signs (ptosis, ocular deviation) to diagnose oculomotor nerve palsy and dilute pilocarpine is not used in this setting.

The early chronic phase (weeks to 2–4 months postdenervation) of a tonic pupil is distinctive for regeneration of nerves and reinnervation of end organs. Accommodative fibers often properly reinnervate the ciliary muscle leading to recovery of accommodation (i.e., patients can read again). However some accommodative fibers are misguided in their regeneration and aberrantly reinnervate the sphincter pupillae. Pupil fibers, however, do not seem to find their way back to the sphincter pupillae. This is presumably related to the bias in the ciliary ganglion in which 97% of the postganglionic parasympathetic neurons innervate the ciliary muscle rather than the sphincter pupillae. Clinically, this aberrant regeneration and reinnervation is observed as a LND of the pupillary constriction (Video 9.4). The pupillary constriction to near effort (accommodation) is slow and redilates very slowly after removal of the near effort. This is one reason for the moniker "tonic." This slow, sustained response may also be seen in the ciliary muscle, leading to unequal accommodation between the two eyes, especially noticeable when the person shifts focus from near tasks, like reading or computer work, to distant vision.

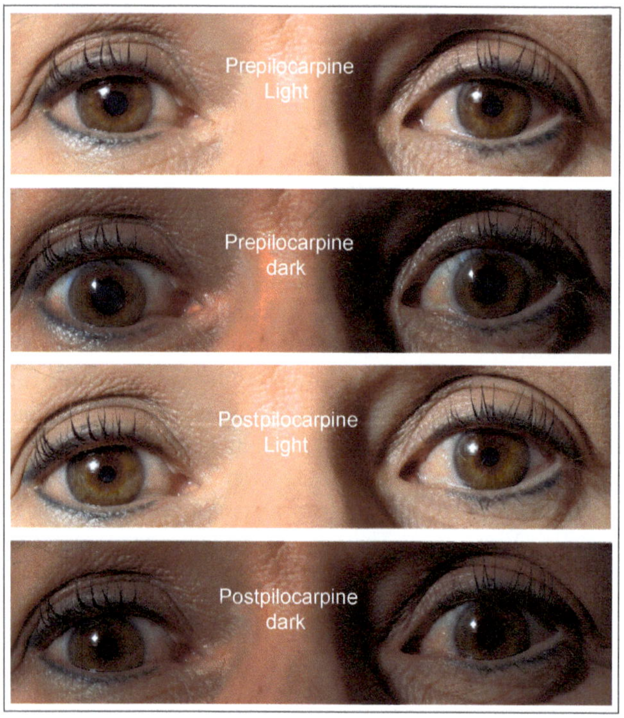

Fig. 9.10. Dilute pilocarpine (0.125%) test demonstrating denervation supersensitivity of a right tonic pupil. Top image: Pupils viewed in room light conditions (before pilocarpine). Note the anisocoria with the right pupil larger than the left pupil. Second image: Pupils viewed in dim light conditions (before pilocarpine). Note that the anisocoria is no longer evident, and the pupils have equal size, as this is the result of primary sympathetic action by darkness. Third image: Pupils viewed in room light 45 min after pilocarpine. Note that the right pupil is now smaller than the left pupil, showing reversal of the anisocoria, thus confirming denervation supersensitivity of the right pupil. The left pupil has not reacted to diluted pilocarpine, as expected. Bottom image: Pupils viewed in dim light 45 min after pilocarpine. Note that the reversed anisocoria persists in dim light conditions.

Finally, in the late chronic phase, when accommodative fibers have extensively re-innervated various parts of the sphincter pupillae, their continuous volley of neural signals to the sphincter causes the pupil to remain in a constant state of constriction known as a "little old Adie's pupil" (Fig. 9.11). In this state, the pupil is small even in darkness. When focusing on close objects, this small pupil may get even smaller due to LND.

While a tonic pupil is most frequently idiopathic, other secondary causes of unilateral mydriasis should be considered (Xu et al., 2022). Parasympathetic damage to the ciliary ganglion may be caused by infectious processes like syphilis, herpes zoster, and Lyme disease, traumatic or surgical affections of the orbit and autoimmune and inflammatory disorders such as Guillain-Barré syndrome, Miller Fisher syndrome, sarcoidosis, and Sjögren syndrome. Bilateral parasympathetic denervation is seen with diabetes mellitus, paraneoplastic syndromes, amyloidosis, autonomic dysfunction affections such as the rare disease called Ross syndrome or the familial dysautonomia syndrome (also called Riley–Day syndrome) but do not demonstrate aberrant re-innervation. Please refer to the autonomic neuropathy section for further details. Other conditions leading to a uni- or bilaterally enlarged pupil with overlapping features of tonic pupil include oculomotor nerve palsy, anticholinergic drug overdose, Argyll Robertson pupil, and congenital mydriasis (see Kardon and Bergamin for further details; Kardon and Bergamin, 2005).

Oculomotor nerve palsy

Oculomotor nerve palsy is another cause of oculoparasympathetic denervation. The classical clinical presentation features upper lid ptosis, ophthalmoplegia with a "down and out" eye, and mydriasis. It is exceptional that unilateral mydriasis is the sole manifestation of an oculomotor nerve palsy. However, careful examination typically reveals a very partial ophthalmoplegia or minimal ptosis that accompanies the pupil anomaly and thus confirms subtle oculomotor nerve palsy. That being said, rare case reports of unilateral and truly isolated pupillary dilation have been reported as consequence of oculomotor

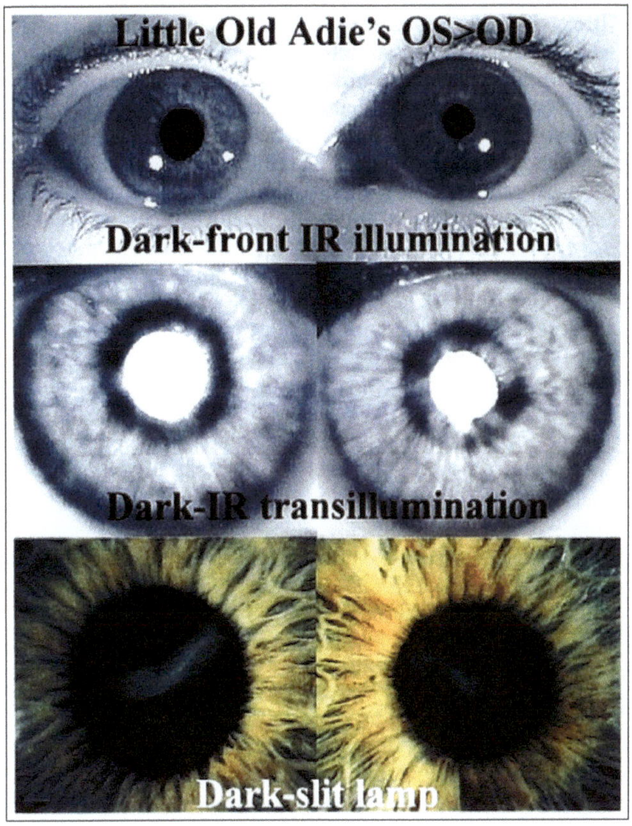

Fig. 9.11. Photos (color and infrared) of bilateral chronic tonic pupils ("little old Adie") in darkness. The eyes are artificially brought closer by use of prisms. The chronic tonic pupils stay abnormally small in darkness due to continuous firing by accommodative fibers which extensively and aberrantly re-innervate the sphincter pupillae. As the sphincter pupillae contract it and the surrounding peri-pupillary iris will appear denser (black) on infrared transillumination. The lighter portions around the pupil correspond to noncontractile segments. Top: Adie's tonic pupil in the left eye is smaller than the right eye in darkness. Middle: The more tonically constricted segments seem denser when the lights are out (adjacent, normally innervated sphincter segments are inhibited and, hence, translucent), and the pattern of the most tonic segments can only be distinguished with infrared transillumination and not with frontal illumination or with the slit lamp. Both pupils are smaller than normal in darkness. The right sphincter shows diffuse increase in density, except at the 9:00 clock-hour position, indicating re-innervation by accommodative fibers. In the left iris sphincter, the increase in density is more clearly segmental (sectorial) and the pupil is smaller in size, suggesting a more tonic contraction in some of the re-innervated segments. Botton: This pattern cannot be discerned in the dynamic slit-lamp examination. The white opacity in the pupillary space is a cortical cataract present in both eyes (Kardon et al., 1998). Reprinted with permission from Kardon RH, Corbett JJ, Thompson HS (1998) Segmental denervation and reinnervation of the iris sphincter as shown by infrared videographic transillumination. Ophthalmology 105: 313–321.

nerve pathology, including vascular conflicts between the oculomotor nerve and the posterior cerebral artery, an arachnoid cyst or a neuroenteric cyst of the oculomotor nerve (Werner et al., 2005; Albayram et al., 2006; Ashker et al., 2008; Chronopoulos et al., 2023). In the comatose patient, clinical examination cannot detect subtle ophthalmoplegia or ptosis; thus, an unresponsive mydriasis must be considered a sign of oculomotor nerve compromise and warrants imaging to rule out aneurysm of the posterior communicating artery, uncal herniation (in this case, the dilated pupil is named Hutchinson pupil), midbrain infarction, or skull base meningitis, among others.

SYMPATHETIC INNERVATION OF THE PUPIL

Anatomy of the sympathetic pathway

The sympathetic pathway to the head and eye is classically described as a three-neuron pathway in which the first-order neuron originates primarily in the hypothalamus, second-order neuron in the lower cervical-upper thoracic spinal cord, and the third-order neuron in the superior cervical ganglion (Fig. 9.12). Details of this pathway are described in the following paragraphs.

Darkness is a potent excitatory stimulus of the oculosympathetic pathway, leading to observable pupillary dilation in 3–5 seconds. Light has both an inhibitory and excitatory effect on sympathetic pathways. The better understood is light-inhibition of sympathetic outflow which occurs simultaneously with light-activation of the parasympathetic pathway, resulting in the well-known pupil light reflex (McDougal and Gamlin, 2015). Histopathologic evidence from isolated human iris dilator has shown functional innervation by adrenergic excitatory as well as cholinergic inhibitory fibers (Yoshitomi et al., 1985). Light has a small excitatory effect on sympathetic discharge, which evokes a small pupillary dilation but this is masked by the more obvious pupillary constriction of the light reflex. A variety of physiologic and psychologic variables such as arousal, attention, cognitive performance, mood, pain, and others also excite the oculosympathetic pathway but such pupil responses are small and best observed using pupillometry (Szabadi, 2018).

The major central sympathetic nuclei are the paraventricular nucleus (PVN) in the hypothalamus, the locus coeruleus, and dorsal raphe nucleus in the brainstem, and collectively, they comprise the first-order (central or premotor) sympathetic neurons destined to the head and eye. The locus coeruleus is an important source of norepinephrine to the forebrain and is a primary mediator of the pupillary dilation seen during mental arousal and cognitive processing. In the hypothalamus, the PVN receives input from suprachiasmatic nucleus, the master circadian regulator, to which temporal light information is directly relayed by retinal MGCs. Thus, the PVN exhibits circadian rhythmicity and may exert a light-mediated excitatory effect on sympathetic preganglionic neurons mediating pupil dilation.

The axons of the central, first-order neurons descend the brainstem in parallel right and left tracts which gradually diverge to the lateral aspect of the medulla oblongata and then continue into the cervical spinal cord to synapse at the intermediolateral (IML) cell column of the cervicothoracic spinal cord at levels C8-T2. The neurons forming the ciliospinal nucleus of Budge–Waller of the IML comprise the second-order (or preganglionic) sympathetic neurons,

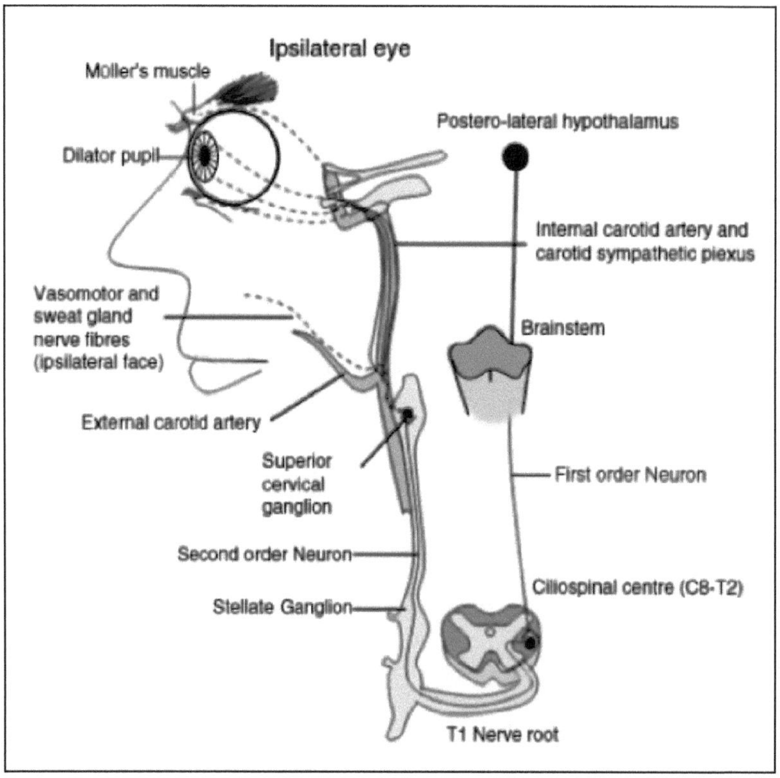

Fig. 9.12. Schematic diagram of the sympathetic nervous system pathway to the eye and face (Kong et al., 2007). Reprinted with permission from Kong YX, Wright G, Pesudovs K, et al. (2007) Horner syndrome. Clin Exp Optom 90: 336–344.

which will innervate the head and eye. Their axons, alongside somatic motor neuros, exit the spinal cord through the ventral roots but after entering the spinal nerves, the sympathetic axons move anteriorly in order to ascend the neck in the cervical sympathetic chain (Kawasaki, 2005). Some sympathetic fibers, particularly from T1, continue into the brachial plexus to regulate vasomotor function of the arm and hand. Sympathetic fibers from T1–T4 pass through the mediastinum to join the vagal parasympathetic fibers, forming the cardiac plexus. The preganglionic sympathetic axons synapse at the superior cervical ganglion located at the level of the bifurcation of the common carotid artery. In the retro- and paraparotid region, ascending sympathetic fibers in the carotid space are located near the lower cranial nerves (IX–XII) after their exit through the jugular foramen (Fig. 9.13).

The third-order neurons, also called postganglionic sympathetic neurons, of the superior cervical ganglion supply vasomotor and sudomotor function to the head and neck as well as pupillodilator and secretomotor functions in the eye and also tarsal muscle function of the eyelids. The postganglionic sympathetic axons form a plexus surrounding carotid arteries (Kawasaki, 2005). Vasomotor and sudomotor fibers primarily follow the external carotid artery, whereas oculosympathetic fibers travel with the internal carotid artery. In the retropharyngeal space, the sympathetic chain lies in proximity to other nerves, notably the superior laryngeal, phrenic, vagus, and glossopharyngeal nerves. The oculosympathetic fibers re-enter the cranial space via the carotid canal and continue with the internal carotid artery into the cavernous sinus. Thereafter, they are conducted via the abducens nerve and the ophthalmic division of the trigeminal nerve (V_1) into the orbital space. Sympathetic fibers to the superior tarsal muscle of the upper eyelid are carried in the lacrimal branch of V_1, whereas the nasociliary branch transmits sympathetic fibers to the inferior tarsal muscle of the lower eyelid via the infratrochlear nerve and to the dilator muscle of the iris via the two long ciliary nerves (nasal and temporal). Some sympathetic fibers pass through the ciliary ganglion without synapsing and reach the eye via short ciliary nerves. Sympathetic fibers surrounding the ophthalmic artery and entering the orbit via the optic canal are mostly involved in regulation of ocular and orbital blood flow but do not innervate the iris or the tarsal muscles of the eyelids (Thakker et al., 2008).

Clinical manifestations of a sympathetic deficit to the eye and head

Horner syndrome

The sympathetic denervation of the eye and head is called Claude Bernard–Horner syndrome or simply Horner syndrome. The commonly cited triad of clinical signs is: eyelid ptosis, pupillary myosis, and facial anhidrosis. Postganglionic sympathetic fibers provide very limited sudomotor function to the face; thus, it is very rare for such patients to report asymmetry of facial sweating. However, patients with central or

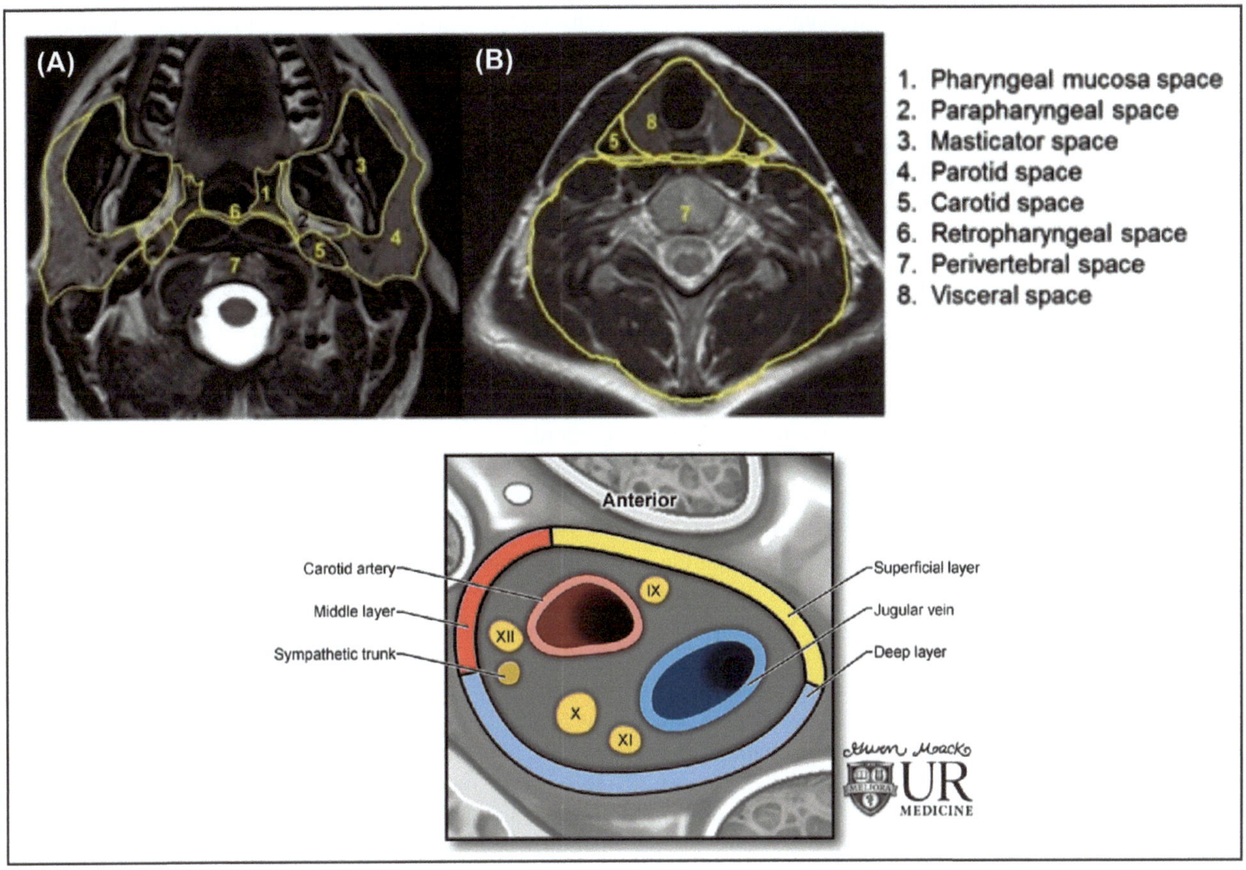

Fig. 9.13. Illustration of the retroparotid space (cross-sectional view), with a magnified schema of the left carotid space showing the position of sympathetic fibers. Top: Axial T2-weighted images showing the deep spaces of the suprahyoid neck at the level of the oropharynx (**A**) and the infrahyoid neck at the level of the thyroid (**B**) which are demarcated by the deep fascia of the neck. The retropharyngeal space is a potential space between the perivertebral space and the pharyngeal mucosa space above and the visceral space below the level of the hyoid bone (Tshering Vogel and Thoeny, 2016). Bottom: The carotid space. Illustration demonstrating the contents and configuration of the left carotid space, including cranial nerves IX, X, XI, and XII (Chengazi and Bhatt, 2019).(A). Tshering Vogel DW, Thoeny HC (2016) Cross-sectional imaging in cancers of the head and neck: how we review and report. Cancer Imaging 16: 20. Available at: https://cancerimagingjournal.biomedcentral.com/articles/10.1186/s40644–016-0075–3/figures/1. Open Access. Reprinted under the terms of the Creative Commons Attribution 4.0 International License (http://creativecommons.org/licenses/by/4.0/), which permits unrestricted use, distribution, and reproduction in any medium. (B) Chengazi HU, Bhatt AA (2019) Pathology of the carotid space. Insights Imaging 10: 21. Available at: https://insightsimaging.springeropen.com/articles/10.1186/s13244-019-0704-z/figures/1. Open Access. Reprinted under the terms of the Creative Commons Attribution 4.0 International License (http://creativecommons.org/licenses/by/4.0/), which permits unrestricted use, distribution, and reproduction in any medium.

preganglionic sympathetic lesions may have notable loss of sweating and vasodilation of the hemiface, hemithorax, and/or hemibody. In our experience at a tertiary care center, the most common clinical presentation of Horner syndrome is a combination of anisocoria and eyelid ptosis on the side of the smaller pupil. The eyelid ptosis of Horner syndrome is usually mild (about 1 mm) and may be frankly absent in 12%–14% of patients (Maloney et al., 1980). For this reason, a new anisocoria can be the sole manifestation of a lesion of the oculosympathetic pathway. Conversely, not everyone with myosis without or with ipsilateral ptosis has a Horner syndrome (Martin, 2007). For these reasons, the diagnosis of Horner syndrome often requires pharmacologic confirmation of sympathetic denervation of the dilator pupillae (described below).

Some signs of oculosympathetic denervation are notable only in the acute stage. For example, the sudden loss of sympathetic autoregulation leads to local vasodilation of ocular structures, and the patient with an acute Horner syndrome may report a sudden conjunctival redness or eyelid puffiness. Iris heterochromia is, however, indicative of a congenital or a very long-standing sympathetic denervation. Finally, the ptosis and myosis of Horner syndrome may be transient and have been reported in association with cluster headache, tension pneumothorax, lumbar epidural anesthesia, cervical cord syrinx, and even intrapetrous carotid dissection, among others. The different clinical manifestations of a sympathetic deficit to the eye and head are detailed in the paragraphs below. Clinically, the eponymic syndrome refers to a sympathetic deficit which involves at least the pupil and/or eyelid.

The anisocoria of Horner syndrome is due to paresis of the dilator pupillae leading to a smaller pupil that dilates poorly.

Anisocoria is the most constant sign of an oculosympathetic defect, present in 98% of patients with Horner syndrome but degree varies, depending on factors such as the completeness of the injury, the patient's state of alertness, and the brightness of the room light (Maloney et al., 1980). For example, sleepiness diminishes premotor sympathetic activation and as the pupils decrease in size, the anisocoria becomes less apparent. A bright room light favors parasympathetic activation and also inhibits sympathetic activation, even if the patient is not sleepy, and the resulting pupillary constriction will minimize, even abolish, the anisocoria due to Horner syndrome. Thus, the anisocoria from sympathetic denervation is best observed in darkness when the patient is alert (Fig. 9.14). Addition of a psychosensory stimulus, e.g., a loud or sudden noise which stimulates sympathetic activity, can enhance the anisocoria in Horner syndrome.

On rare occasion, a patient with Horner syndrome might report that the myotic pupil is sometimes briefly larger than the contralateral pupil and that the anisocoria appears reversed. This transient dilation of a Horner pupil is due to hypersensitivity of the dilator pupillae muscle to circulating catecholamine and adrenergic substances and thus occurs during moments of acute stress, physical exertion, or emotional excitement. In patients with Horner syndrome from cervical cord trauma, the ipsilateral pupil may alternate between relative myosis and mydriasis due to alternating hypofunction and hyperfunction of the damaged sympathetic fibers.

In addition to a smaller resting size of the pupil, sympathetic denervation impairs the dynamic dilation movement of the pupil. Sudden darkness or strong psychosensory stimulus, such as a loud noise or a pinch of the skin, activates a burst of sympathetic discharge while reciprocally inhibiting the EW, and as consequence, the normal pupil actively dilates. Lacking sympathetic activity, a Horner pupil passively dilates from inhibition of parasympathetic tone and relaxation of the sphincter pupillae. In the patient with a unilateral Horner syndrome, the difference between the active rapid dilation of the normal pupil (within 3–5 seconds) and the passive slow dilation taking 15–30 seconds of the Horner pupil can be seen by watching both pupils simultaneously in darkness after abruptly turning off the room light. This impaired pupillary dilation of Horner syndrome is called "dilation lag" and is specific to sympathetic denervation. A mechanically restricted myosis, such as postinflammatory adhesions in the iris, also causes poor pupillary dilation in darkness but this will be immediately observable in the first few seconds, and there is not the continued slow dilation in later seconds, which is pathognomonic of sympathetic denervation. Unfortunately, pupillary dilation lag in the clinical setting is observable in only half of patients with Horner syndrome.

The upper and lower eyelids have smooth tarsal muscles (Müller muscle) innervated by sympathetic fibers. The superior tarsal muscle arises from the undersurface of the levator palpebrae superioris muscle and controls the resting position of the upper lid of the nonclosed eye. It has a maximum excursion of 3 mm so the superior tarsal muscle contributes to but is not the major muscle for superior eyelid elevation. This explains why the ptosis from ocular sympathetic denervation tends to be mild, on the order of 1–2 mm (Tomelleri et al., 2007). In one study of 450 patients, only 88% of patients demonstrated clinically evident upper lid ptosis (Maloney et al., 1980). The inferior tarsal muscle is very small and thin but there is no compensatory voluntary muscle as with the upper eyelid. Paresis of the inferior tarsal muscle leads to a rise in the resting position of the lower lid, a sign known by various names such as lower lid ptosis, inverse ptosis, reverse ptosis, or upside-down ptosis (Fig. 9.15). Lower lid ptosis is a subtle finding and is clinically evident in only two-thirds of patients with Horner syndrome (Nielsen, 1983). However, the loss of tonus in both tarsal muscles narrows the palpebral fissure and can give the

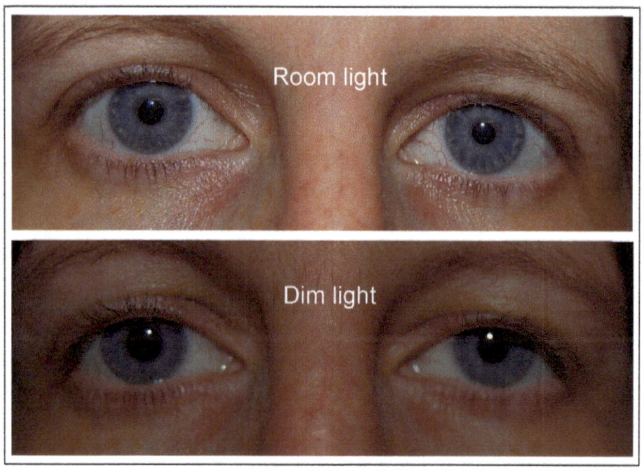

Fig. 9.14. Patient with a left Horner syndrome without ptosis, mimicking physiologic anisocoria. Top: In room light, there is a barely visible anisocoria and a minimal asymmetry in the palpebral fissures. There is no objective ptosis. Bottom: In dim light, a small but visible anisocoria of less than 1 mm can be observed. This anisocoria was pharmacologically confirmed to be a left Horner syndrome.

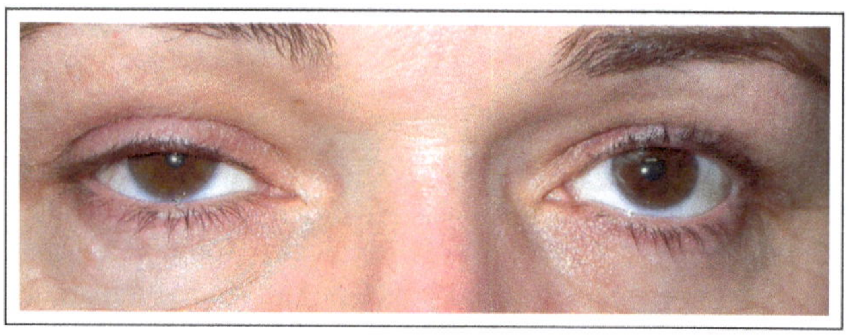

Fig. 9.15. Patient with a right Horner syndrome demonstrating upper and lower lid (inverse) ptosis. The upper lid ptosis on the right side is evident. The lower lid ptosis is best seen by aligning the inferior corneal border of the right eye to the lower lid margin until the inferior scleral space is effaced. The opposite eye, however, shows an observable scleral space inferiorly, indicating that the lower lid of the right eye has a higher resting position. Because there is upper and lower lid ptosis, the palpebral fissure is notably reduced and the eye appears to be smaller and retracted but this is a pseudo-enophthalmia, because exophthalmometry confirmed symmetric position of the globes in the orbit.

eye a sunken appearance and the false impression of enophthalmos.

Sympathetic vasodilatory and sudomotor fibers to the face exit the spinal cord primarily through the second and third thoracic (T2–3) roots (Lance et al., 1988). Sympathetic fibers to the eye exit spinal cord at lower cervical and first thoracic segments (C8–T1) (Turco and Farber, 1996). At the carotid bifurcation, some sudomotor and vasomotor fibers supplying the medial forehead and the side of nose travel with postganglionic oculosympathetic fibers along the internal carotid artery.

Patients with a postganglionic Horner syndrome typically have asymptomatic loss of sudomotor function on the forehead, but patients with a preganglionic Horner syndrome can have hemifacial loss of flushing and sweating (pale and dry skin), which is particularly noticeable during exercise or on hot summer days or in babies, during strenuous crying. This is called the harlequin sign (Lance et al., 1988) (Fig. 9.16) and in association with ipsilateral myosis and ptosis, indicates a lesion of sympathetic preganglionic fibers from C8 to T3. In the largest published series of patients (n=39) with harlequin sign by Bremner, two-thirds had abnormal pupils, and surprisingly, it was most often an ipsilateral postganglionic Horner syndrome (Bremner and Smith, 2008).

Iris color is primarily due to the amount and distribution of melanin pigment present in the melanocytes of the superficial anterior layer of the iris (Sturm and Frudakis, 2004). Sympathetic innervation plays a role in the tyrosinase activity and the development of melanin pigment in iris melanocytes. When sympathetic denervation of the eye occurs during the congenital or early postnatal period, the eye typically develops less iris pigmentation, and iris heterochromia is evident after several months when the normally innervated eye has developed its adult color (Weinstein et al., 1980). In young rabbits, surgical removal of cervical sympathetic fibers postinfancy led to rapid reduction of tyrosinase activity in the choroid and iris and an observable change in iris color as early as 3 weeks after the denervation (Laties, 1974). There is suggestion that the melanin content and iris color in humans is also subject to variation and change after the neonatal and infantile period and indeed, acquired iris heterochromia has been reported in occasional adults with noncongenital Horner syndrome in which the sympathetically denervated eye has a lighter iris color.

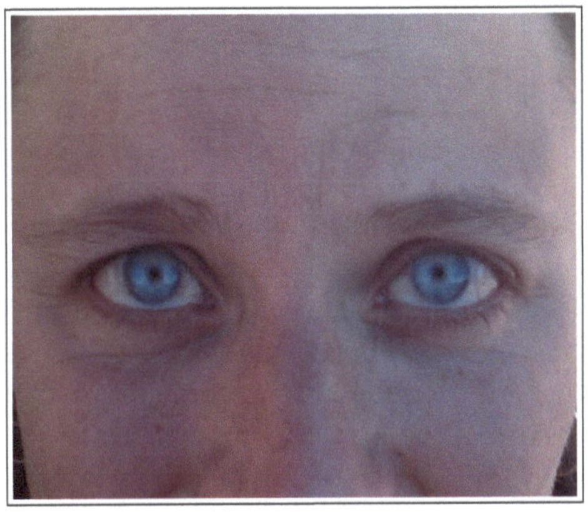

Fig. 9.16. Patient with Horner syndrome and harlequin sign on the left side. The patient has just finished some exercise outdoors. Notice that the right side of her face is flushed but the left side is less so, especially across the forehead, which is notably pale. She also reports decreased sweating of the left face (not shown here). The pupils are constricted because the photo was taken outdoors in sunlight and no anisocoria is observed. However, pharmacologic testing (apraclonidine drops) had confirmed denervation adrenergic supersensitivity on the left side.

Other effects of sympathetic denervation

Sympathetic innervation of the cornea appears to assist maintenance of normal corneal endothelial physiology but sympathetic denervation does not lead to symptomatic corneal surface changes.

Sympathetic fibers innervate the vasculature to the lacrimal gland, and thus play a role in the production and secretion of tears. However, ablation of the superior cervical ganglion does not alter the tear composition. The lipid layer of tears, secreted by meibomian glands, is hormonally regulated with minor neural influence.

As parasympathetic and local peptide pathways dominate the regulation of lacrimation, it is not clear if oculosympathetic denervation results in a symptomatic dry eye syndrome.

Straight (or less curly) hair on the side of Horner syndrome may be seen in congenital cases, perhaps related to inaction of arrectores pilorum muscle (Wang et al., 2012b).

The parotid gland is innervated by parasympathetic and sympathetic nerves carrying secretomotor function. Complete loss of sympathetic innervation to the parotid gland can lead to a painful condition called "first bite syndrome" and typically follows parapharyngeal space surgery with ablation or removal of the cervical sympathetic chain or superior cervical ganglion. Most patients have an ipsilateral Horner syndrome. Upon the first bite of a meal, there is excruciating pain and cramping in the parotid region, which subsides in a matter of seconds and does not recur with subsequent bites. However, the pain invariably returns with the first bite of the next meal (Kawashima et al., 2008). The mechanism is likely related to supersensitivity of myoepithelial cells surrounding salivary ducts. Upon stimulation by food, parasympathetic neurotransmitters in the parasympathetic fibers are released in the parotid and cross-stimulate the sympathetic receptors, causing overaction of the myoepithelial cells and local spasm and pain.

Pharmacologic testing of Horner syndrome

When an anisocoria more apparent in darkness is accompanied by harlequin sign or hemi-anhidrosis or pupillary dilation lag, a clinical diagnosis of Horner syndrome is made with high certitude. However, these distinctive signs of sympathetic deficit are often not present in Horner syndrome, and pharmacologic testing is often necessary for a definitive diagnosis.

Topical cocaine has been a time-honored test of Horner syndrome. In the normal state, terminal sympathetic fibers are continuously releasing norepinephrine into the neuromuscular synapse. Cocaine blocks the enzyme needed to reuptake norepinephrine, thus allowing norepinephrine to accumulate at the neuromuscular junction synapse and saturate postsynaptic adrenergic receptors of effector cells. In the eye, the normal effect of topical cocaine is pupillary dilation, conjunctival blanching, and eyelid retraction. Interruption anywhere along the sympathetic pathway to the eye diminishes or even arrests release of norepinephrine into the terminal neuromuscular synapse, and the action of cocaine is rendered ineffective.

The test is performed by putting two or three drops of cocaine ophthalmic solution (between 4% and 10%) into each eye and then measuring the anisocoria 45–60 minutes later. A Horner pupil will stay small, while the normal pupil will have dilated, resulting in an increase of the baseline anisocoria (Fig. 9.17). Because a partial denervation may permit the Horner pupil to dilate partially to cocaine, a simple rule of thumb is that a persistent postcocaine anisocoria of 1 mm or more is diagnostic for Horner syndrome (Kardon et al., 1990). A valid test requires that the normal (control) eye shows pupillary dilation of 2 mm or more as evidence that cocaine has sufficiently entered and acted upon the eye. In persons with dark irides, cocaine may not dilate the normal pupil and thus, test results cannot be reliably interpreted. Two common errors leading to a false-positive results are failure to put at least two drops in each eye and failure to wait at least 45 minutes before examining the postcocaine anisocoria. Finally, the cocaine test is avoided in patients using other eye drops or wearing contact lenses regularly, because these factors may cause a differential absorption of cocaine between the two eyes.

Apraclonidine (0.5% or 1%) is currently the most popular pharmacologic agent for diagnosing Horner syndrome. Apraclonidine is a relatively selective α2-adrenergic receptor agonist. α2 receptors are predominantly found on presynaptic sympathetic nerve terminals and act to inhibit norepinephrine release. However, apraclonidine also has a weak agonist action on postsynaptic α1 receptors, akin to a dilute adrenergic substance. In a normal eye, this weak α1 action is negligible and the more dominant presynaptic α2 action may slightly decrease (about 0.5 mm) the baseline pupil size. In a sympathetically denervated eye which has developed postsynaptic receptor supersensitivity, the iris dilator pupillae will “over-react” to the weak α1 agonist property of apraclonidine and overcome the α2 action, resulting in pupillary dilation (Fig. 9.17). The apraclonidine test is performed by putting one drop of in each eye and then observing the anisocoria 45–60 minutes later. A Horner pupil will dilate but the normal pupil will not, such that the anisocoria will appear resolved or even reversed, as compared to baseline. Tarsal muscle supersensitivity with resolution of ptosis also occurs with apraclonidine, but this finding has not been validated as a diagnostic criterion for Horner syndrome.

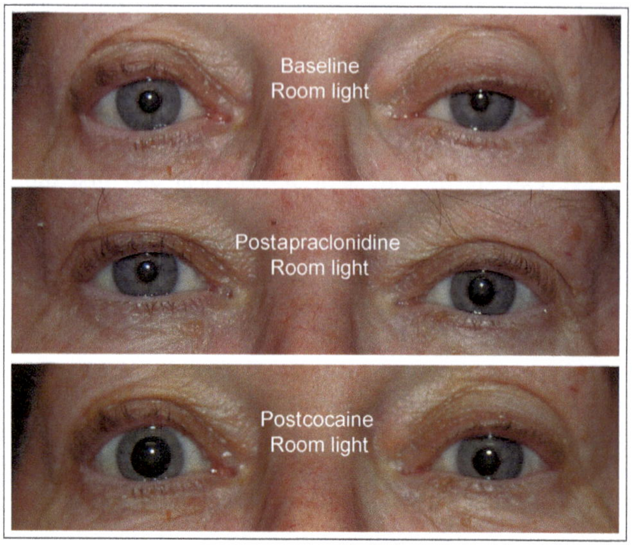

Fig. 9.17. Pharmacologic testing for Horner syndrome using apraclonidine and cocaine drops. Top: The patient presents an anisocoria and left upper lid ptosis. The pupils are briskly reactive to light and the anisocoria becomes more apparent in darkness (not shown). A Horner syndrome is suspected. Middle: 45 min after one drop of 1% apraclonidine had been placed in each eye, the smaller left pupil has dilated and is now the larger pupil, resulting in a reversal of the baseline anisocoria. Additionally, the left upper lid ptosis has resolved. The right pupil has no dilation to apraclonidine which is expected. These results confirm sympathetic denervation supersensitivity on the left side, hence a Horner syndrome. Bottom: The patient was tested with topical 4% cocaine drops (48 h after the apraclonidine test). Forty-five minutes after three drops of cocaine into each eye, the baseline anisocoria increased. This is because the normal right pupil has dilated by nearly 3 mm to cocaine, as expected, and the Horner pupil has dilated little to cocaine. A small dilation can still happen in cases of partial denervation. However, as the postcocaine anisocoria is greater than 1 mm, this result confirms a Horner syndrome on the left side. Note that the upper lid ptosis of Horner syndrome has not completely retracted, while the normal upper lid (right ride) shows the expected lid retraction to cocaine drops.

A recent study has shown greater sensitivity of apraclonidine over cocaine (93% vs 40%) for detecting Horner syndrome. In addition, apraclonidine can be used in patients who have iris damage in the normal eye and in patients suspected to have bilateral Horner syndrome, because any observable increase in pupil size (as measured in dim light) from apraclonidine is considered a positive test result for sympathetic denervation (Bremner, 2019). Because the effect of apraclonidine, like dilute pilocarpine, is not influenced by asymmetric corneal absorption, it can be used on the same day as contact lens wearers. In infants under 12 months of age, early reports of bradycardia and respiratory depression following apraclonidine limits its use in babies but more recent studies suggest significant serious adverse effects are rare (Eldib et al., 2021). We generally avoid apraclonidine to confirm Horner in infants under 12 months and have used cocaine (4%) ophthalmic solution in infants as young as 2 months without significant side effects.

The main limitation of apraclonidine is that it is not a valid test in the setting of an acute sympathetic denervation. This is true to for all pharmacologic tests predicated on denervation supersensitivity as a mechanism of action, because supersensitivity typically requires 5–8 days to develop. Cocaine is, thus, the recommended choice for pharmacologic confirmation of an acute Horner syndrome.

Dilute phenylephrine (1% or less) has been used in the past to detect sympathetic denervation. However, a recent study found that dilator pupillae supersensitivity to 1% phenylephrine was observed in 81% of postganglionic Horner syndrome but not at all in preganglionic or central ones, so dilute phenylephrine is not recommended for diagnostic use (Danesh-Meyer et al., 2004).

Causes of Horner syndrome

The circuitous pathway of descending and ascending sympathetic fibers from brain to thorax to eye exposes these fibers to a variety of pathologies in a variety of locations. Thus, a Horner syndrome has many possible causes (Table 9.1). In adults, commonly recognized causes of Horner syndrome are stroke, trauma, carotid artery dissection, tumor, and cluster headache. Recent literature has highlighted neck pathologies which can injure the preganglionic sympathetic fibers and/or the superior cervical ganglion. Thyroid pathology, benign or malignant, can cause Horner syndrome. Eagle syndrome, in which an abnormally elongated styloid process can injure cranial nerves V, VII, IX, X and XII, has been reported to cause Horner syndrome. Iatrogenic Horner syndrome can complicate any interventional procedure of the neck including neck catheters, carotid endarterectomy, surgery to thyroid or parathyroid glands, cervical spine fusion surgery, and even cervical facet joint corticosteroid injection.

Well-known clinical syndromes are recognized by specific localizing findings accompanying the Horner syndrome. These include Wallenberg syndrome (crossed sensory deficit, ipsilateral ataxia, impaired taste, nystagmus localizing to the lateral medulla), Pancoast syndrome (arm pain, weakness and atrophy localizing to the supraclavicular fossa), Villaret syndrome (lower cranial nerve palsies localizing to the retroparotid space in the neck), Rowland Payne syndrome (paralysis of vocal cord and hemidiaphragm localizing to carotid sheath area at the level of sixth cervical vertebra), and Parkinson sign (Horner syndrome with ipsilateral abducens nerve palsy localizing to the cavernous sinus). Despite localizing clinical signs and technical advances in imaging, a structural lesion is still not defined in more than 40%–70% of adult patients with pharmacologically confirmed Horner syndrome (Sadaka et al., 2017). In patients with an isolated Horner syndrome, the yield of imaging is ever lower at 20% (Beebe et al., 2017; Sabbagh et al., 2020). However, the long-term prognosis of these patients with

Table 9.1

Causes of Horner syndrome in adults.

Central neuron	Preganglionic neuron	Postganglionic neuron
Hypothalamic mass	Cervical spine trauma	Cluster headache
Thalamo-mesencephalic lesion (stroke, mass, demyelination)	Cervical rib	Superior cervical ganglionectomy
Cervicothoracic trauma	Low cervical or upper thoracic disc herniation	Elongated styloid process
Syringomyelia	Spinal root cyst	Calcified stylohyoid ligament
Intramedullary tumor	Clavicular fracture	Retroparotid pathology (abscess, tumor, vascular malformation)
Spinal cord AVM	Brachial plexus trauma	Tonsillectomy
Myelitis	Apical lung mass	Carotid artery dissection or aneurysm
	Pneumothorax	Carotid arteritis
	Mediastinal/intrathoracic tumor	Skull base fracture
	Subclavian artery aneurysm	Skull base lesions (tumor, infection, inflammation, osteomyelitis)
	Jugular vein thrombosis	Cavernous sinus pathology (tumor, aneurysm, inflammation)
	Thyroid mass (tumor, goiter)	Carotid-cavernous sinus fistula
	Neck procedure or surgery (jugular cannulation, chest tube placement, thoracic surgery, thyroid surgery, cervical spine steroid injection)	
	Epidural anesthesia	
	Aspergillosis infection of neck	
	Retroparotid abscess	
	Deep neck lymphadenopathy	
	Sarcoidosis	

"idiopathic" isolated Horner syndrome is benign and later pathology does not appear, even up to 20 years after diagnosis (Bellego et al., 2022).

In a series of 73 pediatric patients, 42% had congenital Horner syndrome in which clinical signs appeared in the first 6–12 months of life (Jeffery et al., 1998). In children with noncongenital Horner syndrome without history of preceding trauma or surgical intervention, various tumors and pathologies must be considered, including neuroblastoma, rhabdomyosarcoma, schwannoma, paraganglioma, and brainstem vascular malformation. Neuroblastoma remains the most common tumor associated with acquired pediatric Horner syndrome with an incidence risk estimated at 1 per 10 cases (Barrea et al., 2016). While birth trauma is a common cause of congenital Horner syndrome, occasional cases are due to congenital structural abnormalities such as internal carotid agenesis, ectopic thymus in the neck, lymphatic malformation, and even congenital neuroblastoma. For this reason, some clinicians strongly advocate that all children with Horner syndrome, even congenital with history of birth trauma, undergo magnetic resonance (MR) imaging of the brain, neck, and chest with angiographic sequences of the neck and chest. Others have questioned the need for imaging when an isolated Horner syndrome is congenital or following trauma (accidental or iatrogenic). In such cases, it is still important to carefully assess accompanying clinical signs before deciding on the type and extent of investigation.

Hyperfunction of oculosympathetic nerves

Unilateral and transient sympathetic hyperfunction with mydriasis, eyelid retraction, and facial sweating can be an idiopathic phenomenon in neurologically intact persons (Pourfour du Petit syndrome). Bilateral excessive sympathetic tone resulting in large pupils with a normal or mildly diminished pupil light reflex is sometimes observed in healthy anxious young persons (Bumke pupils). Episodic focal pupillary distortion may result from focal iris dilator spasm due to spontaneous sympathetic discharge (Udry et al., 2019). Known as tadpole pupil, it has high association with Horner syndrome.

PUPILLARY DYSFUNCTION IN AUTONOMIC NEUROPATHIES

Autonomic neuropathies are characterized by damage to small unmyelinated or thinly myelinated autonomic nerves which leads to diverse clinical manifestations, particularly in cardiovascular, gastrointestinal, urogenital, and sudomotor systems. Pupillary dysfunction would be an expected manifestation of generalized autonomic neuropathy but little is actually known. The asymptomatic and bilateral nature of pupillary dysfunction in autonomic neuropathies contributes to under-reporting and under-detection of pupillary dysfunction in these domains. While pupillometry may demonstrate abnormalities in the dynamics of pupillary responses, there are no standardized stimulus protocols and outcome parameters to establish diagnostic criteria. A simpler bedside evaluation of sympathetic and parasympathetic integrity to the iris muscles is pharmacologic testing using dilute receptor agonists to detect end-organ denervation supersensitivity (eye drop tests). For sympathetic function, pupillary dilation in response to one drop of either apraclonidine (0.5% or 1%) confirms denervation. On the parasympathetic side, pupillary constriction to dilute pilocarpine (0.125% or less) confirms denervation. Pharmacologic testing for unilateral sympathetic or parasympathetic pupil disorders recommends instillation of the test agent in both eyes, as one eye serves as control. In autonomic neuropathy with bilateral pupillary dysfunction, the test agent is instilled in only one eye and normal saline is instilled in the opposite eye.

In some conditions, loss of sympathetic discharge is the dominant autonomic dysfunction, e.g., multiple system atrophy and hereditary sensory autonomic neuropathy type III. Sjogren syndrome seems to predominantly affect parasympathetic pathways. Most autonomic neuropathies (diabetes, amyloidosis, acute dysautonomia, paraneoplastic neuropathy) show abnormalities of sympathetic and parasympathetic systems but not always to equal degree.

Autonomic neuropathy is a well-known complication of diabetes. The most common effect of diabetes on the pupil is a decrease in its resting size (in room light and in darkness), and this finding is related to disease duration. Sympathetic dysfunction, present in half of diabetic patients, may be a major contributor to the smaller resting pupil size. As disease progresses, parasympathetic pupillary dysfunction is evident in changes of the light reflex, including a prolonged latency and reduced amplitude of pupillary constriction. Pupillary dysfunction in diabetes is correlated to diabetic cardiovascular autonomic neuropathy which signals an increased myocardial mortality risk in diabetic patients. Thus, pharmacologic testing of pupils may be one way to screen for developing autonomic neuropathy in diabetic patients.

Another autonomic neuropathy that involves the pupils in up to one-quarter of patients is the Lambert–Eaton myasthenic syndrome (LEMS). LEMS may be idiopathic or associated with tumor, typically a small cell lung cancer. Patients with LEMS produce autoantibodies against P/Q-type voltage-gated calcium channels, which impair release of acetylcholine from the presynaptic membrane of motor nerve terminals. Proximal muscle weakness, diplopia, and autonomic symptoms including sluggish pupil light reflexes and tonic pupils are common clinical manifestations.

CONCLUSION

In conclusion, the pupil can point to a variety of locations of disease and damage. For example, the pupil becomes deformed and dysfunctional in the face of ocular pathology involving the iris. The pupil light reflex is a marker of retinal and optic nerve integrity, revealing an RAPD in the eye with disease or in the worse eye in the event of bilateral damage. In addition, careful clinical and pharmacologic examination of the pupil relays information about parasympathetic and sympathetic function to eye, giving a global assessment of the status of the autonomic nervous system.

ACKNOWLEDGMENTS

The authors wish to thank Yann Leuba and Claudio Fernandes for editing and preparing the figures and videos used to illustrate this chapter.

Supplementary material related to this chapter can be found on online at https://doi.org/10.1016/B978-0-443-15736-3.00001-9.

References

Albayram S, Ozer H, Sarici A et al. (2006). Unilateral mydriasis without ophthalmoplegia – a sign of neurovascular compression? Case report. Neurosurgery 58: E582–E583.

Ashker L, Weinstein JM, Dias M et al. (2008). Arachnoid cyst causing third cranial nerve palsy manifesting as isolated internal ophthalmoplegia and iris cholinergic supersensitivity. J Neuroophthalmol 28: 192–197.

Barnerssoi M, May PJ, Horn AK (2017). GABAergic innervation of the ciliary ganglion in macaque monkeys – a light and electron microscopic study. J Comp Neurol 525: 1517–1531.

Barrea C, Vigouroux T, Karam J et al. (2016). Horner syndrome in children: a clinical condition with serious underlying disease. Neuropediatrics 47: 268–272.

Beebe JD, Kardon RH, Thurtell MJ (2017). The yield of diagnostic imaging in patients with isolated Horner syndrome. Neurol Clin 35: 145–151.

Bellego C, Borruat FX, Kawasaki A (2022). Long-term prognosis of patients with idiopathic Horner syndrome. J Neurol 269: 2781–2783.

Berson DM, Castrucci AM, Provencio I (2010). Morphology and mosaics of melanopsin-expressing retinal ganglion cell types in mice. J Comp Neurol 518: 2405–2422.

Bremner F (2019). Apraclonidine is better than cocaine for detection of Horner syndrome. Front Neurol 10: 55.

Bremner F, Smith S (2008). Pupillographic findings in 39 consecutive cases of harlequin syndrome. J Neuroophthalmol 28: 171–177.

Chengazi HU, Bhatt AA (2019). Pathology of the carotid space. Insights Imaging 10: 21.

Chronopoulos A, Consigli A, Heim J et al. (2023). Isolated internal ophthalmoplegia from posterior cerebral artery neurovascular conflict. Case Rep Ophthalmol 14: 104–110.

Cohen DN, Zakov ZN (1975). The diagnosis of Adie's pupil using 0.0625% pilocarpine solution. Am J Ophthalmol 79: 883–885.

Cox TA (1992). Pupillary escape. Neurology 42: 1271–1273.

Cox TA, Drewes CP (1984). Contraction anisocoria resulting from half-field illumination. Am J Ophthalmol 97: 577–582.

Curcio CA, Sloan KR, Packer JR et al. (1987). Distribution of cones in human and monkey retina: individual variability and radial asymmetry. Science 236: 579–582.

Dacey DM, Liao HW, Peterson BB et al. (2005). Melanopsin-expressing ganglion cells in primate retina signal colour and irradiance and project to the LGN. Nature 433: 749–754.

Danesh-Meyer HV, Savino P, Sergott R (2004). The correlation of phenylephrine 1% with hydroxyamphetamine 1% in Horner's syndrome. Br J Ophthalmol 88: 592–593.

Davis-Silberman N, Ashery-Padan R (2008). Iris development in vertebrates; genetic and molecular considerations. Brain Res 1192: 17–28.

De Maria M, Iannetta D, Moramarco A et al. (2019). Iridocorneal endothelial syndrome in a patient with keratoconus – a case report. BMC Ophthalmol 19: 221.

Do MTH (2019). Melanopsin and the intrinsically photosensitive retinal ganglion cells: biophysics to behavior. Neuron 104: 205–226.

Donadon M, Santoro MM (2021). The origin and mechanisms of smooth muscle cell development in vertebrates. Development 148.

Eldib AA, Patil P, Nischal KK et al. (2021). Safety of apraclonidine eye drops in diagnosis of Horner syndrome in an outpatient pediatric ophthalmology clinic. J AAPOS 25: 336.e1–336.e4.

Fischbach BV, Trout KL, Lewis J et al. (2005). WAGR syndrome: a clinical review of 54 cases. Pediatrics 116: 984–988.

Forrester JV, Dick AD, McMenamin PG et al. (2021). Chapter 1: Anatomy of the eye and orbit. The eye. 6th edition. Elsevier, pp. 1–105.

Gabella G (1974). The sphincter pupillae of the guinea-pig: structure of muscle cells, intercellular relations and density of innervation. Proc R Soc Lond B Biol Sci 186: 369–386.

Haladaj R (2020). Anatomical variations of the ciliary ganglion with an emphasis on the location in the orbit. Anat Sci Int 95: 258–264.

Hattar S, Lucas RJ, Mrosovsky N et al. (2003). Melanopsin and rod-cone photoreceptive systems account for all major accessory visual functions in mice. Nature 424: 76–81.

Hingorani M, Hanson I, Van Heyningen V (2012). Aniridia. Eur J Hum Genet 20: 1011–1017.

Horn AK, Schulze C, Radtke-Schuller S (2009). The Edinger-Westphal nucleus represents different functional cell groups in different species. Ann N Y Acad Sci 1164: 45–50.

Idrees F, Vaideanu D, Fraser SG et al. (2006). A review of anterior segment dysgeneses. Surv Ophthalmol 51: 213–231.

Jeffery AR, Ellis FJ, Repka MX et al. (1998). Pediatric Horner syndrome. J AAPOS 2: 159–167.

Jones R, Vrbova G (1974). Two factors responsible for the development of denervation hypersensitivity. J Physiol 236: 517–538.

Joyce DS, Feigl B, Kerr G et al. (2018). Melanopsin-mediated pupil function is impaired in Parkinson's disease. Sci Rep 8: 7796.

Kaeser PF, Kawasaki A (2010). Disorders of pupillary structure and function. Neurol Clin 28: 657–677.

Kalish JM, Doros L, Helman LJ et al. (2017). Surveillance recommendations for children with overgrowth syndromes and predisposition to Wilms tumors and hepatoblastoma. Clin Cancer Res 23: e115–e122.

Kardon R, Bergamin O (2005). Chapter 37 Adie's Pupile, In: Levin LA, Anthony AC (Eds). Neuro-ophthalmology. The Practical Guid. New York: Thieme publishers, pp. 1–11.

Kardon RH, Corbett JJ, Thompson HS (1998). Segmental denervation and reinnervation of the iris sphincter as shown by infrared videographic transillumination. Ophthalmology 105: 313–321.

Kardon RH, Denison CE, Brown CK et al. (1990). Critical evaluation of the cocaine test in the diagnosis of Horner's syndrome. Arch Ophthalmol 108: 384–387.

Kardon R, Kawasaki A, Miller NR (2006). Origin of the relative afferent pupillary defect in optic tract lesions. Ophthalmology 113: 1345–1353.

Kawasaki A (2005). Chapter 16: Disordes of pupillary function, accomodation and lacrimation, In: Miller NR, Newmann NJ (Eds). Walsh and Hoyt's Clinical Neuro-Ophthalmology. 6th edition. Philadelphia: Lippincott, Williams and Wilkins publishers, pp. 739–805.

Kawasaki AK (2014). Diagnostic approach to pupillary abnormalities. Continuum (Minneap Minn) 20: 1008–1022.

Kawashima Y, Sumi T, Sugimoto T et al. (2008). First-bite syndrome: a review of 29 patients with parapharyngeal space tumor. Auris Nasus Larynx 35: 109–113.

Kivela T, Fuchs U, Tarkkanen A (1992). Cytoskeleton in neuroectodermally derived epithelial and muscle cells of the human iris and ciliary body. J Histochem Cytochem 40: 1517–1526.

Klooster J, Vrensen GF (1998). New indirect pathways subserving the pupillary light reflex: projections of the accessory oculomotor nuclei and the periaqueductal gray to the Edinger-Westphal nucleus and the thoracic spinal cord in rats. Anat Embryol (Berl) 198: 123–132.

Kong YX, Wright G, Pesudovs K et al. (2007). Horner syndrome. Clin Exp Optom 90: 336–344.

Kozicz T, Bittencourt JC, May PJ et al. (2011). The Edinger-Westphal nucleus: a historical, structural, and functional perspective on a dichotomous terminology. J Comp Neurol 519: 1413–1434.

Kuder T, Kuchinka J, Nowak E et al. (2003). A comparative study on the morphology and topography of the ciliary ganglion in midday gerbil (*Meriones meridianus*) and turtle (*Agrionemys horsfieldii*). Ann Anat 185: 351–357.

Lai YL (1972a). The development of the dilator muscle in the iris of the albino rat. Exp Eye Res 14: 203–207.

Lai YL (1972b). The development of the sphincter muscle in the iris of the albino rat. Exp Eye Res 14: 196–202.

Lam BL, Thompson HS, Corbett JJ (1987). The prevalence of simple anisocoria. Am J Ophthalmol 104: 69–73.

Lam BL, Thompson HS, Walls RC (1996). Effect of light on the prevalence of simple anisocoria. Ophthalmology 103: 790–793.

Lance JW, Drummond PD, Gandevia SC et al. (1988). Harlequin syndrome: the sudden onset of unilateral flushing and sweating. J Neurol Neurosurg Psychiatry 51: 635–642.

Laties AM (1974). Ocular melanin and the adrenergic innervation to the eye. Trans Am Ophthalmol Soc 72: 560–605.

Leavitt JA, Wayman LL, Hodge DO et al. (2002). Pupillary response to four concentrations of pilocarpine in normal subjects: application to testing for Adie tonic pupil. Am J Ophthalmol 133: 333–336.

Lingam G, Sen AC, Lingam V et al. (2021). Ocular coloboma – a comprehensive review for the clinician. Eye (Lond) 35: 2086–2109.

Liu EK, Suson KD (2020). Syndromic Wilms tumor: a review of predisposing conditions, surveillance and treatment. Transl Androl Urol 9: 2370–2381.

Loewenfeld IE (1977). "Simple central" anisocoria: a common condition, seldom recognized. Trans Sect Ophthalmol Am Acad Ophthalmol Otolaryngol 83: 832–839.

Lowenstein O (1955). Benign postinfectious disorder of anterior midbrain; alternating contraction anisocoria, combined with general fatigue and peripheral neuritis. AMA Arch Neurol Psychiatry 73: 302–308.

Magalhaes OA, Kronbauer CL, Muller EG et al. (2016). Update and review of Urrets-Zavalia syndrome. Arq Bras Oftalmol 79: 202–204.

Maloney WF, Younge BR, Moyer NJ (1980). Evaluation of the causes and accuracy of pharmacologic localization in Horner's syndrome. Am J Ophthalmol 90: 394–402.

Martin TJ (2007). Horner's syndrome, pseudo-Horner's syndrome, and simple anisocoria. Curr Neurol Neurosci Rep 7: 397–406.

May PJ, Bohlen MO, Perkins E et al. (2021). Superior colliculus projections to target populations in the supraoculomotor area of the macaque monkey. Vis Neurosci 38.

May PJ, Gamlin PD, Warren S (2022). A novel tectal/pretectal population of premotor lens accommodation neurons. Invest Ophthalmol Vis Sci 63: 35.

May PJ, Sun W, Wright NF et al. (2020). Pupillary light reflex circuits in the macaque monkey: the preganglionic Edinger-Westphal nucleus. Brain Struct Funct 225: 403–425.

Mcdougal DH, Gamlin PD (2010). The influence of intrinsically-photosensitive retinal ganglion cells on the spectral sensitivity and response dynamics of the human pupillary light reflex. Vis Res 50: 72–87.

McDougal DH, Gamlin PD (2015). Autonomic control of the eye. Compr Physiol 5: 439–473.

Mitchelson F (2012). Muscarinic receptor agonists and antagonists: effects on ocular function. Handb Exp Pharmacol 208: 263–298.

Moura AL, Nagy BV, Morgia LA et al. (2013). The pupil light reflex in Leber's hereditary optic neuropathy: evidence for preservation of melanopsin-expressing retinal ganglion cells. Invest Ophthalmol Vis Sci 54: 4471–4477.

Nielsen PJ (1983). Upside down ptosis in patients with Horner's syndrome. Acta Ophthalmol (Copenh) 61: 952–957.

Ryabinin AE, Tsivkovskaia NO, Ryabinin SA (2005). Urocortin 1-containing neurons in the human Edinger-Westphal nucleus. Neuroscience 134: 1317–1323.

Sabbagh MA, De Lott LB, Trobe JD (2020). Causes of Horner syndrome: a study of 318 patients. J Neuroophthalmol 40: 362–369.

Sadaka A, Schockman SL, Golnik KC (2017). Evaluation of Horner syndrome in the MRI era. J Neuroophthalmol 37: 268–272.

Schwinn DA, Afshari NA (2006). alpha(1)-Adrenergic receptor antagonists and the iris: new mechanistic insights into floppy iris syndrome. Surv Ophthalmol 51: 501–512.

Silva C, McNaughton N (2019). Are periaqueductal gray and dorsal raphe the foundation of appetitive and aversive control? A comprehensive review. Prog Neurobiol 177: 33–72.

Steward O (2014). Degenerative changes and reactive growth responses of neurons following denervation and axotomy: historical concepts and their modern embodiments, In: Michael Selzer SC, Leonardo C, Gert K (Eds). et al. Textbook of neural repair and rehabilitation. Cambridge: Cambridge University Press.

Sturm RA, Frudakis TN (2004). Eye colour: portals into pigmentation genes and ancestry. Trends Genet 20: 327–332.

Szabadi E (2018). Functional organization of the sympathetic pathways controlling the pupil: light-inhibited and light-stimulated pathways. Front Neurol 9: 1069.

Tamura T, Smelser GK (1973). Development of the sphincter and dilator muscles of the iris. Arch Ophthalmol 89: 332–339.

Thakker MM, Huang J, Possin DE et al. (2008). Human orbital sympathetic nerve pathways. Ophthalmic Plast Reconstr Surg 24: 360–366.

Thompson HS, Kardon RH (2006). The Argyll Robertson pupil. J Neuroophthalmol 26: 134–138.

Tomelleri G, Vattemi G, Filosto M et al. (2007). Eyelid ptosis from sympathetic nerve dysfunction mistaken as myopathy: a simple test to identify this condition. J Neurol Neurosurg Psychiatry 78: 632–634.

Tshering Vogel DW, Thoeny HC (2016). Cross-sectional imaging in cancers of the head and neck: how we review and report. Cancer Imaging, 16: 20.

Turco GR, Farber NE (1996). Postoperative autonomic deficit: a case of harlequin syndrome. Anesthesiology 85: 1197–1199.

Udry M, Kardon RH, Sadun F et al. (2019). The Tadpole pupil: case series with review of the literature and new considerations. Front Neurol 10: 846.

Wang CA, Boehnke SE, White BJ et al. (2012a). Microstimulation of the monkey superior colliculus induces pupil dilation without evoking saccades. J Neurosci 32: 3629–3636.

Wang FM, Wertenbaker C, Cho H et al. (2012b). Unilateral straight hair and congenital Horner syndrome. J Neuroophthalmol 32: 132–134.

Warwick R (1954). The ocular parasympathetic nerve supply and its mesencephalic sources. J Anat 88: 71–93.

Weinstein JM, Zweifel TJ, Thompson HS (1980). Congenital Horner's syndrome. Arch Ophthalmol 98: 1074–1078.

Werner M, Bhatti MT, Vaishnav H et al. (2005). Isolated anisocoria from an endodermal cyst of the third cranial nerve mimicking an Adie's tonic pupil. J Pediatr Ophthalmol Strabismus 42: 176–179.

Wu F, Zhao Y, Zhang H (2022). Ocular autonomic nervous system: an update from anatomy to physiological functions. Vision (Basel) 6: 6.

Xu SY, Song MM, Li L et al. (2022). Adie's pupil: a diagnostic challenge for the physician. Med Sci Monit 28: e934657.

Yamaguchi K (2017). Development of the human oculomotor nuclear complex: centrally-projecting Edinger-Westphal nucleus. Neurosci Lett 646: 8–14.

Yilmaz I, Perente I, Saracoglu B et al. (2016). Changes in pupil size following panretinal retinal photocoagulation: conventional laser vs pattern scan laser (PASCAL). Eye (Lond) 30: 1359–1364.

Yoo YJ, Hwang JM, Yang HK (2021). Dilute pilocarpine test for diagnosis of Adie's tonic pupil. Sci Rep 11: 10089.

Yoshitomi T, Ito Y, Inomata H (1985). Adrenergic excitatory and cholinergic inhibitory innervations in the human iris dilator. Exp Eye Res 40: 453–459.

Younge BR, Buski ZJ (1976). Tonic pupil: a simple screening test. Can J Ophthalmol 11: 295–299.

Handbook of Clinical Neurology, Vol. 216 (3rd series)
The Brainstem
L.T. Grinberg and A.D. Korczyn
DOI: https://doi.org/10.1016/B978-0-443-15736-3.00008-1

Chapter 10

Neurodevelopmental brainstem syndromes

DAVID GOMEZ-ANDRES, MANUEL ROIG-QUILS, AND FRANCINA MUNELL*

Pediatric Neurology Department, University Hospital Vall Hebron, Barcelona, Spain

Abstract

Prenatal development of the brainstem is a meticulously orchestrated process, reliant on the coordinated activity of numerous genes and environmental cues. Disruption of this process, whether by genetic mutations or acquired insults during fetal development, can lead to brainstem dysfunction, characterized by a spectrum of clinical manifestations including cranial nerve palsies, cognitive and behavioral deficits, sensory and motor impairments, dysautonomia, and respiratory insufficiency.

This chapter is dedicated to disorders primarily involving the brainstem and/or cranial nerves. It provides insights into the prenatal development of the brainstem, elucidates the clinical presentations associated with developmental disruptions, and offers an updated perspective on the genetic and acquired brainstem disorders.

The umbrella term "brainstem dysgenesis" is employed herein to encompass all disorders stemming from aberrant brainstem development during embryogenesis. Given the phenotypic heterogeneity observed in brainstem dysgenesis, the identification of characteristic phenotypic patterns linked to acquired insults or inherited conditions is paramount. Such recognition facilitates accurate diagnosis, genetic counseling, prognostication, and the formulation of tailored care and therapeutic strategies.

INTRODUCTION

Disrupted brainstem development often results in intricate and occasionally unique clinical manifestations (Munell et al., 2018). Recent advancements in genetic testing, notably whole-exome sequencing, have significantly enriched our comprehension of brainstem developmental disorders. This surge in genetic insights has highlighted the limitations of conventional diagnostic methods, which predominantly relied on clinical observations and phenotypic classifications. Historically, many of these disorders were clustered and delineated based on observable symptoms, often bearing eponymous names after the researchers or clinicians who initially characterized them. However, dependence on such eponyms can obscure the underlying genetic intricacies and foster confusion, particularly in cases where different disorders exhibit overlapping phenotypes.

The integration of genetic data should not diminish the significance of meticulous clinical evaluation and the rigorous application of diagnostic criteria. On the contrary, genetic investigations have emphasized the indispensability of deep phenotyping, characterized by detailed and comprehensive clinical assessments conducted by expert clinicians. This approach systematically records all facets of a patient condition, enhancing comprehension of the disorder, guiding diagnostic procedures, and facilitating accurate genetic counseling. Moreover, it lays the groundwork for precise prognostication and the formulation of tailored care and targeted therapeutic interventions. Insufficient phenotyping may lead to misdiagnosis and hinder patient access to appropriate care.

"Brainstem malformation" is a radiologic term for abnormal shape, size, or position of the brainstem which results from an abnormal developmental process as identified in imaging studies. Although useful, it is still limited due to the lack of inclusion of clinical phenotype in its definition. "Brainstem dysgenesis" (BSD) is a generic, umbrella term that was first proposed in 2003 to describe all patients with congenital dysfunction of cranial nerves and decreased muscle tone due to prenatal injuries or to disordered development of the brainstem during embryonic growth (Roig et al., 2003). BSD has an advantage over previous eponyms in that it provides a broader approach for this uncommon group of patients. The term "dysgenesis" is used in multiple fields of medicine to indicate the anomalous, partial, or total genesis of one organ and result in missing or underdeveloped components. "Congenital cranial dysinnervation disorders" has been also proposed for congenital, nonprogressive abnormalities or complete absence of one or more cranial nerves with primary or secondary muscle dysinnervation (Ferreira et al., 2011;

*Correspondence to: Francina Munell, MD, PhD, Pediatric Neurology Department, University Hospital Vall Hebron, Passeig Vall d'Hebron 119-129, 08035 Barcelona, Spain. Tel: +34 934 893 156; E-mail: francina.munell@vallhebron.cat.

Whitman, 2021). This term defines only one part of clinical manifestations that could arise from abnormal brainstem development, and we believe that this could be misleading. "Moebius syndrome" is another conflicting term. The first consideration is that as a syndrome, a significant number of particular diseases are encompassed. The second is that it has a clear and limited definition with strict criteria: congenital facial palsy + unilateral or bilateral adduction eye limitation (Carta et al., 2011; MacKinnon et al., 2014). It does not include isolated congenital facial palsy, isolated eye motility problems, or patients with congenital facial palsy and strabismus or vertical eye motility limitations, but could include other manifestations from cranial nerves, brainstem, or extra-neurologic such as limb abnormalities or cleft palate. "Atypic Moebius syndrome," "Moebius-like disorders," or "Moebius-plus syndrome" are terms that should not be used, as they have proved to be confusing in clinical practice.

Cases of BSD can be inherited or sporadic. Sporadic cases can have genetic (mainly, de novo autosomal dominant mutations) or nongenetic causes, such as early ischemic phenomena during embryogenesis or exposure to teratogens to the brainstem particularly prone to (Sarnat et al., 2002). Clinical history and neurologic examination are essential to define the phenotype and orientate the workup toward a genetic study.

EMBRYOLOGY

The brainstem comprises the midbrain (the mesencephalon) and the hindbrain (the rhombencephalon). During development, it is segmentally organized. In the early phases of central nervous system development, the mesencephalon divides into two temporarily present segments called mesomeres and the rhombencephalon into 12 rhombomeres. The cerebellum arises from the rhombomere 0 (isthmic) and 1. These and the rest of the segments with contributions from the flanking neural crest cells will form the reticular formation, the cranial nerves III–XII, and the complex circuitry within the brainstem. They will also guide the formation of the ascending and descending pathways that go through the brainstem. For a detailed description of these processes, we recommend a recent review (Ten Donkelaar et al., 2023).

The first step in brainstem formation is pattern formation and segmentation. There are two patterning phenomena in this process The anterior-posterior pattern controls the formation of the brainstem from rostral to caudal, such as the division of the rhombencephalon into a rostral metencephalon (pons/cerebellum) and a caudal myelencephalon (medulla oblongata). The dorsoventral patterning divides the mesencephalon into the basis (ventral), tum (central), and tectum (dorsal), and further divides the rostral metencephalon into the pons (ventrally) and cerebellum (dorsally).

Anteroposterior patterning is not completely understood but is driven by a complex genetic regulation derived from gradients of secreted growth factors and expressed transcriptional factors. Two critical structures are the diencephalic–mesencephalic junction and the isthmic organizer in the midbrain–hindbrain boundary. This drives the acquisition of individual identities of mesomeres and rhombomeres by the definition of overlapping or nested domains of expression of morphogenetic genes such as Hox genes in rhombomeres.

These signals will contribute to the formation of different progenitor zones that are also organized metamerically in the dorsoventral axis. Specific progenitor zones with specific transcriptional expression (specifically, from genes members of the Nkx and Phox families) will participate in the determination of the progenitors for specialized motor neurons, interneurons, and projecting neurons of the brainstem. Some neurons will migrate to be condensed in the brainstem motor nuclei. For instance, facial motoneurons migrate caudalward giving rise to the internal genu of the facial nerve and the particular relationship with the abducens nuclei. Cranial nerve ganglia are derived from the cranial neural crest cells and the cranial neural placode with proximal and distal subsets regulated by differential gene expression patterns. The established patterning of the brainstem will influence the formation of circuits and pathways that need to be established.

The brainstem during embryogenesis is particularly sensitive to destructive processes. The destructive process that occurs in a rapidly developing tissue may cause both a derangement of subsequent development and injury to an already-developed structure; the less vigorous reactive cellular response to tissue destruction during early development makes it difficult to distinguish genetically determined lesions from destructive lesions (Volpe, 2009).

Brainstem malformations are related to embryologic mechanisms. Classifying brainstem malformations according to the pathogenic mechanism is interesting to provide a conceptual framework and the etiology behind the different syndromes (Table 10.1).

CLINICAL PRESENTATION OF BRAINSTEM DYSGENESIS

Pediatricians, neonatologists, and child neurologists are often faced with neonates or infants with facial diplegia, feeding difficulties, dysphagia, muscle hypotonia, respiratory distress, and joint contractures, sometimes associated with other craniofacial malformations. The causes for this constellation of clinical signs and symptoms are manifold. Congenital tumors, vascular malformations, severe perinatal trauma or hypoxic-ischemic events, and restricted prenatal disruptive lesions or malformations of the cerebral hemispheres or the brainstem, as well as congenital neuromuscular disorders (NMDs), are known to produce the aforementioned clinical picture during the neonatal period or infancy. Clinical suspicion of a brainstem disorder relies heavily on the appropriate interpretation of clinical findings during the neonatal period, and it is difficult to achieve unless it is suspected as early as possible. Once hypotonia subsides and the other presenting symptoms improve, facial diplegia, feeding difficulties, poor handling of oral secretions, dysarthria, and signs of pyramidal tract involvement, including a brisk jaw jerk, are the remaining symptoms. Therefore, it is conceivable that patients in whom the diagnosis of BSD was missed in early infancy will be diagnosed with

Table 10.1

Mechanisms and types of brainstem malformations (Doherty et al., 2013; Munell et al., 2018).

Mechanisms	Type	Genetic	Acquired
Malformation secondary to anteroposterior and dorsoventral patterning	Abnormal anteroposterior patterning: diencephalic–mesencephalic junction dysplasia	PCDH12 L1CAM VRK1 GSX2	Vascular disruption Fetal alcohol syndrome Others
	Abnormal formation of rhombomeres 0 and 1: brainstem disconnection syndrome	Without clear evidence	
	Abnormal formation of caudal rhombomeres	HOXA1 syndromes • Athabaskan brainstem dysgenesis syndrome • Bosley–Salih–Alorainy syndrome HOXB1 syndrome	
	Abnormal dorsoventral patterning: rhombencephalosynapsis	Several genes	Some
Global malformations of the brainstem associated with later development phases	Abnormal signaling processes: Dandy–Walker malformation	Several genes	Some
	Abnormal neuronal and glial proliferation: autism spectrum disorders and other neurodevelopmental disorders	Multiple genes (deletion 22q11.2 as most common)	Toxic
	Abnormal neuronal migration; malformation of cortical development with brainstem involvement	Multiple genes (tubulinopathies as most common)	Rare
	Molar tooth-type dysplasias with defects in ciliary proteins	Joubert syndrome and Meckel–Gruber syndrome	No
Localized malformations associated with later development	Problems in axonal guidance	Hereditary ROBO3	No
	Problems in the signaling of specific rhombomeres	Pontine tegmental cap dysplasia and other cap dysplasias	No
	Specific disruption of some neuronal populations (congenital cranial dysinnervation syndromes)	• Congenital fibrosis of extraocular muscles: KIF21A, PHOX2A, TUBB2B, TUBB3 • Genetic Duane syndrome • Central hypoventilation syndrome • Genetic Moebius syndrome	• Moebius syndrome • Sporadic Duane syndrome • Specific damage of a particular nuclei
Hypoplasia or neurodegeneration of the brainstem	Progressive pontocerebellar hypoplasia	Multiple genes	No
	Nonprogressive pontocerebellar hypoplasia	CASK, RELN, VLDLR	No
	Congenital disorders of glycosylation	Multiple genes	No
	Other early neurodegenerative disorders	DARS2-related disordersWolfram syndrome	No
Temporal immaturity of the circuits	Transitory clinical manifestations	Polygenic contribution	
	Sudden death infant syndrome	Polygenic contribution	

atypic forms of cerebral palsy or congenital suprabulbar paresis when evaluated during childhood (Clark et al., 2010; Munell et al., 2018).

Apart from the overt clinical presentation, many children with this clinical presentation are now detected by fetal imaging due to the presence of polyhydramnios or abnormal ultrasound findings in the fetal central nervous system (Haratz and Lerman-Sagie, 2018).

Neonatal period

In newborns with brainstem dysfunctions (BSD), the most frequent initial clinical symptoms include bilateral VII cranial nerve involvement, muscle hypotonia, and feeding challenges. The cranial nerves most often affected, in descending order of frequency, are the 7th, 9th, 10th, 5th, 12th, 6th, 4th, and 3rd. The reduced muscle tone observed in these patients typically coincides with evidence of pyramidal tract involvement, such as hyperreflexia, pronounced jaw jerk, and Achilles clonus, likely due to simultaneous involvement of the corticospinal and spinocerebellar tracts at the brainstem level (Castilla-Fernández et al., 2013; Munell et al., 2018).

Feeding and swallowing dysfunctions vary based on the involvement of the different parts of the brainstem. In cases with mild facial involvement where swallowing mechanisms remain intact, tube feeding may be stopped once muscle hypotonia improves. However, lesions in the caudal brainstem often result in swallowing difficulties, respiratory issues (as central hypoventilation due to involvement of parafacial respiratory neurons), esophageal motility disorders, and even cardiac arrhythmias, which can sometimes be fatal (Armangue et al., 2012; Castilla-Fernández et al., 2013). Moreover, profuse oral secretions and choking episodes are common in newborns with diffuse BSD, necessitating frequent oropharyngeal suction or, in certain cases, the placement of a continuous nasopharyngeal suction catheter. It is important to recognize that inadequate saliva management can lead to repeated silent microaspiration and subsequent chronic lung damage (Munell et al., 2018). Other associated malformations such as temporomandibular ankylosis, cleft palate, and retromicrognathia, which may or may not be associated with glossoptosis, potentially lead to severe airway complications, which may be difficult to treat requiring alternative methods like laryngeal masks or urgent tracheostomy. Additionally, mandibular growth can be impaired or delayed due to involvement of the trigeminal nerve, resulting in various degrees of micrognathia (Roig Quilis and Pennington, 2011).

Comorbidities and clinical follow-up

Hypotonia tends to improve with time and does not appear to be a significant clinical problem after 3 years of age in most instances. After the first few months of age, facial diplegia, feeding difficulties, poor handling of oral secretions, dysarthria, and signs of pyramidal tract involvement, including a brisk jaw jerk, are the most common clinical findings.

Feeding issues can be life-threatening if not identified and managed properly. In some patients, failure to protect the airway leads to significant aspiration; in milder cases, swallowing disorders might remain undetected or subclinical, often misdiagnosed as reactive bronchitis, leading to recurrent hospital admissions and chronic pulmonary damage (Vernon-Roberts and Sullivan, 2013). Adequate respiratory function is essential for successful oral food intake. Feeding might be facilitated through a nasogastric tube until the airway has been evaluated, and suitable management strategies have been established.

In patients with BSD, abnormal growth and shape of the jaw and palate due to improper innervation of facial and intraoral structures frequently result in progressive tooth misalignment and/or malocclusion. Infants and children with facial diplegia experience not only initial difficulties with sucking and swallowing but also reduced or absent facial expressions and challenges with oral and mimetic communicative functions, leading to significant psychologic impacts. As these patients age, issues such as velopalatine incoordination, facial diplegia, and temporomandibular ankylosis often contribute to speech articulation problems (Roig Quilis and Pennington, 2011). Persistent drooling, common in BSD, arises from difficulties with swallowing and lip closure, causing distress due to social isolation, interference with school or work, and chronic skin irritation that may become infected. While less affected patients may adapt, others improve with age. However, inadequate saliva management can result in repeated silent microaspiration, causing chronic lung injury (Roig Quilis and Pennington, 2011). Tooth decay is notably prevalent among children who are tube-fed or have reduced saliva output due to medication or surgery.

Language development can be delayed by cognitive challenges, but it is usually accompanied by speech articulation issues. The spectrum of speech dysfunction ranges from mild to severe dysarthria with dyspraxic inconsistencies. Primary articulatory challenges are associated with phonemes that require robust labial motor function, such as "P," "B," and "M," or sounds involving tongue-palate friction, like "R," "FR," and "PR." Additionally, the quality of speech production is often further compromised by conductive hearing impairments, frequently a result of chronic serous otitis. This condition stems from Eustachian tube dysfunction, which itself is often caused by secondary deformities of intraoral structures and velo-palatine insufficiency (Roig Quilis and Pennington, 2011).

Cognitive development is altered in many patients. Avoidance of social interaction due to lack of facial expression (Rasing et al., 2024), dysarthria and drooling could have a role, but the role of brainstem as a conduit and modulator of ascending and descending pathways between the brain, the spinal cord, and the cerebellum is also a critical contributor (Venkatraman et al., 2017). Autism spectrum disorders (Seif et al., 2021) and attention deficit disorder with and without hyperactivity (Johnston et al., 2014) have been commonly associated with brainstem and with brainstem disorders (Briegel, 2012). Diagnosis could be complicated and patients could be wrongly diagnosed with emotional and behavioral problems (Briegel, 2012), and parents, teachers, and doctors should pay special attention to bullying and other problems in peer relationships (Strobel and Renner, 2016).

Apart from the oculomotor palsies, ocular motility defects can be more intricate. For instance, congenital oculomotor apraxia, due to lesions in the pathways leading to the horizontal gaze center or the center itself in the paramedian pontine reticular formation, could be detected in examinations in childhood or adulthood (Roig et al., 2003). Abnormal palpebral movement or lack of neural stimulus for the lacrimal glands can impair corneal humidification and ease ulcer formation (Roig Quilis and Pennington, 2011).

Sleep disorders are also a common comorbidity. Common issues include problems with sleep initiation and maintenance, abnormal sleep cycles, and disruptions in Rapid Eye Movement (REM) sleep (Anderson et al., 2007; Roig Quilis and Pennington, 2011). Additionally, brainstem abnormalities may impair autonomic functions during sleep, such as breathing and heart rate regulation, potentially leading to more severe conditions if patients show sleep apnea (Roig Quilis and Pennington, 2011). Specifically, narcolepsy and cataplexy episodes are disabling problems in some patients (Krämer et al., 2014).

DIFFERENTIAL DIAGNOSIS

Before diagnosing BSD, certain conditions must be excluded. Based on clinical considerations (Roig Quilis and Pennington, 2011), we propose organizing the differential diagnosis into three groups:

1. Diseases primarily causing congenital facial palsy, either in isolation or in combination with other symptoms (e.g., cleft palate and micrognathia in Carey–Fineman–Ziter syndrome, tongue fissuring with lip swelling in Melkersson–Rosenthal syndrome, and microtia, facial hypoplasia, or cleft palate in syndromes such as deletion 22q11.2 and CHARGE syndrome).
2. Diseases primarily causing oromotor disorders, including NMDs (e.g., congenital myopathies, congenital myasthenic syndromes, myotonic dystrophy), syndromes characterized by head and neck malformations (e.g., oromandibular-limb hypogenesis syndrome, glossopalatine ankylosis, Pierre Robin sequence, and deletion 22q11.2), and bilateral hemispheric cortical or subcortical lesions.
3. Multiple conditions associated with neurocognitive manifestations, congenital hypotonia, and strabismus that exhibit abnormal brainstem findings on magnetic resonance imaging (MRI).

IMAGING

Conventional MRI remains the cornerstone for diagnosing brainstem malformations due to its superior soft tissue contrast and spatial resolution. Conventional MRI sequences, including T1-weighted, T2-weighted, and fluid-attenuated inversion recovery (FLAIR), are fundamental for initial assessment and identification associated with cranial nerve malformations and for the selection of the appropriate follow-up studies (Poretti et al., 2016; Severino and Huisman, 2019).

Advanced MRI techniques have further refined our understanding of these conditions. High-resolution three-dimensional sequences using constructive interference in steady state (CISS), fast imaging employing steady-state acquisition (FIESTA), or driven equilibrium (DRIVE) sequences offer isotropic voxel representation, which enhances the anatomic delineation of malformations, particularly in complex regions like the cerebellum, the junctions between different brainstem segments, or the visualization of the cranial nerves (Severino and Huisman, 2019). Additionally, diffusion tensor imaging (DTI) provides invaluable information on the microstructural integrity and connectivity of neural pathways. DTI is particularly useful in identifying aberrant fiber tractography and particular patterns that correlate with specific genetic mutations and clinical presentations (Arrigoni et al., 2017). Magnetic resonance (MR) angiography and MR venography are essential in the evaluation of brainstem malformations, as these often involve vascular anomalies (Severino and Huisman, 2019).

Positron emission tomography and functional MRI have been employed to explore the metabolic and functional aspects of brainstem malformations, but their use has been more limited (Severino et al., 2020). Ultra-high-field MRI (e.g., 7 Tesla) offers an increased signal-to-noise ratio and enhanced resolution, allowing for more precise imaging of brainstem structures (Okada et al., 2022). This level of detail is theoretically beneficial in assessing the integrity of small structures like cranial nerve nuclei and their connections, but its clinical use is still to be defined.

Increased understanding of evolving fetal imaging phenotype of these disorders, integration of multiple imaging approaches in pre- and postnatal imaging and contributions for machine learning, and more detailed knowledge about etiologic factors are likely to be the main game-changing factors in the imaging approach to these defects.

OTHER ANCILLARY TESTS

Electrophysiologic studies (nerve conduction studies, electromyography (EMG), blink reflex, brainstem auditory evoked potentials (BAEPs), sleep electroencephalogram (EEG) with polysomnographic recording) are difficult to perform and analyze in neonates or infants; nonetheless, they are crucial for identifying the brainstem origin of this disorder and to exclude other NMDs (Muncll ct al., 2018). Whereas the blink reflex test is the most sensitive for detection of BSD and is abnormal in most patients, BAEPs have the power to localize the lesion at the brainstem but have lower sensitivity (Munell et al., 2018).

GENETIC STUDIES

Under the suspicion of a genetic disorder and following algorithms from other brain malformations (Oegema et al., 2020), whole-exome sequencing (preferentially, under trio schema) may be the first genetic test to perform, as many of the diseases are produced by single nucleotide variants. We will continue with the array comparative genomic hybridization (CGH) and karyotype, if the results are negative. If available and in the future, whole-genome sequencing may substitute whole-exome sequencing, as many noncoding variants are emerging in this particular field (Tenney et al., 2023; Jurgens et al., 2024).

SPECIFIC DISORDERS

Anterior-posterior patterning defects

Disconnection syndromes

This term describes a set of congenital brain malformations resulting from the failure of the development of specific segments of the midbrain and the hindbrain (segmental dysgenesis), leading to the absence or discontinuity of these brain regions (Jissendi-Tchofo et al., 2015). This syndrome is thought to arise from vascular disruptions (Vekemans et al., 2022), toxic exposure (particularly, alcohol) (Yigazu et al., 2014), or impaired genetic expressions local to the affected areas. Mutation in the homeobox gene En2 has been suggested (Sarnat et al., 2002), but not definitively linked. Affected neonates typically exhibit severe symptoms including central hypotonia, cranial neuropathies, spastic quadriparesis, hyperreflexia, and respiratory drive failure, leading to an almost universally fatal outcome. Conventional MRI shows an overall shortening of the brainstem, with certain segments appearing elongated while others are shortened. This variability depends on the location of the disconnection. The mesencephalon is often enlarged with a midline cleft extending to the pons, and the medulla appears thick and dysmorphic. The cerebellum exhibits varying degrees of malformation. Additional imaging such as Constructive Interference in Steady State (CISS) or DTI can help visualize the thin midline cord linking the upper and lower segments of the brainstem, which is typically found in neuropathologic analyses (Severino and Huisman, 2019).

Diencephalic-mesencephalic junction dysplasia

Diencephalic–mesencephalic junction dysplasia (DMJD) is a neurologic disorder characterized by congenital brain malformations at the junction between the diencephalon and mesencephalon. It usually presents as ventriculomegaly in fetal imaging and could be early defined in prenatal studies. DMJD presents distinct morphologic abnormalities, categorized into three types based on imaging features: Type A exhibits a "butterfly" sign indicating hypothalamus–midbrain union; Type B shows partial thalamus–midbrain union; and the newly defined Type C, which demonstrates complete or near-complete midbrain–thalamic continuity. Affected individuals typically face severe developmental delays, cognitive impairments, and various neurologic deficits. Genetic studies have linked DMJD to mutations in the PCDH12 gene (Guemez-Gamboa et al., 2018) but it is a genetic heterogeneous disorder and mutations in L1CAM and biallelic VRK1 and GSX2 variants have been also described (Lawrence et al., 2021).

HOXA1-deficiency syndromes

Homozygous mutations in the HOXA1 gene result in variable phenotypes (Bosley et al., 2008). In Native Americans, it causes Athabaskan brainstem dysgenesis syndrome (ABDS), characterized by gaze palsy, sensorineural deafness, central hypoventilation, and cognitive deficits. Another related disorder, Bosley–Salih–Alorainy syndrome (BSAS), affects individuals in the Middle East, marked by similar eye and ear abnormalities but typically without central hypoventilation. Both conditions include features like delayed motor development, some heart defects, and potential cerebral vascular anomalies. However, ABDS displays more severe hindbrain malformations compared to BSAS. Imaging and genetic studies have confirmed these distinctions and pointed out that these syndromes might be influenced by additional genetic modifiers. Inheritance of these disorders is autosomal recessive with variations in expressivity and penetrance, suggesting that even isolated symptoms like congenital heart disease or Duane retraction syndrome could be manifestations of HOXA1 mutations.

Congenital cranial dysinnervation disorders, Moebius syndrome, and related disorders

Congenital cranial dysinnervation disorders are featured by congenital abnormal eye, eyelid, and/or facial movement disorders, which can occur in combination with deficits in other cranial nerves (Whitman, 2021). Five groups of disorders represent different clinical presentations and different etiologic orientations (Table 10.2) that are formed by the combination or not of congenital disorders of vertical ocular motility, horizontal ocular motility, and/or facial weakness (Ten Donkelaar et al., 2023).

Congenital fibrosis of extraocular muscles (CFEOMs) are an ocular congenital cranial nerve disorder characterized by eyelid ptosis and nonprogressive paralytic strabismus, with limitations in upward gaze and varying restrictions in downward and horizontal eye movements. These disorders are produced mainly by involvement of IIIrd cranial nerve and affected individuals typically exhibit hypoplasia of the superior division of the oculomotor nerve, impacting the levator palpebrae superioris and superior rectus muscles, which results in ptosis and limited upward gaze. In some cases, there are also targeting failures in the oculomotor nerve's inferior division and the abducens nerve, restricting downward and horizontal gaze. Several entities are KIF21A-related and TUBB3-related, the most common (Puri et al., 2023).

There are different types of TUBB3 syndromes that are related to specific mutations. Some mutations are linked to a CFEOM phenotype, others to malformations of cortical development, and in rare cases, to neurodevelopmental phenotypes without CFEOM or malformation of cortical development. Cases of CFEOM are often accompanied by more extensive neurologic and developmental issues such as a range of manifestations including anosmia, hypogonadotropic hypogonadism, optic nerve hypoplasia, facial weakness, swallowing dysfunctions, vocal cord paralysis, progressive polyneuropathy, joint contractures, cyclic vomiting, cardiac arrhythmias, and

Table 10.2

Classification and clinical and radiologic phenotypes of the main congenital cranial dysinnervation disorders.

Group	Entity	Cause	Manifestations that are commonly present	Manifestations that can be associated	Magnetic resonance imaging findings
Predominant vertical ocular motility defects = congenital fibrosis of extraocular muscle (Vivian, 2020; Puri et al., 2023)	KIF21A-related CFEOM (Puri et al. 2023)	Heterozygous missense mutations in KIF21A gene (R954W accounts for > 75%)	Isolated bilateral CFEOM Aberrant residual eye movements (synergistic convergence on attempted upgaze, and a compensatory chin-up head position)	Mild and transitory hypotonia and gross motor delays	• Hypoplasia of cranial nerve III +/– VI • Hypoplasia of extraocular musculature
	PHOX2A-related CFEOM (Nakano et al., 2001)	Biallelic mutations in PHOX2A gene	Bilateral cranial nerve III and IV paresis	• Poorly reactive pupils • Retinal dystrophy	Limited information
	Tukel syndrome (Tukel, 2005)	Suggesting autosomal recessive inheritance linked to 21qter	Variable ophthalmoparesis + limb abnormalities	Abnormal head position	Variable hypoplasia of extraocular muscles
	TUBB3-CFEOM (Puri et al. 2023)	Heterozygous missense mutations in TUBB3 gene	• Isolated unilateral or bilateral CFEOM • Other cranial nerves are usually involved	• Progressive polyneuropathy • Joint contractures • Cyclic vomiting • Cardiac arrhythmias • Cognitive and social impairments	• Hypoplasia or dysplasic corpus callosum or anterior commissure • Dysplasic cerebellar vermis • Basal ganglia abnormalities
Predominant vertical ocular motility defects +/ facial palsy					
Predominant horizontal ocular motility defects + congenital facial palsy	Moebius syndrome (Pedersen et al., 2017; Herrera et al., 2019; López Gutierrez et al., 2024)	• Genetic mutations (rare) • Exposition to teratogens • Factors predisposing to ischemia (hyperthermia, intrauterine trauma, or others)	• Facial weakness • Failure of ocular abduction	Association with swallowing and breathing problemsAssociation with malformations of the head and limbs	• Bilateral absence of the facial colliculus in the pontine tegmentum (horseshoe shape on axial images) • Pons hypoplasia (pontomesencephalic ratio of < 2) • Ventriculomegaly • Dysplasia of corpus callosum • Thalamic fusion • Increased number of arachnoid cysts • Periventricular nodular heterotopia

(*Continued*)

Table 10.2 (Continued)

Group	Entity	Cause	Manifestations that are commonly present	Manifestations that can be associated	Magnetic resonance imaging findings
Predominant horizontal ocular motility defects	Duane restriction (Barry et al., 1993)	• Syndromic (genes): • HOXA1 (AR) • MAFB (AD) • CHN1 (AD) • Duane-radial ray syndrome – Okihiro syndrome: SALL4 (AR) • Nonsyndromic: • Sporadic or CHN1 (AD)	Duane abnormalities	• Other ophthalmologic problems • MAFB: congenital sensorineural hearing loss • SALL4: anomalies in limb	Absence of abducens
	Horizontal gaze paralysis with progressive scoliosis (Pinero-Pinto et al., 2020; Zaka et al., 2021)	Biallelic mutations in ROBO3 or DCC genes	• Lack of horizontal smooth pursuit, saccades, and optokinetic nystagmus with normal vertical movements and convergence • Thoracolumbar scoliosis • Normal cognition	Pendular nystagmus	• Absence of the facial colliculi and medullary dysplasia (butterfly-like appearance) • Absence of major pontine crossing fiber tracts and no decussation of superior cerebellar peduncles in DTI • DCC: agenesis of corpus callosum
Congenital facial palsy	Hereditary congenital facial palsy type 1 (Tenney et al., 2023)	Single nucleotide variants or duplications in noncoding neuron-specific GATA2 regulatory region (autosomal dominant; some de novo)	Bilateral congenital facial weakness	Variable intellectual and psychiatric conditions, craniofacial dysmorphisms	Thin or not visible facial nerves + normal brain and extraocular muscles
	Hereditary facial palsy type 3 (Vogel et al., 2016)	Recessive loss-of-function variants in HOXB1 gene	Bilateral congenital facial weakness + midface retrusion	• Sensorineural hearing loss (3 out of 4) • Esotropia (40%) • Upturned nasal tip with smooth philtrum (70%) • External ear malformations	Bilateral absence of the facial nerve and bilateral abnormal tapering of the basal turn of the cochlea

*CFEOM*congenital fibrosis of extraocular muscles, AR autosomic recessive, AD autosomic dominant

Source: Clinical neuroembryology. Cham: Springer International Publishing. pp. 445–521. https://doi.org/10.1007/978-3-031-26098-8_7.Modified from Ten Donkelaar HJ, Fritzsch B, Cruysberg JRM, et al. (2023) Development and developmental disorders of the brain stem. In: Clinical neuroembryology. Cham: Springer International Publishing. pp. 445–521. https://doi.org/10.1007/978-3-031-26098-8_7.

cognitive and social impairments. In addition to extraocular muscle abnormalities, MRI of syndromic cases often reveal hypoplasia or absence of other cranial nerves, abnormal anterior commissure, corpus callosum or other white matter tracts, dysplastic cerebellar vermis, and basal ganglia malformation. Only in some rare cases, there is a combination of CFEOM with malformation of cortical development (Puri et al., 2023).

Moebius syndrome (MBS), a congenital cranial dysinnervation disorder, arises from a combination of facial and abducens motoneuron loss, due to destructive lesion (hemorrhagic, hypoxic-ischemic, or others) or other cause that significantly impairs the fifth rhombomere. primarily by facial and abducens nerve paralysis. As this damage is not usually restricted to motor nuclei, these patients normally have pyramidal tract symptoms or other cranial nerve manifestations. The presence of a vertical eye motility problem should be considered as an exclusion criterion for MBS and drive suspicion toward a TUBB3 syndrome (MacKinnon et al., 2014). Facial palsy, which is usually bilateral but incomplete, gives rise to altered abilities of facial sensation, speech, feeding, and expression and can further result in exposure keratopathy (Munell et al., 2018). Issues originating due to other cranial nerve palsies can further create problems related to swallowing and breathing as a result of palatopharyngeal dysfunction, inadequate cough reflex, glottic spasm, and laryngeal paralysis (Roig Quilis, 2005). Other head malformations such as cleft lip and palate, micrognathia, maxillary underdevelopment, hypertelorism, aglossia/microglossia, and dental problems have also been noted (Webb et al., 2021). The limb anomalies range from hypoplasia to aplasia of the digits of upper and lower limbs, brachydactyly, syndactyly, ectrodactyly, and several other orthopedic conditions such as talipes equinovarus, ankylosis, and scoliosis (Pedersen et al., 2017; Munell et al., 2018). Occasionally, there is an absence or hypoplasia of the pectoralis major as part of the Poland syndrome (Pedersen et al., 2017). MBS etiology complex, involving multiple factors that contribute to its pathogenesis, is often categorized into vascular, genetic, and teratogenic theories that are related to each other. Exposure to teratogens like misoprostol, cocaine, thalidomide, and ergotamine during pregnancy has been linked to MBS. These substances can cause ischemic damage to the developing fetal brainstem. The vascular theory suggests that MBS results from disruptions in blood supply during critical periods of fetal development, particularly affecting the brainstem watershed zone in which facial and abducens nerves reside. The concept of the subclavian artery supply disruption sequence was used to encompass the association of MBS with Poland syndrome or Klippel–Feil syndrome. Several genetic mutations are associated with rare cases of MBS, including those in the PLXND1 and REV3L genes. PLXND1 has an essential role in axon guidance and angiogenesis during the embryogenesis of the brainstem joining the genetic theory with the vascular one (López Gutierrez et al., 2024).

Duane retraction syndrome is featured by (1) congenital ocular abduction deficiency with variable limitation of eye adduction and (2) retraction of the eye with adduction due to aberrant third nerve innervation and cocontraction of the lateral rectus muscle. Most cases are sporadic but there are rare genetic causes.

Central hypoventilation syndrome

Congenital central hypoventilation syndrome (CCHS) is featured by abnormal control of respiration with normal ventilation while awake and hypoventilation during sleep. The most common form is related to mutations in PHOX2B (mostly polyalanine repeat mutations) (Bachetti and Ceccherini, 2020). Some patients associate Hirschsprung disease. There are other disorders with central hypoventilation and autonomic dysfunction: CCHS Type 2 is associated with biallelic mutations in MYO1H (Spielmann et al., 2017) and CCHS Type 3 is associated with biallelic mutations in LBX1 (Hernandez-Miranda et al., 2018). Central hypoventilation syndrome could also be related to neurodevelopmental disorders such as Rett syndrome (Tascini et al., 2022).

Cap dysplasias

It is a group of disorders suspected to result from impaired axonal guidance or abnormal neuronal migration: (1) pontine tegmental cap dysplasia (Barth et al., 2007), (2) anterior mesencephalic cap dysplasia (Arrigoni et al., 2017), and (3) medullary tegmental cap dysplasia (Gafner et al., 2023). Each disorder has a variable genetic background and variable prognosis.

Other brainstem neurodevelopmental disorders

Apart from the embryologic disorders, there are other disorders that have a prenatal disorder or other malformations in which the brainstem involvement is part of a constellation of other malformation in the central nervous system. We collect the most relevant features of these disorders in Table 10.3.

PROGNOSIS AND THERAPIES

The prognosis of BSD is contingent upon the etiology and extent of the disruptive lesion or abnormal developmental process affecting the brainstem. In nongenetic cases, outcomes often surpass initial clinical manifestations (Munell et al., 2018). Notably, some neonates may exhibit a dysmaturative, reversible brainstem disorder with an excellent prognosis (Frank, 1966; Leroy-Malherbe et al., 1994).

Following a BSD diagnosis, an interdisciplinary team approach encompassing developmental, neurologic, behavioral, speech and language, and cognitive domains becomes imperative (Munell et al. 2018).

The life expectancy of individuals with BSD is influenced by the severity of respiratory insufficiency, heart rate irregularities, the presence of dysphagia, and the extent of gastroesophageal reflux. Evaluating each patient on an individual basis is crucial for establishing a tailored treatment regimen. Early intervention programs should start promptly soon diagnosis, targeting the following functions: speech, hearing, vision, swallowing, nutrition, breathing, mobility, cognition, and behavior (Table 10.4) (Roig Quilis and Pennington, 2011; Munell et al. 2018).

Table 10.3

Other relevant brainstem disorders.

Disease type	Related genes	Clinical presentation	Brainstem magnetic resonance imaging (MRI) features	Cerebellum MRI features	MRI of cerebral cortex, white matter, and ventricles
Progressive pontocerebellar hypoplasia (van Dijk et al., 2018)	TSEN54, VRK1, RARS2, EXOSC3, and others	Developmental delays, movement disorders, often severe	Atrophy or hypoplasia, less pontine prominence	Atrophy with more severe involvement of hemispheres than vermis, possible cyst formation	Possible cerebral atrophy and delayed myelination
CASK-related pontocerebellar hypoplasia (Moog and Kutsche, 1993)	CASK	Microcephaly, ataxia, nystagmus, severe cognitive impairment	Hypoplasia, severe involvement similar to Pontocerebellar hypoplasia type 2	Severe hypoplasia, symmetric involvement of hemispheres and vermis	Severe intellectual disability, dysarthria, seizures
RELN-related hypoplasia and VLDLR-related hypoplasia (Valence et al., 2016)	RELN VLDLR	Severe developmental disabilities, microcephaly, seizures	Hypoplasia with less pontine prominence compared to other types	Severe cerebellar hypoplasia with more severe involvement of the vermis	Mild frontal pachygyria, significant cerebral involvement
α-Dystroglycanopathies (Smith et al., 2005)	Multiple genes involved in dystroglycan glycosylation	Muscle weakness, intellectual disability, eye abnormalities	Brainstem kinking, ventral pontine cleft	Cerebellar dysplasia with cysts, dysplastic tectum	Cobblestone cortex, diffuse periventricular white matter changes
Tubulinopathies (Maillard et al., 2023)	TUBA1A, TUBA8, TUBB2B, TUBB5	Intellectual disability, cerebral palsy, seizures, postnatal microcephaly	Asymmetric midbrain and pons, dysmorphic tectum	Dysplasia, cerebellar hypoplasia, and sometimes asymmetric involvement (Van De Weghe et al., 2022)	Lissencephaly, polymicrogyria, and dysmorphic basal ganglia
Joubert syndrome	More than 30 genes, e.g., TMEM67, AHI1	Hypotonia, ataxia, ocular motor apraxia, neonatal breathing problems, intellectual disability	Dysmorphic tectum and midbrain, thickening, and elongation of midbrain	Vermis hypoplasia, molar tooth sign	Often normal; some cases show hippocampal malrotation, callosal dysgenesis

Table 10.4

Manifestation-based clinical management in brainstem dysgenesis.

Function	Specific aims	Intervention
Speech	Develop speech, language, and communication	Alternative/augmentative methods of communication
Hearing	Improve communication	Consider implant devices
Cognition and behavior	Improve attention, social engagement, and confidence	Early interventional programs Skills training Psychologic support Consider pharmacologic therapy for Attention deficit hyperactivity disorder (ADHD)
Eye movements and closure	Improve eye motion and sight and avoid corneal ulcers	Artificial eye drops Surgical replacement with amniotic membrane for corneal ulcers Consider surgery for strabismus and tarsorrhaphy for impaired blinking
Swallowing	Achieve safe and efficient oral feeding Decrease drooling Lowering broncho-aspiration rate	Postural counseling Food texture modification, Thickener agents Nasogastric tube Gastrostomy (consider fundoplication if associated gastroesophageal reflux) Consider anticholinergic agents (scopolamine, glycopyrrolate) and Botulinum toxin injections
Nutrition	Achieve efficient eating and prevent failure to thrive Maintain vitamin levels Avoid constipation	Parental counseling Nutritional supplements Dietary treatment of constipation (hydration, fiber) Consider laxatives and prokinetic agents Consider cisapride if cyclic vomiting (frequent response in TUBB3 mutations, personal communication)
Breathing and sleep	Identify and prevent sleep disorders	Physiotherapy Ventilatory assistance Consider surgery for obstructive apnea (tonsillectomy and adenoidectomy) Consider cataplexy therapies
Motility	Prevent or reduce joint ankyloses or deformitiesAchieve maximal limb motor independence	Physiotherapy Orthoses Consider surgery for joint deformities Consider surgery for facial palsy

FUTURE IN BRAINSTEM DYSGENESIS

One of the primary challenges in BSD is the ability to predict prognosis accurately and provide precise genetic counseling along with appropriate multidisciplinary and expert care. Achieving this objective necessitates enhancing our understanding of the disorder and identifying the underlying cause, whether acquired or inherited, as early as feasible. This entails possessing profound knowledge for conducting precise clinical characterization and employing the most appropriate diagnostic technologies.

Therefore, it is imperative to focus on identifying the associations between each specific disorder and its phenotypic traits and to disseminate comprehensive phenotyping data within the international scientific community. This data will facilitate the construction of the natural history profiles of these disorders. Furthermore, combined with advancements in the scientific understanding of these conditions, it will aid in the development and testing of novel therapeutic approaches.

References

Anderson K, Shneerson J, Smith I (2007). Mobius syndrome in association with the REM sleep behaviour disorder. J Neurol Neurosurg Psychiatry 78: 659–660.

Armangue T, Macaya A, Vazquez E et al. (2012). Central hypoventilation and brainstem dysgenesis. Pediatr Neurol 46: 257–259.

Arrigoni F, Romaniello R, Peruzzo D et al. (2017). Anterior mesencephalic cap dysplasia: novel brain stem malformative features associated with Joubert syndrome. Am J Neuroradiol 38: 2385–2390.

Bachetti T, Ceccherini I (2020). Causative and common *PHOX2B* variants define a broad phenotypic spectrum. Clin Genet 97: 103–113.

Barry BJ, Whitman MC, Hunter DG et al. Duane syndrome. In: Adam MP, Feldman J, Mirzaa GM et al. (eds.) GeneReviews®. Seattle, WA: University of Washington.

Barth PG, Majoie CB, Caan MWA et al. (2007). Pontine tegmental cap dysplasia: a novel brain malformation with a defect in axonal guidance. Brain 130: 2258–2266.

Bosley TM, Alorainy IA, Salih MA et al. (2008). The clinical spectrum of homozygous *HOXA1* mutations. Am J Med Genet A 146A: 1235–1240.

Briegel W (2012). Self-perception of children and adolescents with Möbius sequence. Res Dev Disabil 33: 54–59.

Carta A, Mora P, Neri A et al. (2011). Ophthalmologic and systemic features in Möbius syndrome an Italian case series. Ophthalmology 118: 1518–1523.

Castilla-Fernández Y, Boix H, Macaya A et al. (2013). Brainstem dysgenesis during the neonatal period: diagnosis and management. J Perinat Med 41: 445–453.

Clark M, Harris R, Jolleff N et al. (2010). Worster-Drought syndrome: poorly recognized despite severe and persistent difficulties with feeding and speech. Dev Med Child Neurol 52: 27–32.

Doherty D, Millen KJ, Barkovich AJ (2013). Midbrain and hindbrain malformations: advances in clinical diagnosis, imaging, and genetics. Lancet Neurol 12: 381–393.

Ferreira RM, Amaral LLF, Gonçalves MVM et al. (2011). Imaging findings in congenital cranial dysinnervation disorders. Top Magn Reson Imaging 22: 283–294.

Frank MM (1966). Transient pharyngeal incoordination in the newborn. Arch Pediatr Adolesc Med 111: 178.

Gafner M, Garel C, Leibovitz Z et al. (2023). Medullary tegmental cap dysplasia: fetal and postnatal presentations of a unique brainstem malformation. Am J Neuroradiol 44: 334–340.

Guemez-Gamboa A, Çağlayan AO, Stanley V et al. (2018). Loss of *Protocadherin-12* leads to diencephalic-mesencephalic junction dysplasia syndrome. Ann Neurol 84: 638–647.

Haratz KK, Lerman-Sagie T (2018). Prenatal diagnosis of brainstem anomalies. Eur J Paediatr Neurol 22: 1016–1026.

Hernandez-Miranda LR, Ibrahim DM, Ruffault P-L et al. (2018). Mutation in LBX1/Lbx1 precludes transcription factor cooperativity and causes congenital hypoventilation in humans and mice. Proc Natl Acad Sci U S A 115: 13021–13026.

Herrera DA, Ruge NO, Florez MM et al. (2019). Neuroimaging findings in Moebius sequence. Am J Neuroradiol 40: 862–865.

Jissendi-Tchofo P, Severino M, Nguema-Edzang B et al. (2015). Update on neuroimaging phenotypes of mid-hindbrain malformations. Neuroradiology 57: 113–138.

Johnston BA, Mwangi B, Matthews K et al. (2014). Brainstem abnormalities in attention deficit hyperactivity disorder support high accuracy individual diagnostic classification. Hum Brain Mapp 35: 5179–5189.

Jurgens JA, Barry BJ, Chan W-M et al. (2024). Expanding the genetics and phenotypes of ocular congenital cranial dysinnervation disorders. medRxiv.

Krämer S, Goldammer U, Sindern E (2014). Möbius-syndrom und narkolepsie. Nervenarzt 85: 1569–1572.

Lawrence AK, Whitehead MT, Kruszka P et al. (2021). Prenatal diagnosis of diencephalic-mesencephalic junction dysplasia: fetal magnetic resonance imaging phenotypes, genetic diagnoses, and outcomes. Prenat Diagn 41: 778–790.

Leroy-Malherbe V, Husson B, Landrieu P et al. (1994). Syndrome of delayed maturation of the sucking-swallowing reflex. Arch. Pediatr 1: 20–25.

López Gutierrez D, Luna López I, Medina Mata BA et al. (2024). Physiopathologic bases of Moebius syndrome: combining genetic, vascular, and teratogenic theories. Pediatr Neurol 153: 1–10.

MacKinnon S, Oystreck DT, Andrews C et al. (2014). Diagnostic distinctions and genetic analysis of patients diagnosed with Moebius syndrome. Ophthalmology 121: 1461–1468.

Maillard C, Roux CJ, Charbit-Henrion F et al. (2023). Tubulin mutations in human neurodevelopmental disorders. Semin Cell Dev Biol 137: 87–95.

Moog U, Kutsche K (1993). CASK disorders. In: Adam MP, Feldman J, Mirzaa GM et al. (eds.) GeneReviews®. Seattle, WA: University of Washington.

Munell F, Tormos MA, Roig-Quilis M (2018). Brainstem dysgenesis: beyond Moebius syndrome. Rev Neurol 66: 241–250.

Nakano M, Yamada K, Fain J et al. (2001). Homozygous mutations in ARIX(PHOX2A) result in congenital fibrosis of the extraocular muscles type 2. Nat Genet 29: 315–320.

Oegema R, Barakat TS, Wilke M et al. (2020). International consensus recommendations on the diagnostic work-up for malformations of cortical development. Nat Rev Neurol 16: 618–635.

Okada T, Fujimoto K, Fushimi Y et al. (2022). Neuroimaging at 7 Tesla: a pictorial narrative review. Quant Imaging Med Surg 12: 3406–3435.

Pedersen LK, Maimburg RD, Hertz JM et al. (2017). Moebius sequence – a multidisciplinary clinical approach. Orphanet J Rare Dis 12: 4.

Pinero-Pinto E, Pérez-Cabezas V, Tous-Rivera C et al. (2020). Mutation in ROBO3 gene in patients with horizontal gaze palsy with progressive scoliosis syndrome: a systematic review. Int J Environ Res Public Health 17: 4467.

Poretti A, Boltshauser E, Huisman TAGM (2016). Cerebellar and brainstem malformations. Neuroimaging Clin N Am 26: 341–357.

Puri D, Barry BJ, Engle EC (2023). TUBB3 and KIF21A in neurodevelopment and disease. Front Neurosci 17: 1226181.

Rasing NB, Van De Geest-Buit W, Chan OYA et al. (2024). Psychosocial functioning in patients with altered facial expression: a scoping review in five neurological diseases. Disabil Rehabil 46: 3772–3791.

Roig M, Gratacòs M, Vazquez E et al. (2003). Brainstem dysgenesis: report of five patients with congenital hypotonia, multiple cranial nerve involvement, and ocular motor apraxia. Dev Med Child Neurol 45: 489–493.

Roig Quilis M (2005). Disgenesia troncoencefálica: los síndromes de Möbius, Cogan y Pierre Robin en revisión. An Pediatr 62: 346–351.

Roig Quilis M, Pennington L (eds.) (2011). Oromotor disorders in childhood. Viguera, Barcelona.

Sarnat HB, Benjamin DR, Siebert JR et al. (2002). Agenesis of the mesencephalon and metencephalon with cerebellar hypoplasia: putative mutation in the *EN2* gene—report of 2 cases in early infancy. Pediatr Dev Pathol 5: 54–68.

Seif A, Shea C, Schmid S et al. (2021). A systematic review of brainstem contributions to autism spectrum disorder. Front Integr Neurosci 15: 760116.

Severino M, Geraldo AF, Utz N et al. (2020). Definitions and classification of malformations of cortical development: practical guidelines. Brain 143: 2874–2894.

Severino M, Huisman TAGM (2019). Posterior fossa malformations. Neuroimaging Clin N Am 29: 367–383.

Smith AS, Levine D, Barnes PD et al. (2005). Magnetic resonance imaging of the kinked fetal brain stem: a sign of severe dysgenesis. J Ultrasound Med 24: 1697–1709.

Spielmann M, Hernandez-Miranda LR, Ceccherini I et al. (2017). Mutations in *MYO1H* cause a recessive form of central hypoventilation with autonomic dysfunction. J Med Genet 54: 754–761.

Strobel L, Renner G (2016). Quality of life and adjustment in children and adolescents with Moebius syndrome: evidence for specific impairments in social functioning. Res Dev Disabil 53–54: 178–188.

Tascini G, Dell'Isola GB, Mencaroni E et al. (2022). Sleep disorders in Rett syndrome and Rett-related disorders: a narrative review. Front Neurol 13: 817195.

Ten Donkelaar HJ, Fritzsch B, Cruysberg JRM et al. (2023). Development and developmental disorders of the brain stem. Clinical neuroembryology. Cham: Springer International Publishing, pp. 445–521.

Tenney AP, Di Gioia SA, Webb BD et al. (2023). Noncoding variants alter GATA2 expression in rhombomere 4 motor neurons and cause dominant hereditary congenital facial paresis. Nat Genet 55: 1149–1163.

Tukel T (2005). A new syndrome, congenital extraocular muscle fibrosis with ulnar hand anomalies, maps to chromosome 21qter. J Med Genet 42: 408–415.

Valence S, Garel C, Barth M et al. (2016). *RELN* and *VLDLR* mutations underlie two distinguishable clinico-radiological phenotypes. Clin Genet 90: 545–549.

Van De Weghe JC, Gomez A, Doherty D (2022). The Joubert-Meckel-nephronophthisis spectrum of ciliopathies. Annu Rev Genomics Hum Genet 23: 301–329.

van Dijk T, Baas F, Barth PG et al. (2018). What's new in pontocerebellar hypoplasia? An update on genes and subtypes. Orphanet J Rare Dis 13: 92.

Vekemans M, Maurice P, Lachtar M et al. (2022). Additional evidence for the vascular disruption defect hypothesis in a novel case of brainstem disconnection syndrome. Birth Defects Res 114: 1298–1306.

Venkatraman A, Edlow BL, Immordino-Yang MH (2017). The brainstem in emotion: a review. Front Neuroanat 11: 15.

Vernon-Roberts A, Sullivan PB (2013). Fundoplication versus post-operative medication for gastro-oesophageal reflux in children with neurological impairment undergoing gastrostomy. Cochrane Database Syst Rev 2013: CD006151.

Vivian AJ (2020). Congenital fibrosis of the extra-ocular muscles (CFEOM) and the cranial dysinnervation disorders. Eye 34: 251–255.

Vogel M, Velleuer E, Schmidt-Jiménez LF et al. (2016). Homozygous *HOXB1* loss-of-function mutation in a large family with hereditary congenital facial paresis. Am J Med Genet A 170: 1813–1819.

Volpe JJ (2009). Brain injury in premature infants: a complex amalgam of destructive and developmental disturbances. Lancet Neurol 8: 110–124.

Webb BD, Manoli I, Engle EC et al. (2021). A framework for the evaluation of patients with congenital facial weakness. Orphanet J Rare Dis 16: 158.

Whitman MC (2021). Axonal growth abnormalities underlying ocular cranial nerve disorders. Annu Rev Vis Sci 7: 827–850.

Yigazu P, Kalra V, Altinok D (2014). Brainstem disconnection in a late preterm neonate with classic features of fetal alcohol syndrome. Pediatr Neurol 51: 745–746.

Zaka A, Shahzad S, Rao HZ et al. (2021). A novel homozygous frameshift mutation in the *DCC* gene in a Pakistani family with autosomal recessive horizontal gaze palsy with progressive scoliosis-2 with impaired intellectual development. Am J Med Genet A 185: 355–361.

Handbook of Clinical Neurology, Vol. 216 (3rd series)
The Brainstem
L.T. Grinberg and A.D. Korczyn
https://doi.org/10.1016/B978-0-443-15736-3.00020-2

Chapter 11

Intrinsic pontine tumors: Pathology of brainstem tumors and exploration of diffuse intrinsic pontine glioma

RYAN MOON[1,*], SABINE MUELLER[2], AND PASCALE VARLET[3,*]

[1]*Department of Neurology, University of California San Francisco, San Francisco, CA, United States*

[2]*Department of Neurology, Neurosurgery and Pediatrics, University of California San Francisco, San Francisco, CA, United States*

[3]*Department of Neuropathology, GHU Paris, Psychiatry and Neurosciences, Sainte-Anne Hospital, Paris, France*

Abstract

Over the past decade, advances in brainstem biopsies and the discovery of the histone H3 mutation have revealed a new spectrum of brainstem tumors. In the brainstem, diffuse midline gliomas (DMGs) are most common in the pediatric population, followed by less aggressive tumors like pilocytic astrocytomas. The 2016 WHO central nervous system (CNS) classification introduced diagnostic criteria for DMGs, which have since evolved into five subtypes. These new subtypes present challenges for histomolecular diagnosis in biopsies. Recent data suggest these subtypes differ in clinical outcomes and prognosis, and most likely will require different therapies highlighting the need for tissue for diagnostic purposes for brainstem lesions.

This chapter will first cover an overview of different brainstem pathologies, followed by a discussion on imaging, treatment standards, and the evolving role of biopsies in DMG including diffuse intrinsic pontine gliomas.

PATHOLOGY OF BRAINSTEM TUMORS

Knowledge of the various distinct brainstem tumoral types has progressed enormously over the last 10 years, thanks to the widespread use of biopsy explorations in this location, which was previously considered unnecessary. As a direct consequence, in addition to the contemporary discovery of mutations of a new driver Histone H3K27 in midline tumors, the knowledge of the pathology of these tumors has improved, with the discovery of several subtypes of H3K27-altered gliomas and the description of others, previously unknown.

Overview of intrinsic brainstem tumors

HGGs non-DIPG: high-grade glioma (HGG) with MYCN-amplification, radio-induced glioma, and diffuse IDH-mutant gliomas

High-grade pontine gliomas, without H3K27me3-loss, are exceptional and correspond to high-grade gliomas with MYCN-amplification (HGGs MYCN), radio-induced gliomas (RIG), and diffuse isocitrate dehydrogenase (IDH)-mutant gliomas. Although the relative proportion of these three subtypes is unknown, they depend on the age of the patient: the HGG MYCN subtype is a pediatric tumor occurring at a median age of 3 years (Tauziède-Espariat et al., 2019), contrarily to the two other subgroups, which occur at the median age of 15 years for RIG and in adulthood for diffuse IDH-mutant gliomas (DeSisto et al., 2021; Aboubakr et al., 2023).

Pontine HGGs MYCN present distinctive morphologic features, including areas of infiltration and nodules of poorly differentiated, spindle, or epithelioid cells. The existence of large cells with distinct nucleoli is quite characteristic (Tauziède-Espariat et al., 2019). The immunophenotype is also quite particular: Olig2-negative, focally glial fibrillary acidic protein (GFAP)-positive, and nuclear accumulation of p53 (Tauziède-Espariat et al., 2019). Two molecular subtypes exist; the most frequent pontine subtype is the co-amplification of MYCN and ID2, or in a mutually exclusive fashion, a *P53*

*Correspondence to: Ryan Moon, Department of Neurology, University of California, San Francisco, San Francisco, CA, United States. E-mail: ryan.moon@ucsf.edu; Pascale Varlet, Department of Neuropathology, GHU Paris, Psychiatry and Neurosciences, Sainte-Anne Hospital, Paris, France. E-mail: P.VARLET@ghu-paris.fr.

mutation in the spectrum of Li–Fraumeni syndrome (Buczkowicz et al., 2014b; Tauziède-Espariat et al., 2023). The prognosis of these tumors is worse than that of the diffuse intrinsic pontine glioma (DIPG) H3.3K27-mutant (Tauziède-Espariat et al., 2019).

Pontine RIGs generally occur after an average of 8 years from completion of radiotherapy for various posterior fossa tumors (Klimo et al., 2016). The histologic characteristics of RIG are not well known; however, from a molecular point of view, they are characterized by a distinct cluster of DNA methylation (Ped*RTK1*), recurrent copy-number alterations of chromosome (Ch.) 1p loss/1q gain, Ch. 13q and Ch. 14q loss, and *PDGFRA* amplification (DeSisto et al., 2021). The prognosis of these tumors is also worse than the DIPG H3.3K27-mutant subtype (Klimo et al., 2016).

The majority of diffuse IDH-mutant gliomas in pontine location are astrocytomas, as oligodendroglioma IDH-mutant, 1p/19q-codeleted according to a strict molecular description, is limited to solitary case reports (Hodges et al., 2015; Aboubakr et al., 2023). Histologic aspects of these infiltrative tumors are not different from their supratentorial counterparts but are molecularly enriched by non-R132H variants and IDH2 mutations (Banan et al., 2020; Aboubakr et al., 2023). These tumors, which are more common in adults, are still less frequent than DIPG H3.3K27-mutant, even at this age range (Wang et al., 2022). Their prognosis is better than that of the DIPG H3.3K27-mutant with a median overall survival (OS) of 11 months (Banan et al., 2020; Wang et al., 2022).

Benign gliomas: pilocytic astrocytoma, angiocentric glioma with MYB/MYBL1-alteration, and astroblastoma, MN1-altered

Benign pontine gliomas are less frequent than malignant gliomas in this particular location, but represent a major differential diagnosis given their different prognosis and treatment. The relative proportion of these three tumor types is not really known, mainly because their precise histomolecular description is recent. However, age at diagnosis is a good element of orientation because MYB/MYBL1 tumors occur in young children, while astroblastomas occur in adolescents/young adults.

Pilocytic astrocytomas represent the most common glioma in children, and the brainstem is one of the preferred sites after the cerebellum (Dangouloff-Ros et al., 2019; Ryall et al., 2020). Their histologic appearance does not differ from that observed in other locations, but their molecular characteristics are significantly different, with enrichment for FGFR1 alterations and, more generally, non-KIAA1549::BRAF fusions (Ryall et al., 2020).

MYB/MYBL1-altered gliomas comprise two distinct tumor types according to the current WHO central nervous system (CNS) tumor classification: angiocentric glioma (AG) and diffuse astrocytoma, MYB/MYBL1-altered. AGs are mainly located in the cerebral cortex, presenting a nice phenotype/genotype correlation, which is less evident in the diffuse astrocytoma, MYB/MYBL1-altered (Reisz et al., 2023). Both have rarely been described in pontine location (D'Aronco et al., 2017; Chiang et al., 2020; Reisz et al., 2023; Trkova et al., 2023). The AG histologic appearance is identical to that of tumors observed in a cortical location; however, the characteristic angiocentric appearance could be more difficult to observe under stereotaxic biopsy conditions. Histologically, small bipolar cells are radially arranged around the vessels. Myxoid and microcysts can explain the significant intratumoral T1-weighted high intensity (Kurokawa et al., 2022). Unlike all other infiltrating glial tumors encountered in the pons, these tumors are Olig2-negative and can show epithelial membrane antigen (EMA) dot positivity (Reisz et al., 2023) without loss of H3K27 trimethylation. AG mostly presents with a fusion of *MYB* transcription factor with the tumor suppressor gene *QKI*, whereas the rearrangements of MYB/MYBL1 are more varied in diffuse astrocytomas with MYB/MYBL1-alterations. Favorable therapy response has been observed after radiation therapy only or after high-dose chemotherapy (Trkova et al., 2023). Making this differential diagnosis between AG and DIPG is thus key to patient management and therefore reinforces the role of stereotaxic biopsy and histomolecular analysis in pontine tumors.

The diagnosis of astroblastoma, MN1-altered, has gained a lot in precision since the introduction of histomolecular diagnosis. We now know that astroblastomas with MN1::BEND2 fusion characteristically occur in young girls (average age 12 years) in supratentorial locations, whereas the rarest astroblastomas with EWSR1::BEND2 fusion occur in midline locations (brainstem, spine). The histologic appearance is less characteristic in these cases, with few astroblastic figures and GFAP-, S100-, and Olig2-positive cells (Smith-Cohn et al., 2021; Lucas et al., 2022).

Malignant embryonal tumors (ETMR, ATRT)

The spectrum heterogeneity of pontine tumors, particularly in children, has become apparent, especially since the widespread use of biopsy exploration. Brainstem embryonal tumors, however, are much rarer than in the cerebellar location or than brainstem glial tumors (Klimo et al., 2016). The relative proportion of these embryonal tumors, atypical teratoid rhabdoid tumors (ATRT) and embryonal tumor with multi-layered rosettes (ETMR), is not known, as their descriptions are limited to a small series or solitary case reports (Udaka et al., 2013; Sato et al., 2015). According to the Surveillance, Epidemiology, and End Results (SEER) register, brainstem location accounts for 12% of ATRT cases (Lau et al., 2015), and from a large series of 97 ETMRs, 17% were located in the brainstem (Korshunov et al., 2014). The histologic and molecular characteristics of these two tumoral types are not peculiar in this specific location and justify the systematic INI1 and LIN28A immunostaining, in addition to the immunohistochemical panel, when analyzing a brainstem tumor in an infant (Dangouloff-Ros et al., 2019).

Glio-neuronal tumors of the brainstem

Ganglioglioma is the glio-neuronal tumor (GNT) type that dominates in this location, bearing in mind that many of these heterogeneous GNTs are classically cortical and epileptogenic. Gangliogliomas therefore represent an exception within this large GNT group, which can be observed in the midline (thalamus, brainstem, and cervico-medullary junction) (Janjua et al., 2017; Oushy et al., 2020). Their mixed glio-neuronal histologic aspects do not differ in this location, and the percentage of cases with BRAF V600E mutation seems discretely higher (50%) than in other locations (Janjua et al., 2017). A less favorable prognosis has been described, although the reason(s) for this cannot always be determined: difficulty in performing a complete resection, appearance of respiratory failure or swallowing disorders, and also the existence in 10% of midline BRAF V600E-mutant gangliogliomas of an associated H3K27 mutation (Pagès et al., 2018), which could correspond to a new type of diffuse midline glioma (DMG) with H3K27 and BRAF/FGFR1 co-alteration (Janjua et al., 2017). New data are needed to better assess whether or not there is an overlap between these particular brainstem gangliogliomas and DMGs with double H3K27 and MAP.

Diffuse intrinsic pontine glioma pathology

The histopathologic hallmark of DIPG is a diffusely infiltrative glial pattern, without any particular perivascular or perineuronal tropism. They display a broad spectrum of histologic grading criteria and morphologic features, mainly composed of small and monomorphic infiltrative cells but may incorporate giant, polymorphous, poorly differentiated, and epithelioid cells (Solomon et al., 2016). About 10% of DIPGs lack mitotic features, microvascular proliferation, or necrosis (Buczkowicz et al., 2014a). These criteria are not necessary for the DIPG diagnosis, as they are not independent predictors of poor overall survival (Buczkowicz et al., 2014b; Castel et al., 2015; Louis et al., 2018). DIPGs are considered grade 4, regardless of the presence of necrosis or neo-angiogenesis.

Clinical data (age and outcome) and distinct molecular patterns (methylation profiling, gene expression, and cooperating mutations) suggest the presence of five distinct DIPG subtypes: (1) H3.3K27-mutant, (2) H3.1/3.2K27-mutant, (3) H3-WT DIPG with EZHIP (EZH inhibitory protein) overexpression, (4) DIPG *EGFR*-mutant, and (5) DIPG H3K27 and BRAF/FGFR1 co-altered. No significant histologic differences have been noted among these five different subtypes, except for the H3K27 and mitogen-activated protein kinase (MAPK) co-altered, as they present microcalcifications and/or a solid tumor component and/or an associated neuronal morphotype in a subset of cases (Auffret et al., 2024). The H3.3K27-mutant subtype largely predominates in the pontine location, characterized by subtype-specific cooperating mutations such as *TP53*, *PPM1D*, *ATM* mutations, or PDGFRa amplification, and a worse prognosis (median OS of 10 months) (Buczkowicz et al., 2014a; Castel et al., 2015). The H3.1/3.2K27-mutant subtype is restricted to the pons, in younger patients (mean patient age: 5 years) and with *ACVR1* co-driver mutation (Castel et al., 2015). These two distinct subtypes present two distinct methylation clusters. In rare H3-WT DIPGs with EZHIP overexpression (<5% of DIPG), EZHIP overexpression mimics the H3K27 "oncohistone" mutation, impairing EZH2 methyltransferase activity, repressing Polycomb Repressive Complex 2 (PRC2) activation and thereby generating a global H3K27me3 loss (Castel et al., 2020). This subtype, lacking an H3K27 mutation, presents alterations in genes frequently involved in other DIPG such as *ACVR1*, *TP53*, *PPM1D*, and a methylation cluster close to H3.1K27-mutant tumors (Castel et al., 2020). The overall survival of patients with H3-WT DIPGs with EZHIP overexpression is similar to that of H3.1/3.2K27-mutant subtype, which in turn is significantly higher than that of H3.3K27-mutant patients (median OS of 15 months) (Castel et al., 2020). The more recently described subtype, called H3K27 and BRAF/FGFR1 co-altered, constitutes a unique methylation cluster distinct from other DMG subtypes and various low- or high-grade MAPKinase-altered CNS tumors (Auffret et al., 2024). They harbor particular radiologic and histopathologic aspects (calcifications, solid compact tumoral component) and a longer overall survival (median OS > 3 years) (Auffret et al., 2024).

PATHOGENESIS

The recurrent oncohistone driver mutation defining this entity is a substitution, in one of the histone H3 isoforms (in canonical H3.1/H3.2 or in noncanonical H3.3 variants), of a lysine at position 27 by methionine (K27) or rarely an isoleucine (I), resulting in a generalized loss of the repressive H3K27 trimethylation (H3K27me3) mark through various mechanisms including inhibition of PRC2 methyltransferase activity (Schwartzentruber et al., 2012; Wu et al., 2012; Lewis et al., 2013). Despite representing only a small proportion of the total cellular H3 pool, H3K27 results in widespread loss of H3K27me3 (Lewis et al., 2013). As PRC2 promotes cellular differentiation through H3K27me3, H3K27 causes developmental arrest, with distinct progenitor populations remaining blocked in an immature precursor state, resembling to oligodendrocyte precursor cells (Filbin et al., 2018). Note that differences between the DNA-based numbering (H3K28) and amino acid numbering used in the study of histones (H3K27M), could coexist in literature but are to be considered equivalent.

Somatic heterozygous H3K27 mutations are generally found to be associated with collaborating mutations, invariably maintained during the disease (Nikbakht et al., 2016). These target canonical cancer-associated pathways, are observed with a subtype-specific enrichment, such as the p53 pathway (*TP53*, *PPM1D*, *ATM*) predominantly in H3.3K27 cases, and PI3K/MAPK (*PIK3CA*, *PIK3R1*, *PTEN*) largely in the H3.1/H3.2K27 subtype. Mutations in *ACVR1* are found in ~25% cases (Buczkowicz et al., 2014b). They are almost entirely restricted to the pons, with identical base changes to those found in the germline of patients with rare autosomal dominant disorder, Fibrodysplasia Ossificans Progressiva; however, these patients do not have a predisposition for DIPG (Han et al., 2018). Genetic amplifications of growth factor receptors involved in brain development are common and target PDGFRA (hindbrain, diencephalon, telencephalon), FGFR1 (diencephalon), and ACVR1 (hindbrain). H3K27M mutations are mutually exclusive with IDH or G34R/V histone H3 gene mutations. CDKN2A/CDKN2B deletions or MGMT promoter methylation represents rare events in DMGs (Mackay et al., 2017).

CXorf67 gene (chromosome X open reading frame 67) codes for a small interactor protein EZHIP, which is physiologically expressed during early development, to block PRC2. EZHIP directly binds to EZH2 (enhancer of zeste homolog 2), a core component of PRC2, which normally acts as a methyltransferase. In H3-wild-type tumors (< 10%), the EZHIP overexpression mimics the H3K27M "oncohistone," abolishing the EZH2 methyltransferase activity, generating a global H3K27me3 loss and a derepression of PRC2 target genes, resulting in the maintenance of the tumor cells in an undifferentiated state (Castel et al., 2020; Filbin et al., 2018; Hübner et al., 2019).

Epidermal growth factor receptor (EGFR)-mutant is genetically defined by alterations in the *EGFR* gene, mostly small in-frame duplication/insertion in exon 20, or amplification or missense mutation in exon 7 (Mondal et al., 2020; Sievers et al., 2021).

IMMUNOPROFILE

The loss of an H3K27me3 immunoexpression is common to all five DIPG subtypes and is very useful for the diagnosis, if performed with appropriate negative and positive controls. The combination of a loss of H3K27me3 immunoexpression and the use of H3K27M-mutant specific immunohistochemistry, is highly effective for the diagnosis of H3K27-mutant DIPG, with positive nuclear staining in neoplastic cells even if infiltrative tumor cells are few (Venneti et al., 2014; Solomon et al., 2016). However, no H3K27-mutant staining is observed in DIPG with H3K27i mutation, EZHIP overexpression, or EGFR-mutant. In the latter two tumors, the loss of nuclear trimethylation H327me3 immunoexpression is associated with an EZHIP overexpression (Antin et al., 2020; Castel et al. 2020; Ajuyah et al., 2023).

DIPG is a highly infiltrative glial Olig2-positive tumor, and neuronal stainings highlight the infiltrative neuropil and axonal network. GFAP expression is variable. Sox10 expression is also positive except in the DMG EGFR-mutant subtype (Aboubakr et al., 2024). Nuclear accumulation of P53 suggests an underlying collaborating *TP53* mutation, which has been shown to be associated with radioresistance in H3K27-mutant DIPG (Werbrouck et al., 2019). ATRX loss is seen in a subset of DIPG, which could correspond to the rare DIPGs H3K27 and MAPK co-altered subtype (Auffret et al., 2024) (Table 11.1).

Table 11.1

Summary of pathologic and molecular features of intrinsic brainstem tumors.

	Diagnosis	Morphologic/histologic/location	Immunophenotype/molecular
High-grade glioma (HGG) non-DIPG (diffuse intrinsic pontine glioma) without H3K27me3 loss	HGGs with MYCN-amplification	Areas of infiltration and nodules of poorly differentiated, spindle, or epithelioid cells and the existence of large cells with distinct nucleoli (Tauziède-Espariat et al., 2019).	Olig2-negative, focally GFAP-positive, and nuclear accumulation of p53 (Tauziède-Espariat et al., 2019). The most frequent pontine subtype is the co-amplification of MYCN and ID2 or a *P53* mutation in the spectrum of Li–Fraumeni syndrome (Buczkowicz et al., 2014b; Tauziède-Espariat et al., 2023).
	Radio-induced gliomas	Characterized by a distinct cluster of DNA methylation (Ped*RTK1*), recurrent copy-number alterations of chromosome (Ch.) 1p loss/1q gain, Ch. 13q and Ch. 14q loss, and *PDGFRA* amplification (DeSisto et al., 2021).	
	Diffuse IDH-mutant gliomas	The majority of diffuse IDH-mutant gliomas in pontine location are astrocytomas, as oligodendroglioma IDH-mutant, 1p/19q-codeleted according to a strict molecular description, is limited to solitary case reports (Hodges et al., 2015; Aboubakr et al., 2023).	Molecularly enriched by non-R132H variants and IDH2 mutations (Banan et al., 2020; Aboubakr et al., 2023).
Benign gliomas	Pilocytic astrocytoma	Brainstem is one of the preferred sites after the cerebellum (Dangouloff-Ros et al., 2019; Ryall et al., 2020).	Presents with an enrichment in FGFR1 alterations and more generally non-KIAA1549::BRAF fusions (Ryall et al., 2020).
	Angiocentric glioma with MYB/MYBL1-alteration	Histologically, small bipolar cells are radially arranged around the vessels. Myxoid and microcysts can explain the significant intratumoral T1-weighted high intensity (Kurokawa et al., 2022).	Olig2-negative and can show EMA dot positivity (Reisz et al., 2023) without loss of H3K27 trimethylation. It presents with a fusion of *MYB* transcription factor with the tumor suppressor gene *QKI*.

	Astroblastoma, MN1-altered	The histologic appearance is less characteristic in these cases, with few astroblastic figures and GFAP-, S100-, and Olig2-positive cells (Smith-Cohn et al., 2021; Lucas et al., 2022).	MN1::BEND2 fusion occurs characteristically in young girls (average age 12 years) in a supratentorial location, whereas the rarest astroblastomas with EWSR1::BEND2 fusion occur in midline locations (brainstem, spine).
Malignant embryonal tumors	Atypical teratoid rhabdoid tumors (ATRT) Embryonal tumor with multilayered rosettes (ETMR)	Brainstem location accounts for 12% of ATRT cases (Lau et al., 2015), and from a large series of 97 ETMRs, 17% were in the brainstem (Korshunov et al., 2014).	The histologic and molecular characteristics of these two tumoral types justify the systematic INI1 and LIN28A immunostaining addition to the immunohistochemical panel when analyzing a brainstem tumor in an infant (Dangouloff-Ros et al., 2019).
Glio-neuronal tumors	Ganglioglioma	The percentage of cases with BRAF V600E mutation seems discretely higher (50%) in the midline than in other locations (Janjua et al., 2017). Ten percent of midline BRAF V600E-mutant gangliogliomas of an associated H3K27 mutation (Pagès et al., 2018), which could correspond to a new type of DMG with H3K27 and BRAF/FGFR1 co-alteration (Janjua et al., 2017).	
DIPG		The histopathologic hallmark of DIPG is a diffusely infiltrative glial pattern, without any particular perivascular or perineuronal tropism. They display small and monomorphic infiltrative cells but may incorporate giant, polymorphous, poorly differentiated, and epithelioid cells (Solomon et al., 2016). Around 10% of DIPGs lack mitotic features, microvascular proliferation, or necrosis (Buczkowicz et al., 2014a).	Five distinct DIPG subtypes: (1) H3.3K27-mutant, (2) H3.1/3.2K27-mutant, (3) H3-WT DIPG with EZHIP overexpression, (4) DIPG *EGFR*-mutant, and (5) DIPG H3K27 and BRAF/FGFR1 co-altered. The loss of an H3K27me3 immunoexpression is common to all five DIPG subtypes. No H3K27-mutant staining is observed in DIPG with H3K27i mutation, EZHIP overexpression, or EGFR-mutant. In the latter two tumors, the loss of nuclear trimethylation H327me3 immunoexpression is associated with an EZHIP overexpression (Antin et al., 2020; Castel et al., 2020; Ajuyah et al., 2023).

CLINICAL OVERVIEW OF DIFFUSE INTRINSIC PONTINE GLIOMA

DMG arises from glial cells found in the spinal cord, pons, midbrain, cerebellum, and thalamus (Tosi and Souweidane, 2024). In the rest of this chapter, we focus on DMGs found in the pons, also known as diffuse intrinsic pontine glioma (DIPG) (Barkley and Hauptman, 2022). With an estimated annual incidence of 200–300 DIPG cases per year in the United States, these tumors are the second most common pediatric brain malignancy (Tosi and Souweidane, 2024). Although DIPGs were reclassified by the World Health Organization in 2021 as a *diffuse midline glioma tumor with the H3K27M-altered*, DIPG is a term that still has significance and is often used in the clinical setting and more precisely reflects an imaging diagnosis rather than a distinct pathologic entity (Vitanza and Monje, 2019; Barkley and Hauptman, 2022; Vallero et al., 2023).

The location and biology of DIPGs make it one of the most challenging to treat brain tumors. DIPG has a median age of diagnosis of 6–8 years and a median survival of 8–11 months (Hargrave et al., 2006; Warren, 2020). Despite the many years of research dedicated to DIPG treatments, the survival outcomes in DIPG remain poor. In recent years, biopsies have led to molecular, genetic, and epigenetic findings in DIPG, giving hope through targeted therapies and clinical trials.

CLINICAL PRESENTATION

DIPG is most encountered in children younger than 10 years of age, with a median age of diagnosis of 6–8 years (Hargrave et al., 2006; Hoffman et al., 2018; Warren, 2020). Hoffman et al. in 2018 found that in over 50% of cases, patients diagnosed with DIPG presented with one or more of the following symptoms: cranial nerve palsies, pyramidal tract, or cerebellar signs. These three symptoms constitute the "classic" triad associated with DIPG (Vanan and Eisenstat, 2015). Cranial nerve palsy can manifest as facial weakness and asymmetry due to damage to cranial nerve VII and disconjugate gaze and diplopia possibly due to cranial nerve VI dysfunction (Schroeder et al., 2014; Johung and Monje, 2017). Pyramidal tract dysfunction can manifest as hyperreflexia, extremity weakness, and Babinski signs possibly because of disruption of long motor tracts through the pons by the tumor. Cerebellar signs such as ataxia, dysmetria, or dysarthria might be present due to impact of the tumor on cerebellopontine connections (Schroeder et al., 2014; Johung and Monje, 2017). In less than 10% of patients with DIPG, the tumor may cause obstructive hydrocephalus (Schroeder et al., 2014). At the time of progression, DIPG often disseminates throughout the brain and spine. Spinal dissemination at the time of diagnosis may occur, but its overall incidence remains unclear due to the lack of consistent spinal magnetic resonance imaging (MRI) assessments in all subjects.

Adult diffuse intrinsic pontine glioma

Adult DIPGs are less common, and adults with the H3K27M-altered tumor may be more commonly found in nonbrainstem structures, such as the thalamus, spinal cord, cerebellum, or other midline locations (Schulte et al., 2020). Adult brainstem gliomas account for 1%–2% of intracranial gliomas, while around 20% of primary pediatric brain tumors are DIPGs (Hu et al., 2016; Schulte et al., 2020). Although it has been suggested that adult DIPGs may have better outcomes than pediatric DIPGs, this is an area that requires further investigation (Meyronet et al., 2017; Kleinschmidt-DeMasters Mulcahy Levy, 2018; Schulte et al., 2020).

HISTORICAL VIEW ON DIFFUSE INTRINSIC PONTINE GLIOMA

One of the first intricately discussed cases of a pons tumor comes from a case report by Mary Putnam Jacobi from 1888. In her report titled "Case of a Probable Tumor of the Pons," she describes a patient as having headaches, trouble balancing and walking, weakness in the extremities, odd eye movements in the form of divergent strabismus, and drawling speech. These are all symptoms that are now known to be associated with DIPG. Three months later, the patient passed away (Jacobi, 1889).

As conveyed by her immensely detailed report of these cases, Dr. Jacobi seemed particularly fascinated by this unique type of tumor. Upon autopsy by Dr. Henry Hun, Dr. Jacobi and Dr. Hun describe the way the tumor infiltrated and encroached upon the pons. The appearance of the tumor was so intertwined with the pons that it appeared to be a hypertrophied pons (Jacobi, 1889). Dr. Jacobi and Dr. Hun's description closely mirrors that of what is now referred to as DIPG (Warren, 2020).

The location of the tumor has posed challenges for treatment since its discovery. Dr. Henry Cushing, often noted as the father of neurosurgery, wrote about tumors in the pons in a 1911 letter to Dr. Henry Christian, joking "I would like to see the day when somebody would be appointed surgeon somewhere who had no hands, for the operative part is the least part of the work" (Cushing, 2004; Kuzan-Fischer and Souweidane, 2019). Dr. Cushing foresaw the challenges of logistically operating on this tumor. In 1926, Dr. Wilfred Harris documented the first DIPG, describing this tumor as having infiltrative growth. He labeled this tumor a "pontine glioma" (Harris and Newcomb, 1926). The location of the tumor in the pons, in addition to the diffuse, intrinsic, infiltrative nature of DIPG, renders it nonresectable (Warren, 2020). Donald Matson wrote in 1969 for Neurosurgery of Infancy and Childhood, "regardless of specific histology, brainstem gliomas must be classified as malignant tumors since their location in itself renders them inoperable" (Matson and Ingraham, 1969; Kuzan-Fischer and Souweidane, 2019).

The first documented stereotactic biopsy of the brainstem was reported in 1978 (Gleason et al., 1978). Because of the limitations of CT imaging in diagnosis, stereotactic biopsies would be an important tool to diagnose tumors of the midbrain and pons. Soon thereafter, Reigal et al. in 1979 published a paper titled "Biopsy of pediatric brainstem tumors" in which three of the patients had lesions located in the pons (Reigel et al., 1979). CT scan-guided stereotactic surgery was believed to be the safest and most reliable method of obtaining a diagnosis for tumors of the pons (Coffey and Lunsford, 1985). However, the paradigm of brainstem biopsy would shift with the advent of MRI. In recent years, however, biopsy has been reintroduced into clinical practice and is now increasingly considered the standard of care for these patients to ensure the correct pathology classification (Kuzan-Fischer and Souweidane, 2019; Coleman et al., 2023).

IMAGING

MRI gave radiologists the ability to parse out the origin, location, diffuse versus focal, absence or presence of cysts, extent and direction, hemorrhage, hydrocephalus, and necrosis of a given lesion in the brainstem (Barkovich et al., 1990; Fischbein et al., 1996). In their review, Barkovich et al. found that diffuse pontine gliomas with these radiographic features had a poorer response to therapy than other groups, a finding that would hold true for the coming decades (Barkovich et al., 1990; Hargrave et al., 2006).

Radiologists are now able to diagnose DIPG using MRI techniques. On T2-weighted sequences, DIPG is typically homogeneously (72%) or heterogeneously (28%) hyperintense (Hargrave et al., 2008; Dang and Phillips, 2017; Warren, 2020) (Fig. 11.1). On T1-weighted imaging, the lesion appears mostly hypo- (92%) and isointense (8%) (Hargrave et al., 2008; Warren, 2020) (Fig. 11.2). With contrast, DIPG typically shows minimal enhancement and involves more than half of the pons (Tosi and Souweidane, 2023). Given this advancement in imaging techniques, the necessity of a biopsy to acquire a diagnosis and prognosis was diminished (Hargrave et al., 2006). In fact, Hargrave et al.'s paper articulates how a consensus was formed against the biopsy of a DIPG because the risks outweighed the benefits (Hargrave et al., 2006).

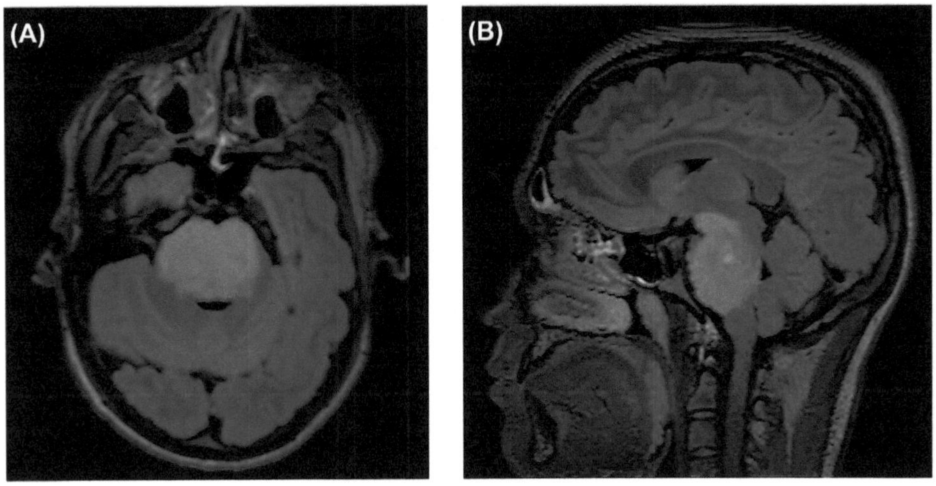

Figure 11.1. Typical T1-weighted imaging characteristics of diffuse intrinsic pontine glioma are predominantly hypointense (92%) and less commonly isointense (8%) (Hargrave et al., 2008; Warren, 2020). (**A**) T1-weighted imaging. (**B**) Image demonstrating a lesion which appears iso to mildly hypointense.

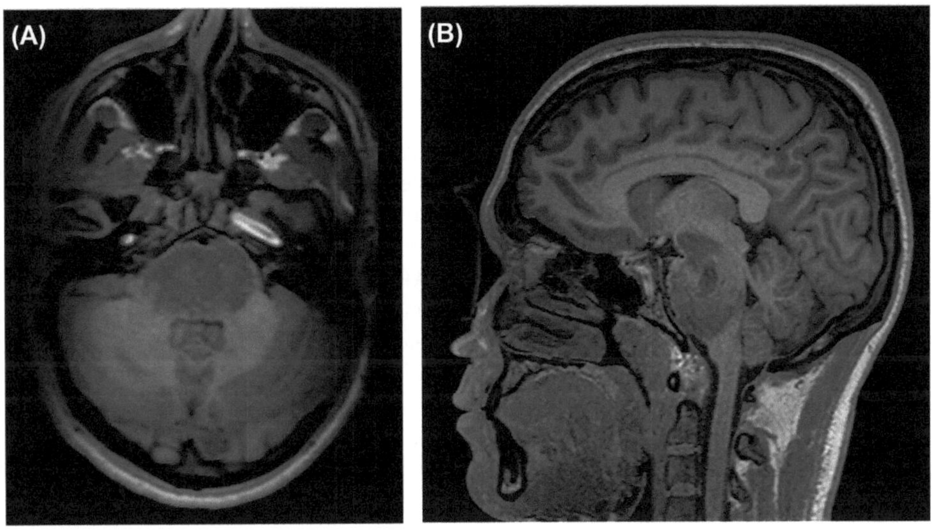

Figure 11.2. Typical T2-weighted imaging characteristics of diffuse intrinsic pontine glioma are homogeneously (72%) or heterogeneously (28%) hyperintense (Hargrave et al., 2008; Dang and Phillips, 2017; Warren, 2020). (**A**) T2 FLAIR demonstrates a heterogeneously FLAIR hyperintense mass. (**B**) Mass located in the pons with extension into the midbrain.

BIOPSY

Despite decades of hundreds of clinical trials, the prognosis for patients with DIPG remains poor, underscoring the urgent need for novel therapeutic strategies, which are now being informed by recent advances in molecular profiling. As molecular profiling of tumors advanced, the call to understand DIPG from a molecular perspective increased (Williams et al., 2020; Kline et al., 2022). However, in order to study this tumor, tissue samples were required. Thus, this reluctance to biopsy has slowly dissipated in the last decade. Multiple studies have found the biopsy of DIPG to be safe, feasible, and with low morbidity (Cage et al., 2013; Puget et al., 2015; Hamisch et al., 2017; Gupta et al., 2018; Williams et al., 2020). Puget et al., in their prospective analysis of 130 children biopsied with DIPG, observed transient morbidity in the form of worsening of neurologic deficit in 3.9% (5) of the cohort, zero mortalities, and a diagnostic yield of 100% (Puget et al., 2015). In their paper, they stated "Stereotactic biopsies of DIPG can be considered as a safe procedure in well-trained neurosurgical teams and could be incorporated in protocols," echoing the findings of Walker et al. (2013), Cage et al. (2013), Kickingereder et al. (2013), Hamisch et al. (2017), Gupta et al. (2018), and Puget et al., (2015). Although biopsies still carry risks, the biopsy of DIPG was recommended in international consensus for the continued biologic study of this tumor to understand and find targeted therapies (Walker et al., 2013). These samples from biopsies in combination with more advanced tools have allowed for pathologic, molecular, genetic, and epigenetic studies of the tumor including the defining H3K27M mutation of DIPG (Castel et al., 2015; Warren, 2020). Conventional therapy has failed thus far; biopsies and pathologic analyses of this tissue may be the key to improved outcomes for children diagnosed with pontine tumors, especially in the context that DIPGs consist of different subtypes.

CHEMOTHERAPY

In DIPG, the addition of chemotherapy has not improved outcomes compared to radiation alone (Hargrave et al., 2006; Robison and Kieran, 2014; Williams, 2020). The following list is some of the chemotherapies and combinations of chemotherapies that have been attempted: lomustine and vincristine; 5-fluorouracil and lomustine; cisplatin and cyclophosphamide; cisplatin and etoposide, vincristine, and cyclophosphamide; temozolomide; vincristine and etoposide; temozolomide and cis-retinoic acid; temozolomide and thalidomide; methotrexate; cisplatin, etoposide, vincristine, and ifosfamide; temozolomide, gefitinib, tipifarnib, and nimotuzumab; PEGylated interferon α-2b; bevacizumab; irinotecan/cetuximab; busulfan/thiotepa; and tamoxifen (Bouffet et al., 2000; Chassot et al., 2012; Warren, 2020; Tosi and Souweidane, 2023). None of these combinations has improved OS (Warren, 2020; Tosi and Souweidane, 2023). Given the negative outcome data from a Children's Oncology Group study on the addition of temozolomide to the treatment regimen for DIPG, it is often omitted in the United States (Cohen et al., 2011). It remained the standard of care in Europe, in part, because of data from International DIPG registries (Szychot et al., 2025). There remains debate, however, because the registries' data may be biased due to an ill-defined H3K27-altered DMG subgroup and non-randomized data (Szychot et al., 2025). In the 2025 European standard clinical practice recommendations for pediatric high-grade gliomas, palliative irradiation is noted as the standard of care (Szychot et al., 2025).

RADIATION THERAPY

Focal irradiation has been the only effective treatment to temporarily extend the length of survival (Halperin, 1985; Hargrave et al., 2006; Hoffman et al., 2018; Tinkle et al., 2020; Warren, 2020). Patients with DIPG who chose to forego radiation had a median OS of 4.6 months (140 days), whereas those who underwent radiation had a median OS of 8–11 months (Langmoen et al., 1991; Hargrave et al., 2006; Warren, 2020). Thus, radiation treatment has been and continues to be the standard of treatment for DIPG. The standard regimen using focal photon radiation therapy administers 54–60 Gy in 180–200 cGy fractions over around 6 weeks, 5 days per week (Hargrave et al., 2006; Gallitto et al., 2019; Warren, 2020; Kim and Suh, 2023). Doses greater than 60 Gy may be associated with toxicity without much benefit (Warren, 2020; Kim and Suh 2023). Doses less than 50 Gy may not be sufficient to elicit a response (Warren, 2020; Kim and Suh, 2023). Hyperfractionated regimens and proton therapy have not improved outcomes thus far (Gallitto et al., 2019; Kim and Suh, 2023). Hypofractionation, however, appears to be effective, showing similar outcomes to conventional fractionation in early studies, while also reducing the need for daily anesthesia during radiation therapy for these children, and is now in some centers the standard approach (Zaghloul et al., 2022; Mankuzhy et al., 2024). Re-irradiation has been shown to extend survival without significantly increasing toxicity (Cacciotti et al., 2021; Chavaz et al., 2022; Mankuzhy et al., 2023). Thus, re-irradiation is being integrated more frequently upon occurrence and has become a new standard of care at the time of progression at some institutions. In summary, focal radiation therapy is considered the standard of care for DIPG with developing evidence for re-irradiation and hypofractionated strategies.

INVESTIGATIONAL APPROACHES FOR THE TREATMENT OF DIFFUSE INTRINSIC PONTINE GLIOMA

There is tremendous interest in harnessing immunotherapy strategies for children with DIPG. In recent years, clinical trials have increasingly focused on assessing various immunotherapy approaches, including vaccines, chimeric antigen receptor (CAR) T cells, oncolytic viruses, and checkpoint inhibitors (Foster et al., 2023; Lin et al., 2024). This growing body of research is well reviewed in several articles. Notably, there is specific interest in CAR T cell strategies for DIPG, driven by promising preclinical data showing that CAR T cell therapy can cure mice with DIPG (Mount et al., 2018; Thomas et al., 2023).

However, these trials are predominantly in the phase 1 stage, with initial efforts focused primarily on safety. Further understanding is needed to elucidate why some patients benefit from these therapies while others do not. The integration of routine cerebrospinal fluid and tumor tissue collection, coupled with advanced technologies such as single-cell analysis and proteomics, is expected to provide significant insights and enable more precise stratification of patients. Further, as outlined above, DIPG consists of a heterogenous group of tumors (Mackay et al., 2017). Therefore, several groups have aimed to target the canonical histone mutation as well as some of the partnering mutations via several strategies including vaccine strategies or histone deacetylase inhibitors for the H3 mutations as well as other targeted inhibitors for other alterations (Lin et al., 2019; Mueller et al., 2020; Vitanza et al., 2021; Grassl et al., 2023a; Grassl et al., 2023b). Other strategies include targeting specific co-mutation that frequently occurs in the H3K27M-altered DMGs, such as PIK3CA, MAPK pathway alteration, or ACVR1, to name a few (Cao et al., 2020; Fortin et al.,

2020; Carvalho et al., 2022; Izquierdo et al., 2022; Dalle Ore et al., 2023; Jackson et al., 2023; Duchatel et al., 2024). Others have investigated metabolic targeting using imipridones such as ONC201, and initial studies show promising biomarker response as well as overall survival benefit (Chi et al., 2019; Venneti et al., 2023; Odia et al., 2024). Further emphasis is placed on optimizing strategies for effective drug delivery and enhancing penetration across the blood–brain barrier. These approaches include convection-enhanced delivery and, more recently, techniques utilizing focused ultrasound (Souweidane et al., 2018; Himes et al., 2019; Szychot et al., 2021; Martinez et al., 2023; Mueller et al., 2023; Parekh et al., 2023; Arms et al., 2024; Pandit-Taskar et al., 2024).

The prevailing consensus among neuro-oncology experts is that a multifaceted therapeutic strategy is essential to significantly improve the currently poor prognosis associated with these tumors. Furthermore, it is recognized that outcome data must be evaluated in the context of the tumor's molecular signature, anatomic location, and the patient's age.

CONCLUSION

Despite significant advances in elucidating the molecular and genetic underpinnings of brainstem tumors including DIPG, there has been minimal progress in improving clinical outcomes for these patients. The complexity of brainstem tumor pathology, compounded by the integration of molecular characteristics into the broader classification schema, underscores the increasing importance of tissue-based diagnostics—particularly in the context of evaluating novel therapeutic strategies in clinical trials. While ongoing trials are investigating diverse treatment modalities—especially for DIPG—there remains cautious optimism for the future. Over the last 15 years, researchers have made considerable strides in deciphering the intricate biology which ultimately will lead to better therapies.

References

Aboubakr O, Métais A, Hasty L et al. (2023). Brainstem oligodendroglioma, IDH-mutant, and 1P/19Q-codeleted: a potential diagnostic pitfall. J Neuropathol Exp Neurol 82: 958–962.

Aboubakr O, Métais A, Maillard J et al. (2024). Utility of combining OLIG2 and SOX10 IHC expression in CNS tumours: promising biomarkers for subtyping paediatric- and adult-type gliomas. Histopathology 84: 893–899.

Ajuyah P, Mayoh C, Lau L et al. (2023). Histone H3-wild type diffuse midline gliomas with H3K27me3 loss are a distinct entity with exclusive EGFR or ACVR1 mutation and differential methylation of homeobox genes. Sci Rep 13: 3775.

Antin C, Tauziède-Espariat A, Debily M-A et al. (2020). EZHIP is a specific diagnostic biomarker for posterior fossa ependymomas, group PFA and diffuse midline gliomas H3-WT with EZHIP overexpression. Acta Neuropathol Commun 8: 183.

Arms L, Duchatel R, Jackson E et al. (2024). Current status and advances to improving drug delivery in diffuse intrinsic pontine glioma. J Control Release 370: 835–865.

Auffret L, Ajlil Y, Tauziède-Espariat A et al. (2024). A new subtype of diffuse midline glioma, H3 K27 and BRAF/FGFR1 co-altered: a clinico-radiological and histomolecular characterisation. Acta Neuropathol 147: 2.

Banan R, Stichel D, Bleck A et al. (2020). Infratentorial IDH-mutant astrocytoma is a distinct subtype. Acta Neuropathol 140: 569–581.

Barkley A, Hauptman JS (2022). Pediatric brainstem tumors. Pediatric neurosurgery for clinicians. Cham: Springer International Publishing, pp. 327–344.

Barkovich AJ, Krischer J, Kun LE et al. (1990). Brain stem gliomas: a classification system based on magnetic resonance imaging. Pediatr Neurosurg 16: 73–83.

Bouffet E, Raquin M, Doz F et al. (2000). Radiotherapy followed by high dose busulfan and thiotepa : a prospective assessment of high dose chemotherapy in children with diffuse pontine gliomas. Cancer 88: 685–692.

Buczkowicz P, Bartels U, Bouffet E et al. (2014a). Histopathological spectrum of paediatric diffuse intrinsic pontine glioma: diagnostic and therapeutic implications. Acta Neuropathol 128: 573–581.

Buczkowicz P, Hoeman C, Rakopoulos P et al. (2014b). Genomic analysis of diffuse intrinsic pontine gliomas identifies three molecular subgroups and recurrent activating ACVR1 mutations. Nat Genet 46: 451–456.

Cacciotti C, Liu K, Haas-Kogan D et al. (2021). Reirradiation practices for children with diffuse intrinsic pontine glioma. Neurooncol Pract 8: 68–74.

Cage T, Samagh S, Mueller S et al. (2013). Feasibility, safety, and indications for surgical biopsy of intrinsic brainstem tumors in children. Childs Nerv Syst 29: 1313–1319.

Cao H, Jin M, Gao M et al. (2020). Differential kinase activity of ACVR1 G328V and R206H mutations with implications to possible TβRI cross-talk in diffuse intrinsic pontine glioma. Sci Rep 10: 6140.

Carvalho D, Richardson P, Olaciregui N et al. (2022). Repurposing Vandetanib plus Everolimus for the treatment of ACVR1-mutant diffuse intrinsic pontine glioma. Cancer Discov 12: 416–431.

Castel D, Kergrohen T, Tauziède-Espariat A et al. (2020). Histone H3 wild-type DIPG/DMG overexpressing EZHIP extend the spectrum diffuse midline gliomas with PRC2 inhibition beyond H3-K27M mutation. Acta Neuropathol 139: 1109–1113.

Castel D, Philippe C, Calmon R et al. (2015). Histone H3F3A and HIST1H3B K27M mutations define two subgroups of diffuse intrinsic pontine gliomas with different prognosis and phenotypes. Acta Neuropathol 130: 815–827.

Chassot A, Canale S, Varlet P et al. (2012). Radiotherapy with concurrent and adjuvant temozolomide in children with newly diagnosed diffuse intrinsic pontine glioma. Neuro-Oncology 106: 399–407.

Chavaz L, Janssens G, Bolle S et al. (2022). Neurological symptom improvement after re-irradiation in patients with diffuse intrinsic pontine glioma: a retrospective analysis of the SIOP-E-HGG/DIPG Project. Front Oncol 12:926196.

Chi A, Tarapore R, Hall M et al. (2019). Pediatric and adult H3 K27M-mutant diffuse midline glioma treated with the selective DRD2 antagonist ONC201. Neuro-Oncology 145: 97–105.

Chiang J, Diaz A, Makepeace L et al. (2020). Clinical, imaging, and molecular analysis of pediatric pontine tumors lacking characteristic imaging features of DIPG. Acta Neuropathol Commun 8: 57.

Coffey RJ, Lunsford DL (1985). Stereotactic surgery for mass lesions of the midbrain and pons. Neurosurgery 17: 12–18.

Cohen K, Heideman R, Zhou T et al. (2011). Temozolomide in the treatment of children with newly diagnosed diffuse intrinsic pontine gliomas: a report from the Children's Oncology Group. Neuro-Oncology 13: 410–416.

Coleman C, Chen K, Lu A et al. (2023). Interdisciplinary care of children with diffuse midline glioma. Neoplasia 35: 100851.

Cushing H (2004). No hands [Letter from 1911]. BMJ 329: 374.

Dalle Ore C, Coleman C, Gupta N et al. (2023). Advances and clinical trials update in the treatment of diffuse intrinsic pontine gliomas. Pediatr Neurosurg 58: 259–266.

Dang M, Phillip C (2017). Pediatric brain tumors. Continuum 23: 1727–1757.

Dangouloff-Ros V, Tauziède-Espariat A, Roux C-J et al. (2019). CT and multimodal MR imaging features of embryonal tumors with multilayered rosettes in children. AJNR 40: 732–736.

D'Aronco L, Rouleau C, Gayden T et al. (2017). Brainstem angiocentric gliomas with MYB–QKI rearrangements. Acta Neuropathol 134: 667–669.

DeSisto J, Lucas JJ, Xu K et al. (2021). Comprehensive molecular characterization of pediatric radiation-induced high-grade glioma. Nat Commun 12: 5531.

Duchatel R, Jackson E, Parackal S et al. (2024). PI3K/mTOR is a therapeutically targetable genetic dependency in diffuse intrinsic pontine glioma. J Clin Investig 134: e170329.

Filbin M, Tirosh I, Hovestadt V et al. (2018). Developmental and oncogenic programs in H3 K27M gliomas dissected by single-cell RNA-seq. Science 360: 331–335.

Fischbein NJ, Prados MD, Wara W, Russo C, Edwards MSB, Barkovich J (1996). Radiologic classification of brain stem tumors: correlation of magnetic resonance imaging appearance with clinical outcome. Pediatr Neurosurg 24: 9–23.

Fortin J, Tian R, Zarrabi I et al. (2020). Mutant ACVR1 arrests glial cell differentiation to drive tumorigenesis in pediatric gliomas. Cancer Cell 37: 308–323.

Foster J, Alonso M, Sayour E et al. (2023). Translational considerations for immunotherapy clinical trials in pediatric neuro-oncology. Neoplasia 42: 100909.

Gallitto M, Lazarev S, Wasserman I et al. (2019). Role of radiation therapy in the management of diffuse intrinsic pontine glioma: a systematic review. Adv Radiat Oncol 4: 520–531.

Gleason CA, Wise BL, Feinstein B (1978). Stereotactic localization (with computerized tomographic scanning), biopsy, and radiofrequency treatment of deep brain lesions. Neurosurgery 2: 217–222.

Grassl N, Poschke I, Lindner K (2023a). A H3K27M-targeted vaccine in adults with diffuse midline glioma. Nat Med 29: 2586–2592.

Grassl N, Sahm K, Süße H et al. (2023b). INTERCEPT H3: a multicenter phase I peptide vaccine trial for the treatment of H3-mutated diffuse midline gliomas. Neurol Res Pract 5: 55.

Gupta N, Goumnerova LC, Manley P et al. (2018). Prospective feasibility and safety assessment of surgical biopsy for patients with newly diagnosed diffuse intrinsic pontine glioma. Neuro-Oncology 20: 1547–1555.

Halperin EC (1985). Pediatric brain stem tumors: patterns of treatment failure and their implications for radiotherapy. Int J Radiat Oncol Biol Phys 11: 1293–1298.

Hamisch C, Kickingereder P, Fischer M et al. (2017). Update on the diagnostic value and safety of stereotactic biopsy for pediatric brainstem tumors: a systematic review and meta-analysis of 735 cases. J Neurosurg Pediatr 20: 261–268.

Han HJ, Jain P, Resnick AC (2018). Shared ACVR1 mutations in FOP and DIPG: opportunities and challenges in extending biological and clinical implications across rare diseases. Bone 109: 91–100.

Hargrave D, Bartels U, Bouffet E (2006). Diffuse brainstem glioma in children: critical review of clinical trials. Lancet Oncol 7: 241–248.

Hargrave D, Chuang N, Bouffet E (2008). Conventional MRI cannot predict survival in childhood diffuse intrinsic pontine glioma. Neuro-Oncology 86: 313–319.

Harris W, Newcomb WD (1926). A case of pontine glioma, with special reference to the paths of gustatory sensation. J R Soc Med 19: 1–5.

Himes BT, Zhang L, Daniels DJ (2019). Treatment strategies in diffuse midline gliomas with the H3K27M mutation: the role of convection-enhanced delivery in overcoming anatomic challenges. Front Oncol 931: 31.

Hodges SD, Malafronte P, Gilhooly J et al. (2015). Rare brainstem oligodendroglioma in an adult patient: presentation, molecular characteristics and treatment response. J Neurol Sci 355: 209–210.

Hoffman LM, Veldhuijzen van Zanten SEM, Colditz N et al. (2018). Clinical, radiologic, pathologic, and molecular characteristics of long-term survivors of diffuse intrinsic pontine glioma (DIPG): a collaborative report from the International and European Society for Pediatric Oncology DIPG Registries. J Clin Oncol 36: 1963–1972.

Hu J, Western S, Kesari S (2016). Brainstem glioma in adults. Front Oncol 6: 180.

Hübner JM, Müller T, Papageorgiou DN et al. (2019). EZHIP/CXorf67 mimics K27M mutated oncohistones and functions as an intrinsic inhibitor of PRC2 function in aggressive posterior fossa ependymoma. Neuro-Oncology 21: 878–889.

Izquierdo E, Carvalho DM, Mackay A et al. (2022). DIPG harbors alterations targetable by MEK inhibitors, with acquired resistance mechanisms overcome by combinatorial inhibition. Cancer Discov 12: 712–729.

Jackson ER, Duchatel RJ, Staudt DE et al. (2023). ONC201 in combination with paxalisib for the treatment of H3K27-altered diffuse midline glioma. Cancer Res 83: 2421–2437.

Jacobi MP (1889). Case of probable tumor of the pons. J Nerv Ment Dis 14: 115–129.

Janjua MB, Ivasyk I, Pisapia DJ et al. (2017). Ganglioglioma of brain stem and cervicomedullary junction: a 50 years review of literature. J Clin Neurosci 44: 34–46.

Johung TB, Monje M (2017). Diffuse intrinsic pontine glioma: new pathophysiological insights and emerging therapeutic targets. Curr Neuropharmacol 15: 88–97.

Kickingereder P, Willeit P, Simon T et al. (2013). Diagnostic value and safety of stereotactic biopsy for brainstem tumors: a systematic review and meta-analysis of 1480 cases. Neurosurgery 72: 873–882.

Kim HJ, Suh CO (2023). Radiotherapy for diffuse intrinsic pontine glioma: insufficient but indispensable. Brain Tumor Res Treat 11: 79–85.

Kleinschmidt-DeMasters BK, Mulcahy Levy JM (2018). H3 K27M-mutant gliomas in adults vs. children share similar histological features and adverse prognosis. Clin Neuropathol 37: 53–63.

Klimo PJ, Nesvick CL, Broniscer A et al. (2016). Malignant brainstem tumors in children, excluding diffuse intrinsic pontine gliomas. J Neurosurg Pediat 17: 57–65.

Kline C, Jain P, Kilburn L et al. (2022). Upfront biology-guided therapy in diffuse intrinsic pontine glioma: therapeutic, molecular, and biomarker outcomes from PNOC003. Clin Cancer Res 28: 3965–3978.

Korshunov A, Sturm D, Ryzhova M et al. (2014). Embryonal tumor with abundant neuropil and true rosettes (ETANTR), ependymoblastoma, and medulloepithelioma share molecular similarity and comprise a single clinicopathological entity. Acta Neuropathol 128: 279–289.

Kurokawa R, Baba A, Emile P et al. (2022). Neuroimaging features of angiocentric glioma: a case series and systematic review. J Neuroimaging 32: 389–399.

Kuzan-Fischer CM, Souweidane MM (2019). The intersect of neurosurgery with diffuse intrinsic pontine glioma: JNSPG 75th Anniversary Invited Review Article. J Neurosurg Pediatr 24: 611–621.

Langmoen IA, Lundar T, Storm-Mathisen I et al. (1991). Management of pediatric pontine gliomas. Childs Nerv Syst 7: 13–15.

Lau CS, Mahendraraj K, Chamberlain RS (2015). Atypical teratoid rhabdoid tumors: a population-based clinical outcomes study involving 174 patients from the Surveillance, Epidemiology, and End Results database (1973-2010). Cancer Manag Res 7: 301–309.

Lewis PW, Müller MM, Koletsky MS et al. (2013). Inhibition of PRC2 activity by a gain-of-function H3 mutation found in pediatric glioblastoma. Science 340: 857–861.

Lin C, Smith C, Rutka J (2024). Current immunotherapeutic approaches to diffuse intrinsic pontine glioma. Front Genet 15: 1349612.

Lin GL, Wilson KM, Ceribelli M et al. (2019). Therapeutic strategies for diffuse midline glioma from high-throughput combination drug screening. Sci Transl Med 11: eaaw0064.

Louis DN, Giannini C, Capper D et al. (2018). cIMPACT-NOW update 2: diagnostic clarifications for diffuse midline glioma, H3 K27M-mutant and diffuse astrocytoma/anaplastic astrocytoma, IDH-mutant. Acta Neuropathol 135: 639–642.

Lucas CG, Gupta R, Wu J et al. (2022). EWSR1-BEND2 fusion defines an epigenetically distinct subtype of astroblastoma. Acta Neuropathol 143: 109–113.

Mackay A, Burford A, Carvalho D et al. (2017). Integrated molecular meta-analysis of 1,000 pediatric high-grade and diffuse intrinsic pontine glioma. Cancer Cell 32: 520–537.

Mankuzhy NP, Tringale KR, Dunkel IJ et al. (2024). Hypofractionated re-irradiation for diffuse intrinsic pontine glioma. Pediatr Blood Cancer 71:e30929.

Mankuzhy NP, Tringale KR, Farouk Sait S et al. (2023). Re-irradiation for diffuse intrinsic pontine glioma with 3 Gy per fraction. Int J Radiat Oncol Biol Phys 117: e530.

Martinez P, Nault G, Steiner J et al. (2023). MRI-guided focused ultrasound blood–brain barrier opening increases drug delivery and efficacy in a diffuse midline glioma mouse model. Neuro-Oncol Adv 5: vdad111.

Matson DD, Ingraham FD (eds.) (1969). Neurosurgery of Infancy and Childhood. Second edition. Springfield, IL: Thomas.

Meyronet D, Esteban-Mader M, Bonnet C et al. (2017). Characteristics of H3 K27M-mutant gliomas in adults. Neuro-Oncology 19: 1127–1134.

Mondal G, Lee JC, Ravindranathan A et al. (2020). Pediatric bithalamic gliomas have a distinct epigenetic signature and frequent EGFR exon 20 insertions resulting in potential sensitivity to targeted kinase inhibition. Acta Neuropathol 139: 1071–1088.

Mount CW, Majzner RG, Sundaresh S et al. (2018). Potent antitumor efficacy of anti-GD2 CAR T cells in H3-K27M^{+} diffuse midline gliomas. Nat Med 24: 572–579.

Mueller S, Kline C, Stoller S et al. (2023). PNOC015: repeated convection-enhanced delivery of MTX110 (aqueous panobinostat) in children with newly diagnosed diffuse intrinsic pontine glioma. Neuro-Oncology 25: 2074–2086.

Mueller S, Taitt JM, Villanueva-Meyer JE et al. (2020). Mass cytometry detects H3.3K27M-specific vaccine responses in diffuse midline glioma. J Clin Investig 130: 6325–6337.

Nikbakht H, Panditharatna E, Mikael LG et al. (2016). Spatial and temporal homogeneity of driver mutations in diffuse intrinsic pontine glioma. Nat Commun 7: 11185.

Odia Y, Hall MD, Cloughesy TF et al. (2024). Selective DRD2 antagonist and ClpP agonist ONC201 in a recurrent non-midline H3 K27M-mutant glioma cohort. Neuro-Oncology 26: S165–S172.

Oushy S, Perry A, Graffeo CS et al. (2020). Pediatric ganglioglioma of the brainstem and cervicomedullary junction: a retrospective cohort study. J Neurosurg Pediatr 25: 30–36.

Pagès M, Beccaria K, Boddaert N et al. (2018). Co-occurrence of histone H3 K27M and BRAF V600E mutations in paediatric midline grade I ganglioglioma. Brain Pathol 28: 103–111.

Pandit-Taskar N, Zanzonico PB, Grkovski M et al. (2024). Theranostic intratumoral convection-enhanced delivery of ^{124}I-Omburtamab in patients with diffuse intrinsic pontine glioma: pharmacokinetics and lesion dosimetry. J Nucl Med 65: 1364–1370.

Parekh K, LeBlang S, Nazarian J et al. (2023). Past, present and future of Focused Ultrasound as an adjunct or complement to DIPG/DMG therapy: a consensus of the 2021 FUSF DIPG meeting. Neoplasia 37: 100876.

Puget S, Beccaria K, Blauwblomme T et al. (2015). Biopsy in a series of 130 pediatric diffuse intrinsic Pontine gliomas. Childs Nerv Syst 31: 1773–1780.

Reigel DH, Scarff TB, Woodford JE (1979). Biopsy of pediatric brain stem tumors. Childs Brain 5: 329–340.

Reisz Z, Radics BL, Nemes P et al. (2023). Case report: brainstem angiocentric glioma presenting in a toddler child-diagnostic and therapeutic challenges. Pathol Oncol Res 29: 1611231.

Robison NJ, Kieran MW (2014). Diffuse intrinsic pontine glioma: a reassessment. Neuro-Oncology 119: 7–15.

Ryall S, Zapotocky M, Fukuoka K et al. (2020). Integrated molecular and clinical analysis of 1,000 pediatric low-grade gliomas. Cancer Cell 37: 569–583.

Sato H, Terakawa Y, Tsuyuguchi N et al. (2015). Embryonal tumor with abundant neuropil and true rosettes in the brainstem: case report. J Neurosurg Pediatr 16: 291–295.

Schroeder KM, Hoeman CM, Becher OJ (2014). Children are not just little adults: recent advances in understanding of diffuse intrinsic pontine glioma biology. Pediatr Res 75: 205–209.

Schulte JD, Buerki RA, Lapointe S et al. (2020). Clinical, radiologic, and genetic characteristics of histone H3 K27M-mutant diffuse midline gliomas in adults. Neuro-Oncol Adv 2: vdaa142.

Schwartzentruber J, Korshunov A, Liu XY et al. (2012). Driver mutations in histone H3.3 and chromatin remodelling genes in paediatric glioblastoma. Nature 482: 226–231.

Sievers P, Sill M, Schrimpf D et al. (2021). A subset of pediatric-type thalamic gliomas share a distinct DNA methylation profile, H3K27me3 loss and frequent alteration of EGFR. Neuro-Oncology 23: 34–43.

Smith-Cohn MA, Abdullaev Z, Aldape KD et al. (2021). Molecular clarification of brainstem astroblastoma with EWSR1-BEND2 fusion in a 38-year-old man. Free Neuropathol 2: 16.

Solomon DA, Wood MD, Tihan T et al. (2016). Diffuse midline gliomas with histone H3-K27M mutation: a series of 47 cases assessing the spectrum of morphologic variation and associated genetic alterations. Brain Pathol 26: 569–580.

Souweidane MM, Kramer K, Pandit-Taskar N et al. (2018). Convection-enhanced delivery for diffuse intrinsic pontine glioma: a single-centre, dose-escalation, phase 1 trial. Lancet Oncol 19: 1040–1050.

Szychot E, Giraud G, Hargrave D et al. (2025). European standard clinical practice recommendations for paediatric high-grade gliomas. EJC Paediatric Oncology 5: 100210.

Szychot E, Walker D, Collins P et al. (2021). Clinical experience of convection-enhanced delivery (CED) of carboplatin and sodium valproate into the pons for the treatment of diffuse intrinsic pontine glioma (DIPG) in children and young adults after radiotherapy. Int J Clin Oncol 26: 647–658.

Tauziède-Espariat A, Debily MA, Castel D et al. (2019). An integrative radiological, histopathological and molecular analysis of pediatric pontine histone-wildtype glioma with MYCN amplification (HGG-MYCN). Acta Neuropathol Commun 7: 87.

Tauziède-Espariat A, Uro-Coste E, Nicaise Y et al. (2023). Refinement of diagnostic criteria for pediatric-type diffuse high-grade glioma, IDH- and H3-wildtype, MYCN-subtype including histopathology, TP53, MYCN and ID2 status. Acta Neuropathol Commun 11: 1–170.

Thomas BC, Staudt DE, Douglas AM et al. (2023). CAR T cell therapies for diffuse midline glioma. Trends Cancer 9: 791–804.

Tinkle CL, Simone B, Chiang J et al. (2020). Defining optimal target volumes of conformal radiation therapy for diffuse intrinsic pontine glioma. Int J Radiat Oncol Biol Phys 106: 838–847.

Tosi U, Souweidane M (2023). Fifty years of DIPG: looking at the future with hope. Childs Nerv Syst 39: 2675–2686.

Tosi U, Souweidane M (2024). Diffuse midline gliomas: challenges and new strategies in a changing clinical landscape. Cancers 16: 219.

Trkova K, Sumerauer D, Krskova L et al. (2023). DIPG-like MYB-altered diffuse astrocytoma with durable response to intensive chemotherapy. Childs Nerv Syst 39: 2509–2513.

Udaka YT, Shayan K, Chuang NA et al. (2013) Atypical presentation of atypical teratoid rhabdoid tumor in a child. Case Rep Oncol Med 2013: 815923.

Vallero SG, Bertero L, Morana G et al. (2023). Pediatric diffuse midline glioma h3k27-altered: a complex clinical and biological landscape behind a neatly defined tumor type. Front Oncol 12.

Vanan MI, Eisenstat DD (2015). DIPG in children – what can we learn from the past? Front Oncol 5: 237.

Venneti S, Kawakibi AR, Ji S et al. (2023). Clinical efficacy of ONC201 in H3K27M-mutant diffuse midline gliomas is driven by disruption of integrated metabolic and epigenetic pathways. Cancer Discov 13: 2370–2393.

Venneti S, Santi M, Felicella MM et al. (2014). A sensitive and specific histopathologic prognostic marker for H3F3A K27M mutant pediatric glioblastomas. Acta Neuropathol 128: 743–753.

Vitanza NA, Biery MC, Myers C et al. (2021). Optimal therapeutic targeting by HDAC inhibition in biopsy-derived treatment-naïve diffuse midline glioma models. Neuro-Oncology 23: 376–386.

Vitanza NA, Monje M (2019). Diffuse intrinsic pontine glioma: from diagnosis to next-generation clinical trials. Curr Treat Options Neurol 21: 37.

Walker DA, Liu J, Kieran M et al. (2013). A multi-disciplinary consensus statement concerning surgical approaches to low-grade, high-grade astrocytomas and diffuse intrinsic pontine gliomas in childhood (CPN Paris 2011) using the Delphi method. Neuro-Oncology 15: 462–468.

Wang Y, Pan C, Xie M et al. (2022). Adult diffuse intrinsic pontine glioma: clinical, radiological, pathological, molecular features, and treatments of 96 patients. J Neurosurg 137: 1628–1638.

Warren KE (2020). Radiation and chemotherapy for brainstem tumors. In: Jallo G, Noureldine M, Shimony N (eds.). Brainstem tumors. Cjam: Springer.

Werbrouck C, Evangelista CCS, Lobón-Iglesias MJ et al. (2019). TP53 pathway alterations drive radioresistance in diffuse intrinsic pontine gliomas (DIPG). Clin Cancer Res 25: 6788–6800.

Williams JR, Young CC, Vitanza NA et al. (2020). Progress in diffuse intrinsic pontine glioma: advocating for stereotactic biopsy in the standard of care. Neurosurg Focus 48: E4.

Wu G, Broniscer A, McEachron TA et al. (2012). Somatic histone H3 alterations in pediatric diffuse intrinsic pontine gliomas and non-brainstem glioblastomas. Nat Genet 44: 251–253.

Zaghloul MS, Nasr A, Tolba M et al. (2022). Hypofractionated radiation therapy for diffuse intrinsic pontine glioma: a noninferiority randomized study including 253 children. Int J Radiat Oncol Biol Phys 113: 360–368.

Further reading

Mathisen JR, Giunta F, Marini G et al. (1987). Transcerebellar biopsy in the posterior fossa: 12 years experience. Surg Neurol 28: 100–104.

Ostrom QT, Gittleman H, Truitt G et al. (2018). CBTRUS statistical report: primary brain and other central nervous system tumors diagnosed in the United States in 2011-2015. Neuro-Oncology 20: iv1–iv86.

Handbook of Clinical Neurology, Vol. 216 (3rd series)
The Brainstem
L.T. Grinberg and A.D. Korczyn
DOI: https://doi.org/10.1016/B978-0-443-15736-3.00010-X

Chapter 12

Autoimmune maladies of the brainstem

KATSUICHI MIYAMOTO[1] AND SUSUMU KUSUNOKI[2,*]

[1]*Department of Neurology, Wakayama Medical University, Wakayama, Japan*

[2]*Department of Neurology, Faculty of Medicine, Kindai University, Osaka, Japan*

Abstract

There are numerous autoimmune diseases that present with brainstem involvement including multiple sclerosis, neuromyelitis optica spectrum disorders, myelin oligodendrocyte glycoprotein antibody-associated disease, acute disseminated encephalomyelitis, Bickerstaff brainstem encephalitis, chronic lymphocytic inflammation with pontine perivascular enhancement responsive to steroids (CLIPPERS), autoimmune encephalitis/encephalopathy, systemic lupus erythematosus, Sjögren's syndrome, neuro-sarcoidosis, and neuro-Behçet's disease. Symptoms of these diseases depend on the mechanisms, site and extent of the lesions. MRI images and autoantibodies in the blood are useful in diagnosing these diseases. Immunotherapy is the common treatment, though the response to treatment varies from disease to disease. In recent years, biologics have been applied for maintenance therapy.

INTRODUCTION

Many diseases cause brainstem involvement through autoimmune mechanisms. They are called as autoimmune disorders because the laboratory data, such as autoantibodies, suggest immunologic pathogenesis and immunotherapies are effective. In addition to diseases that are primary to the brainstem, there are also diseases in which the brainstem is involved as a manifestation of systemic neuroimmunologic diseases. There are various pathologic conditions, and symptoms include impaired consciousness, ataxia, diplopia, abnormal sensation in the face and oral cavity, facial paralysis, hearing loss, dysphagia, and weakness in the limb muscles. Diagnosis is made by CT and MRI imaging, blood tests, spinal fluid analysis, electroencephalography, and electrophysiologic studies. Because autoimmune maladies of the brainstem often respond to immunotherapy, early diagnosis and treatment are important. A summary of the representative diseases is shown in Table 12.1.

MULTIPLE SCLEROSIS

Overview

Multiple sclerosis (MS) is a chronic inflammatory demyelinating disease of the central nervous system (CNS) characterized by lesions of dissemination in time and dissemination in space. The diagnosis is usually confirmed by a detailed history and neurologic examination showing time and spaces of dissemination of lesions, and by exclusion of other diseases (Compston and Coles, 2008). The lesions are mainly located around the lateral ventricles of the cerebrum, but brainstem lesions are also frequently seen. They can sometimes be the initial symptom of MS (Habek, 2013).

Pathophysiology

Although the detailed pathogenesis of MS is still unclear, the presence of lymphocyte and macrophage infiltration in the lesions suggests that demyelination is caused by inflammation through an autoimmune mechanism. It is also assumed that genetic factors, such as racial differences, are involved, and that environmental factors such as sunlight hours and viral infections influence the pathogenesis of the disease.

Symptoms

Symptoms depend on the site of the lesion. Symptoms of brainstem lesions include diplopia, abnormal facial and oral sensations, trigeminal neuralgia, and dysphagia.

*Correspondence to: Susumu Kusunoki, Department of Neurology, Kindai University Faculty of Medicine, Osaka, Japan. E-mail: kusunoki@med.kindai.ac.jp.

Table 12.1

Summary of autoimmune maladies affecting brainstem.

	MS	NMOSD	MOGAD	ADEM	GFAP ab-associated disease	BBE
Clinical characteristics	DIT and DIS	Area postrema syndrome, optic neuritis (chiasma), LETM	Optic neuritis, LETM, encephalitis	Onset after infection or vaccination	Meningoencephalitis, psychiatric symptoms, seizures	Ophthalmoplegia, ataxia, impaired consciousness
Laboratory data	Oligoclonal IgG band	AQP4 ab	MOG ab	–	GFAP ab	GQ1b ab
Treatment in the acute phase	Steroid pulse, PE	Steroid pulse, IVIg, PE	Steroid pulse, IVIg, PE	Steroid pulse, IVIg, PE	steroid pulse, IVIg, PE	IVIg, PE, steroids
Maintenance therapy	IFNβ, GA, DMF, fingolimod, siponimod, natalizumab, ofatumumab, etc.	Steroids, ISD, rituximab, inebilizumab, satralizumab, eculizumab, ravulizumab, etc.	Steroids, ISD, rituximab, etc.	In case of recurrence: steroids, ISD	Steroids, ISD, rituximab etc.	–

ab, antibody; *ADEM*, acute disseminated encephalomyelitis; *AQP4*, aquaporin 4; *BBE*, Bickerstaff brainstem encephalitis; *DIT*, dissemination in time; *DIS*, dissemination in space; *DMF*, dimethyl fumarate; *GA*, glatiramer acetate; *GFAP*, glial fibrillary acidic protein; *IFNβ*, interferon β; *ISD*, immunosuppressive drugs; *IVIg*, immunoglobulin therapy; *LETM*, longitudinally extensive transverse myelitis; *MOG*, myelin oligodendrocyte glycoprotein; *MOGAD*, myelin oligodendrocyte glycoprotein antibody-associated disease; *MS*, multiple sclerosis; *NMOSD*, neuromyelitis optica spectrum disorders; *PE*, plasma exchange.

Diagnosis

McDonald's diagnostic criteria are often used after excluding other diseases (Thompson et al., 2018). Cerebrospinal fluid (CSF) findings are often normal, but during the exacerbation phase, a mild increase in the number of cells, mainly lymphocytes, and total protein level may be seen. In the presence of active foci, an elevated IgG index and oligoclonal IgG bands may be seen (Tintore et al., 2010). MRI images show T2 high-signal plaques in the white matter (Fig. 12.1). Auditory brainstem response is a useful tool to investigate brainstem lesions in MS.

Treatment

Treatment of MS consists of treatment during the acute exacerbation phase, relapse prevention and progression prevention, symptomatic treatment during the acute and chronic phases, and rehabilitation. In the acute phase, high-dose intravenous steroid therapy (steroid pulse therapy) and plasma exchange are administered. Disease-modifying drugs approved for prevention of recurrence include interferon β, glatiramer acetate, dimethyl fumarate, fingolimod, siponimod, natalizumab, and ofatumumab (Gholamzad et al., 2019; Hauser and Cree, 2020).

Prognosis

MS has a long course with repeated relapses and remissions. The early stages of the disease are usually relapsing-remitting, but as the disease lengthens, it shifts to a secondary progressive form, in which physical disability progresses even in the absence of relapses.

NEUROMYELITIS OPTICA SPECTRUM DISORDERS

Overview

Neuromyelitis optica spectrum disorder (NMOSD) is a CNS autoimmune disease caused by anti-aquaporin 4 (AQP4) antibodies. It was previously considered a subtype of MS, but is now classified as a separate and independent disease (Lennon et al., 2005). The optic nerves and spinal cord are mainly injured, but intracranial lesions are also common, with brainstem lesions present in one-third of NMOSD cases.

Pathogenesis

In the pathogenesis of NMOSD, anti-AQP4 antibodies cause complement-mediated injury to astrocytes in the CNS, with demyelination occurring secondarily (Ratelade et al., 2013).

Symptoms

Symptoms vary depending on the site of injury, but NMOSD tends to be more severe than MS. Brainstem lesions are characterized by a high frequency of area postrema syndrome,

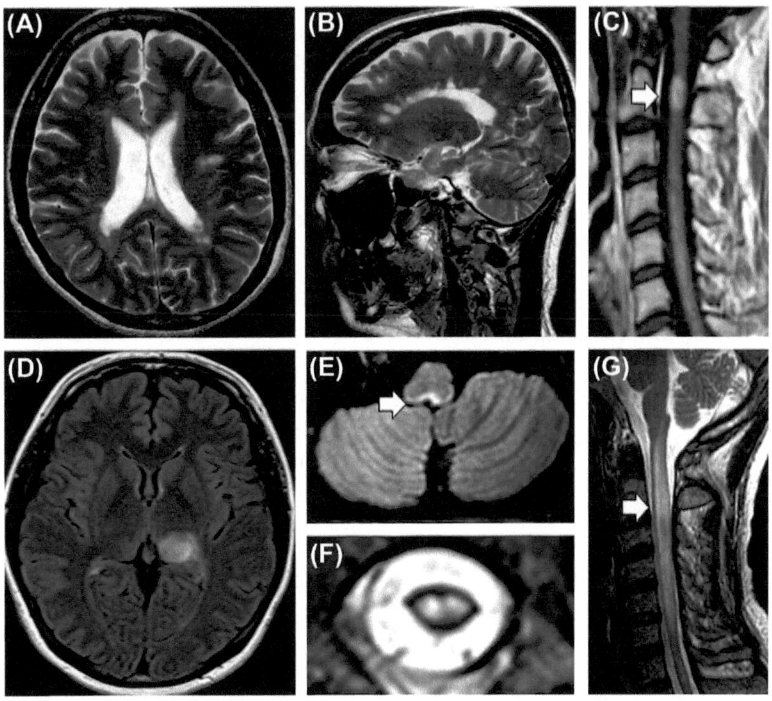

Fig. 12.1. MRI images of (**A–C**) multiple sclerosis (MS) and (**D–G**) neuromyelitis optica spectrum disorder (NMOSD). Cerebral white matter lesions in MS plaques are characterized by oval-shaped foci with a long axis perpendicular to the ventricles around the lateral ventricles (**A** and **B**). Brain lesions in NMOSD are less often white matter and are more often found in the mesencephalon, hypothalamus, and area postrema of the medulla oblongata (**D** and **E**). Spinal cord lesions are short in MS (**C**), but NMOSD is characterized by longitudinally extensive spinal cord lesions extending over three vertebrae in the central region (**F** and **G**).

refractory stuttering, nausea, and vomiting (Kremer et al., 2014). These symptoms may be the first manifestation of the disease (Apiwattanakul et al., 2010).

Diagnosis

International consensus diagnostic criteria for NMOSD are used (Wingerchuk et al., 2015). A positive serum anti-AQP4 antibody can be used to make the diagnosis if symptomatic CNS lesions are identified. If anti-AQP4 antibody is negative, the diagnosis should be made with caution because of the possibility of other diseases. CSF findings include increased cell counts with lymphocyte predominance and increased protein levels; unlike MS, oligoclonal IgG bands are rarely seen. Longitudinally extensive spinal cord lesions extending over three vertebrae are observed (Fig. 12.1). Lesions are also seen in the mesencephalon, hypothalamus, periventricular area of the fourth ventricle, periventricular area of the third ventricle, and area postrema of the medulla oblongata.

Treatment

Acute treatment is the same as for MS, including steroid pulses and plasma exchange. Oral steroids and immunosuppressive drugs have been used for relapse prevention, but in recent years, highly effective biologic agents have been used in an increasing number of cases (Pittock et al., 2019). Agents that suppress the B-cell system involved in antibody production and those that suppress the complement system are now available. Rituximab (an anti-CD20 antibody), inebilizumab (an anti-CD19 antibody), satralizumab (an anti-IL-6 (interleukin-6) receptor antibody), and eculizumab and ravulizumab (complement C5 inhibitors) are used.

Prognosis

Since the advent of biologic agents, the number of relapse-free cases has increased significantly; it is important to start relapse prevention as soon as possible, since NMOSD is considered to have no decline in physical function without relapse.

MYELIN OLIGODENDROCYTE GLYCOPROTEIN ANTIBODY-ASSOCIATED DISEASE

Overview

Myelin oligodendrocyte glycoprotein antibody-associated disease (MOGAD) is a new CNS autoimmune disease that was established as a part of CNS demyelinating diseases previously diagnosed as MS, acute disseminated encephalomyelitis (ADEM), or neuromyelitis optica, when anti-MOG antibodies were found in serum or spinal fluid (Banwell et al., 2023). MOGAD is a disease characterized by the presence of autoantibodies against myelin oligodendrocyte glycoprotein (MOG), one of the myelin sheath proteins of the CNS. MOGAD is a group of diseases with a variety of symptoms including optic neuritis, myelitis, brainstem encephalitis, and cerebral encephalitis. The age of onset ranges from infancy to adulthood, and ADEM is the most common clinical form of childhood-onset MOGAD.

Pathogenesis

The pathogenesis of MOGAD is thought to involve an autoimmune mechanism of pathogenesis by anti-MOG antibodies (Reindl and Waters, 2019). Genetic predisposition similar to other autoantibody-associated diseases is predicted, but the complications with other autoimmune diseases, tumor associations, and association with specific HLA haplotypes are unknown.

Symptoms

Symptoms vary according to the type of disease and the location of demyelinating findings (Hegen and Reindl, 2020). ADEM is more common in children and presents with impaired consciousness, abnormal behavior, fever, seizures, movement disorders, and ataxia. Transverse myelitis and/or brainstem encephalitis can cause motor and sensory deficits and vesicorectal disturbances, depending on the level of involvement.

Diagnosis

The International MOGAD Panel proposed diagnostic criteria for MOGAD in 2023 (Banwell et al., 2023). MRI scans show demyelinating findings with contrast effects in the affected regions including brainstem. CSF examination often shows elevated cell counts and protein levels. Identification of anti-MOG antibodies in serum or spinal fluid supports the diagnosis of MOGAD.

Treatment

Steroid pulses are administered in the acute phase, with a better response to treatment than in NMOSD, but plasma exchange is used in refractory cases. Since about half of the patients relapse, oral steroids are administered as prophylaxis, and the dose is reduced over several months while monitoring for relapses. In cases of frequent recurrence, immunosuppressive drugs and periodic immunoglobulin therapy should be administered.

Prognosis

The pathogenesis of MOGAD is variable, with some cases having a monophasic course without recurrence. In children, a negative conversion of anti-MOG antibodies has been reported to be a significant predictor of a monophasic course.

ACUTE DISSEMINATED ENCEPHALOMYELITIS

Overview

ADEM is an acute-onset, monophasic course, disseminated encephalomyelitis affecting the CNS. It often develops after infection or vaccination. ADEM is a heterogeneous disease, including its clinical form, and it is sometimes difficult to differentiate it from MS, neuromyelitis optica, or infectious encephalomyelitis (Hardy et al., 2016).

Pathogenesis

Viral infection or vaccination triggers an autoimmune mechanism that causes sporadic demyelination and inflammation in the CNS. Pathologically, the main lesions are small perivascular demyelinating foci and cellular infiltrates in the white matter of the CNS.

Symptoms

Acute onset of disease occurs within a few days to 4 weeks after viral infection or vaccination. It often begins with headache, fever, and vomiting, and is often accompanied by impaired consciousness and convulsions (Pellegrino et al., 2013). Neurologic symptoms are seen corresponding to the site of the disorder. Brainstem and cerebellar symptoms include cerebellar ataxia, myoclonus, nystagmus, eye movement disorders, and bulbar palsy.

Diagnosis

Although there are no internationally standardized clinical diagnostic criteria, the diagnostic criteria for pediatric ADEM proposed by the International Pediatric MS Study Group require the presence of neurologic symptoms of multifocal CNS origin in addition to encephalopathy (altered consciousness and behavioral changes) that cannot be explained by fever (Krupp et al., 2013). The diagnosis is made in conjunction with spinal fluid examination and CT/MRI findings. CT shows extensive low-absorption areas in the cerebral white matter, and MRI shows multiple high-signal areas with unclear boundaries from subcortical to deep white matter on T2-weighted images. The most common areas are around the lateral ventricles, brainstem, cerebellum, optic nerves, and spinal cord. The lesions are often contrast-enhanced.

Treatment

Although there is no treatment that has been proven effective in large-scale clinical trials, steroid pulse therapy has been used in many cases. High-dose immunoglobulin therapy and plasma exchange are effective in some cases.

Prognosis

The course of the disease is mostly monophasic, and recurrence is rare, but anti-MOG antibody-positive cases may be included among the recurrent cases.

GLIAL FIBRILLARY ACIDIC PROTEIN ANTIBODY-ASSOCIATED DISEASE

Overview

Glial fibrillary acidic protein (GFAP) antibody-associated disease is a newly proposed immune-mediated neurologic disease. Although the pathomechanism of the disease has not been fully elucidated, a perivascular T-cell infiltrate has been reported, suggesting the involvement of antigen-specific T cells. Meningoencephalitis or myelitis is the most common manifestation of this disease, which is associated with impaired consciousness, dysuria, dyskinesia, meningism, and cognitive dysfunction. CSF analysis shows mononuclear cell proliferation and anti-GFAP antibodies are detected in the serum (Ding et al., 2020).

Pathogenesis

The pathogenesis of GFAP antibody-associated disease has not been fully elucidated, but antigen-specific CD8-positive T cells, cytokines, and chemokines are thought to play an important role (Kimura et al., 2021; Shu et al., 2018). Since GFAP is an intracellular antigen, it is unlikely that it is directly involved in the pathogenesis, but this issue remains to be clarified. In cases complicated with anti-*N*-methyl-D-aspartate receptor (NMDAR) or AQP4 antibodies, it has been speculated that astrocytes may be damaged by the inflammatory process caused by the direct action of the antibodies, resulting in a secondary immune response against GFAP antigen (Flanagan et al., 2017).

Symptoms

The median age of patients is in their 50s, and males are more frequently affected. Patients present with a variety of symptoms including impaired consciousness, dysuria, dyskinesia (tremor, myoclonus, ataxia), meningism, pyramidal tract signs, and cognitive dysfunction. Some patients may present with optic papillary edema, respiratory disturbance requiring the use of a respirator, and area postrema syndrome.

Diagnosis

CSF in most patients shows an increased number of mononuclear cells and elevated protein levels, and oligoclonal IgG bands are positive in approximately 70% of patients. Head MRI shows multiple lesions in the cerebral white matter, basal ganglia, and brainstem. Spinal cord MRI shows long spinal cord lesions extending over three vertebrae, sometimes with abnormal intramedullary or intrathecal contrast effects.

Treatment

Immunotherapy is effective, and steroid therapy is the mainstay of treatment, but there are refractory cases and relapsed cases. Other treatments include intravenous immunoglobulin and plasma exchange.

Prognosis

Steroid therapy is effective in most cases, but relapses may occur, and some cases are refractory to steroid therapy. The prognosis is likely to be poor in elderly patients and those who start treatment too late. Sequelae often include dysuria and cognitive dysfunction.

CHRONIC LYMPHOCYTIC INFLAMMATION WITH PONTINE PERIVASCULAR ENHANCEMENT RESPONSIVE TO STEROIDS

Overview

Chronic lymphocytic inflammation with pontine perivascular enhancement responsive to steroids (CLIPPERS) presents with clinical symptoms suggesting damage to the surrounding tissues. Head MRI shows scattered gadolinium punctate enhancement in the brainstem, cerebellum, and spinal cord. Pathologically, the disease is characterized by a marked perivascular lymphocytic infiltrate and responds well to immunotherapy such as steroids but is prone to relapse after early reduction of steroid dosage (Dudesek et al., 2014).

Pathogenesis

The pathogenesis of the disease is not autoantibody-mediated. There is no vasculitis or clear demyelination, but a prominent perivascular infiltrate of nonneoplastic lymphocytes with a predominance of T cells, mainly in the pons. Infiltration of activated macrophages is also prominent, but no Langhans giant cells or infectious pathogens are seen. A case of hemophagocytic lymphohistiocytosis diagnosed by genetic testing shows a pathology similar to that of CLIPPERS (Tobin et al., 2017).

Symptoms

Because the lesions are characterized by foci centered on brainstem and cerebellar symptoms such as ataxic gait, diplopia, facial dysesthesia, dysarthria, and vertigo are frequent in the initial symptoms, cognitive dysfunction and spinal cord disorder may occur as the lesions spread beyond the brainstem (Pittock et al., 2010).

Diagnosis

There is subacute pontine and cerebellar dysfunction that responds well to steroid therapy, with homogeneous gadolinium-enhancing nodules (<3 mm in diameter with no ring enhancement or occupying effect) seen on MRI images, mainly in the pons and cerebellum. The foci extend peripherally, but become less frequent distal to the pons. Enhancement is uniform and even, with a low degree of central enhancement and no so-called ring enhancement. Cases with imaging findings similar to CLIPPERS have been reported in patients with anti-GFAP or anti-MOG antibody-positive disease (Obeidat et al., 2021). CSF examination reveals mildly elevated protein and increased cell counts, and oligoclonal IgG bands may be positive.

Treatment

Clinical symptoms and imaging findings improve markedly with steroid treatment. However, many cases relapse after discontinuation or reduction of steroid medication, and concomitant use of immunosuppressive drugs is required for long-term steroid reduction (Taieb et al., 2017).

Prognosis

Patients with recurrent relapses may be diagnosed with malignant lymphoma or hemophagocytic lymphohistiocytosis during the course of the disease. Hemophagocytic lymphohistiocytosis gene mutations should be confirmed, and hematopoietic stem cell transplantation may be an option if such mutations are confirmed.

AUTOIMMUNE ENCEPHALITIS/ ENCEPHALOPATHY

Overview

Autoimmune encephalitis/encephalopathy presents acutely or subacutely with memory impairment, psychiatric symptoms, somnolence, personality changes, epileptic seizures, and impaired consciousness, which may lead to coma. The disease should be suspected in the absence of infectious signs during the course of the disease.

Pathogenesis

In addition to damage to brain tissue caused by acute encephalitis/encephalopathy, autoimmune abnormalities against multiple brain tissue antigens are thought to be involved. Several antibodies have been identified, including anti-NMDAR antibody, anti-LGI1 (leucine-rich glioma-inactivated 1) antibody, and other autoantibodies against neuronal cell surface structures. These autoantibodies against cell surface structures are thought to be involved in the pathogenesis. In addition, involvement of unknown antineuronal antibodies and paraneoplastic causes have been suggested (Dalmau, 2016).

Symptoms

Most of the patients with autoimmune encephalitis/encephalopathy present with impaired consciousness, cognitive dysfunction, and epileptic seizures in the acute phase, which possibly lead to coma and death. The target antigens that have been reported to be associated with brainstem encephalitis include LGI1, NMDAR, Ri, Ma1, Ma2, IgLON5, and DPPX (Fuseya et al., 2020). Even after recovery, sequelae such as cognitive impairment, motor dysfunction, and epilepsy remain, depending on the site of brain damage.

Diagnosis

The Graus criteria for the diagnosis of autoimmune encephalitis are based on clinical symptoms and findings of various tests (MRI, CSF, EEG, and antineural antibodies) (Graus et al., 2016). As the disease often develops paraneoplastically, tumor screening with imaging studies and tumor markers is also necessary.

Treatment

Early initiation of immunotherapy can improve prognosis, shorten disease duration, and reduce the risk of recurrence. Immunotherapy should be started as early as possible in cases of strong clinical suspicion of autoimmune encephalitis, even if antibody results are not known. In paraneoplastic cases, treatment should be given against the tumor. Immunotherapy includes steroid pulse therapy, high-dose intravenous immunoglobulin, and plasma exchange. In treatment-resistant cases, rituximab and intravenous cyclophosphamide have been advocated (Abboud et al., 2021). There is no evidence for maintenance therapy, but immunomodulatory therapies (steroids, immunosuppressive drugs), symptomatic therapy, surgical treatment, and rehabilitation are used.

Prognosis

Although some patients have a good prognosis after successful treatment in the acute phase of the disease, they may still have cognitive and motor deficits after recovery from the acute phase, and may develop epilepsy that is drug-resistant or have a prolonged course. Encephalitis and encephalopathy involving antibodies that target cell surface antigens are more likely to respond to immunotherapy.

BICKERSTAFF BRAINSTEM ENCEPHALITIS

Overview

Bickerstaff brainstem encephalitis (BBE) is an autoimmune disease characterized by ophthalmoplegia, ataxia, and disturbance of consciousness, with the brainstem as the primary site of involvement. Three cases were reported by Bickerstaff and Cloake in 1951: "a central nervous system disease with disturbance of consciousness, paralysis of external eye muscles, and ataxia as core symptoms accompanied by impaired consciousness" (Bickerstaff and Cloake, 1951). Symptoms peak within 4 weeks, followed by a gradual recovery. The disease occurs in all age groups, but is more common in young people, especially those in their 30s.

Pathophysiology

Since limb weakness is seen during the course of the disease in about half of the cases, and serum IgG anti-GQ1b antibodies are found in about 70% of cases, as in Fisher syndrome (FS), an autoimmune mechanism induced by infection has been postulated, as in Guillain–Barré syndrome (Chiba et al., 1993; Kusunoki et al., 2021). BBE shares many common findings with FS and is considered to be a spectrum of the same disease (Odaka et al., 2001). Thus, BBE can be considered a subtype of FS complicated by central lesions. The nerve root is the boundary between the blood–nerve barrier and the blood–brain barrier, and inflammation in this region may result in spillover of inflammation to the CNS, where anti-GQ1b antibodies may enter the CNS and immunologically attack the GQ1b epitope. It was reported that BBE sera, but not FS sera, destroyed blood–brain barrier (Saito et al., 2013). That may explain the CNS involvement of BBE.

Symptoms

In addition to ophthalmoplegia and ataxia, symptoms of CNS disturbance such as impaired consciousness are observed. About half of the patients develop muscle weakness in the extremities during the course of the disease, about 60% have oropharyngeal paralysis, and about 30% have autonomic nervous system symptoms such as arrhythmias. About 20% of patients require ventilator management, and about 60% of patients are unable to walk with assistance.

Diagnosis

The following criteria is reported to diagnose BBE (Koga et al., 2012). It is based on the four points described below.

Definite: All of (1), (2), and (4) are met.

Probable: Both (1) and (4) are met, or (2) (3) and (4) are met.

(1) All of the following three signs are acutely progressive, peaking within 4 weeks of onset, and showing a tendency to lighten within 3 months: 1) bilateral external ophthalmoplegia (symmetry is the rule, but minor differences between right and left eye muscles are acceptable), 2) ataxia, and 3) decreased level of consciousness.
(2) Positive blood IgG-type GQ1b antibody.
(3) Some of the clinical features in (1) are not matched (multiple items are acceptable): 1) difficulty in evaluation of ataxia due to muscle weakness and decreased level of consciousness, 2) no tendency to remission can be confirmed, 3) a severe left-right difference in external ophthalmoplegia, and 4) no decrease in the level of consciousness, but there are findings suggestive of long tract signs.
(4) The following diseases can be excluded from various tests (CSF, imaging studies, etc.) Wernicke's encephalopathy, cerebrovascular disease, MS, neuromyelitis optica, ADEM, neuro-Behçet's disease, neuro-sweet disease, pituitary apoplexy, viral brainstem encephalitis,

myasthenia gravis, brainstem tumor lesions, vasculitis, botulinism, and Hashimoto's encephalopathy.

Treatment

There is no established effective treatment, but intravenous immunoglobulin, plasma exchange, and corticosteroids are used. Systemic management, such as ventilator and cardiovascular management, may be necessary.

Prognosis

About 10% of patients are unable to walk on their own at 1 year after onset of disease. Many patients have long-term sequelae such as motor and sensory deficits. Anti-GQ1b antibody-positive cases have clinical features more similar to FS, with a faster recovery, milder CSF findings, and milder MRI findings (Tables 12.2 and 12.3) (Yoshikawa et al., 2020).

NEURO-BEHÇET'S DISEASE

Overview

Behçet's disease is an inflammatory disease of unknown cause characterized by recurrent oral aphthae, vulvar ulcers, mucocutaneous lesions, and ocular lesions. Secondary symptoms include arthritis, epididymitis, ulcers in the ileocecal region, vascular lesions, and CNS lesions. A total of 10%–20% of patients with Beichet's disease present with CNS symptoms (Borhani-Haghighi et al., 2020).

Pathogenesis

In neuro-Behçet's disease, pathologically, brainstem lesions are seen in most cases, most often ventral to the pons to midbrain mesencephalon, and recurrent inflammation, mainly of capillaries to small veins, is the predominant form of lesion formation. Demyelinating lesions are often present, but they are mainly secondary to inflammation and have blurred boundaries.

Symptoms

There are two main types of symptoms: an acute form that develops acutely as meningitis or brainstem encephalitis, and a chronic progressive form that has a poor prognosis with neurologic symptoms such as hemiparesis, cerebellar symptoms, pyramidal tract symptoms, and psychiatric symptoms such as dementia. Neurologic findings include impaired consciousness, convulsions, eye movement disorder, cognitive

Table 12.2

Clinical features of patients with Bickerstaff brainstem encephalitis.

	Anti-GQ1b antibody-positive (*n*=73)	Anti-GQ1b antibody-negative (*n*=10)	*p*-Value
Age (Median (range))	40 (15–80)	50 (19–84)	n.s.
Sex (male/female)	43/30	5/5	n.s.
Preceding infection, *n* (%)	65 (89)	5 (50)	<0.01
Respiratory infection	51 (70)	2 (20)	<0.01
Gastrointestinal infection	11 (15)	0 (0)	n.s.
Ataxia, *n* (%)	68 (93)	10 (100)	n.s.
Cerebellar ataxia	39 (53)	6 (60)	n.s.
Sensory ataxia	9 (12)	0 (0)	n.s.
Undeterminable	20 (27)	4 (40)	n.s.
Muscle weakness, *n* (%)	45 (62)	6 (60)	n.s.
Tendon reflex, *n* (%)			
Brisk	17 (23)	3 (30)	n.s.
Normal	13 (18)	2 (20)	n.s.
Decreased	14 (19)	2 (20)	n.s.
Absent	29 (40)	3 (30)	n.s.
Pathological reflex, *n* (%)	35 (48)	2 (20)	n.s.
Sensory disturbance, *n* (%)	41 (56)	1 (10)	<0.01
Dysesthesia	32 (44)	0 (0)	<0.01
Superficial sense impairment	8 (11)	0 (0)	n.s.
Pain	2 (3)	1 (10)	n.s.
Deep sense impairment	8 (11)	0 (0)	n.s.
Autonomic dysfunction, *n* (%)	26 (36)	4 (40)	n.s.
Mechanical ventilation, *n* (%)	15 (21)	2 (20)	n.s.

n.s., not significant.

Source: Cited from Yoshikawa K, Kuwahara M, Morikawa M, et al. (2020) Bickerstaff brainstem encephalitis with or without anti-GQ1b antibody. Neurol Neuroimmunol Neuroinflamm 7: e889.

Table 12.3

Cerebrospinal fluid and radiologic findings in patients with Bickerstaff brainstem encephalitis.

	Anti-GQ1b antibody-positive (n=73)	Anti-GQ1b antibody-negative (n=10)	p-Value
Duration (d), median (range)	3 (1–23)	5.5 (1–22)	n.s.
Pleocytosis (> 5/μL), *n* (%)	34/71 (48)	6/9 (67)	n.s.
Median (range)	12.5 (5.3–90)	75.9 (11–251)	< 0.01
Elevated protein (≥45 mg/dL), *n* (%)	20/71 (28)	8/9 (89)	< 0.01
Median (range)	63.5 (47–132)	159 (59–381)	< 0.01
Brain MRI abnormal findings, *n* (%)	6/73 (8)	5/10 (50)	< 0.01

Duration, days from onset to conduct of lumbar puncture; *n.s.*, not significant.

Source: Cited from Yoshikawa K, Kuwahara M, Morikawa M, et al. (2020) Bickerstaff brainstem encephalitis with or without anti-GQ1b antibody. Neurol Neuroimmunol Neuroinflamm 7: e889.

decline, cerebellar ataxia, hemiplegia, quadriplegia, sensory disturbance, and various other symptoms (Akman-Demir et al., 1999).

Diagnosis

After the diagnosis of Behcet's disease is made, MRI images of the acute form show foci in the pons, midbrain, thalamus, and other areas. In the chronic progressive form, MRI images show progressive atrophy of the brainstem, cerebellum, and cerebrum. CSF findings are characterized by increased cell counts, elevated protein levels, and persistently elevated IL-6 levels, which may reflect chronic inflammation in the CNS.

Treatment

Steroid pulse therapy is administered during the acute exacerbation phase, and thereafter, oral steroids are started, and the dose is reduced while confirming efficacy. If continued treatment is necessary, immunosuppressive agents or infliximab, a TNF inhibitor, should be considered. In the chronic progressive form, immunosuppressive agents, methotrexate, and anti-TNFα agents are used.

Prognosis

The acute form of the disease responds well to steroids but may progress to the chronic progressive form with recurrent relapses. The chronic progressive form is considered to have an unfavorable prognosis because the neurologic symptoms may continue to recur in the acute form or progress subclinically (Kidd et al., 1999).

OTHER DISEASES

Narcolepsy Type 1 with cataplexy and autoimmune diabetes insipidus are also thought to be caused by an autoimmune mechanism that results in damage to brainstem function (Xu et al., 2024). Narcolepsy is a central hypersomnolence caused by a dysfunction of the sleep/wake center, and can be divided into two types: type 1 with and type 2 without emotional desensitization attack. Type 1 narcolepsy causes seizures of emotional weakness that occur when emotions such as joy, anger, and sadness are aroused. Consciousness and memory are usually preserved during these attacks. Type 1 narcolepsy is deficient in the neuropeptide hypocretin and is thought to result from the destruction of neurons in the lateral hypothalamus by an autoimmune mechanism.

Renal enuresis is a condition in which the kidneys are unable to produce antidiuretic hormone due to genetic abnormalities, electrolyte abnormalities, or drugs, whereas idiopathic central enuresis is thought to involve an autoimmune mechanism and is called as autoimmune diabetes insipidus.

CONCLUSIONS

Brainstem is affected with various autoimmune diseases. In addition to the abovementioned diseases, such diseases as systemic lupus erythematosus and Sjögren's syndrome, neurosarcoidosis can also cause lesions in brainstem. While immunotherapies are commonly used for the autoimmune maladies in brainstem, the response to them depends on the pathogenetic mechanism of each disease. Molecular targeted therapies are becoming to be used for the treatment of neuroimmunologic diseases, including those causing brainstem lesions. Future investigation of the real-world data is needed to evaluate their effectiveness.

DECLARATION OF INTEREST

The authors have no conflicts of interest to declare.

References

Abboud H, Probasco JC, Irani S et al. (2021). Autoimmune encephalitis: proposed best practice recommendations for diagnosis and acute management. J Neurol Neurosurg Psychiatry 92: 757–768.

Akman-Demir G, Serdaroglu P, Tasçi B (1999). Clinical patterns of neurological involvement in Behçet's disease: evaluation of 200 patients. The Neuro-Behçet Study Group. Brain 122: 2171–2182.

Apiwattanakul M, Popescu BF, Matiello M et al. (2010). Intractable vomiting as the initial presentation of neuromyelitis optica. Ann Neurol 68: 757–761.

Banwell B, Bennett J, Marignier R et al. (2023). Diagnosis of myelin oligodendrocyte glycoprotein antibody-associated disease: International MOGAD Panel proposed criteria. Lancet Neurol 22: 268–282.

Bickerstaff ER, Cloake PCP (1951). Mesencephalitis and rhombencephalitis. Br Med J 2: 77–81.

Borhani-Haghighi A, Kardeh B, Banerjee S et al. (2020). Neuro-Behcet's disease: an update on diagnosis, differential diagnoses, and treatment. Mult Scler Relat Disord 39: 101906.

Chiba A, Kusunoki S, Obata H et al. (1993). Serum anti-GQ1b IgG antibody is associated with ophthalmoplegia in Miller Fisher syndrome and Guillain-Barré. Neurology 43: 1911–1917.

Compston A, Coles A (2008). Multiple sclerosis. Lancet 372: 1502–1517.

Dalmau J (2016). NMDA receptor encephalitis and other antibody-mediated disorders of the synapse: the 2016 Cotzias Lecture. Neurology 87: 2471–2482.

Ding J, Ren K, Wu J et al. (2020). Overlapping syndrome of MOG-IgG associated disease and autoimmune GFAP astrocytopathy. J Neurol 267: 2589–2593.

Dudesek A, Rimmele F, Tesar S et al. (2014). CLIPPERS: chronic lymphocytic inflammation with pontine perivascular enhancement responsive to steroids. Review of an increasingly recognized entity within the spectrum of inflammatory central nervous system disorders. Clin Exp Immunol 175: 385–396.

Flanagan EP, Hinson SR, Lennon VA et al. (2017). Glial fibrillary acidic protein immunoglobulin G as biomarker of autoimmune astrocytopathy: analysis of 102 patients. Ann Neurol 81: 298–309.

Fuseya K, Kimura A, Yoshikura N et al. (2020). Corticobasal syndrome in a patient with anti-IgLON5 antibodies. Mov Disord Clin Pract 7: 557–559.

Gholamzad M, Ebtekar M, Ardestani MS et al. (2019). A comprehensive review on the treatment approaches of multiple sclerosis: currently and in the future. Inflamm Res 68: 25–38.

Graus F, Titulaer MJ, Balu R et al. (2016). A clinical approach to diagnosis of autoimmune encephalitis. Lancet Neurol 15: 391–404.

Habek M (2013). Evaluation of brainstem involvement in multiple sclerosis. Expert Rev Neurother 13: 299–311.

Hardy TA, Reddel SW, Barnett MH et al. (2016). Atypical inflammatory demyelinating syndromes of the CNS. Lancet Neurol 15: 967–981.

Hauser SL, Cree BAC (2020). Treatment of multiple sclerosis: a review. Am J Med 133: 1380–1390.

Hegen H, Reindl M (2020). Recent developments in MOG-IgG associated neurological disorders. Adv Neurol Disord 13: 1756286420945135.

Kidd D, Steuer A, Denman AM et al. (1999). Neurological complications in Behçet's syndrome. Brain 122: 2183–2194.

Kimura A, Kato S, Takekoshi A et al. (2021). Autoimmune glial fibrillary acidic protein astrocytopathy resembling isolated central nervous system lymphomatoid granulomatosis. J Neuroimmunol 361: 577748.

Koga M, Kusunoki S, Kaida K et al. (2012). Nationwide survey of patients in Japan with Bickerstaff brainstem encephalitis: epidemiological and clinical characteristics. J Neurol Neurosurg Psychiatry 83: 1210–1215.

Kremer L, Mealy M, Jacob A et al. (2014). Brainstem manifestations in neuromyelitis optica: a multicenter study of 258 patients. Mult Scler 20: 843–847.

Krupp LB, Tardieu M, Amato MP et al. (2013). International Pediatric Multiple Sclerosis Study Group criteria for pediatric multiple sclerosis and immune-mediated central nervous system demyelinating disorders: revisions to the 2007 definitions. Mult Scler 19: 1261–1267.

Kusunoki S, Willison HJ, Jacobs BC (2021). Antiglycolipid antibodies in Guillain-Barré and Fisher syndromes: discovery, current status and future perspective. J Neurol Neurosurg Psychiatry 92: 311–318.

Lennon VA, Kryzer TJ, Pittock SJ et al. (2005). IgG marker of optic-spinal multiple sclerosis binds to the aquaporin-4 water channel. J Exp Med 202: 473–477.

Obeidat AZ, Block AN, Hooshmand SI (2021). "Peppering the Pons": CLIPPERS or myelin oligodendrocyte glycoprotein associated disease? Mult Scler Relat Disord 51: 102874.

Odaka M, Yuki N, Hirata K (2001). Anti-GQ1b IgG antibody syndrome: clinical and immunological range. J Neurol Neurosurg Psychiatry 70: 50–55.

Pellegrino P, Carnovale C, Perrone V et al. (2013). Acute disseminated encephalomyelitis onset: evaluation based on vaccine adverse events reporting systems. PLoS One 8: e77766.

Pittock SJ, Debruyne J, Krecke KN et al. (2010). Chronic lymphocytic inflammation with pontine perivascular enhancement responsive to steroids (CLIPPERS). Brain 133: 2626–2634.

Pittock SJ, Berthele A, Fujihara K et al. (2019). Eculizumab in aquaporin-4-positive neuromyelitis optica spectrum disorder. N Engl J Med 381: 614–625.

Ratelade J, Asavapanumas N, Ritchie AM et al. (2013). Involvement of antibody-dependent cell-mediated cytotoxicity in inflammatory demyelination in a mouse model of neuromyelitis optica. Acta Neuropathol 126: 699–709.

Reindl M, Waters P (2019). Myelin oligodendrocyte glycoprotein antibodies in neurological disease. Nat Rev Neurol 15: 89–102.

Saito K, Shimizu F, Koga M et al. (2013). Blood-brain barrier destruction determines Fisher/Bickerstaff clinical phenotypes: an in vitro study. J Neurol Neurosurg Psychiatry 84: 756–765.

Shu Y, Long Y, Chang Y et al. (2018). Brain Immunohistopathology in a Patient with autoimmune glial fibrillary acidic protein astrocytopathy. Neuroimmunomodulation 25: 1–6.

Taieb G, Allou T, Labauge P et al. (2017). Therapeutic approaches in CLIPPERS. Curr Treat Options Neurol 19: 17.

Thompson AJ, Banwell BL, Barkhof F et al. (2018). Diagnosis of multiple sclerosis: 2017 revisions of the McDonald criteria. Lancet Neurol 17: 162–173.

Tintore M, Rovira A, Arrambide G et al. (2010). Brainstem lesions in clinically isolated syndromes. Neurology 75: 1933–1938.

Tobin WO, Guo Y, Krecke KN et al. (2017). Diagnostic criteria for chronic lymphocytic inflammation with pontine perivascular enhancement responsive to steroids (CLIPPERS). Brain 140: 2415–2425.

Xu W, Ding W, Zhang Y et al. (2024). The role of T cells in the pathogenesis of narcolepsy type 1: a narrative review. Int J Mol Sci 25: 11914.

Yoshikawa K, Kuwahara M, Morikawa M et al. (2020). Bickerstaff brainstem encephalitis with or without anti-GQ1b antibody. Neurol Neuroimmunol Neuroinflamm 7: e889.

Wingerchuk DM, Banwell B, Bennett JL et al. (2015). International consensus diagnostic criteria for neuromyelitis optica spectrum disorders. Neurology 85: 177–189.

Handbook of Clinical Neurology, Vol. 216 (3rd series)
The Brainstem
L.T. Grinberg and A.D. Korczyn
https://doi.org/10.1016/B978-0-443-15736-3.00005-6

Chapter 13

Dysarthria

PAMELA M. ENDERBY[1] AND ANGELA MORGAN[2,*]

[1]*School of Health and Related Research, University of Sheffield, Sheffield, United Kingdom*

[2]*Murdoch Children's Research Institute, Melbourne, VIC, Australia*

Abstract

The ability to communicate pervades every aspect of life, reflecting one's personality to oneself and to others, and is critical in forming and sustaining relationships, impacting education, employment, and social engagement. This chapter provides information on dysarthria, given the role of the brainstem in articulation, intonation, prosody, and fluency. The different types of dysarthria associated with different lesions are described and methods of assessment and treatment are outlined, emphasizing the importance of a holistic approach to the management of this often overlooked, but profoundly disabling disorder.

INTRODUCTION

Dysarthria is a motor *speech* disorder caused by disturbance of muscular control which can be classified according to the underlying neuropathology. Dysarthria occurs due to weakness, paralysis, imprecision, or incoordination of the muscles of articulation. As a result, dysarthria is associated with disturbances of respiration, laryngeal function, airflow direction, and articulation, resulting in slurred and abnormal sounding speech which may be difficult to understand. It is not surprising that patients with dysarthria are also likely to have some degree of dysphagia and drooling, given the common innervation. Further, children with neurodevelopmental disorders, such as rare genetic conditions, may also present with apraxia, alongside dysarthria, in the young developing system (Kaspi et al., 2023).

The importance of the brainstem in speech is demonstrated by patients with locked-in syndrome which is most often caused by substantial damage to the pons. The pons contains the neuronal pathways between the cerebrum, spinal cord, and cerebellum. In locked-in syndrome, there is an interruption of motor fibers running from the gray matter to the body's muscles, and damage to the centers in the brainstem important for facial control and speaking, resulting in absent speech (anarthria), in the presence of preserved cognition (Farr et al., 2021).

*Dysphasi*a (also called aphasia) is an impairment of *language*. Language is the process in which thoughts and ideas become spoken, but also includes other higher-level cognitive modes of comprehension and expression such as reading, writing, and nonverbal communication. It involves both the selection or understanding of words (semantics) and the formulation or processing of appropriate sentences or phrases (syntax). In the pediatric field, language disturbances are more commonly known as language disorders, rather than aphasia (Bishop et al., 2017).

Dyspraxia of speech results from dysfunction of the cerebral hemisphere, especially the parietal lobe or nerve pathways that connect these lobes to other parts of the brain, such as the frontal and/or temporal lobes. It is different to dysphasia given that it is a motor speech disorder rather than a language disorder. It is different to dysarthria in that it is the inability to carry out learned purposeful movements despite good motor, sensory, and coordination functions. A person with speech dyspraxia is unable to initiate purposeful speech automatically and accurately with correct prosody. Oral dyspraxia may also be present with speech dyspraxia. It is easier to conceptualize true praxis in the context of oral dyspraxia, where an individual may be able to automatically lick a stamp or ice cream but when asked to stick out their tongue as if licking, they would not be able to complete the movement when under this volitional control. In children, acquired and persistent dyspraxia is extremely rare. However, there is a neurodevelopmental form of the condition, known as childhood apraxia of speech, which is commonly genetic. As many as one in three children have a genetic diagnosis explanatory for their speech presentation (Eising et al., 2019; Hildebrand et al., 2020; Kaspi et al., 2023).

The terms "anarthria," "aphasia," and "apraxia" mean a total absence of the ability to form speech or language, but they are often used when the terms *dysarthria*, *dysphasia*, and *dyspraxia* would be more correct.

*Correspondence to: Angela Morgan, PhD, Professor of Speech Pathology, University of Melbourne, 550 Swanston Street, Melbourne, Parkville VIC 3052, Australia. E-mail: amor@unimelb.edu.au.

There is often some overlap of these conditions, but a person who has pure dysarthria without dysphasia would be able to comprehend, read, and write as normal and make meaningful gestures, provided that the associated necessary motor pathways were intact.

Dysphagia is a disorder of eating, drinking, or swallowing that has a functional impact at mealtimes, effects nutrition and growth, and is associated with aspiration pneumonia. Like dysarthria, dysphagia can also result from damage of the motor cortex or brainstem, resulting in reduced neurophysiologic control of the oral, pharyngeal, laryngeal, or esophageal aspects of eating and drinking.

Within this chapter, we will be concentrating on the issues of dysarthria, with some reference to dysphagia. However, we will not be deliberating on dysphasia or apraxia, given that these conditions are more typically associated with cortical damage and as such, are beyond the focus of this book.

SIGNS AND SYMPTOMS

Dysarthria can affect one or more of the following five systems that speech involves:

1. *Respiration*: Exhalation moves air across the adducted vocal cords, creating sounds that the mouth and nose shape into words. In speech production, the duration of inspiration decreases, and velocity of exhalation increases and is modified according to the cognitive requirements. Respiration is the power behind speech production.
2. *Phonation*: This system uses airflow from the lungs plus vocal cord vibrations to produce vocalization which may be whispered, voiced, breathy, falsetto, etc. A column of air pressure is moved toward the vocal folds by the coordinated action of the diaphragm, abdominal and chest muscles. Movement of the arytenoids during expiration results in fine alterations in vocal fold tension causing vibration of the vocal folds.
3. *Resonance*: This refers to the quality/timbre and intensity of speech sounds that the vocal tract produces. The vocal tract enhances harmonics of the phonation based on the size and/or shape of the supralaryngeal cavities, the pharynx, and the oral and nasal cavities. An important distinction for resonance includes oral/nasal contrasts and hence altered resonance often manifests as hyper or hyponasality.
4. *Articulation*: This term means shaping sounds with the lips, teeth, and tongue into recognizable words conforming to the practice of a particular language and involves forming precise and accurate vowels, diphthongs, consonants, and blends.
5. *Prosody*: The rhythm, stress, and intonation of speech that give words and phrases additional meaning.

These five speech systems work together, meaning that impairment in one system can impinge on the others.

People who have dysarthria often have slurred speech with distorted consonants and vowel sounds. Additionally, they may have an abnormally quiet or loud speaking voice, a monotonous tone, a rough, scratchy or hoarse phonation or imbalanced nasality of tone. Some patients will have a vocal tremor or speak too quickly or slowly.

People with dysarthria may experience additional oral and facial symptoms including tremors or involuntary movements of the jaw, tongue or lips; abnormal sensitivity of the gag reflex; and weakness and muscle wasting of oral musculature (Table 13.1).

Table 13.1

Incidence and prevalence of dysarthria in adults in the United Kingdom and the United States (American Speech Language and Hearing Association, 2021; Enderby, 2013).

Selected neurologic condition	Dysarthria within the disorder	
	Incidence	Prevalence
Traumatic brain injury	33% (Enderby, 2012)	10%–65% (Jayaraman and Das, 2023)
Parkinson disease	50%–90%[a] (Enderby, 2012)	70%–100% (Jayaraman and Das, 2023)
Stroke	20%	50% diminishing as recovery (Enderby, 2012)
Motor neuron disease	Not known	More than 80% of people with MND will experience communication difficulties at some point in the disease trajectory. (Enderby and John, 2019)
Multiple sclerosis	25% (Jayaraman and Das, 2023)	50% at some stage during the continuity of their disease (Jayaraman and Das, 2023)
Progressive neurologic disorder	Not known	30%–80% (Enderby, 2012)

[a] Range given, with increasing incidence as the disease progresses.

Source: dysarthria. In: Barnes MP, Good DC (eds.) Handbook of clinical neurology, Vol. 110. Elsevier, pp. 273–281. https://doi.org/10.1016/B978-0-444-52901-5.00022-8; American Speech Language and Hearing Association (2021). Dysarthria in adults. Available at: https://www.asha.org/practice-portal/clinical-topics/dysarthria-in-adults/#collapse_1.From Enderby P (2013) Chapter 22 – Disorders of communication: dysarthria. In: Barnes MP, Good DC (eds.) Handbook of clinical neurology, Vol. 110. Elsevier, pp. 273–281. https://doi.org/10.1016/B978-0-444-52901-5.00022-8; American Speech Language and Hearing Association (2021). Dysarthria in adults. Available at: https://www.asha.org/practice-portal/clinical-topics/dysarthria-in-adults/#collapse_1.

Table 13.2

Prevalence of dysarthria in pediatric clinical cohorts.

Selected neurologic condition	Prevalence of dysarthria within the condition	Reference
Cerebral palsy	78%	Mei et al. (2020)
Neuromuscular diseases (e.g., muscular dystrophy, congenital myopathy)	30%	Kooi-van Es et al. (2020)
Moderate/severe traumatic brain injury	20%–67%	Morgan et al. (2010)
Stroke	74%	Liégeois et al. (2019)
Posterior fossa tumor	30%	Mei and Morgan. (2011)

Dysarthria is infrequently seen in children with mild traumatic brain injury (Morgan et al., 2010).

INCIDENCE AND PREVALENCE OF DYSARTHRIA

Dysarthria is associated with neurologic disease and disorder. See Table 13.2 for incidence and prevalence.

DYSARTHRIA TYPES

There are six major types of dysarthria: *flaccid dysarthria* associated with lower motor neuron impairment, *spastic dysarthria* associated with damaged upper motor neurons linked to the motor areas of the cerebral cortex, *ataxic* dysarthria primarily caused by cerebellar dysfunction, and *hyperkinetic* dysarthria and *hypokinetic* dysarthria, which are related to disorders of the extrapyramidal system. The sixth is generally termed a *mixed* dysarthria and is associated with damage in more than one area, resulting in speech characteristics of at least two groups. The features of the speech disturbance of these six major types of dysarthria are distinctive and can assist with diagnosis.

There is also a further division of spastic dysarthria, known as unilateral upper motor neuron dysarthria (Duffy, 2012). Children with neurodevelopmental conditions may also show a single subtype of dysarthria, but most commonly there is a mixed dysarthria profile in line with the neuropathology, which is typically more bilateral and diffuse in nature (Liégeois and Morgan, 2012).

Flaccid dysarthria

There may be specific parts of the oropharyngeal musculature affected, identified by eliciting syllables of "ka," "la," and "ma," which involves CN IX/X, XII, and VII, respectively.

The hallmark of flaccid dysarthria is difficulty pronouncing consonants. Damage to the peripheral nervous system (PNS) is responsible for this type of dysarthria. The PNS connects the brain and spinal cord to the rest of the body.

Flaccid dysarthria can result from neurosurgery, a tumor in the brainstem, or traumatic brain injury.

Fasciculations in these muscles, but especially the tongue, is associated with motor neuron disease/amyotrophic lateral sclerosis.

Spastic dysarthria

Spastic dysarthria is caused by damage to the motor neurons in the central nervous system (CNS) which includes the brain and spinal cord. It is the most common type of dysarthria. It is typically caused by bilateral dysfunction of the corticobulbar pathways innervating the cranial nerves mentioned above. This type of dysarthria is often seen in those with bilateral strokes, demyelinating disease, for example, MS, or motor neurone disease/amyotrophic lateral sclerosis. Features can include a strained hoarse voice with slow and imprecise articulation.

There is also a subtype of spastic dysarthria, referred to as unilateral motor neuron spastic dysarthria. In this condition, dysarthria results from a unilateral cortical lesion, but the speech symptoms are similar to those noted in individuals with bilateral spastic dysarthria.

Ataxic dysarthria

Ataxic dysarthria is associated with "scanning" speech with irregular intonation, pitch, volume, and speed. The voice is frequently loud, and speech is arhythmic/uncoordinated. This type of dysarthria can occur if a person sustains damage to the cerebellum which is the part of the brain responsible for receiving sensory information and regulating movement. Children with cerebral palsy frequently have ataxic dysarthria.

Hypokinetic dysarthria

A malfunction in the brain's extrapyramidal system causes hypokinetic dysarthria characterized by a quiet, breathy, and monotone voice with difficulty in initiating speech demonstrated by stuttering and difficulty in pronouncing blended consonants. This dysarthria is likely to demonstrate rigidity, tremors, and muscle spasms along with reduced movements in the face and articulators.

Hyperkinetic dysarthria

Hyperkinetic movement disorders, sometimes called dyskinesia, result in excessive, often repetitive, involuntary movements that intrude upon the normal flow of motor activity and occur as the result of damage to parts of the brain that are collectively termed the basal ganglia. The basal ganglia play a role in various functions, including involuntary muscle movement.

Patients with hyperkinetic dysarthria often find speaking tiring and have slow slurred speech with a shaky voice and poor breath control when speaking.

Damage to the basal ganglia can develop as a result of neurodegenerative diseases resulting in hyper or hypokinetic dysarthria, such as Parkinson disease and Huntington disease.

ASSESSMENT OF DYSARTHRIA

Auditory-perceptual analysis of speech disorders has been seen as accurate and comprehensive but is recognized as being time-consuming and requiring special expertise. Wannberg et al. (2016) investigated whether a simplified rating scale and/or checklist of oral motor and speech performance could be reliably used to confirm the presence of dysarthria and identify the diagnostically salient aspects of speech production whilst saving time and being reliable.

Their study indicated that a general assessment of dysarthria could be made with high reliability to identify problem indicating severity and type of dysarthria. Yet the authors stressed that such checklists should be complemented with an additional assessment of overall intelligibility. Two frequently used checklists are detailed below:

Robertson dysarthria profile

The Robertson dysarthria profile (RDP) revised in 1982 and 1994 is used to evaluate adults with acquired dysarthria. A further study using the RDP was carried out with patients with dysarthria associated with Parkinsonism and multiple sclerosis by Snowden (1995). Each section of the RDP (revised) profile has been analyzed to evaluate whether certain aspects of the scale would identify subtypes of dysarthria caused by a certain pathology or trauma. Findings supported this hypothesis.

Frenchay dysarthria assessment

The Frenchay dysarthria assessment – second edition (FDA-2) aims to be a quick and simple to administer tool (Enderby and Palmer, 2008). The FDA-2 has been shown to be accurate, with good psychometric properties. The patient/client is rated on several simple performance tasks related to speech function. It is divided into eight sections:

1. Reflexes – Ratings for cough, swallow, and dribble/drool
2. Respiration – Ratings at rest and in speech
3. Lips – Ratings for at rest, spread, seal, alternate, and in speech
4. Palate – Ratings for fluids, maintenance, and in speech
5. Laryngeal – Ratings for time, pitch, volume, and in speech
6. Tongue – Ratings for at rest, protrusion, elevation, lateral, alternate, an in speech
7. Intelligibility – Ratings for words, sentences, and conversation
8. Influencing Factors – Includes hearing, sight, teeth, language, mood, posture, rate (words per minute), and sensation.

The test results can be recorded on the FDA-2 Rating Form, indicating both the client's strengths and weaknesses in speech production. Normative data are reported for adults without dysarthria as well as patients with specific dysarthria associated with confirmed medical diagnoses. Separate tables enable the speech therapist to compare individual results with those of known dysarthric groups. This assessment has been translated into several different languages including Japanese, Hebrew, Swedish, Italian, French, German, and Arabic.

Assessment of dysarthria – more broadly

It is helpful to consider the model of the International Classification of Functioning, Disability and Health (ICF) as a framework for describing assessment of dysarthria (World Health Organization, 2001). The ICF has become standard language for rehabilitation and is a conceptual basis for the definition and measurement of health and disability. The ICF was approved for use by the World Health Assembly in 2001, after extensive testing across the world involving people with disabilities and people from a range of relevant disciplines. It moves clinical practice from considering a physiologic disorder alone to the broader consideration of its impact on life and well-being.

The assessments mentioned above primarily focus on the disorder/impairment of speech and includes intelligibility (activity)

Hustad, Beukelman, and Yorkston were ahead of their time in 1998 to draw attention to the impact of dysarthria at the societal level (Hustad et al., 1998). They developed an assessment to indicate the overall degree of success/difficulty a speaker has in specific real-world speaking situations focusing on "functional" assessment of persons with dysarthria and its impact on life, that is, social participation.

In the ICF, functioning and disability are multidimensional concepts, relating to:

- the body functions and structures and impairments (disorders) at the level of the body – in this case, the level and type of dysarthria (would include assessment such as the Robertson and the FDA-2);
- the activities, independence, and functioning and the activity limitations they experience – in this case, intelligibility and quality of speech as covered in the Robertson and the FDA-2 assessments;
- the social participation or involvement of people in all areas of life, for example, recreation, work, education, etc.

THERAPY OUTCOME MEASURE

The Therapy Outcome Measures (TOMs) for Rehabilitation Professionals was designed to be a simple, reliable, cross-disciplinary and cross-client group method of gathering psychometrically robust information on a broad spectrum of issues to assist in identifying rehabilitation goals and evaluating the impact of therapy which frequently considers a broad range of issues other than the speech impairment alone (Enderby et al., 1995; Enderby and John, 1999; Enderby et al., 2000; Enderby et al., 2003; Enderby et al., 2013). The TOM was based on a review of patient notes which indicated that therapy aims to

- improve the disorder,
- reduce dependence (in dysarthria – ability to communicate),
- improving social participation, and
- impacting on the well-being of the individual and carer.

The procedure for using the TOM requires that the health-care professional follows their usual assessment procedures for the individual referred for treatment, that is, standardized tests, observation, report, and consideration of medical and social history. Thus, no additional work/assessments or tests are required but rather, the TOM allows for a summary of the holistic impact of the disorder on the individual and their carer. It is a summarising tool in this regard, bringing no additional clinical burden.

The four TOM dimensions (impairment, activity, participation, and well-being) are rated on an ordinal rating scale, with 0 representing the severe end of the scale and 5 representing normal for a human being given the age, sex, and culture. Half points increase the scale to 11 points. The integers are defined with a semantic operational code that identifies the severity of the difficulty experienced on each dimension allowing the assessor to indicate whether the person is slightly better or worse than the descriptor, resulting in an 11-point scale. The ratings are entered at the start and end of a rehabilitation/enablement intervention/episode of care.

MEDICAL DIAGNOSES ASSOCIATED WITH DYSARTHRIA

The Royal College of Speech and Language Therapists (SLTs) has developed a national online outcome tool (ROOT) for the use of SLTs in the United Kingdom to collect data on all patients that they see. The database facilitates the collection of demographic information, ICD-10, information on speech and language disorder and TOMs (Enderby et al., 2013; Enderby and John, 2019). The reports generated by the ROOT can be used by SLTs to inform clinical decision-making, drive quality improvement, and demonstrate the impact of speech and language therapy. This real-world data provides contextual information on clinical practice and referral patterns in the United Kingdom.

As of July 2023, there were 4336 records associated with referrals to speech and language therapy for management of dysarthria. The 22 most frequently referred medical conditions associated with dysarthria are detailed in Table 13.3. A further 97 different ICD-10 codes were associated with referrals of individual patients with dysarthria, illustrating the broad range of medical conditions giving rise to this disorder of speech.

Table 13.3

Medical Diagnoses associated with referrals to speech and language therapy for dysarthria as recorded on ROOT database.

Primary medical diagnosis – ICD-10 codes	Count of primary medical diagnosis
Stroke, not specified as hemorrhage or infarction	1153
Parkinson disease	587
Cerebral infarction	423
Motor neuron disease	106
Cerebrovascular disease, unspecified	56
Malignant neoplasm of other and ill-defined sites in the lip, oral cavity, and pharynx	56
Multiple sclerosis	42
Intracerebral hemorrhage	30
Person with feared complaint in whom no diagnosis is made	25
Cerebral infarction, unspecified	24
Diffuse brain injury	19
Malaise and fatigue	19
Dysphagia	18
Other cerebrovascular diseases	17
Pneumonia, organism unspecified	17
Subarachnoid hemorrhage	17
Intracranial injury	11
Malignant neoplasm of other and ill-defined sites: head, face, and neck	11
Late-onset cerebellar ataxia	11
Malignant neoplasm of brain	11
Subdural hemorrhage (acute) (nontraumatic)	10
Myasthenia gravis	9

SPEECH AND LANGUAGE THERAPY FOR DYSARTHRIA

Dysarthria is a frequent symptom of many neurologic conditions and is commonly associated with progressive neurologic disease. It has a profound effect upon the patient and their families, as communication is integrally related with expressing personality and social relationships.

Speech and language therapy aims to encourage the person to use the speech that they are capable of more effectively by increasing the range and consistency of sound production and teaching strategies for improving intelligibility and communicative effectiveness. The therapist will teach methods of speaking which are less tiring and more successful. Appropriate augmentative and alternative communication (AAC) systems/devices may be introduced as required. Of importance is the support the therapist gives to the individual and their families encouraging resilience and improving confidence in social settings.

Reviewing the evidence of the effectiveness of speech language therapy for people with dysarthria has proved problematic given the heterogeneous nature of the underlying causes of dysarthria in children, adolescents, and adults which is compounded further by the severity of the disorders and the range of different speech and language therapy approaches and interventions.

Much of the research that has taken place in the field of dysarthria treatment has focused on adult presentations of dysarthria as a symptom of progressive disorders associated with Parkinson disease with less attention paid to longstanding, nonprogressive dysarthria often experienced by people with cerebral palsy or following a stroke or traumatic brain injury (Enderby and Emerson, 1995).

A Cochrane review published in 2002 concluded that at that time, there was no evidence to support or refute the effectiveness of speech and language therapy interventions for adult dysarthria following nonprogressive brain damage and the authors called for good quality research in this area (Sellars et al., 2002). A similar conclusion was expressed in the Cochrane review of speech therapy for children with dysarthria acquired before three years of age (Pennington et al., 2016) and for older children with acquired pathologies (Morgan and Vogel, 2009). This is despite thousands of papers being identified in the initial searches for both the adult and pediatric-based reviews indicating that research is being undertaken in the field of speech and language therapy for persons with dysarthria. However, meta-analysis is problematic due to the many different pathologic causes of dysarthria, making it difficult for researchers to recruit enough patients with similar types of motor speech disorder in order to assess the efficacy of therapy. Furthermore, the broad and individual psychosocial impact of communication difficulties leads to different approaches being identified as appropriate for some patients and not others. For example, in some settings, consideration of the use of AAC would be favored whilst in others there may be physical, resource, or relationship issues precluding this as an appropriate way forward.

Additionally, it is not surprising that a patient with spastic dysarthria associated with a progressive disease would require a different approach to intervention to one with a similar level and type of dysarthria secondary to an acute but nonprogressive disorder. Consideration of the persons personality, character, and communication needs would assist the therapist in identifying the appropriate way forward to achieve their goals. Some patients would find computerized therapy to improve articulation proficiency, volume modulation, and speed of speech acceptable, motivating, and effective, whilst other may find this too challenging and intimidating.

In a further practically oriented systematic review, Palmer et al. (2007) summarized different treatment techniques for speech and language therapists to use with adult patients with stable dysarthria. These authors concluded that there was a strong indication of the success of the Lee Silverman technique which may be due to the required combination of goals, verbal encouragement, and intensive treatment delivery (four 1-hour treatment sessions per week for 4 weeks) (Fox et al., 2006). The other approach supported by evidence was related to rate reduction (slowing down speech rate) which improved articulation accuracy and intelligibility.

Over the last two decades there has been increasing interest in developing and evaluating computerized approaches for the treatment of dysarthria. Technologic advances have led to more sophisticated and engaging programs which encourage practice and give consistent feedback. Motor learning principles suggest that to be effective, treatment needs to involve frequent, consistent practice with accurate and prompt feedback. One study found that patients being treated with a computerized approach spent 37% more time practicing than those involved in traditional therapy (Palmer et al., 2007). But it is important for the therapist to consider the aspects of speech that need to be modified and to match these with the computerized therapy approach. Some programs will aim to modify changes in volume, pitch, and speed, and others may encourage other motor speech features. Being able to personalize the program to meet the patient's needs as well as interests should be considered carefully.

ASSISTIVE AND AUGMENTATIVE COMMUNICATION

Technology is now used to assist communication of those who have no difficulty speaking as well as those who have some difficulty in communicating intelligibly. Communication devices and systems can support, extend, and replace natural verbal interaction.

A broad range of devices are available to those who are unable to communicate verbally and can be controlled by minimal physical movement. It is beyond the scope of this chapter to provide a list of AAC devices but there are certain issues that should be considered:

1. A communication aid should not only be considered for those who have no speech at all. A device may assist an

individual with even quite mild dysarthria in certain situations, for example, noisy environment or if they are tired. This may help them to extend their social participation.
2. Providing a communication aid is not sufficient on its own. The individual needs to be taught how to use all its functions and should be encouraged to use the aid frequently. Those around the patient need to understand and be part of the therapy to encourage access and use.
3. Encouraging children who have dysarthria to use a communication aid will not reduce their likelihood of improving their speech and will encourage development of language (vocabulary, grammatical structures, etc.) and facilitate social development.
4. The chosen AAC device or a combination of devices should be available to the user to improve communication in all settings.

Consideration should be given to the backup of an AAC device in the event that it may breakdown. This may be a nontechnical approach, for example, letter board or picture system.

EMOTIONAL IMPACT OF DYSARTHRIA

Brady et al. (2011) and colleagues remind us that there is complex association between the severity of an individual's dysarthria and its impact on their lives more broadly. They report findings from interviews with 24 individuals with dysarthria following stroke and suggest that, in some cases, the coping strategies adopted by the participants could be seen to further exacerbate isolation "dysarthria transcends the physiologic impairment to impact upon individuals' social participation, which is key to the process of rehabilitation."

Studies have shown that listeners make negative attributions toward people with communication impairments. Fox and Pring (2005) found that doctors judged the cognitive competence of persons with dysarthria more negatively than should have been the case, and that this was associated with the severity of the dysarthria.

CONCLUSION

The motor speech disorder, dysarthria, is associated with a broad range of neurologic conditions and has diagnostic importance. Speech and Language Therapists have an important role to play in the diagnosis and treatment. Intervention aims to assist with improving the quality and ease of communication and needs to consider the impact of this condition on social participation, employment, education, and relationships. There are an increasing number of assistive technologic devices to overcome communication restrictions and which can be integrated into rehabilitation programs.

References

American Speech Language and Hearing Association (2021) Dysarthria in adults. https://www.asha.org/practice-portal/clinical-topics/dysarthria-in-adults/#collapse_1.

Bishop DVM, Snowling MJ, Thompson PA et al. (2017). Phase 2 of CATALISE: a multinational and multidisciplinary Delphi consensus study of problems with language development: terminology. J Child Psychol Psychiatry 58: 1068–1080.

Brady MC, Clark AM, Dickson S et al. (2011). The impact of stroke-related dysarthria on social participation and implications for rehabilitation. Disabil Rehabil 33: 178–186.

Duffy JR (2012). Motor speech disorders: substrates, differential diagnosis, and management. Elsevier Health Sciences.

Eising E, Carrion-Castillo A, Vino A et al. (2019). A set of regulatory genes co-expressed in embryonic human brain is implicated in disrupted speech development. Mol Psychiatry 24: 1065–1078.

Enderby P (2012). Communication disorders and dysphagia. Pathy's Principles and Practice of Geriatric Medicine. pp. 1691–699.

Enderby P (2013). Chapter 22 – Disorders of communication: dysarthria. Barnes MP. Good DC. Handbook of clinical neurology, Vol. 110. Elsevier, pp. 273–281.

Enderby P, Emerson J (1995). Does speech and language therapy work. Wiley.

Enderby P, John A (2019). Therapy outcome measure user guide and scales. J&R Press Ltd.

Enderby P, John A, Hughes A et al. (2000). Benchmarking in rehabilitation: comparing physiotherapy services. Clin Perform Qual Health Care 8: 86–92.

Enderby P, John A, Hughes A et al. (2003). Using benchmarking data for assessing performance in occupational therapy. Clin Gov Int J 8: 290–295.

Enderby P, John A, Petheram B (2013). Therapy outcome measures for rehabilitation professionals: speech and language therapy, physiotherapy, occupational therapy. John Wiley & Sons.

Enderby P, John A, Sloan M (1995). Outcome measurement in speech and language therapy: results of a pilot study. Int J Lang Commun Disord 30: 70.

Enderby P, Palmer R (2008). Frenchay dysarthria assessment. 2nd edition. Pro-Ed.

Enderby PM, John A (1999). Therapy outcome measures in speech and language therapy: comparing performance between different providers. Int J Lang Commun Disord 34: 417–429.

Farr E, Altonji K, Harvey RL (2021). Locked-in syndrome: practical rehabilitation management. PM R 13: 1418–1428.

Fox A, Pring T (2005). The cognitive competence of speakers with acquired dysarthria: judgements by doctors and speech and language therapists. Disabil Rehabil 27: 1399–1403.

Fox CM, Ramig LO, Ciucci MR et al. (2006). The science and practice of LSVT/LOUD: neural plasticity-principled approach to treating individuals with Parkinson disease and other neurological disorders. Semin Speech Lang 27: 283–299.

Hildebrand MS, Jackson VE, Scerri TS et al. (2020). Severe childhood speech disorder: gene discovery highlights transcriptional dysregulation. Neurology 94: e2148–e2167.

Hustad KC, Beukelman DR, Yorkston KM (1998). Functional outcome assessment in dysarthria. Semin Speech Lang 19: 291–302.

Jayaraman D, Das J (2023). In: StatPearls [Internet]. StatPearls Publishing; 2025 Jan. https://www.ncbi.nlm.nih.gov/books/NBK592453/.

Kaspi A, Hildebrand MS, Jackson VE et al. (2023). Genetic aetiologies for childhood speech disorder: novel pathways co-expressed during brain development. Mol Psychiatry 28: 1647–1663.

Kooi-van Es M, Erasmus CE, de Swart BJM et al. (2020). Dysphagia and dysarthria in children with neuromuscular diseases, a prevalence study. J Neuromuscul Dis 7: 287–295.

Liégeois FJ, Mei C, Pigdon L et al. (2019). Speech and language impairments after childhood arterial ischemic stroke: does hemisphere matter? Pediatr Neurol 92: 55–59.

Liégeois FJ, Morgan AT (2012). Neural bases of childhood speech disorders: lateralization and plasticity for speech functions during development. Neurosci Biobehav Rev 36: 439–458.

Mei C, Morgan AT (2011). Incidence of mutism, dysarthria and dysphagia associated with childhood posterior fossa tumour. Childs Nerv Syst 27: 1129–1136.

Mei C, Reilly S, Bickerton M et al. (2020). Speech in children with cerebral palsy. Dev Med Child Neurol 62: 1374–1382.

Morgan AT, Mageandran SD, Mei C (2010). Incidence and clinical presentation of dysarthria and dysphagia in the acute setting following paediatric traumatic brain injury. Child Care Health Dev 36: 44–53.

Morgan AT, Vogel AP (2009). A Cochrane review of treatment for dysarthria following acquired brain injury in children and adolescents. Eur J Phys Rehabil Med 45: 197–204.

Palmer R, Enderby P, Hawley M (2007). Addressing the needs of speakers with longstanding dysarthria: computerized and traditional therapy compared. Int J Lang Commun Disord 42: 61–79.

Pennington L, Parker NK, Kelly H et al. (2016). Speech therapy for children with dysarthria acquired before three years of age. Cochrane Database Syst Rev 7: CD006937.

Sellars C, Hughes T, Langhorne P (2002). Speech and language therapy for dysarthria due to nonprogressive brain damage: a systematic Cochrane review. Clin Rehabil 16: 61–68.

Snowden C (1995). A study to investigate the value of the Revised Robertson Dysarthria Profile to differentially diagnose the motor speech characteristics of Parkinson's Disease and Multiple Sclerosis [Dissertation, Manchester Metropolitan University].

Wannberg P, Schalling E, Hartelius L (2016). Perceptual assessment of dysarthria: comparison of a general and a detailed assessment protocol. Logoped Phoniatr Vocol 41: 159–167.

World Health Organization (2001). International classification of functioning, disability and health (ICF). Geneva: World Health Organization.

Handbook of Clinical Neurology, Vol. 216 (3rd series)
The Brainstem
L.T. Grinberg and A.D. Korczyn
https://doi.org/10.1016/B978-0-443-15736-3.00017-2

Chapter 14

The brainstem in neurodegenerative diseases

LEA TENENHOLZ GRINBERG*

Departments of Laboratory Medicine, Pathology and Neurosciences, Mayo Clinic, Jacksonville, FL, United States

Abstract

The brainstem, despite its modest size relative to the cerebral cortex, is critically involved in the pathology and clinical manifestations of numerous neurodegenerative diseases (NDDs). Historically, research on NDDs such as Alzheimer disease, Lewy body disease, and frontotemporal lobar degeneration predominantly adopted a corticocentric perspective. However, emerging neuropathologic evidence underscores the brainstem's essential role, with early pathologic changes often predating cortical involvement. This chapter highlights salient points regarding the pathology and clinicopathologic correlations of brainstem involvement across major NDDs, emphasizing the chronology of disease progression. Key mechanisms, including protein misfolding and aggregation, selective neuronal vulnerability, and neurotransmitter dysfunction, are explored. Clinical correlations illustrate how early brainstem pathology significantly contributes to prodromal symptoms and helps define distinct clinical phenotypes, such as autonomic dysfunction, sleep disturbances, and mood disorders. Recognizing the chronologic order and specific nuclei affected in the brainstem broadens our understanding of disease progression, highlighting opportunities for targeted interventions at earlier disease stages.

INTRODUCTION

Neurodegenerative diseases (NDDs) encompass a broad spectrum of disorders characterized by the progressive loss of neuronal structure or function, ultimately culminating in neuronal death. These conditions pose a significant global health challenge as populations grow older and the prevalence of chronic illnesses increases (Alzheimer's Disease Facts and Figures, 2021). Although individual NDDs vary widely in their specific causes and clinical presentations, they share several fundamental mechanisms, including protein misfolding and aggregation, progressive neuronal loss, selective neuronal vulnerability, stereotypical pathologic spread, and neuroinflammation (Kovacs, 2019). From both patient and public health perspectives, the decline in cognitive or motor abilities is profoundly distressing and debilitating, frequently necessitating long-term care and substantial healthcare resources (Alzheimer's Disease Facts and Figures, 2021).

Among the most prominent neurodegenerative conditions are Alzheimer disease (AD), Lewy body disease (LBD, encompassing dementia with Lewy bodies (DLB) and Parkinson disease dementia [PDD]), limbic-predominant age-related TAR DNA-binding protein 43 (TDP-43) encephalopathy (LATE), frontotemporal lobar degeneration (FTLD, typically associated with either tau or TDP-43 deposits), multiple system atrophy (MSA), and Huntington disease (HD; Suemoto et al., 2017; Karanth et al., 2020). Despite considerable research progress, the precise causes of many neurodegenerative disorders remain only partially understood. Genetic predispositions undoubtedly play a role, as evidenced by familial forms of HD and certain cases of amyotrophic lateral sclerosis (ALS), frontotemporal FTLD, and LBD (Nalls et al., 2014; Greaves and Rohrer, 2019; Kunkle et al., 2019). However, most cases appear sporadic and are influenced by a complex interplay of genetic risk factors and environmental triggers. Aging is the most significant known risk factor for many of these disorders, as the cellular and molecular changes accompanying advancing age – such as mitochondrial decline, oxidative damage, and altered protein homeostasis – can lay the groundwork for disease onset.

The brainstem serves as a vital hub for numerous essential functions. Despite its relatively small size compared to the cerebral cortex, the brainstem is extensively connected to neuronal circuits

*Correspondence to: Lea Tenenholz Grinberg, MD, PhD, Mayo Clinic, 4500 San Pablo Road South, Jacksonville, FL 32224, United States. Tel: +1-415-502-7174; E-mail: grinberg.lea@mayo.edu.

crucial for both motor and sensory processing. In recent years, growing evidence has underscored the brainstem's pivotal involvement in a range of NDDs, highlighting its key role not only in symptom development but also in disease pathogenesis. Mechanistically, brainstem involvement draws attention to specific cellular vulnerabilities, protein misfolding and aggregation, and neurotransmitter system dysfunction. For instance, α-synuclein pathology in LBD and AD suggests that certain brainstem regions may be uniquely susceptible to harmful protein accumulation (Table 14.1). Additionally, disrupted oxidative metabolism, mitochondrial dysfunction, and neuroinflammatory processes within brainstem neurons can exacerbate disease progression, especially because some brainstem nuclei are part of the neuromodulatory subcortical system (Ehrenberg et al., 2023) and/or modulate neuroinflammatory response. Appreciating the brainstem's contributions to neurodegenerative processes is thus essential for constructing more comprehensive disease models and, ultimately, for designing new therapeutic strategies.

Clinically, recognizing brainstem involvement is critical for accurate diagnosis and developing individualized management plans. Emerging imaging techniques, such as advanced MRI protocols, nuclear medicine scans, and objective sleep measurements, are increasingly allowing the detection of subtle brainstem abnormalities. Furthermore, targeted treatments – whether pharmacologic, surgical (e.g., deep brain stimulation), or lifestyle-based – must factor in the brainstem's extensive regulatory functions to produce meaningful therapeutic benefits.

This chapter outlines current knowledge on the human brainstem's involvement in NDDs, exploring the underlying pathophysiologic mechanisms and the clinical manifestations linked to disease-specific brainstem dysfunction. It aims to provide a comprehensive perspective on the brainstem's critical role in the onset and progression of these debilitating conditions.

FROM A CORTICO-CENTRIC VIEW TO UNDERSTANDING BRAINSTEM INVOLVEMENT IN NEURODEGENERATIVE DISEASES

For much of the 20th century, a deeply ingrained "cortico-centric" perspective dominated the study of NDDs (Parvizi, 2009). Researchers and clinicians tended to focus on the cerebral cortex as the principal site of pathology, largely because many hallmark symptoms of conditions like AD and FTLD manifest through cortical dysfunction – such as memory loss, language deficits, and changes in executive function. This cortex-first approach shaped diagnostic criteria, therapeutic strategies, and our broader conceptual framework for understanding neurodegeneration. Subcortical structures, including the brainstem, were often regarded as secondary players (Ehrenberg et al., 2023).

This perspective stemmed, in part, from historical biases and methodological constraints. Nineteenth-century scientists, under the influence of hierarchical models of evolution and society, conceptualized the brain as a vertical neuroaxis: the further "up" one went, the more "advanced" the functions. As a result, the field overlooked compelling evidence that many NDDs begin subcortically. For instance, although it had long been noted that tau pathology in AD consistently originates in specific brainstem nuclei (Stratmann et al., 2016) before spreading cortically, these observations did not receive the attention they deserved, overshadowed by an emphasis on hippocampal and cortical tau aggregates.

A major turning point came with the advent of refined immunohistochemical techniques in the past 30 years. By tagging proteins such as tau, α-synuclein, and TDP-43, immunohistochemistry helped to identify pathologic inclusions in earlier and more diverse brain regions. This approach revealed that, in many NDDs, subcortical structures – including brainstem nuclei – harbor

Table 14.1

Brainstem nuclei involvement across neurodegenerative conditions.

Neurodegenerative disease	Early affected nuclei	Primary symptoms	Pathologic hallmark proteins
Lewy body disease	DMV, LC, SN	GI symptoms, sleep disorders	α-Synuclein
Alzheimer disease	LC, DRN	Mood disturbances, sleep issues	Tau, β-amyloid
PSP	SN, superior colliculus	Gaze palsy, early falls	4-repeat tau
MSA	Medulla, pontomedullary reticular formation	Autonomic failure, ataxia	α-synuclein (GCIs)
Huntington disease	EBR, superior olive, raphe nuclei	Balance, ocular motor issues	Huntingtin
Spinocerebellar ataxias	Pontine nuclei, inferior olive	Coordination deficits, ataxia	Polyglutamine expansions
FTLD-TDP	Superior colliculus, substantia nigra	Oculomotor deficits, Parkinsonism	TDP-43
ALS	Hypoglossal nucleus, pre-Bötzinger complex	Bulbar dysfunction, respiratory failure	TDP-43, SOD1, FUS

ALS, Amyotrophic lateral sclerosis; *DMV*, dorsal motor nucleus of the vagus nerve; *DRN*, dorsal raphe nucleus; *EBR*, Edinger–Westphal nucleus; *FTLD-TDP*, frontotemporal lobar degeneration with TDP-43 pathology; *FUS*, fused in sarcoma protein; *GCIs*, glial cytoplasmic inclusions; *GI*, gastrointestinal; *LC*, locus coeruleus; *MSA*, multiple system atrophy; *PSP*, progressive supranuclear palsy; *SN*, substantia nigra; *SOD1*, superoxide dismutase 1; *TDP-43*, TAR DNA-binding protein 43.

pathologic changes years before overt cortical degeneration is evident (Braak et al., 2003a; Simic et al., 2009; Kovacs et al., 2020). In LBD, for example, α-synuclein inclusions often appear in the dorsal motor nucleus of the vagus nerve (DMV) in the medulla and the locus coeruleus (LC) in the pons long before classic motor or cognitive symptoms emerge (Seidel et al., 2015). Meanwhile, in AD, refined Braak staging preconizes that tau aggregates begin in the brainstem long before cortical involvement (Braak et al., 2011).

The subsequent rise of advanced neuroimaging techniques, such as high-resolution structural MRI, diffusion tensor imaging, and novel PET tracers, has reinforced this shift in perspective as subtle brainstem changes that previously went undetected can be detected (Betts et al., 2019; Bueicheku et al., 2024). Clinical correlations – ranging from autonomic dysfunction and sleep disturbances to early mood and cognitive alterations – have further highlighted the brainstem's integrative role in NDDs (Benarroch et al., 2004, 2006; Boeve et al., 2007; Dickson et al., 2010; Rub et al., 2016; Saper and Fuller, 2017; Lew et al., 2021).

BRAINSTEM INVOLVEMENT IN THE MAIN NEURODEGENERATIVE DISEASES

Lewy body disease

LBD is an umbrella term that includes clinically defined Parkinson disease (PD), PDD, and DLB. These conditions share the pathologic hallmark of α-synuclein accumulations within neuronal inclusions known as Lewy bodies and Lewy neurites.

Historically, research on LBD focused on cortical and midbrain pathology – particularly dopaminergic cell loss in the substantia nigra (SN). However, it is widely accepted now that α-synuclein pathology often originates within the lower brainstem (Braak et al., 2003b), highlighting that early dysfunction in these nuclei can precede the classic motor and cognitive symptoms commonly associated with PD, PDD, or DLB, the so-called nonmotor symptoms (Grinberg et al., 2010; Jellinger, 2015; van Wamelen et al., 2022). Understanding this chronology of brainstem involvement is essential for unraveling the pathogenesis, refining early diagnostic strategies, and developing targeted therapies for LBD.

The most influential model describing the progression of α-synuclein pathology through the central nervous system (CNS) comes from Braak et al. (2003b). Their staging framework for PD (and related synucleinopathies) maps out how pathologic aggregates move from the lower brainstem upward, and it is divided into stages 0–6:

- **Stage 1** – Pathology begins in the olfactory structures and the lower brainstem. The DMV in the medulla oblongata is especially noteworthy for exhibiting early α-synuclein inclusions (Lewy bodies and neurites). The DMV is intimately connected with parasympathetic control of the gastrointestinal tract and various autonomic functions. Gastrointestinal symptoms – such as constipation, delayed gastric emptying, or nausea – often appear years or even decades before the hallmark motor signs (Jost and Klingelhoefer, 2022). The pathologic insult to the DMV and its projecting pathways helps explain why these autonomic disturbances are not mere "side symptoms" but can be fundamental indicators of prodromal disease. At Stage 1, α-synuclein inclusions may also appear in the olfactory bulb in isolation or concomitant with lower brainstem pathology (Hawkes et al., 2007; Borghammer et al., 2022).
- **Stage 2** – α-synuclein lesions expand to the LC in the pons and magnocellular regions of the reticular formation. As the LC is the primary source of noradrenergic input to much of the brain, which modulates several functions, early LC pathology correlates with several nonmotor manifestations, ranging from sleep disturbances to mood disorders. Indeed, depression and anxiety are remarkably common in LBD, and their onset can precede motor deficits by years (Tsai and Gopalakrishna, 2022).
- **Stage 3** – Pathology extends further into the midbrain, targeting the substantia nigra pars compacta (SNc) (Dickson, 2018). This stage precedes the onset of the classic motor symptoms (resting tremor, rigidity, bradykinesia) commonly associated with PD and usually manifests when a substantial portion of nigral neurons die. Moreover, involvement of the pedunculopontine nucleus (PPN) can contribute to gait problems and postural instability. Additionally, pathology often spreads to the raphe nuclei, which produce serotonin, affecting mood, pain modulation, and various autonomic functions. Widespread involvement of these core brainstem neurotransmitter systems (dopaminergic, noradrenergic, serotonergic, and cholinergic) creates a complex clinical profile that includes autonomic, motor, and psychiatric manifestations, underscoring why LBD are so heterogeneous.

Certainly, individual patient variability exists, and some may diverge from this neat staging system (Attems et al., 2021). Nevertheless, while Braak's model was originally formulated to describe PD, the general pattern of α-synuclein spread in other Lewy body disorders (including DLB) shows similarities, with initial brainstem involvement preceding extensive cortical pathology (Halliday et al., 2008; Jellinger and Korczyn, 2018). Still, the fundamental observation remains: key nuclei in the lower brainstem are among the earliest sites of α-synuclein deposition.

Rapid eye movement (REM) sleep behavior disorder (RBD) is among the most conspicuous and telling prodromal features of LBD (Boeve, 2013). It involves dream-enactment behaviors due to a failure of the normal atonia that should occur in REM sleep. The pathology in pontine structures, including the sublaterodorsal nucleus and ventral medulla, disrupts the networks responsible for suppressing muscle activity during dreams (Grinberg et al., 2010; Stefani et al., 2025). Longitudinal studies show that many individuals with idiopathic RBD eventually develop a synucleinopathy (most frequently LBD). Thus, these early pontine changes can presage a future diagnosis by years and offer a potential window for early intervention (Sobreira-Neto et al., 2023).

The recognition that α-synuclein pathology often commences in the brainstem has shifted clinical thinking toward earlier, more holistic diagnostic criteria. Nonmotor symptoms now play a larger role in identifying prodromal and early-stage LBD (Non-Motor Parkinson's Disease, 2022). Polysomnography for RBD detection, specialized autonomic function tests, and advanced neuroimaging (e.g., neuromelanin-sensitive MRI of the LC, diffusion tensor imaging, and PET tracers specific for α-synuclein or related pathways) can all be employed to detect subtle brainstem abnormalities before motor symptoms dominate. Therapeutically, this evolving understanding points to the need for interventions that go beyond the dopaminergic system. Targeting noradrenergic, cholinergic, or serotonergic pathways might help manage the wide array of nonmotor symptoms, especially in the early stages. Moreover, experimental strategies aiming to curb α-synuclein misfolding or propagation (e.g., immunotherapies) could be most effective if administered before significant neuronal loss occurs, potentially during the subclinical phase when pathology is confined largely to brainstem nuclei.

Alzheimer disease

AD is widely recognized as the most common cause of dementia. Clinically, most cases are characterized by progressive memory impairment, along with a decline in executive function, language abilities, and other cognitive domains. A smaller subset of patients presents with early language, behavioral, motor, or visual symptoms before clear memory loss emerges. These are classified as atypical forms of AD.

Neuropathologically, AD is defined by two hallmark lesions: extracellular β-amyloid plaques and intracellular neurofibrillary tangles (NFTs) composed of hyperphosphorylated tau. Notably, the progression of these pathologies follows distinct anatomic trajectories. β-Amyloid accumulation begins in the neocortex and subsequently involves the allocortex, basal ganglia, brainstem, and cerebellum (Thal et al., 2002). In contrast, NFTs typically originate in subcortical nuclei of the neuromodulatory subcortical systems (NSS), particularly in regions such as the LC, before spreading to the allocortex, most notably the hippocampus and entorhinal cortex, and then to the neocortex (Braak and Braak, 1991; Ehrenberg et al., 2023).

Historically, both research and clinical focus have centered on cortical and limbic areas, especially the allocortex, as the earliest and most functionally relevant regions affected in AD (Parvizi, 2009). However, growing evidence suggests that tau pathology may originate outside these canonical areas, particularly within the NSS, which play critical roles in regulating mood, arousal, sleep, and autonomic functions (Ehrenberg et al., 2023).

Although the symptoms linked to early involvement of these subcortical structures can be less overt than the hallmark memory deficits of later stages, their presence may be key to understanding the earliest phases of AD. Involvement of the brainstem and other NSS components has been associated with noncognitive symptoms such as mood disturbances, sleep–wake disruptions, and autonomic dysfunction – features that often precede or accompany the cognitive decline. Recognizing these early changes is crucial for redefining the timeline of disease progression and for developing strategies aimed at early diagnosis and intervention.

Heiko Braak and Eva Braak developed a classic staging scheme for NFTs in AD, primarily emphasizing medial temporal lobe and cortical progression (Braak and Braak, 1991). Yet, subsequent investigators by other authors and reexamination of the Braak's collection noted that certain subcortical regions, particularly the NSS, often harbor early tau pathology, sometimes prior to or alongside the earliest detectable changes in the entorhinal cortex (Rub et al., 2016). This observation has led to refined or alternative staging proposals that better account for the possibility that AD-associated pathology (especially tau-related changes) can begin in the brainstem (Simic et al., 2009). It is unclear if these nuclei develop NFT at the same time or if there is a chronology there. The most vulnerable brainstem nuclei to NFT are part of the isodendritic core, such as the dorsal raphe nucleus (DRN) and LC, and belong to the NSS (Grinberg et al., 2011). Conversely, cranial nerves (CNs) are spared in AD. Neuropathologic studies have shown that, in a subset of individuals, AD-NFT pathology occurs in the brainstem in the absence of β-amyloid deposition. Since the current diagnostic criteria for AD require the presence of β-amyloid pathology, there has been ongoing debate as to whether these individuals would have eventually developed β-amyloid pathology had they lived longer, or whether they represent a distinct condition. To address this, a working group introduced the term *Primary Age-Related Tauopathy* (PART) to classify these cases and facilitate further study (Crary et al., 2014; Duyckaerts et al., 2015). Evidence to date indicates that individuals lacking β-amyloid pathology tend to exhibit milder clinical symptoms, particularly those related to cortical dysfunction, consistent with the observation that β-amyloid accelerates the spread of tau pathology. Interestingly, cryo-electron microscopy studies have revealed that the structure of NFTs is similar in cases with and without β-amyloid, suggesting that both may represent manifestations of a shared pathologic process (Shi et al., 2021).

Traditionally, evidence for early brainstem pathology in AD was obtained postmortem. However, new imaging strategies are helping detect subtle brainstem changes in living patients. High-resolution MRI protocols (including neuromelanin-sensitive imaging) can reveal changes in the LC. Recent studies have used neuromelanin-sensitive MRI to demonstrate that LC signal reductions correlate with disease severity and cognitive measures (Betts et al., 2019; Bueicheku et al., 2024). Similarly, functional imaging may highlight disruptions in arousal or stress-response circuits mediated by the LC and other brainstem nuclei.

Earliest sites

The **LC**, located in the pons, is the brain's principal noradrenergic nucleus, projecting widely to cortical and subcortical structures. Noradrenergic signaling modulates attention, arousal, stress responses, and aspects of memory consolidation. Several postmortem studies have consistently shown that the LC frequently contains tangles of hyperphosphorylated tau in very early stages – sometimes even in individuals without overt cognitive decline (Theofilas et al., 2017) (Arendt et al., 2015). This aligns with a growing body of work suggesting that subclinical degeneration in the LC may contribute to subtle mood disturbances, disruptions in alertness, and stress reactivity long before more classic AD symptoms arise. LC dysfunction accelerates AD through multiple mechanisms (Van Egroo et al., 2024). As the brain's primary norepinephrine (NE) source, LC degeneration reduces NE, impairing neuroprotection, increasing oxidative stress, and promoting neuroinflammation. Additionally, LC dysfunction exacerbates amyloid accumulation by failing to suppress microglial overactivation in experimental models (Heneka et al., 2006). These findings suggest that preserving LC integrity could be a therapeutic strategy to slow AD progression.

DRN and other brainstem targets: Serotonergic neurons located in the DRN also show tau deposition in early or prodromal stages of AD (Grinberg et al., 2009). Serotonin plays a crucial role in mood, appetite, and sleep–wake regulation. Consequently, involvement of the DRN may help explain common neuropsychiatric symptoms, such as depression, anxiety, and sleep disturbances, that often accompany or precede cognitive decline in AD (Pierson et al., 2025). Experimental overexpression of mutant tau in the DRN of mice induced depressive-like behaviors and serotonergic dysfunction, suggesting that tau pathology in the DRN can lead to neuropsychiatric symptoms associated with AD (Pierson et al., 2025).

Additionally, **other cholinergic and noncholinergic groups** in the pontine and midbrain areas, such as the ventral tegmental area (VTA) and periaqueductal gray, can show early pathology. While less studied than the LC or DRN, their dysfunction can give rise to varied symptoms that are not always recognized as part of the AD spectrum, underscoring the importance of a broader diagnostic lens.

Clinical correlates of brainstem involvement in Alzheimer disease

Neuropsychiatric symptoms

Depression, anxiety, irritability, apathy, and other mood or behavioral changes occur frequently in people with AD, sometimes manifesting in mild cognitive impairment or even earlier. Such symptoms can relate to the progressive loss of monoaminergic neurons that modulate emotion and behavior. This mechanism suggests that addressing subcortical pathology might mitigate or at least slow certain neuropsychiatric aspects of AD. A clinicopathologic study with over 400 cases in the AD pathologic spectrum and excluding pathologies suggested an increased risk of agitation, depression, nocturnal behavior, and appetite changes preceding cortical involvement by AD pathology (Ehrenberg et al., 2023).

Sleep–wake cycle disturbances

Sleep disturbances are increasingly recognized as an early marker of AD pathophysiology, with chronic insomnia, fragmented sleep, and changes in circadian rhythm preceding more overt cognitive deficits. The brainstem serves as a central hub for regulating sleep–wake transitions. Of importance, key regulators of wakefulness and alertness, such as the LC and DRN, are early affected by tau pathology (Lew et al., 2021). Thus, even early AD stages can lead to dysregulated sleep–wake cycles (Oh et al., 2022). Additionally, poor sleep quality may exacerbate amyloid or tau accumulation, creating a vicious cycle that accelerates disease progression.

In sum. recognition that AD pathology extends to the brainstem in its earliest stages broadens the range of potential therapeutic targets. While most current treatments aim to modulate cortical or hippocampal neurotransmission (e.g., cholinesterase inhibitors, NMDA receptor antagonists), emerging strategies may focus on preventing or slowing the spread of tau pathology from the LC, DRN, or other brainstem nuclei to cortical regions (Fig. 14.1).

Tauopathies

Tau proteins are central to a range of neurodegenerative disorders known as "tauopathies." Unlike AD, which also involves β-amyloid pathology, *primary tauopathies* feature tau dysfunction as the principal driver of neuronal injury (Kovacs et al., 2022). These disorders include progressive supranuclear palsy (PSP), corticobasal degeneration (CBD), Pick's disease (PiD), and certain forms of genetic FTLD with MAPT mutation. Despite differences in clinical profiles – ranging from parkinsonian features to cognitive and language impairments, tau inclusion distribution, and tau ultrastructure – the brainstem is usually affected (Irwin, 2016).

Tau normally stabilizes neuronal microtubules, ensuring proper axonal transport. In tauopathies, various pathologic processes – such as hyperphosphorylation and misfolding – cause tau to aggregate into filamentous inclusions. These appear in neurons (NFTs), astrocytes (tufted astrocytes), or oligodendrocytes (coiled bodies), damaging cellular function and ultimately leading to neuron loss. Different primary tauopathies often exhibit unique tau isoform profiles: some predominantly accumulate 3-repeat (3R) tau, others 4-repeat (4R) tau, and still others contain mixtures of these isoforms. They also have different ultrastructural folds which impact the pattern of tau distribution and selective neuronal vulnerability, making it critical to examine each tauopathy independently (Kovacs et al., 2022).

Progressive supranuclear palsy

PSP is the most prevalent 4-repeat tauopathy. PSP is not a singular clinical-pathologic entity but rather a spectrum disorder comprising multiple subtypes, each with varying degrees and patterns of brain involvement. The classical Richardson's syndrome (PSP-RS) is characterized by early postural instability, vertical gaze palsy, and pronounced midbrain atrophy, making the brainstem – particularly the SN, tectum and pontine tegmentum – a central site of

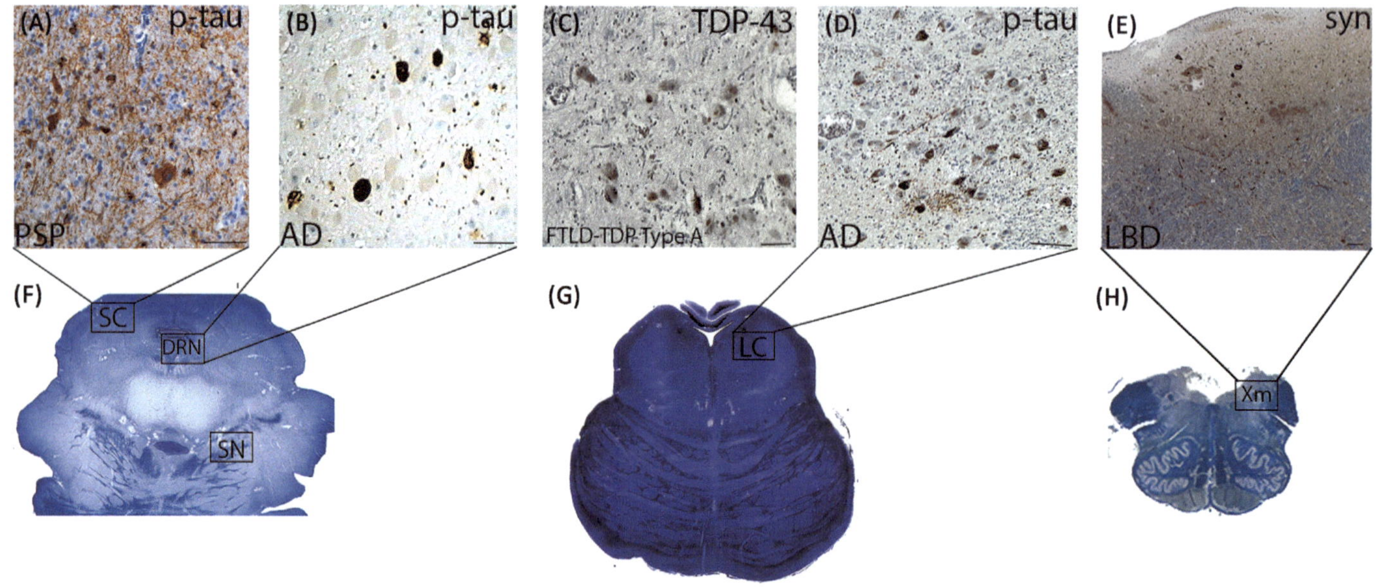

Fig. 14.1. Neurodegenerative proteinopathies affecting distinct brainstem nuclei. Representative immunohistochemistry for disease-defining proteins in selected brainstem regions. (A) Phospho-tau (p-tau) accumulation in the superior colliculus (SC) in progressive supranuclear palsy (PSP). (B) p-tau–positive neurofibrillary tangles in the dorsal raphe nucleus (DRN) in Alzheimer's disease (AD). (C) Phosphorylated TDP-43 inclusions in the midbrain tegmentum in frontotemporal lobar degeneration with TDP-43 type A pathology (FTLD-TDP type A). (D) p-tau pathology in the locus coeruleus (LC) in early stage AD. (E) α-synuclein (syn) pathology in the dorsal motor nucleus of the vagus (Xm) in Lewy body disease (LBD). (F–H) Corresponding Nissl-stained sections illustrating the anatomic localization of sampled nuclei. Scale bars: 10 μm.

pathology. Other subtypes, such as PSP-parkinsonism and PSP with speech/language disorder, display more asymmetric or regionally confined tau pathology, with some sparing of brainstem structures early in the disease. Nevertheless, neuropathologic studies consistently demonstrate that brainstem degeneration is a unifying feature across PSP variants (Kovacs et al., 2020).

PSP-RS, the most common PSP subtype, prominently involves the midbrain from the early stages (Kovacs et al., 2020). Atrophy of the midbrain tegmentum near the superior colliculi produces the classic "hummingbird sign" or "penguin sign" on sagittal MRI. Tau inclusions (often in the form of tufted astrocytes or globous neuronal inclusions) can appear in the superior colliculus, oculomotor nucleus, SN, pontine reticular formation, the PPN, the raphe nuclei, and the LC (Roemer et al., 2022). The degeneration of these nuclei leads to a broad constellation of motor and nonmotor symptoms:

Eye movement abnormalities

A distinctive hallmark of PSP-RS is the difficulty with voluntary vertical gaze, particularly downward gaze (Garbutt et al., 2008). Over time, patients may lose voluntary control of eye movements altogether, although reflexive eye movements (e.g., via the vestibulo-ocular reflex) can remain relatively intact until later stages. These supranuclear eye movement issues reflect pathology in the midbrain structures that coordinate vertical gaze, notably the superior colliculus, oculomotor nucleus, rostral interstitial nucleus of the medial longitudinal fasciculus (riMLF), and the interstitial nucleus of Cajal.

Sleep disorders

Hyper-insomnia in PSP is a result of a disbalance between severe and early involvement of sleep-promoting nuclei with the relative presentation of wake-promoting nuclei (Kaalund et al., 2020; Oh et al., 2022; Rastegar-Pouyani et al., 2025). New evidence suggests that non-REM sleep-promoting neurons also reside in the brainstem, and future research should examine these areas in PSP.

Other brainstem-related clinicopathologic correlations

1. Substantia nigra (SN)– Severe neuronal loss in the SN leads to akinesia, rigidity, and bradykinesia, contributing to PSP's parkinsonism, though unresponsive to levodopa (Morris et al., 1999; Dickson et al., 2007).
2. Pendiculopontine nucleus (PPN) – Degeneration of this cholinergic wake-promoting nucleus is linked to postural instability, early falls, and gait freezing, distinguishing PSP from PD. Also the PPN is critical for generating and modulating REM sleep.
3. Dorsal raphe nucleus (DRN) – Loss of serotonergic neurons contributes to apathy, depression, and sleep disturbances, common in PSP.
4. Medullary reticular formation – Involvement of these circuits leads to bulbar dysfunction, including dysarthria and dysphagia, worsening disease progression.

Corticobasal degeneration

CBD, another 4 R tauopathy, often presents with asymmetric rigidity, dystonia, cortical sensory loss, and the so-called "alien limb" phenomenon. Many patients exhibit a parkinsonian syndrome that overlaps with PSP or other atypical parkinsonian disorders, leading to diagnostic challenges. While CBD is traditionally associated with frontoparietal cortical atrophy and basal ganglia pathology, brainstem structures – including the SN pars compacta – are also significantly affected. In fact, the SN is often severely degenerated in CBD, resulting in parkinsonism (Dickson et al., 2011). Tau inclusions may also appear in pontine nuclei, but less prominently than in PSP. Sleep disorders seem to be less frequent and severe than in PSP and LBD, another distinguishable point (Oh et al., 2022).

Pick's disease

PiD is characterized by round aggregates known as Pick bodies, predominantly composed of 3R tau isoforms. Clinically, it often presents as a behavioral-variant frontotemporal dementia (bvFTD) or nonfluent primary progressive aphasia. While the cortex is disproportionately involved, subcortical regions can be affected to a lesser degree. Braak and Braak noted conspicuous pathology in precerebellar nuclei, which receive major input from the cerebral cortex, such as pontine gray, the arcuate nucleus, the pontobulbar body, and the paramedian reticular nucleus (Braak and Braak, 1998). As opposed to PSP and CBD, the SN is usually spared.

Familial FTLD-tau variants

FTLD-tau caused by mutations in the MAPT gene often presents with a spectrum of clinical and pathologic features, including prominent brainstem involvement, depending on the mutation. For instance, PSP-like pathology is found in several of these mutations, such as R5L, N279K, S285R, S303S, and S305S (Forrest et al., 2018). Within the brainstem, several nuclei are notably affected in several of these mutations, including the SN, LC, raphe nuclei, Pontine nuclei, inferior olivary nucleus, and periaqueductal gray region (Ghetti et al., 2015).

TDP-43 proteinopathies

TDP-43 proteinopathies includes neurodegenerative conditions with pathologic accumulation of TDP-43 in neuronal and glial cytoplasm. TDP-43 proteinopathies are classified as FTLD types A, B, C, D, and E or the age-related limbic-predominant type (LATE-NC) (Mackenzie et al., 2011; Nelson et al., 2019). TDP-43 inclusions are also the main molecular signature of ALS (Cairns et al., 2007). While cortical involvement (particularly in FTLD) and spinal motor neuron loss (in ALS) are well recognized, the brainstem is frequently implicated in TDP-43 pathophysiology, especially motor, preocular, and precerebellar nuclei (Grinberg et al., 2011). LATE-NC is a relatively newly recognized entity and less is known about brainstem involvement in this condition.

Though traditionally viewed as a cortical disorder, FTLD with TDP-43 pathology (FTLD-TDP) also affects the brainstem in a distinct, subtype-specific manner. Among the most consistently affected regions is the superior colliculus, a midbrain structure involved in saccadic eye movements and visuomotor coordination (Grinberg et al., 2011). While overt oculomotor dysfunction is not a hallmark of FTLD-TDP, subclinical abnormalities in saccadic initiation, smooth pursuit, and visual attention may reflect this involvement – especially in patients with bvFTD who exhibit gaze apraxia (Munoz and Everling, 2004).

Other preocular premotor nuclei, such as the raphe interpositus, reticulotegmental nucleus, and the riMLF, show variable TDP-43 pathology across subtypes. These nuclei are critical for vertical gaze control, and while frank vertical gaze palsy is uncommon in FTLD-TDP, subtle impairments in upgaze or downgaze saccades have been observed on quantitative oculomotor testing, especially in patients with overlapping corticobasal syndrome features. Involvement of precerebellar nuclei, such as the pontine nuclei and inferior olivary complex, supports the presence of subtle cerebellar motor signs, including gait instability, limb incoordination, or dysmetria – symptoms sometimes attributed to overlapping motor phenotypes or misclassified as ALS-Plus syndromes. This precerebellar degeneration may also contribute to the spastic dysarthria and dysphagia seen in FTLD-TDP cases with bulbar involvement. Perhaps most compelling is the observation of TDP-43 pathology in the SN across all FTLD-TDP subtypes. This dopaminergic midbrain region is essential for basal ganglia function, and its degeneration aligns with extrapyramidal features, including parkinsonism, bradykinesia, and rigidity, which are occasionally observed in FTLD-TDP. These motor signs may be misdiagnosed as PD in early stages, delaying appropriate recognition of the underlying FTLD-TDP spectrum.

The involvement of nonmotor brainstem nuclei also carries significant clinical implications (Geser et al., 2009). The DRN, a key serotonergic center, shows TDP-43 accumulation in many cases, possibly contributing to neuropsychiatric manifestations such as depression, anxiety, impulsivity, or emotional blunting – frequent but under-recognized features of FTLD-TDP (Brettschneider et al., 2014). In contrast, the LC, a noradrenergic nucleus implicated in arousal, attention, and autonomic regulation, tends to be relatively spared. Altogether, these findings reinforce that brainstem involvement in FTLD-TDP is integral to its pathophysiology. Recognizing these correlations could improve diagnostic accuracy, support biomarker development, and inform targeted symptom management, especially in cases that deviate from classical frontotemporal presentations.

Amyotrophic lateral sclerosis

ALS is a progressive and fatal NDD that primarily targets motor neurons. While much of the research and clinical focus has historically centered on degeneration of cortical and spinal motor neurons, involvement of the brainstem is equally critical, especially in relation to bulbar dysfunction and respiratory failure, which are major determinants of symptom burden and survival. The molecular pathology of ALS is heterogeneous. Mutations in superoxide dismutase 1 (SOD1) and Fused in Sarcoma (FUS) define distinct subtypes of the disease, but the most common pathologic hallmark involves cytoplasmic aggregation of TDP-43. For decades, ALS-TDP and FTLD-TDP have been recognized as sharing a common molecular spectrum. However, there remains ongoing debate about whether these diseases originate from similar or distinct anatomic sites (Liscic et al., 2008; Brettschneider et al., 2013).

While brainstem pathology varies across ALS subtypes (Piao et al., 2003; Bede et al., 2019), most forms share overlapping patterns of involvement.

Motor nuclei in the rostral brainstem, such as those controlling eye movement (CN III, IV, and VI), are typically spared in ALS. The trigeminal motor nucleus (CN V) is also usually preserved or only mildly affected. In contrast, caudal cranial nerve motor nuclei are significantly impacted. The hypoglossal nucleus (CN XII), located in the medulla, is often affected early, leading to tongue atrophy, fasciculations, and dysarthria. Involvement of the glossopharyngeal and vagus nuclei (CN IX, X) results in swallowing dysfunction, dysphagia, and choking episodes, particularly in bulbar-onset ALS. The facial nerve nucleus (CN VII) is commonly affected, resulting in facial weakness, reduced emotional expression, difficulty sealing the lips, and sialorrhea. In bulbar-onset ALS, facial involvement often occurs early; in spinal-onset ALS, it appears later in the disease course.

As ALS progresses, respiratory dysfunction becomes a leading cause of morbidity and mortality. Beyond motor neuron loss in the phrenic and intercostal motor pools, the disease affects key medullary respiratory centers, such as the pre-Bötzinger complex, which is essential for generating respiratory rhythm. The pontomedullary reticular formation may also be involved, contributing to breathing irregularities. In some cases, subtle involvement of autonomic brainstem centers may lead to heart rate **or** blood pressure variability. The red nucleus and precerebellar nuclei have also been described as involved in ALS-FTD, a similarity with FTLD-TDP (Brettschneider et al., 2013).

Diagnosing ALS with prominent brainstem involvement typically relies on a combination of clinical examination, including assessment of speech and swallowing, and electromyography to detect signs of bulbar denervation. MRI and other imaging tools are useful to exclude structural lesions but are not usually diagnostic for ALS itself.

In conclusion, brainstem degeneration in ALS is central to many of the disease's most debilitating and life-threatening symptoms. Early involvement of cranial nerve nuclei and respiratory centers underscores the importance of timely diagnosis and intervention. Recognizing these features early is crucial to preserving function, maintaining quality of life, and potentially prolonging survival in affected individuals.

Multiple system atrophy

MSA is a rare, progressive neurodegenerative disorder defined by the presence of α-synuclein-positive glial cytoplasmic inclusions (GCIs – Papp–Lantos bodies) primarily affecting oligodendrocytes, classifying it within the spectrum of synucleinopathies (Jellinger, 2014). Neuronal inclusions and loss are also present. Clinically, MSA features a combination of autonomic dysfunction, parkinsonism, cerebellar ataxia, and pyramidal signs, and it has been classically divided into two main subtypes: MSA with predominant parkinsonism (MSA-P) and MSA with predominant cerebellar features (MSA-C). Although MSA affects multiple regions of the CNS, the brainstem plays a particularly critical role in the manifestation of clinical symptoms and disease progression (Dickson, 2012).

In MSA, the most prominently affected brainstem nuclei include those modulating central autonomic, nigral, and olivopontocerebellar systems. Postmortem human studies consistently reveal significant neuronal loss and gliosis in these regions, often correlating with the clinical symptoms experienced by patients during life (Wenning et al., 1997; Ozawa et al., 2004). Early autonomic dysfunction in MSA – manifesting as orthostatic hypotension, urogenital disturbances, and respiratory irregularities – reflects degeneration in medullary structures such as the nucleus tractus solitarius and the ventrolateral medulla (Benarroch et al., 1998; Ozawa, 2007). These regions are integral to baroreflex and chemoreflex pathways. Furthermore, involvement of the pontomedullary reticular formation and pre-Bötzinger complex likely underlies the frequent and life-threatening respiratory complications, such as inspiratory stridor and central sleep apnea (Schwarzacher et al., 2011). Neuropathologic studies underscore the severe atrophy of the pons and medulla in MSA, particularly in MSA-C, where degeneration of the pontocerebellar fibers contributes to cerebellar ataxia (Ozawa et al., 2004). In MSA-P, degeneration of the SN parallels the parkinsonian features observed clinically. Importantly, imaging studies such as MRI frequently reveal brainstem atrophy – especially the "hot cross bun" sign in the pons – as a strong predictor of MSA, reflecting loss of transverse pontocerebellar fibers and longitudinal pontine raphe neurons.

The presence of GCIs within the brainstem oligodendrocytes has been associated with local demyelination and axonal degeneration, potentially impairing neural network integrity in autonomic and motor circuits. Human postmortem analyses have further clarified the progression of pathology in MSA, suggesting that brainstem involvement often precedes more widespread neurodegeneration, reinforcing the hypothesis that early targeting of these nuclei might offer a window for therapeutic intervention (Jellinger, 2014). Thus, the clinicopathologic correlation between brainstem degeneration and symptomatology in MSA is both robust and clinically informative.

In conclusion, the brainstem is a central pathologic target in MSA, with its degeneration underpinning many of the disease's hallmark clinical features. Continued research utilizing human brain tissues remains essential for elucidating the mechanisms of brainstem dysfunction in MSA and for guiding the development of early diagnostic markers and therapeutic strategies.

Huntington disease

HD is an autosomal dominant neurodegenerative disorder caused by an expanded CAG repeat in the HTT gene. It commonly manifests in mid-adult life with motor dysfunction (often chorea), psychiatric disturbances, and progressive cognitive impairment. Although the striatum (caudate and putamen) has long been considered the primary pathologic site, growing evidence indicates that brainstem structures also undergo early and progressive changes. Additionally, the presence of respiratory irregularities, mood and sleep disorders, and ocular motor abnormalities in HD suggests that brainstem pathology may influence diverse autonomic and sensorimotor functions, underscoring the need for more targeted investigation. Most of the few anatomic studies on brainstem involvement were led by the late Udo Rub and colleagues, whose work highlighted under-recognized clinical features of HD (Rüb et al., 2015).

In a pivotal study of eight genetically confirmed cases, Rub et al. (2014) documented consistent neuronal loss in the midbrain, pons, and medullary regions. The most affected nuclei were the Edinger–Westphal, superior olive, and raphe interpositus nucleus, each contributing to horizontal saccades, balance, or rhythmic modulation, respectively. Additionally, mild neuronal depletion in the SN – encompassing its dopaminergic pars compacta and GABAergic pars reticulata – may underlie parkinsonian features such as bradykinesia and rigidity in advanced HD. Neuronal loss also appeared in the precerebellar pontine nucleus, reticulotegmental nucleus of the pons, and the lateral and medial vestibular nuclei. Since the vestibular nuclei and superior olive are crucial for equilibrium and spatial orientation, their degeneration could exacerbate postural and gait disturbances. Interestingly, the degree of brainstem pathology did not correlate with CAG repeat length, age at onset, disease duration, or striatal atrophy, suggesting a partially independent disease pathway.

In the same study, immunoreactive intranuclear inclusions of mutant huntingtin were detected in nuclei that seemed relatively preserved, implying a broader molecular impact even before overt neuronal loss. p62-positive axonal inclusions were also identified in nearly all major brainstem fiber tracts, underscoring axonal pathology's role in HD progression. Another study points to subtle cellular changes preceding measurable neuron loss: Jahanshahi et al. (2013) found in rat models and human HD that, despite normal neuron counts in the DRN (B6 region in rats), there was a higher proportion of dopaminergic cells. Because the dorsal raphe is predominantly serotonergic, this shift could influence mood disturbances and perhaps chorea, given that dopaminergic pathways have been linked to hyperkinetic movements. Intriguingly, a transcranial sonography investigation in HD patients – some with depressive symptoms and others without – revealed a relationship between depression severity and pathologic raphe echogenicity (Krogias et al., 2011).

In summary, recognizing brainstem pathology is vital to understanding HD in its entirety and discovering new therapeutic targets. Integrating insights from both striatal and brainstem involvement is key to develop more comprehensive strategies for treating this complex disorder.

Spinocerebellar ataxias

Spinocerebellar ataxias (SCAs) are a group of hereditary neurodegenerative disorders in most cases, primarily affecting the cerebellum and its connections (Klockgether et al., 2019). However, substantial evidence shows that brainstem structures, from cranial nerve nuclei in the medulla to pontine and midbrain pathways, are deeply involved in disease processes. These subcortical changes are key to understanding the complex presentation of SCAs, which often includes coordination deficits (ataxia), bulbar symptoms, and autonomic dysfunction (Rub et al., 2013).

SCAs comprise dozens of subtypes, each linked to specific mutations:

- **Polyglutamine (PolyQ) expansions** (SCA1, SCA2, SCA3, SCA6, SCA7, SCA17): Caused by expanded CAG repeats coding for elongated polyQ tracts in different proteins. The resultant misfolded proteins aggregate within neurons, disrupting normal function. It is usually of adult onset.
- **Non-PolyQ mutations** (SCA8, SCA10, and others) arise from diverse mechanisms, ranging from large untranslated repeat expansions to point mutations. Regardless of subtype, neuronal integrity in both the cerebellum and the brainstem can be severely affected.

Though "spinocerebellar" emphasizes the cerebellum and spinal tracts, mounting data confirm that brainstem involvement is ubiquitous in many SCAs, often preceding or paralleling cerebellar degeneration. The pons and medulla, in particular, harbor critical relay stations for sensory and motor circuits that interface closely with the cerebellum. Dysfunction here intensifies clinical ataxia and adds features like bulbar palsies, oculomotor disturbances, and autonomic issues. Thus, neurologic examinations in SCAS should always scrutinize bulbar and oculomotor functions. MRI can reveal atrophy in the cerebellar vermis and hemispheres and the pons. Genetic testing remains essential for definitive identification of the specific SCA mutation, but correlating clinical brainstem signs with imaging helps refine the differential.

Patterns of brainstem pathology by SCA subtype and clinicopathologic correlations

1. **SCA1**
   - Notable for atrophy in the pontine nuclei and inferior olive, but also displaying neuronal loss in nuclei associated with sensory systems, PPN, SN, and ventral tegmental area (Rub et al., 2013). The Purkinje cell layer degenerates in tandem with brainstem nuclei, exacerbating gait ataxia and bulbar dysfunction (e.g., dysarthria). Clinically, SCA1 is considered a "cerebellar plus'' syndrome with variable expression of brainstem-related symptoms.
2. **SCA2**
   - Neuropathologic involvement resembles SCA1, as does the clinical picture, which may involve Parkinsonian

early saccade slowing and abnormal eye movements, and autonomic dysfunction.

3. **SCA3 (Machado–Joseph disease)**
   - Known as the most common dominant ataxia worldwide. A range of brainstem symptoms, such as ophthalmoplegia, facial myokymia, and bulbar weakness, reflect widespread pathology in cranial nerve nuclei and reticular formation, which is similar to SCA1 and SCA2.
4. **SCA6**
   - Primarily a calcium channelopathy with prominent cerebellar involvement, yet subtle brainstem changes (such as SN degeneration) do occur.
5. **SCA7**
   - Distinguished by retinal degeneration leading to vision loss, but the pons and medulla can also develop tau or polyQ protein aggregates, similarly to SCAs 1, 2, and 3. Dysphagia and speech difficulties often arise from bulbar pathology, aggravating the disease burden.
6. **Non-PolyQ SCAs**
   - Subtypes like SCA8, SCA10, etc., can exhibit variable involvement of the olivopontocerebellar system, highlighting the significance of these subcortical sites in shaping clinical outcomes.

CONCLUSIONS

In summary, the brainstem emerges as a crucial but often overlooked site in the pathology and clinical evolution of various NDDs. By reviewing salient pathologic findings and highlighting clinicopathologic correlations, this chapter emphasizes that brainstem involvement typically occurs early, often preceding or paralleling cortical pathology. Understanding the chronology and selective vulnerability of specific brainstem nuclei provides critical insights into disease pathogenesis and symptom progression. Recognizing the significance of these early pathologic events not only enhances diagnostic accuracy and clinical management but also opens avenues for earlier and more targeted therapeutic interventions. Continued exploration of brainstem pathology thus promises to refine our comprehension of NDD mechanisms and improve patient care outcomes.

References

Alzheimer's Disease Facts and Figures. (2021). Alzheimers Dement; 17: 327-406.

Arendt T, Bruckner MK, Morawski M et al. (2015). Early neurone loss in Alzheimer's disease: cortical or subcortical? Acta Neuropathol Commun 3: 10.

Attems J, Toledo JB, Walker L et al. (2021). Neuropathological consensus criteria for the evaluation of Lewy pathology in post-mortem brains: a multi-centre study. Acta Neuropathol 141: 159–172.

Bede P, Chipika RH, Finegan E et al. (2019). Brainstem pathology in amyotrophic lateral sclerosis and primary lateral sclerosis: a longitudinal neuroimaging study. Neuroimage Clin 24: 102054.

Benarroch EE, Schmeichel AM, Low PA et al. (2004). Involvement of medullary serotonergic groups in multiple system atrophy. Ann Neurol 55: 418–422.

Benarroch EE, Schmeichel AM, Sandroni P et al. (2006). Involvement of vagal autonomic nuclei in multiple system atrophy and Lewy body disease. Neurology 66: 378–383.

Benarroch EE, Smithson IL, Low PA et al. (1998). Depletion of catecholaminergic neurons of the rostral ventrolateral medulla in multiple systems atrophy with autonomic failure. Ann Neurol 43: 156–163.

Betts MJ, Kirilina E, Otaduy MCG et al. (2019). Locus coeruleus imaging as a biomarker for noradrenergic dysfunction in neurodegenerative diseases. Brain 142: 2558–2571.

Boeve BF (2013). Idiopathic REM sleep behaviour disorder in the development of Parkinson's disease. Lancet Neurol 12: 469–482.

Boeve BF, Dickson DW, Olson EJ et al. (2007). Insights into REM sleep behavior disorder pathophysiology in brainstem-predominant Lewy body disease. Sleep Med 8: 60–64.

Borghammer P, Just MK, Horsager J et al. (2022). A postmortem study suggests a revision of the dual-hit hypothesis of Parkinson's disease. NPJ Parkinsons Dis 8: 166.

Braak H, Braak E (1991). Neuropathological stageing of Alzheimer-related changes. Acta Neuropathol 82: 239–259.

Braak H, Braak E (1998). Involvement of precerebellar nuclei in Pick's disease. Exp Neurol 153: 351–365.

Braak H, Del Tredici K, Rub U et al. (2003a). Staging of brain pathology related to sporadic Parkinson's disease. Neurobiol Aging 24: 197–211.

Braak H, Del Tredici K, Rüb U et al. (2003b). Staging of brain pathology related to sporadic Parkinson's disease. Neurobiol Aging 24: 197–211.

Braak H, Thal DR, Ghebremedhin E et al. (2011). Stages of the pathologic process in Alzheimer disease: age categories from 1 to 100 years. J Neuropathol Exp Neurol 70: 960–969.

Brettschneider J, Del Tredici K, Irwin DJ et al. (2014). Sequential distribution of pTDP-43 pathology in behavioral variant frontotemporal dementia (bvFTD). Acta Neuropathol 127: 423–439.

Brettschneider J, Del Tredici K, Toledo JB et al. (2013). Stages of pTDP-43 pathology in amyotrophic lateral sclerosis. Ann Neurol 74: 20–38.

Bueicheku E, Diez I, Kim CM et al. (2024). Spatiotemporal patterns of locus coeruleus integrity predict cortical tau and cognition. Nat Aging 4: 625–637.

Cairns NJ, Bigio EH, Mackenzie IR et al. (2007). Neuropathologic diagnostic and nosologic criteria for frontotemporal lobar degeneration: consensus of the Consortium for Frontotemporal Lobar Degeneration. Acta Neuropathol 114: 5–22.

Crary JF, Trojanowski JQ, Schneider JA et al. (2014). Primary age-related tauopathy (PART): a common pathology associated with human aging. Acta Neuropathol 128: 755–766.

Dickson DW (2012). Parkinson's disease and parkinsonism: neuropathology. Cold Spring Harb Perspect Med 2: a009258.

Dickson DW (2018). Neuropathology of Parkinson disease. Parkinsonism Relat Disord 46: S30–S33.

Dickson DW, Ahmed Z, Algom AA et al. (2010). Neuropathology of variants of progressive supranuclear palsy. Curr Opin Neurol 23: 394–400.

Dickson DW, Hauw J-J, Agid Y et al. (2011). Progressive supranuclear palsy and corticobasal degeneration. Neurodegeneration: the molecular pathology of dementia and movement disorders. Wiley-Blackwell, pp. 135–155.

Dickson DW, Rademakers R, Hutton ML (2007). Progressive supranuclear palsy: pathology and genetics. Brain Pathol 17: 74–82.
Duyckaerts C, Braak H, Brion JP et al. (2015). PART is part of Alzheimer disease. Acta Neuropathol 129: 749–756.
Ehrenberg AJ, Kelberman MA, Liu KY et al. (2023). Priorities for research on neuromodulatory subcortical systems in Alzheimer's disease: position paper from the NSS PIA of ISTAART. Alzheimers Dement 19: 2182–2196.
Forrest SL, Kril JJ, Stevens CH et al. (2018). Retiring the term FTDP-17 as MAPT mutations are genetic forms of sporadic frontotemporal tauopathies. Brain 141: 521–534.
Garbutt S, Matlin A, Hellmuth J et al. (2008). Oculomotor function in frontotemporal lobar degeneration, related disorders and Alzheimer's disease. Brain 131: 1268–1281.
Geser F, Martinez-Lage M, Robinson J et al. (2009). Clinical and pathological continuum of multisystem TDP-43 proteinopathies. Arch Neurol 66: 180–189.
Ghetti B, Oblak AL, Boeve BF et al. (2015). Invited review: Frontotemporal dementia caused by microtubule-associated protein tau gene (MAPT) mutations: a chameleon for neuropathology and neuroimaging. Neuropathol Appl Neurobiol 41: 24–46.
Greaves CV, Rohrer JD (2019). An update on genetic frontotemporal dementia. J Neurol 266: 2075–2086.
Grinberg LT, Rub U, Ferretti RE et al. (2009). The dorsal raphe nucleus shows phospho-tau neurofibrillary changes before the transentorhinal region in Alzheimer's disease. A precocious onset? Neuropathol Appl Neurobiol 35: 406–416.
Grinberg LT, Rueb U, Alho AT et al. (2010). Brainstem pathology and non-motor symptoms in PD. J Neurol Sci 289: 81–88.
Grinberg LT, Rueb U, Heinsen H (2011). Brainstem: neglected locus in neurodegenerative diseases. Front Neurol 2: 42.
Halliday G, Hely M, Reid W et al. (2008). The progression of pathology in longitudinally followed patients with Parkinson's disease. Acta Neuropathol 115: 409–415.
Hawkes CH, Del Tredici K, Braak H (2007). Parkinson's disease: a dual-hit hypothesis. Neuropathol Appl Neurobiol 33: 599–614.
Heneka MT, Ramanathan M, Jacobs AH et al. (2006). Locus coeruleus degeneration promotes Alzheimer pathogenesis in amyloid precursor protein 23 transgenic mice. J Neurosci 26: 1343–1354.
Irwin DJ (2016). Tauopathies as clinicopathological entities. Parkinsonism Relat Disord 22: S29–S33.
Jahanshahi A, Vlamings R, van Roon-Mom WMC et al. (2013). Changes in brainstem serotonergic and dopaminergic cell populations in experimental and clinical Huntington's disease. Neuroscience 238: 71–81.
Jellinger KA (2014). Neuropathology of multiple system atrophy: new thoughts about pathogenesis. Mov Disord 29: 1720–1741.
Jellinger KA (2015). Neuropathobiology of non-motor symptoms in Parkinson disease. J Neural Transm (Vienna) 122: 1429–1440.
Jellinger KA, Korczyn AD (2018). Are dementia with Lewy bodies and Parkinson's disease dementia the same disease? BMC Med 16: 34.
Jost WH, Klingelhoefer L (2022). Gastrointestinal disturbances in Parkinson's disease including the management of sialorrhea. In: Gálvez-Jiménez N, Korczyn AD, Lugo-Sanchez R (eds.). Nonmotor Parkinson's disease. Cambridge: Cambridge University Press, pp. 145–154.
Kaalund SS, Passamonti L, Allinson KSJ et al. (2020). Locus coeruleus pathology in progressive supranuclear palsy, and its relation to disease severity. Acta Neuropathol Commun 8: 11.
Karanth S, Nelson PT, Katsumata Y et al. (2020). Prevalence and clinical phenotype of quadruple misfolded proteins in older adults. JAMA Neurol 77: 1299–1307.
Klockgether T, Mariotti C, Paulson HL (2019). Spinocerebellar ataxia. Nat Rev Dis Primers 5: 24.
Kovacs GG (2019). Molecular pathology of neurodegenerative diseases: principles and practice. J Clin Pathol 72: 725–735.
Kovacs GG, Ghetti B, Goedert M (2022). Classification of diseases with accumulation of Tau protein. Neuropathol Appl Neurobiol 48: e12792.
Kovacs GG, Lukic MJ, Irwin DJ et al. (2020). Distribution patterns of tau pathology in progressive supranuclear palsy. Acta Neuropathol 140: 99–119.
Krogias C, Strassburger K, Eyding J et al. (2011). Depression in patients with Huntington disease correlates with alterations of the brain stem raphe depicted by transcranial sonography. J Psychiatry Neurosci 36: 187–194.
Kunkle BW, Grenier-Boley B, Sims R et al. (2019). Genetic meta-analysis of diagnosed Alzheimer's disease identifies new risk loci and implicates Abeta, tau, immunity and lipid processing. Nat Genet 51: 414–430.
Lew CH, Petersen C, Neylan TC et al. (2021). Tau-driven degeneration of sleep- and wake-regulating neurons in Alzheimer's disease. Sleep Med Rev 60: 101541.
Liscic RM, Grinberg LT, Zidar J et al. (2008). ALS and FTLD: two faces of TDP-43 proteinopathy. Eur J Neurol 15: 772–780.
Mackenzie IR, Neumann M, Baborie A et al. (2011). A harmonized classification system for FTLD-TDP pathology. Acta Neuropathol 122: 111–113.
Morris HR, Wood NW, Lees AJ (1999). Progressive supranuclear palsy (Steele-Richardson-Olszewski disease). Postgrad Med J 75: 579–584.
Munoz DP, Everling S (2004). Look away: the anti-saccade task and the voluntary control of eye movement. Nat Rev Neurosci 5: 218–228.
Nalls MA, Pankratz N, Lill CM et al. (2014). Large-scale meta-analysis of genome-wide association data identifies six new risk loci for Parkinson's disease. Nat Genet 46: 989–993.
Nelson PT, Dickson DW, Trojanowski JQ et al. (2019). Limbic-predominant age-related TDP-43 encephalopathy (LATE): consensus working group report. Brain 142: 1503–1527.
Non-Motor Parkinson's Disease (2002). In: Gálvez-Jiménez N, Korczyn AD, Lugo-Sanchez R (eds.). Non-motor Parkinson's disease. Cambridge: Cambridge University Press, pp. i-ii.
Oh JY, Walsh CM, Ranasinghe K et al. (2022). Subcortical neuronal correlates of sleep in neurodegenerative diseases. JAMA Neurol 79.
Ozawa T (2007). Morphological substrate of autonomic failure and neurohormonal dysfunction in multiple system atrophy: impact on determining phenotype spectrum. Acta Neuropathol 114: 201–211.
Ozawa T, Paviour D, Quinn NP et al. (2004). The spectrum of pathological involvement of the striatonigral and olivopontocerebellar systems in multiple system atrophy: clinicopathological correlations. Brain 127: 2657–2671.
Parvizi J (2009). Corticocentric myopia: old bias in new cognitive sciences. Trends Cogn Sci 13: 354–359.
Piao Y-S, Wakabayashi K, Kakita A et al. (2003). Neuropathology with clinical correlations of sporadic amyotrophic lateral sclerosis: 102 autopsy cases examined between 1962 and 2000. Brain Pathol 13: 10–22.
Pierson SR, Fiock KL, Wang R et al. (2025). Tau pathology in the dorsal raphe may be a prodromal indicator of Alzheimer's disease. Mol Psychiatry 30: 532–546.
Rastegar-Pouyani S, Lew C, Pereira FL et al. (2025). Tau-associated neuronal loss in the intermediate nucleus of the human

hypothalamus (VLPO Analog): unveiling the basis of NREM sleep dysfunction in PSP and Alzheimer's disease. bioRxiv.

Roemer SF, Grinberg LT, Crary JF et al. (2022). Rainwater Charitable Foundation criteria for the neuropathologic diagnosis of progressive supranuclear palsy. Acta Neuropathol 144: 603–614.

Rub U, Hentschel M, Stratmann K et al. (2014). Huntington's disease (HD): degeneration of select nuclei, widespread occurrence of neuronal nuclear and axonal inclusions in the brainstem. Brain Pathol 24: 247–260.

Rub U, Schols L, Paulson H et al. (2013). Clinical features, neurogenetics and neuropathology of the polyglutamine spinocerebellar ataxias type 1, 2, 3, 6 and 7. Prog Neurobiol 104: 38–66.

Rub U, Stratmann K, Heinsen H et al. (2016). The brainstem tau cytoskeletal pathology of Alzheimer's disease: a brief historical overview and description of its anatomical distribution pattern, evolutional features, pathogenetic and clinical relevance. Curr Alzheimer Res 13: 1178–1197.

Rüb U, Vonsattel J, Heinsen H et al. (2015). Widespread brainstem neurodegeneration in Huntington's disease (HD). The neuropathology of Huntington's disease: classical findings, recent developments and correlation to functional neuroanatomy. Advances in anatomy, embryology and cell biology. Springer, pp. 83–90.

Saper CB, Fuller PM (2017). Wake-sleep circuitry: an overview. Curr Opin Neurobiol 44: 186–192.

Schwarzacher SW, Rub U, Deller T (2011). Neuroanatomical characteristics of the human pre-Botzinger complex and its involvement in neurodegenerative brainstem diseases. Brain 134: 24–35.

Seidel K, Mahlke J, Siswanto S et al. (2015). The brainstem pathologies of Parkinson's disease and dementia with Lewy bodies. Brain Pathol 25: 121–135.

Shi Y, Zhang W, Yang Y et al. (2021). Structure-based classification of tauopathies. Nature 598: 359–363.

Simic G, Stanic G, Mladinov M et al. (2009). Does Alzheimer's disease begin in the brainstem? Neuropathol Appl Neurobiol 35: 532–554.

Sobreira-Neto MA, Stelzer FG, Gitai LLG et al. (2023). REM sleep behavior disorder: update on diagnosis and management. Arq Neuropsiquiatr 81: 1179–1194.

Stefani A, Antelmi E, Arnaldi D et al. (2025). From mechanisms to future therapy: a synopsis of isolated REM sleep behavior disorder as early synuclein-related disease. Mol Neurodegener 20: 19.

Stratmann K, Heinsen H, Korf HW et al. (2016). Precortical phase of Alzheimer's disease (AD)-related tau cytoskeletal pathology. Brain Pathol 26: 371–386.

Suemoto CK, Ferretti-Rebustini RE, Rodriguez RD et al. (2017). Neuropathological diagnoses and clinical correlates in older adults in Brazil: a cross-sectional study. PLoS Med 14: e1002267.

Thal DR, Rüb U, Orantes M et al. (2002). Phases of Aβ-deposition in the human brain and its relevance for the development of AD. Neurology 58: 1791–1800.

Theofilas P, Ehrenberg AJ, Dunlop S et al. (2017). Locus coeruleus volume and cell population changes during Alzheimer's disease progression: a stereological study in human postmortem brains with potential implication for early-stage biomarker discovery. Alzheimers Dementia 13: 236–246.

Tsai P-H, Gopalakrishna G (2022). Mood disorders in Parkinson's disease. In: Gálvez-Jiménez N, Korczyn AD, Lugo-Sanchez R (eds.). Non-motor Parkinson's disease. Cambridge: Cambridge University Press, pp. 91–98.

Van Egroo M, van Someren EJW, Grinberg LT et al. (2024). Associations of 24-hour rest-activity rhythm fragmentation, cognitive decline, and postmortem locus coeruleus hypopigmentation in Alzheimer's disease. Ann Neurol 95: 653–664.

van Wamelen D, Leta V, Chaudhuri KR (2022). Parkinson's disease: an overview of the non-motor symptomatology. In: Gálvez-Jiménez N, Korczyn AD, Lugo-Sanchez R (eds.). Non-motor Parkinson's disease. Cambridge: Cambridge University Press, pp. 3–17.

Wenning GK, Tison F, Ben Shlomo Y et al. (1997). Multiple system atrophy: a review of 203 pathologically proven cases. Mov Disord 12: 133–147.

Handbook of Clinical Neurology, Vol. 216 (3rd series)
The Brainstem
L.T. Grinberg and A.D. Korczyn
DOI: https://doi.org/10.1016/B978-0-443-15736-3.00011-1

Chapter 15

Brainstem vascular syndromes

JORGE MONCAYO-GAETE[1] AND JULIEN BOGOUSSLAVSKY[2,*]

[1]*School of Medicine. Universidad de las Américas, Quito, Ecuador*

[2]*Genolier Swiss Medical Network Neurocenter, Clinique Valmont, Glion/Montreux, Switzerland*

Abstract

The brainstem is a heterogeneous assembly of well-organized nuclei and pathways that play a critical role in motor, sensory, arousal, and autonomic functions. Because of its extensive reciprocal connections with the cortex, basal ganglia, and cerebellum, the brainstem is involved in generating, controlling, and regulating all eye movements. Brainstem vascular supply, which is especially complex in the midbrain, comes entirely from the vertebrobasilar system. Brainstem infarcts are infrequent, and hemorrhages are even less common.

The pons is the section most often affected by ischemic or hemorrhagic stroke. A wide spectrum of clinical motor, sensory, eye movement, or behavioral manifestations – some with devastating effects – may occur in brainstem infarcts and hemorrhages. Neuro-ophthalmologic signs, particularly in the vertical plane and sometimes associated with the most complex eye movement patterns observed in clinical neurology, may predominate in midbrain stroke. Pontine stroke is characterized by horizontal eye movement disorders and motor, sensory, and cerebellar manifestations, either isolated or in combination. Sensory and cerebellar deficits and eye movement disorders due to central vestibular pathway involvement are the main clinical features of medullary stroke; autonomic dysfunction may sometimes occur. Occlusive large artery disease is the leading infarct etiology in almost all brainstem vascular territories.

INTRODUCTION

Relatively small and yet highly interconnected, the brainstem is one of the phylogenetically oldest and most anatomically and functionally complex structures of the central nervous system (CNS). It is made up of three structures along its rostro-caudal axis – the midbrain, the pons, and the medulla oblongata – and is strategically placed in the anterior portion of the posterior cranial fossa, nestled between the clivus and the cerebellum.

The brainstem has important and extensive connections with the cerebral cortex, basal ganglia, diencephalon, cerebellum, and spinal cord. Long white matter tracts run downward through the brainstem, participating in motor control (movement, posture, tone, locomotion), and upward, conveying epicritic and protopathic sensory information. The brainstem also contains nuclei that receive, integrate, and relay information to and from the cortex, cerebellum, and spinal cord, as well as all the cranial nerve nuclei and connections except those of the olfactory and optic nerves. Some brainstem cell groups have widespread projections throughout the CNS and are involved in sensory, motor, autonomic, behavioral, affective, and cognitive functions (Vanderah and Gould, 2021).

The brainstem is a main neural hub regulating a myriad of functions essential for life, including breathing, blood pressure, temperature, pain, sleep, and consciousness. In addition, a complex and highly sophisticated brainstem neural network participates along with the cerebral cortex, basal ganglia, and cerebellum in generating, controlling, and regulating all functional classes of eye movements (Horn and Leigh, 2011; Bae et al., 2013).

Consequently, since the brainstem houses critical nuclei and pathways in a densely packed arrangement, any ischemic or hemorrhagic lesion will produce a wide spectrum of clinical manifestations, some with devastating effects.

The vertebrobasilar (VB) system, or posterior circulation, serves as the arterial supply to the brainstem. The posterior circulation is composed of two vertebral arteries (VA), three paired cerebellar arteries, the basilar artery (BA), and two posterior cerebral arteries (PCAs). Besides the brainstem, this vascular system also supplies the spinal cord, the cerebellum, the inferior aspects of the temporal lobe, and the lateral and medial aspects of the occipital lobe (Vanderah and Gould, 2021).

Brainstem vascular supply, especially that of the midbrain, is quite complex. Compared to the anterior circulation, the

*Correspondence to: Julien Bogousslavsky, MD, Swiss Medical Network Neurocenter, Clinique Valmont, Route de Valmont 19, 1823 Glion/Montreux, Switzerland. Tel: 41 21 962 37 00; E-mail: jbogousslavsky@cliniquevalmont.ch.

posterior circulation arteries present many anatomic variations in terms of their origin, diameter, presence, and the course of the main arteries and their branches – all of which may be clinically significant (Park et al., 2007; Songur et al., 2008; Mitsumura et al., 2016).

According to hospital-based stroke registries, about 10% of all brain infarcts involve at least one brainstem vascular territory (Bogousslavsky et al., 1988a; Kumral et al., 1998; Vemmos et al., 2000; Paciaroni et al., 2003). The pons is the most frequent location of brainstem ischemic stroke, followed by the medulla oblongata and the midbrain (Bogousslavsky et al., 1988a; Paciaroni et al., 2003).

Brainstem hemorrhages are by far less common than brainstem infarcts; hospital-based stroke registries showed brainstem hemorrhage rates of between 2% and 9% of all intracerebral hemorrhages (Tatu et al., 2000; Karelis et al., 2021). Again, the pons is the most common location among brainstem hemorrhages, while midbrain and medullary hemorrhages are unusual.

Diagnosis of brainstem stroke may be challenging, particularly in cases of minute infarcts or hemorrhages (Arch et al., 2016). The clinical manifestations of posterior circulation infarcts, particularly in the brainstem, may be atypical or overlap with those of anterior circulation strokes (Tao et al., 2012).

This chapter describes the vascular supply of the brainstem and the clinical manifestations, etiologies, and prognosis of infarcts according to their main arterial territories in the three main subdivisions of the brainstem. The clinical manifestations, etiologies, and prognosis of hemorrhages according to their internal organization are also addressed.

MIDBRAIN VASCULAR SYNDROMES

The midbrain, or mesencephalon, is the smallest and uppermost structure of the brainstem and occupies the tentorial notch Traditionally, its superior boundary is an imaginary line between the mamillary bodies and the posterior commissure (PC), while its inferior limit is formed by the trochlear nucleus and the decussation of the superior cerebellar peduncles (SCP). (Bogousslavsky et al., 1994; Vanderah and Gould, 2021).

MIDBRAIN VASCULARIZATION

The most complex vascular supply of the brainstem is in the midbrain and derives mainly from the first two segments of the PCA (whose contribution increases caudo-rostrally), the upper portion of the BA, the superior cerebellar artery (SCA), and the anterior choroidal artery. In addition, branches issuing from the posterior choroidal artery (PChoA) can contribute to perfusing the midbrain (Bogousslavsky et al., 1994). As a result of the contribution from several vascular sources and of individual arterial variations, an overlap between the arterial territories of the BA, PCA, and SCA is not uncommon. The extent of this overlap will depend on its midbrain level.

Perforating branches and short circumferential SCA branches mainly perfuse the latero-dorsal surface of the lowest two-thirds of the midbrain (Bogousslavsky et al., 1994). The BA usually contributes minimally to midbrain irrigation. The upper BA can give off direct perforators supplying the ventral paramedian region (Saeki et al., 1977; Tulleken and Luiten, 1987; Vogels et al., 2021). In 50 dissected brains, Saeki and Rhoton observed an average of eight perforators emerging mostly from the dorsal surface of the BA just before its bifurcation into PCAs (Saeki and Rhoton, 1977). These perforators entered the midbrain at the midline. Nevertheless, Caruso et al. and Djulejić et al. did not observe any perforating arteries perfusing the midbrain from the tip of the BA (Caruso et al., 1990; Djulejić et al., 2015). Rather, they found perforators in the BA irrigating the upper paramedian pons and the lower part of the posterior perforated substance (Caruso et al., 1990; Djulejić et al., 2015).

The collicular artery (CA) is a small artery principally supplying the peduncle, the lateral tectum, and both colliculi, as it branches from the first segment of the PCA (Duvernoy, 1999). According to Foix and Hillemand, the CA is also known as the quadrigeminal or pedunculo-quadrigeminal artery, the latter denomination probably better describing its vascular territory (Foix and Hillemand, 1925; Duvernoy, 1999).

Most perforating arteries supplying the upper paramedian midbrain arise from the first two segments of the PCA. The thalamoperforating arteries often stem from precommunicating segment (also called the mesencephalic artery or communicating BA) of the PCA. Two types of thalamoperforating arteries, the diencephalic and the mesencephalic perforators, have been recognized (Marinković and Gibo, 1994). An average of 2.1 mesencephalic perforators with a mean diameter of 255 μm supply the posterior perforated substance and the medial aspect of the cerebral peduncle (Pedroza et al., 1986; Kaya et al., 2010; Djulejić et al., 2015). The diencephalic perforators supply the thalamus and the mesodiencephalic junction along with perforating branches issuing from the postcommunicating segment of the PCA (Pedroza et al., 1986; Djulejić et al., 2015).

Among other branches, the postcommunicating segment of the PCA sends out the peduncular perforating arteries, which perfuse the corticospinal and the corticobulbar pathways, the substantia nigra, the red nucleus, and other structures of the mesencephalic tegmentum (Zeal and Rhoton 1978; Giotta Lucifero et al., 2021).

The PChoA usually originates as a single trunk from the postcommunicating segment of the PCA and is made up of medial and lateral branches. The medial branch gives off numerous smaller anterolateral, lateral, and posterior branches. The anterolateral and lateral branches mainly feed the lower lateral part of the crus cerebri, while the posterior branches supply the peduncle and the midbrain tegmentum (Vinas et al., 1995; Duvernoy, 1999; Kaya et al., 2010). Occasionally, medial branches of the anterior choroidal artery may supply the upper paramedian midbrain (Viader et al., 1984; Helgason, 1988).

Based on anatomic studies initially by Gillian and Lazorthes and subsequently by Duvernoy, the most widely recognized vascular territories are the anteromedial (paramedian), anterolateral, lateral, and dorsal (Gillilan, 1964; Lazorthes et al., 1976; Duvernoy, 1999).

MIDBRAIN INFARCT EPIDEMIOLOGY

Despite the crucial role of the midbrain in vertical gaze movements, there are few patient series featuring pure midbrain infarctions (Bogousslavsky et al., 1994; Martin et al., 1998; Kumral et al., 2002a; Kim and Kim 2005; Ogawa et al., 2012; Gilberti et al., 2014). Midbrain infarcts have been described more frequently in mesodiencephalic infarct or posterior circulation infarct series (Gomez et al., 1990; Bogousslavsky et al., 1993; Lee et al., 2006).

Midbrain infarcts are uncommon; incidence rates range between 0.6% and 2.3% of all brain infarcts (Bogousslavsky et al., 1994; Martin ct al., 1998; Kumral et al., 2002a; Kim and Kim, 2005; Gilberti et al., 2014). Midbrain infarction represents between 0.6% and 8% of all posterior circulation ischemic strokes (Bogousslavsky et al., 1994; Martin et al., 1998; Kumral et al., 2002a; Kim and Kim, 2005). Midbrain infarcts more often occur with concomitant infarction mainly involving the neighboring thalamus or pons or, more rarely, the temporal and occipital lobes (Martin et al., 1998; Kumral et al., 2002a; Kim and Kim, 2005). In a series of midbrain infarcts included in the New England Medical Center Posterior Circulation Registry, Martin et al. found only three patients with midbrain-restricted ischemia, while 36 patients had midbrain infarcts associated with diencephalic or pontine ischemic stroke (Martin et al., 1998). Accordingly, they stated that it is 10 times more likely that midbrain ischemia will be associated with infarcts in other VB territories than simply confined to the midbrain (Martin et al., 1998).

The most common midbrain infarct locations are the paramedian territory, with nearly two-thirds of the total, and the anterolateral territory, with around one-fourth. Infarctions in the lateral and, particularly, the dorsal territory are quite uncommon (Bogousslavsky et al., 1994; Kim and Kim, 2005; Ogawa et al., 2012; Gilberti et al., 2014).

The third (oculomotor) nerve palsy – often associated with other clinical manifestations – is considered the clinical hallmark of midbrain infarctions since the original descriptions of midbrain infarcts in the second half of the 18th century. Nevertheless, clinical diagnosis of mesencephalic infarcts may often be challenging and easily overlooked, particularly when other unusual neuro-ophthalmologic signs are present or when clinical manifestations are atypical or overlap with clinical signs of other concomitant posterior circulation infarctions. Consequently, midbrain infarctions are sometimes found or confirmed mainly after brain magnetic resonance imaging (MRI) in the setting of suspected acute posterior circulation ischemia (Zhang et al., 2017; Zhou et al., 2021).

PARAMEDIAN MIDBRAIN INFARCTIONS

Paramedian (anteromedial) infarcts predominate in ischemia, whether confined to the midbrain or otherwise. Up to 70% of patients with middle and upper paramedian midbrain infarctions present one or more of a broad spectrum of neuro-ophthalmologic manifestations. These signs are commonly associated with other clinical manifestations that reflect midbrain or posterior circulation involvement and are rarely isolated (Bogousslavsky and Regli, 1983a; Bogousslavsky and Meienberg, 1987; Bogousslavsky, 1989). Moreover, such neuro-ophthalmologic signs may sometimes be present at the same time, giving rise to the most complex eye-movement patterns observed in clinical neurology.

Supranuclear (prenuclear) vertical gaze palsy is a prominent clinical finding in upper paramedian midbrain infarction. Henoch (1864), Wernicke (1876), and Gee (1890) were the first to describe upward gaze, upward and downward gaze, and downward gaze palsies, respectively (Bender, 1980). In addition, Parinaud (1883) was the first to suggest the existence of a vertical gaze center anatomically distinct from the third oculomotor nucleus (Parinaud, 1883). Subsequently, the neural substrate of vertical gaze palsies was deduced from clinical descriptions, which initially sometimes lacked pathologic confirmation (Parinaud, 1883; Spiller, 1905). More recently, neuropathologic correlations and anatomic or electrophysiologic studies in experimental animal models confirm the neural substrate (Moschovakis et al., 1991a, 1991b).

It has now been clearly established that the rostral interstitial nucleus of the medial longitudinal fascicle (riMLF), the interstitial nucleus of Cajal (INC), and the PC are the critical structures housed in the tegmental rostral midbrain responsible for generating vertical eye movements (Bhidayasiri et al., 2000; Horn and Leigh, 2011).

Supranuclear conjugate vertical gaze palsies may result from occlusion of SCA or PCA perforators, resulting in infarction in the PC or periaqueductal gray matter (Bogousslavsky and Meienberg, 1987). Supranuclear gaze palsy may also occur if the posterior paramedian thalamic-subthalamic artery and the paramedian mesencephalic perforating artery stem from a single pedicle, which may give rise to a butterfly-shaped infarction with paramedian upper midbrain and medial thalamus involvement (Percheron, 1976; Castaigne et al., 1981; Bogousslavsky et al., 1988b).

Selective upward gaze palsy is the most frequent vertical gaze disorder, followed by combined upward and downward gaze palsy, and then selective downward gaze palsy (Moncayo and Bogousslavsky, 2009). Selective upward gaze palsy has been considered a revealing clinical sign for midbrain ischemia. Vertical gaze disorders mainly affect saccadic movements, while smooth pursuit movements are variably affected, and vestibulo-ocular reflexes (VOR) are usually unaffected (Bogousslavsky and Meienberg, 1987; Bogousslavsky, 1989). Combined gaze palsy and selective upward gaze or downward gaze palsy mostly occur with bilateral paramedian upper midbrain infarcts affecting the riMLF on both sides (Bogousslavsky, 1989; Bogousslavsky et al., 1994). All these vertical gaze disorders may result from unilateral infarcts as well, though more rarely (Hommel and Bogousslavsky, 1991).

Combined vertical gaze palsy may result from unilateral paramedian ischemia. In such cases, the ischemia damages the riMLF, the PC, the INC and PC nuclei, both the riMLF and INC, or the central mesencephalic reticular formation

(Ranalli et al., 1988; Bogousslavsky et al., 1990a; Alemdar et al., 2006; Fujioka and Nakajima, 2008). Therefore, a unilateral anatomic lesion situated in the riMLF or in the pretectal region may produce a bilateral functional impairment of the pathways mediating upward vertical gaze (Bogousslavsky and Meienberg, 1987; Hommel and Bogousslavsky, 1991).

Selective downward gaze palsy is often due to bilateral ischemic lesions on the riMLF that damage its medio-caudal region (Trojanowski and Wray, 1980; Pierrot-Deseilligny et al., 1982). Restricted downward gaze paralysis may also result from bilateral ischemia that affects the dorsolateral periaqueductal gray matter, the PC and its nuclei while sparing both riMLF (Jacobs et al., 1985). Downward gaze palsy may also result from bilateral lesions at the thalamomesencephalic junction near the riMLF that result from the occlusion of a single posterior thalamo-subthalamic artery (Halmagyi et al., 1978).

Besides supranuclear conjugate vertical gaze disorders, rostral midbrain infarctions may be associated with supranuclear disconjugate vertical gaze disorders, which are less common clinical findings but present with more complex patterns that may be overlooked on routine clinical examination. Monocular elevation paresis (MEP), vertical one-and-a-half syndrome (OAAHS), crossed paralysis, see-saw nystagmus (SSN), and skew deviation are the main manifestations of disconjugate vertical gaze disorders. These uncommon disorders have been reported by Hommel and Bogousslavsky in three patients with vertical OAAHS and two patients with MEP among 11 patients with unilateral midbrain infarctions and vertical gaze disorders (Hommel and Bogousslavsky, 1991).

Double elevator palsy, or MEP, is an asymmetrical upward gaze disorder presenting with a combined monocular palsy of both elevator muscles (superior rectus and inferior oblique) either in abduction or adduction. Double elevator palsy in the affected eye involves saccadic and smooth pursuit movements, VOR, and upward optokinetic nystagmus (Hommel and Bogousslavsky, 1991; Thömke et al., 1992). Contralateral and ipsilateral ischemia has been documented on brain MRI mainly in the upper midbrain (Hommel and Bogousslavsky, 1991; Thömke et al., 1992). In contralateral ischemia, the lesion is thought to affect decussated riMLF fibers before they reach the oculomotor nucleus complex, while ipsilateral ischemia may affect fibers immediately after they leave the riMLF and before crossing in the PC (Hommel and Bogousslavsky, 1991; Thömke et al., 1992).

SSN is a very unusual and peculiar form of vertical disconjugate nystagmus. It is characterized by both conjugate torsional movement and disconjugate vertical movement. Thalamic hematomas extending to the mesodiencephalic regions account for most SSN vascular cases (Halmagyi et al., 1994). Upper paramedian midbrain infarction is seldom a cause of SSN (Kanter et al., 1987; Halmagyi et al., 1994). Ischemia at the INC level may affect the central otolithic connections, though the pathophysiologic mechanism of the SSN still remains unclear.

Crossed paralysis is another uncommon disconjugate downward gaze disorder and may result from unilateral upper midbrain infarctions extending into the ipsilateral thalamus (Wiest et al., 1996). It is characterized by ipsilateral downward gaze palsy with concomitant contralateral MEP. In such cases, ischemia may affect the ipsilateral and contralateral fibers issuing from the riMLF that, respectively, mediate downward and upward gaze (Wiest et al., 1996).

The vertical OAAHS is another supranuclear disconjugate vertical gaze disorder due to bilateral or unilateral upper midbrain infarcts. Two patterns of this complex eye disorder have been described. The first manifests with simultaneous bilateral upward gaze palsy and unilateral downward gaze palsy, while the second presents with simultaneous bilateral downward gaze palsy and unilateral upward gaze palsy (Bogousslavsky and Regli, 1984; Deleu et al., 1989; Hommel and Bogousslavsky, 1991). This disconjugate vertical eye disorder may be explained by the spatial segregation in the riMLF premotor motoneurons for vertical gaze movements according to their specific motoneuron targets in the oculomotor and trochlear nuclei. In both types of vertical OAAHS, the bilateral vertical gaze palsy is attributed to ischemia selectively affecting the riMLF premotor fibers crossing the PC that mediate downward gaze (a bilateral lesion) or upward gaze (a unilateral lesion), while the unilateral vertical gaze palsy is attributed to ischemia affecting the riMLF premotor fibers before or after their decussation in the PC en route to the oculomotor complex nucleus or the trochlear nucleus (Bogousslavsky, 1989; Hommel and Bogousslavsky, 1991).

Unilateral or bilateral ischemic lesions at the level of the thalamomesencephalic junction may also produce a broad variety of vertical gaze disorders. Pseudo-abducens palsy, described initially by Fisher in 1965, is a probably underrecognized rare condition invariably associated with other neuro-ophthalmologic signs due to thalamomesencephalic infarcts or top of the basilar infarctions (Fisher, 1965a; Caplan, 1980; Pullicino et al., 2000). Pullicino et al. reported seven patients with pseudo-abducens palsy, all of whom had associated upward gaze palsy, while six had concomitant downgaze palsy, four exhibited a fascicular third oculomotor palsy, and five had signs of increased convergence activity (unilateral esotropia or convergence retraction nystagmus (CRN)). Bilateral pseudo-abducens palsy due to extensive ischemia affecting the thalamus, mesodiencephalic junction, and midbrain was more common than unilateral pseudo-abducens palsy. One patient had a small infarct in the INC with some extension to ipsilateral riMLF. The anatomic substrate has not yet been completely identified, although it seems that INC and riMLF may be critical to producing this supranuclear abduction deficit (Pullicino et al., 2000). In addition, damage to descending thalamic vergence pathways in the subthalamus before or after their decussation at the level of the mesodiencephalic junction may also explain pseudo-abducens palsy. The vergence pathways project to vergence-sensitive neurons in the supra-oculomotor area located dorsally and laterally to the oculomotor third nerve nucleus. Damage to vergence-sensitive neurons or their outputs to the ipsilateral lateral rectus nucleus and medial rectus subnuclei may also be involved in this ocular abduction impairment (Pullicino et al., 2000).

CRN is also considered a landmark sign of upper midbrain lesion. Nevertheless, it is rarely observed in posterior thalamus or pretectal-region infarcts or hemorrhages affecting the PC (Luda et al., 1992; Pullicino et al., 2000; Scheitler et al., 2021). It is deemed a vergence disorder and is not a true nystagmus because it lacks a slow phase (Rambold et al., 2001). CRN is always associated with other clinical signs of excessive convergence drive or vertical gaze disorders (Jouvent et al., 2005).

Peduncular hallucinosis (PH) is a rare stroke manifestation characterized by vivid, dynamic, colorful, and well-formed transient visual hallucinations following lesions located in the midbrain, pons, thalamus, or basal ganglia. In a seminal article, Lhermitte described a woman in her 70s with initial clinical manifestations similar to Claude's syndrome, who a couple of weeks later reported visual hallucinations (Lhermitte, 1922). Based on the clinical findings, it was at that time assumed that a pontomesencephalic lesion was associated with PH (Lhermitte, 1922). Five years later, Van Bogaert was the first to confirm a midbrain lesion at autopsy of a patient with PH and concomitant Weber syndrome (Van Bogaert, 1927). PH may rarely manifest in unilateral or bilateral paramedian midbrain ischemia or midbrain hemorrhages, almost always with other signs of tegmental dysfunction (Geller and Bellur, 1987; McKee et al., 1990).

Paramedian middle midbrain infarcts almost invariably present with a nuclear third nerve palsy, which is characterized by ipsilateral oculomotor dysfunction along with eye elevation paresis on the opposite side due to a lesion in the superior rectus subnucleus, which sends axons across the midline to subserve the contralateral superior rectus muscle. Contralateral ptosis may be a prominent finding when the common caudal subnucleus subserving the upper lid elevation on both sides is involved. Conversely, bilateral pupillary involvement may occur if the lesion affects the single rostral visceral nucleus of Edinger–Westphal. In such a case, lid elevation may be spared (Pierrot-Deseilligny et al., 1981a; Bogousslavsky and Regli, 1983a; Kobayashi et al., 1986; Bogousslavsky, 1989). A complete or partial involvement of the subnuclei may occur in minute infarcts depending on the columnar and dorso-ventral anatomic arrangement of the oculomotor nuclear complex. In these cases, a third nuclear involvement, either isolated or associated with other clinical signs, may occur (Bogousslavsky et al., 1994; Kumral et al., 2002a). The main clinical findings of midbrain infarctions are summarized in Table 15.1.

Paramedian middle midbrain infarcts and, less commonly, small hemorrhages may affect the intramesencephalic (intra-axial) segment of the third oculomotor nerve and are usually associated with contralateral motor deficits, giving rise to classic crossed midbrain syndromes described in the second half of the 19th century (Table 15.2). The development of neuroanatomy over the past two centuries has led to the description of more than 20 crossed brainstem syndromes (CBS), often named after their first describers. CBS are typically characterized by ipsilateral paralysis of one or more cranial nerves along with contralateral signs of long-tract involvement. It appears that some of these classical CBS are theoretically possible but are very uncommonly observed in clinical practice, while others were based mainly on detailed clinical descriptions but lacked pathologic confirmation. Also, a

Table 15.1

Main clinical manifestations of midbrain infarctions.

Midbrain territory	Clinical manifestations	Structure damaged
Upper paramedian territory	Supranuclear conjugate gaze disorders Supranuclear disconjugate gaze disorders	riMLF, posterior commissure, interstitial nucleus of Cajal
Middle paramedian territory	Ipsilateral nuclear third nerve palsy Fascicular third neve palsy associated with crossed hemiparesis, contralateral, or contralateral hemiataxia	Oculomotor nerve nucleus Intra-axial segment of the third oculomotor nerve Corticospinal and corticobulbar tracts Upper or lower red nucleus Brachium conjunctivum
Caudal paramedian territory	Internuclear ophthalmoplegia Bilateral limb ataxia Abnormal movements: tremor, dystonia, or parkinsonism Superior oblique palsy on the contralateral eye	Medial longitudinal fasciculus SCP decussation Medial portion of the substantia nigra Trochlear nucleus or trochlear nerve fascicle
Anterolateral middle midbrain territory	Fascicular third nerve palsy associated with crossed hemiplegia, contralateral hemiataxia, and abnormal movements	Intramesencephalic segment of the third oculomotor nerve Corticospinal and corticobulbar tracts Upper or lower red nucleus Brachium conjunctivum
Lateral territory	Restricted acral (cheiro-oral and cheiro-oral-pedal) sensory deficits	Medial and trigeminal lemniscus

riMLF, rostral interstitial nucleus of the medial longitudinal fascicle; *SCP*, superior cerebellar peduncle.

Table 15.2

Crossed brainstem syndromes associated with infarcts and hemorrhages.

Syndrome (year)	Clinical features	Structures damaged
Weber (1863)	Oculomotor fascicular palsy and contralateral hemiplegia	Oculomotor nucleus, cerebral peduncle, substantia nigra
Nothnagel (1879)	Ipsilateral or bilateral Oculomotor palsy, ipsilateral Limb ataxia	Oculomotor nucleus (dorsomedial), superior cerebellar peduncle
Parinaud (1883) Dorsal midbrain syndrome	Vertical gaze palsy; vergence dysfunction (convergence spasm or paresis); light-near dissociation of the pupils; eyelid retraction; convergence-retraction nystagmus; skew deviation	Pineal gland and rostral midbrain
Benedikt (1889)	Oculomotor fascicular palsy. Contralateral hemiparesis and ataxia, choreoathetosis	Oculomotor nucleus, red nucleus, superior cerebellar peduncle
Claude (1912)	Fascicular third nerve palsy and contralateral cerebellar ataxia	Oculomotor nucleus, red nucleus, and superior cerebellar peduncle
Millard-Gubler (1856)	Ipsilateral facial nerve palsy and contralateral hemiplegia	Dorsolateral pontine tegmentum
Foville (1858)	Ipsilateral horizonal gaze palsy and facial nerve palsy; contralateral hemihypoesthesia and hemiparesis	Medial longitudinal fasciculus, paramedian reticular formation, nuclei of the abducens, and facial nerves
Raymond (1896)	Ipsilateral abducens palsy; contralateral facial paresis and hemiparesis	Ventral and medial pons
Raymond–Céstan (1903)	Ipsilateral internuclear ophthalmoplegia; contralateral hemiparesis and hemihypoesthesia	Ventral pontine tegmentum and pontine basis: medial longitudinal fascicle, medial lemniscus, and corticospinal tract
Gasperini (1912)	Ipsilateral horizontal gaze palsy, facial hypoalgesia, peripheral facial palsy, and deafness; contralateral hemiparesis and hemihypoesthesia	Abducens nucleus, facial nerve, lateral lemniscus: medial lemniscus, ventral trigeminothalamic tract, and corticospinal tract
Marie-Foix (1913)	Ipsilateral abducens nucleus palsy, hemiataxia, Horner's syndrome; contralateral hemiparesis and hemihypoesthesia	Rostral lateral pons; middle cerebellar peduncle, corticospinal tract
Avellis (1891)	Ipsilateral palatolaryngeal paresis; contralateral hemiparesis and/or hemihypesthesia	Nucleus ambiguus, ventral trigeminothalamic tract; medial lemniscus and corticospinal tract
Reinhold (1894)	Ipsilateral facial hypoesthesia, Horner syndrome, vail paresis, vocal cord paresis and hemiataxia; contralateral hemiparesis and hemihypoesthesia	Hemimedullary syndrome (medial paramedian, lateral and dorsal areas)
Babinski–Nageotte (1902)	Ipsilateral facial hypoesthesia, Horner syndrome, vail paresis, vocal cord paresis and hemiataxia; contralateral hemiparesis and hemihypoesthesia	Complete lateral medullary syndrome and incomplete medial medullary (ipsilateral hypoglossal palsy absent)
Céstan–Chenais (1903)	Ipsilateral Horner's syndrome, vocal cord paralysis, paralysis of soft palate and posterior wall of pharynx; contralateral hemiparesis and hemihypoesthesia	Lateral part of medulla oblongata; ipsilateral cerebellar ataxia absent due to spare of posterior spinocerebellar tract

number of these syndromes were not originally described in patients with brainstem stroke and may be caused by non-vascular pathologies. Moreover, the clinical picture of CBS may sometimes overlap, clinical findings may be imprecise and not very consistent with the original description, and crossed clinical manifestations in a stroke patient may not fit any of the classical CBS. In 10% of 189 patients with acute brainstem infarction on brain MRI, clinical signs were consistent with one of the CBS, most commonly in the pons and medulla oblongata. Wallenberg syndrome was the most common brainstem syndrome, followed by Babinski–Nageotte and Raymond–Céstan syndromes, while Weber and Claude syndromes were the least common (Marx and Thömke, 2009).

Paramedian infarcts located at the caudal midbrain either unilaterally or bilaterally are not common. Occlusion of the inferior paramedian arteries in the interpeduncular fossa results in Wernekinck commissure ischemia, which may selectively affect the decussation of both the SCP, as well as the frontopontine tract, the medial portion of the substantia nigra, the MLF, the central tegmental tract, and the trochlear nucleus (Voogd and van Baarsen, 2014). The outstanding clinical signs are limb tetra-ataxia, gait ataxia, and truncal ataxia, which are rarely isolated and more often seen in association with severe dysarthria, eye movement disorders – internuclear ophthalmoplegia (INO) or wall-eye bilateral internuclear ophthalmoplegia (WEBINO) – and involuntary movements – tremor or palatal myoclonus

(Mossuto-Agatiello, 2006; Zhou et al., 2017). A lesion of the Guillain–Mollaret triangle may sometimes produce an inferior olivary degeneration or a palatal myoclonus in a delayed form after a Wernekinck syndrome. Unilateral ischemia may affect the dentatorubrothalamic tract before and after it crosses the midline (Mossuto-Agatiello, 2006; Zhou et al., 2017).

Superior oblique palsy in the eye contralateral to a restricted caudal midbrain infarct may very rarely occur because of damage to the trochlear nucleus or the adjacent trochlear nerve fascicle before it decussates (Thömke and Ringel, 1999; Lee et al., 2010; Chen et al., 2021).

A tremor resulting from midbrain stroke is an uncommon manifestation. Holmes tremor (HT) is a high-amplitude, low-frequency tremor (< 4.5 Hz) affecting predominantly the upper limb, followed by the upper and lower limb on the same side and, rarely, the lower limb (Raina et al., 2016). Resting and kinetic components are almost always present, and a postural component is usually present as well. The pathophysiologic substrate of this rare tremor lies in the combined involvement of the nigrostriatal dopaminergic system and the dento-rubro-olivary pathway or the cerebellar-thalamo-cortical tract. HT has also been referred to as rubral or mesencephalic tremor, though both terms can be misleading since various lesions located outside the midbrain can also produce this type of tremor, and experimental red nucleus lesions do not necessarily cause tremor (Raina et al., 2016). HT arises from a mild midbrain ischemic or hemorrhagic stroke, and is usually associated with hemiparesis, ataxia, or sensory deficit. HT, which may respond to levodopa, emerges around 4 weeks after stroke though longer delays have been reported (Sato et al., 2008; Raina et al., 2016).

Dystonia and parkinsonism are very rare as acute or delayed manifestations of caudal paramedian infarcts. These lesions may affect the cerebellar-motor loop, bringing on dystonia, or the contralateral substantia nigra, leading to parkinsonism. These two movement disorders may be the only clinical signs of caudal paramedian infarcts but are more often associated with cerebellar signs or dysarthria (Akyol et al., 2006; Choi et al., 2015; Robles, 2016).

ANTEROLATERAL MIDBRAIN INFARCTIONS

Small anterolateral middle midbrain infarcts may present with fascicular third nerve palsy, either as a sole manifestation or, more often, associated with contralateral hemiparesis, hemiataxia, or involuntary movements (Breen et al., 1991; Kim et al., 1993; Kim and Kim, 2005; Miura et al., 2012). Several topographic models of third nerve fascicle arrangement have been proposed to explain this single or multiple selective fascicle involvement by midbrain ischemia anywhere along the intramesencephalic fascicular route (Castro et al., 1990; Ksiazek et al., 1994; Vitošević et al., 2013).

LATERAL MIDBRAIN INFARCTS

Restricted acral (cheiro-oral and cheiro-oral-pedal) sensory deficits – whether associated or not with hemiataxia and hemiparesis – are due to damage to the medial and trigeminal lemniscus and comprise the main clinical findings of lateral infarcts at the middle or caudal mesencephalon level (Bogousslavsky et al., 1994; Kim and Kim, 2005) (Table 15.1).

DORSAL TERRITORY INFARCTS

Infarcts exclusively involving this territory are exceedingly uncommon due to concomitant involvement of the cerebellar territory of the SCA. The clinical findings in this territory may be isolated trochlear nerve palsy, ipsilateral Horner's syndrome, isolated hemihypesthesia, and contralateral ataxia (Tsivgoulis et al., 2006; Lee et al., 2010).

BILATERAL OR COMBINED MIDBRAIN INFARCTS

Bilateral infarcts are very unusual events (Kumral et al., 2002a; Kim and Kim, 2005). When they occur, they often affect both paramedian territories at the upper level, with ischemia usually extending rostrally to the thalamus. The main clinical findings are different patterns of vertical gaze palsy. Classical locked-in syndrome may result from infarcts extensively affecting both cerebral peduncles while sparing the pontine basis and thalamus (Karp and Hurtig, 1974; Dehaene and Dom, 1982; Park et al., 1997). Total locked-in midbrain syndrome may result from infarcts extending across the tegmentum and the peduncles with additional infarcts in the thalamus and cerebellum (Fujimoto et al., 2012). Bilateral ventral midbrain infarcts may rarely manifest with pseudobulbar palsy and tetraparesis. In such cases, vertical and horizontal eye movements are preserved (Zhou et al., 2017). Pseudobulbar palsy with tetra-ataxia may also occur, though not with motor deficit or gaze disorder (Fu et al., 2019). Chen et al. reported acute tetraplegia and altered level of consciousness as the main clinical findings in 14 patients with bilateral peduncular ischemia and simultaneous pontine and cerebellar infarcts (Chen et al., 2015).

In the rare cases of combined anteromedial and anterolateral midbrain infarcts, oculomotor signs with hemiparesis and hemiataxia are the outstanding clinical features (Kim and Kim, 2005).

MIDBRAIN INFARCT ETIOLOGY

Large artery atherosclerotic occlusive disease (occlusion or stenosis greater than 50% of VA, BA, or PCA) was the leading cause in 34 (35.5%) of 104 patients with infarcts restricted to the midbrain. These patients were included in six series of midbrain infarcts, four of which comprised 92 patients with restricted midbrain infarcts, while the patients in the other two series had concomitant infarcts in the posterior circulation (Bogousslavsky et al., 1994; Martin et al., 1998; Kumral et al., 2002a; Kim and Kim, 2005; Ogawa et al., 2012; Gilberti et al., 2014). Small vessel disease and cardioembolic source were the presumed etiologies in 33.6% and 9.6% of these 104 patients, respectively. Etiology could not be determined in 20% of these cases (Bogousslavsky et al., 1994; Martin et al., 1998; Kumral et al., 2002a; Kim and Kim, 2005; Ogawa et al., 2012; Gilberti et al., 2014). Large artery atherosclerotic occlusive disease and cardioembolism were also the leading etiologies for midbrain infarcts in the two series comprising concomitant posterior circulation infarcts (Martin et al., 1998; Kumral et al., 2002a). Large artery atherosclerotic occlusive disease was the predominant etiology in bilateral peduncular infarcts (Chen et al., 2015).

PROGNOSIS

Pure midbrain infarcts often have a favorable outcome, with 50%–90% of pure midbrain infarction patients achieving independence during follow-up (Martin et al., 1998; Kumral et al., 2002a; Kim and Kim, 2005; Gilberti et al., 2014). Prognosis is not as favorable when midbrain infarcts coexist with other infarcts in the posterior circulation. In such cases, most patients remain dependent (with a modified Rankin score from 3 to 5). Death is rare in pure midbrain infarction, while up to 20% of patients with midbrain and concomitant infarcts in the posterior circulation die (Martin et al., 1998; Kumral et al., 2002a). Bilateral peduncular ischemic stroke, particularly associated with other posterior circulation infarcts, have a dismal prognosis: most patients die or end up with moderate to severe disability (Chen et al., 2015).

MIDBRAIN HEMORRHAGES

Nontraumatic spontaneous midbrain hemorrhages are unusual (Weisberg, 1986a; Fingerote et al., 1990). Most often, midbrain hemorrhages are described as part of large hemorrhages extending rostrally to the thalamus or caudally to the pons. Hemorrhages restricted to the midbrain often involve the tegmentum. Neuro-ophthalmologic findings predominate (Weisberg, 1986a; Fingerote et al., 1990; Link et al., 1993). The clinical manifestations may be elaborate, particularly when bleeding or associated edema involve the neighboring structures, or when obstructive hydrocephalus occurs due to intraventricular extension (Caplan and Goodwin, 1982; Weisberg, 1986a; Fingerote et al., 1990).

Small peduncular hemorrhages may be characterized by complete or partial oculomotor nuclear complex or fascicular third nerve palsy, associated or not with contralateral hemiparesis (Han et al., 1989; Gaymard et al., 1990; Mizushima and Seki, 2002). Scanning dysarthria with upward gaze palsy and gait instability is an exceedingly rare manifestation (Perez and Nuñez, 2008). Tegmental hemorrhages may be characterized predominantly by uni- or bilateral INO, total bilateral ophthalmoplegia, pure sensory stroke (PSS), hemiparkinsonism, and cerebellar signs (Durward et al.,1982; Azouvi et al., 1989; Han et al., 1989; Abe and Yanagihara, 1996; Worthington and Halmagyi, 1996). A unilateral HT may develop some months after the midbrain hemorrhage (Raina et al., 2007). Transient retrograde amnesia together with third nerve palsy and contralateral hemiparesis, is a rare occurrence in tegmental midbrain hemorrhages (Mehler and Ragone, 1988). The classic crossed midbrain syndromes (Weber, Nothnagel, Benedikt, and Claude) have also been reported in midbrain hemorrhages (Weisberg, 1986a; Link et al., 1993).

Complete or partial dorsal midbrain syndrome (Parinaud's syndrome) may be a prominent clinical manifestation of tectal hemorrhages (Sand et al., 1986; Han et al., 1989; Turkes, 2017). Dorsal midbrain hemorrhages may uncommonly manifest by involuntary movements (bilateral asterixis and ipsilateral upper limb dystonia), bilateral oblique superior paresis, or sudden monoaural tinnitus (Esteban Muñoz et al., 1996; Stimmer et al., 2009).

Regarding midbrain hemorrhage etiology, vascular malformations account for up to 40% of cases (Durward et al., 1982; Link et al., 1993). Although arterial hypertension is the leading cause of hemorrhages in most other brain locations, it is responsible for only around 25% of those in the midbrain. However, etiology remains undetermined in around 33% of patients. Functional outcome in isolated midbrain hemorrhages is usually favorable, though recovery from gaze disorders may be incomplete (Weisberg, 1986a; Link et al., 1993). Death is not currently a common occurrence in midbrain hemorrhages (Weisberg, 1986a).

PONTINE VASCULAR SYNDROMES

The largest part of the brainstem is the pons or metencephalon, located in its middle portion. It is the structure most often affected by both ischemic and hemorrhagic stroke.

The arterial supply to the pons comes mainly from the BA and its branches, which serve the pontine basis and most of the tegmentum (Bassetti et al., 1996; Duvernoy, 1999). The pons has four distinct arterial territories. The first is the anteromedial territory, which is irrigated by small penetrators from the BA that supply the paramedian basis and the anterior portion of the tegmentum. The corticospinal tract and the exiting fibers of the abducens nuclei (AN) are fed by these penetrators. The second is the anterolateral territory, which is supplied by short circumferential arteries arising from the BA that supply the lateral portion of the corticospinal tract and parts of the medial tegmentum (Bassetti et al., 1996; Duvernoy, 1999). The third is the lateral territory, which is vascularized by long circumferential arteries from the proximal part (the anterior pontine segment) of the antero inferior cerebellar artery (AICA) to the lower lateral pons (at the facial colliculus level). The rostral lateral pons is supplied at the level of the principal sensory and motor trigeminal nuclei by a series of perforating arteries from the first two segments of the SCA. According to Hardy, these are long circumflex perforating branches and terminate mainly in the tegmentum at the junction between the SCP and the middle cerebellar peduncles (Hardy et al., 1980). These arteries nourish most of the nuclei and white matter tracts of the pons. The fourth is the dorsal territory, the smallest pontine arterial territory, which is fed by perforating branches issuing from the SCA toward the SCP, the mesencephalic nucleus of the trigeminal nerve, and the locus coeruleus (Bassetti et al., 1996; Duvernoy, 1999).

EPIDEMIOLOGY OF PONTINE INFARCTS

Pontine infarcts are the most common vascular lesions among brainstem strokes (Bogousslavsky et al., 1993; Paciaroni et al., 2003). A large hospital-based posterior circulation stroke registry showed that pontine infarcts, which are in the "middle" territory according to its rostro-caudal vascular territory classification, occur most often with concomitant infarcts mainly affecting the cerebellar and PCA artery territories (Caplan et al., 2005). Infarcts confined to the pons account for 2.6%–7.5% of all ischemic strokes (Kumral et al., 2002b; Erro et al., 2005; Vemmos et al., 2005; Kobayashi et al., 2014). On the other hand, restricted pontine infarcts account for around a third of all posterior circulation infarctions (Bassetti et al., 1996; Kataoka et al., 1997; Kumral et al., 2002b; Vemmos et al., 2005; Xia et al., 2017). Surprisingly, Ha et al. found that 70% of 241 patients with MRI-proven pontine infarction had ischemia restricted to the pons, while the remaining patients had pontine infarcts combined with infarcts in the cerebellar and PCA artery territories (Ha et al., 2022).

Anteromedial, anterolateral, lateral, and dorsal are the four major pontine vascular territories recognized in patients with proven neuroimaging infarcts (Bassetti et al., 1996; Tatu et al., 1996; Kumral et al., 2002b). Combined territory infarcts may also occur (Bassetti et al., 1996; Kumral et al., 2002b). In a series of 30 patients with MRI-proven pontine infarcts, the rostral and caudal pontine levels were involved in 27% and 23% of cases, respectively; 10% of cases showed ischemia in the middle pons, and 40% of infarcts involved at least two pontine levels in the rostro-caudal dimension (Vlašković et al., 2022a).

In series of restricted pontine ischemia, the anteromedial and the anterolateral territories are the most common topographic patterns, followed by the lateral territory. Most isolated pontine infarcts are unilateral, and nearly 10% of them are bilateral (Bassetti et al., 1996; Kataoka et al., 1997; Kumral et al., 2002b; Kobayashi et al., 2014; Huang et al., 2019).

ANTEROMEDIAL PONTINE INFARCTS

The frequency of anteromedial pontine territory infarcts ranges between 30% and 70% (Bassetti et al., 1996; Kumral et al., 2002b; Kobayashi et al., 2014; Vlašković et al., 2022a). Pure motor hemiparesis (PMH) is a prominent clinical sign in paramedian pontine infarcts. These infarcts may be deeply located in the basis or may extend to the basal surface (Toyoda et al., 1994; Kim et al., 1995a; Kataoka et al., 1997; Erro et al., 2005; Kobayashi et al., 2014). In such instances, a moderate to severe facio-brachio-crural hemiparesis is commonly found, often associated with dysarthria (Fisher and Curry, 1965b; Bassetti et al., 1996; Kataoka et al., 1997; Kumral et al., 2002b; Schmahmann et al., 2004; Kobayashi et al., 2014). Sometimes, hemiparesis is worse in the arm, though very rarely in the leg (Nighoghossian et al., 1993; Schneider and Gautier, 1994; Kim et al., 1995a; Kataoka et al., 1997). Motor weakness predominates in the distal part of affected limbs (Kim et al., 1995a; Bassetti et al., 1996). Isolated brachial monoparesis may occur (Song et al., 2007).

A history of previous vertigo, transient gait ataxia, and dysarthria may be useful clinical manifestations to differentiate pontine motor deficit from motor deficit due to an internal capsule lesion (Nighoghossian et al., 1993). PMH is caused by occlusion of the BA-paramedian branches supplying the corticospinal tract in the pontine basis. Ischemia may extend over the tegmentum in about 50% of cases, leading to horizontal eye movement disorders, sensory abnormalities, and ataxia, alone or in combination (Kim et al., 1995a; Arboix et al., 2004; Schmahmann et al., 2004).

Small infarcts in the pontine basis may be present, though uncommonly, with the classical lacunar syndromes (LS) described almost 60 years ago by Fisher et al.: ataxic hemiparesis, PSS, and dysarthria-clumsy hand syndrome (Fisher and Cole, 1965c; Fisher, 1965d, 1967a). The latter syndrome occurs particularly when ischemia involves the ventral surface of the upper pons (Kataoka et al., 1997). In addition, an increasing number of clinical variants of classical LS have been described over the years in patients with small anteromedial pontine infarcts (Kim et al., 1995a; Urban et al., 2001; Schmahmann et al., 2004).

Paramedian tegmental infarcts lying just ventral to the fourth ventricle may manifest as body lateropulsion in the absence of vertigo and limb weakness, which is most often associated with INO and is rarely an isolated manifestation (Yi et al., 2007). The direction of body fall is contralateral to the lesion and is almost always associated with contraversive tilting of the subjective visual vertical in the same direction of body lateropulsion (Yi et al., 2007).

Paramedian dorsal infarcts may produce predominant or isolated lemniscal sensory deficits in a cheiro-oral, perioral, or bilateral facial distribution (Kim et al., 1998a). Moreover, dysesthesias restricted to cheek, jaw, shoulder, breast, and abdomen together with epicritic sensory deficit in these body areas may exceptionally occur in paramedian dorsal infarcts (Shintani et al., 1994). The main clinical findings of pontine infarctions are described in Table 15.3.

Transient focal neurologic deficits and emotional expression disorders, often under-recognized and diagnostically challenging, may rarely precede a paramedian pontine infarction. Short-lasting, recurrent, stereotyped hemiparesis, sensory dysfunction, ataxia, dysarthria, or ophthalmoplegia (termed "pontine warning syndrome") may rarely herald paramedian pontine ischemia due to basilar branch disease (Saposnik et al., 2008; Muengtaweepongsa et al., 2010). A particular type of intense transient pain manifesting as a burning sensation ("salt and pepper") in the eyes, nose, or face may also herald a paramedian pontine infarction (Caplan and Gorelick, 1983). A prodromal pathologic laughter ("fou rire prodromique") may precede uni- or bilateral paramedian pontine infarcts (Wali, 1993; Tei and Sakamoto, 1997; Assal et al., 2000). It has been speculated that a loss of tonic control of laughter occurs due to interruption of tonic innervation by ischemia of the faciorespiratory center in the lower brainstem (Wali, 1993).

Table 15.3

Main clinical manifestations of pontine infarctions.

Pontine territory	Clinical manifestation	Structure damaged
Anteromedial territory	Pure motor hemiparesis	Corticospinal tract
	Dysarthria	Corticobulbar tract
	Ataxic hemiparesis	Medial lemniscus
	DCHS	
	Restricted cheiro-oral sensory deficits	
Anterolateral territory	Pure motor hemiparesis, Ataxic hemiparesis, DCHS	Corticospinal tract
	Conjugate gaze palsy	Corticobulbar tract
	OAAHS	PPRF, abducens nucleus or MLF
	Hemisensory deficits	Medial lemniscus
Lateral territory	Conjugate or disconjugate horizontal gaze disorders	PPRF, abducens nucleus or MLF
	Hemisensory loss or restricted acral sensory syndromes	Medial lemniscus or spinothalamic tract
	Limb ataxia	Efferent cerebellar pathways

DCHS, dysarthria-clumsy hand syndrome; *MLF*, medial longitudinal fasciculus; *OAAHS*, one-and-a-half syndrome; *PPRF*, paramedian pontine reticular formation.

ANTEROLATERAL PONTINE INFARCTS

In isolated pontine infarct series, anterolateral infarctions represented between 14% and 25% of all ischemic lesions (Bassetti et al., 1996; Kumral et al., 2002b; Kobayashi et al., 2014; Huang et al., 2019). The lower and middle pons are usually damaged, and the most predominant manifestation is a classical LS or its variants (Bassetti et al., 1996). Lemniscal sensory loss, saccadic intrusion into pursuit eye movements, conjugate gaze palsy, and OAAHS are observed in around 50% of cases (Bassetti et al., 1996; Kumral et al., 2002b).

LATERAL PONTINE INFARCTS

The lateral (tegmental) territory makes up between 3% and 30% of all infarcts confined to the pons (Bassetti et al.,1996; Kumral et al., 2002b; Kobayashi et al., 2014; Huang et al., 2019). Ischemia primarily located in the pontine tegmentum or extending over this area from the basis may lead to a wide constellation of predominantly horizontal gaze disorders (Bronstein et al., 1990a, 1990b; Bassetti et al., 1996; Kumral et al., 2002b). Such disorders more often occur together with other signs of tegmental dysfunction and, less commonly, as isolated clinical findings. A third to a half of all patients with lateral territory pontine infarcts may present conjugate or disconjugate horizontal gaze disorders (Bassetti et al., 1996; Kobayashi et al., 2014). The AN at the caudal pons is located near the midline beneath the facial colliculi at the floor of the fourth ventricle. The paramedian pontine reticular formation (PPRF) is located rostrally and ventrally to the AN. The PPRF, the adjacent AN, the MLF, and the rostral portion of the vestibular nucleus are the main structures mediating saccadic and smooth pursuit movements at the horizontal-axis level (Leigh and Zee, 2015). The PPRF contains the premotor neurons for horizontal saccades to both sides and the omnipause neurons. The PPRF sends projections to the adjacent AN, the prepositus hypoglossal nucleus, the medial vestibular nucleus, and the motor neurons of the cervical muscles activated on gaze direction changes (Leigh and Zee, 2015). The AN is the final pathway for all horizontal eye movements and houses the motoneurons for the rectus lateral muscle and the interneurons joining the MLF crossing the midline to the contralateral medial rectus motoneurons in the motor ocular nuclei complex (Moncayo and Bogousslavsky, 2009; Leigh and Zee, 2015).

Selective lesions of the AN due to small infarcts or hemorrhages are uncommon, but when they occur, all ipsilateral horizontal movements (saccadic, smooth pursuit, and optokinetic and vestibular eye movements) are impaired, while vertical axis and convergence movements are relatively spared (Bennett and Savill, 1889; Bogousslavsky and Meienberg, 1987; Bronstein et al., 1990b; Miller et al., 2002). AN lesions are more often associated with ipsilateral facial palsy due to involvement of the adjacent genu of the facial nerve (Bogousslavsky and Meienberg, 1987).

The PPRF corresponds to the medial portions of the nucleus reticularis magnus celullaris and lies in the pontine tegmentum near the midline. It extends rostrally from the AN toward the brachium conjuctivum and trochlear nucleus. The PPRF is the last supranuclear relay involved in horizontal gaze movements (Moncayo and Bogousslavsky, 2009; Bae et al., 2013). Projections from cortical frontal eye fields relay into the colliculus superior and then provide signals to the excitatory burst neurons (EBNs) of the PPRF for ipsilateral saccades. The PPRF also participates in smooth pursuit movements and the nystagmus slow phase. The EBNs send projections to both neuron populations lying in the ipsilateral AN. In addition, the PPRF sends projections to the riMLF, thus contributing to vertical eye movements (Leigh and Zee, 2015). The extent of PPRF lesion determines the complete or partial alteration of horizontal gaze movements. Ipsilateral saccadic horizontal gaze impairment occurs with restricted PPRF lesion, while ipsilateral horizontal gaze movements with vestibular stimuli are preserved. In the acute stroke phase, contralateral conjugate gaze deviation to the lesion may be observed (Bogousslavsky and Meienberg, 1987).

A partial PPRF lesion produces only ipsilateral saccade slowness, while smooth pursuit movements and VOR may be preserved, particularly if the lesion affects the PPRF rostral portion (Bogousslavsky and Meienberg, 1987). Horizontal smooth pursuit movements and optokinetic nystagmus to the side of the lesion may be selectively impaired. In such a case, ischemia and hemorrhages restricted to the basal pons affect its nuclei, mainly the dorsolateral and lateral groups and, to a lesser extent, the dorsomedial group. These pontine nuclei are critical structures mediating smooth pursuit signals between the cerebral cortex and the cerebellum (Gaymard et al., 1993; Bassetti et al., 1996; Kumral et al., 2002b).

PPRF, AN, and MLF lesions result in the OAAHS, which was described in the 1960s by Miller Fisher in a 79-year-old woman with a caudal pons infarct (Fisher, 1967b). Kataoka noted OAAHS in 8% of 49 patients with paramedian pontine infarcts, all of whom had a tegmental lesion (Kataoka et al., 1997). In his series of 150 patients with pontine infarcts, Kumral reported OAAHS in 8% and 17% of patients with anterolateral and tegmental infarcts, respectively (Kumral et al., 2002b). This uncommon neuro-ophthalmologic manifestation of unilateral tegmental pontine damage is rarely an isolated feature of ischemic or hemorrhagic pontine stroke, most often occurring with other cranial nerve palsies, skew deviation, and motor or sensory deficits (Wall and Wray, 1983; Hitchings et al., 1988; De Seze et al., 1999). OAAHS is characterized by an ipsilateral horizontal gaze paresis due to PPRF or AN lesion, with concomitant INO due to MLF lesion just after it crosses the midline. In this instance, the contralateral eye can only abduct and nystagmus in abduction may sometimes occur (Pierrot-Deseilligny et al., 1981b). In acute stroke, a paralytic or nonparalytic exotropia in the eye contralateral to the lesion may be noted (Bogousslavsky and Regli, 1983b; Bogousslavsky and Meienberg, 1987).

Over the last few years, the clinical spectrum of OAAHS has expanded. Consequently, new uncommon variants – named arithmetically according to the sum of the cranial nerves involved – have been reported (Eggenberger, 1998; Cummins et al., 2011; Rosini et al., 2013; Dube et al., 2019; Li et al., 2019, 2020).

Unilateral ischemic or hemorrhagic stroke damaging the MLF results in INO (Nakajima 1983; Kim 2004). The MLF is the final common pathway for saccadic and smooth pursuit movements, as well as VOR in both horizontal and vertical planes (Frohman et al., 2008). The first reports of this disconjugate horizontal gaze disorder occurred in the first quarter of the 20th century. At that time, Lhermitte coined the term INO, which refers to paresis of adduction for conjugate horizontal movement on the side of lesion, combined in most of cases with a dissociated jerking nystagmus in the abducting eye (Paton, 1921; Freeman, 1922; Smith and Cogan, 1959; Crane et al., 1983). The basis for this nystagmus has not been fully clarified, although it has been hypothesized that it may be an adaptative response to overcome the contralateral medial rectus weakness (Zee et al., 1987). All horizontal movements (saccadic, smooth pursuit, and optokinetic following phase and quick phase of nystagmus) are affected in unilateral INO. Adduction with convergence movements is usually preserved despite the impossibility of voluntary adduction (known as dissociation of convergence or posterior INO) (Cogan, 1970; Bogousslavsky and Meienberg, 1987). Sometimes vertical gaze–evoked nystagmus may be seen. Bassetti reported INO in nearly a third of patients with lateral pontine infarctions (Bassetti et al., 1996). In the Kim series, 46% of 30 patients with INO had pontine infarcts, most in the rostral pons (Kim, 2004). INO was isolated or more often associated with perioral, hand, or foot paresthesias, mild dysarthria, or gait ataxia (Kim, 2004). Skew deviation was present in 40% of patients with INO. Ipsilateral hypertropia was common in patients with INO, reflecting the imbalance of crossed otolith inputs traveling in the MLF (Kim, 2004). Moreover, about two-thirds of patients with INO also had contralateral exotropia in forward gaze (Kim, 2004).

In a series of 120 INO patients (87 unilateral and 33 bilateral), 76% and 5% of cases were due to infarction or hemorrhage in the brainstem, respectively (Zwergal et al., 2008). Nearly all unilateral INO patients and only about a third of bilateral INO patients also had at least one component (subjective visual vertical, ocular torsion, or skew deviation) of the ocular tilt reaction (OTR) on the opposite side of stroke. In these cases, most strokes were located in the paramedian upper pontine tegmentum between the AN and the oculomotor nuclei. OTR is a perceptual and ocular motor dysfunction in the roll plane attributed to asymmetric damage to crossed graviceptive pathways from vertical semicircular canals and otolith organs ascending in the MLF toward the contralateral INC in the midbrain (Zwergal et al., 2008; Dieterich and Brandt, 2015). Bilateral INO has mainly been attributed to multiple sclerosis and less frequently to brainstem vascular lesions. Bilateral INO due to vascular lesions has been described in a third of patients with INO, although the anatomic location of the stroke was only confirmed pathologically in two patients in an older series, while in a more recent series the type and location of stroke were not mentioned (Gonyea, 1974; Bolaños et al., 2004). In another series, bilateral INO was found in only two of 30 isolated INO patients, both of whom had a dorsal pontine infarct (Kim, 2004). The wall-eyed bilateral INO (WEBINO syndrome) is a much less common bilateral INO variant. The cardinal signs of WEBINO are bilateral adduction impairment, nystagmus of the abducting eye, and primary gaze exotropia (Chen and Lin, 2007; Kim et al., 2008; Uzawa et al., 2020). There is still controversy regarding the anatomic site of the responsible lesion. WEBINO syndrome has been reported with both midbrain and pontine paramedian tegmental infarcts and tiny hemorrhages. The pathologic substrate mainly proposed is a bilateral MLF lesion between the trochlear nucleus and the AN with no extraocular muscle nuclei involvement. Unlike bilateral INO, convergence is usually impaired in WEBINO syndrome and often manifests with a variety of vertical gaze disorders as well (Chen and Lin, 2007; Wu et al., 2015). Additionally, WEBINO may manifest with other signs of tegmental dysfunction (Im et al., 2020). In extremely rare instances, a tiny infarct confined to the tegmental pons and affecting the MLF may produce a unilateral INO together with ipsilateral exotropia (WEMINO syndrome) (Johnston and Sharpe, 1994; Ikeda and Okamoto, 2002; Jeon et al., 2005).

Divergence paralysis, initially described by Parinaud in 1883, is rarely reported in unilateral caudal tegmental pontine infarcts or small upper pontine hemorrhages extending to the midbrain (Lee et al., 1987; Tsuda et al., 2006). It therefore is a rare cause of acquired uncrossed horizontal diplopia at far distance with no deviation at near distance (Bender and Savistky, 1940). The existence of a specific anatomic divergence center is still debated; nevertheless, it appears that the nucleus reticularis tegmenti pontis plays a key role in divergence eye movements (Rambold et al., 2004; Tsuda et al., 2006).

Sensory disturbance is a common clinical manifestation of tegmental pontine infarcts (Bassetti et al., 1996; Kumral et al., 2002b). Crural, brachio-crural, or facio-brachio-crural distribution of sensory loss may occur due to a lemniscal or spinothalamic lesion (Bassetti et al., 1996; Helgason and Wilbur, 1991). Restricted acral sensory syndromes may also be present, though more rarely. In such cases, a cheiro-pedal pattern of sensory deficit predominates (Kumral et al., 2002b) (Table 15.3).

Bilateral limb cerebellar ataxia with no other tegmental signs is an unusual finding of upper tegmental pontine infarcts (Lee and Cho, 2003). Damage to crossed and uncrossed efferent cerebellar pathways traveling in the PCS has been postulated. Occasionally, crossed brainstem syndromes may occur but are often not consistent with the classical clinical descriptions of pontine ischemic stroke syndromes (Table 15.2).

OTHER PONTINE INFARCTIONS

Dorsal territory infarcts account for up to 10% of all isolated pontine infarcts (Kobayashi et al., 2014). Sensory disturbances followed by oculomotor dysfunction are the leading clinical manifestations, while pyramidal signs are the least common (Kobayashi et al., 2014).

Around 10% of pontine infarcts affect both paramedian territories (Bassetti et al., 1996; Kataoka et al., 1997; Kumral et al., 2002b). Their notable clinical manifestations are acute pseudobulbar palsy and locked-in syndrome (Bassetti et al., 1996; Kumral et al., 2002b). Multiple territory infarcts (anteromedial and lateral or tegmental and dorsal) represent fewer than 5% of pontine infarcts. The predominant clinical manifestation is sensory-motor deficit in a facio-brachio-crural distribution (Kumral et al., 2002b).

PONTINE INFARCTION ETIOLOGY AND PROGNOSIS

Basilar artery branch disease (BABD), confirmed by necropsy or by angiography, is the leading cause of anteromedian and anterolateral infarctions (Kubik and Adams, 1946; Fisher and Caplan, 1971; Fisher, 1977; Bogousslavsky et al., 1993; Bassetti et al., 1996; Kumral et al., 2002b; Erro et al., 2005; Kobayashi et al., 2014; Huang et al., 2019). This pathologic mechanism described by Fisher and Caplan 50 years ago is characterized by occlusion of the ostium by an intramural atheroma of the perforating branches of the BA (Fisher and Caplan, 1971). Large artery occlusive disease (LAOD) in the VB system and small artery disease (SAD) are the most common causes of pontine infarctions after BABD in almost all series (Bassetti et al., 1996; Kumral et al., 2002b; Vemmos et al., 2005). SAD is particularly frequent as the etiology in lateral tegmental infarcts. Cardioembolism is not a common cause of infarctions restricted to the pons but becomes an important one when ischemia affects multiple territories of the posterior circulation (Bassetti et al., 1996). The cause of up to 10% of pontine infarcts may remain unknown (Kumral et al., 2002b).

Patients with paramedian pontine infarcts have the worst functional prognosis, likely because ischemia is larger compared to other pontine territories and consequently affects the corticospinal tract, while tegmental infarcts and anterolateral infarcts usually have better functional outcomes (Bassetti et al., 1996; Kumral et al., 2002b; Kobayashi et al., 2014). In addition, upper pontine infarcts have a better prognosis than infarcts in the caudal pons. The risk of ischemic stroke recurrence is highest in patients with pontine infarcts due to SAD or BABD (Bassetti et al., 1996; Kumral et al., 2002b).

PONTINE HEMORRHAGES

Hemorrhages in the pons are less common than infarcts. Nontraumatic pontine hemorrhages account for about 2%–11.5% of all intracranial hemorrhages (Nakajima, 1983; Tatu et al., 2000; Meguro et al., 2015). The clinical and imaging classifications for pontine hemorrhages are massive, bilateral tegmental, basal-tegmental, and small unilateral tegmental (Chung and Park, 1992). Of these, the bilateral tegmental and basal-tegmental are the most common types (Goto et al., 1980; Weisberg, 1986b; Murata et al., 1999; Rabinstein et al., 2004; Wessels et al., 2004).

In general, the clinical manifestations of pontine hemorrhages are quite similar to those of ischemic lesions, particularly in the case of hemorrhages affecting the tegmentum or both the tegmentum and the basis. About 20% of pontine hemorrhages belong to the massive type (Chung and Park, 1992; Murata et al., 1999; Rabinstein et al., 2004). Bleeding often occurs at the junction between the basis and the tegmentum at the middle pons level due to rupture of large paramedian perforators of the BA and may extend to the midbrain and, rarely, to the thalamus in 50% of cases (Goto et al., 1980; Nakajima, 1983; Wijdicks and St Louis, 1997). Bleeding usually spreads to the adjacent fourth ventricle, as well, though not commonly to the medulla oblongata. The most common clinical findings are loss of consciousness, respiratory abnormalities (Cheyne–Stokes or inspiratory gaps of apneustic respiration), tetraplegia, bilateral small (sometimes pinpoint) pupils reacting to light, paralysis of all conjugate horizontal eye movements, and ocular bobbing (Fisher, 1964; Weisberg, 1986b). The latter, in its classical form, consists of spontaneous, abrupt, conjugate, downward jerks of both eyes followed by a slow return to the mid position

(Fisher, 1964). Since its first description by Fisher in 1959, this unusual clinical sign denoting extensive pontine damage has been described in large pontine hemorrhages and infarctions (Fisher, 1964). Prognosis is dismal; mortality is the highest among pontine hemorrhages, and survivors end up severely disabled (Wijdicks et al., 1997).

The unilateral basal-tegmental type makes up nearly 40% of all pontine hemorrhages (Chung and Park, 1992). Hemorrhages of this type are often large, and bleeding may extend over the opposite side, the fourth ventricle, or rostrally to the midbrain (Goto et al., 1980). Bilateral tegmental hemorrhages are less common. Bleeding restricted to the basis or basotegmental junction may manifest as LS. The clinical manifestations of these types of hemorrhages are similar and include disturbance of consciousness, cranial nerve palsies, bilateral or unilateral (ipsilateral to the lesion) horizontal gaze palsies, hemiparesis, or quadriparesis (Weisberg, 1986b). More rarely, an unusual sensorial deficit called two-glove-and-one-sock syndrome may be a cardinal manifestation in small bilateral tegmental hemorrhages (Wu and Pai, 2012). Outcome is poor, particularly in bilateral tegmental hemorrhages.

Frequency of pontine hemorrhages limited to the tegmentum ranges between 9% and 66% (Chung and Park, 1992; Murata et al., 1999; Rabinstein et al., 2004). Small hemorrhages may present with LS. In other instances, the prominent features may be horizontal gaze disorders (ipsilateral gaze palsy, INO, OAAHS, homolateral sixth nerve palsy, and ocular bobbing), and pupillary abnormalities (ipsilateral small pupil reacting to light). Additionally, ipsi- or contralateral cerebellar limb ataxia, contralateral isolated hemiparesis, contralateral hemisensory loss (lemniscal deficit), or isolated trigeminal sensory deficit may occur (Caplan and Goodwin, 1982; Holtzman et al., 1987). Prognosis is favorable; nearly all patients survive with minimal or no disability (Chung and Park, 1992; Wijdicks et al., 1997). Arterial hypertension is the leading etiology in primary pontine hemorrhages, while cavernous malformations are most common in younger, nonhypertensive patients (Rabinstein et al., 2004).

MEDULLA OBLONGATA VASCULAR SYNDROMES

The lowermost segment of the brainstem, the medulla, houses the hypoglossal, vagus, glossopharyngeal, and spinal trigeminal nerve nuclei. Despite its short length, the medulla also contains critical centers that regulate cardiac, vasomotor, respiratory, autonomic, and vomiting functions. In addition, the vestibular, perihypoglossal, and inferior olivary nuclei, as well as ascending tracts that play a critical role in the horizontal and vertical gaze-holding mechanisms, reside in the medulla (Vanderah and Gould, 2021).

MEDULLA OBLONGATA VASCULAR SUPPLY

The intracranial VA (ICVA), posterior inferior cerebellar artery (PICA), anterior spinal artery (ASA), and posterior spinal arteries (PSA) are the main contributors to the medullary vascular supply. The proximal ICVA gives off small branches to irrigate the anteromedial surface of the upper third of the medulla (Duvernoy, 1999; Tatu et al., 1996). Branches issuing from the ASA penetrate in the anteromedial sulcus and perfuse the caudal two-thirds of the anteromedial region. The anteromedial arteries supply the hypoglossal nucleus, the MLF, the medial lemniscus, the medial portion of the pyramidal tract and its decussation, and the internal arcuate fibers (Akar et al., 1995; Tatu et al., 1996; Duvernoy 1999). The short anterolateral arteries, described by Stopford in 1916, stem from the ASA and irrigate the lateral aspect of the pyramidal tract and the medial part of the inferior olive (Stopford 1916a, 1916b). Sometimes, the anterolateral territory is supplied by the lateral medullary arteries (Duvernoy, 1999).

The lateral portion of the medulla is fed by a large number of small arteries in the entire lateral medullary fossa. The following rami may be recognized according to their relation to the IX and X cranial nerve roots: the inferior rami branching from the PICA, the middle rami branching from VA, and the superior rami branching from the proximal BA and, to a lesser extent, from the AICA that irrigate the upper lateral medulla. The lateral group of medullary arteries supply the spinothalamic and spinocerebellar tracts, the lateral reticular formation, and the IX and X cranial nerve nuclei. The dorsal medulla is irrigated by perforating branches from the PICA and the PSA. Thus, the dorsal medullary arteries irrigate the vestibular nuclei, the restiform body, and the gracilis and cuneatus nuclei (Bassetti et al., 1997; Duvernoy, 1999).

MEDULLARY INFARCT EPIDEMIOLOGY

Despite the growing use of high-resolution brain MRI in diagnosing acute brain ischemia, initial brain MRI with diffusion weighted imaging (DWI) pulse sequences may not reveal ischemia in patients with medullary infarction (MI) (Bogousslavsky et al., 1986; Shono et al., 2010; Fukuoka et al., 2012; Akimoto et al., 2017). One series found initial false-negative DWI in half of the patients with small infarcts in the dorsal territory (Lee et al., 2015). Medullary infarcts are the least common of brainstem or posterior circulation infarctions, accounting for up to 2.6% of all brain infarcts (Toyoda et al., 1996; Bassetti et al., 1997; Shono et al., 2010; Akimoto et al., 2017; Makita et al., 2019). A recent series of MRI-proven MI, however, showed these infarcts accounting for 28% of all brainstem infarcts (Dogan et al., 2020).

Infarcts may be restricted to the medulla or may also occur with ischemia involving the pons or, more rarely, the occipital lobes (Akimoto et al., 2017; Hong et al., 2018; Kang et al., 2018; Vlašković et al., 2022b). In a large series of lateral medullary infarcts (LMI), concomitant infarcts involving the whole PICA territory, or its lateral or medial branches, occurred in a third of patients. Less commonly, LMI also occurred with infarcts in the AICA territory (Kang et al., 2018; Vlašković et al., 2022b).

The most widely recognized medullary infarcts are those involving the medial (MMI) and the lateral (LMI) territories.

Unilateral infarcts in the latero-dorsal territory outnumber those located in the medial territory, while hemimedullary infarcts (Babinski–Nageotte syndrome) and bilateral MMI are less common (Escourolle et al., 1976; Kim et al., 1994b; Kameda et al., 2004; Hong et al., 2018, Makita et al., 2019; Dogan et al., 2020; Tao et al., 2021).

Based on the rostro-caudal medulla division, the upper third of the medulla, which is adjacent to the pontomedullary junction, is nearly always involved in MMI (Bassetti et al., 1997; Kameda et al., 2004; Akimoto et al., 2017; Hong et al., 2018; Makita et al., 2019; Dogan et al., 2020; Tao et al., 2021). The medullary blood supply from ASA diminishes caudorostrally until it is replaced by medial branches stemming from the VA, which may explain the predominance of ischemia at this level in MMI (Stopford, 1916a, 1916b). The middle and the rostral third of the medulla are the most frequently affected in LMI (Kameda et al., 2004; Hong et al., 2018; Makita et al., 2019). The lower medulla is rarely affected by ischemia (Makita et al., 2019).

MEDIAL MEDULLARY INFARCTS

The clinical picture of an infarct in the anterior and medial surface of the medulla oblongata was first described by William Spiller early in the 20th century (Spiller, 1908). Subsequently, Déjerine described the classic triad characterized by ipsilateral hypoglossal nerve palsy together with contralateral paresis and decreased positions and vibration sense in the upper and lower limbs (Déjerine, 1914). In a series of 30 MMI patients, no patient met the classic clinical triad described by Déjerine, while in other clinical MMI series, the frequency of this triad ranges between 4% and 90% of patients (Sawada et al., 1990; Toyoda et al., 1996; Bassetti et al., 1997; Kumral et al., 2002c; Shono et al., 2010; Akimoto et al., 2017; Vlašković et al., 2022b). The variable extent of ischemia in the ventromedial medulla may likely explain the variability in the frequency of the Déjerine triad. A contralateral hemiparesis of variable severity but sparing the face is the most constant clinical manifestation of MMI (Toyoda et al., 1996; Bassetti et al., 1997; Shono et al., 2010; Akimoto et al., 2017). A contralateral flaccid brachio-crural paresis often associated with increased muscle reflexes may in rare cases be the only MMI clinical manifestation. In such cases, ischemia predominantly affects the medullary pyramid (Chokroverty et al., 1975; Leestma and Noronha, 1976; Ropper et al., 1979). A contralateral monoparesis or tetraparesis may also occur (Kim et al., 1995b; Bassetti et al., 1997; Kim and Han, 2009; Akimoto et al., 2017). In addition, the hemiparesis may be ipsilateral due to a caudal extent of ischemia impinging on the crossed corticospinal tract (Bassetti et al., 1997).

Ipsilateral supranuclear facial paresis is far less common than contralateral supranuclear facial paresis, which can be found in up to 70% of patients when ischemia extends rostrally affecting uncrossed corticobulbar fibers running toward the contralateral facial nucleus (Kim et al., 1994a; Toyoda et al., 1996; Bassetti et al., 1997; Terao et al., 1997; Kameda et al., 2004; Kim and Han, 2009; Akimoto et al., 2017).

Loss of the contralateral position and vibration sense is the second most common clinical manifestation of MMI. In such cases, the infarct covers the medullary middle ventrodorsal aspect. A loss of spinothalamic sensation is less common (Kim and Han, 2009; Shono et al., 2010; Akimoto et al., 2017). Numbness in the contralateral arm and leg due to medial lemniscal involvement is a complaint reported by half of MMI patients (Bassetti et al., 1997; Kim and Han, 2009; Akimoto et al., 2017). Restricted sensory deficits may occur exceptionally if the lesion selectively affects the somatotopic representation in the medial lemniscus or the ascending sensory trigeminal tract. In some cases, pseudo-syringomyelia, pseudo-spinal, dermatomal, and perioral deficits may be seen (Kim et al., 1998a; Lee et al., 2001; Ryman and Ju, 2013).

Hypoglossal nerve palsy on the side of the infarct is the least common element of the triad, occurring in 15%–70% of patients. This palsy is invariably associated with hemiparesis (Ho and Meyer, 1981; Toyoda et al., 1996; Bassetti et al., 1997; Kameda et al., 2004; Kim and Han, 2009; Tao et al., 2021). When the hypoglossal nerve is spared, ischemia neither affects the dorsally located hypoglossal nucleus nor the hypoglossal nerve fibers running laterally and then exiting the medulla through the anterolateral sulcus. The main clinical manifestations of medullary infarcts are summarized in Table 15.4.

Vertigo and mild limb ataxia are found in 50% and 30% of MMI patients, respectively (Kameda et al., 2004; Shono et al., 2010; Tao et al., 2021). Dysphagia and dysarthria due to the lesion of corticobulbar fibers innervating the nucleus ambiguus are found, respectively, in 66% and 40% of cases (Kameda et al., 2004; Shono et al., 2010; Akimoto et al., 2017; Tao et al., 2021). However, Kwon et al. reported dysphagia in nearly 80% of unilateral MMI patients, in whom swallowing was assessed by videofluoroscopy (Kwon et al., 2005a). Cervical pain and central respiratory disturbances have rarely been reported (Tao et al., 2021). Ocular manifestations are uncommon and are summarized in Table 15.5.

BILATERAL MEDIAL MEDULLARY INFARCTS

The cardinal features of this uncommon vascular syndrome are flaccid tetraparesis sparing the face, bilateral loss of deep sensation, and bilateral hypoglossal palsy (Kameda et al., 2004; Hu et al., 2022). In addition, severe dysphagia, attributed to bilateral damage to the supranuclear fibers of the nucleus ambiguus in the rostral medulla, is common (Kwon et al., 2005a; Paliwal et al., 2009). Sometimes, the rostral extent of the lesion leads to horizontal gaze palsy because of the involvement of the pontine MLF or the PPRF (Toyoda et al., 1992).

LATERAL MEDULLARY INFARCTIONS

In the late 19th century, Wallenberg provided the first detailed clinical, although nonpathologic, description of LMI, which is

Table 15.4

Main clinical manifestations of medullary infarctions.

Infarction topography	Clinical manifestation	Structure damaged
Medial medullary territory	Ipsilateral hypoglossal nerve palsy	Hypoglossal nerve nucleus
	Contralateral hemiparesis sparing the face	Corticospinal tract
	Loss of vibration and proprioceptive sense	Medial lemniscus
	Dysphagia and dysarthria	Nucleus ambiguus
	Limb ataxia (unusual)	Pontocerebellar fibers and/or proprioceptive sensory loss
	Vertigo, nystagmus and ocular motor dysfunction	Vestibular nuclei, MLF, and NPH
Lateral medullary territory	Ipsilateral facial hypalgesia	Trigeminal spinal nucleus and tract involvement
	Contralateral trunk and limbs hypalgesia	Spinothalamic tract
	Dysphagia	Nucleus solitarius, nucleus ambiguus or its fascicle
	Dysarthria	Nucleus ambiguus or its fascicle
	Gait ataxia or ipsilateral limb ataxia	Inferior cerebellar peduncle, spinocerebellar fibers
	Nystagmus	Inferior vestibular nucleus
	Horner's syndrome	Central sympathetic tract
	Hiccup	Reticular system
	Respiratory failure	Respiratory central pattern generators

MLF, medial longitudinal fasciculus; *NPH*, nucleus prepositus hypoglossi.

Table 15.5

Ocular manifestations in medial and lateral medullary infarctions.

Clinical manifestation	Medial medullary infarction	Lateral medullary infarction
Nystagmus	Uncommon Horizontal gaze nystagmus and primary position nystagmus. Both nystagmus beat to the side of lesion Upbeat nystagmus in primary position or more marked in up-gaze (perihypoglossal nucleus lesion)	Common. Horizontal or mixed horizontal-torsional spontaneous nystagmus in primary position (usually beats away from the lesion) Gaze-evoked nystagmus (more prominent when looking to the contralateral side) Head-shaking nystagmus
Ocular tilt reaction (Triad of skew deviation, head tilt, and ocular torsion)	Exceptional in rostral infarcts; contralesional	Common, partial or total Skew deviation with ipsilateral hypotropia; head tilt toward the side of the lesion, ipsilateral cyclodeviation, and ipsilateral deviation of subjective visual vertical
Ocular lateropulsion of Barré	Infrequent, contrapulsion	Common. Ipsilateral associated with axial lateropulsion Climbing fibers damage before or after decussation
Horizontal saccades	Ipsilateral hypometria; contralesional hypermetria	Ipsilateral hypermetria; contralateral hypometria Inhibition of the ipsilateral fastigial nucleus
Horner's syndrome	15% of patients	90% of patients

perhaps the most well-known of all brainstem stroke syndromes. Dizziness, imbalance, severe vertigo, nausea, and vomiting are common early clinical manifestations (Kim, 2003; Kameda et al., 2004; Tao et al., 2021).

Sensory deficits are nearly always present in LMI and are major, long-lasting sequelae (Kim and Choi-Kwon, 1999; Kim, 2003). A selective ipsilateral loss of the spinothalamic sensation in the face and in the contralateral trunk and limbs is the classical sensory deficit pattern (Stopford classification type I). This pattern is attributed to lesion of the homolateral, uncrossed, descending tract, and the nucleus of the trigeminal nerve, as well as the crossed lateral spinothalamic tract, respectively. However, such a pattern only occurs in 25% of LMI (Kim et al., 1997).

Sensory deficits in LMI are diverse in their manifestations and may be seen as often as the classical pattern (Currier et al., 1961; Vuilleumier et al., 1995; Kim et al., 1997; Hongo et al., 2014). A contralateral spinothalamic deficit involving the face and the upper part of the body (Stopford classification type II) may be seen. This pattern results from damage of the crossed ventral trigeminothalamic tract and the most medial part of the spinothalamic tract, which carries pain and temperature information from the cervical segment of the body (Kim et al., 1997). Large medial lateral infarcts cause bilateral facial and contralateral trunk and limb hypoalgesia (Stopford classification type III). In such cases, both the descending and ascending trigeminal tracts and the spinothalamic tract are affected (Kim et al., 1997). Loss of the senses of pain and temperature in the entire contralateral hemibody (Stopford classification type IV), reported for the first time by Vuadens and Bogousslavsky, is due to crossed ventral trigeminothalamic tract and spinothalamic tract involvement. In this sensory deficit, the lesion predominantly affects the retro-olivary area in the ventromedial tegmentum (Vuadens and Bogousslavsky, 1998). Additionally, spinothalamic loss restricted to the contralateral limbs and trunk, with no facial sensory deficit, is another pattern found in about 20% of patients (Kim et al., 1997). With infarcts located more dorsally, the sensory deficit may be limited on rare occasions to the ipsilateral face, sparing the contralateral hemibody (Nakamura et al., 1996; Kim et al., 1997). Dissociation of spinothalamic deficit may also occur (Kim, 1998b). An ipsilateral lemniscal sensory deficit may rarely be observed when ischemia extends caudally involving the ascending or crossed dorsal column tracts (Kim et al., 1994b; Brochier et al., 1999). Ipsilateral loss of lemniscal modalities and a contralateral spinothalamic impairment restricted to the distal part of both arms is hardly ever observed (Cerrato et al., 2000). Ipsilateral numbness, mainly in the fingers, may occasionally be reported when the lesion extends dorsomedially involving the homolateral dorsal column or the crossed lemniscal fibers (Kim, 2001).

Dysphagia is a rarely isolated, nearly invariable finding in LMI and is usually accompanied by concurrent dysarthria and, less frequently, by dysphonia (Kim et al., 1994b; Lee e al., 1999; Kim, 2003; Kameda et al., 2004; Kwon et al., 2005a; Tao et al., 2021). It is more frequent and severe with rostral LMI than caudal infarcts and is often associated with hyolaryngeal excursion problems (Kwon et al., 2005a). Dysphagia may be transient or long-lasting, requiring early nasogastric tube feeding or percutaneous endoscopic gastrostomy. It may also be associated with aspiration pneumonia, increased mortality, and longer hospitalization. The central pattern generator of swallowing, which includes the solitary nucleus, the nucleus ambiguus, and the surrounding reticular structure, is affected in dysphagia (Sacco et al., 1993; Jean, 2001) (Table 15.4).

Though rarely, LMI may exhibit a variety of bizarre visual disturbances. These are often transient and recurrent and may erroneously be interpreted as being of psychiatric origin. Sensations of body and environmental tilt (the surroundings tilting to one side or even upside and down). These perceptual symptoms have been found in connection with otolith pathway lesions in the peripheral vestibular organ, the brainstem, and the parietal cortex (Estañol and Lopez-Rios, 1982; Tilikete et al., 2001). A variety of oculomotor abnormalities exist in LMI (Table 15.5).

Severe gait ataxia and, less often, ipsilateral limb ataxia are also frequently observed (Vuilleumier et al., 1995). Ataxia affects the inferior cerebellar peduncle, spinocerebellar fibers, or cerebellum (Marx et al., 2008). Ipsilateral axial lateropulsion, often associated with other manifestations of LMI, may be related to spinocerebellar pathway interruption and restiform body involvement (Thömke et al., 2005).

Hiccup occurs early in LMI and in up to 36% of cases. It is usually transient and associated with respiratory center lesion in the dorsolateral region of the middle or caudal medulla (Vuilleumier et al., 1995; Park et al., 2005). Occipital headache or posterior cervical pain, both ipsilateral to the lesion, are reported mainly in LMI due to cervical VA dissection (Kameda et al., 2004).

Respiratory failure due to damage of respiratory central pattern generators including the dorsal respiratory group (tractus solitary nucleus) and the ventral respiratory group (ambiguus and retroambiguus nuclei) may be expected in LMI (Benarroch, 2007; Prabhakar et al., 2020). Fatal overt respiratory failure is uncommon (Bogousslavsky et al., 1990b; Kim, 2003; Terao et al., 2004; Makita et al., 2019). Furthermore, diverse sleep breathing disorders manifesting with obstructive or central sleep apnea, central alveolar hypoventilation (Ondine's curse), or periodic breathing may occur (Oku and Okada, 2008; Pavšič et al., 2020; Prabhakar et al., 2020; Pavšič et al., 2021). Sometimes these disorders may develop acutely, subacutely, or some months after infarction and may go unnoticed (Oya et al., 2001; Mendoza and Latorre, 2013; Mon and Tamaki, 2018).

Sinus arrythmia, sick sinus syndrome, blood pressure variability, paroxysmal hypertension or orthostatic hypotension, and syncope reflecting autonomic dysregulation may occur in small caudal LMI and are sometimes missed in the acute and subacute setting (Phillips et al., 2000; Lassman and Mayer, 2005; Koay and Dewan, 2013; Huynh et al., 2018; Lee et al., 2021; Alloush et al., 2022).

Opalski syndrome is an unusual LMI variant that manifests with concomitant mild to severe ipsilateral hemiparesis, hyperreflexia, and Babinski's sign (Opalski, 1946). It is still debated whether hemiparesis is due to the caudal extent of ischemia affecting the ipsilesional crossed corticospinal fibers or if a hypotonic dysregulation of motor loops is the pathologic substrate (Currier et al., 1961; Kim, 2003; Nakamura et al., 2010). Other features of this syndrome include Horner's syndrome, ipsilateral cerebellar signs, ipsilateral facial loss of pain and temperature sensation, and contralateral loss of pain and temperature sensation in the limbs (Hermann et al., 2009; Kon et al., 2017).

OTHER VASCULAR MEDULLARY SYNDROMES

The bilateral medial medullary and hemimedullary syndromes are infrequent. Besides these, other very rarely found medullary syndromes (Babinski–Nageotte, Avellis, Céstan–Chenais, Reinhold) have been reported in medullary stroke (Table 15.2).

MEDULLARY INFARCTION ETIOLOGY

LAOD involving the VA is the main cause in unilateral and bilateral MMI followed by SAD (Escourolle et al., 1976; Kim, 2003; Pongmoragot et al., 2013; Akimoto et al., 2017; Hong et al.,

2018; Makita et al., 2019; Tao et al., 2021; Hu et al., 2022). VA dissection is not uncommon in MMI and is the cause in up to 25% of cases, while cardioembolism is a rare occurrence in both unilateral and bilateral MMI (Sacco et al., 1993; Vuilleumier et al., 1995; Kim et al., 1998c; Kim, 2003; Kameda et al., 2004; Kang et al., 2018; Dogan et al., 2020; Tao et al., 2021). The cause remains undetermined in 9%–20% of MMI cases (Akimoto et al., 2017; Dogan et al., 2020; Tao et al., 2021).

Atherosclerotic VA disease, either in the proximal or distal VA segment, is the leading etiology in LMI (Kim et al., 1998c; Kameda et al., 2004; Fukuoka et al., 2012; Hong et al., 2018; Dogan et al., 2020). This infarction was initially associated with PICA occlusion; nevertheless, in their seminal work in 1961, Fisher et al. demonstrated arterial occlusion in 14 of 16 LMI patients, 12 of whom had VA occlusion, while the rest showed occlusion restricted to the PICA (Fisher et al., 1961). As in MMI, SAD is the second most common cause of LMI. VA dissection, either spontaneous or traumatic, is a prominent LMI etiology, particularly in younger patients, who are usually free of stroke risk factors (Kim, 2003; Kang et al., 2018; Yu et al., 2022). VA dissection is documented in up to 33% of LMI cases (Sacco et al., 1993; Vuilleumier et al., 1995; Kim et al., 1998c; Kim, 2003; Kameda et al., 2004; Kang et al., 2018; Dogan et al., 2020; Tao et al., 2021). Embolism from a cardiac source is uncommon and is often associated with infarcts in the caudal medulla's dorsal territory (Vuilleumier et al., 1995; Kim et al., 1998c; Kim, 2003). Simultaneous cerebellar infarcts in the PICA territory occur in around 40% of patients whose lesions are almost exclusively located in the dorsal medulla, while such an occurrence is not observed when ischemia affects the lateral medulla (Vuilleumier et al., 1995; Kameda et al., 2004). The cause goes unidentified in around 15% of LMI cases, usually when the lesion affects the rostral medulla (Kim, 2003; Tao et al., 2021).

MEDULLARY INFARCT PROGNOSIS

In unilateral MMI series, fatalities are uncommon and around 70% of patients have a good outcome (modified Rankin Scale score ≤2) at discharge and at three months after infarct onset (Shono et al., 2010; Fukuoka et al., 2012; Akimoto et al., 2017; Tao et al., 2021). On the other hand, the outcome in bilateral MMI is not as good, with 62% of patients ending up dependent for activities of daily living three months after the index infarct (Pongmoragot et al., 2013). Bilateral MMI was associated with higher fatality rates (Paliwal et al., 2009). In a recent systematic review of 38 patients with MRI-proven bilateral MMI, 24% of the patients died (Pongmoragot et al., 2013).

Like MMI, LMI has a good outcome, with about 72% of patients exhibiting a modified Rankin Scale score ≤2 at discharge and at three months after infarct onset (Fukuoka et al., 2012; Tao et al., 2021). In the long-term follow-up, 36% of LMI patients had a poor functional outcome (modified Rankin Scale ≥3). Recurrence is seen more often in association with cerebellar infarcts in the PICA territory (Kang et al., 2018).

MEDULLARY HEMORRHAGES

Primary medullary hemorrhages are extremely uncommon and are most often due to caudal extension of pontine hemorrhages. Basal or tegmental unilateral medullary hemorrhages are the most common and are rarely located in the dorsolateral aspect of the medulla. Clinical manifestations are quite similar to those of ischemic strokes, the most common being horizontal nystagmus, ipsilateral facial sensory deficit, limb and gait ataxia, dysarthria, and dysphagia (Barinagarrementeria and Cantu, 1994). Contralateral hemiparesis and ipsilateral hypoglossal palsy are found with dorsolateral hemorrhages as the result of spreading of bleeding to the ventral and medial medulla. Hypertension and ruptured vascular malformations may account for some causes, but the etiology of medullary hemorrhages most often remains unidentified (Neumann et al., 1985; Biller et al., 1986; Barinagarrementeria and Cantu, 1994; Kwon et al., 2005b; Kumral et al., 2020).

References

Abe K, Yanagihara T (1996). Hemiparkinsonism following haemorrhage in the contralateral substantia nigra. Neuroradiology 38: S67–S69.

Akar ZC, Dujovny M, Gómez-Tortosa E et al. (1995). Microvascular anatomy of the anterior surface of the medulla oblongata and olive. J Neurosurg 82: 97–105.

Akimoto T, Ogawa K, Morita A et al. (2017). Clinical study of 27 patients with medial medullary infarction. J Stroke Cerebrovasc Dis 26: 2223–2231.

Akyol A, Akyildiz UO, Tataroglu C (2006). Vascular Parkinsonism: a case of lacunar infarction localized to mesencephalic substantia nigra. Parkinsonism Rel Disord 12: 459–461.

Alemdar M, Kamaci S, Budak F (2006). Unilateral midbrain infarction causing upward and downward gaze palsy. J Neuroophthalmol 26: 173–176.

Alloush TK, Alloush AT, Sami M et al. (2022). Sinus arrest following acute lateral medullary infarction. Neurol Sci 43: 6555–6559.

Arboix A, Bell Y, García-Eroles L et al. (2004). Clinical study of 35 patients with dysarthria-clumsy hand syndrome. J Neurol Neurosurg Psychiatry 75: 231–234.

Arch AE, Weisman DC, Coca S et al. (2016). Missed ischemic stroke diagnosis in the emergency department by emergency medicine and neurology services. Stroke 47: 668–673.

Assal F, Valenza N, Landis T et al. (2000). Clinicoanatomical correlates of a Fou rire prodromique in a pontine infarction. J Neurol Neurosurg Psychiatry 69: 697–698.

Azouvi P, Tougeron A, Hussonois C et al. (1989). Pure sensory stroke due to midbrain haemorrhage limited to the spinothalamic pathway. J Neurol Neurosurg Psychiatry 52: 1427–1428.

Bae YJ, Kim JH, Choi BS et al. (2013). Brainstem pathways for horizontal eye movement: pathologic correlation with MR imaging. Radiographics 33: 47–59.

Barinagarrementeria F, Cantu C (1994). Primary medullary hemorrhage. Report of four cases and review of the literature. Stroke 25: 1684–1687.

Bassetti C, Bogousslavsky J, Barth A et al. (1996). Isolated infarcts of the pons. Neurology 46: 165–175.
Bassetti C, Bogousslavsky J, Mattle H et al. (1997). Medial medullary stroke: report of seven patients and review of the literature. Neurology 48: 882–890.
Benarroch EE (2007). Brainstem respiratory chemosensitivity: new insights and clinical implications. Neurology 68: 2140–2143.
Bender MB, Savistky N (1940). Paralysis of divergence. Arch Ophthalmol 23: 1046–1051.
Bender MB (1980). Brain control of conjugate horizontal and vertical eye movements: a survey of the structural and functional correlates. Brain 103: 23–69.
Bennett H, Savill TH (1889). A case of permanent conjugate deviation of the eyes and head, the result of a lesion limited to the sixth nucleus; with remarks on associated lateral movements of the eyeballs, and rotation of the head and neck. Brain 12: 102–116.
Bhidayasiri R, Plant GT, Leigh RJ (2000). A hypothetical scheme for the brainstem control of vertical gaze. Neurology 54: 1985–1993.
Biller J, Gentry LR, Adams HP Jr et al. (1986). Spontaneous hemorrhage in the medulla oblongata: clinical MR correlations. J Comput Assist Tomogr 10: 303–306.
Bogousslavsky J, Regli F (1983). Nuclear and prenuclear syndromes of the oculomotor nerve. Neuro-Ophthalmology 3: 211–216.
Bogousslavsky J, Regli F (1983). Exotropie pontique paralytique et non paralytique. Rev Neurol (Paris) 139: 219–223.
Bogousslavsky J, Regli F (1984). Upgaze palsy and monocular paresis of downward gaze from ipsilateral thalamo-mesencephalic infarction: a vertical "one-and-a-half" syndrome. J Neurol 231: 43–45.
Bogousslavsky J, Fox AJ, Barnett HJ et al. (1986). Clinico-topographic correlation of small vertebrobasilar infarct using magnetic resonance imaging. Stroke 17: 929–938.
Bogousslavsky J, Meienberg O (1987). Eye-movement disorders in brain-stem and cerebellar stroke. Arch Neurol 44: 141–148.
Bogousslavsky J, Van Melle G, Regli F (1988). The Lausanne Stroke Registry: analysis of 1,000 consecutive patients with first stroke. Stroke 19: 1083–1092.
Bogousslavsky J, Regli F, Uske A (1988). Thalamic infarcts: clinical syndromes, etiology, and prognosis. Neurology 38: 837–848.
Bogousslavsky J (1989). Syndromes oculomoteurs résultant de lésions mésencéphaliques chez l'homme [Oculomotor syndromes resulting from mesencephalic lesions in man]. Rev Neurol (Paris) 145: 546–559.
Bogousslavsky J, Miklossy J, Regli F et al. (1990). Vertical gaze palsy and selective unilateral infarction of the rostral interstitial nucleus of the medial longitudinal fasciculus (riMLF). J Neurol Neurosurg Psychiatry 53: 67–71.
Bogousslavsky J, Khurana R, Deruaz JP et al. (1990). Respiratory failure and unilateral caudal brainstem infarction. Ann Neurol 28: 668–673.
Bogousslavsky J, Regli F, Maeder P et al. (1993). The etiology of posterior circulation infarcts: a prospective study using magnetic resonance imaging and magnetic resonance angiography. Neurology 43: 1528–1533.
Bogousslavsky J, Maeder P, Regli F et al. (1994). Pure midbrain infarction: clinical syndromes, MRI, and etiologic patterns. Neurology 44: 2032–2040.
Bolaños I, Lozano D, Cantú C (2004). Internuclear ophthalmoplegia: causes and long-term follow-up in 65 patients. Acta Neurol Scand 110: 161–165.
Breen LA, Hopf HC, Farris BK et al. (1991). Pupil-sparing oculomotor nerve palsy due to midbrain infarction. Arch Neurol 48: 105–106.
Brochier T, Ceccaldi M, Milandre L et al. (1999). Dorsolateral infarction of the lower medulla: clinical-MRI study. Neurology 52: 190–193.
Bronstein AM, Rudge P, Gresty MA et al. (1990). Abnormalities of horizontal gaze. Clinical, oculographic and magnetic resonance imaging findings. II. Gaze palsy and internuclear ophthalmoplegia. J Neurol Neurosurg Psychiatry 53: 200–207.
Bronstein AM, Morris J, Du Boulay G et al. (1990). Abnormalities of horizontal gaze. Clinical, oculographic and magnetic resonance imaging findings. I. Abducens palsy. J Neurol Neurosurg Psychiatry 53: 194–199.
Caplan LR (1980). "Top of the basilar" syndrome. Neurology 30: 72–79.
Caplan LR, Goodwin JA (1982). Lateral tegmental brainstem hemorrhages. Neurology 32: 252–260.
Caplan L, Gorelick P (1983). Salt and pepper on the face pain in acute brainstem ischemia. Ann Neurol 13: 344–345.
Caplan L, Chung CS, Wityk R et al. (2005). New England Medical Center Posterior Circulation Stroke Registry: I. Methods, data base, distribution of brain lesions, stroke mechanisms, and outcomes. J Clin Neurol 1: 14–30.
Caruso G, Vincentelli F, Giudicelli G et al. (1990). Perforating branches of the basilar bifurcation. J Neurosurg 73: 259–265.
Castaigne P, Lhermitte F, Buge A et al. (1981). Paramedian thalamic and midbrain infarct: clinical and neuropathological study. Ann Neurol 10: 127–148.
Castro O, Johnson LN, Mamourian AC (1990). Isolated inferior oblique paresis from brainstem infarction: perspective on oculomotor fascicle organization in the ventral midbrain tegmentum. Arch Neurol 47: 235–237.
Cerrato P, Imperiale D, Bergui M et al. (2000). Restricted dissociated sensory loss in a patient with a lateral medullary syndrome: a clinical-MRI study. Stroke 31: 3064–3066.
Chen CM, Lin SH (2007). Wall-eyed bilateral internuclear ophthalmoplegia from lesions at different levels in the brainstem. J Neuroophthalmol 27: 9–15.
Chen W, Yi T, Chen Y et al. (2015). Assessment of bilateral cerebral peduncular infarction: magnetic resonance imaging, clinical features, and prognosis. J Neurol Sci 357: 131–135.
Chen WC, Li YS, Huang P (2021). Isolated trochlear palsy as the only presentation of midbrain infarction: a case report. J Int Med Res 49: 3000605211008292.
Choi HY, Jung YJ, Shin HW (2015). Multifocal dystonia as a manifestation of acute midbrain infarction. J Neurol Sci 356: 217–218.
Chokroverty S, Rubino FA, Haller C (1975). Pure motor hemiplegia due to pyramidal infarction. Arch Neurol 32: 647–648.
Chung CS, Park CH (1992). Primary pontine hemorrhage: a new CT classification. Neurology 42: 830–834.
Cogan DG (1970). Internuclear ophthalmoplegia, typical and atypical. Arch Ophthalmol 84: 583–589.
Crane TB, Yee RD, Baloh RW et al. (1983). Analysis of characteristic eye movement abnormalities in internuclear ophthalmoplegia. Arch Ophthalmol 101: 206–210.
Cummins G, O'Hare A, Dunne R et al. (2011). "Sixteen and a half": a novel pontine neuro-ophthalmological syndrome. J Neurol 258: 1347–1348.
Currier RD, Giles CL, DeJong RN (1961). Some comments on Wallenberg's lateral medullary syndrome. Neurology 11: 778–791.
Dehaene I, Dom R (1982). A mesencephalic locked-in syndrome. J Neurol 227: 255–259.

Déjerine J (1914). Sémiologie des affections du système nerveux. Paris: Masson.
Deleu D, Buisseret T, Ebinger G (1989). Vertical one-and-a-half syndrome. Supranuclear downgaze paralysis with monocular elevation palsy. Arch Neurol 46: 1361–1363.
De Seze J, Lucas C, Leclerc X et al. (1999). One-and-a-half syndrome in pontine infarcts: MRI correlates. Neuroradiology 41: 666–669.
Dieterich M, Brandt T (2015). The bilateral central vestibular system: its pathways, functions, and disorders. Ann N Y Acad Sci 1343: 10–26.
Djulejić V, Marinković S, Milić V et al. (2015). Common features of the cerebral perforating arteries and their clinical significance. Acta Neurochir (Wien) 157: 743–754.
Dogan SN, Bayrak AH, Yazgu R (2020). Topographic evaluation of medullary infarcts from the radiologist's point of view. Neuroradiology 62: 947–953.
Dube M, Dani R, Dubey A et al. (2019). Twenty-and-a-half syndrome: a case report. J Med Case Rep 13: 35.
Durward QJ, Barnett HJ, Barr HW (1982). Presentation and management of mesencephalic hematoma. Report of two cases. J Neurosurg 56: 123–127.
Duvernoy HM (1999). Human brain stem vessels. 2nd edition. Berlin/Heidelberg/New York: Springer-Verlag.
Eggenberger E (1998). Eight-and-a-half syndrome: one-and-a-half syndrome plus cranial nerve VII palsy. J Neuroophthalmol 18: 114–116.
Erro ME, Gállego J, Herrera M et al. (2005). Isolated pontine infarcts: etiopathogenic mechanisms. Eur J Neurol 12: 984–988.
Escourolle R, Hauw JJ, Agopian PD et al. (1976). Les infarctus bulbaires. Etude des lésions vasculaires dans 26 observations. J Neurol Sci 28: 103–113.
Estañol B, Lopez-Rios G (1982). Neuro-otology of the lateral medullary infarct syndrome. Arch Neurol 39: 176–179.
Esteban Muñoz J, Tolosa E, Saiz A et al. (1996). Upper-limb dystonia secondary to a midbrain hemorrhage. Mov Disord 11: 96–99.
Fingerote RJ, Shuaib A, Brownell AK (1990). Spontaneous midbrain hemorrhage. South Med J 83: 280–282.
Fisher CM, Karnes WE, Kubi CS (1961). Lateral medullary infarction-the pattern of vascular occlusion. J Neuropathol Exp Neurol 20: 323–379.
Fisher CM (1964). Ocular bobbing. Arch Neurol 11: 543–546.
Fisher C (1965). Acute hypertensive cerebellar hemorrhage. J Nerv Ment Dis 140: 38–57.
Fisher CM, Curry HB (1965). Pure motor hemiplegia of vascular origin. Arch Neurol 13: 30–44.
Fisher CM, Cole M (1965). Homolateral ataxia and crural paresis: a vascular syndrome. J Neurol Neurosurg Psychiatry 28: 48–55.
Fisher CM (1965). Pure sensory stroke involving face, arm, and leg. Neurology 15: 76–80.
Fisher CM (1967). A lacunar stroke. The dysarthria-clumsy hand syndrome. Neurology 17: 614–617.
Fisher CM (1967). Some neuro-ophthalmological observations. J Neurol Neurosurg Psychiatry 30: 383–392.
Fisher CM, Caplan LR (1971). Basilar artery branch occlusion: a cause of pontine infarction. Neurology 21: 900–905.
Fisher CM (1977). Bilateral occlusion of basilar artery branches. J Neurol Neurosurg Psychiatry 40: 1182–1189.
Foix C, Hillemand P (1925). Les artères de l'axe encéphalique jusqu'au diencéphale inclusivement. Rev Neurol 2: 705–739.
Freeman W (1922). Paralysis of associated lateral movements of the eyes: a symptom of intrapontine lesion. Arch Neurol Psychiatry 7: 454–487.
Frohman TC, Galetta S, Fox R et al. (2008). Pearls & oysters: the medial longitudinal fasciculus in ocular motor physiology. Neurology 70: e57–e67.
Fu X, Li H, Tian X et al. (2019). Rare presentation of an isolated bilateral cerebral peduncular infarction: a case report and review of the literature. Medicine (Baltimore) 98: e17665.
Fujimoto Y, Ohnishi Y, Wakayama A, et al. (2012). Transient total mesencephalic locked-in syndrome after bilateral ptosis due to basilar artery thrombosis. J Stroke Cerebrovasc Dis. 21: 909.e7-e8.
Fujioka S, Nakajima M (2008). Combined up- and downgaze palsy associated temporarily with a unilateral midbrain infarction. Eur Neurol 59: 272–273.
Fukuoka T, Takeda H, Dembo T et al. (2012). Clinical review of 37 patients with medullary infarction. J Stroke Cerebrovasc Dis 21: 594–599.
Gaymard B, Larmande P, de Toffol B et al. (1990). Reversible nuclear oculomotor nerve paralysis. Caused by a primary mesencephalic hemorrhage. Eur Neurol 30: 128–131.
Gaymard B, Pierrot-Deseilligny C, Rivaud S et al. (1993). Smooth pursuit eye movement deficits after pontine nuclei lesions in humans. J Neurol Neurosurg Psychiatry 56: 799–807.
Geller TJ, Bellur SN (1987). Peduncular hallucinosis: magnetic resonance imaging confirmation of mesencephalic infarction during life. Ann Neurol 21: 602–604.
Gilberti N, Gamba M, Costa A et al. (2014). Pure midbrain ischemia and hypoplastic vertebrobasilar circulation. Neurol Sci 35: 259–263.
Gillilan LA (1964). The correlation of the blood supply to the human brain stem with clinical brain stem lesions. J Neuropathol Exp Neurol 23: 78–108.
Giotta Lucifero A, Baldoncini M, Bruno N et al. (2021). Microsurgical neurovascular anatomy of the brain: the posterior circulation (Part II). Acta Biomed 92: e2021413.
Gomez CR, Saul RF, Selhorst JB et al. (1990). Meso-diencephalic infarction: a not so rare form of stroke. Ital J Neurol Sci 11: 551–557.
Gonyea EF (1974). Bilateral internuclear ophthalmoplegia. Association with occlusive cerebrovascular disease. Arch Neurol 31: 168–173.
Goto N, Kaneko M, Hosaka Y et al. (1980). Primary pontine hemorrhage: clinicopathological correlations. Stroke 11: 84–90.
Ha SH, Ryu JC, Bae JH et al. (2022). Isolated pontine infarction versus pontine plus infarction: prevalence, pathogenic mechanism, and outcomes. J Neurol 269: 4375–4382.
Halmagyi GM, Evans WA, Hallinan JM (1978). Failure of downward gaze: the site and nature of the lesion. Arch Neurol 35: 22–26.
Halmagyi GM, Aw ST, Dehaene I et al. (1994). Jerk-waveform see-saw nystagmus due to unilateral meso-diencephalic lesion. Brain 117: 789–803.
Han SH, Roh JK, Myung HJ (1989). Mesencephalic hemorrhage – a report of 3 cases. J Korean Med Sci 4: 1–5.
Hardy DG, Peace DA, Rhoton AL Jr (1980). Microsurgical anatomy of the superior cerebellar artery. Neurosurgery 6: 10–28.
Helgason CM (1988). A new view of anterior choroidal artery territory infarction. J Neurol 235: 387–391.
Helgason CM, Wilbur AC (1991). Basilar branch pontine infarction with prominent sensory signs. Stroke 22: 1129–1136.

Hermann DM, Jung HH, Bassetti CL (2009). Lateral medullary infarct with alternating and dissociated sensorimotor deficits: Opalski's syndrome revisited. Eur J Neurol 16: e72–e74.
Hitchings L, Crum A, Troost BT (1988). Isolated one-and-a-half syndrome from focal brainstem hypertensive hemorrhage: precise localization with MRI. Neurology 38: 1501–1502.
Ho KL, Meyer KR (1981). The medial medullary syndrome. Arch Neurol 38: 385–387.
Holtzman RN, Zablozki V, Yang WC et al. (1987). Lateral pontine tegmental hemorrhage presenting as isolated trigeminal sensory neuropathy. Neurology 37: 704–706.
Hommel M, Bogousslavsky J (1991). The spectrum of vertical gaze palsy following unilateral brainstem stroke. Neurology 41: 1229–1234.
Hong YH, Zhou LX, Yao M et al. (2018). Lesion topography and its correlation with etiology in medullary infarction: analysis from a multi-center stroke study in China. Front Neurol 9: 813.
Hongo H, Tanaka Y, Shimada Y et al. (2014). Dermatomal sensory manifestations in lateral medullary infarction. J Stroke Cerebrovasc Dis 231: 182–183.
Horn AK, Leigh RJ (2011). The anatomy and physiology of the ocular motor system. Handb Clin Neurol 102: 21–69.
Hu F, Nie H, Bai Y et al. (2022). The stroke mechanism, clinical presentation, and radiological feature of bilateral medial medullary infarction. Neurol Sci 43: 6881–6888.
Huang J, Qiu Z, Zhou P et al. (2019). Topographic location of unisolated pontine infarction. BMC Neurol 1: 186.
Huynh TR, Decker B, Fries TJ et al. (2018). Lateral medullary infarction with cardiovascular autonomic dysfunction: an unusual presentation with review of the literature. Clin Auton Res 28: 569–576.
Ikeda Y, Okamoto K (2002). Lesion responsible for WEMINO syndrome confirmed by magnetic resonance imaging. J Neurol Neurosurg Psychiatry 73: 204–205.
Im K, Ju H, Lee M et al. (2020). A case report of wall-eyed bilateral internuclear ophthalmoplegia with bilateral facial palsy. J Stroke Cerebrovasc Dis 29: 105139.
Jacobs L, Heffner RR Jr, Newman RP (1985). Selective paralysis of downward gaze caused by bilateral lesions of the mesencephalic periaqueductal gray matter. Neurology 35: 516–521.
Jean A (2001). Brain stem control of swallowing: neuronal network and cellular mechanisms. Physiol Rev 81: 929–969.
Jeon SB, Chung SJ, Ahn H et al. (2005). Wall-eyed monocular internuclear ophthalmoplegia (WEMINO) with contraversive ocular tilt reaction. J Clin Neurol 1: 101–103.
Johnston J, Sharpe J (1994). The WEMINO syndrome – wall-eyed monocular internuclear ophthalmoplegia: an oculographic and neuropathologic characterization. Neurology 44: A311.
Jouvent E, Benistry S, Fenelon G et al. (2005). Nystagmus de convergence et paralysie de la verticalité du regard d'origine vasculaire. Rev Neurol 161: 595.
Kameda W, Kawanami T, Kurita K et al. (2004). Lateral and medial medullary infarction: a comparative analysis of 214 patients. Stroke 35: 694–699.
Kang HG, Kim BJ, Lee SH et al. (2018). Lateral medullary infarction with or without extra-lateral medullary lesions: what is the difference? Cerebrovasc Dis 45: 132–140.
Kanter DS, Ruff RL, Leigh RJ et al. (1987). See-saw nystagmus and brainstem infarction: MRI findings. Neuroophthalmology 7: 279–283.
Karelis G, Micule M, Klavina E (2021). The Riga East University Hospital Stroke Registry – an analysis of 4915 consecutive patients with acute stroke. Medicina (Kaunas) 57: 632.
Karp JS, Hurtig HI (1974). "Locked-in" state with bilateral midbrain infarcts. Arch Neurol 30: 176–178.
Kataoka S, Hori A, Shirakawa T et al. (1997). Paramedian pontine infarction. Neurological/topographical correlation. Stroke 28: 809–815.
Kaya AH, Dagcinar A, Ulu MO et al. (2010). The perforating branches of the P1 segment of the posterior cerebral artery. J Clin Neurosci 17: 80–84.
Kim JS, Kang JK, Lee SA et al. (1993). Isolated or predominant ocular motor nerve palsy as a manifestation of brain stem stroke. Stroke 24: 581–586.
Kim JS, Lee MC, Kim HG et al. (1994). Isolated trigeminal sensory change due to pontine hemorrhage. Clin Neurol Neurosurg 96: 168–169.
Kim JS, Lee JH, Suh DC et al. (1994). Spectrum of lateral medullary syndrome: correlation between clinical findings and magnetic resonance imaging in 33 subjects. Stroke 25: 1405–1410.
Kim JS, Lee JH, Im JH et al. (1995). Syndromes of pontine base infarction. A clinical-radiological correlation study. Stroke 26: 950–955.
Kim JS, Kim HG, Chung CS (1995). Medial medullary syndrome. Report of 18 new patients and a review of the literature. Stroke 26: 1548–1552.
Kim JS, Lee JH, Lee MC (1997). Patterns of sensory dysfunction in lateral medullary infarction. Clinical-MRI correlation. Neurology 49: 1557–1563.
Kim JS, Koh JY, Lee JH (1998). Medial medullary infarction with restricted sensory symptom. Eur Neurol 39: 174–177.
Kim JS (1998). Dissociation of spinothalamic sensation in lateral medullary infarction. Neurology 51: 1774.
Kim JS, Lee JH, Choi CG (1998). Patterns of lateral medullary infarction: vascular lesion-magnetic resonance imaging correlation of 34 cases. Stroke 29: 645–652.
Kim JS, Choi-Kwon S (1999). Sensory sequelae of medullary infarction: differences between lateral and medial medullary syndrome. Stroke 30: 2697–2703.
Kim JS (2001). Sensory symptoms in ipsilateral limbs/body due to lateral medullary infarction. Neurology 57: 1230–1234.
Kim JS (2003). Pure lateral medullary infarction: clinical-radiological correlation of 130 acute, consecutive patients. Brain 126: 1864–1872.
Kim JS (2004). Internuclear ophthalmoplegia as an isolated or predominant symptom of brainstem infarction. Neurology 62: 1491–1496.
Kim JS, Kim J (2005). Pure midbrain infarction: clinical, radiologic, and pathophysiologic findings. Neurology 64: 1227–1232.
Kim JS, Jeong SH, Oh YM et al. (2008). Teaching NeuroImage: wall-eyed bilateral internuclear ophthalmoplegia (WEBINO) from midbrain infarction. Neurology 70: e35.
Kim JS, Han YS (2009). Medial medullary infarction: clinical, imaging, and outcome study in 86 consecutive patients. Stroke 40: 3221–3225.
Koay S, Dewan B (2013). An unexpected Holter monitor result: multiple sinus arrests in a patient with lateral medullary syndrome. BMJ Case Rep 2013: bcr2012007783.
Kobayashi J, Ohara T, Minematsu K et al. (2014). Etiological mechanisms of isolated pontine infarcts based on arterial territory involvement. J Neurol Sci 339: 113-117.

Kobayashi S, Mukuno K, Tazaki Y et al. (1986). Oculomotor nerve nuclear complex syndrome: a case with clinicopathological correlation, Neuro-Ophthalmology 6: 55–59.

Kon T, Funamizu Y, Ueno T et al. (2017). Dermatomal sensory manifestations in Opalski syndrome. J Stroke Cerebrovasc Dis 26: e18–e19.

Ksiazek SM, Slamovitz TL, Rosen CE et al. (1994). Fascicular arrangement in partial oculomotor paresis. Am Ophthalmol 118: 97–103.

Kubik CS, Adams RD (1946). Occlusion of the basilar artery; a clinical and pathological study. Brain 69: 73–121.

Kumral E, Ozkaya B, Sagduyu A et al. (1998). The Ege Stroke Registry: a hospital-based study in the Aegean region, Izmir, Turkey. Analysis of 2,000 stroke patients. Cerebrovasc Dis 8: 278–288.

Kumral E, Bayulkem G, Akyol A et al. (2002). Mesencephalic and associated posterior circulation infarcts. Stroke 33: 2224–2231.

Kumral E, Bayülkem G, Evyapan D (2002). Clinical spectrum of pontine infarction. Clinical-MRI correlations. J Neurol 249: 1659–1670.

Kumral E, Afsar N, Kirbas D et al. (2002). Spectrum of medial medullary infarction: clinical and magnetic resonance imaging findings. J Neurol 249: 85–93.

Kwon M, Lee JH, Kim JS (2005). Dysphagia in unilateral medullary infarction: lateral vs medial lesions. Neurology 65: 714–718.

Kwon HM, Park JM, Lee JY (2005). Primary medullary hemorrhage associated with hypertension. J Clin Neurol 1: 177–179.

Kumral E, Bayam FE, Özerol R et al. (2020). Predictors of outcome in patients with medullary hemorrhage. J Stroke Cerebrovasc Dis 29: 105337.

Lassman AB, Mayer SA (2005). Paroxysmal apnea and vasomotor instability following medullary infarction. Arch Neurol 62: 1286–1288.

Lazorthes G, Gouaze A, Salamon G (1976). Vascularisation et circulation de l'encephale. Paris: Masson.

Lee BC, Hwang SH, Chang GY (1999). Isolated dysphagia due to a medullary infarction: a new lacunar syndrome. Eur Neurol 41: 53–54.

Lee H, Cho YW (2003). Bilateral cerebellar ataxia as the sole manifestation of a unilateral rostral pontine tegmental infarct. J Neurol Neurosurg Psychiatry 74: 1445.

Lee JH, Han SJ, Yun YH et al. (2006). Posterior circulation ischemic stroke in Korean population. Eur J Neurol 13: 742–748.

Lee SA, Sunwoo IN, Kim KW (1987). Divergence paralysis due to a small hematoma in the tegmentum of the brainstem. Yonsei Med J 28: 326–328.

Lee SH, Kim DE, Song EC et al. (2001). Sensory dermatomal representation in the medial lemniscus. Arch Neurol 58: 649–651.

Lee SW, Lee H, Yoo J et al. (2021). Sick sinus syndrome combined with wallenberg syndrome: a case report. Brain Neurorehabil 143: e27.

Lee SH, Park SW, Kim BC et al. (2010). Isolated trochlear palsy due to midbrain stroke. Clin Neurol Neurosurg 112: 68–71.

Lee SU, Park SH, Park JJ et al. (2015). Dorsal medullary infarction: distinct syndrome of isolated central vestibulopathy. Stroke 46: 3081–3087.

Leestma JE, Noronha A (1976). Pure motor hemiplegia, medullary pyramid lesion, and olivary hypertrophy. J Neurol Neurosurg Psychiatry 39: 877–884.

Leigh RJ, Zee DS (2015). The neurology of eye movements5th ed. Oxford/New York: Oxford University Press.

Lhermitte J (1922). Syndrome de la calotte du pédoncule cérébral. Les troubles psycho-sensoriels dans les lésions du mésencéphale. Rev Neurol (Paris) 38: 1359–1365.

Li M, Li X, Liu L et al. (2019). A case report of the rare fifteen-and-a-half syndrome. Medicine (Baltimore) 98: e14553.

Li B, Song J, Zhang Y et al. (2020). "Sixteen syndrome" – a new pontine ophthalmo-neurological syndrome within the one-and-a-half syndrome spectrum of disorders: case report and literature review. Quant Imaging Med Surg 10: 518–521.

Link MJ, Bartleson JD, Forbes G et al. (1993). Spontaneous midbrain hemorrhage: report of seven new cases. Surg Neurol 39: 58–65.

Luda E, Sicuro L, Albera R et al. (1992). Retraction-convergence nystagmus: clinical and radiological study of 4 cases of vascular origin. Ital J Neurol Sci 13: 119–124.

Makita N, Yamamoto Y, Nagakane Y et al. (2019). Stroke mechanisms and their correlation with functional outcome in medullary infarction. J Neurol Sci 400: 1–6.

Marinković S, Gibo H (1994). The neurovascular relationships and the blood supply of the oculomotor nerve: the microsurgical anatomy of its cisternal segment. Surg Neurol 42: 505–516.

Martin PJ, Chang HM, Wityk R et al. (1998). Midbrain infarction: associations and aetiologies in the New England Medical Center Posterior Circulation Registry. J Neurol Neursurg Psychiatry 64: 392–395.

Marx JJ, Iannetti GD, Thömke F et al. (2008). Topodiagnostic implications of hemiataxia: an MRI-based brainstem mapping analysis. Neuroimage 39: 1625–1632.

Marx JJ, Thömke F (2009). Classical crossed brain stem syndromes: myth or reality? J Neurol 256: 898–903.

McKee AC, Levine DN, Kowall NW et al. (1990). Peduncular hallucinosis associated with isolated infarction of the substantia nigra pars reticulata. Ann Neurol 27: 500–504.

Meguro T, Kuwahara K, Tomita Y et al. (2015). Primary pontine hemorrhage in the acute stage: clinical features and a proposed new simple scoring system. J Stroke Cerebrovasc Dis 24: 860–865.

Mehler MF, Ragone PS (1988). Primary spontaneous mesencephalic hemorrhage. Can J Neurol Sci 15: 435–438.

Mendoza M, Latorre JG (2013). Pearls and oysters: reversible Ondine's curse in a case of lateral medullary infarction. Neurology 80: e13–e16.

Miller NR, Biousse V, Hwang T et al. (2002). Isolated acquired unilateral horizontal gaze paresis from a putative lesion of the abducens nucleus. J Neuroophthalmol 22: 204–207.

Mitsumura H, Miyagawa S, Komatsu T et al. (2016). Relationship between vertebral artery hypoplasia and posterior circulation ischemia. J Stroke Cerebrovasc Dis 25: 266–269.

Miura K, Nagaoka T, Ikeda K et al. (2012). A case of inferolateral oculomotor fascicular infarction: a review of the clinicoradiological literature. Intern Med 51: 921–924.

Mizushima H, Seki T (2002). Midbrain hemorrhage presenting with oculomotor nerve palsy: case report. Surg Neurol 58: 417–420.

Mon Y, Tamaki C (2018). A case of lateral medullary infarction with severe dysphagia and worsening of respiratory failure in the chronic phase of recovery. Clin Case Rep 6: 1608–1611.

Moncayo J, Bogousslavsky J (2009). Eye movement disorders in posterior circulation stroke. Expert Rev Ophthalmol 4: 259–281.

Moschovakis AK, Scudder CA, Highstein SM (1991). The structure of the primate oculomotor burst generator. I. Medium-lead burst neurons with upward on-directions. J Neurophysiol 65: 203–217.

Moschovakis AK, Scudder CA, Highstein SM et al. (1991). The structure of the primate oculomotor burst generator II. Medium-lead burst neurons with downward on directions. J Neurophysiol 65: 218–229.

Mossuto-Agatiello L (2006). Caudal paramedian midbrain syndrome. Neurology 66: 1668–1671.

Muengtaweepongsa S, Singh NN, Cruz-Flores S (2010). Pontine warning syndrome: case series and review of literature. J Stroke Cerebrovasc Dis 19: 353–356.

Murata Y, Yamaguchi S, Kajikawa H et al. (1999). Relationship between the clinical manifestations, computed tomographic findings and the outcome in 80 patients with primary pontine hemorrhage. J Neurol Sci 167: 107–111.

Nakajima K (1983). Clinicopathological study of pontine hemorrhage. Stroke 14: 485–493.

Nakamura K, Yamamoto T, Yamashita M (1996). Small medullary infarction presenting as painful trigeminal sensory neuropathy. J Neurol Neurosurg Psychiatry 61: 138.

Nakamura S, Kitami M, Furukawa Y (2010). Opalski syndrome: ipsilateral hemiplegia due to a lateral-medullary infarction. Neurology 75: 1658.

Neumann PE, Mehler MF, Horoupian DS (1985). Primary medullary hypertensive hemorrhage. Neurology 35: 925–928.

Nighoghossian N, Ryvlin P, Trouillas P et al. (1993). Pontine versus capsular pure motor hemiparesis. Neurology 43: 2197–2201.

Ogawa K, Suzuki Y, Oishi M et al. (2012). Clinical study of twenty-one patients with pure midbrain infarction. Eur Neurol 67: 81–89.

Oku Y, Okada M (2008). Periodic breathing and dysphagia associated with a localized lateral medullary infarction. Respirology 13: 608–610.

Opalski A (1946). Un nouveau syndrome sous-bulbaire partiell' artere vertebro-spinal posterieure. Paris Med 1: 214–220.

Oya S, Tsutsumi K, Yonekura I et al. (2001). Delayed central respiratory dysfunction after Wallenberg's syndrome – case report. Neurol Med Chir (Tokyo) 41: 502–504.

Paciaroni M, Silvestrelli G, Caso V et al. (2003). Neurovascular territory involved in different etiological subtypes of ischemic stroke in the Perugia Stroke Registry. Eur J Neurol 10: 361–365.

Paliwal VK, Kalita J, Misra UK (2009). Dysphagia in a patient with bilateral medial medullary infarcts. Dysphagia 24: 349–353.

Parinaud H (1883). Paralysie des movements associés des yeux. Arch Neurol 5: 145–172.

Park SA, Sohn YH, Kim WC (1997). Locked-in syndrome with bilateral peduncular infarct. J Neuroimaging 7: 126–128.

Park MH, Kim BJ, Koh SB et al. (2005). Lesional location of lateral medullary infarction presenting hiccups (singultus). J Neurol Neurosurg Psychiatry 76: 95–98.

Park JH, Kim JM, Roh JK (2007). Hypoplastic vertebral artery: frequency and associations with ischaemic stroke territory. J Neurol Neurosurg Psychiatry 78: 954–958.

Paton L (1921). Ocular palsies. Br J Ophthalmol 5: 250–271.

Pavšič K, Pretnar-Oblak J, Bajrović FF et al. (2020). Prospective study of sleep-disordered breathing in 28 patients with acute unilateral lateral medullary infarction. Sleep Breath 24: 1557–1563.

Pavšič K, Pretnar-Oblak J, Bajrović FF et al. (2021). Breathing patterns in relation to sleep stages in acute unilateral lateral medullary infarction: an exploratory study. Respir Physiol Neurobiol 285: 103592.

Pedroza A, Dujovny M, Ausman JI et al. (1986). Microvascular anatomy of the interpeduncular fossa. J Neurosurg 64: 484–493.

Percheron G (1976). Les artères du thalamus humain. II. Artères et territoires thalamiques paramédians de l'artère basilaire communicante. Rev Neurol (Paris) 132: 309–324.

Perez FJ, Nuñez CT (2008). Scanning dysarthria secondary to spontaneous midbrain hemorrhage: case report and review of literature. Eur Neurol 60: 89–91.

Phillips AM, Jardine DL, Parkin PJ et al. (2000). Brain stem stroke causing baroreflex failure and paroxysmal hypertension. Stroke 31: 1997–2001.

Pierrot-Deseilligny C, Schaison M, Bousser MG et al. (1981). Syndrome nucléaire du nerf moteur oculaire commun: à propos de deux observations cliniques. Rev Neurol 137: 217–222.

Pierrot-Deseilligny C, Chain F, Serdaru M et al. (1981). The 'one-and-a-half' syndrome. Electro-oculographic analyses of five cases with deductions about the physiological mechanisms of lateral gaze. Brain 104: 665–699.

Pierrot-Deseilligny CH, Chain F, Gray F et al. (1982). Parinaud's syndrome: electro-oculographic and anatomical analyses of six vascular cases with deductions about vertical gaze organization in the premotor structures. Brain 105: 667–696.

Pongmoragot J, Parthasarathy S, Selchen D et al. (2013). Bilateral medial medullary infarction: a systematic review. J Stroke Cerebrovasc Dis 22: 775–780.

Prabhakar A, Sivadasan A, Shaikh A et al. (2020). Network localization of central hypoventilation syndrome in lateral medullary infarction. J Neuroimaging 30: 875–881.

Pullicino P, Lincoff N, Truax BT (2000). Abnormal vergence with upper brainstem infarcts: pseudoabducens palsy. Neurology 55: 352–358.

Rabinstein AA, Tisch SH, McClelland RL et al. (2004). Cause is the main predictor of outcome in patients with pontine hemorrhage. Cerebrovasc Dis 17: 66–71.

Raina GB, Velez M, Pardal MF et al. (2007). Holmes tremor secondary to brainstem hemorrhage responsive to levodopa: report of 2 cases. Clin Neuropharmacol 30: 95–100.

Raina GB, Cersosimo MG, Folgar SS et al. (2016). Holmes tremor: clinical description, lesion localization, and treatment in a series of 29 cases. Neurology 86: 931–938.

Rambold H, Kömpf D, Helmchen (2001). Convergence retraction nystagmus: a disorder of vergence? Ann Neurol 50: 677–681.

Rambold H, Neumann G, Helmchen C (2004). Vergence deficits in pontine lesions. Neurology 62: 1850–1853.

Ranalli PJ, Sharpe JA, Fletcher WA (1988). Palsy of upward and downward saccadic, pursuit, and vestibular movements with a unilateral midbrain lesion: pathophysiologic correlations. Neurology 38: 114–122.

Robles LA (2016). Pure hemiparkinsonism secondary to contralateral lacunar stroke in the substantia nigra. J Stroke Cerebrovasc Dis 25: e20–e21.

Ropper AH, Fisher CM, Kleinman GM (1979). Pyramidal infarction in the medulla: a cause of pure motor hemiplegia sparing the face. Neurology 29: 91–95.

Rosini F, Pretegiani E, Guideri F et al. (2013). Eight and a half syndrome with hemiparesis and hemihypesthesia: the nine syndrome. J Stroke Cerebrovasc Dis 22: e637–e638.

Ryman DC, Ju YE (2013). Medial medullary infarction with central sparing. Neurology 81: 2054–2055.

Sacco RL, Freddo L, Bello JA et al. (1993). Wallenberg's lateral medullary syndrome. Clinical-magnetic resonance imaging correlations. Arch Neurol 50: 609–614.

Saeki N, Rhoton AL Jr (1977). Microsurgical anatomy of the upper basilar artery and the posterior circle of Willis. J Neurosurg 46: 563–578.

Sand JJ, Biller J, Corbett JJ et al. (1986). Partial dorsal mesencephalic hemorrhages: report of three cases. Neurology 36: 529–533.

Saposnik G, Noel de Tilly L, Caplan LR (2008). Pontine warning syndrome. Arch Neurol 65: 1375–1377.

Sato S, Toyoda K, Kawase K et al. (2008). A caudal mesencephalic infarct presenting only with tetra-ataxia and tremor. Cerebrovasc Dis 25: 187189.

Sawada H, Seriu N, Udaka F et al. (1990). Magnetic resonance imaging of medial medullary infarction. Stroke 21: 963–966.

Scheitler KM, Mustafa R, Wijdicks EFM (2021). Lid and convergence retraction nystagmus in thalamic-midbrain hematoma. Neurocrit Care 34: 691–692.

Schmahmann JD, Ko R, MacMore J (2004). The human basis pontis: motor syndromes and topographic organization. Brain 127: 1269–1291.

Schneider R, Gautier JC (1994). Leg weakness due to stroke. Site of lesions, weakness patterns and causes. Brain 117: 347–354.

Shintani S, Tsuruoka S, Shiigai T (1994). Pure sensory stroke caused by a pontine infarct. Clinical, radiological, and physiological features in four patients. Stroke 25: 1512–1515.

Shono Y, Koga M, Toyoda K et al. (2010). Medial medullary infarction identified by diffusion-weighted magnetic resonance imaging. Cerebrovasc Dis 30: 519–524.

Smith JL, Cogan DG (1959). Internuclear ophthalmoplegia: a review of fifty-eight cases. Arch Ophthalmol 61: 687–694.

Song IU, Kim JS, Kim YI et al. (2007). Brachial monoparesis as an isolated manifestation of a paramedian pontine ischemic lesion. Clin Neurol Neurosurg 109: 376–378.

Songur A, Gonul Y, Ozen OA et al. (2008). Variations in the intracranial vertebrobasilar system. Surg Radiol Anat 30: 257–264.

Spiller WG (1905). The importance in clinical diagnosis of paralysis of associated movements of the eyeballs, especially of upward and downward associated movements. J Nerv Ment Dis 32: 497–539.

Spiller WG (1908). The symptom-complex of a lesion of the upper most portion of the anterior spinal and adjoining portion of the vertebral arteries. J Nerv Ment Dis 35: 775–778.

Stimmer H, Borrmann A, Löer C et al. (2009). Monaural tinnitus from a contralateral inferior colliculus hemorrhage. Audiol Neurootol 14: 35–38.

Stopford JSB (1916a). The arteries of the pons and medulla oblongata. Part 1. J Anat Physiol 50: 131–164.

Stopford JSB (1916b). The arteries of the pons and medulla oblongata. Part 2. J Anat Physiol 50: 255–280.

Tao WD, Liu M, Fisher M et al. (2012). Posterior versus anterior circulation infarction: how different are the neurological deficits? Stroke 43: 2060–2065.

Tao LS, Lin JJ, Zou M et al. (2021). A comparative analysis of 375 patients with lateral and medial medullary infarction. Brain Behav 11: e2224.

Tatu L, Moulin T, Bogousslavsky J et al. (1996). Arterial territories of human brain: brainstem and cerebellum. Neurology 47: 1125–1135.

Tatu L, Moulin T, El Mohamad R et al. (2000). Primary intracerebral hemorrhages in the Besançon Stroke Registry. Initial clinical and CT findings, early course and 30-day outcome in 350 patients. Eur Neurol 43: 209–214.

Tei H, Sakamoto Y (1997). Pontine infarction due to basilar artery stenosis presenting as pathological laughter. Neuroradiology 39: 190–191.

Terao S, Miura N, Osano Y et al. (2004). Rapidly progressive fatal respiratory failure (Ondine's curse) in the lateral medullary syndrome. J Stroke Cerebrovasc Dis 13: 41–44.

Terao S, Takatsu S, Izumi M et al. (1997). Central facial weakness due to medial medullary infarction: the course of facial corticobulbar fibres. J Neurol, Neurosurg, Psychiatry 633: 391–393.

Thömke F, Hopf H (1992). Acquired monocular elevation paresis: an asymmetric upgaze palsy. Brain 115: 1901–1910.

Thömke F, Marx JJ, Iannetti GD et al. (2005). A topodiagnostic investigation on body lateropulsion in medullary infarcts. Neurology 64: 716–718.

Thömke F, Ringel K (1999). Isolated superior oblique palsies with brainstem lesions. Neurology 53: 1126–1127.

Tilikete C, Rode G, Nighoghossian N et al. (2001). Séméiologie otolithique dans le syndrome de Wallenberg. Rev Neurol (Paris) 157: 198–208.

Toyoda K, Hasegawa Y, Yonehara T et al. (1992). Bilateral medial medullary infarction with oculomotor disorders. Stroke 23: 1657–1659.

Toyoda K, Saku Y, Ibayashi S et al. (1994). Pontine infarction extending to the basal surface. Stroke 25: 2171–2178.

Toyoda K, Imamura T, Saku Y et al. (1996). Medial medullary infarction: analyses of eleven patients. Neurology 47: 1141–1147.

Trojanowski JQ, Wray SH (1980). Vertical gaze ophthalmoplegia: selective paralysis of downgaze. Neurology 30: 605–610.

Tsivgoulis G, Spengos K, Vassilopoulou S et al. (2006). Isolated dorsal midbrain infarct: an uncommon cause of pure sensory stroke. Cerebrovasc Dis 21: 139–140.

Tsuda H, Ishikawa H, Koga N et al. (2006). Magnetic resonance imaging findings in divergence paralysis. Neuro-Ophthalmology 30: 59–62.

Tulleken CA, Luiten ML (1987). The basilar artery bifurcation: microscopical anatomy. Acta Neurochir (Wien) 85: 50–55.

Turkes M (2017). Primary spontaneous mesencephalic hemorrhage. Acta Neurol Belg 117: 311–312.

Urban PP, Wicht S, Vukurevic G et al. (2001). Dysarthria in acute ischemic stroke: lesion topography, clinicoradiologic correlation, and etiology. Neurology 56: 1021–1027.

Uzawa A, Takeda Y, Kuwabara S (2020). Wall-eyed bilateral internuclear ophthalmoplegia by ischemic stroke: case report and literature review. Neurologist 25: 82–84.

Van Bogaert L (1927). L'hallucinose pédonculaire. Rev Neurol 1: 608–617.

Vanderah TW, Gould DJ (2021). Nolte's the human brain. 8th edition. Philadelphia, PA: Elsevier Inc.

Vemmos KN, Takis CE, Georgilis K et al. (2000). The Athens Stroke Registry: results of a five-year hospital-based study. Cerebrovasc Dis 10: 133–141.

Vemmos KN, Spengos K, Tsivgoulis G et al. (2005). Aetiopathogenesis and long-term outcome of isolated pontine infarcts. J Neurol 252: 212–217.

Viader F, Masson M, Marion MH et al. (1984). Infarctus cérébral dans le territoire de l'artère choroidienne antérieure avec trouble oculomoteur. Rev Neurol (Paris) 140: 668–670.

Vinas FC, Lopez F, Dujovny M (1995). Microsurgical anatomy of the posterior choroidal arteries. Neurol Res 17: 334–344.

Vitošević Z, Marinković S, Cetković M et al. (2013). Intramesencephalic course of the oculomotor nerve fibers: microanatomy and possible clinical significance. Anat Sci Int 88: 70–82.

Vlašković T, Brkić BG, Stević Z et al. (2022). Anatomic and MRI bases for pontine infarctions with patients presentation. J Stroke Cerebrovasc Dis 31: 106613.

Vlašković T, Brkić BG, Stević Z et al. (2022). Anatomic and MRI bases for medullary infarctions with patients' presentation. J Stroke Cerebrovasc Dis 31: 106730.

Vogels V, Dammers R, van Bilsen M et al. (2021). Deep cerebral perforators: anatomical distribution and clinical symptoms: an overview. Stroke 52: e660–e674.

Voogd J, van Baarsen K (2014). The horseshoe-shaped commissure of Wernekinck or the decussation of the brachium conjunctivum methodological changes in the 1840s. Cerebellum 13: 113–120.
Vuadens P, Bogousslavsky J (1998). Face-arm-trunk-leg sensory loss limited to the contralateral side in lateral medullary infarction: a new variant. J Neurol Neurosurg Psychiatry 65: 255–257.
Vuilleumier P, Bogousslavsky J, Regli F (1995). Infarction of the lower brainstem. Clinical, aetiological and MRI-topographical correlations. Brain 118: 1013–1125.
Wali GM (1993). "Fou rire prodromique" heralding a brainstem stroke. J Neurol Neurosurg Psychiatry 56: 209–210.
Wall M, Wray S (1983). The one-and-a-half syndrome-a unilateral disorder of the pontine tegmentum: a study of 20 cases and review of the literature. Neurology 33: 971–980.
Weisberg LA (1986). Mesencephalic hemorrhages: clinical and computed tomographic correlations. Neurology 36: 713–716.
Weisberg LA (1986). Primary pontine haemorrhage: clinical and computed tomographic correlations. J Neurol Neurosurg Psychiatry 49: 346–352.
Wessels T, Möller-Hartmann W, Noth J et al. (2004). CT findings and clinical features as markers for patient outcome in primary pontine hemorrhage. AJNR Am J Neuroradiol 25: 257–260.
Wiest G, Baumgartner C, Schnider P et al. (1996). Monocular elevation paresis and contralateral downgaze paresis from unilateral mesodiencephalic infarction. J Neurol Neurosurg Psychiatry 60: 579–581.
Wijdicks EF, St Louis E (1997). Clinical profiles predictive of outcome in pontine hemorrhage. Neurology 49: 1342–1346.
Worthington JM, Halmagyi GM (1996). Bilateral total ophthalmoplegia due to midbrain hematoma. Neurology 46: 1176–1177.
Wu YJ, Pai MC (2012). Two gloves and one sock syndrome after a pontine hemorrhage. Neurologist 18: 196–198.
Wu YT, Cafiero-Chin M, Marques C (2015). Wall-eyed bilateral internuclear ophthalmoplegia: review of pathogenesis, diagnosis, prognosis and management. Clin Exp Optom 98: 25–30.
Xia C, Chen HS, Wu SW et al. (2017). Etiology of isolated pontine infarctions: a study based on high-resolution MRI and brain small vessel disease scores. BMC Neurol 17: 216.
Yi HA, Kim HA, Lee H et al. (2007). Body lateropulsion as an isolated or predominant symptom of a pontine infarction. J Neurol Neurosurg Psychiatry 78: 372–374.
Yu C, Zhu Z, Li S et al. (2022). Clinical and radiological features of medullary infarction caused by spontaneous vertebral artery dissection. Stroke Vasc Neurol 7: 245–250.
Zeal AA, Rhoton ALJr, (1978). Microsurgical anatomy of the posterior cerebral artery. J Neurosurg 48: 534–559.
Zee DS, Hain TC, Carl JR (1987). Abduction nystagmus in internuclear ophthalmoplegia. Ann Neurol 21: 383–388.
Zhang Y, Wang L, He M (2017). Isolated INO as a presentation of midbrain paramedian area lacunar infarction in patients with diabetes. J Clin Neurosci 45: 261–264.
Zhou C, He Y, Chao Z et al. (2017). Wernekink commissure syndrome secondary to bilateral caudal paramedian midbrain infarction presenting with a unique "Heart or V" appearance sign: case report and review of the literature. Front Neurol 3: 376.
Zhou C, Xu Z, Huang B et al. (2021). Caudal paramedian midbrain infarction: a clinical study of imaging, clinical features and stroke mechanisms. Acta Neurol Belg 121: 443–450.
Zwergal A, Cnyrim C, Arbusow V et al. (2008). Unilateral INO is associated with ocular tilt reaction in pontomesencephalic lesions: INO plus. Neurology 71: 590–593.

Handbook of Clinical Neurology, Vol. 216 (3rd series)
The Brainstem
L.T. Grinberg and A.D. Korczyn
https://doi.org/10.1016/B978-0-443-15736-3.00022-6

Chapter 16

Traumatic brainstem injury

WELLINGSON SILVA PAIVA[1,*], RAPHAEL BERTANI[1], AND SÁVIO BATISTA[2]

[1]*Division of Neurosurgery, Hospital das Clínicas, University of São Paulo Medical School, São Paulo, Brazil*

[2]*Medical School, Federal University of Rio de Janeiro, Rio de Janeiro, Brazil*

Abstract

Traumatic brainstem injuries represent some of the most severe and life-threatening forms of head trauma, demanding a thorough understanding of their profound clinical implications. These injuries assume critical significance given the brainstem's role in controlling vital autonomic functions. Brainstem damage may be observed in many cases of head injury, occurring as primary injury or as hemorrhagic and ischemic complications from brain herniation and intracranial hypertension. This pattern is commonly seen as a secondary event in cases such as extra-axial or subdural hemorrhage, or intracerebral bleeding following head injury.

In addition to hemorrhagic injuries and ischemia resulting from increased intracranial hypertension, traumatic brainstem injury may cause a significant impact on brain tissue as a form of diffuse axonal injury. Poor prognosis has been a common feature following severe traumatic brain injury, particularly in patients with brainstem involvement. In this chapter, we provide a critical discussion of the pathophysiology, diagnosis, and prognosis of these injuries.

DEFINITION

Traumatic brainstem injury (TBSI) occurs when the brainstem, a vital part of the central nervous system connecting the brain to the spinal cord, sustains damage from traumatic events such as accidents, falls, or sports-related injuries (Se et al., 2009). This injury involves damage to the medulla oblongata, pons, and midbrain, each of which is responsible for essential functions, including regulation of respiration and heart rate, sensory relay, and motor control (Hilario et al., 2012).

Brainstem damage may be observed in many cases of head injury, occurring as primary injury or as hemorrhagic and ischemic complications from brain herniation and intracranial hypertension. This is commonly seen as a secondary event in cases such as extra-axial or subdural hemorrhage, or intracerebral bleeding following head injury (Sandhu et al., 2019).

CLINICAL IMPORTANCE AND IMPACT ON PATIENTS' QUALITY OF LIFE

Traumatic brainstem injuries stand out as some of the most severe and life-threatening forms of head trauma, demanding a thorough understanding of their profound clinical implications. These injuries assume critical significance when we consider the brainstem's role in controlling autonomic functions. The brainstem serves as the command center for critical bodily processes such as breathing and heart rate regulation. Neglecting an injury in this area can precipitate dire consequences, including respiratory failure, cardiac arrhythmias, and, in the gravest cases, death, given that these functions are indispensable for sustaining life (Izzy et al., 2017).

Furthermore, exploring the ramifications of brainstem damage unveils a cascade of severe clinical effects. Such injuries can lead to a spectrum of alterations in consciousness, ranging from profound coma to impaired awareness, profoundly impacting a patient's overall quality of life and independence. Within the intricate architecture of the brainstem, vital motor and sensory pathways are meticulously housed, and any injury here can result in paralysis, weakness, or sensory deficits, significantly impeding a patient's ability to perform even the most basic daily activities. Specific injuries to the brainstem can also disrupt functions such as swallowing and speech, necessitating an extended period of rehabilitation and the development of adaptive strategies (Al-Sarraj et al., 2012).

Moreover, traumatic brainstem injuries extend their reach into the cognitive and emotional domains, often resulting in complex disturbances that reverberate through a patient's life. These disturbances can affect relationships, employability, and

[*]Correspondence to: Wellingson Paiva, MD, PhD, Divison of Neurosurgery, Hospital das Clínicas, University of São Paulo, São Paulo, SP, Brazil. E-mail: wellingson.paiva@hc.fm.usp.br

overall well-being, adding layers of complexity to the patient's recovery journey. These cognitive and behavioral changes frequently render patients heavily reliant on caregivers for their daily needs, imposing substantial burdens on families and healthcare systems (Barman et al., 2016).

Understanding the prognosis for recovery in cases of traumatic brainstem injuries requires a nuanced perspective. The potential for recovery can vary widely, contingent on factors such as the extent of the injury and the timeliness of intervention. Rehabilitation assumes an indispensable role in this context, encompassing a comprehensive approach that includes physical therapy to regain motor function, speech therapy to address communication challenges, and occupational therapy to reestablish the ability to perform daily activities. This multifaceted rehabilitation approach not only aids in restoring functionality but also serves as a cornerstone in enhancing the patient's overall quality of life, offering hope in the face of this formidable challenge (Kim, 2012).

TYPES OF TRAUMA

Brainstem injuries can emanate from diverse traumatic incidents, each characterized by unique circumstances and injury patterns. Understanding the common types of trauma that lead to brainstem injuries is crucial for prevention and treatment (Rad et al., 2009).

Traffic accidents represent a significant mechanism of TBSI. Motor vehicle collisions, motorcycle accidents, and pedestrian incidents are frequent culprits. The rapid deceleration and sheer force of impact can inflict severe, often irreversible, damage to the brainstem. Similarly, accidental falls from heights, including slips, trips, and tumbles, can result in brainstem injuries. The angle and force of impact are critical variables that determine the severity and extent of the damage. Contact sports such as football, martial arts, and others present inherent risks of brainstem injuries given the potential for head impacts and concussions (Andrade et al., 2009).

Physical assaults and acts of violence can lead to traumatic brainstem injuries, especially when the head is deliberately targeted. The force and location of the impact are critical factors in the resulting damage (Andrade et al., 2009). Injuries involving penetration, such as gunshot or stab wounds to the head, can directly compromise the structural integrity of the brainstem, leading to immediate and severe consequences. Blast injuries, commonly encountered in military combat or industrial accidents, can also result in brainstem injuries. The shockwave generated by the blast exerts a significant impact on the brain, potentially harming the brainstem (Cury et al., 2017).

Rad and colleagues, in an autopsy study, found brainstem injury in 39 out of 200 severe traumatic brain injury patients. Autopsy findings revealed the following types of brainstem injuries: 17 patients had pontine contusion, 10 had medullary contusion, 9 had midbrain contusion, 6 had pontomedullary junction tearing, and 1 had cervicomedullary junction tearing (Rad et al., 2009).

Kim, analyzing 2,695 head-injured patients, found 136 TBSI cases and described that in the evaluation of TBSI on outcome, the variables of imaging findings, type of injury, and age were revealed as three critical factors to have an unfavorable effect on early stage outcome as expressed by Glasgow Coma Scale (GCS) score (Kim, 2012).

INJURY MECHANISMS

Understanding how traumatic incidents cause brainstem injuries is crucial. These injuries can be categorized into several types, each shedding light on how and why they occur (Paiva et al., 2012).

In deceleration injuries, such as those seen in traffic accidents, the brainstem is at risk when a vehicle suddenly decelerates upon impact. This abrupt change in motion can cause the brainstem to collide with the rigid, protective bony structures inside the skull. This collision can result in the tearing or stretching of delicate neural structures within the brainstem, leading to injury. This highlights the brainstem's vulnerability to rapid shifts in movement and force (Vieira et al., 2024).

Penetrating trauma, such as gunshot wounds or stab wounds, directly breaches the protective cranial vault, causing direct damage to the brainstem. The severity of these injuries will vary according to factors such as the trajectory and velocity of the penetrating object. This can lead to localized damage or more extensive trauma. In craniocerebral gunshot injuries, the energy from the penetrating bullet can cause damage to brain tissue both along the bullet's path and in areas beyond it, known as cavitation injury. This damage is due to shearing, compressing, and stretching forces exerted during the bullet's passage. Magnetic resonance imaging (MRI) scans can reveal these areas of damage, which may appear as "enlightened" or "translucent" zones around the bullet's trajectory (Cook et al., 2024) (Fig. 16.1).

Primary TBSI due to initial trauma of the head is distinguished from secondary brainstem lesions resulting from shift and distortion of the brainstem by raised intracranial pressure (ICP) after injury. A significantly higher incidence of hematoma is observed in hemorrhagic cases, and the most common site of hemorrhage is either the pons alone or in association with the midbrain, thalamus, and hypothalamus. Secondary midbrain lesions are paramedian, with hemorrhage or necrosis in a distorted midbrain. In contrast, primary lesions are present in an undistorted brainstem and are typically lateral, tegmental, and often hemorrhagic (Fig. 16.4) (Mannion et al., 2007).

In cases of rotational forces, often seen in sports-related injuries or falls, the brainstem can be twisted within the confines of the skull. These rotational forces can disrupt neural pathways, blood vessels, and other critical structures, leading to brainstem injury. Compression and shearing forces, which can occur in various traumatic scenarios including falls and assaults,

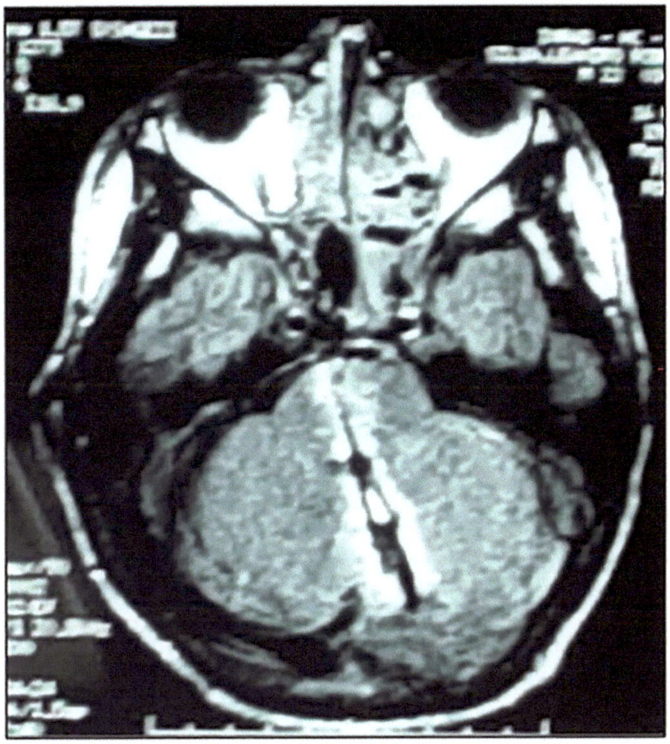

Fig. 16.1. Magnetic resonance image showing a transcranial brainstem lesion seen as hyperintense area transversing the dorsal pons and middle cerebellar peduncle.

can subject the brainstem to compression against the skull's bony ridges or cause shearing of nerve fibers. These forces can lead to injury by compressing vital structures or damaging neural pathways (Izzy et al., 2017).

Diffuse axonal injury (DAI) may involve the brainstem. DAI is conventionally graded according to location: Grade 1 (lesions confined to the subcortical lobar white matter or cerebellum only), Grade 2 (lesions in the corpus callosum, with or without lesions in the lobar white matter), and Grade 3 (traumatic lesions in the brainstem in areas typical of DAI) (Sandhu et al., 2019).

Finally, blast injuries introduce unique risks to the brainstem. In cases of explosions or blasts, the shockwave generated can transmit damaging forces to the brainstem. The rapid change in atmospheric pressure can stretch and compress brain tissues, resulting in injury. Comprehending these intricate injury mechanisms unravels the delicate nature of the brainstem and the diverse ways in which traumatic incidents can lead to significant injury. This knowledge is critical for both prevention and effective treatment of brainstem injuries (Yamamura et al., 2022).

SIGNS AND SYMPTOMS

The brainstem is a vital region of the central nervous system that regulates numerous critical functions. When lesions or injuries occur in this area, they can lead to a wide range of signs and symptoms, often reflecting the brainstem's role in controlling various bodily processes. Common clinical manifestations associated with brainstem lesions are described below (Bhatoe, 1999).

Brainstem lesions can result in muscle weakness or even paralysis, affecting one or both sides of the body depending on the lesion's location and extent. Damage to the brainstem can impair coordination and balance, leading to difficulties walking or performing fine-motor tasks (Sandhu et al., 2019).

Respiratory disorders may be observed, as brainstem lesions can disrupt respiratory control, leading to abnormal breathing patterns, including rapid, shallow breathing or irregular breathing rhythms. Severe brainstem injury can result in respiratory failure, where the individual is unable to breathe without mechanical support (Hongo et al., 2025).

Brainstem lesions often cause alterations in consciousness, ranging from confusion and disorientation to coma. In some cases, particularly with extensive brainstem damage, individuals may fall into a comatose state, where they are unresponsive to external stimuli (Sandhu et al., 2019).

Disorders of consciousness and cranial nerves

The brainstem houses the cranial nerve nuclei, and various cranial nerve abnormalities can be observed following injury. For example, lesions can affect eye movements, leading to symptoms such as double vision (diplopia) or difficulty focusing. Pupillary abnormalities may also occur. Damage to the facial nerve nuclei can result in facial weakness or complete paralysis on one or both sides of the face. Issues with hearing, vertigo (a spinning sensation), or imbalance may arise if cranial nerves controlling these functions are affected. Brainstem lesions can impair swallowing, leading to choking or aspiration of food or liquids. Impaired control of the vocal cords and tongue can result in slurred speech or difficulty articulating words (Nourallah et al., 2018).

Sensory deficits, such as numbness, tingling, or loss of sensation, may occur in the face or body. Some individuals may experience pain or abnormal sensations in areas served by affected sensory pathways. Beyond these, autonomic functions such as heart rate, blood pressure, and body temperature regulation may be disrupted, leading to symptoms such as dizziness, blood pressure fluctuations, or abnormal heart rhythms (Sandhu et al., 2019).

DIAGNOSIS

Clinical and neurologic evaluation

Clinical and neurologic evaluations are the initial steps in diagnosing brainstem lesions resulting from traumatic injury. Healthcare providers conduct these evaluations and encompass a detailed medical history (gathering information about the traumatic event, loss of consciousness, and any neurologic symptoms) as well as a comprehensive physical examination (assessing motor function, reflexes, cranial nerve function, sensory perception, and changes in consciousness) (Kim, 2012).

These evaluations help establish a baseline of neurologic function and identify any immediate life-threatening issues that require urgent intervention. Clinical and neurologic evaluations are essential in the assessment of traumatic brainstem injuries as they provide critical information about the patient's clinical status and neurologic deficits. They guide healthcare providers in selecting the most appropriate diagnostic tests, ensuring timely and accurate diagnosis. Clinical assessments help determine the severity of the injury, monitor the patient's progress, and guide treatment decisions. Continuous neurologic evaluation throughout the course of care is essential to track changes in the patient's condition, detect complications, and assess response to treatment (Hilario et al., 2012).

Computed tomography

Computed tomography (CT) scans are often the first-line imaging studies used to assess traumatic brainstem injuries due to their rapid acquisition and ability to detect acute bleeding, fractures, or gross structural abnormalities (Fig. 16.2). CT scans are particularly valuable in emergencies to rule out life-threatening intracranial hemorrhages. However, CT scans have limitations, including insufficient resolution to identify injuries to the posterior fossa and brainstem or to detect DAI that may lead to severe cortical dysfunction. There exists a significant population of severely head-injured patients whose outcome is much worse than might be predicted from their initial clinical and CT findings, and it is widely acknowledged that patients can suffer severe and life-threatening Traumatic brain injury (TBI) without displaying significant abnormality on CT – the so-called diffuse brain injury (Mannion et al., 2007).

Magnetic resonance imaging

MRI is the preferred imaging modality for evaluating brainstem lesions in greater detail, especially when there is suspicion of more subtle or complex injuries. The use of acute MRI to investigate severe TBI offers clear advantages over the use of CT alone, particularly concerning the visualization of posterior fossa structures and diffuse brain injury (Mannion et al., 2007).

MRI offers superior contrast and resolution compared to CT, making it ideal for visualizing soft tissue and detecting lesions such as contusions, edema, or small hemorrhages (Fig. 16.1) (Firsching et al., 2001). T2-weighted MRI offers superior visualization of the brainstem and has been increasingly employed to evaluate stabilized TBI patients in acute settings. Advanced MRI sequences, such as diffusion-weighted imaging and susceptibility-weighted imaging, can provide valuable insights into the extent and nature of brainstem injuries (Fig. 16.3) (Mannion et al., 2007).

This imaging modality can detect injuries often associated with poor outcomes, such as edema, ischemia, hemorrhage, and infarction in the brainstem. However, MRI provides valuable anatomic insights; its role as a prognostic tool for TBSI remains undefined, necessitating further research to establish its efficacy in predicting patient outcomes (Mannion et al., 2007) (Fig. 16.4).

Firsching and colleagues developed an MRI score to classify brainstem injury by its rostrocaudal and mediolateral extent in the midbrain and pons, independent of the pattern or degree of supratentorial injury, with notable results (Firsching et al., 2002). Other studies have reported on the use of MRI in severe TBI; however, the timing of MRI after TBI has varied from a few days up to one month, and the strength of association between the presence of brainstem injury and poor clinical outcome has varied significantly (Paterakis et al., 2000; Firsching et al., 2001; Aguas et al., 2005). In part, this has depended on the anatomy and extent of brainstem injury. Still, there is also debate over the significance of the accompanying supratentorial injury and a generally poor understanding of the underlying mechanisms of brainstem injury.

QUANTITATIVE MRI: DIFFUSION TENSOR TRACTOGRAPHY (DTT)

The human brainstem remains one of the most challenging regions to explore with DTI due to the difficulty of accurately delineating small nuclei and complex fiber connections in this area. Distortions and artifacts are common in DTI acquisition due to the increased echo time required to incorporate diffusion gradients. For this reason, manual DTI tractography was used in some studies aimed at analyzing the complex anatomy of the brainstem (Jones et al., 2013). Recently, some studies (Yeh et al., 2018; Adil et al., 2021) have described strategies for

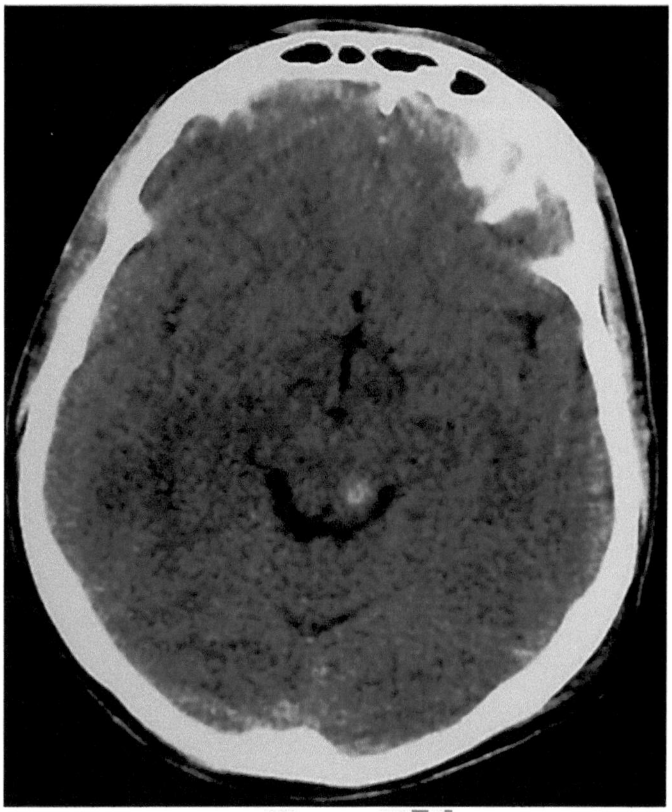

Fig. 16.2. Computed tomography scan of skull showing brainstem hemorrhage seen as a hypodense circumscribed lesion at the left superior colliculus of the midbrain.

automated white matter analyses. This strategy could potentially use individual tractography by transforming imaging into an an atlas of fiber clusters/landmarks from standard space. While these studies of landmarks or fiber clusters are promising, the actual use of these concept will largely depend on the avaliability and environment of the end user

Despite these difficulties, DTI tractography has been used to improve our understanding of longitudinal TBI effects upon the connectome as a function of age (Irimia et al., 2017; Trotter et al., 2015), recovery in the late phase of trauma (Ewing-Cobbs et al., 2006), and trauma mechanism (Petrie et al., 2014). However, despite the value of DTI studies to characterize TBI sequelae within the connectome, the utility and interpretation of DTI measures in TBI remain equivocal (Van Horn et al., 2017). Serial changes in white matter DTI metrics in moderate TBI have been correlated with neurocognitive function (Hunter et al., 2019). The new DTI and other quantitative MRI modalities need further research to define their actual value as outcome predictors and as novel task-based imaging biomarkers.

Electrophysiologic tests

Electrophysiologic studies, including electromyography and nerve conduction studies, may be used to assess motor and sensory function, particularly when motor deficits are present (Sandhu et al., 2019).

Cerebrospinal fluid analysis

In some instances, a lumbar puncture may be performed to analyze cerebrospinal fluid (CSF), primarily to rule out infection or bleeding within the central nervous system (Zhang et al., 2024).

Angiography

Angiography techniques, such as cerebral angiography or magnetic resonance angiography, can evaluate the vascular structures within and around the brainstem, helping to identify vascular injuries or abnormalities.

TREATMENT

Initial patient care

When managing patients with brainstem injuries, the initial approach is critical and focuses on immediate stabilization and life support measures (Carney et al., 2017). Stabilization may involve securing the patient's airway, ensuring adequate

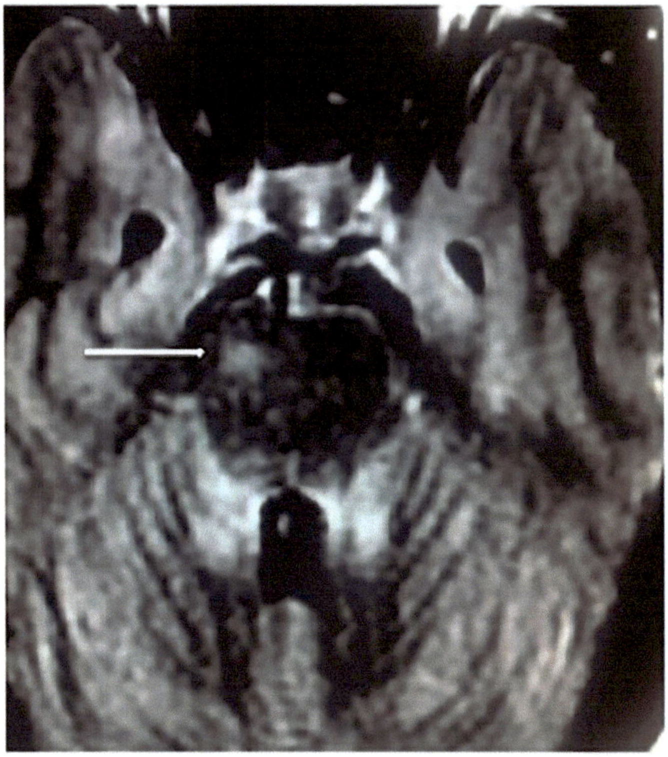

Fig. 16.3. Magnetic resonance image in FLAIR sequence showing a brainstem lesion seen as a hyperintense circumscribed lesion at the right pontine tegmentum (white arrow).

ventilation and oxygenation, and addressing any life-threatening issues such as bleeding or elevated ICP. Patients often require close monitoring in an intensive care unit (ICU) to ensure optimal physiologic parameters, including blood pressure, heart rate, and ICP (Carney et al., 2017).

Surgical interventions

Surgical options may be necessary, depending on the nature and severity of the brainstem injury (Amorim et al., 2014).

Surgical Decompression: In cases of severe brainstem injuries with significant swelling or mass effect, surgical decompression may be performed to relieve pressure within the cranial vault through decompressive hemicraniectomy. This procedure involves removing a portion of the skull to create space for the swollen brain tissue, reducing the risk of further damage. Decompressive hemicraniectomies (DHCs) are surgical interventions primarily used to manage refractory intracranial hypertension, a condition associated with higher mortality rates. However, there is no consensus on the optimal timing and methodology for this surgical intervention. The DECRA trial indicated that while DHCs effectively reduced ICP and ICU length of stay, they were associated with more unfavorable outcomes (Cooper et al., 2011). The RESCUEICP study also showed increased survival but at the cost of higher disability rates (Hutchinson et al., 2016). These studies mainly focused on DHC as a last-resort treatment rather than a preemptive measure. In contrast, some military-based studies have advocated for early, aggressive cranial decompression, particularly in cases of penetrating injuries (Hawryluk et al., 2020).

The effectiveness of DHC varies depending on the context. Studies conducted during the Iraq and Afghanistan wars recommended aggressive early intervention, while others based on civilian scenarios have been more cautious. One study reported more favorable outcomes in gunshot–head wound patients treated with DHC, suggesting its potential as a standard care option. However, most existing research does not consider primary DHC, which is increasingly being used, nor does it focus on treating or anticipating brain herniation. Furthermore, the generalizability of military-based studies to civilian settings is limited due to differences in injury mechanisms and patient health. Therefore, the role of DHC – whether as a primary intervention or a last resort – remains an area requiring more comprehensive research (Tian et al., 2021; Bertani et al., 2023). A study suggests that higher preoperative Glasgow Coma Score and shorter injury-to-surgery time may significantly improve patient outcomes (Tian et al., 2021).

Hematoma Evacuation: If a hematoma (blood clot) is compressing the brainstem, especially an acute subdural or epidural hematoma, surgical evacuation may be required to relieve pressure and prevent further injury (Amorim et al., 2014).

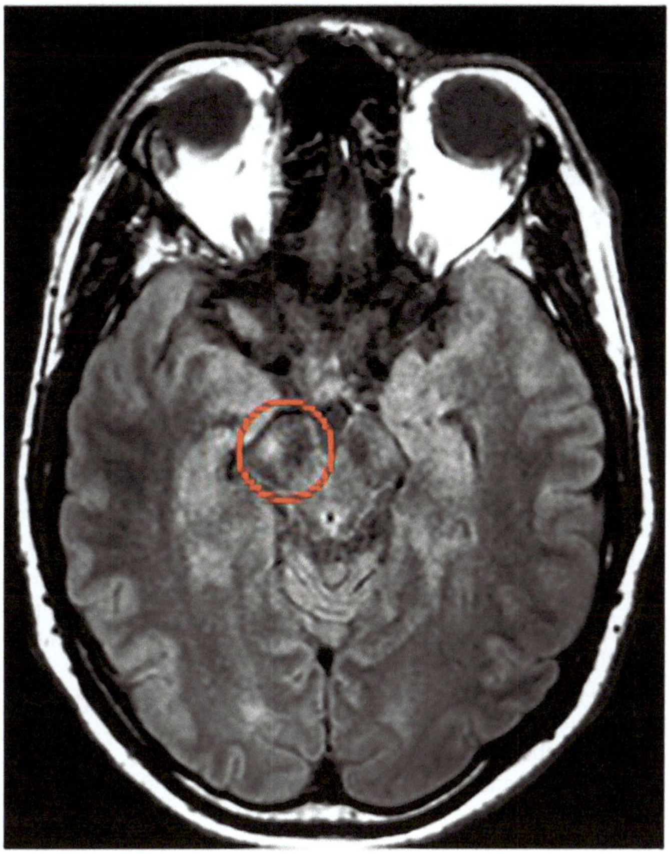

Fig. 16.4. Magnetic resonance image showing a brainstem injury seen as a hyperintense circumscribed lesion at the right cerebral peduncle of the midbrain (red circle).

Rehabilitation and long-term care

Rehabilitation is a critical component of the treatment plan for patients with brainstem injuries, aiming to maximize functional recovery and improve the patient's quality of life (Mostert et al., 2022). Patients often require physical therapy to address motor deficits, improve muscle strength, and enhance mobility. For individuals with speech and swallowing difficulties, speech therapy is essential to regain communication skills and ensure safe swallowing. Occupational therapists work with patients to regain independence in activities of daily living, such as dressing, grooming, and eating. For those with respiratory impairments, respiratory therapy helps optimize lung function and maintain proper breathing (Mostert et al., 2022).

Depending on the level of disability, patients may require assistive devices such as wheelchairs, communication aids, or adaptive equipment to enhance their independence. Patients with brainstem injuries may require long-term follow-up care and monitoring to address any complications or evolving needs, such as managing spasticity or adjusting medications. Brainstem injuries can have a profound emotional and psychologic impact on patients and their families. Psychosocial support and counseling are often integral parts of long-term care (Cassinat et al., 2024).

Disorders of consciousness (DoC) are medical conditions characterized by prolonged unconsciousness or reduced consciousness following severe TBI. The American Academy of Neurology and the American Congress of Rehabilitation Medicine have issued guidelines for managing prolonged DoC, which include the use of structured, standardized assessment methods. For prognostic purposes, clinicians may use imaging techniques such as MRI. Specifically, the presence of injuries in the corona radiata, dorsolateral upper brainstem, or corpus callosal regions on MRI scans 6–8 weeks postinjury is indicative of a poor prognosis. Amantadine is the only pharmacologic treatment recommended to accelerate functional recovery in patients with traumatic unresponsive wakefulness syndrome or minimally conscious state (MCS) (Russell et al., 2024).

Current clinical guidelines, specifically the 2018 American Academy of Neurology (AAN) update, assign Amantadine a "moderate quality" evidence rating for accelerating recovery in traumatic DoC, a recommendation anchored by the 2012 Giacino et al.'s trial (Giacino et al., 2012; Giacino et al., 2018a). The relationship between the drug and the assessment is critical. Because Amantadine hastens functional recovery rather than resolving the condition immediately, the Coma Recovery Scale-Revised (CRS-R) is the essential instrument for capturing these granular improvements. Without the CRS-R's sensitivity to subtle signs of conscious behavior (which standard exams often miss), the clinical efficacy of Amantadine cannot be reliably tracked or validated, making the two standards – pharmacologic and diagnostic – interdependent in practice (Giacino et al., 2018b; Kondziella et al., 2020). Importantly, recent evidence suggests that monitoring basic vital signs (blood pressure, heart rate, respiration) may serve as a complementary bedside tool, as fluctuations in these parameters correlate with consciousness recovery or deterioration in primary brainstem hemorrhage patients, providing clinicians with immediate prognostic data alongside formal neurobehavioral scales (Zuo et al., 2024).

Nonpharmacologic interventions to improve consciousness include various brain stimulation techniques and sensory stimulation programs. Transcranial direct current stimulation has shown some clinical effectiveness in patients with MCS, although the research is inconclusive and often involves small sample sizes. Sensory stimulation programs such as familiar auditory stimulation training also offer potential benefits but lack sufficient evidence. Despite existing guidelines and emerging treatments, the field requires more robust, large-scale research to establish effective protocols for managing DoC (Li et al., 2015).

COMPLICATIONS

Brainstem injuries can be associated with cerebral edema, which is the swelling of brain tissue. This can lead to increased ICP, potentially causing further damage. Monitoring and managing cerebral edema are crucial to prevent complications. When surgical interventions are necessary, patients are at risk of developing infections, including surgical site infections, meningitis, or brain abscesses. Prompt treatment with antibiotics or surgical drainage may be necessary. Because the brainstem regulates respiration, patients with brainstem injuries may develop respiratory complications, including pneumonia, atelectasis, or ventilator-associated pneumonia. Adequate respiratory care and monitoring are essential to prevent these issues (Rabinstein et al., 2023).

Depending on the location and extent of the brainstem injury, specific neurologic syndromes may develop. Locked-in syndrome occurs when patients are conscious and aware but have minimal or no voluntary muscle control except for eye movements. Wallenberg syndrome (lateral medullary syndrome) results from a lesion in the lateral medulla and is characterized by facial weakness, difficulty swallowing, and sensory changes. Brainstem injuries often lead to dysphagia, making it challenging for patients to swallow safely. This can increase the risk of aspiration pneumonia and malnutrition if not addressed through rehabilitation and dietary modifications (León-Carrión et al., 2002).

Duret brainstem hemorrhage

Secondary brainstem hemorrhage associated with intracranial hypertension and transtentorial herniation is called Duret hemorrhage. Duret hemorrhages are hemorrhagic areas typically located on the ventral and paramedian aspects of the upper brainstem (midbrain and pons), representing secondary brainstem hemorrhages that occur in patients with descending transtentorial herniation (Edlow et al., 2019).

The pathophysiology of Duret hemorrhage remains under discussion, as to whether it is of arterial origin (stretching and laceration of the perforating pontine branches of the basilar artery) or venous (thrombosis and venous infarction). Beucler and colleagues, in a meta-analysis including 32 patients, described that Duret brainstem hemorrhage was caused by sudden downward displacement of the upper brainstem secondary to supratentorial intracranial hypertension (91%), intracranial hypotension (6%), or mechanical traction (3%). Brainstem focal symptoms and decompressive craniectomy were potential favorable prognostic factors, whereas an age greater than 50 years showed a trend toward a poor prognosis (Beucler et al., 2023).

The presence of Duret hemorrhage has been invariably regarded as a poor prognostic sign. However, by itself, it should not trigger a decision to withdraw care. Aggressive intracranial monitoring and treatment of a Duret hemorrhage arising secondary to cerebral herniation may enable a good recovery in selected patients after severe TBI (Stiver et al., 2009).

PROGNOSIS

The relationship between brainstem injury and clinical outcome is unclear and has varied significantly. The prognosis for individuals with brainstem injuries varies widely based on several factors (Chew et al., 2012).

The extent and severity of the brainstem injury play a significant role in determining prognosis; severe injuries often result in more profound and lasting deficits. The specific location of the lesion within the brainstem can affect the clinical presentation and prognosis; injuries to vital centers for respiration, consciousness, or cranial nerve nuclei may have a poorer prognosis. Prompt and appropriate medical and surgical interventions can improve outcomes, and early detection and management of complications such as cerebral edema or infections are crucial. The extent and success of rehabilitation efforts significantly influence prognosis, with comprehensive programs including physical, speech, and occupational therapy helping patients regain function and independence. The patient's age and general health also affect prognosis; younger, healthier individuals tend to have better recovery prospects. A strong support system, including family and caregivers, can positively impact a patient's rehabilitation and long-term outlook. Proactive management of complications and ongoing medical monitoring can mitigate potential long-term issues.

In the context of TBI prognosis, a significant study offers valuable insights into outcomes associated with structural brainstem lesions. The research indicates that if the volume of these lesions does not exceed 1 cm^3, patients do not experience worsened outcomes at six months postinjury (O'Phelan, 2021). This finding brings to light the possibility of favorable outcomes in patients with early imaging evidence of structural brainstem lesions. Furthermore, the study enhances prognostic modeling by incorporating brainstem findings from CT scans, thereby contributing to the development of more accurate future prognostic tools (Williams et al., 2021).

TRACK-TBI: lesion morphology matters

The Transforming Research and Clinical Knowledge - Traumatic brain injury (TRACK-TBI) collaboration provided crucial data by stratifying brainstem injuries by lesion type, rather than relying solely on GCS score (Williams et al., 2021). Two distinct outcome trajectories were identified: patients with primary traumatic axonal injury (TAI) of the brainstem achieved a mean six-month glasgow outcome scale extended (GOSE) score of 4.0 (moderate disability with functional independence in many domains). In contrast, patients with Duret hemorrhage or contusion achieved a mean GOSE of 1.4 (severe disability or death). This difference was statistically significant ($p = 0.002$).

The biologic basis for this difference lies in the injury: Duret hemorrhage is a secondary injury (from herniation/displacement) reflecting irreversible damage, whereas primary TAI involves axonal shearing, which, although serious, allows for neuroplasticity and functional reorganization over time. The conclusion is that incorporating lesion type alongside GCS significantly improved discrimination for this subgroup. Generalized prognostic models that fail to account for brainstem pathology subtypes will misclassify patients, leading to inappropriate counseling and possibly biased treatment. The performance of the IMPACT and CRASH prognostic models has been evaluated in contemporary multicenter cohorts, showing good discrimination but potential for improvement when incorporating additional imaging and biomarker data (Yuh et al., 2021; Dijkland et al., 2022; Gravesteijn et al., 2024).

Therapeutic nihilism and early prognostication

Inaccurate or premature prognostication can introduce biased decisions regarding life-sustaining treatment. Therapeutic nihilism refers to inappropriately pessimistic expectations regarding patient recovery and is common, often derived from anchoring decisions to historical data rather than individualized injury patterns (Hemphill et al., 2009; Hawryluk and Bhattacharjee, 2018). Up to 70% of deaths in severe TBI result from the withdrawal of life-sustaining treatment (WLST), rather than from the injury itself (Turgeon et al., 2011). These withdrawal decisions often occur within 72 hours postinjury, a period characterized by maximum clinical uncertainty and ongoing evolution of secondary injury (van Veen et al., 2021).

This creates a problematic cycle: clinicians predict a worse outcome based on historical rates (availability bias), recommend early WLST, and the subsequent death validates the original pessimistic prediction, thus inflating mortality. Contemporary practice should aim for a shift toward provisional and refined prognostication rather than definitive early predictions. For example, the 72-hour threshold for end-of-life discussions remains unvalidated (Letsinger et al., 2017; Dijkland et al., 2020).

Severe brainstem injury requires adequate time to resolve transient secondary injuries (edema, shock) before irreversible decisions are made. A robust approach should incorporate iterative assessment: serial neurologic examination, refined lesion characterization on high-resolution imaging (TAI versus hemorrhage), and longitudinal biomarker trends. The clinical goal should be to distinguish authentic biologic futility (no realistic possibility of meaningful recovery) from discomfort with outcome uncertainty and prognostic bias.

The prognosis of these patients is often dismal because of the critical location. Some studies have shown that the mortality of patients with TBSI ranges from 67% to 83% (Se et al., 2009). However, these studies are not without limitations. Firstly, research has relied on noncontrast CT scans conducted within the first 48 hours postadmission. While CT scans are standard in clinical care, MRI could provide more detailed characterization of lesions, and the potential to incorporate MRI into future studies could further refine prognostic models. Secondly, the impact of withdrawing life-sustaining therapies (WLST) on study outcomes cannot be overlooked. WLST is a leading cause of death in severe TBI cases, and in one study, 7 out of 29 patients died due to WLST before the six-month outcome measure was reached (Min and Shin, 2022). This factor may disproportionately affect the prognosis for patients with brainstem hemorrhages and raises questions about the potential for therapeutic nihilism based on possibly inaccurate prognostic predictions (Williams et al., 2021).

FUTURE RESEARCH AND ADVANCES

Advanced imaging techniques

Traumatic brain injury is associated with various vascular and metabolic changes that can lead to altered cerebral blood flow (CBF), ischemia, and infarction. Various imaging techniques have been employed to study these changes, including single-photon emission computed tomography (SPECT), positron emission tomography (PET), and arterial spin-labeling (ASL), among others. Each technique has its own advantages and limitations; for example, SPECT and dynamic susceptibility contrast (DSC) allow only for qualitative comparisons, while PET and ASL offer quantitative accuracy. Notably, reduced cerebral blood volume as measured by DSC MRI has been correlated with worse clinical outcomes, even in patients with otherwise normal MRIs (Grassi et al., 2021). ASL, which does not require intravenous contrast, has shown decreased regional CBF in the thalamus correlating with neurocognitive impairment. However, these studies are generally underpowered, and more research is needed to establish the prognostic value of these techniques.

Functional MRI (fMRI) uses blood–oxygen level dependent (BOLD) contrast to study the relationships among CBF, metabolism, and neuronal activity. While fMRI offers high spatial resolution, its temporal resolution is limited. Despite significant advances in TBI imaging, further validation is needed to transition these techniques from research to routine clinical use. This is particularly true for conditions like mild TBI and chronic traumatic encephalopathy, which often show no abnormalities on conventional imaging (Chew et al., 2012). As new therapeutic options for TBI emerge, advanced imaging techniques like DTI, magnetic resonance spectroscopy (Mrs), perfusion weighted imaging (PWI), and fMRI are likely to play a critical role in patient selection and treatment evaluation (Lee, 2020).

Neuroinflammation and immunomodulation

Investigating the intricate role of neuroinflammation in brainstem injuries is a key research focus. Researchers are exploring novel immunomodulatory approaches and anti-inflammatory agents to mitigate the harmful effects of inflammation and promote neural repair (Dadas et al., 2018).

The neuroinflammatory response to TBI is a complex interplay of pro- and anti-inflammatory cytokines and chemokines, primarily released by microglia. Key proinflammatory cytokines, such as interleukin-1 beta (IL-1β) and tumor necrosis factor alpha (TNF-α), are elevated shortly after TBI and contribute to brain damage. However, the role of TNF-α is complex: it acts as an inflammatory mediator in the acute phase and a neurotrophic factor in the delayed phase. Anti-inflammatory cytokines like IL-10 and TGFβ1 also play a role, often mitigating the damaging effects of proinflammatory cytokines. Sustained neuroinflammation could potentially hinder neuroprotection and neuroregeneration (Su et al., 2022).

An example of treatment is ACT001, which serves as an antineuroinflammatory agent by enhancing the integrity of the blood–brain barrier (BBB) and reducing microglial activation post-TBI. It operates via the protein kinase B/nuclear factor kappa B/NLR family pyrin domain containing 3 (AKT/NFκB/NLRP3) pathway, inhibiting AKT phosphorylation, which, in turn, restrains NFκB nuclear translocation and downregulates NLRP3 inflammasome activation (Cai et al., 2022). This results in a reduced neuroinflammatory response, thereby offering a targeted strategy to mitigate the secondary injuries commonly associated with TBI. This has been demonstrated only in rats to date (Su et al., 2022).

Neuroprotective strategies

TBI involves a two-step process: primary and secondary injury mechanisms. Primary injury occurs at the time of trauma, causing immediate mechanical damage to neurons, axons, glia, and blood vessels due to shearing, tearing, or stretching. Secondary injury mechanisms evolve over minutes to months after the initial trauma and are characterized by neurochemical, metabolic, and cellular changes (Andrade et al., 2009). These secondary mechanisms are responsible for many of the neurologic deficits observed posttrauma and offer a window for therapeutic intervention. Secondary injury involves a range of factors, including ionic imbalance, neurotransmitter release, mitochondrial dysfunction, and inflammatory responses, among others. These mechanisms can lead to further complications like cerebral edema and increased ICP, contributing to neuronal cell death. Neuroprotection strategies aim to prevent or mitigate the development of secondary injury (Loane et al., 2015).

Several neuroprotective strategies have been investigated. Progesterone, a neurosteroid, has demonstrated neuroprotective effects in experimental models of TBI, spinal cord injury, and stroke; however, its effectiveness has been questioned, and further clinical trials are needed. Ciclosporin, an immunosuppressant drug, stabilizes mitochondrial function and has shown promise in reducing axonal damage and improving metabolic markers in TBI patients. Diketopiperazines have shown neuroprotective properties in both in vitro and animal TBI studies, with a clinically relevant therapeutic window. SUR1-regulated NCCa-ATP channel inhibitors, like glibenclamide, have been effective in reducing edema and secondary hemorrhage in TBI models. Cell cycle inhibitors such as flavopiridol and roscovitine have shown potent neuroprotective effects in experimental TBI models by inhibiting cell cycle activation, thereby reducing inflammation and neuronal cell death. Statins, commonly used for cholesterol management, have shown promise in reducing inflammation and improving neuronal survival in TBI models; they also have a large therapeutic window, making them well suited for clinical trials.

The translation of neuroprotective treatments from animal models to human TBI has been largely unsuccessful. Several factors contribute to this translational gap, including methodologic differences between preclinical and clinical studies, inadequate understanding of secondary injury mechanisms, and insufficient preclinical testing across diverse models (Xu et al., 2024).

Biomarkers for diagnosis and prognosis

Several biomarkers can be found in CSF or blood, and more recently in saliva. Examples include proteins such as S100B and neuron-specific enolase, which can provide insights into the extent of brain damage and the integrity of the BBB. However, the field faces challenges such as the variable origins of these biomarkers, which can also be found outside the central nervous system, and the need for more reliable tools to predict outcomes like posttraumatic epilepsy (PTE) or to guide treatment selection (Drieu et al., 2022). Standardization of biomarker measurement methods and cutoff values across institutions remains an important consideration for clinical implementation (Manley et al., 2025).

Serum biomarkers in brainstem injury

Circulating biomarkers are emerging as more objective tools for assessing TBI severity. In the context of brainstem trauma, blood-based proteins provide quantifiable measures of injury burden and aid in prognostication. Two distinct classes of biomarkers are now clinically relevant: acute triage markers (GFAP and UCH-L1) and markers of chronic axonal damage (neurofilament light chain, NfL).

GFAP (glial fibrillary acidic protein) and UCH-L1 (ubiquitin C-terminal hydrolase-L1): Food and Drug Administration **(FDA)-Cleared Triage Markers**. GFAP, a cytoskeletal protein from astrocytes, and UCH-L1, derived from neuronal cytoplasm, are FDA-approved for clinical use and are utilized in rapid point-of-care platforms. Their primary function is to identify patients who can safely avoid CT imaging (Bazarian et al., 2018). In mild to moderate TBI, both markers demonstrate high sensitivity and a negative predictive value that approaches 1.00 within the first 12 hours postinjury (Papa et al., 2016). Beyond ruling out injury, GFAP and UCH-L1 levels correlate with overall injury severity (Okonkwo et al., 2020; Welch et al., 2025). Hyperacute measurements (within 2 hours) show significantly higher biomarker levels in CT-positive patients than in CT-negative patients. Combined assessment improves diagnostic discrimination (area under the curve (AUC) 0.94) (Papa et al., 2012).

NfL: Quantifying Axonal Burden. NfL is the most sensitive serum biomarker for TAI, a pathologic hallmark of severe brainstem trauma (Shahim et al., 2020). Its presence in peripheral blood indicates axonal breakdown, a critical distinction in brainstem injuries where TAI to the reticular activating system and major motor pathways directly determines neurologic outcome. Elevated NfL appears to correlate with poor long-term functional outcomes (Shahim et al., 2022). It is a marker of white matter damage and is validated by its concordance with imaging modalities such as diffusion tensor imaging (DTI). The timing of elevation distinguishes clinical roles: GFAP/UCH-L1 peak acutely (hours), providing immediate injury detection for triage, while NfL rises later and persists, reflecting the true extent of ongoing axonal degeneration (Czeiter et al., 2020). Chronically elevated NfL suggests progressive neurodegeneration and potential risk for chronic sequelae, making it valuable for long-term rehabilitation planning and prognostication beyond the acute phase (Graham et al., 2021).

As a practical consideration for the clinical use of biomarkers, the lack of standardization may be problematic if institutions do not set cutoff values in common agreement (Maas et al., 2022).

Neuroregeneration and stem cell therapies

Researchers are exploring innovative approaches to stimulate neural regeneration in the brainstem. Stem cell-based therapies, tissue engineering, and gene editing techniques are under investigation to promote axonal regrowth and functional recovery. Various types of stem cells – mesenchymal stem cells, neural stem cells, multipotent adult progenitor cells, and endothelial progenitor cells – have demonstrated potential in ameliorating neurologic impairments post-TBI (Loane et al., 2015). These stem cells operate through diverse mechanisms, including anti-inflammatory actions, angiogenesis promotion, and neuroprotection, thereby addressing the multifaceted challenges posed by TBI.

Imai and colleagues investigated a new treatment for TBI using genetically modified induced pluripotent stem cells (iPSCs). These iPSCs were engineered to express a specific enzyme-prodrug gene, reducing the risk of tumorigenesis, a major hurdle in stem cell therapies. The modified stem cells improved motor functions and reduced brain injury in a TBI mouse model, demonstrating their potential for safe and effective use in regenerative medicine (Imai et al., 2023). While clinical applications are still in the nascent stages, stem cell therapies hold potential to improve patient outcomes.

Telemedicine and remote monitoring

A review study indicated that follow-up technologies (FUTs) for managing TBI are still in the early stages of development, particularly in terms of large-scale patient studies, formal technology assessments, and applicability in low-income settings (Smith et al., 2022). Despite these limitations, FUTs hold promise for enhancing clinicians' understanding of patient well-being postdischarge by offering more frequent and precise data collection, aligning with the World Health Organization's goals for outcome data and injury surveillance. Future research should focus on empirically evaluating the utility, feasibility, and cost-effectiveness of these technologies, as well as understanding the challenges of implementation and adherence from the perspectives of patients, caregivers, and healthcare providers (Smith et al., 2022).

Genomic and precision medicine

Advancements in genomics and personalized medicine are being harnessed to tailor treatments for brainstem injuries based on individual genetic profiles. Precision medicine approaches may optimize therapeutic strategies. A study in rats revealed significant DAI in the brainstem, accompanied by notable molecular alterations. Specifically, 579 genes, 70 proteins, and 183 metabolites exhibited significant changes, primarily related to neural signal regulation, inflammation, neuroprotection, and immune response (Su et al., 2023). Furthermore, comprehensive analysis indicated that these molecular changes were intricately linked to neuroactive ligand–receptor interactions, with particular emphasis on G protein-coupled receptor (GPCR)-regulated pathways. Notably, gamma-aminobutyric acid (GABA's) corresponding receptors (GABA-A, GABA-B, GABA-C) and transporter GAT were inhibited to varying degrees, suggesting a complex interplay of molecular mechanisms following TBSI. Detailed molecular mapping may provide a foundation for targeted therapeutic interventions, potentially revolutionizing the treatment of traumatic brain injuries by modulating specific genes, proteins, and metabolic pathways (Su et al., 2023).

Patient-specific models and simulations

Computational models and simulations are becoming increasingly sophisticated, allowing researchers to create patient-specific models of brain injuries. These models can simulate injury scenarios and predict outcomes based on individualized data. For example, in one study, a dynamic nonlinear mathematical model based on the Michaelis–Menten equation was applied to data from 156 acquired brain injury patients and showed an accuracy rate of 85%, sensitivity of 90.6%, and specificity of 62.5% in predicting positive or negative Glasgow outcome scores (GOS) at discharge. The model's predictive power improves significantly 86 days postinjury, offering clinicians an interesting tool for tailoring interventions based on predicted outcomes (Panunzi et al., 2023).

CONCLUSION

TBSI encompasses a heterogeneous spectrum of pathology – from primary TAI with potential for functional recovery (mean GOSE 4.0) to secondary Duret hemorrhage portending severe disability or death (mean GOSE 1.4). Prognosis hinges not merely on injury presence but on lesion morphology, volume threshold (< 1 cm³ as a favorable cutoff), and timing of assessment. Critically, up to 70% of deaths in severe TBI result from WLST rather than from the injury itself, often within 72 hours – a window characterized by maximum clinical uncertainty. This increases the risk of therapeutic nihilism and self-fulfilling prophecies when early, definitive prognostication drives irreversible decisions.

This synthesis is critical to mention because it challenges the traditional conflation of "brainstem injury" with "futility." By stratifying outcomes according to injury mechanism (primary TAI versus secondary herniation) and integrating contemporary biomarker data (GFAP/UCH-L1 for acute triage, neurofilament light for axonal burden), clinicians can move beyond binary prognostic statements toward individualized, iterative assessment. Recognizing that the CRS-R and Amantadine function as interdependent standards – pharmacologic intervention requiring sensitive outcome measurement – further refines rehabilitation pathways for patients with prolonged disorders of consciousness.

Future investigations should prioritize prospective multi-center cohorts with standardized high-resolution MRI (including DTI) acquired at uniform time points, paired with serial biomarker sampling, to refine lesion-specific prognostic models. Studies examining the impact of delayed prognostication protocols on WLST rates and long-term outcomes are essential to quantify the contribution of therapeutic nihilism to observed mortality. Finally, randomized trials of neuroprotective and neuroregenerative strategies such as stem cell therapies, immunomodulation, and targeted molecular interventions that require brainstem-specific outcome measures to determine whether the translational gap seen in broader TBI research can be bridged for this uniquely vulnerable population approach to management and care.

References

Adil SM, Calabrese E, Charalambous LT et al. (2021). A high-resolution interactive atlas of the human brainstem using magnetic resonance imaging. Neuroimage 237: 118135.

Aguas J, Begue R, Diez J (2005). Brainstem injury diagnosed by MRI. An epidemiologic and prognostic reappraisal. Neurocirugia (Astur) 16: 14–20.

Al-Sarraj S, Fegan-Earl A, Ugbade A et al. (2012). Focal traumatic brain stem injury is a rare type of head injury resulting from assault: a forensic neuropathological study. J Forensic Leg Med 19: 144–151.

Amorim RL, Stiver SI, Paiva WS et al. (2014). Treatment of traumatic acute posterior fossa subdural hematoma: report of four cases with systematic review and management algorithm. Acta Neurochir (Wien) 156: 199–206.

Andrade AF, Paiva WS, Amorim RL et al. (2009). The pathophysiological mechanisms following traumatic brain injury. Rev Assoc Med Bras 55: 75–81.

Barman A, Chatterjee A, Bhide R (2016). Cognitive impairment and rehabilitation strategies after traumatic brain injury. Indian J Psychol Med 38: 172–181.

Bazarian JJ, Biberthaler P, Welch RD et al. (2018). Serum GFAP and UCH-L1 for prediction of absence of intracranial injuries on head CT (ALERT-TBI): a multicentre observational study. Lancet Neurol 17: 782–789.

Bertani R, Koester SW, Perret C et al. (2023). Decompressive hemicraniectomies as a damage control approach for multilobar firearm projectile injuries: a single-center experience. World Neurosurg 169: e96–e101.

Beucler N, Cungi PJ, Dagain A (2023). Duret brainstem hemorrhage after transtentorial descending brain herniation: a systematic review and meta-analysis. World Neurosurg 173: 251–262.e4.

Bhatoe HS (1999). Primary brainstem injury: benign course and improved survival. Acta Neurochir (Wien) 141: 515–519.

Cai L, Gong Q, Qi L et al. (2022). ACT001 attenuates microglia-mediated neuroinflammation after traumatic brain injury via inhibiting AKT/NFκB/NLRP3 pathway. Cell Commun Signal 20: 56.

Carney N, Totten AM, O'Reilly C et al. (2017). Guidelines for the management of severe traumatic brain injury, fourth edition. Neurosurgery 80: 6–15.

Cassinat J, Nygaard J, Hoggard C et al. (2024). Predictors of mortality and rehabilitation location in adults with prolonged coma following traumatic brain injury. PM R 16: 1064–1071.

Chew BG, Spearman CM, Quigley MR et al. (2012). The prognostic significance of traumatic brainstem injury detected on T2-weighted MRI. J Neurosurg 117: 722–728.

Cook R, Zima L, Khazaal J et al. (2024). Low-velocity penetrating brain injury: a review of the literature and illustrative case. Brain Inj 31: 1–7.

Cooper DJ, Rosenfeld JV, Murray L et al. (2011). Decompressive craniectomy in diffuse traumatic brain injury. N Engl J Med 364: 1493–1502.

Cury RG, Barbosa ER, Freitas C et al. (2017). Holmes tremor secondary to a stabbing lesion in the midbrain. Tremor Other Hyperkinet Mov (N Y) 7: 523.

Czeiter E, Amrein K, Gravesteijn BY et al. (2020). Blood biomarkers on admission in acute traumatic brain injury: relations to severity, CT findings and care path in the CENTER-TBI study. EBioMedicine 56: 102785.

Dadas A, Washington J, Diaz-Arrastia R et al. (2018). Biomarkers in traumatic brain injury (TBI): a review. Neuropsychiatr Dis Treat 14: 2989–3000.

Dijkland SA, Foks KA, Polinder S et al. (2020). Prognosis in moderate and severe traumatic brain injury: a systematic review of contemporary models and validation studies. J Neurotrauma 37: 1–13.

Dijkland SA, Gravesteijn BY, Lingsma HF (2022). Prognostic models for traumatic brain injury have good discrimination but poor overall model performance for predicting mortality and unfavorable outcomes. Neurocrit Care 37: 273–283.

Drieu A, Lanquetin A, Prunotto P et al. (2022). Persistent neuroinflammation and behavioural deficits after single mild traumatic brain injury. J Cereb Blood Flow Metab 42: 2216–2229.

Edlow BL, Threlkeld ZD, Fehnel KP et al. (2019). Recovery of functional independence after traumatic transtentorial herniation with Duret hemorrhages. Front Neurol 10: 1077.

Ewing-Cobbs L, Prasad MR, Kramer L et al. (2006). Late intellectual and academic outcomes following traumatic brain injury sustained during early childhood. J Neurosurg 105: 287–296.

Firsching R, Woischneck D, Klein S et al. (2001). Classification of severe head injury based on magnetic resonance imaging. Acta Neurochir (Wien) 143: 263–271.

Firsching R, Woischneck D, Klein S et al. (2002). Classification of severe head injury based on MRI. Neurol Res 24: 145–146.

Giacino JT, Katz DI, Schiff ND et al. (2018a). Practice guideline update recommendations summary: disorders of consciousness. Neurology 91: 450–460.

Giacino JT, Katz DI, Schiff ND et al. (2018b). Comprehensive systematic review update summary: disorders of consciousness. Arch Phys Med Rehabil 99: 1699–1709.

Giacino JT, Whyte J, Bagiella E et al. (2012). Placebo-controlled trial of Amantadine for severe traumatic brain injury. N Engl J Med 366: 819–826.

Graham NSN, Zimmerman KA, Moro F et al. (2021). Axonal marker neurofilament light predicts long-term outcomes and progressive neurodegeneration after traumatic brain injury. Sci Transl Med 13: eabg9922.

Grassi DC, Zaninotto AL, Feltrin FS et al. (2021). Dynamic changes in white matter following traumatic brain injury and how diffuse axonal injury relates to cognitive domain. Brain Inj 35: 275–284.

Gravesteijn BY, Sewalt CA, Stocchetti N et al. (2024). Performance of IMPACT and CRASH prognostic models for traumatic brain injury in a contemporary multicenter cohort: a TRACK-TBI study. J Neurosurg 140: 1165–1174.

Hawryluk GWJ, Bhattacharjee S (2018). Should we have a guard against therapeutic nihilism for patients with severe traumatic brain injury?. Neural Regen Res 13: 406–407.

Hawryluk GWJ, Rubiano AM, Totten AM et al. (2020). Guidelines for the management of severe traumatic brain injury: 2020 update of the decompressive craniectomy recommendations. Neurosurgery 87: 427–434.

Hemphill JC 3rd, White DB (2009). Clinical nihilism in neuroemergencies. Emerg Med Clin North Am 27: 27–37.

Hilario A, Ramos A, Millan JM et al. (2012). Severe traumatic head injury: prognostic value of brain stem injuries detected at MRI. AJNR Am J Neuroradiol 33: 1925–1931.

Hongo S, Shimizu H, Saji E et al. (2025). Acute respiratory failure caused by brainstem demyelinating lesions in an older patient with an atypical relapsing autoimmune disorder. Neuropathology 45: 3–12.

Hunter LE, Lubin N, Glassman NR et al. (2019). Comparing region of interest versus voxel-wise diffusion tensor imaging analytic methods in mild and moderate traumatic brain injury: a systematic review and meta-analysis. J Neurotrauma 36: 1222–1230.

Hutchinson PJ, Kolias AG, Timofeev IS et al. (2016). Trial of decompressive craniectomy for traumatic intracranial hypertension. N Engl J Med 375: 1119–1130.

Imai R, Tamura R, Yo M et al. (2023). Neuroprotective effects of genome-edited human iPS cell-derived neural stem/progenitor cells on traumatic brain injury. Stem Cells 41: 603–616.

Irimia A, Goh S-YM, Wade AC et al. (2017). Traumatic Brain Injury Severity, Neuropathophysiology, and Clinical Outcome: Insights from Multimodal Neuroimaging. Front Neurol 8: 530.

Izzy S, Mazwi NL, Martinez S et al. (2017). Revisiting Grade 3 diffuse axonal injury: not all brainstem microbleeds are prognostically equal. Neurocrit Care 27: 199–207.

Jones DK, Knösche TR, Turner R (2013). White matter integrity, fiber count, and other fallacies: the do's and don'ts of diffusion MRI. Neuroimage 73: 239–254.

Kim HJ (2012). The prognostic factors related to traumatic brain stem injury. J Korean Neurosurg Soc 51: 24–30.

Kondziella D, Bender A, Diserens K et al. (2020). European Academy of Neurology guideline on the diagnosis of coma and other disorders of consciousness. Eur J Neurol 27: 741–756.

Lee AL (2020). Advanced imaging of traumatic brain injury. Korean J Neurotrauma 16: 3–17.

Letsinger J, Rommel C, Hirschi R et al. (2017). The aggressiveness of neurotrauma practitioners and the influence of the IMPACT prognostic calculator. PLoS One 12: e0183552.

León-Carrión J, van Eeckhout P, Domínguez-Morales MR et al. (2002). The locked-in syndrome: a syndrome looking for a therapy. Brain Inj 16: 571–582.

Li S, Zaninotto AL, Neville IS et al. (2015). Clinical utility of brain stimulation modalities following traumatic brain injury: current evidence. Neuropsychiatr Dis Treat 11: 1573–1586.

Loane DJ, Stoica BA, Faden AI (2015). Neuroprotection for traumatic brain injury. Handb Clin Neurol 127: 343–366.

Maas AIR, Menon DK, Manley GT et al. (2022). Traumatic brain injury: progress and challenges in prevention, clinical care, and research. Lancet Neurol 21: 1004–1060.

Manley GT, Dams-O'Connor K, Alosco ML et al. (2025). NIH-NINDS TBI Classification and Nomenclature Initiative. A new characterisation of acute traumatic brain injury: the NIH-NINDS TBI Classification and Nomenclature Initiative. Lancet Neurol 24: 512–523.

Mannion RJ, Cross J, Bradley P et al. (2007). Mechanism-based MRI classification of traumatic brainstem injury and its relationship to outcome. J Neurotrauma 24: 128–135.

Min JH, Shin I (2022). Treatment and rehabilitation for traumatic brain injury: current update. Brain Neurorehabil 15: e14.

Mostert CQB, Singh RD, Gerritsen M et al. (2022). Long-term outcome after severe traumatic brain injury: a systematic literature review. Acta Neurochir (Wien) 164: 599–613.

Nourallah B, Menon DK, Zeiler FA (2018). Midline shift is unrelated to subjective pupillary reactivity assessment on admission in moderate and severe traumatic brain injury. Neurocrit Care 29: 203–213.

Okonkwo DO, Puffer RC, Puccio AM et al. (2020). Point-of-care platform blood biomarker testing of glial fibrillary acidic protein

versus S100 calcium-binding protein B for prediction of traumatic brain injuries: a TRACK-TBI study. J Neurotrauma 37: 2460–2467.
O'Phelan KH (2021). Not always a nail in the coffin! Brainstem lesions after traumatic brain injury. Neurocrit Care 35: 306–307.
Paiva WS, de Andrade AF, Amorim RL et al. (2012). Brainstem injury by penetrating head trauma with a knife. Br J Neurosurg 26: 779–781.
Panunzi S, Lucca LF, De Tanti A et al. (2023). Modeling outcome trajectories in patients with acquired brain injury using a non-linear dynamic evolution approach. Sci Rep 13: 1–15.
Papa L, Brophy GM, Welch RD et al. (2016). Time course and diagnostic accuracy of glial and neuronal blood biomarkers GFAP and UCH-L1 in a large cohort of trauma patients with and without mild traumatic brain injury. JAMA Neurol 73: 551–560.
Papa L, Lewis LM, Falk JL et al. (2012). Elevated levels of serum glial fibrillary acidic protein breakdown products in mild and moderate traumatic brain injury are associated with intracranial lesions and neurosurgical intervention. Ann Emerg Med 59: 471–483.
Paterakis K, Karantanas AH, Komnos A et al. (2000). Outcome of Patients with Diffuse Axonal Injury: The Significance and Prognostic Value of MRI in the Acute Phase. J Trauma: Injury, Infection, Crit Care 49(6): 1071–1075.
Petrie EC, Cross DJ, Yarnykh VL et al. (2014). Neuroimaging, behavioral, and psychological sequelae of repetitive combined blast/impact mild traumatic brain injury in Iraq and Afghanistan war veterans. J Neurotrauma 31(5): 425–436.
Rabinstein AA, Cinotti R, Bösel J (2023). Liberation from mechanical ventilation and tracheostomy practice in traumatic brain injury. Neurocrit Care 38: 439–446.
Rad BS, Zeinali MS, Bavil MS et al. (2009). Autopsy findings of brainstem in head trauma in comparison with CT scan findings in brain trauma ward in Tabriz, Iran. Acta Medica Iran 47: 409–414.
Russell ME, Hammond FM, Murtaugh B (2024). Prognosis and enhancement of recovery in disorders of consciousness. NeuroRehabilitation 54: 43–59.
Sandhu S, Soule E, Fiester P et al. (2019). Brainstem diffuse axonal injury and consciousness. J Clin Imaging Sci 9: 32.
Se YB, Kim CH, Bak KH et al. (2009). Traumatic brainstem hemorrhage presenting with hemiparesis. J Korean Neurosurg Soc 45: 176–178.
Shahim P, Politis A, van der Merwe A et al. (2020). Neurofilament light as a biomarker in traumatic brain injury. Neurology 95: e610–e622.
Shahim P, Zetterberg H, Blennow K (2022). Neurofilament protein and antineurofilament antibodies following traumatic brain injury. Expert Rev Neurother 22: 309–322.
Smith BG, Tumpa S, Mantle O et al. (2022). Remote follow-up technologies in traumatic brain injury: a scoping review. J Neurotrauma 39: 1289–1317.
Stiver SI, Gean AD, Manley GT (2009). Survival with good outcome after cerebral herniation and Duret hemorrhage caused by traumatic brain injury. J Neurosurg 110: 1242–1246.
Su Q, Chen Q, Li Z et al. (2022). Multi-omics analysis reveals GABAergic dysfunction after traumatic brainstem injury in rats. Front Neurosci 16: 1003300.
Su Q, Chen Q, Li Z et al. (2023). Molecular mapping of traumatic brainstem injury pathways. Front Neurosci 17: 1105234.
Tian R, Dong J, Liu W et al. (2021). Prognostic analysis of emergency decompressive craniectomy for patients with severe traumatic brain injury with bilateral fixed dilated pupils. World Neurosurg 146: e1307–e1317.
Trotter BB, Robinson ME, Milberg WP et al. (2015). Military blast exposure, ageing and white matter integrity. Brain 138(Pt 8): 2278–2292.
Turgeon AF, Lauzier F, Simard JF et al. (2011). Mortality associated with withdrawal of life-sustaining therapy for patients with severe traumatic brain injury: a Canadian multicentre cohort study. CMAJ 183: 1581–1588.
Van Horn JD, Bhattrai A, Irimia A (2017). Multimodal imaging of neurometabolic pathology due to traumatic brain injury. Trends Neurosci 40(1): 39–59.
van Veen E, van der Jagt M, Citerio G et al. (2021). Occurrence and timing of withdrawal of life-sustaining measures in traumatic brain injury patients: a CENTER-TBI study. Intensive Care Med 47: 1115–1129.
Vieira RCA, Pipek LZ, Oliveira DV et al. (2024). The relationship between injury characteristics and post-traumatic recovery after diffuse axonal injury. Biomedicines 12: 311.
Welch RD, Bazarian JJ, Chen JY et al. (2025). A high-performance core laboratory GFAP/UCH-L1 test for the prediction of intracranial injury after mild traumatic brain injury. Am J Emerg Med 89: 129–134.
Williams JR, Nieblas-Bedolla E, Feroze A et al. (2021). Prognostic value of hemorrhagic brainstem injury on early computed tomography: a TRACK-TBI study. Neurocrit Care 35: 335–346.
Xu X, He L, Li M et al. (2024). Neuroinflammation plays a potential role in the medulla oblongata after moderate traumatic brain injury in mice as revealed by nontargeted metabonomics analysis. J Neurotrauma 41: e2026–e2038.
Yamamura K, Kiriu N, Tomura S et al. (2022). The cause of acute lethality of mice exposed to a laser-induced shock wave to the brainstem. Sci Rep 12: 9490.
Yeh FC, Panesar S, Fernandes D et al. (2018). Population-averaged atlas of the macroscale human structural connectome and its network topology. Neuroimage 178: 57–68.
Yuh EL, Jain S, Sun X et al. (2021). Pathological Computed Tomography Features Associated With Adverse Outcomes After Mild Traumatic Brain Injury: A TRACK-TBI Study With External Validation in CENTER-TBI. JAMA Neurol 78: 1137–1148.
Zhang Y, Li Z, Wang H et al. (2024). Molecular biomarkers of diffuse axonal injury: recent advances and future perspectives. Expert Rev Mol Diagn 24: 39–47.
Zuo S, Feng Y, Sun J et al. (2024). The assessment of consciousness status in primary brainstem hemorrhage (PBH) patients can be achieved by monitoring changes in basic vital signs. Geriatr Nurs 59: 498–506.

Section III

Monitoring and intervention

Handbook of Clinical Neurology, Vol. 216 (3rd series)
The Brainstem
L.T. Grinberg and A.D. Korczyn
DOI: https://doi.org/10.1016/B978-0-443-15736-3.00002-0

Chapter 17

Computed tomography and magnetic resonance imaging of the brainstem

DIOGO GOULART CORRÊA[1,2,*] AND TOMÁS FREDDI[3]

[1]*Department of Radiology, Clínica de Diagnóstico por Imagem (CDPI)/DASA, Rio de Janeiro, Brazil*

[2]*Department of Diagnostic Imaging, Rio de Janeiro State University, Rio de Janeiro, Brazil*

[3]*Department of Radiology, Hospital do Coração, São Paulo, Brazil*

Abstract

The brainstem is a fundamental central nervous system component, essential for life. It comprises the midbrain, pons, and medulla oblongata. It is connected superiorly to the diencephalon, inferiorly to the spinal cord, and posteriorly to the cerebellum. The brainstem is pivotal in permitting transit for ascending and descending tracts, integrative functions, such as consciousness and respiratory cycle, and transmitting action potentials to cranial nerves. Lesions in this region can lead to diverse neurologic deficits. Imaging the brainstem presents challenges due to the proximity of critical structures and potential artifacts. While computed tomography and ultrasound have limitations, magnetic resonance imaging (MRI) is preferred, particularly at higher field strengths, offering superior anatomic details. The compact nature of the brainstem poses challenges in distinguishing its nuclei on MRI. Despite technical hurdles, adequate MRI protocols, including T1- and T2-weighted imaging, fluid-attenuated inversion recovery, susceptibility-weighted imaging, and diffusion-weighted imaging, enable a comprehensive evaluation. Additionally, diffusion tensor imaging with fiber tractography provides qualitative and quantitative insights into white matter tracts. This chapter reviews the normal anatomy of the brainstem through imaging, emphasizing MRI, and addresses common diseases affecting this crucial region.

INTRODUCTION

The brainstem is the most primitive part of the brain, which together with the upper cervical spinal cord, are the only central nervous system structure essential for life (Sciacca et al., 2019; Shepherd and Hoch, 2022). It is the lower extension of the brain, connecting superiorly with the diencephalon (thalamus, hypothalamus, epithalamus, and subthalamus), inferiorly to the spinal cord, and posteriorly to the cerebellum (Sciacca et al., 2019). It is located in the posterior cranial fossa, grossly composed of three transverse regions: midbrain or mesencephalon which is the most rostral portion; pons located in the middle part; and medulla oblongata which is the most caudal segment, which connects with the spinal cord (Querol-Pascual, 2010).

The brainstem has three main functions: (1) to provide transit for ascending and descending tracts and nuclei, leading connections to and from the cerebrum, cerebellum, and spinal cord; (2) to play a role in integrative functions, such as level of consciousness, the sleep–wake cycle, muscles tone, and respiratory and cardiac cycles; and (3) to send or receive action potentials of the cranial nerves, composed of sensory fibers terminating in the brainstem nuclei and motor fibers originating in the brainstem nuclei (Smith and DeMyer, 2003; Querol-Pascual, 2010). This brain region contains all the cranial nerve nuclei except for the olfactory (I), optic (II), and the spinal rootlets of the accessory (XI) nerves (Smith and DeMyer, 2003; Sciacca et al., 2019). Thus, even a small and isolated lesion in the brainstem can manifest with devastating neurologic deficits (Querol-Pascual, 2010).

Imaging the brainstem can be challenging due to the variety of lesions that can affect this region and because several different structures with essential functions are next to each other in such a small volume (Guzmán-De-Villoria et al., 2010a). The brainstem can be imaged by several methods, such as computed tomography (CT), magnetic resonance imaging (MRI), ultrasound (US), digital subtraction angiography (DSA), and positron emission tomography. However, brainstem evaluation by CT is limited by bone-hardening artifacts, whereas US windows limit the sonography evaluation. Brain MRI is the most sensitive and specific imaging technique

*Correspondence to: Diogo Goulart Corrêa, MD, PhD, Department of Radiology, Clínica de Diagnóstico por Imagem (CDPI)/DASA, Avenida das Américas, 4666, Barra da Tijuca, Rio de Janeiro, RJ, 22640-102, Brazil. Tel: +55 (21) 99384-3617; E-mail: diogogoulartcorrea@yahoo.com.br.

for evaluating the anatomy and diagnosing disorders of the brainstem, although there are technical challenges in evaluating the brainstem using MRI (Gass et al., 2000).

This chapter aims to review the brainstem's normal anatomy on imaging exams, with emphasis on MRI, and highlight the main diseases that affect it.

IMAGING TECHNIQUE

Differently from the cerebral hemispheres, the brainstem has a highly compacted arrangement, with many small nuclei and small-diameter white matter tracts. Therefore, imaging methods lack spatial resolution to detect all the functional organization of the brainstem. Most brainstem nuclei are indistinguishable on MRI images because of the similar signal intensity, especially at 1.5- or 3-Tesla (T). Furthermore, the brainstem moves more than the cerebrum during the imaging procedure due to a complex interaction between the cerebrospinal fluid (CSF) flow, overlapping respiratory movements, and cardiac pulsations. Also, the intracranial posterior fossa, which contains the brainstem, is adjacent to bone, air spaces, and ligamentous structures, creating magnetic field inhomogeneities that degrade some MRI sequences, including higher magnetic fields (Shepherd and Hoch, 2022). Higher field–strength scanners are recommended to image the brainstem, since they provide an increased signal-to-noise ratio, allowing better anatomic details. Therefore, 3-T MRI is preferred to 1.5-T MRI when spatial and contrast resolution is crucial. Although 7-T MRI is promising for further gains, it remains investigational (Casselman et al., 2008).

Current MRI techniques allow us to evaluate some of the internal structures of the brainstem, making it possible to locate some of the intrinsic structures (Angeles Fernández-Gil et al., 2010). A standard MRI protocol to evaluate the brainstem includes multiplanar T1-weighted imaging before and after intravenous gadolinium injection, T2-weighted imaging, fluid-attenuated inversion recovery (FLAIR), susceptibility-weighted imaging (SWI), and diffusion-weighted imaging (DWI), including diffusion tensor imaging (DTI). Perfusion-weighted imaging and proton magnetic resonance (^{1}H-MR) spectroscopy may be added depending on the visualized lesions. The inclusion of a steady-state free precession sequence, such as constructive interference in steady state or fast imaging employing steady-state acquisition, is essential to evaluate the subarachnoid space, including the cisternal segments of the cranial nerves, and conditions affecting the CSF circulation, such as inside the cerebral aqueduct (Sarma et al., 2021). DTI with fiber tractography is an advanced imaging technique that provides qualitative and quantitative information about the micro-architecture of white matter tracts based on their location and direction, helpful in investigating the microstructural neuroarchitecture of brain tumors or brain malformations (Spalice et al., 2010; Poretti et al., 2013). Table 17.1 shows a suggested MRI protocol for patients with diseases affecting the brainstem.

Assessment of the brainstem using CT is very limited (Gass et al., 2000), beyond the bone-hardening artifacts, which generally project on the brainstem. Therefore, CT can only identify the surface anatomy of the brainstem and is unsuitable to identify any of the internal structures. Also, CT is not an adequate method to study the cisternal segments of the cranial nerves.

ANATOMY

A detailed description of the anatomy of the brainstem is beyond the objective of this chapter. Therefore, we will emphasize the anatomy identifiable on a clinical brain MRI.

Midbrain

The midbrain is the upper portion of the brainstem, connected with the forebrain, pons, and cerebellum. The cerebral or Sylvius aqueduct passes through it, draining the CSF from the third to the fourth ventricle, surrounded by the periaqueductal gray matter, which has a role in pain modulation and control of

Table 17.1

Suggested magnetic resonance imaging protocol for evaluating the brainstem.

MRI sequences	Plane
T1-weighted imaging	Sagittal
T2-weighted imaging	Axial and coronal
Fluid-attenuated inversion recovery	Axial
Diffusion-weighted imaging	Axial
Susceptibility-weighted imaging	Axial
T1-weighted imaging after intravenous gadolinium injection	Axial
Optional sequences	Plane
Diffusion tensor imaging (especially for patients with congenital mal formations)	Axial
Steady-state free precession sequence, such as three-dimensional constructive interference in steady state or fast imaging employing steady-state acquisition (to evaluate the cisternal segments of cranial nerves)	Axial
Perfusion-weighted imaging (to evaluate brain tumors)	Axial
T1-weighted neuromelanin-sensitive (for suspected cases of Parkinson disease)	Axial

emotional responses (Sciacca et al., 2019). In general, the longitudinal length of the ventral pons should be approximately twice that of the midbrain, from the ventral midbrain–pons junction to the third ventricle. Also, the longitudinal length of the midbrain should be roughly the same as that of the medulla, from the ventral pontomedullary to the obex. The posterior margin of the brainstem, from the caudal sylvian aqueduct to the obex, is a straight line (Fig. 17.1).

The ventral midbrain comprises the cerebral peduncles (crus cerebri), divided by a deep depression in the midline, the interpeduncular fossa. The corticospinal, corticopontine, and corticobulbar tracts pass through the cerebral peduncles. The corticospinal tract originates from the precentral gyrus, responsible for the voluntary movements of the limbs and trunk, occupying the middle three-fifths of the cerebral peduncle (Fig. 17.2). The corticopontine tract carries impulses from the frontal and parieto-occipital cortices to the pontine nuclei, which are transmitted to the contralateral cerebellum through the pontocerebellar tracts within the middle cerebellar peduncles, responsible for the coordination of planned movements. The corticobulbar tract, also known as the corticonuclear tract, innervates the motor cranial nerves nuclei, supplying upper motor innervation to the head and face (Ruchalski and Hathout, 2012; Sciacca et al., 2019).

The midbrain tegmentum is ventral to the cerebral aqueduct and dorsal to the cerebral peduncles. It contains the nuclei of the oculomotor (third) and trochlear (fourth) nerves, white matter tracts, and gray matter. Among the gray matter stand out the substantia nigra, which functions in motor control and reward pathways, and the red nucleus, which is part of the extrapyramidal system, with a role in motor coordination (Bae et al., 2013). On a specific 3-T high-resolution SWI sequence is possible to identify the "swallow tail" sign, which corresponds to a hyperintensity within the posterolateral substantia nigra, surrounded anterolaterally and medially by low signal intensity (Fig. 17.3). The linear signal hyperintensity of the "swallow tail" sign is credited to the normal nigrosome-1 (Frohman et al., 2008; Schwarz et al., 2014). The white matter tracts passing through the midbrain tegmentum and the length of the brainstem include the medial longitudinal fasciculus (a paired white matter tract located in the dorsomedial pontine and midbrain tegmentum, which connects the abducens nerve to the contralateral oculomotor nerve and allows concerted eye movements and conjugate gaze) (Frohman et al., 2008; Bae et al., 2013), medial lemniscus (which carries proprioceptive, vibratory, and touch–pressure sensations), lateral lemniscus (carrying auditory information to the auditory cortex), central tegmental tract (which contains ascending fibers connecting the rostral nucleus solitarius to the thalamus and cortical taste area), spinothalamic tract (carrying touch, pain, and temperature information) and the rubro-olivary tract (with

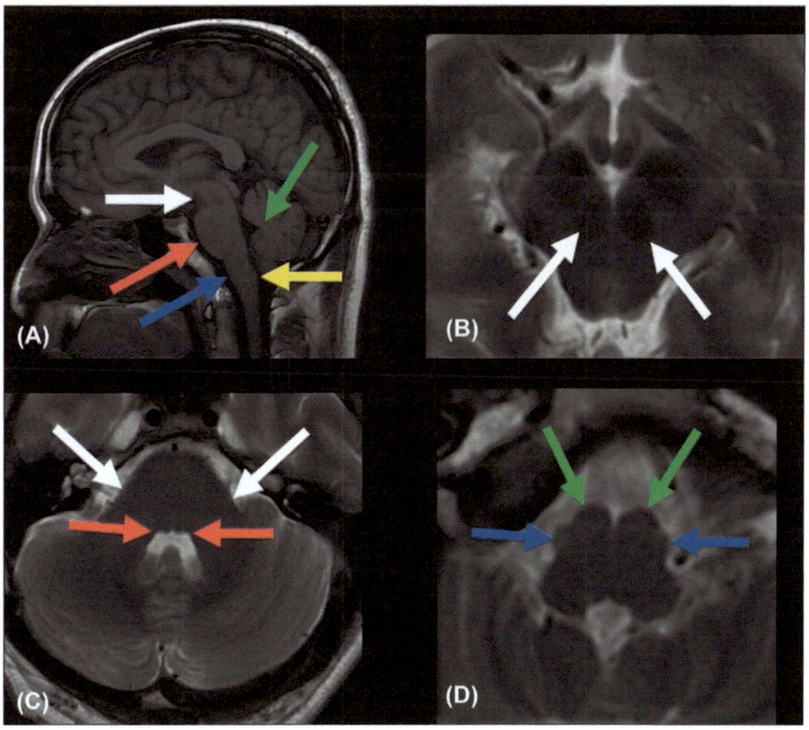

Fig. 17.1. Normal anatomy of the brainstem. (**A**) T1-weighted imaging in the midline of the sagittal plane, demonstrating the midbrain (white arrow in **A**), pons (red arrow in **A**), and medulla (blue arrow in **A**). Also note the normal fastigium of the fourth ventricle (green arrow in **A**) and the posterior margin of the brainstem, which is a straight line from the caudal sylvian aqueduct to the obex (yellow arrow in **A**). T2-weighted imaging in the midbrain axial plane at the superior colliculi level (**B**) demonstrates the red nuclei (arrows in **B**). T2-weighted imaging in the axial plane of the pons demonstrating the middle cerebellar peduncles (white arrows in **C**). Also, note the facial colliculi (red arrows in **C**). T2-weighted imaging in the axial plane of the medulla demonstrating the pyramids (green arrows in **D**) and the olives (blue arrows in **D**).

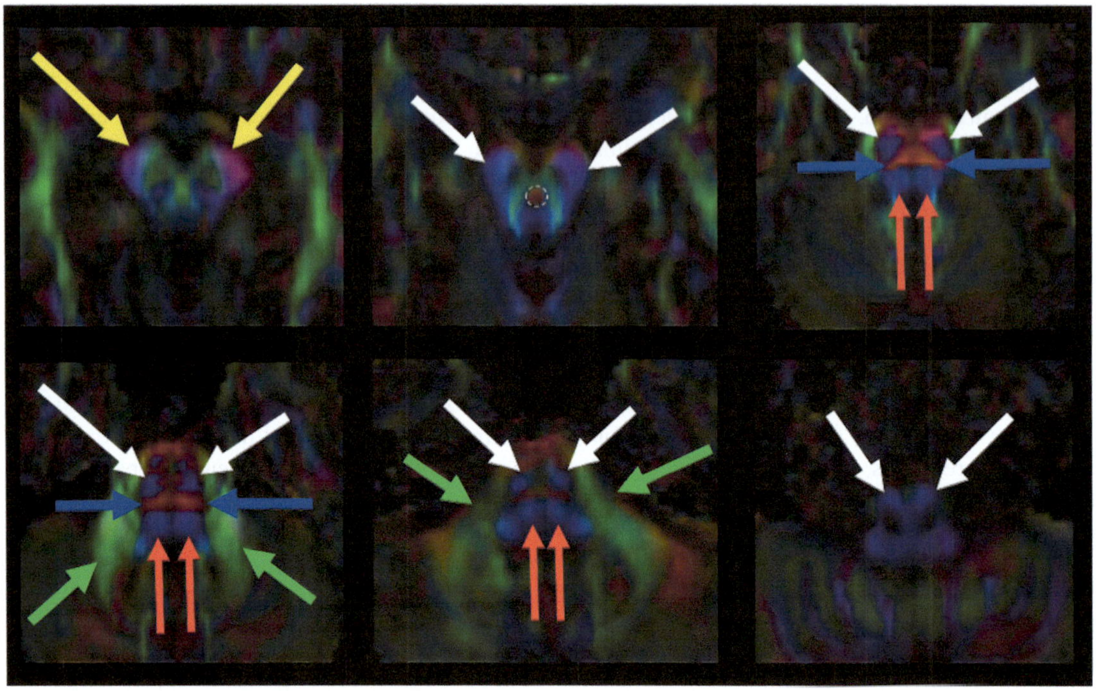

Fig. 17.2. Diffusion tensor imaging color-coded map of a healthy person in the axial plane. Locations of white matter tracts are assigned on color maps. The direction of the main fiber tracts is represented by red (right–left), green (anterior–posterior), and blue (superior–inferior). Several fibers on color maps are annotated based on anatomic knowledge. Yellow arrows: cerebral peduncles. White arrows: corticospinal tracts. Blue arrows: Transverse pontine fibers. Red arrows: central tegmental tracts. Green arrows: middle cerebellar peduncles. Dashed circle: superior cerebellar peduncles decussation.

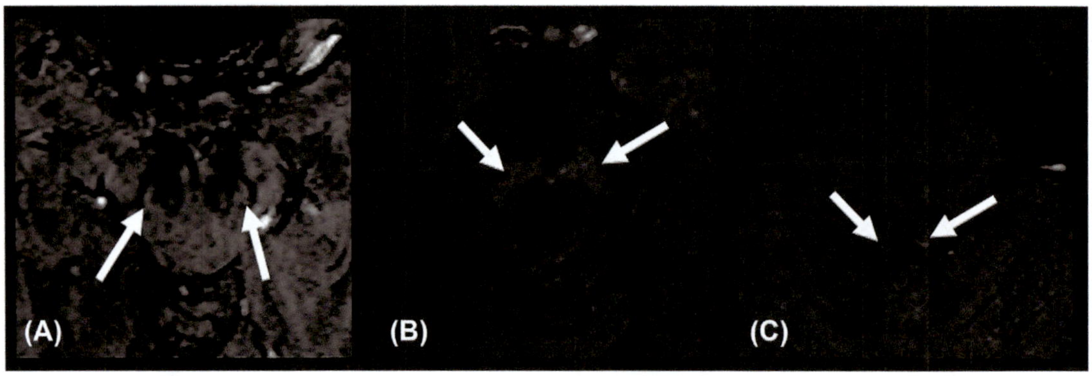

Fig. 17.3. Normal substantia nigra at 3-T MRI. High-resolution susceptibility-weighted imaging demonstrates hyperintensity within the posterolateral substantia nigra (arrows in **A**), surrounded anterolaterally and medially by low signal intensity ("swallow tail" sign). The T1-weighted neuromelanin-sensitive sequence demonstrates hyperintense signals in the substantia nigra (arrows in **B**) and locus coeruleus (arrows in **C**), regions containing this pigment.

descending fibers from the red nucleus to the inferior olivary nucleus) (Sciacca et al., 2019).

The main oculomotor nuclear complex (composed of five motor nuclei) is located in the midbrain, at the level of the superior colliculi, in a paramedian position, ventral to the cerebral aqueduct, partially in the periaqueductal gray matter, and dorsal to the medial longitudinal fasciculus and red nucleus (Fig. 17.4). From the main oculomotor nuclear complex originate the somatic efferent fibers of the third cranial nerve, which innervate the superior rectus, medial rectus, inferior rectus, inferior oblique, and levator palpebrae superioris muscles, being responsible for most of the eye movements. Immediately dorsal and medial to the main oculomotor nuclear complex lies the Edinger–Westphal nucleus, part of the parasympathetic system responsible for the accommodation and light reflex. The oculomotor nerves emerge from the oculomotor sulcus in the ventral midbrain, at the medial border of the cerebral peduncles, to enter into the interpeduncular cistern (Mark, 1998; Corrêa et al., 2022).

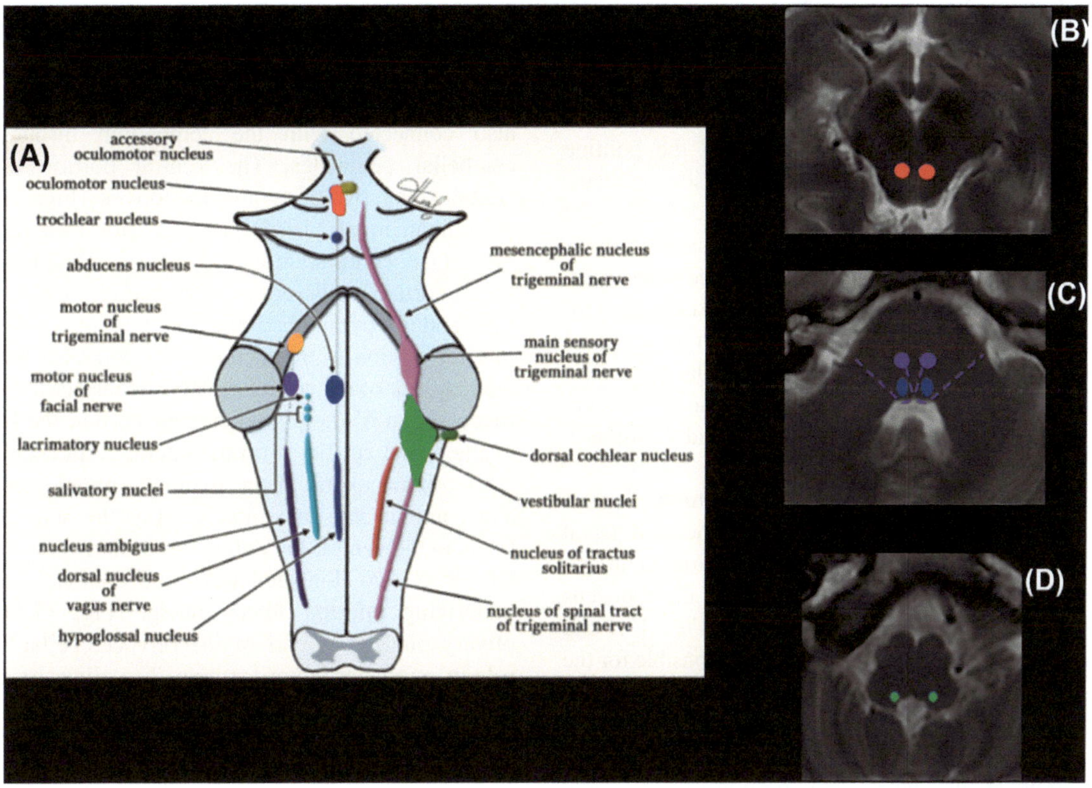

Fig. 17.4. Normal anatomy of the brainstem. The drawing of the brainstem nuclei in the longitudinal view (**A**) demonstrates the motor nuclei on the left and the sensory nuclei on the right of the drawing. Although MRI does not demonstrate the cranial nerves nuclei, the oculomotor nerve nuclear complex is located in a paramedian position, ventral to the cerebral aqueduct, partially in the periaqueductal gray matter, at the level of the superior colliculi (red circles in **B**). Also note the facial motor nuclei (purple circles in **C**) and the course of the facial nerve fibers within the brainstem (purple dashed lines in **C**) curving laterally around the abducens nucleus (blue circles in **C**), forming the facial colliculi in the floor of the fourth ventricle. At the level of the medulla, note the pyramids and olives and part of the vestibular nuclei (green circles in **D**).

The trochlear nerve nucleus lies in the midbrain tegmentum's dorsal surface, at the inferior colliculus level, just caudal to the oculomotor nucleus, ventral to the cerebral aqueduct. The intra-axial segment of the trochlear nerves crosses the midline before exiting the dorsal surface of the brainstem, inferior to the inferior colliculus. This nerve is responsible for the motor innervation of the superior oblique muscle (Freddi, 2022).

The tectum is the dorsal portion of the midbrain, posterior to the cerebral aqueduct. Its posterior surface has two paired eminences: the superior and inferior colliculi. The superior colliculi have roles in the visual pathway and eye movement control, projecting fibers to the lateral geniculate nucleus of the thalamus. The inferior colliculi are part of the auditory pathway, projecting fibers to the medial geniculate nucleus of the thalamus, which relays auditory information to the primary auditory cortex (Angeles Fernández-Gil et al., 2010).

Pons

The pons is situated between the midbrain and the medulla. Its name is due to the appearance of its anterior surface in the axial plane, resembling a bridge. The pons also connects the brain to the cerebellum through the middle cerebellar peduncles. The pons' anterior portion is convex and presents a shallow groove (the basilar groove) in which the basilar artery is lodged. The medullopontine sulcus is located between the pons and the medulla, where the roots of abducens, facial, and vestibulocochlear nerves emerge from medial to lateral, respectively. The posterior surface of the pons forms the superior half of the floor of the fourth ventricle (Angeles Fernández-Gil et al., 2010; Sciacca et al., 2019).

Longitudinal fibers from the corticospinal, corticobulbar, and corticopontine tracts pass through the ventral portion of the pons. The pontine tegmentum contains the nuclei of the trigeminal (fifth cranial nerve), abducens (sixth), facial (seventh), and vestibulocochlear (eighth) cranial nerves, as well as white matter tracts, such as the medial longitudinal fasciculus, medial lemniscus, lateral lemniscus, spinothalamic tract, central tegmental tract, and trapezoid body (part of the auditory pathway) (Angeles Fernández-Gil et al., 2010).

The trigeminal nerve is composed of sensory and motor fibers. The sensory portion of the trigeminal nerve presents three nuclei: (1) the main sensory nucleus, located in the pons; (2) the spinal nucleus, which is the inferior continuation of the main sensory nucleus, extending from the lower pons through the medulla to the level of C2–C3, where it becomes continuous with the dorsal horn of the spinal cord; and (3) the mesencephalic nucleus, which is the superior extension of the main sensory nucleus, through the midbrain, situated along the anterolateral portion of the periaqueductal gray matter, near the superior cerebellar peduncle, ventral to the inferior colliculi (Borges and Casselman, 2010). The motor nucleus of the trigeminal nerve lies on the lateral surface of the pontine tegmentum, medial and anterior to the main sensory nucleus. The main sensory nucleus receives fibers for touch sensations and pressure on the head and face. The spinal nucleus receives fibers responsible for pain and temperature of the head and face. The mesencephalic nucleus receives fibers responsible for proprioceptive information from the masticator muscles, hard palate, teeth, periodontium, and temporomandibular joint. The motor nucleus of the trigeminal nerve provides motor innervation to the muscles of mastication (Woolfall and Coulthard, 2001; Freddi et al., 2022).

The abducens nerve nuclei are located adjacent to the floor of the fourth ventricle. On MRI in the axial plane, the facial colliculus (a small lump in the floor of the fourth ventricle, produced by the facial nerve axons looping around the abducens nerve nucleus) is a mark of the abducens nucleus (Figs. 17.1 and 17.4). The abducens nerve is purely motor and responsible for the lateral rectus muscle motor innervation (Lucio et al., 2022).

The facial nerve presents sensory, motor, and parasympathetic fibers. The main motor nucleus is located deep in the reticular formation, at the ventrolateral portion of the caudal pontine tegmentum, next to the spinal nucleus of the trigeminal nerve. The fibers course upward and curve laterally around the abducens nucleus (internal genu of the facial nerve), forming the facial colliculus in the floor of the fourth ventricle and going slightly ventrolaterally to exit the brainstem in the medullopontine sulcus at the cerebellopontine angle (Fig. 17.4) (Al-Noury and Lotfy, 2011). The motor fibers of the facial nerve are responsible for the innervation of the facial muscles. The sensory and parasympathetic fibers of the facial nerve emerge from the brainstem through the nervus intermedius of Wrisberg, in the medullopontine sulcus, between the facial nerve proper and the vestibulocochlear nerve. In the internal auditory canal, the nervus intermedius forms a unique nerve with the motor segment of the facial nerve. The parasympathetic nuclei of the facial nerve are the superior salivatory and lacrimal nuclei, posterolateral to the main motor nucleus, in the inferior pons. They are responsible for the motor parasympathetic fibers of the submandibular, sublingual, and lacrimal glands. The sensory nucleus of the facial nerve is the upper part of the nucleus of the tractus solitarius, which is located posterolaterally to the facial nerve motor nucleus. It receives taste inputs from the anterior two-thirds of the tongue via the chorda tympani nerve (Sarma et al., 2021; Ottaiano et al., 2023).

The vestibulocochlear nerve is purely sensitive. It has six nuclei in the lower portion of the pons (two cochlear nuclei – dorsal and ventral; and four vestibular nuclei – superior, inferior, lateral, and medial) along the lateral floor of the fourth ventricle. The cochlear portion of this nerve is responsible for transmitting afferent information related to hearing, whereas the vestibular portion of this nerve is related to afferent fibers responsible for balance (Corrêa et al., 2023a).

Medulla

The medulla or medulla oblongata is the most inferior portion of the brainstem. It connects the pons with the spinal cord. It is also connected with the cerebellum through the inferior cerebellar peduncles. The ventral portion of the medulla contains the pyramids and the olives (Figs. 17.1 and 17.4) (Flannigan et al., 1985). The anterior median fissure of the medulla is continuous with the anterior median fissure of the spinal cord. This fissure is in the anterior midline, between the right and left pyramids containing the corticospinal tracts. In the inferior portion of the medulla, the pyramids become thin, and the majority of the corticospinal tract fascicles cross the midline (decussation of the pyramids) and go into the lateral funiculus of the spinal cord to form the lateral corticospinal tract. A few corticospinal axons in the pyramids remain on the same side and continue into the spinal cord as the anterior corticospinal tracts in the ventral funiculus. Posterolateral to the pyramids are the olives, two elongated and rounded elevations shaped by the underlying inferior olivary nuclei (Fig. 17.1). The inferior olivary nucleus is part of the olivocerebellar system, with a role in cerebellar motor learning. The olives are separated from the pyramids by the preolivary groove. Posterior to the olives are located in the postolivary grooves. The inferior cerebellar peduncles are posterior to the postolivary grooves (Angeles Fernández-Gil et al., 2010; Sciacca et al., 2019).

The tegmentum contains the nuclei of the glossopharyngeal (ninth), vagus (tenth), accessory (eleventh), and hypoglossal (twelfth) cranial nerves. The medial longitudinal fasciculus, medial lemniscus, spinothalamic, central tegmental, and spinocerebellar tracts also pass through the medullary tegmentum. The obex is located in the posterior surface of the medulla as a thin band of neural tissue, extending transversely between the lateral nuclei of the area postrema (Quisling et al., 1993). In the sagittal plane, the obex forms the dorsal roof of the central canal aperture, and its tip defines the inferior margin of the foramen of Magendie. The aperture of the central canal and the tip of the obex can be seen on conventional midsagittal MRI (Milhorat et al., 1993). Immediately caudal and laterally to the obex lie the gracile tubercles, separated by the posterior median sulcus, produced by the underlying gracile nucleus. Lateral to the gracile tubercle is the cuneate tubercle, produced by the underlying cuneate nucleus (Quisling et al., 1993; Angeles Fernández-Gil et al., 2010).

The glossopharyngeal and vagus nerves are sensory and motor. Also, these nerves share some of their nuclei and functions. Both nerves have three nuclei: (1) the main motor nucleus, (2) the sensory nucleus, and (3) the parasympathetic nucleus (Angeles Fernández-Gil et al., 2010). The motor nucleus of the glossopharyngeal nerve is the superior portion of the ambiguus nucleus, which lies deep in the reticular formation of the medulla, supplying the stylopharyngeus muscle (Lucio and Freddi, 2023). The main sensitive nucleus of the glossopharyngeal nerve is the nucleus of the tractus solitarius, which receives gustatory afferent fibers posterior one-third of the tongue and afferent impulses from baroreceptors of the carotid sinus via this nerve.

Although the visceral afferent fibers, responsible for the general sensory innervation of the posterior third of the tongue, uvula, pharynx, auditory tube, and external auditory canal travel through the glossopharyngeal nerve, their fibers synapse with the spinal nucleus of the trigeminal nerve. The parasympathetic nucleus of the glossopharyngeal nerve is the inferior salivary nucleus, whose postganglionic fibers innervate the parotid gland (Policeni and Smoker, 2008; Angeles Fernández-Gil et al., 2010).

The motor nucleus of the vagus nerve is the inferior portion of the ambiguus nucleus, innervating the soft palate, pharyngeal constrictor muscle, intrinsic muscles of the larynx, and palatoglossus muscle of the tongue (Soldatos et al., 2014; Lucio and Freddi, 2023). The solitary tract nucleus also receives gustatory fibers from the epiglottis through the vagus nerve. Sensory information from the tympanic membrane, external auditory canal, external ear, and meninges of the posterior fossa also travel through the vagus nerve, but terminate in the trigeminal spinal nucleus. The parasympathetic fibers of the vagus nerve originate from the dorsal motor nucleus of the vagus nerve, which lies lateral to the hypoglossal nucleus, providing innervation to the involuntary muscles of the heart, bronchi, esophagus, stomach, liver, pancreas, and intestines (Traylor and Branstetter, 2022; Lucio and Freddi, 2023).

The accessory nerve is purely motor, innervating the sternocleidomastoid and trapezius muscles. A cranial and spinal rootlets union forms it. The medullary motor fibers originate in the nucleus ambiguus and the spinal motor fibers in the spinal nucleus of the accessory nerve. The glossopharyngeal, vagus, and cranial rootlets of the accessory nerves emerge from the lateral medulla in the postolivary groove, posteriorly to the olive (Lucio and Freddi, 2023).

The hypoglossal nerve is purely motor, supplying all the intrinsic muscles of the tongue, the styloglossus, hyoglossus, and genioglossus muscles. The hypoglossal nucleus is immediately beneath the floor of the inferior portion of the fourth ventricle. The hypoglossal nucleus produces a bulge into the floor of the fourth ventricle, named hypoglossal eminence. The hypoglossal nerve rootlets emerge from the medulla in the preolivary groove anteriorly to the olive (de Sousa Costa et al., 2023).

CONGENITAL MALFORMATIONS

Congenital malformations are nonprogressive, congenital morphologic abnormalities of an organ or body part, caused by intrinsic alterations in the cellular and molecular pathways involved in the organogenesis developmental program, secondary to gene mutations, teratogens, or combined effects (Hennekam et al., 2013). The complex development of the posterior fossa, including the cerebellum and brainstem and the high number of involved genes, results in a broad spectrum of malformations. Generally, the diagnostic criteria for posterior fossa malformations are based on neuroimaging findings, usually classified into three gross patterns: (1) predominantly cerebellar, (2) predominantly brainstem, and (3) cerebellar and brainstem malformations (Poretti et al., 2016).

Predominantly brainstem malformations

Pontine tegmental cap dysplasia

Pontine tegmental cap dysplasia is a rare sporadic brainstem congenital malformation with an unknown genotype and no familial recurrence. Patients usually present with hearing loss, facial paralysis, trigeminal anesthesia, difficulty in swallowing, and ataxia. Vertebral segmentation anomalies, rib malformations, and congenital heart defects can also be seen. Cognitive development varies, ranging from mild delay to severe disability (Barth et al., 2007).

Neuroimaging findings are pathognomonic, characterized by a flattened ventral pons, vaulted pontine tegmentum, with a rounded bump (the cap aspect) or an angular protrusion with partial absence of the middle cerebellar peduncles and absence of the inferior cerebellar peduncles. Vermian hypoplasia, molar tooth sign-like aspect of the pontomesencephalic junction, and absent inferior olivary prominence are found in some patients (Fig. 17.5) (Jissendi-Tchofo et al., 2009). Duplicated internal auditory canals and hypoplastic cranial nerves may be present (Desai et al., 2011). DTI may show the absence of the transverse pontine fibers and the presence of a dorsal transverse axonal band at the level of the cap in the dorsal pons. The basal pons is connected to the cerebellar hemispheres through misoriented fibers in the middle cerebellar peduncles. The decussation of the superior cerebellar peduncles is missing (Barth et al., 2007; Jissendi-Tchofo et al., 2009).

Horizontal gaze palsy with progressive scoliosis

Horizontal gaze palsy with progressive scoliosis is also a rare brainstem congenital malformation secondary to an autosomal recessive mutation in the ROBO3 gene, which encodes a receptor required for axonal guidance (Bosley et al., 2005). Affected patients present with progressive scoliosis and congenital absence of horizontal eye movements, but preservation

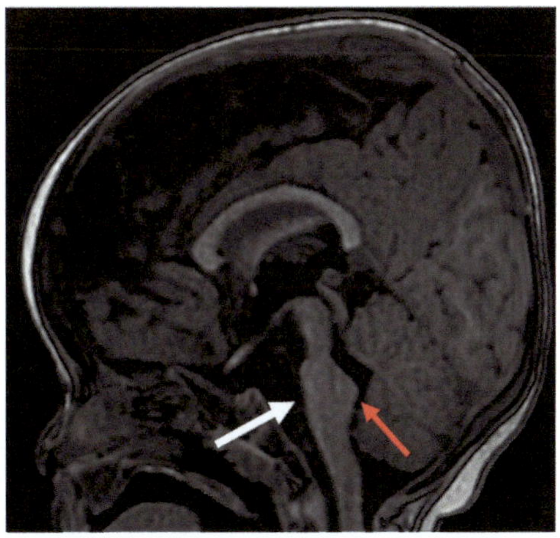

Fig. 17.5. A 1-year-old boy with pontine tegmental cap dysplasia. Brain MRI showed a flattened ventral pons (white arrow), associated with a posteriorly vaulted pontine tegmentum (the cap aspect – red arrow), on sagittal T1-weighted imaging.

of vertical gaze and convergence. Cognitive functions are typically preserved (Severino and Huisman, 2019).

A butterfly-shaped medulla characterizes neuroimaging due to the missing prominence of the gracile and cuneate nuclei, pontine hypoplasia with a dorsal midline cleft, absence of the facial colliculi in the dorsal pons, and prominence of inferior olivary nuclei (Fig. 17.6) (Rossi et al., 2004). DTI may demonstrate the absence of the decussation of the corticospinal tracts in the medulla, superior cerebellar peduncles, and pontine sensory tracts, which may result in a lack of integration of proprioceptive stimuli, which is hypothesized to be related with the development of scoliosis. Tractography of the corpus callosum is normal (Haller et al., 2008; Wahl and Mukherjee, 2009).

Diencephalic–mesencephalic junction dysplasia

Diencephalic–mesencephalic junction dysplasia presents with an abnormal dorsoventral enlargement and abnormal butterfly-like contour of the midbrain on axial sections. Clinical presentation includes microcephaly, severe cognitive impairment, axial hypotonia, and spastic tetraparesis. Additional imaging features are pontocerebellar hypoplasia, ventriculomegaly, calcifications, abnormalities of the corticospinal tracts, hypoplasia or agenesis of the corpus callosum and olfactory bulbs, and agenesis, hypoplasia, or fusion of the basal ganglia (Zaki et al., 2012; Severino and Huisman, 2019).

This malformation can be classified as Type A, which is characterized by complete hypothalamic-mesencephalic fusion, with dorsoventral enlargement and abnormal contour of the midbrain on axial planes, with a possible midbrain ventral cleft or linear T2-hyperintensities; and Type B, which is characterized by incomplete thalamic-mesencephalic cleavage on sagittal plane, with parenchymal bands connecting the interthalamic adhesion with the midbrain, variably associated with brainstem and/or vermian hypoplasia.

Patients with Type B diencephalic–mesencephalic junction dysplasia are less neurologically impaired, usually with a mild developmental delay (Severino et al., 2016).

Brainstem disconnection

Brainstem disconnection is defined by the nearly complete absence of a brainstem segment, with intact rostral and caudal portions connected only by a thin cord of tissue. Both malformative and disruptive mechanisms have been proposed in its pathophysiology. The prognosis is poor, and most patients die within the first 2 months after birth (Jurkiewicz et al., 2010). Typically, patients are symptomatic at birth, with absent or weak suck and swallowing, central respiratory insufficiency, increased or decreased muscle tone, reduced or poor visual fixation, seizures, and unstable body temperature. Imaging findings are pathognomonic, with the absence of a brainstem portion, with thin strands connecting the remaining portions, usually associated with cerebellar hypoplasia. The basilar artery is also absent in most patients. Congenital cardiac abnormalities, hydronephrosis, and vertebral body anomalies are also common (Poretti et al., 2007; Jurkiewicz et al., 2010).

Cerebellar and brainstem malformations

Pontocerebellar hypoplasia

Pontocerebellar hypoplasia is used as a descriptive term, referring to a reduction in the volumes of both pons and cerebellum. However, this term embraces a heterogeneous group of prenatal-onset neurodegenerative disorders, mainly affecting the pons and cerebellum (van Dijk et al., 2018). Currently, there are 17 different types listed in Online Mendelian Inheritance in Man (OMIM, 2023). Characteristic MRI features are (1) cerebellar hypoplasia with superimposed atrophy, (2) absence or significant reduction of the pontine prominence, (3) normal or slightly reduced posterior fossa size, and (4) variable cerebral involvement, including atrophy and delayed myelination (Fig. 17.7) (Severino and Huisman, 2019). The description of the different types of pontocerebellar hypoplasia is beyond the scope of this chapter. The interested reader should go to specific texts (Rüsch et al., 2020).

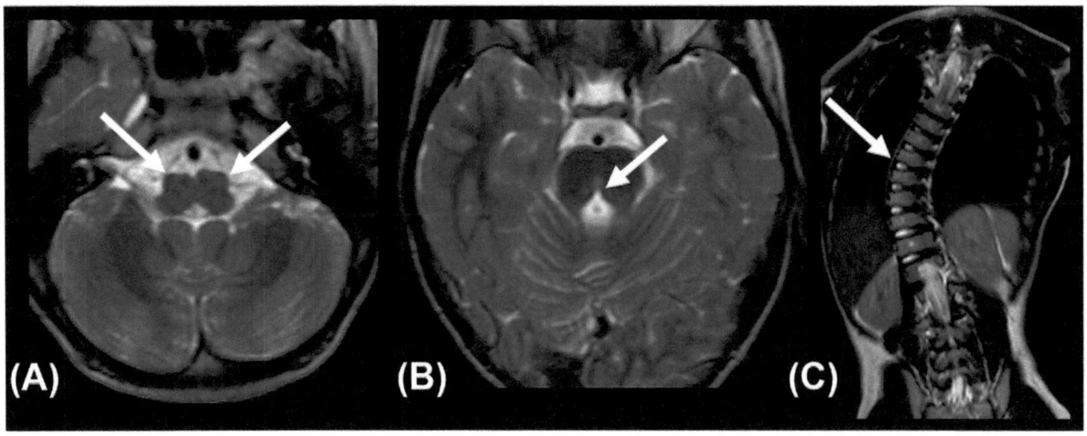

Fig. 17.6. A 4-year-old boy with horizontal gaze palsy with progressive scoliosis. Brain MRI demonstrated a butterfly-shaped medulla on axial T2-weighted imaging (arrow in **A**) and pontine hypoplasia with a dorsal midline cleft (arrow in **B**). Dorsal spinal MRI demonstrated a scoliosis (arrow in **C**) on coronal T2-weighted imaging.

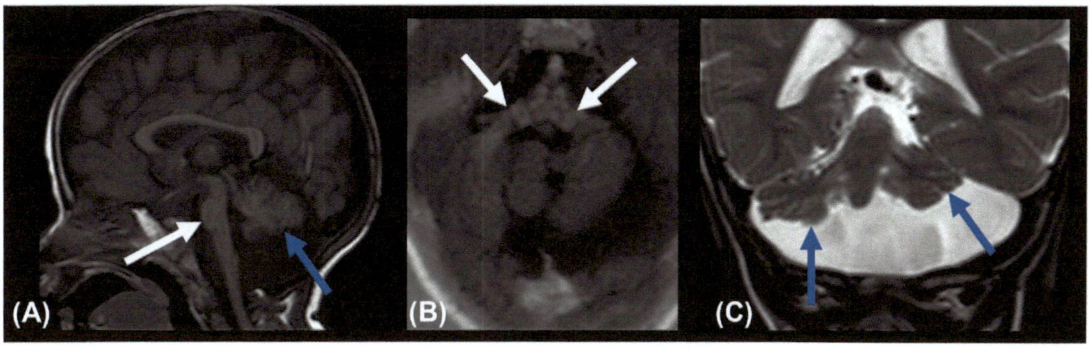

Fig. 17.7. A 3-year-old girl with pontocerebellar hypoplasia. Brain MRI showed a reduction of the ventral pontine prominence on T1-weighted imaging in the sagittal plane (white arrow in **A**), associated with reduced volume of the middle cerebellar peduncles in the axial plane (arrows in **B**), as well as cerebellar hypoplasia, seen on T2-weighted imaging in the coronal plane (arrows in **C**) and T1-weighted imaging (blue arrow in **A**). Also, note the normal posterior fossa size.

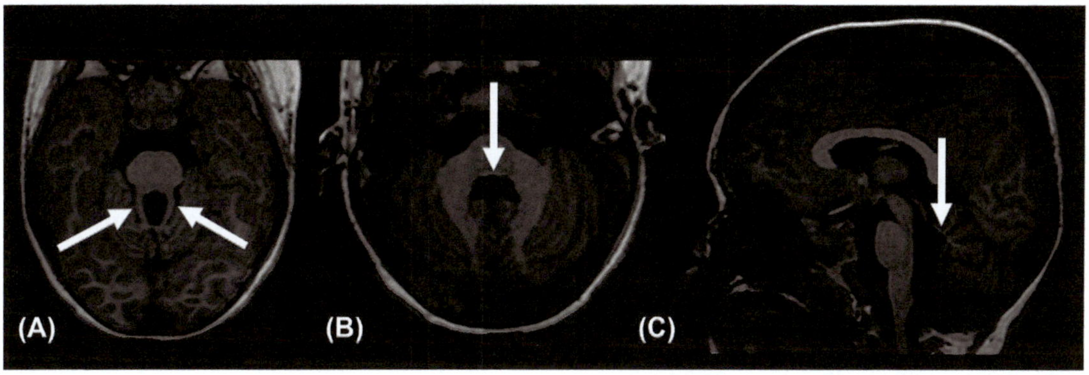

Fig. 17.8. A 2-year-old boy with Joubert syndrome. Brain MRI demonstrated thickened and elongated superior cerebellar peduncles, resulting in a "molar tooth" appearance on T1-weighted imaging in the axial plane (arrows in **A**). Note the elongated fourth ventricle. At an inferior level, the fourth ventricle has a bat-wing appearance (arrow in **B**). In the sagittal plane, the fastigium appears elongated and rounded (arrow in **C**). The vermis was hypoplastic and clefted.

The morphologic pattern of the pons and cerebellum hypoplasia is not specific to pontocerebellar hypoplasia. There are several disorders with pontocerebellar hypoplasia-like imaging aspects not listed in OMIM as pontocerebellar hypoplasia. It has been demonstrated in several conditions, such as in extreme prematurity, neurometabolic diseases (e.g., congenital disorder of glycosylation type 1A), and patients with mutations in the calcium/calmodulin-dependent serine protein kinase (CASK), RELN, and Very Low Density Lipoprotein Receptor (VLDLR) genes (Rüsch et al., 2020).

Molar tooth malformation

Molar tooth malformation presents with a characteristic hypoplastic cerebellar vermis with thickened, elongated, and horizontal superior cerebellar peduncles, resulting in a "molar tooth" appearance, on axial MRI or CT images at the lower midbrain level (Jissendi-Tchofo et al., 2015). In an inferior level, the hypoplastic cerebellar vermis, with the cerebellar hemispheres set side by side, gives the fourth ventricle the appearance of a bat with the wings outstretched. Molar tooth malformation is mandatory for diagnosing Joubert syndrome and related disorders (Fig. 17.8). This group of disorders presents with hypotonia or ataxia, developmental delay, ocular motor apraxia, and breathing abnormalities, associated with involvement of other organs such as the retina, kidneys, liver, and the orofacial region (Parisi, 2009). DTI may demonstrate an absence of both normal transverse fibers in the central cerebellar vermis, superior cerebellar peduncles decussation, and corticospinal tracts decussation, in addition to the lateral location of dysmorphic deep cerebellar nuclei and hypoplastic medial lemnisci (Jissendi-Tchofo et al., 2015). Some patients present additional asymmetrical cerebral peduncles, due to unilateral absence of the corticospinal tract in its usual location and anterior displacement of the transverse pontine fibers, characterizing the "decayed molar tooth" sign (Scortegagna et al., 2024).

INFLAMMATORY DISEASES

Any inflammatory disease can potentially affect the brainstem, such as Sjögren syndrome, lupus erythematosus, and sarcoidosis, among others, often associated with supratentorial lesions, with or without contrast enhancement. This section will address the imaging aspect of the main inflammatory diseases, including demyelinating diseases that affect the brainstem.

Multiple sclerosis

Multiple sclerosis commonly affects the brainstem. Typically, these lesions have a significant clinical correlation due to the scarce functional resilience of white matter tracts and gray matter nuclei in the brainstem (Quattrocchi et al., 2010).

Multiple sclerosis lesions present with hyperintense signal on T2-weighted imaging, most commonly distributed in the pons, followed by the midbrain and the medulla.

The infratentorial compartment is one of the four locations considered in the dissemination in space criteria for the diagnosis of multiple sclerosis (Fig. 17.9). Identification of the typical supratentorial multiple sclerosis lesions, specifically in the periventricular white matter and the callososeptal interface, and cortico/subcortical region, as well as in the spinal cord, is essential for the diagnosis (Thompson et al., 2018). Furthermore, the presence of brainstem lesions increases the risk of conversion to multiple sclerosis in patients with clinically isolated syndromes (Guzmán-De-Villoria et al., 2010a).

The multiple sclerosis lesions are typically located in the periphery of the pons and in the intraparenchymal tracts of the trigeminal nerves. The middle cerebellar peduncles are also frequently involved, more commonly than the cerebellar hemispheres (Weidauer et al., 2021). Lesions in the brainstem may also present gadolinium enhancement when active, similarly to supratentorial lesions. However, they are less likely to evolve to pronounced hypointense signals on T1-weighted imaging ("black holes") (Guzmán-De-Villoria et al., 2010a).

Neuromyelitis optica spectrum disorder

Neuromyelitis optica spectrum disorder (NMOSD) is an inflammatory demyelinating disease of the central nervous system. It is characterized by severe attacks of optic neuritis, typically bilateral, affecting the optic chiasm, and longitudinally extensive transverse myelitis, typically affecting three or more spinal cord levels (Wingerchuk et al., 2015). In addition, hyperintense lesions on T2-weighted imaging/FLAIR, with or without gadolinium enhancement in the brain may also be seen. Area postrema syndrome, with episodes of otherwise unexplained hiccups or nausea and vomiting, and acute brainstem syndrome are core clinical characteristics present in the International Consensus Diagnostic Criteria for NMOSD. Brain lesions are frequently located in areas associated with high aquaporin-4 expression, such as periependymal areas surrounding the ventricular system. Therefore, in the brainstem, midbrain lesions surrounding the cerebral aqueduct and dorsal brainstem lesions adjacent to the fourth ventricle, including the area postrema, are characteristic (Fig. 17.10). Also, a rostral extension of cervical myelitis into the brainstem is common (Fig. 17.11) (Wingerchuk et al., 2015; Renard et al., 2022).

Myelin oligodendrocyte glycoprotein antibody-associated disorder

Myelin oligodendrocyte glycoprotein antibody-associated disorder (MOGAD) is a distinct nosologic entity considered an oligodendropathy, with specific diagnostic and therapeutic criteria (Banwell et al., 2023). The most common feature is optic neuritis, usually affecting a long segment of the anterior portion of the optic nerves, while the chiasm and retrochiasmatic portions are spared. Optic perineuritis is a specific and often present finding characterized by perineural enhancement on MRI.

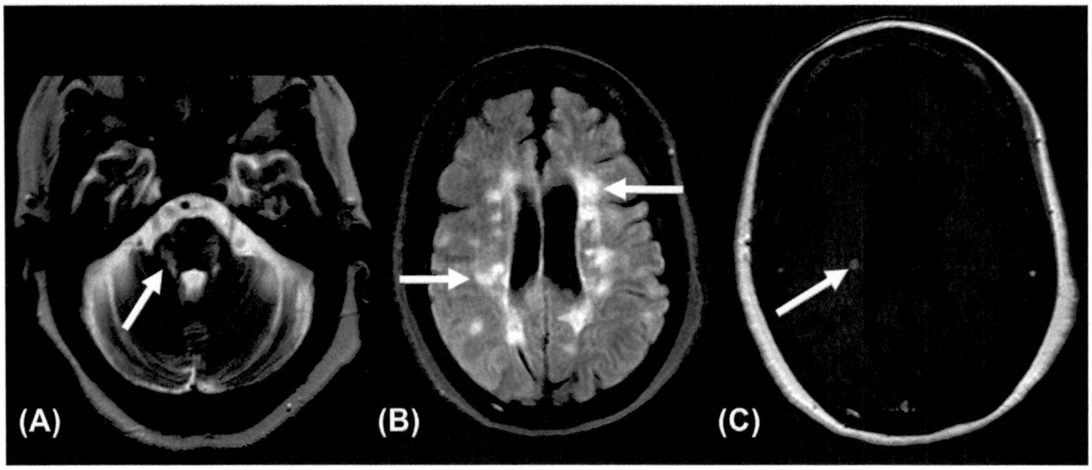

Fig. 17.9. A 44-year-old woman with multiple sclerosis. Brain MRI demonstrated multiple hyperintense lesions in the pons, including in the entry zone of the right trigeminal nerve root, seen on axial T2-weighted imaging (arrow in **A**). Note the typical periventricular supratentorial lesions on axial fluid-attenuated inversion recovery (arrows in **B**) and that one lesion presented gadolinium enhancement, due to active inflammatory demyelination (arrow in **C**).

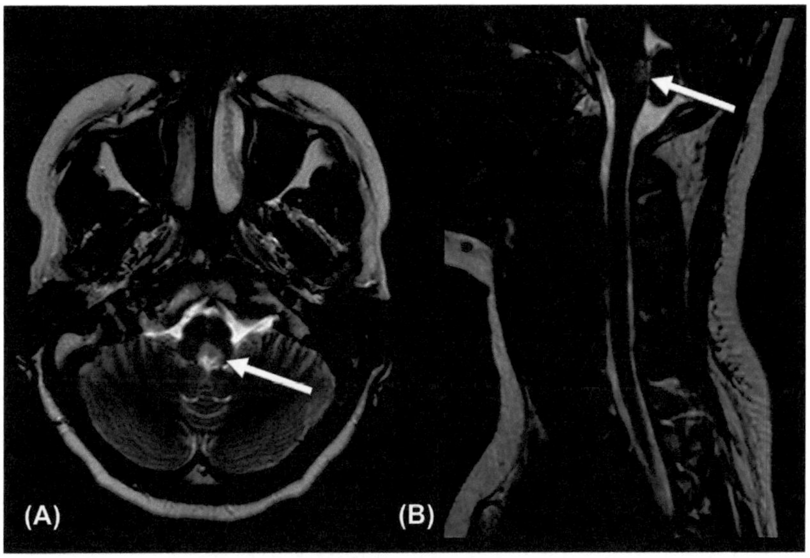

Fig. 17.10. A 40-year-old woman with neuromyelitis optica spectrum disorder presented with vomiting and uncontrolled hiccups. Brain MRI revealed a lesion with hyperintense signal on T2-weighted imaging in the area postrema in the axial (arrow in **A**) and sagittal planes (arrow in **B**). The patient also had positive anti-aquaporin-4 antibodies in the serum.

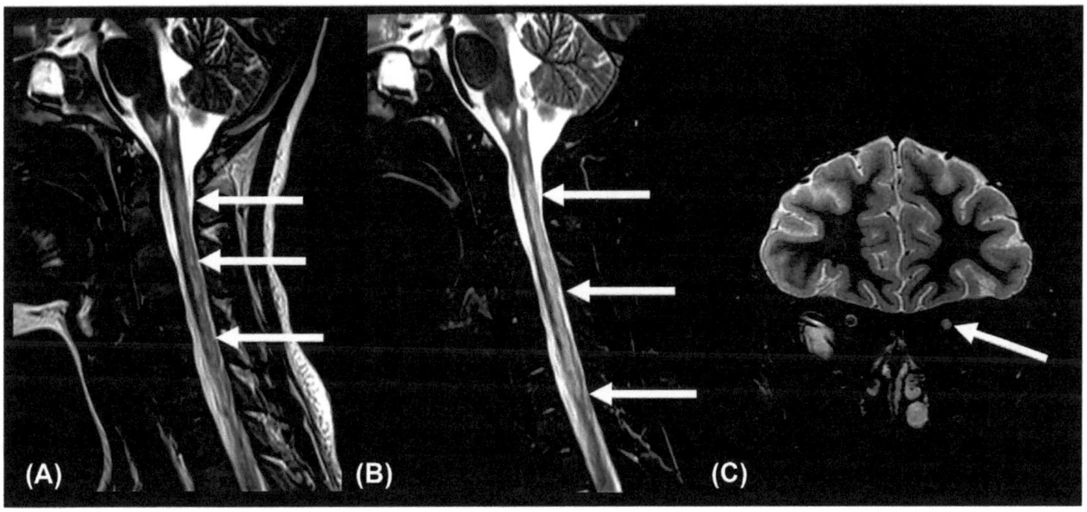

Fig. 17.11. A 20-year-old woman with neuromyelitis optica spectrum disorder. Brain MRI demonstrated longitudinally extensive myelitis, with rostral extension into the brainstem, including the area postrema, seen on sagittal T2-weighted imaging (arrows in **A**) andShort Tau Inversion Recovery (STIR) (arrows in **B**). The patient also had optic neuritis in the left optic nerve (arrow in **C**). The patient had positive anti-aquaporin-4 antibodies in the serum.

Typically, MOGAD myelitis involves the conus medullaris, which is considered specific for the diagnosis (Dutra et al., 2018). Although less common, it can be associated with cerebral cortical encephalitis, brainstem, or cerebellar symptoms.

MOGAD may present similar brainstem lesions compared with NMOSD. The imaging appearance of brainstem lesions does not reliably distinguish MOGAD patients from those with NMOSD (Banwell et al., 2023).

Acute disseminated encephalomyelitis

In general, acute disseminated encephalomyelitis (ADEM) is a monophasic and multifocal inflammatory demyelinating disease associated with encephalopathy and a typical temporal relationship with infection or vaccination that occurred 1–3 weeks before the clinical onset (Pohl et al., 2016; Tardieu et al., 2016). Lesions vary from multifocal punctate to large flocculent or pseudo-tumoral lesions. Gadolinium enhancement can be absent, punctate, ring, or incomplete ring in all or nearly all lesions. The imaging aspect of brainstem lesions partially overlaps with multiple sclerosis, and these two entities can be indistinguishable on MRI. Still, ADEM has a higher frequency of midbrain lesions, and lesions tend to be more symmetrical and bilateral than multiple sclerosis (Fig. 17.12). Also, ADEM affects the corpus callosum and periventricular white matter less

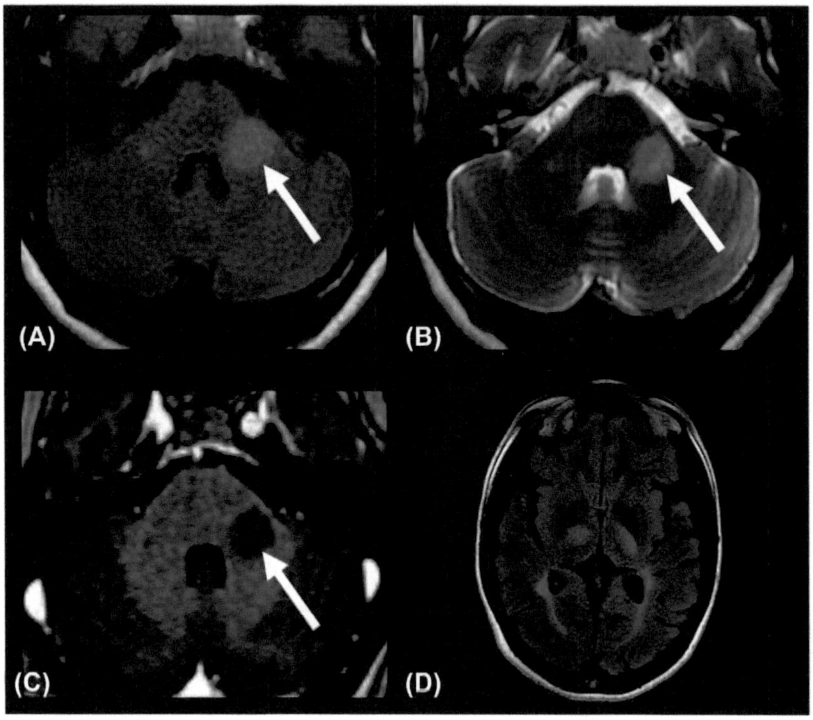

Fig. 17.12. A 10-year-old girl with a recent history of respiratory infection and encephalopathy after two weeks due to acute disseminated encephalomyelitis. Brain MRI revealed a hyperintense lesion on fluid-attenuated inversion recovery (arrow in **A**) and T2-weighted imaging (arrow in **B**) in the right pons/middle cerebellar peduncle, with no gadolinium enhancement (arrow in **C**). There were also relatively symmetrical lesions in periventricular white matter and nucleocapsular regions (**D**).

frequently than multiple sclerosis, and thalamic and basal ganglia lesions are more frequent in ADEM (Quattrocchi et al., 2016).

Chronic lymphocytic inflammation with pontine perivascular enhancement responsive to steroids

Chronic lymphocytic inflammation with pontine perivascular enhancement responsive to steroids (CLIPPERS) is an inflammatory syndrome that predominantly compromises the pons. It has characteristic lesions on neuroimaging consisting of multiple small, punctuate and curvilinear gadolinium-enhancing foci centered in the pons and brachium pontis, variably extending to the midbrain and medulla. Diagnostic criteria include (1) homogenous, gadolinium-enhancing nodules without ring enhancement or mass effect predominating in the pons and cerebellum, measuring < 3 mm in diameter and a pattern consistent with perivascular involvement, (2) marked improvement after corticosteroid treatment, (3) signal abnormality on T2-weighted imaging not significantly exceeding the size of the area with gadolinium enhancement (suggestive of absence of vasogenic edema), and (4) spinal cord lesions with similar T2 and gadolinium-enhancing lesions (Tobin et al., 2017).

Behçet's disease

Behçet's disease is a relapsing and multisystem vasculitis, which can affect any part of the central nervous system, with a propensity to affect the brainstem (Borhani-Haghighi et al., 2020). Acute and subacute lesions present with hyperintense signal on T2-weighted MRI, usually with gadolinium enhancement. The most commonly affected regions are the mesodiencephalic junction (Fig. 17.13) and the pons/medulla transition. Chronically, the lesions tend to decrease in size or resolve completely, but the brainstem may become atrophic (Koçer et al., 1999; Lee et al., 2001).

INFECTIOUS DISEASES

Rhombencephalitis refers to inflammation in the pons, medulla, and cerebellum. Although the midbrain is not a component of the rhombencephalon, as it is continuous with the pons, it is often involved and included as a part of rhombencephalitis (Wasenko et al., 2002). The etiology is often difficult to determine, but infections are the most common cause. The two most frequent agents causing infectious rhombencephalitis are *Listeria monocytogenes* and *Mycobacterium tuberculosis*, among bacteria, and enterovirus 71, herpes simplex virus 1 and 2, Epstein–Barr virus, and human herpes virus 6, among viruses. However, the infectious agent is often not identified (Wasenko et al., 2002; Sotoudeh et al., 2021).

Although the imaging findings depend on the infectious agent and the disease stage, infectious rhombencephalitis usually presents with lesions with hyperintense signal on T2-weighted imaging with or without gadolinium enhancement and restricted diffusion. Associated leptomeningeal gadolinium

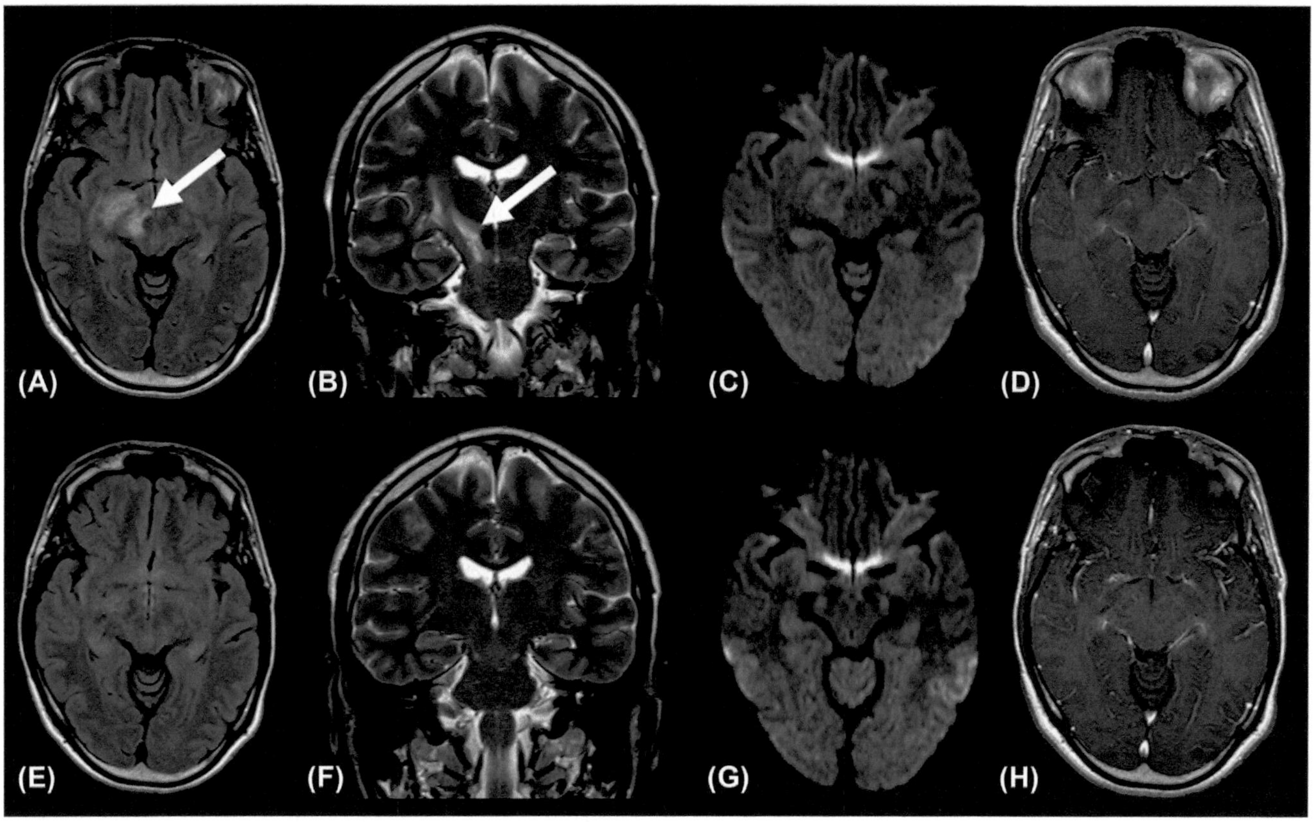

Fig. 17.13. A 30-year-old man with a history of recurrent oral and genital ulcers and Behçet's disease diagnosis. Brain MRI showed a lesion in the right portion of the brainstem, extending through the mesodiencephalic junction, affecting the midbrain, cerebral peduncle, and right basal ganglia, with the hyperintense signal on fluid-attenuated inversion recovery (FLAIR) (arrow in **A**) and T2-weighted imaging (arrow in **B**), with no restricted diffusion (**C**) and no gadolinium enhancement (**D**). A follow-up MRI 3 months after treatment demonstrated complete resolution of the lesion on FLAIR (**E**), T2-weighted imaging (**F**), diffusion-weighted imaging (**G**), and T1-weighted imaging with gadolinium (**H**).

enhancement seen on T1-weighted imaging or postcontrast FLAIR can also be seen (Corrêa et al., 2022; Renard et al., 2022).

Other imaging features of infectious diseases are granulomas, characterized as a hypodense lesion with irregular margins on CT and as a lesion with predominant hypointense signal on T1- and T2-weighted imaging, with peripheral or nodular gadolinium enhancement on MRI (Fig. 17.14). Abscesses are characterized as expansive lesions with a central hypointense signal on T1-weighted imaging (but slightly hyperintense related to CSF), a central hyperintense signal on T2-weighted imaging and FLAIR, and a thin peripheral capsule that has a hypointense signal on T2-weighted imaging. The peripheral capsule also presents with gadolinium enhancement (Aiken, 2010). The central portion of the lesion has restricted diffusion, which is related to the pus inside the cavity.

Furthermore, brainstem infarction may result from a secondary vasculitis to infections, such as tuberculosis, syphilis, and mucormycosis, among others (Corrêa et al., 2023b; Nunes et al., 2024). It is essential to consider that the etiology of brainstem encephalitis remains unknown in about 30% of cases despite extensive clinical investigation (Fig. 17.15) (Sotoudeh et al., 2021).

METABOLIC DISEASES

Posterior reversible encephalopathy syndrome

Posterior reversible encephalopathy syndrome (PRES) is a clinicoradiologic syndrome that most commonly affects the cortex and subcortical white matter of the parieto-occipital regions and is typically reversible. However, despite its name, the frontal lobes, the brainstem and basal ganglia can also be affected (Bartynski, 2008). Also, although most cases are reversible, 19% of patients die and there is a varying degree of long-term impairments, including epilepsy, and motor deficits have been reported in 44% of patients, and even recurrences may be seen. The exact pathophysiology of PRES remains unknown, but several hypotheses have been proposed, mainly focusing on blood flow dysregulation due to abnormal cerebrovascular tone and endothelial dysfunction. Other mechanisms, such as immune system dysregulation, may also play a role (Bartynski, 2008; Battal and Castillo, 2024).

In PRES, the most common imaging finding is bilateral and symmetric vasogenic edema in the cortex and subcortical white matter, centered at the watershed areas. The distribution is not confined to a specific vascular territory and commonly involves

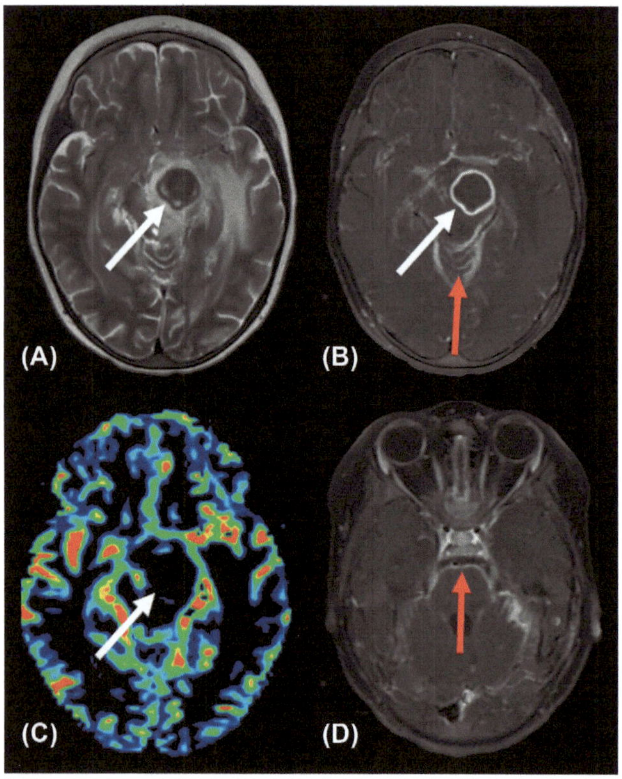

Fig. 17.14. A 40-year-old HIV-positive man presented with reduced level of consciousness and an epileptic seizure due to tuberculous meningitis associated with a granuloma due to a tuberculoma. Brain MRI showed an expansive lesion in the left cerebral peduncle/left thalamus, with hypointense signal on T2-weighted imaging (arrow in **A**), peripheral gadolinium enhancement (arrow in **B**), with peripheral edema, with no signs of hyperperfusion (arrow in **C**). Note the leptomeningeal enhancement at the basal cisterns, notably in the prepontine cistern and the cerebellar folia (red arrows in **B** and **D**).

the posterior parieto-occipital regions (Battal and Castillo, 2024). Beyond the classic parieto-occipital pattern, vasogenic edema around the superior frontal sulcus and central pattern can occur, affecting the brainstem and/or basal ganglia. Pons is the most frequently affected portion of the brainstem in PRES, followed by the midbrain and the medulla (Fig. 17.16) (McKinney et al., 2013; Li et al., 2020).

Furthermore, several atypical/rare patterns have been documented, including holohemispheric and asymmetrical cases. In addition to atypical distributions, some rarer imaging characteristics have also been reported, such as restricted diffusion, hemorrhages, and contrast enhancement, and should not rule out the hypothesis of PRES in the correct clinical context (Battal and Castillo, 2024).

Wernicke encephalopathy

Wernicke encephalopathy occurs secondary to vitamin B1 (thiamine) deficiency. The classical clinical triad is ocular dysfunction (nystagmus, conjugate gaze palsies or ophthalmoplegia), ataxia, and altered mental status, but it is only observed in 30% of patients (Geibprasert et al., 2010). The major mechanisms include inadequate nutritional intake, decreased absorption of vitamin B1, and decreased thiamine utilization in the cells. It is typically related to alcohol abuse but is also related to nonalcoholic patients, associated with hyperemesis gravidarum, cancer, bariatric surgery, hunger strikes, excess soft drink intake in children, depression, Crohn's disease, schizophrenia, anorexia nervosa, ulcerative colitis, and incidental thiamine-deficient infant formula (Oudman et al., 2021). If left untreated, patients may develop Korsakoff psychosis, characterized by memory loss and confabulation. These clinical findings are related to damage to the mammillary bodies, anterior thalamic nuclei, and interruption of the diencephalic-hippocampal circuits. Rapid thiamine repositioning has to be done to prevent the otherwise high mortality rates (Geibprasert et al., 2010). MRI of patients with Wernicke encephalopathy demonstrates symmetric lesions with signal hyperintensity on T2-weighted imaging and FLAIR in the periventricular area, such as the mammillary bodies, dorsomedial thalami, tectal plate, fornix, and periaqueductal region (Fig. 17.17). Less frequently involved areas include the cerebellar dentate nuclei, the cerebellar peduncles, the vermis, the corpus callosum, and the caudate nucleus. Involvement of the red nuclei and substantia nigra in the midbrain or diffuse brainstem lesions can sometimes be observed (Sciacca et al., 2019; Renard et al., 2022).

Osmotic demyelination syndrome

Osmotic demyelination syndrome can occur secondary to rapid osmotic gradient change, such as rapid iatrogenic correction of hyponatremia. It consists of the destruction of the myelin sheaths, diagnosed in patients with pronounced plasma osmolality alterations. Pons is the most commonly affected region, but extrapontine myelinolysis may also occur, affecting the cerebellum, lateral geniculate body, external and extreme capsules, hippocampus, putamen, cerebral cortex/subcortical white matter, thalamus, and caudate, in descending order of frequency.

The lesions are hyperintense on FLAIR and T2-weighted imaging and hypointense on T1-weighted imaging, with restricted diffusion, in the acute phase. Contrast enhancement is rare. The pontine lesion usually involves the transverse pontocerebellar, and pontine midline structures, sparing the corticospinal tracts and ventrolateral aspect of the pons (trident sign) (Fig. 17.18). Imaging alterations may occur within a certain delay after the onset of clinical symptoms, usually characterized by lethargy, quadriplegia, dysarthria, ophthalmoplegia, dysphagia, ataxia, and altered reflexes, generally beginning 2–7 days after the start of electrolyte correction, or may even be seen in the absence of clinical abnormalities. DWI may demonstrate alterations before T1- and T2-weighted imaging (Guzmán-De-Villoria et al., 2010b).

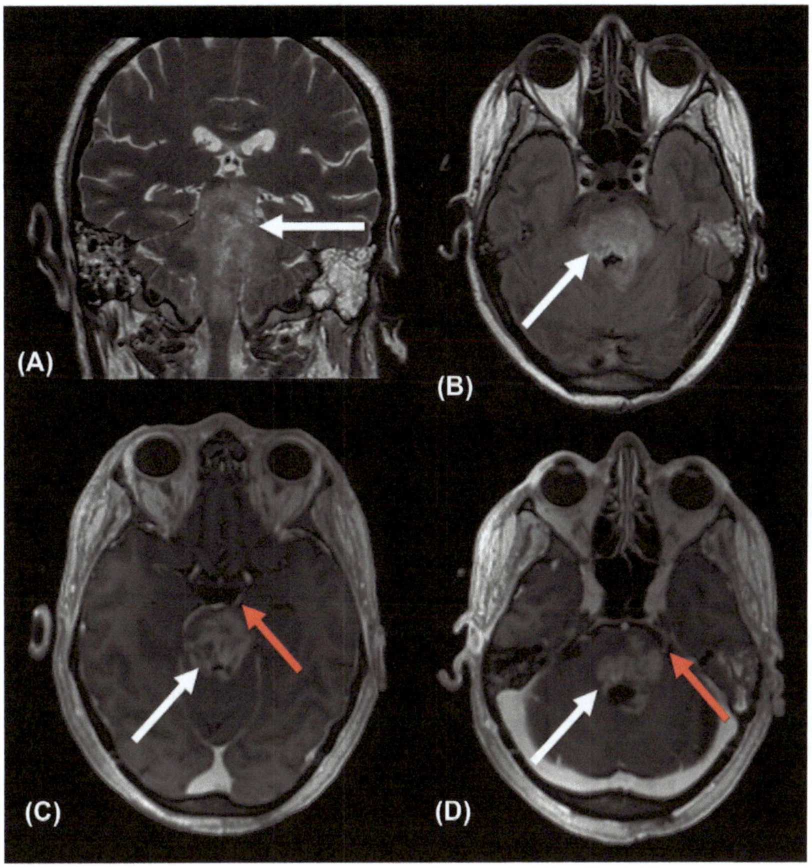

Fig. 17.15. A 26-year-old woman with idiopathic rhombencephalitis presented with multiple cranial nerve palsy and a reduced level of consciousness. Brain MRI revealed a hyperintense signal all along the brainstem on T2-weighted imaging (arrow in **A**) and fluid-attenuated inversion recovery (arrow in **B**), associated with gadolinium enhancement (white arrows in **C** and **D**). Also note the enhancement along the left oculomotor and trigeminal nerves cisternal segments (red arrows in **C** and **D**, respectively). Despite an extensive investigation, the etiology was not determined. Biopsy showed a mononuclear inflammatory process, with no evidence of neoplasm. Cultures of the surgical specimen were negative.

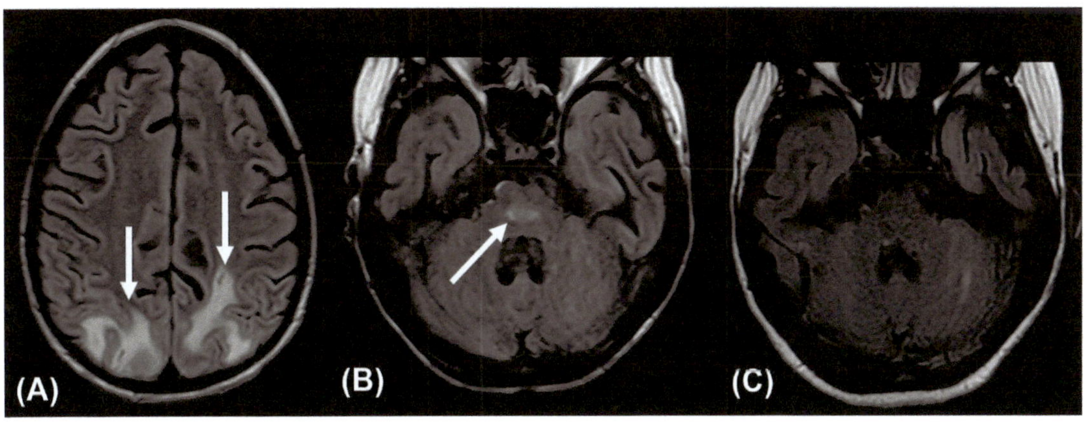

Fig. 17.16. A 32-year-old woman in the puerperium, with preeclampsia, diagnosed with posterior reversible encephalopathy syndrome. Brain MRI demonstrated parieto-occipital vasogenic edema (arrows in **A**) and signal abnormality in the pons (arrow in **B**), seen on fluid-attenuated inversion recovery. After blood pressure control, there was an improvement in the imaging findings (**C**).

Wilson's disease

Wilson's disease, also known as hepatolenticular degeneration, is a rare autosomal recessive inherited disorder caused by ATP7B gene mutations, with a lack of ceruloplasmin, resulting in errors of copper transport and accumulation in the liver and deposition in other organs, including the brain. The diagnosis is usually established by low ceruloplasmin levels and Kayser–Fleischer ring around the iris, associated with increased urine copper and

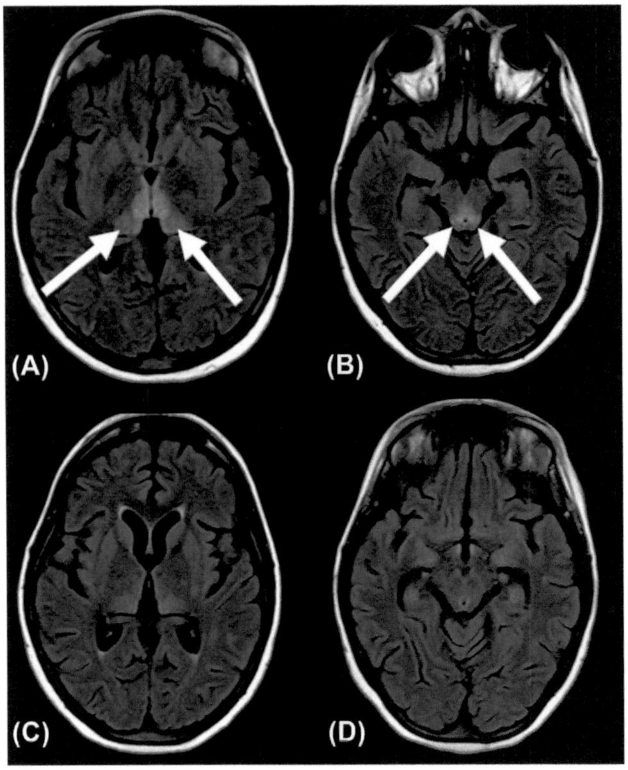

Fig. 17.17. A malnourished, 60-year-old alcoholic man with Wernicke's encephalopathy. Brain MRI revealed symmetric lesions with hyperintense signal on fluid-attenuated inversion recovery in the dorsomedial thalami (arrows in **A**) and periaqueductal area in the midbrain (arrows in **B**). A follow-up MRI 2 months after thiamine reposition demonstrated partial resolution of the lesions (**C** and **D**).

genetic testing. Neurologic symptoms, characterized by extrapyramidal and behavioral symptoms, typically begin in the second or third decade of life (Van Cauter et al., 2020).

Neuroimaging demonstrates bilateral and symmetric deep gray matter signal hyperintensity on T2-weighted imaging and FLAIR. Restricted diffusion may be seen in the early stage. Putamen is the most frequently affected structure, followed by caudate, thalamus, midbrain, pons, cerebral white matter, cortex, medulla, and cerebellum (Sinha et al., 2006). Hyperintensity in the midbrain, with sparing of red nuclei, substantia nigra, and superior colliculi produces the "face of the giant panda" sign (Fig. 17.19), which is virtually pathognomonic for Wilson disease, but is present in a minority of patients (Van Cauter et al., 2020). Bilateral and symmetric hyperintense signal on T1-weighted imaging in the globus pallidus results from manganese deposition. The imaging alterations regress with copper chelating treatment (Ranjan et al., 2015).

NEOPLASMS

Gliomas

Diffuse midline glioma

Brainstem gliomas account for 10%–20% of central nervous system tumors in childhood. The most common primary tumor in the pons is diffuse midline glioma that usually occurs between 5 and 6 years of age, occupying more than 75% of the pons, with a tendency to extend to the cerebellar peduncles and midbrain.

Diffuse midline glioma is considered a central nervous system grade 4 tumor, irrespective of necrosis or microvascular proliferation. Previous studies demonstrated that H3 K27M-mutation, seen in approximately 70%–80% of diffuse midline gliomas, defines a clinically and biologically distinct subgroup, with a poor prognosis, similar to glioblastomas. This led to a newly defined entity in the 2016 World Health Organization (WHO) classification, "diffuse midline glioma, H3 K27M-mutant." However, alternative mechanisms that alter the pathogenic pathway in this tumor have been described since then. Thus, in the 2021 WHO classification, the terminology has been revised to "diffuse midline glioma, H3 K27-altered," with four subtypes of diffuse midline gliomas that are defined by the driving oncohistone alteration: (1) H3.3 p.K28M (K27M)-mutant, (2) H3.1 or 3.2 p.K28M (K27M)-mutant, (3) H3-wildtype with EZHIP overexpression, and (4) EGFR-mutant. Although diffuse midline glioma, H3 K27-altered, is a pediatric-type glioma, it is also frequently diagnosed in young adults and less frequently in older patients. Then, H3 K27M alteration should be tested in diffuse gliomas that involve the midline, including the brainstem, even in older adults (Louis et al., 2021; Park et al., 2023).

Furthermore, these tumors do not occur exclusively in the pons. They can be found in the thalamus, cerebellum, spinal cord, and other midline structures, including the third ventricle, hypothalamus, and pineal region (Johnson et al., 2023).

Diffuse midline glioma is a locally invasive tumor that, because of its location, cannot always be biopsied. Therefore, its diagnosis is often based on imaging aspects. In general, patients with pontine gliomas present with an onset of symptoms related to the brainstem compromise, specifically presenting with a clinical triad of cranial neuropathies, deficits related to long white matter tracts impairment and ataxia (Paldino et al., 2011).

On CT, it appears as a hypodense lesion with an infiltrative appearance, poorly defined limits, and epicenter in the pons. Calcification is rare and may correlate with a better prognosis. On MRI, these tumors appear as an expansive lesion with an infiltrative appearance with heterogeneous hyperintense signal on T2-weighted imaging, determining an increased pons volume. If there is anterior growth of the lesion, the tumor may embrace the basilar artery, which is characteristic. Generally, they present faint, focal gadolinium enhancement (Fig. 17.20) (Johnson et al., 2023; Park et al., 2023).

At the time of diagnosis, usually they do not present with restricted diffusion, hyperperfusion, or a significant increase in choline peak on spectroscopy. However, some tumors can reveal these characteristics along the clinical course or treatment (Paldino et al., 2011).

Low-grade gliomas

Although diffuse midline glioma can occur in the midbrain and medulla, the most common type of primary tumor in these

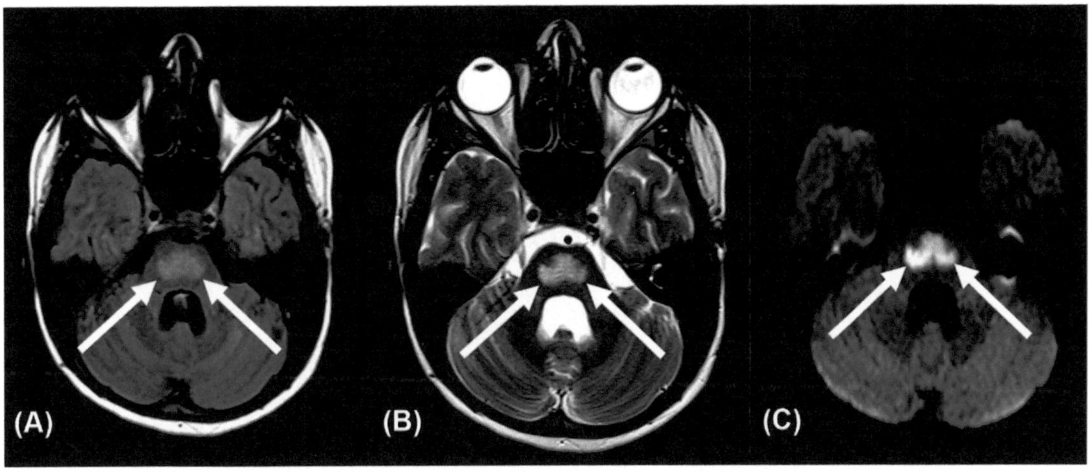

Fig. 17.18. A 13-year-old boy with pontine osmotic demyelination after rapid correction of hyponatremia presented with a reduced level of consciousness. Brain MRI showed hyperintense signal on axial fluid-attenuated inversion recovery (**A**) and T2-weighted imaging (**B**) in the pons, affecting the pontine midline structures (arrows in **A** and **B**), sparing the corticospinal tracts and ventrolateral aspect of the pons (trident sign). The lesion presented restricted diffusion due to cytotoxic edema (arrows in **C**).

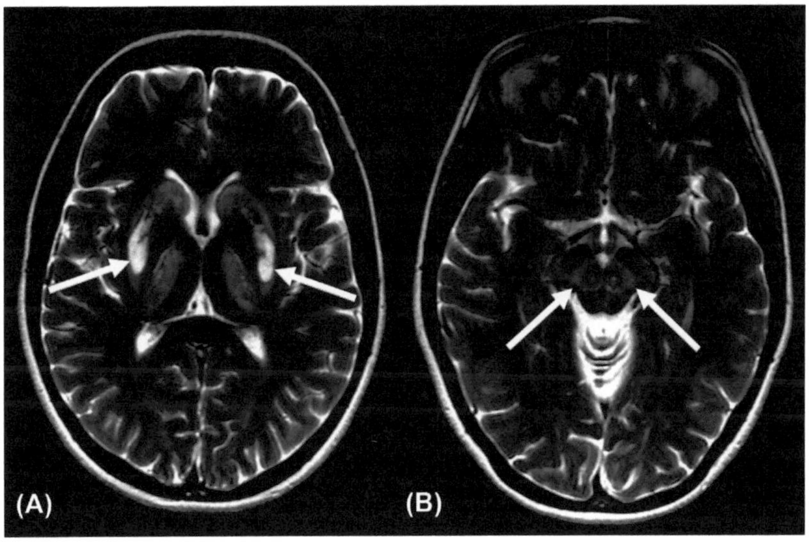

Fig. 17.19. A 36-year-old man with a diagnosis of Wilson disease, associated with liver cirrhosis and Kayser–Fleischer rings. Brain MRI demonstrated a hyperintense signal on T2-weighted imaging in the bilateral putamen, caudate, and internal capsule (arrows in **A**). There was also a hyperintense signal on T2-weighted imaging in the midbrain, sparing the red nuclei, substantia nigra, and superior colliculi, determining the "face of the giant panda" sign (arrows in **B**).

regions is pilocytic astrocytoma. These tumors tend to show a low degree of aggressiveness and may present an exophytic growth. Attention should be given to midbrain tectal glioma, which is a specific midbrain location of these tumors that usually presents with signs and symptoms of increased intracranial pressure resulting from obstruction of the cerebral aqueduct and Parinaud syndrome (characterized by paralysis of upward gaze, absence of pupillary reaction to light, eyelid retraction, and gaze convergence disorders). Like other low-grade gliomas, they present as a hypodense lesion, on CT, and with hypointense signal on T1-weighted imaging and hyperintense signal on FLAIR and T2-weighted imaging, with heterogeneous contrast enhancement, without restricted diffusion or hyperperfusion on MRI (Figs. 17.21 and 17.22) (Igboechi et al., 2013; Ramos et al., 2013).

Ependymoma

Ependymal tumors are a heterogeneous group of neuroepithelial neoplasms, classified based on histologic and molecular characteristics, and location. There are seven types of ependymal cell tumors according to the 2021 WHO classification of tumors of the central nervous system (supratentorial ependymoma, ZFTA fusion–positive; supratentorial ependymoma, YAP1

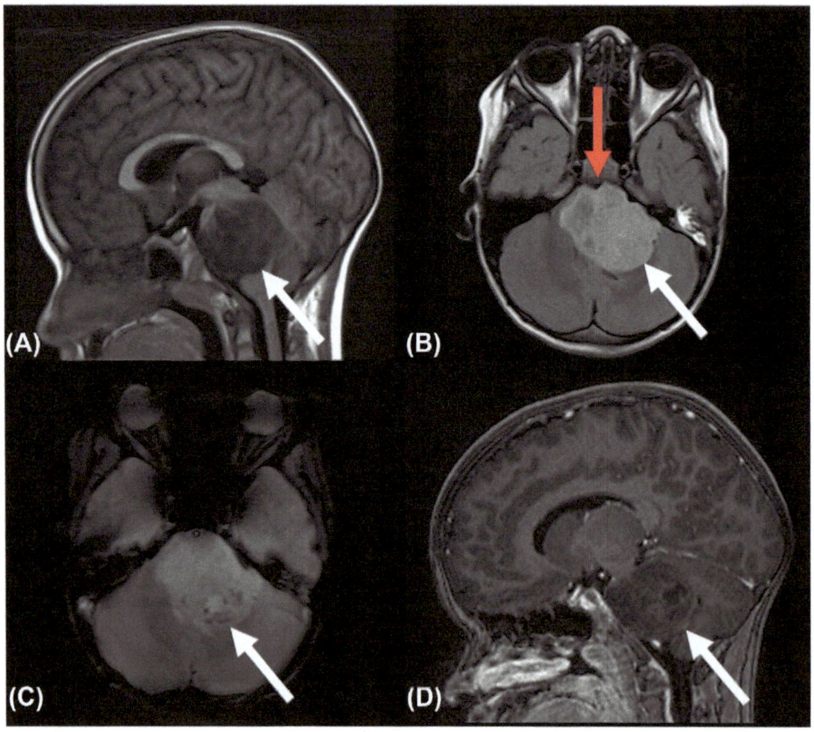

Fig. 17.20. A 7-year-old boy with a pontine diffuse midline glioma. Brain MRI demonstrated an expansive lesion centered in the pons, with hypointense signal on sagittal T1-weighted imaging (arrow in **A**), hyperintense signal on axial fluid-attenuated inversion recovery (white arrow in **B**), with hypointense foci on axial susceptibility-weighted imaging (arrow in **C**), due to micro-hemorrhages, with no gadolinium enhancement (arrow in **D**). Note how the lesion embraces the basilar artery (red arrow in **B**).

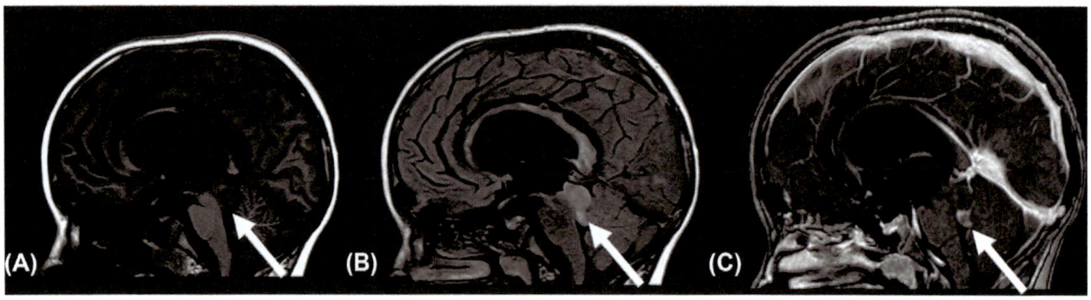

Fig. 17.21. A 9-year-old boy with a tectal glioma. Brain MRI demonstrated an expansive lesion centered in the midbrain tectum, with a hypointense signal on sagittal T1-weighted imaging (arrow in **A**), a hyperintense signal on sagittal fluid-attenuated inversion recovery (arrow in **B**), with partial gadolinium enhancement (arrow in **C**). Note that the tumor obstructed the cerebral aqueduct. The biopsy revealed a pilocytic astrocytoma.

fusion–positive; posterior fossa group A (PFA) ependymoma; posterior fossa group B (PFB) ependymoma; spinal ependymoma, MYCN–amplified; myxopapillary ependymoma; and subependymoma) (McNamara et al. 2022). About two-thirds of all ependymomas occur in the posterior fossa, typically inside the fourth ventricle (Renard et al., 2022).

Posterior fossa ependymomas are subdivided on the basis of methylomic profiling into the two most common subtypes: PFA and PFB ependymomas. PFA ependymomas are seen predominantly in infants, with a worse prognosis. The PFB ependymomas are seen in older children, with a better prognosis (Cohen, 2022). In general, posterior fossa ependymomas present as an irregular expansive lesion, with mild to moderate contrast enhancement, often associated with cysts, calcifications, and intratumoral hemorrhage on imaging, inside the fourth ventricle, extending through the foramina of Luschka and Magendie, associated with hydrocephalus due to mass effect. Restricted diffusion can be seen in areas with anaplasia and hemorrhage (Renard et al., 2022).

Although PFA and PFB ependymomas cannot be distinguished with certainty based solely on imaging appearance, PFA ependymomas usually present as a voluminous hyperdense and calcified mass in the posterior fossa, affecting young children, with mild contrast enhancement, extending through

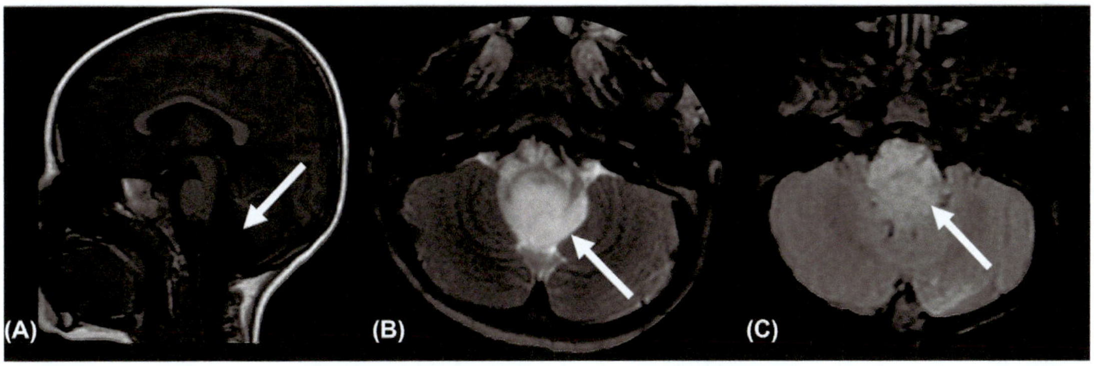

Fig. 17.22. A 15-year-old girl with an exophytic pilocytic astrocytoma in the medulla. Brain MRI demonstrated an exophytic expansive lesion in the posterior portion of the medulla, with a hypointense signal on sagittal T1-weighted imaging (arrow in **A**) and hyperintense signal on axial T2-weighted imaging (arrow in **B**) and fluid-attenuated inversion recovery (arrow in **C**). There was no gadolinium enhancement.

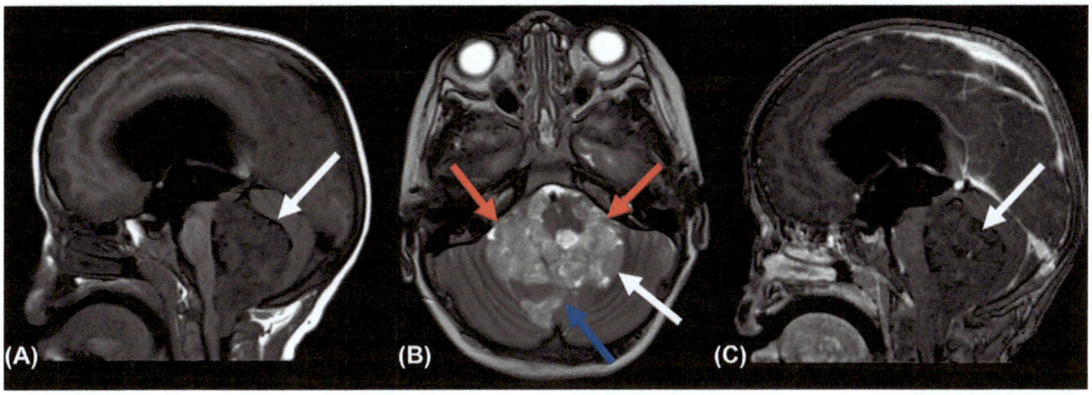

Fig. 17.23. A 2-year-old boy with a posterior fossa group A ependymoma. Brain MRI demonstrated an voluminous expansive lesion inside the fourth ventricle, with hypointense signal on T1-weighted imaging (arrow in **A**), heterogeneous signal on T2-weighted imaging (white arrow in **B**), extending along the foramina of Luschka (red arrows in **B**) and Magendie (blue arrow in **B**), with mild gadolinium enhancement (arrow in **C**).

the fourth ventricle foramina (Fig. 17.23). PFB ependymomas usually present as smaller noncalcified lesions in adolescents and young adults, within the fourth ventricle, minimally extending through the fourth ventricle foramina, with homogeneous contrast enhancement (Leclerc et al., 2024).

Medulloblastoma

Medulloblastomas are embryonal cell tumors arising in the posterior fossa, classified secondary to a combination of molecular and histopathologic features, and all types are WHO grade 4 lesions (Cohen, 2022). The molecularly defined groups are (1) medulloblastoma, Wingless/Int1 (WNT)-activated; (2) medulloblastoma,Sonic Hedgehog (SHH)-activated, TP53-wildtype; (3) medulloblastoma, SHH-activated, TP53-mutant; and (4) medulloblastoma, non-WNT/non-SHH (groups 3 and 4). Histologic subtypes are classic, desmoplastic/nodular, medulloblastoma with extensive nodularity, and large cell. Several associations exist between molecular and histologic subtypes of medulloblastomas (McNamara et al., 2022).

Medulloblastomas generally present as a hyperdense expansive lesion on CT, with or without calcifications. On MRI, these tumors present hypointense signal on T1- and T2-weighted imaging, with restricted diffusion and hyperperfusion. Although a location-based imaging approach was proposed to distinguish the different molecular subgroups of medulloblastoma, this strategy is not definitive. Although, WNT-activated medulloblastomas can occur in all posterior fossa locations, this subtype tends to be localized in the middle cerebellar peduncle/cerebellopontine angle, in close association with the foramen of Luschka, with a broad contact with the brainstem (Fig. 17.24). SHH-activated medulloblastomas are more commonly seen in the cerebellar hemispheres. Medulloblastoma, non-WNT/non-SHH, is the most common subtype, representing 50%–60% of all medulloblastomas, usually occurring in the cerebellar vermis. Group 3 tumors tend to be less well-defined and demonstrate more prominent contrast enhancement than group 4 (Fig. 17.25) (McNamara et al., 2022; Osborn et al., 2022).

Lymphoma

Primary central nervous system lymphomas are rare, comprising only 2% of all brain tumors and 1%–2% of malignant lymphomas. Furthermore, only 3% of primary central nervous system lymphomas affect the brainstem. Lymphoma can occur

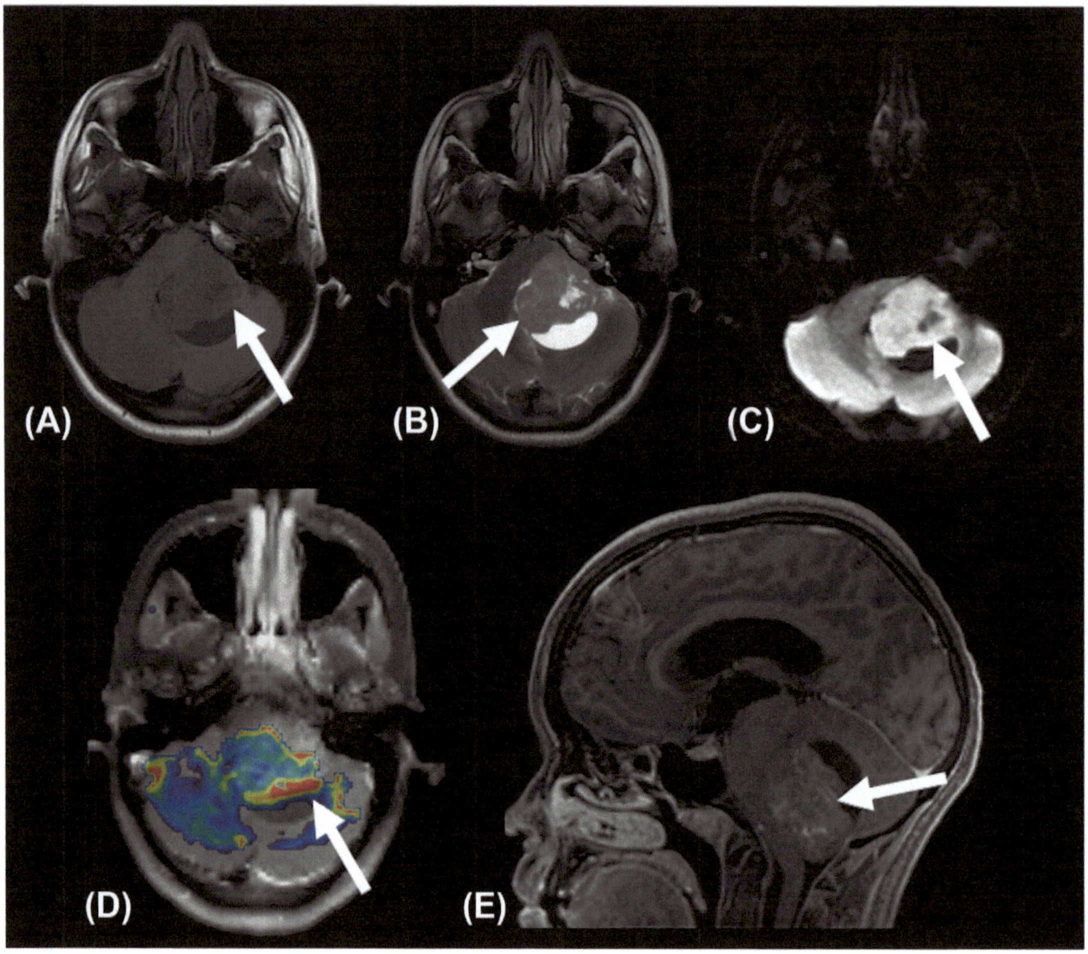

Fig. 17.24. A 14-year-old girl with a WNT-activated medulloblastoma. Brain MRI showed an expansive lesion in the left middle cerebellar peduncle/cerebellopontine angle, with a broad contact with the brainstem, with hypointense signal on T1-weighted imaging (arrow in **A**), isointense signal on T2-weighted imaging (arrow in **B**), restricted diffusion (arrow in **C**), signs of hyperperfusion (arrow in **D**), and mild gadolinium enhancement (arrow in **E**).

at any age, but is more common in the 17th and 18th decades of life. Imaging-based diagnosis is challenging because there are no specific neuroimaging features (Guzmán-De-Villoria et al., 2010b). Typically, lymphoma presents with single or multiple brain lesions, with hypo to isointense signal on T1-, T2-weighted imaging and FLAIR, associated with marked restricted diffusion (due to high cellularity and a high nuclear-to-cytoplasmic ratio), mild increased perfusion, and strong and homogeneous contrast enhancement (Fig. 17.26) (Ribas et al., 2024). However, lymphomas in immunocompromised patients can be multifocal, with necrotic areas and hemorrhagic foci, making the diagnosis difficult. In addition, an important feature is that lymphomas often decrease or even temporarily disappear after steroid treatment (Brandão and Castillo, 2016; Pons-Escoda et al., 2023).

Metastases

Brainstem metastases are often associated with other brain and/or nonbrain metastasis. Only 5% of brain metastases occur in the brainstem. These lesions present with hypointense signal on T1-and hyperintense signal on T2-weighted imaging and FLAIR, with nodular or ring-like contrast enhancement and surrounding edema (Fig. 17.27). Due to melanin or hemorrhages, some metastatic lesions present with hyperintense signal on T1-weighted imaging (Renard et al., 2022).

DEGENERATIVE DISEASES

Alzheimer disease

Alzheimer disease is the most common neurodegenerative disease, corresponding to 50%–70% of all cases of dementia. It is characterized by intracellular phosphorylated-tau accumulation (forming neurofibrillary tangles) and extracellular amyloid-β-proteins deposition (in the form of neuritic plaques), and neuronal loss. Since 2009, it has been demonstrated that the brainstem, especially the dorsal raphe nucleus of the midbrain, is affected by neurofibrillary tangles even before the entorhinal cortex. In addition, the locus coeruleus, substantia nigra, and the nucleus basalis of Meynert are also affected in the early stages of the disease (Grinberg et al., 2011). However, the anatomic

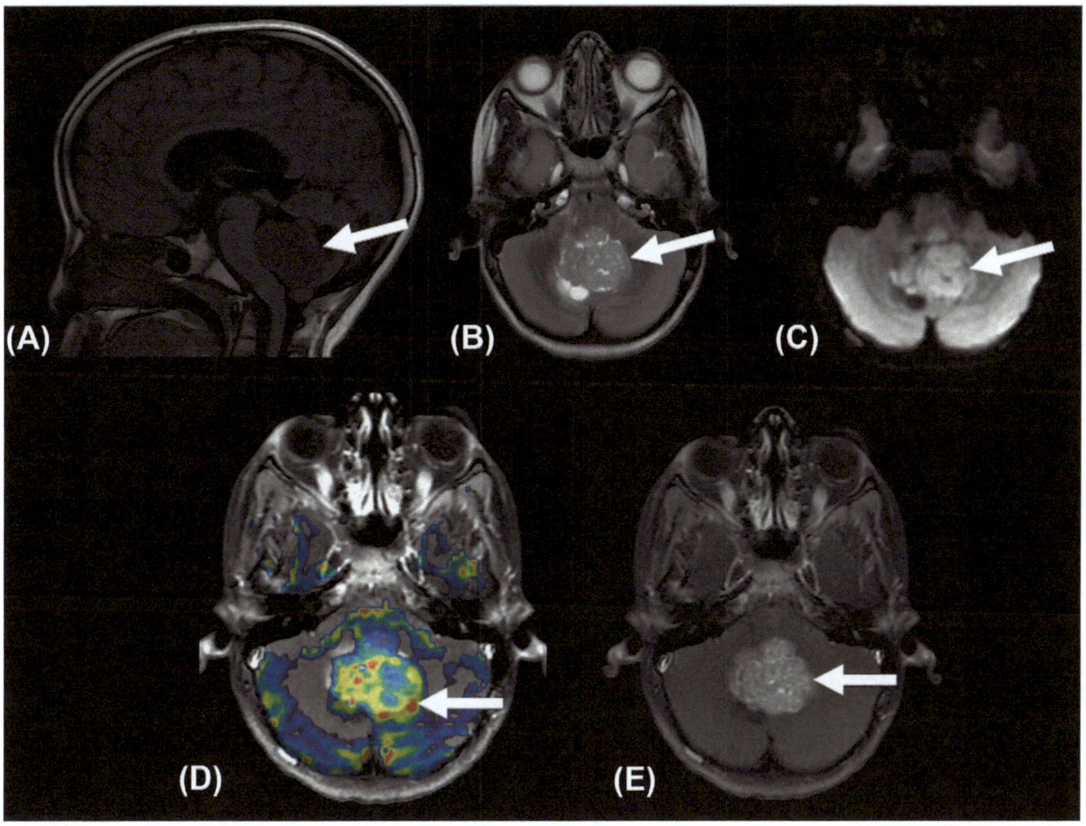

Fig. 17.25. A 7-year-old boy with a group 3 medulloblastoma. Brain MRI demonstrated a mass in the cerebellar vermis, with hypointense signal on T1-weighted imaging (arrow in **A**), isointense signal on T2-weighted imaging (arrow in **B**), restricted diffusion (arrow in **C**), signs of hyperperfusion (arrow in **D**), and gadolinium enhancement (arrow in **E**).

complexity of the brainstem and the lack of resolution to individualize its internal nuclei through the clinically available 1.5-T and 3-T MRI scanners limit the assessment of these structures on imaging (Braun and Van Eldik, 2018).

Patients with Alzheimer disease may present a reduced global brainstem volume, compared with healthy individuals, detected by structural MRI (Lee et al., 2015). Studies using functional MRI have seen a reduction of the default mode network connectivity with the posterior hippocampus, medial cingulo-parieto-occipital regions, and the dorsal raphe nucleus in patients with Alzheimer disease (Zhou et al., 2010) and reduced functional connectivity between the hippocampus and the dorsal raphe nucleus, associated with reduced serotonin transporter in the dorsal raphe nucleus, in patients with mild cognitive impairment (Barrett et al., 2017). In healthy patients, catecholamine synthesis is associated with the neuromelanin production, including the noradrenergic neurons in the locus coeruleus. In general, neuromelanin levels increases with aging and then decline around the seventh decade of life. The paramagnetic properties of neuromelanin can be mapped on high-resolution T1-weighted imaging acquired on 3-T scanners. A significantly reduced neuromelanin signal on T1-wighted imaging has been reported in patients with Alzheimer disease and mild cognitive impairment (Takahashi et al., 2015). However, despite being well defined in studies with small groups of patients, these findings are difficult to reproduce at the individual level, in the daily clinical practice. In the future, use of ultra-high-field MRI may enhance the evaluation of these brainstem nuclei (Braun and Van Eldik, 2018).

Lewy body disease

The brainstem is the earliest vulnerable structure in Lewy body disease, a condition that can express clinically as Parkinson disease or dementia with Lewy body. The dorsal motor nucleus of the vagal nerve in the medulla, which receives inputs from the neurenteric system, is classically considered the first involved nucleus, before the serotoninergic raphe nuclei and noradrenergic locus coeruleus nuclei of the pons. Neurodegeneration then reaches the midbrain, more particularly the dopaminergic substantia nigra pars compacta (SNc) and cholinergic pedunculopontine nuclei. The SNc plays a pivotal role, with classic Parkinson disease motor symptoms that occur when 30% or more of its dopaminergic neurons have disappeared (Arribarat et al., 2020).

The healthy nigrosome-1 can be depicted on high-resolution 3-T MRI on SWI, as the appearance of a "swallow tail" in the dorsolateral substantia nigra, and this feature is lost in Lewy body disease (Fig. 17.28). Visual assessment on SWI has an excellent diagnostic accuracy for the diagnosis of Parkinson

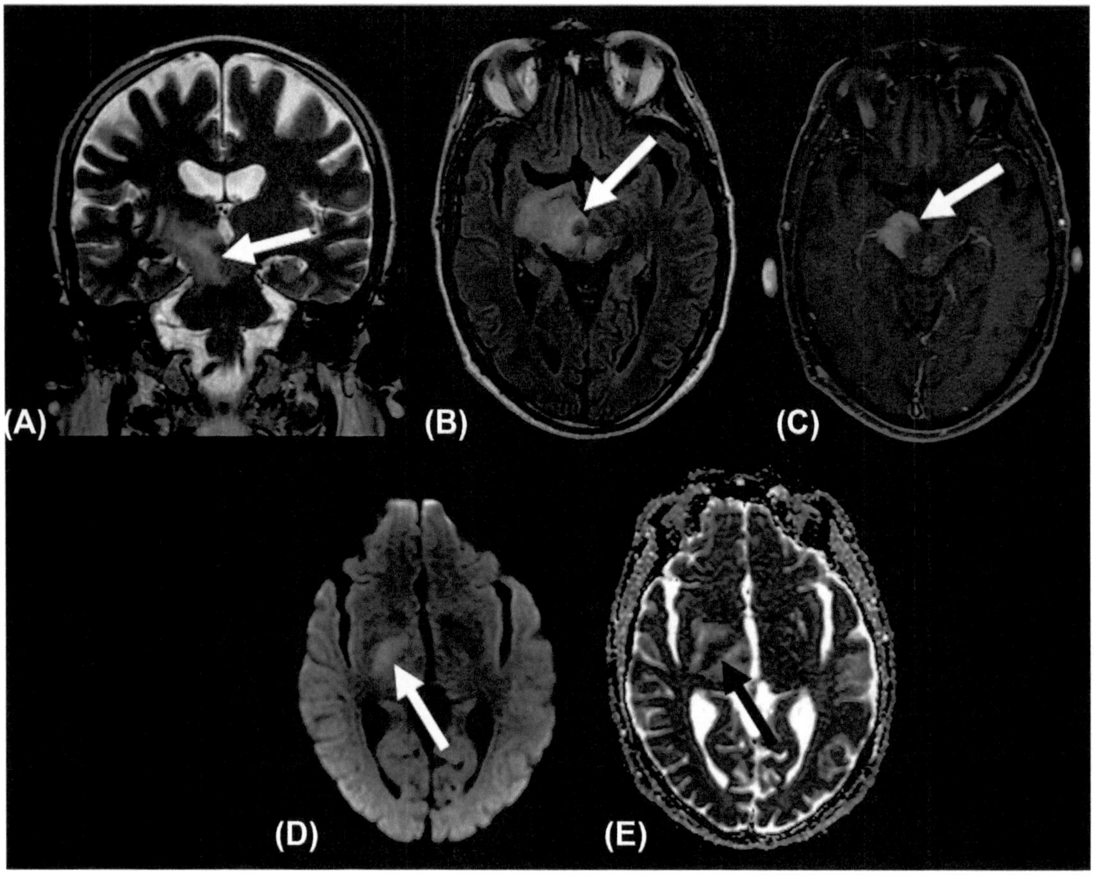

Fig. 17.26. A 70-year-old man with a brainstem lymphoma. Brain MRI demonstrated an infiltrative lesion centered in the right midbrain, extending to the basal ganglia region, with a hyperintense signal on coronal T2-weighted imaging (white arrow in **A**) and axial fluid-attenuated inversion recovery (white arrow in **B**), with gadolinium enhancement (white arrow in **C**), associated with restricted diffusion, as seen on diffusion-weighted imaging (white arrow in **D**) and apparent diffusion coefficient map (black arrow in **E**).

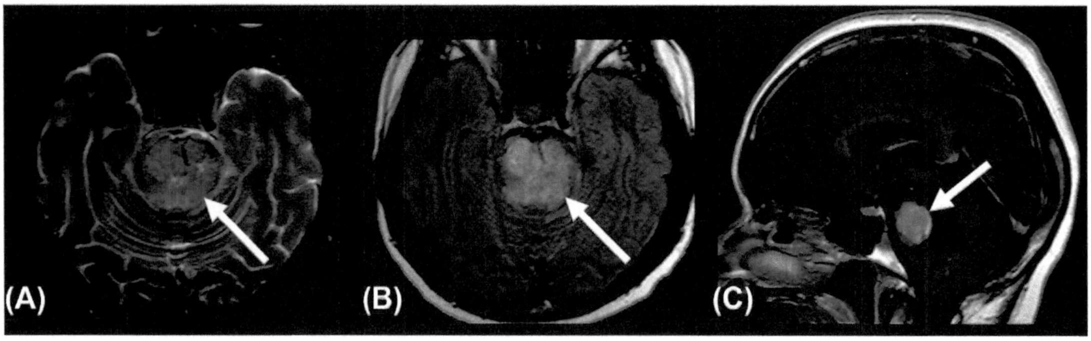

Fig. 17.27. A 68-year-old man with lung cancer and brainstem metastasis. Brain MRI showed a nodular lesion with hyperintense signal on T2-weighted imaging (arrow in **A**) and fluid-attenuated inversion recovery (arrow in **B**), with solid gadolinium enhancement (arrow in **C**) located in the ventral surface of the pons.

disease (Schwarz et al., 2014), but interpretation of the presence or absence of the "swallow tail" sign can be challenging due to nonoptimization of the sequence for this objective, presence of motion artifacts that difficult interpretation, nonavailability of a 3-T scanner and, eventually, the physician lack of experience in the area.

Also, neuromelanin-containing neurons are mainly concentrated in the SNc and locus coeruleus. The MRI signal in neuromelanin-containing neurons is paramagnetic. T1-weighted neuromelanin-sensitive MRI produces hyperintense signals in regions containing this pigment. Studies have confirmed that patients with Parkinson and Lewy body diseases have a significantly reduced neuromelanin signal in the substantia nigra and locus coeruleus on T1-weighted neuromelanin-sensitive sequences (Wang et al., 2018).

Multiple system atrophy

Multiple system atrophy (MSA) is a progressive neurodegenerative disease that clinically courses with parkinsonism, autonomic failure, and cerebellar dysfunction in various combinations. Pathologically, it presents with glial cytoplasmic inclusions and neuronal loss, predominantly in striatonigral and olivopontocerebellar systems. According to The Movement Disorder Society Criteria for the Diagnosis of Multiple System Atrophy, in addition to the core clinical features: autonomic dysfunction (defined as unexplained voiding difficulties with postvoid urinary residual volume ≥100 mL, unexplained urinary urge incontinence, neurogenic orthostatic hypotension) and at least one of (1) poorly L-dopa-responsive parkinsonism or (2) cerebellar syndrome (at least two of gait ataxia, limb ataxia, cerebellar dysarthria, or oculomotor features), at least one MRI feature suggestive of the disease is essential for the diagnosis of clinically established MSA.

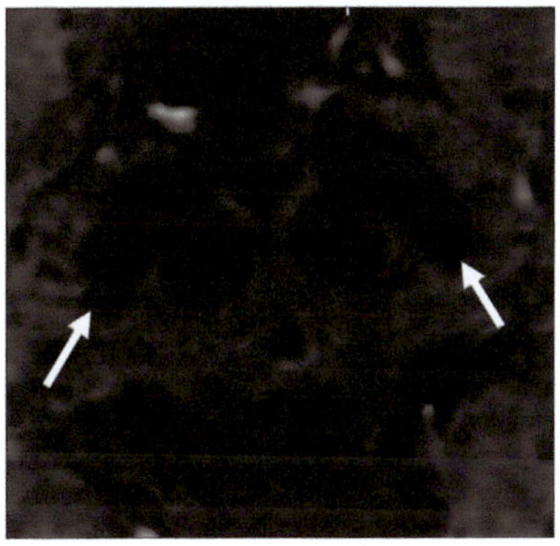

Fig. 17.28. A 67-year-old man with Parkinson disease. Brain 3-T MRI demonstrated the absence of the "swallow tail" sign in the dorsolateral substantia nigra on high-resolution susceptibility-weighted imaging.

MRI features of MSA are atrophy of putamen (with or without a slit-like T2/FLAIR hyperintense putaminal rim, and signal decrease on iron-sensitive sequences, such as SWI), atrophy of the pons, cerebellum and middle cerebellar peduncle, and the "hot cross bun" sign, or increased diffusivity in the putamen or middle cerebellar peduncles (Fig. 17.29). The hot cross bun sign is characterized by a cruciform pontine hyperintensity on T2-weighted imaging and FLAIR, secondary to degeneration of the pontine neurons and transverse pontocerebellar fibers, preserving the pyramidal tract and pontine tegmentum, especially in MSA with predominant cerebellar dysfunction (Wenning et al., 2022).

Progressive supranuclear palsy

Progressive supranuclear palsy (PSP) is a group of tauopathies that encompass a range of clinical phenotypes involving behavioral, language, and movement abnormalities. There are several clinical phenotypes of PSP, with different imaging features, but in most cases, nonspecific (Boxer et al., 2017). Although the objective of this chapter is not to describe in detail the characteristics of each form of PSP presentation, the classical presentation of PSP is Richardson's syndrome (PSP-Richardson's syndrome), which corresponds to 50% of all PSP cases. It is characterized by early postural instability, falls, and supranuclear gaze palsy.

The most representative imaging findings of PSP on MRI are brainstem atrophic changes, characterized by the hummingbird sign or penguin-silhouette sign, due to midbrain atrophy on sagittal images; morning glory sign or Mickey Mouse sign, due to midbrain atrophy, on axial images (Fig. 17.30); atrophy of the pontine tegmentum; atrophy of the superior cerebellar peduncles; atrophy of the frontotemporal lobes and/or corpus callosum; hyperintense signal on T2-weighted imaging in the periaqueductal region, and in the superior cerebellar peduncles, or globus pallidus (Boxer et al., 2017; Sakurai et al., 2017). Other clinical subtypes are less distinctive, contributing to a delay in the diagnosis or misdiagnosis.

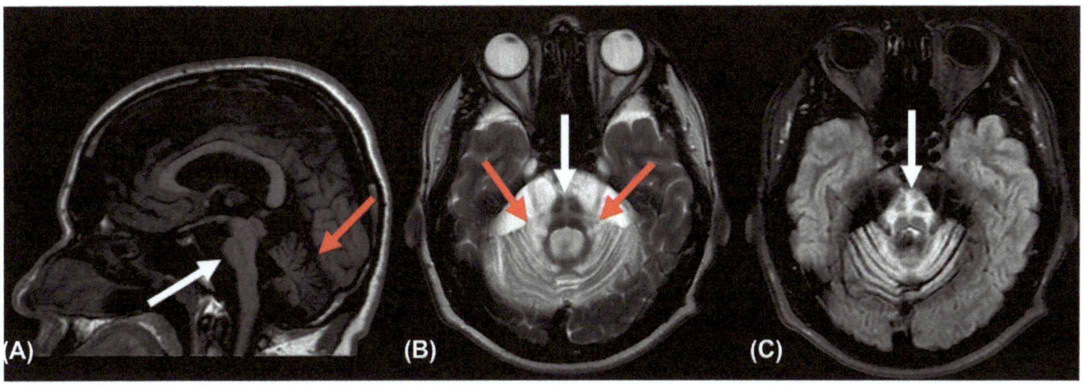

Fig. 17.29. A 44-year-old man with multiple system atrophy with predominant cerebellar dysfunction. Brain MRI showed atrophy of the pons (white arrow in **A**) and cerebellum (red arrow in **A**), seen on sagittal T1-weighted imaging. Axial T2-weighted imaging (**B**) and fluid-attenuated inversion recovery (**C**) demonstrated atrophy of the middle cerebellar peduncles (red arrows in **B**) and a cruciform pontine hyperintensity due to the "hot cross bun" sign (white arrows in **B** and **C**).

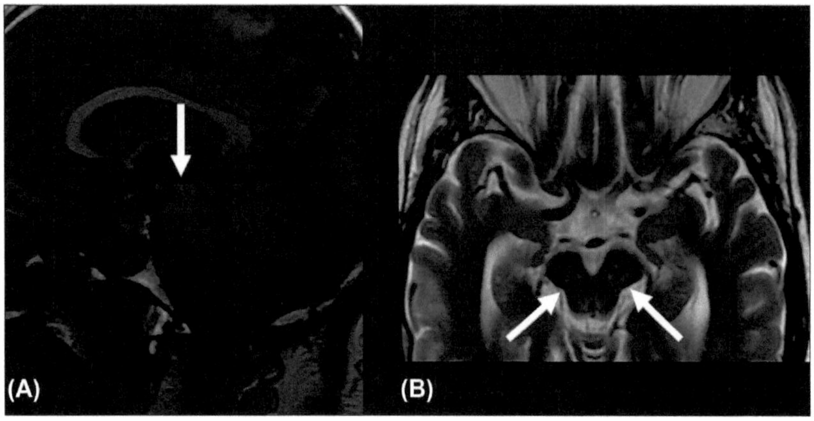

Fig. 17.30. A 65-year-old woman with progressive supranuclear palsy, with eight years of symptoms. Brain MRI demonstrated midbrain atrophy on sagittal T1-weighted imaging (arrow in **A**), with preservation of the pons (hummingbird sign) and on axial T2-weighted imaging (arrows in **B** – morning glory sign).

PSP-parkinsonism corresponds to 20%–35% of PSP cases, the second most common subtype, characterized by asymmetric tremors, ophthalmoplegia, and postural instability. The degree of midbrain atrophy is less severe in this subtype, especially in the early disease stage, and there may be no specific findings on MRI.

PSP-pure akinesia with gait freezing is the third well-described subtype of PSP, clinically characterized by freezing at initiating gait, freezing during walking, writing, and speaking, without rigidity, tremor, dementia, and eye movement abnormality during the early disease. This subtype does not exhibit characteristic atrophic changes on MRI, only nonspecific mild cerebral atrophy and ventricular dilatation. However, an atrophic and triangular-shaped globus pallidus, with hyperintense signal on T2-weighted imaging, can be seen.

PSP-corticobasal syndrome (CBS) presents with asymmetric dystonia, apraxia, alien limb and cortical sensory loss, similar to corticobasal degeneration. Most patients with PSP-CBS eventually develop postural instability, axial rigidity, and eye movement abnormality. This subtype presents variable midbrain atrophy and asymmetric atrophy of the frontal lobe and cerebral peduncle, usually contralateral to the clinically more profoundly affected side.

PSP-progressive nonfluent aphasia presents with progressive speech and language dysfunction, with hesitancy, agrammatism, apraxia of speech, and phonemic errors. On MRI, left-sided frontal lobe atrophy can be seen, with relative sparing of the midbrain.

PSP with cerebellar ataxia is presented with cerebellar ataxia as a prodromal symptom before the main features of PSP. Therefore, these patients might be misdiagnosed with spinocerebellar ataxia. These patients may present with or without cerebellar atrophy in the early stages, in addition to the typical midbrain atrophy (Sakurai et al., 2017).

Spinocerebellar ataxia

The spinocerebellar ataxias are a group of autosomal dominant disorders, clinically characterized by symptoms secondary to cerebellum and cerebellar interconnection degeneration, such as cerebellar ataxia. Extracerebellar findings, such as pyramidal and extrapyramidal symptoms, oculomotor abnormalities, and peripheral neuropathy, often accompany ataxia. Sleep disorders, cognitive impairment, and psychiatric symptoms are also frequent, possibly due to a more widespread neurodegeneration (Klaes et al., 2016).

The most used classification of the spinocerebellar ataxias is to number them according to the order of gene discovery, from 1 to 48. Although genetic testing is increasingly available, it is important to know the specific clinical, laboratorial, and imaging features of each type of spinocerebellar ataxias, especially when genetic testing is not available. However, neuroimaging cannot predict the genotype of a specific type of spinocerebellar ataxia, because of the overlap between symptoms and imaging findings between them, but it is possible to cluster subgroups of spinocerebellar ataxias through imaging aspects, limiting the diagnostic possibilities. The main MRI findings are cortical cerebellar atrophy and spinal cord atrophy, with or without brainstem involvement. None of these imaging features are specific of a type of spinocerebellar ataxia, but together with clinical and epidemiologic information, can guide the genetic diagnosis (Meira et al., 2019). A detailed description of the clinical and imaging features of all spinocerebellar ataxias is also beyond the scope of this chapter, and we recommend to the interested reader to look for specific texts (Klaes et al., 2016; Meira et al., 2019; Cocozza et al., 2021).

In general, pure cerebellar atrophy is seen in spinocerebellar ataxia types 4, 5, 6, 10, 11, 13, 14, 15, 16, 18, 19, 21, 22, 24, 25, 26, 27, 28, 29, 30, 31, 32, 35, 37, 38, 41, 42, 43, 44, and 47. Cerebellar atrophy associated with pontine atrophy is seen in spinocerebellar ataxia types 1, 2, 3, 7, 8, 13, 34, and 40; spinal cord may be associated with cerebellar atrophy in types 3 and 7. The pontine "hot cross bun sign" can occur in types 1, 2, 3, 6, 7, 8, and 34. Calcifications of the dentate nuclei were described in type 20 (Meira et al., 2019).

Spinocerebellar ataxia type 3 (Machado–Joseph disease) is the most common worldwide. It is determined by an abnormal

CAG trinucleotide repeats (52–86) in ATXN3 gene encoding for the Ataxin3 protein. Clinically, it is characterized by ataxia, dysarthria, diplopia, nystagmus, and hyperreflexia, usually beginning between the second and fifth decades of life. Sleep disturbance and deficits in executive functions may also be present. Death usually occurs due to pulmonary complication and cachexia, after a few to 30 years of disease onset. MRI shows variable degrees of spinal cord, pontine, and cerebellar atrophy, mainly involving the vermis, dorsal pons, and dentate nuclei, with fourth ventricle ectasia (Fig. 17.31). The "hot cross bun" sign may also be seen (Cocozza et al., 2021).

Hypertrophic olivary degeneration

Hypertrophic olivary degeneration occurs due to trans-synaptic degeneration caused by the loss of afferent signals for the inferior olivary nucleus in the medulla, secondary to a lesion in the dentato-rubro-olivary pathway, within the Guillain–Mollaret triangle. The Guillain–Mollaret triangle is formed by the red nucleus (in the midbrain), which is connected to the ipsilateral inferior olivary nucleus (in the medulla) and the contralateral dentate nucleus (in the cerebellum). Lesions along this pathway can lead to hypertrophic olivary degeneration, with the most common causes being hemorrhage and ischemic stroke.

MRI usually demonstrates a hyperintense signal on T2-weighted imaging and swelling in the medullary olive (Figs. 17.32 and 17.33). The olivary hyperintensity appears around three weeks after the initial injury, and the hypertrophy appears after 5–15 months. In the long term, the olivary swelling subsides, but T2-hyperintensity may persist (Kitajima et al., 1994).

VASCULAR DISEASES

Ischemic stroke

The main arteries that supply the brainstem are the vertebral, basilar, anterior spinal, posterior inferior cerebellar, anterior inferior cerebellar, superior cerebellar, posterior cerebral, posterior communicating, and anterior choroidal arteries. The collaterals of these arteries are divided into four regions (anteromedial, anterolateral, lateral, and posterior). The size and location of the brainstem ischemic stroke depend on the affected artery. In general, proximal basilar artery occlusion is often caused by a thrombus related to atherosclerosis. Distal basilar artery occlusion is typically caused by a cardiac or artery-to-artery embolism (Burger et al., 2005).

Brainstem ischemic stroke corresponds to approximately 10% of all ischemic strokes. Typically, the patients present with an acute neurologic deficit that varies depending on the affected region. The pons is the most common brainstem portion affected by ischemic strokes, followed by the medulla and midbrain (Guzmán-De-Villoria et al., 2010a).

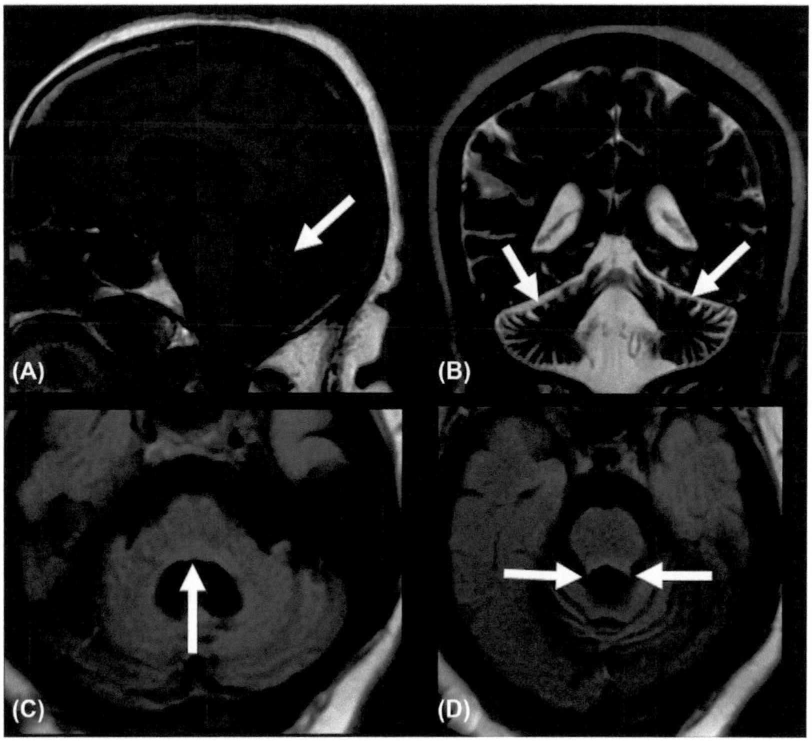

Fig. 17.31. A 45-year-old man with spinocerebellar ataxia type 3 (Machado–Joseph disease). Brain MRI showed cerebellar atrophy and dilatation of the fourth ventricle, on sagittal T1-weighted imaging (arrow in **A**) and coronal T2-weighted imaging (arrows in **B**), as well as atrophic changes of the pontine tegmentum and bilateral superior cerebellar peduncles, on axial fluid-attenuated inversion recovery (arrows in **C** and **D**, respectively).

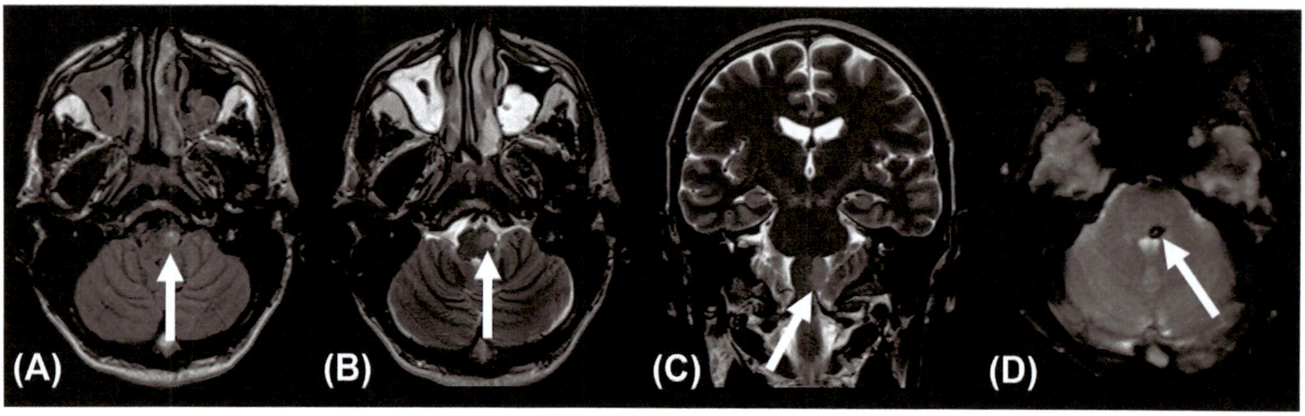

Fig. 17.32. A 45-year-old man with left hypertrophic olivary degeneration due to a previous brainstem hemorrhage. Brain MRI showed swelling and hyperintense signal in the left olive on fluid-attenuated inversion recovery (arrow in **A**) and T2-weighted imaging in the axial (arrow in **B**) and coronal planes (arrow in **C**). Axial susceptibility-weighted imaging demonstrated a hemosiderin deposit in the pontine tegmentum, probably related to a cavernoma (arrow in **D**).

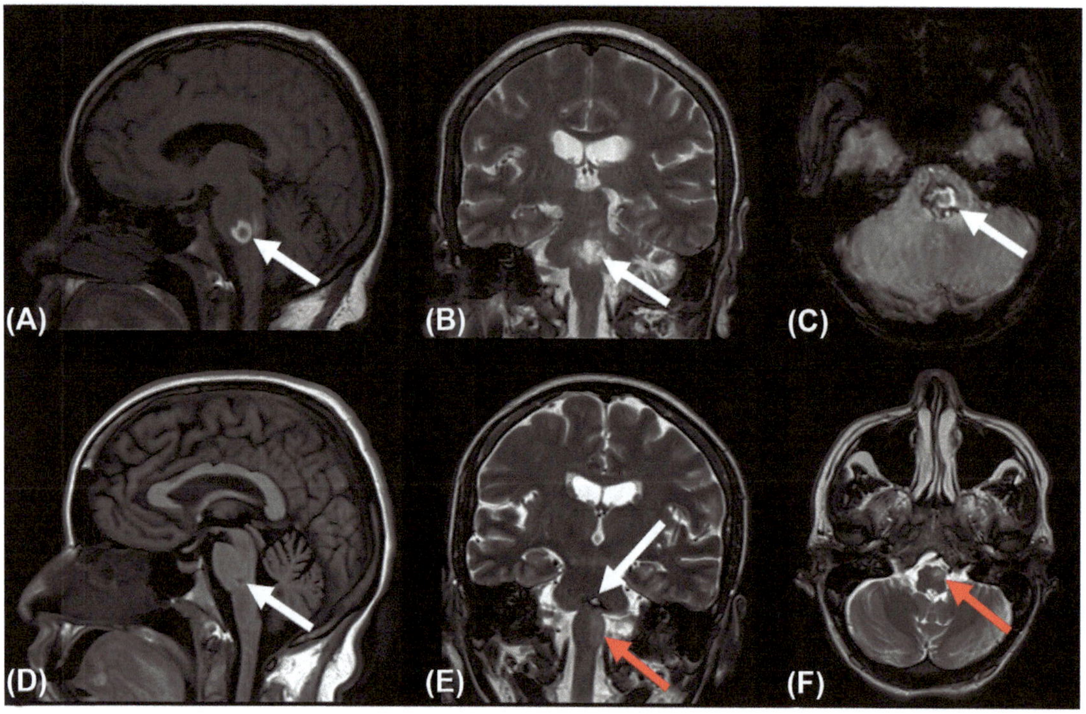

Fig. 17.33. A 50-year-old woman with a pontine hypertensive hemorrhagic stroke. Brain MRI demonstrated an expansive lesion with a heterogeneous signal, predominantly hyperintense on T1- (arrow in **A**) and T2-weighted imaging (arrow in **B**), with a peripheral hypointense signal on susceptibility-weighted imaging (arrow in **C**), compatible with a hematoma. A follow-up MRI 3 months later demonstrated a nearly complete resolution of the hematoma, with a hemosiderin deposit seen on T1- (arrow in **D**) and T2-weighted imaging (white arrow in **E**). The second MRI also demonstrated left hypertrophic olivary degeneration in the coronal (red arrow in **E**) and axial planes (red arrow in **F**).

In the acute clinical scenario, unenhanced CT is frequently normal but can demonstrate a hyperdense artery or brainstem hypodensity. Acute brainstem ischemic stroke is best visualized on MRI on DWI, which can demonstrate restricted diffusion due to cytotoxic edema, as hyperintense signal on DWI and hypointense signal on apparent diffusion coefficient map (Fig. 17.34). T2-weighted imaging and FLAIR can also show signal hyperintensity, but are less sensitive than DWI in the hyperacute phase (Abdelrasoul et al., 2019). Usually, these lesions have well-defined borders and do not extend through the midline, although it should be kept in mind that bilateral ischemic lesions are not uncommon in the brainstem (de Morais et al., 2021). MR-angiography or CT-angiography may demonstrate steno-occlusive lesions of the vertebral-basilar system and its main trunks, but smaller penetrating arteries cannot be visualized in these methods. Perfusion-weighted imaging is unreliable for brainstem stroke assessment (Abdelrasoul et al., 2019).

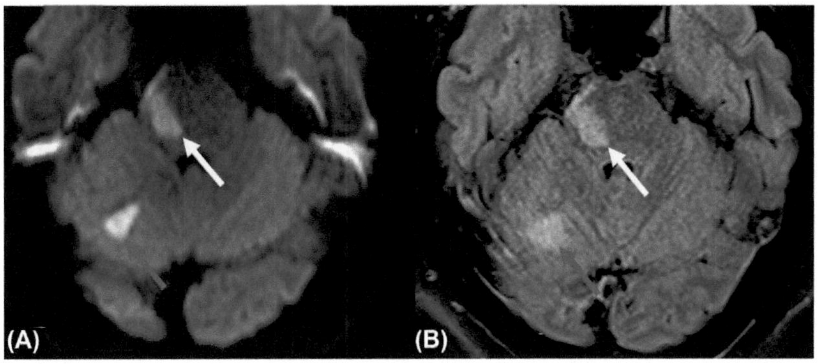

Fig. 17.34. A 55-year-old man with an acute ischemic stroke in the right pons and cerebellar hemisphere. Brain MRI showed a lesion with restricted diffusion (white arrow in **A**), with hyperintense signal on fluid-attenuated inversion recovery (white arrow in **B**), associated with another ischemic lesion in the right cerebellar hemisphere (red arrows in **A** and **B**).

Finally, a chronic supratentorial stroke can lead to a progressive anterograde disintegration after injury to the cell body, a process called Wallerian degeneration, which can sometimes be observed as a T2/FLAIR hyperintensity and atrophy involving the corticospinal, pontocerebellar, and the dentate-rubro-olivary pathways in the brainstem (Renard et al., 2022).

Hemorrhagic stroke

Hemorrhagic stroke in the brainstem is usually associated with arterial hypertension. Five to ten percent of all hemorrhagic strokes occur in the brainstem (Aguilar and Brott, 2011). The pons is the most commonly affected region of the brainstem. Common complications are intraventricular hemorrhage and hydrocephalus due to the proximity of the cerebral aqueduct and fourth ventricle. A hemorrhagic stroke will present as an intraparenchymal hematoma, seen as a round or elliptical well-delineated hyperdense mass lesion, surrounded by edema, on CT; and as mixed hyper/hypointense signal, on T1-, T2-weighted imaging and FLAIR, and prominent hypointense signal on SWI, depending on the timing of imaging (Fig. 17.33). In active hemorrhages, the so-called "spot sign" can be seen inside the hematoma when intravenous contrast is used. This feature is a biomarker for rapid hematoma growth (Renard et al., 2022).

Cavernoma

Cavernoma, or cavernous angioma, is a vascular malformation composed of sinusoidal channels lined by endothelium, without an elastic membrane, muscular tissue, or interposed neural tissue, containing blood in different stages of metabolism. They can be solitary or multiple and occur throughout the central nervous system. The pons is the most common localization in the brainstem. When multiple cavernomas are detected, there may be a familial tendency for the disease. They are also characterized as hidden vascular malformations, which do not communicate with the arteries. Therefore, digital angiography, MR-angiography, and CT-angiography are generally incapable of identifying this lesion and do not play a significant role in the diagnosis. The association with venous angiomas is quite common.

On CT, they can be identified as focal, hyperdense lesions containing small calcifications. However, this aspect is not specific. The imaging method of choice for diagnosis is MRI, which may reveal a characteristic "popcorn" appearance characterized by a lesion with lobulated contour, with a central portion with heterogeneous signal hyperintensity on T1- and T2-weighted images, associated with a peripheral ring of T2 hypointensity (Hegde et al., 2012). Contrast enhancement may or may not be present inside the lesion. Furthermore, the appearance may vary depending on its size and the stage of the metabolic degradation of blood-derived products. Therefore, not all cavernomas present with the classic "popcorn" appearance. SWI is an extremely sensitive MRI technique for detecting bleeding and calcifications. Therefore, cavernomas will show a markedly hypointense signal on SWI ("blooming effect") (Mittal et al., 2009). SWI also allows the identification of other cavernomatous lesions, which can lead to a presumptive diagnosis of family involvement and therefore guide a broader approach to first-degree relatives (Fig. 17.35).

Arteriovenous malformation

Arteriovenous malformation (AVM) is a congenital anomaly of blood vessel development, which consists of a tangle of vascular structures without an intervening capillary network between the arterial and venous beds, leading to arteriovenous shunt. The transition between the arterial and venous beds occurs through the so-called nidus (which is the central portion of the AVM), characterized by a tangle of abnormal vessels within the brain parenchyma that forms the transition between the nourishing artery(ies) and the draining vein(s). The tributary veins of the AVMs become dilated and tortuous due to the pressure to which they are subjected.

On MRI, AVM appears as an agglomerate of serpiginous structures with marked signal hypointensity on T1- and T2-weighted images (known as "flow-voids") (Fig. 17.36). Areas of signal hyperintensity on T2-weighted images in the adjacent brain parenchyma can represent gliosis and/or edema (Vattipally and Bronen, 2006).

The diagnostic criteria for an AVM are the presence of a nidus in the brain parenchyma and early venous drainage. DSA,

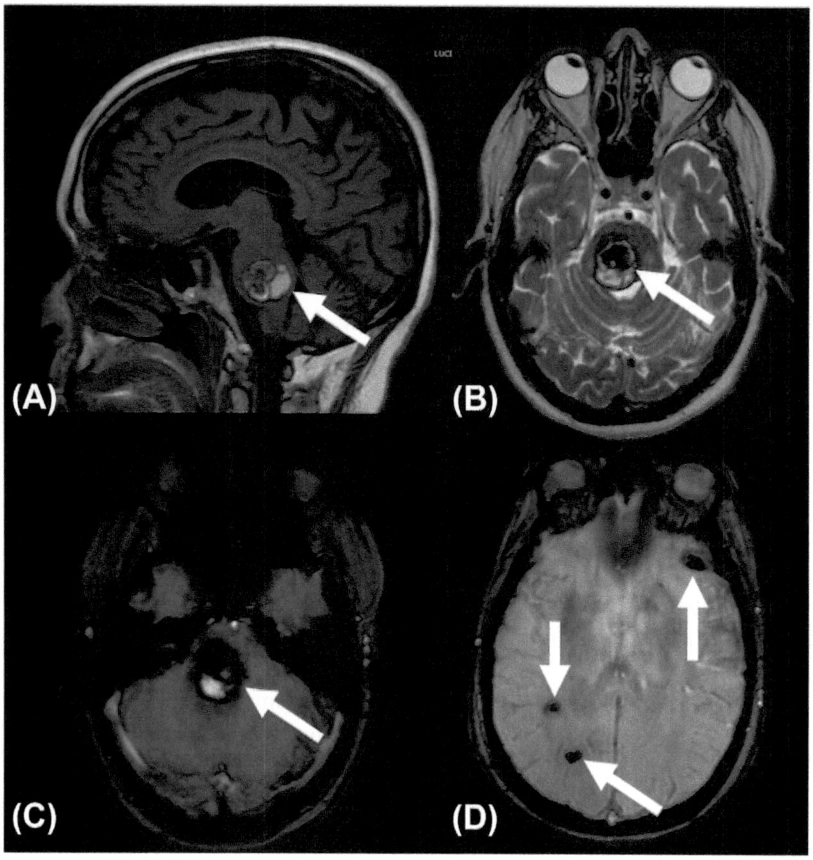

Fig. 17.35. A 26-year-old woman with familial cerebral cavernous malformations. A brain MRI showed a heterogeneous lesion, with a predominant hyperintense signal on T1-weighted imaging (arrow in **A**), predominant hypointense signal on T2-weighted imaging (arrow in **B**), and a marked hypointense signal on susceptibility-weighted imaging (SWI), with a blooming effect (arrow in **C**), suggestive of cavernoma in the pons. SWI also demonstrated other supratentorial cavernomas (arrows in **D**).

however, can be necessary for a better characterization due to better temporal resolution and the ability to investigate distinct vascular territories (Nagae et al., 2016).

TRAUMA

The incidence of traumatic brainstem injury increases with the severity of the trauma. Patients with brainstem traumatic injuries present with loss of consciousness, abnormal cranial nerves reflexes or internuclear ophthalmoplegia, and decerebration, respiratory dysfunction, or dysautonomia in more severe cases.

Primary brainstem trauma often occurs due to contusions of the posterior portions of the midbrain, superior cerebellar peduncles, or pons at the sharp edges of the cerebellar tentorium. These lesions can be hemorrhagic or not (Guzmán-De-Villoria et al., 2010b).

However, the most common cause of brainstem posttraumatic injury is diffuse axonal injury, secondary to indirect shear–strain forces, caused by severe head trauma. In order of increasing severity, the diffuse axonal injury typically involves the subcortical lobar white matter (especially the gray–white matter junction), corpus callosum, and the midbrain. About 20% of diffuse axonal injuries involve the brainstem (Igboechi et al., 2013). This lesion is better seen on DWI, as areas with restricted diffusion, due to cytotoxic edema, and SWI, as foci of microbleeds in the regions mentioned above. Other features of traumatic brain injuries can be seen, such as epidural or subdural hematomas, subarachnoid, or intraventricular hemorrhages, and parenchymal hematomas (Guzmán-De-Villoria et al., 2010b; Sandhu et al., 2019).

Another type of secondary brainstem hemorrhage, which may be associated with trauma, is Duret hemorrhage. The pathophysiology of this brainstem hemorrhage is related with a rapidly expanding supratentorial mass lesion, such as a posttraumatic hematoma or acute cerebral edema, which causes a severe increase in intracranial pressure, resulting in a descending transtentorial herniation (Parizel et al., 2002). Then, the brainstem is pushed inferiorly, foreshortened, and buckled, leading to stretching, spasm, infarction, and hemorrhage of the central perforating arteries, which arise from the relatively immobile basilar artery (Friede and Roessmann, 1966). Venous congestion is other possible mechanism, due to prolonged elevation in intracranial pressure, causing vascular thrombosis within the brainstem, which evolves to hemorrhage (Andeweg, 1999). On imaging, this hemorrhage can be single or multiple, in the midline of the midbrain or pons, near the pontomesencephalic junction, with or without extension to the cerebellar peduncles (Fig. 17.37). Outcome is usually poor (Parizel et al., 2002).

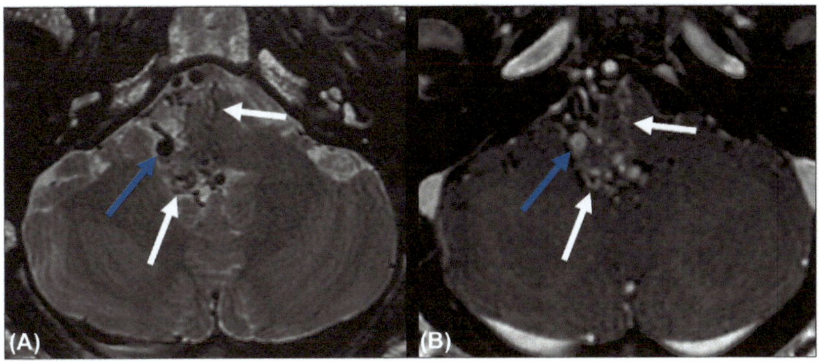

Fig. 17.36. A 35-year-old man with medullary arteriovenous malformation (AVM). Brain MRI revealed serpiginous structures with hypointense signal ("flow-voids") on T2-weighted imaging in the right medulla, which corresponds to the nidus (white arrows in **A**), as well as dilated tributary veins of the AVM (blue arrow in **A**). Note how some of the vessels in the nidus (white arrows in **B**) and tributary veins (blue arrow in **B**) become easily visible on postcontrast T1-weighted imaging.

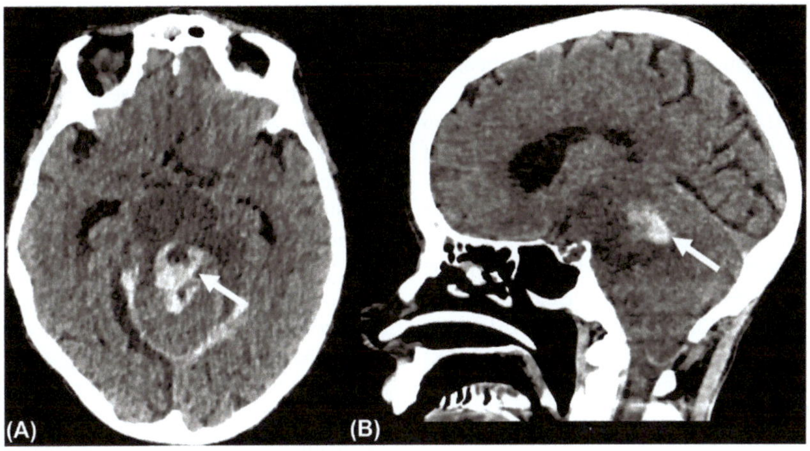

Fig. 17.37. A 30-year-old man with Duret hemorrhage, with a history of car accident. Brain CT showed extensive edema in the midbrain and mesodiencephalic junction, associated with a hematoma in the midline of the midbrain, in the axial (white arrow in **A**) and sagittal planes (white arrow in **B**).

CONCLUSION

The brainstem is a relatively small region that contains important structures that are difficult to access surgically. A wide variety of lesions can affect the brainstem and manifest clinically in relatively similar ways. Therefore, neuroimaging, mainly MRI, is of utmost importance to evaluate the normal anatomy of the brainstem and to determine the diagnosis or at least narrow the differential diagnosis of patients presenting brainstem lesions.

References

Abdelrasoul AA, Elsebaie NA, Gamaleldin OA et al. (2019). Imaging of brain infarctions: beyond the usual territories. J Comput Assist Tomogr 43: 443–451.

Aguilar MI, Brott TG (2011). Update in intracerebral hemorrhage. Neurohospitalist 1: 148–159.

Aiken AH (2010). Central nervous system infection. Neuroimaging Clin N Am 20: 557–580.

Al-Noury K, Lotfy A (2011). Normal and pathological findings for the facial nerve on magnetic resonance imaging. Clin Radiol 66: 701–707.

Andeweg J (1999). Consequences of the anatomy of deep venous outflow from the brain. Neuroradiology 41: 233–241.

Angeles Fernández-Gil M, Palacios-Bote R, Leo-Barahona M et al. (2010). Anatomy of the brainstem: a gaze into the stem of life. Semin Ultrasound CT MR 31: 196–219.

Arribarat G, De Barros A, Péran P (2020). Modern brainstem MRI techniques for the diagnosis of Parkinson's disease and parkinsonisms. Front Neurol 11: 791.

Bae YJ, Kim JH, Choi BS et al. (2013). Brainstem pathways for horizontal eye movement: pathologic correlation with MR imaging. Radiographics 33: 47–59.

Banwell B, Bennett JL, Marignier R et al. (2023). Diagnosis of myelin oligodendrocyte glycoprotein antibody-associated disease: International MOGAD Panel proposed criteria. Lancet Neurol 22: 268–282.

Barrett FS, Workman CI, Sair HI et al. (2017). Association between serotonin denervation and resting-state functional connectivity in mild cognitive impairment. Hum Brain Mapp 38: 3391–3401.

Barth PG, Majoie CB, Caan MW et al. (2007). Pontine tegmental cap dysplasia: a novel brain malformation with a defect in axonal guidance. Brain 130: 2258–2266.

Bartynski WS (2008). Posterior reversible encephalopathy syndrome, part 1: fundamental imaging and clinical features. AJNR Am J Neuroradiol 29: 1036–1042.

Battal B, Castillo M (2024). Imaging of reversible cerebral vasoconstriction syndrome and posterior reversible encephalopathy syndrome. Neuroimaging Clin N Am 34: 129–147.

Borges A, Casselman J (2010). Imaging the trigeminal nerve. Eur J Radiol 74: 323–340.

Borhani-Haghighi A, Kardeh B, Banerjee S et al. (2020). Neuro-Behcet's disease: an update on diagnosis, differential diagnoses, and treatment. Mult Scler Relat Disord 39: 101906.

Bosley TM, Salih MA, Jen JC et al. (2005). Neurologic features of horizontal gaze palsy and progressive scoliosis with mutations in ROBO3. Neurology 64: 1196–1203.

Boxer AL, Yu JT, Golbe LI et al. (2017). Advances in progressive supranuclear palsy: new diagnostic criteria, biomarkers, and therapeutic approaches. Lancet Neurol 16: 552–563.

Brandão LA, Castillo M (2016). Lymphomas – part 1. Neuroimaging Clin N Am 26: 511–536.

Braun DJ, Van Eldik LJ (2018). *In vivo* brainstem imaging in Alzheimer's disease: potential for biomarker development. Front Aging Neurosci 10: 266.

Burger KM, Tuhrim S, Naidich TP (2005). Brainstem vascular stroke anatomy. Neuroimaging Clin N Am 15: 297–324.

Casselman J, Mermuys K, Delanote J et al. (2008). MRI of the cranial nerves – more than meets the eye: technical considerations and advanced anatomy. Neuroimaging Clin N Am 18: 197–231.

Cocozza S, Pontillo G, De Michele G et al. (2021). Conventional MRI findings in hereditary degenerative ataxias: a pictorial review. Neuroradiology 63: 983–999.

Cohen AR (2022). Brain tumors in children. N Engl J Med 386: 1922–1931.

Corrêa DG, Hygino da Cruz LC Jr, Freddi TAL (2022). The oculomotor nerve: anatomy and pathology. Semin Ultrasound CT MR 43: 389–399.

Corrêa DG, Hygino da Cruz LC Jr, Freddi TAL (2023). The vestibulocochlear nerve: anatomy and pathology. Semin Ultrasound CT MR 44: 81–94.

Corrêa DG, Pacheco FT, da Cruz LCH et al. (2023). Intracranial vessel wall magnetic resonance imaging features of infectious vasculitis. Clin Imaging 98: 26–35.

de Morais RQ, de Melo ASA, Corrêa DG (2021). Bilateral medullary infarct: the radiologist's point of view. Neuroradiology 63: 15–16.

Desai NK, Young L, Miranda MA et al. (2011). Pontine tegmental cap dysplasia: the neurotologic perspective. Otolaryngol Head Neck Surg 145: 992–998.

de Sousa Costa R, Ventura N, de Andrade Lourenção Freddi T et al. (2023). The hypoglossal nerve. Semin Ultrasound CT MR 44: 104–114.

Dutra BG, da Rocha AJ, Nunes RH et al. (2018). Neuromyelitis optica spectrum disorders: spectrum of MR imaging findings and their differential diagnosis. Radiographics 38: 169–193.

Flannigan BD, Bradley WG Jr, Mazziotta JC et al. (1985). Magnetic resonance imaging of the brainstem: normal structure and basic functional anatomy. Radiology 154: 375–383.

Freddi TAL (2022). The trochlear nerve: anatomy and pathology. Semin Ultrasound CT MR 43: 400–402.

Freddi TAL, Ottaiano AC, Lucio LL et al. (2022). The trigeminal nerve: anatomy and pathology. Semin Ultrasound CT MR 43: 403–413.

Friede RL, Roessmann U (1966). The pathogenesis of secondary midbrain hemorrhages. Neurology 16: 1210–1216.

Frohman TC, Galetta S, Fox R et al. (2008). Pearls & oysters: the medial longitudinal fasciculus in ocular motor physiology. Neurology 70: e57–e67.

Gass A, Filippi M, Grossman RI (2000). The contribution of MRI in the differential diagnosis of posterior fossa damage. J Neurol Sci 172: S43–S49.

Geibprasert S, Gallucci M, Krings T (2010). Alcohol-induced changes in the brain as assessed by MRI and CT. Eur Radiol 20: 1492–1501.

Grinberg LT, Rueb U, Heinsen H (2011). Brainstem: neglected locus in neurodegenerative diseases. Front Neurol 2: 42.

Guzmán-De-Villoria JA, Fernández-García P, Ferreiro-Argüelles C (2010). Differential diagnosis of T2 hyperintense brainstem lesions: part 1. Focal lesions. Semin Ultrasound CT MR 31: 246–259.

Guzmán-De-Villoria JA, Ferreiro-Argüelles C, Fernández-García P (2010). Differential diagnosis of T2 hyperintense brainstem lesions: part 2. Diffuse lesions. Semin Ultrasound CT MR 31: 260–274.

Haller S, Wetzel SG, Lütschg J (2008). Functional MRI, DTI and neurophysiology in horizontal gaze palsy with progressive scoliosis. Neuroradiology 50: 453–459.

Hegde AN, Mohan S, Lim CC (2012). CNS cavernous haemangioma: "popcorn" in the brain and spinal cord. Clin Radiol 67: 380–388.

Hennekam RC, Biesecker LG, Allanson JE et al. (2013). Elements of morphology: general terms for congenital anomalies. Am J Med Genet A 161A: 2726–2733.

Igboechi C, Vaddiparti A, Sorenson EP et al. (2013). Tectal plate gliomas: a review. Child Nerv Syst 29: 1827–1833.

Jissendi-Tchofo P, Doherty D, McGillivray G et al. (2009). Pontine tegmental cap dysplasia: MR imaging and diffusion tensor imaging features of impaired axonal navigation. AJNR Am J Neuroradiol 30: 113–119.

Jissendi-Tchofo P, Severino M, Nguema-Edzang B et al. (2015). Update on neuroimaging phenotypes of mid-hindbrain malformations. Neuroradiology 57: 113–138.

Johnson DR, Giannini C, Vaubel RA et al. (2023). A radiologist's guide to the 2021 WHO Central Nervous System Tumor Classification: part I – key concepts and the spectrum of diffuse gliomas. Radiology 306: e229036.

Jurkiewicz E, Dobrzańska A, Nowak K et al. (2010). MRI findings in the young infant with brainstem disconnection and extracerebral features. Report of one case and review of the literature. Brain Dev 32: 495–498.

Kitajima M, Korogi Y, Shimomura O et al. (1994). Hypertrophic olivary degeneration: MR imaging and pathologic findings. Radiology 192: 539–543.

Klaes A, Reckziegel E, Franca MC et al. (2016). MR imaging in spinocerebellar ataxias: a systematic review. AJNR Am J Neuroradiol 37: 1405–1412.

Koçer N, Islak C, Siva A et al. (1999). CNS involvement in neuro-Behçet syndrome: an MR study. AJNR Am J Neuroradiol 20: 1015–1024.

Leclerc T, Levy R, Tauziède-Espariat A et al. (2024). Imaging features to distinguish posterior fossa ependymoma subgroups. Eur Radiol 34: 1534–1544.

Lee JH, Ryan J, Andreescu C et al. (2015). Brainstem morphological changes in Alzheimer's disease. Neuroreport 26: 411–415.

Lee SH, Yoon PH, Park SJ et al. (2001). MRI findings in neuro-Behçet's disease. Clin Radiol 56: 485–494.

Li K, Yang Y, Guo D et al. (2020). Clinical and MRI features of posterior reversible encephalopathy syndrome with atypical regions: a descriptive study with a large sample size. Front Neurol 11: 194.

Louis DN, Perry A, Wesseling P et al. (2021). The 2021 WHO Classification of Tumors of the Central Nervous System: a summary. Neuro-Oncology 23: 1231–1251.

Lucio LL, Freddi TAL (2023). Glossopharyngeal, vagus and accessory nerves: anatomy and pathology. Semin Ultrasound CT MR 44: 95–103.

Lucio LL, Freddi TAL, Ottaiano AC (2022). The abducens nerve: anatomy and pathology. Semin Ultrasound CT MR 43: 414–419.

Mark AS (1998). Oculomotor motion disorders: current imaging of cranial nerves 3, 4, and 6. Semin Ultrasound CT MR 19: 240–256.

McKinney AM, Jagadeesan BD, Truwit CL (2013). Central-variant posterior reversible encephalopathy syndrome: brainstem or basal ganglia involvement lacking cortical or subcortical cerebral edema. AJR Am J Roentgenol 201: 631–638.

McNamara C, Mankad K, Thust S et al. (2022). 2021 WHO Classification of Tumours of the Central Nervous System: a review for the neuroradiologist. Neuroradiology 64: 1919–1950.

Meira AT, Arruda WO, Ono SE et al. (2019). Neuroradiological findings in the spinocerebellar ataxias. Tremor Other Hyperkinet Mov (N Y) 9.

Milhorat TH, Miller JI, Johnson WD et al. (1993). Anatomical basis of syringomyelia occurring with hindbrain lesions. Neurosurgery 32: 748–754.

Mittal S, Wu Z, Neelavalli J et al. (2009). Susceptibility-weighted imaging: technical aspects and clinical applications, part 2. AJNR Am J Neuroradiol 30: 232–252.

Nagae LM, Lall N, Dahmoush H et al. (2016). Diagnostic, treatment, and surgical imaging in epilepsy. Clin Imaging 40: 624–636.

Nunes RH, Corrêa DG, Pacheco FT et al. (2024). Neuroimaging of infectious vasculopathy. Neuroimaging Clin N Am 34: 93–111.

OMIM®: an online catalog of human genes and genetic disorders. Available at: < https://omim.org/ > (accessed 3 November 2023).

Osborn AG, Louis DN, Poussaint TY et al. (2022). The 2021 World Health Organization Classification of Tumors of the Central Nervous System: what neuroradiologists need to know. AJNR Am J Neuroradiol 43: 928–937.

Ottaiano AC, Gomez GD, Freddi TAL (2023). The facial nerve: anatomy and pathology. Semin Ultrasound CT MR 44: 71–80.

Oudman E, Wijnia JW, Oey MJ et al. (2021). Wernicke-Korsakoff syndrome despite no alcohol abuse: a summary of systematic reports. J Neurol Sci 426: 117482.

Paldino MJ, Faerber EN, Poussaint TY (2011). Imaging tumors of the pediatric central nervous system. Radiol Clin N Am 49: 589–616.

Parisi MA (2009). Clinical and molecular features of Joubert syndrome and related disorders. Am J Med Genet C Semin Med Genet 151C: 326–340.

Parizel PM, Makkat S, Jorens PG et al. (2002). Brainstem hemorrhage in descending transtentorial herniation (Duret hemorrhage). Intensive Care Med 28: 85–88.

Park YW, Vollmuth P, Foltyn-Dumitru M et al. (2023). The 2021 WHO Classification for Gliomas and Implications on Imaging Diagnosis: part 2 – summary of imaging findings on pediatric-type diffuse high-grade gliomas, pediatric-type diffuse low-grade gliomas, and circumscribed astrocytic gliomas. J Magn Reson Imaging 58: 690–708.

Pohl D, Alper G, Van Haren K et al. (2016). Acute disseminated encephalomyelitis: updates on an inflammatory CNS syndrome. Neurology 87: S38–S45.

Policeni BA, Smoker WR (2008). Pathologic conditions of the lower cranial nerves IX, X, XI, and XII. Neuroimaging Clin N Am 18: 347–368.

Pons-Escoda A, Naval-Baudin P, Velasco R et al. (2023). Imaging of lymphomas involving the CNS: an update – review of the full spectrum of disease with an emphasis on the World Health Organization Classifications of CNS Tumors 2021 and Hematolymphoid Tumors 2022. AJNR Am J Neuroradiol 44: 358–366.

Poretti A, Boltshauser E, Huisman TA (2016). Cerebellar and brainstem malformations. Neuroimaging Clin N Am 26: 341–357.

Poretti A, Boltshauser E, Plecko B (2007). Brainstem disconnection: case report and review of the literature. Neuropediatrics 38: 210–212.

Poretti A, Meoded A, Rossi A et al. (2013). Diffusion tensor imaging and fiber tractography in brain malformations. Pediatr Radiol 43: 28–54.

Quattrocchi CC, Cherubini A, Luccichenti G et al. (2010). Infratentorial lesion volume correlates with sensory functional system in multiple sclerosis patients: a 3.0-Tesla MRI study. Radiol Med 115: 115–124.

Quattrocchi CC, Errante Y, Rossi Espagnet MC et al. (2016). Magnetic resonance imaging differential diagnosis of brainstem lesions in children. World J Radiol 8: 1–20.

Querol-Pascual MR (2010). Clinical approach to brainstem lesions. Semin Ultrasound CT MR 31: 220–229.

Quisling RG, Quisling SG, Mickle JP (1993). Obex/nucleus gracilis position: its role as a marker for the cervicomedullary junction. Pediatr Neurosurg 19: 143–150.

Ramos A, Hilario A, Lagares A et al. (2013). Brainstem gliomas. Semin Ultrasound CT MR 34: 104–112.

Ranjan A, Kalita J, Kumar S et al. (2015). A study of MRI changes in Wilson disease and its correlation with clinical features and outcome. Clin Neurol Neurosurg 138: 31–36.

Renard D, Guillamo JS, Ion I et al. (2022). Brainstem lesions: MRI review of standard morphological sequences. Acta Neurol Belg 122: 597–613.

Ribas GA, De Mori LH, Freddi TAL et al. (2024). Primary central nervous system lymphoma: Imaging features and differential diagnosis. Neuroradiol J. 376: 705–722.

Rossi A, Catala M, Biancheri R et al. (2004). MR imaging of brainstem hypoplasia in horizontal gaze palsy with progressive scoliosis. AJNR Am J Neuroradiol 25: 1046–1048.

Ruchalski K, Hathout GM (2012). A medley of midbrain maladies: a brief review of midbrain anatomy and syndromology for radiologists. Radiol Res Pract 2012: 258524.

Rüsch CT, Bölsterli BK, Kottke R et al. (2020). Pontocerebellar hypoplasia: a pattern recognition approach. Cerebellum (London, England) 19: 569–582.

Sakurai K, Tokumaru AM, Shimoji K et al. (2017). Beyond the midbrain atrophy: wide spectrum of structural MRI finding in cases of pathologically proven progressive supranuclear palsy. Neuroradiology 59: 431–443.

Sandhu S, Soule E, Fiester P et al. (2019). Brainstem diffuse axonal injury and consciousness. J Clin Imaging Sci 9: 32.

Sarma A, Heck JM, Ndolo J et al. (2021). Magnetic resonance imaging of the brainstem in children, part 1: imaging techniques, embryology, anatomy and review of congenital conditions. Pediatr Radiol 51: 172–188.

Schwarz ST, Afzal M, Morgan PS et al. (2014). The 'swallow tail' appearance of the healthy nigrosome – a new accurate test of Parkinson's disease: a case-control and retrospective cross-sectional MRI study at 3T. PLoS One 9: e93814.

Sciacca S, Lynch J, Davagnanam I et al. (2019). Midbrain, pons, and medulla: anatomy and syndromes. Radiographics 39: 1110–1125.

Scortegagna F, Corrêa DG, Pacheco FT et al. (2024). Joubert syndrome with the decaying molar tooth sign: report of 2 cases. Acta Neurol Belg 125: 1481–1484

Severino M, Huisman TAGM (2019). Posterior fossa malformations. Neuroimaging Clin N Am 29: 367–383.

Severino M, Tortora D, Pistorio A et al. (2016). Expanding the spectrum of congenital anomalies of the diencephalic-mesencephalic junction. Neuroradiology 58: 33–44.

Shepherd TM, Hoch MJ (2022). MRI-visible anatomy of the brainstem. Neuroimaging Clin N Am 32: 553–564.

Sinha S, Taly AB, Ravishankar S et al. (2006). Wilson's disease: cranial MRI observations and clinical correlation. Neuroradiology 48: 613–621.

Smith LH, DeMyer WE (2003). Anatomy of the brainstem. Semin Pediatr Neurol 10: 235–240.

Soldatos T, Batra K, Blitz AM et al. (2014). Lower cranial nerves. Neuroimaging Clin N Am 24: 35–47.

Sotoudeh H, Razaei A, Saadatpour Z et al. (2021). Brainstem encephalitis. The role of imaging in diagnosis. Curr Probl Diagn Radiol 50: 946–960.

Spalice A, Nicita F, Papetti L et al. (2010). Usefulness of diffusion tensor imaging and fiber tractography in neurological and neurosurgical pediatric diseases. Child Nerv Syst 26: 995–1002.

Takahashi J, Shibata T, Sasaki M et al. (2015). Detection of changes in the locus coeruleus in patients with mild cognitive impairment and Alzheimer's disease: high-resolution fast spin-echo T1-weighted imaging. Geriatr Gerontol Int 15: 334–340.

Tardieu M, Banwell B, Wolinsky JS et al. (2016). Consensus definitions for pediatric MS and other demyelinating disorders in childhood. Neurology 87: S8–S11.

Thompson AJ, Banwell BL, Barkhof F et al. (2018). Diagnosis of multiple sclerosis: 2017 revisions of the McDonald criteria. Lancet Neurol 17: 162–173.

Tobin WO, Guo Y, Krecke KN et al. (2017). Diagnostic criteria for chronic lymphocytic inflammation with pontine perivascular enhancement responsive to steroids (CLIPPERS). Brain 140: 2415–2425.

Traylor KS, Branstetter BF 4th, (2022). Cranial nerve anatomy. Neuroimaging Clin N Am 32: 565–576.

Van Cauter S, Severino M, Ammendola R et al. (2020). Bilateral lesions of the basal ganglia and thalami (central grey matter) – pictorial review. Neuroradiology 62: 1565–1605.

van Dijk T, Baas F, Barth PG et al. (2018). What's new in pontocerebellar hypoplasia? An update on genes and subtypes. Orphanet J Rare Dis 13: 92.

Vattipally VR, Bronen RA (2006). MR imaging of epilepsy: strategies for successful interpretation. Magn Reson Imaging Clin N Am 14: 225–247.

Wahl M, Mukherjee P (2009). Diffusion imaging of congenital brain malformations. Semin Pediatr Neurol 16: 111–119.

Wang J, Li Y, Huang Z et al. (2018). Neuromelanin-sensitive magnetic resonance imaging features of the substantia nigra and locus coeruleus in de novo Parkinson's disease and its phenotypes. Eur J Neurol 25: 949–e73.

Wasenko JJ, Park BJ, Jubelt B et al. (2002). Magnetic resonance imaging of mesenrhombencephalitis. Clin Imaging 26: 237–242.

Weidauer S, Raab P, Hattingen E (2021). Diagnostic approach in multiple sclerosis with MRI: an update. Clin Imaging 78: 276–285.

Wenning GK, Stankovic I, Vignatelli L et al. (2022). The movement disorder society criteria for the diagnosis of multiple system atrophy. Mov Disord 37: 1131–1148.

Wingerchuk DM, Banwell B, Bennett JL et al. (2015). International consensus diagnostic criteria for neuromyelitis optica spectrum disorders. Neurology 85: 177–189.

Woolfall P, Coulthard A (2001). Pictorial review: Trigeminal nerve: anatomy and pathology. Br J Radiol 74: 458–467.

Zaki MS, Saleem SN, Dobyns WB et al. (2012). Diencephalic-mesencephalic junction dysplasia: a novel recessive brain malformation. Brain 135: 2416–2427.

Zhou J, Greicius MD, Gennatas ED et al. (2010). Divergent network connectivity changes in behavioural variant frontotemporal dementia and Alzheimer's disease. Brain 133: 1352–1367.

Handbook of Clinical Neurology, Vol. 216 (3rd series)
The Brainstem
L.T. Grinberg and A.D. Korczyn
https://doi.org/10.1016/B978-0-443-15736-3.00006-8

Chapter 18

Ultrasound studies of brainstem

BİNNUR ÇELİK[1] AND REZZAK YILMAZ[1,2,]*

[1]*Department of Neurology, Ankara University School of Medicine, Ankara, Turkey*

[2]*Ankara University Brain Research Center, Ankara, Turkey*

Abstract

Transcranial B-mode sonography (TCBS) is a well-established neuroimaging technique that enables the visualization and structural evaluation of certain brain structures. This chapter focuses on the application of TCBS and its potential use in the diagnosis of several neurologic disorders. The first part of this chapter outlines the technical requirements and describes the ultrasound method for an optimal TCBS examination in detail. The second section focuses on the existing evidence in movement disorders, with an emphasis on Parkinson disease and parkinsonism. In the third part of the chapter, the focus will be on the limitations and advantages of the TCBS. Current data indicate that TCBS is a cost-effective and user-friendly imaging technique with high accuracy and interrater reliability for diagnosing Parkinson disease. TCBS also serves as a valuable supplementary tool for distinguishing various movement disorders. Its limitations include potential difficulties with insufficient temporal bone windows in some individuals and the requirement for an experienced operator to achieve accurate results.

INTRODUCTION

Transcranial sonography is the ultrasonographic examination of the structures in the brain. While ultrasound of the brain structures is routinely used in pediatrics where the brain is visualized through the fontanels, examination of adults is also possible using some areas in the head where the skull is thinner, thus allowing visualization. Transcranial ultrasound studies are conducted under two primary imaging modalities. (1) Transcranial color-coded duplex sonography examines the structure and the performance (such as velocity and direction of the blood flow) of the main arteries of the brain to investigate the pathologies related to the field of vascular neurology/stroke. (2) Transcranial B-mode sonography (TCBS) is employed mainly in the field of movement disorders. It examines the structures of the brainstem, ventricles, and basal ganglia, and to a lesser extent other structures such as the insula or medial temporal lobe. This chapter will focus on TCBS.

METHOD OF TRANSCRANIAL B-MODE SONOGRAPHY

Equipment and ultrasound settings

First of all, a high-end ultrasound machine is recommended for the best performance and image resolution. Structures of the brain can be visualized using a phased-array transducer. Here, it should be noted that the shape, imaging depth, and resolution of the transducer play a role in the intended use. As a rule, high-frequency transducers (such as 10–12 MHz) provide better resolution at the expense of image depth and thus are suitable for examining more superficial structures. On the other hand, low-frequency transducers allow the examination of deeper structures. Given that the brainstem lies around 7–9 cm medially from the temporal bone, TCBS requires a low-frequency (2–4 MHz) transducer with adequate penetration.

Basic knowledge of the settings of an ultrasound machine is also necessary for an optimal examination. In most devices, an imaging depth of 15 cm and a dynamic range of 40–60 dB provide the best resolution. Time-gain compensation buttons of the ultrasound (these buttons are mostly located physically on the side of the keyboard, and are adjusted manually) can also be set depending on the depth of the relevant structure. The focus button of the machine (which is usually shown as an arrow and can be multiple) can also be adjusted to better visualize the target structure (Yilmaz and Berg, 2018).

Investigation procedure, imaging planes, and visible structures

The TCBS examination should be performed in a dark and comfortable room. The patient is placed in a supine position,

*Correspondence to: Rezzak Yilmaz, MD, Department of Neurology, Ankara University School of Medicine, Talatpasa Blv. 82, Ankara 06230, Turkey. Tel: +903125082214; E-mail: rezzak.yilmaz@ankara.edu.tr.

and the examiner sits behind the patient's head at the rear of the examination bed. This allows the easiest examination of both (right and left) sides. Then the transducer is placed on the preauricular temporal bone in front of the upper part of the auricle, parallel to the so-called orbitomeatal line (Fig. 18.1A) (Yilmaz and Berg, 2017). This part of the temporal bone is the starting point of the TCBS examination because it is the thinnest area in the reachable skull and provides the best visualization of the brain structures, and thus is called the "bone window." However, as the skull size, shape, and thickness vary in each individual, the location of the bone window may also vary. The transducer should be slowly moved in each direction to determine the best area providing the bone window. One should also make sure that there is enough ultrasound gel on the transducer to achieve sufficient contact between the transducer and the skin. If there is air between the transducer and the skin, the image quality will be poor. Also, the transducer should not be pressed too hard, which may cause a painful examination. The transducer should be held firmly. Pressing harder would not result in a better image.

The terminology in reporting the results of an ultrasound examination is also crucial. For ultrasound, the word "echogenicity" is used, which corresponds to the words "density" or "intensity" of computerized tomography (CT) or magnetic resonance imaging (MRI), respectively. While emerging technologies increasingly offer new postprocessing abilities, a standard B-mode examination displays the structures in the black-and-white spectrum. Within this spectrum, dark structures are reported as "hypoechogenic," while bright structures are "hyperechogenic." The word "normoechogenic" is used for structures that have a mid-echogenicity, that is, those that appear gray-looking. It should always be taken into account that the default echogenicity of a structure can be hypo- or hyper- and that does not necessarily indicate an abnormality. Thus, the "normal" echogenicities of the examined structures should be known to detect abnormalities in TCBS.

When the temporal bone window that enables the best visualization of the structures is found, the examination can begin. In TCBS, the brain structures are predominantly examined in two planes: the mesencephalic plane and the third ventricle plane (Fig. 18.1B).

Mesencephalic plane

When a patient is lying in a supine position, an axial image of the mesencephalic plane can be visualized by placing the transducer at the anterior section of the temporal bone, at the level of the orbitomeatal line (Fig. 18.1A). As mentioned, the transducer should be gently moved back and forth in circular movements to find the area providing the best bone window for the patient. At this plane, the most significant structure is the butterfly-shaped mesencephalon, which appears hypoechogenic (dark) in the middle at approximately 7 cm of penetration depth, surrounded by the hyperechogenic (bright) basal cisterns (Fig. 18.2A). In the mesencephalon, the brainstem raphe (BR) and the substantia nigra (SN) are the most important structures visible in this plane. In healthy individuals, the mesencephalic part of the midline raphe is visualized as a vertical hyperechogenic line which is located in the middle of the mesencephalon (Fig. 18.1B). The midline raphe should be assessed regarding its continuity, that is, an interruption

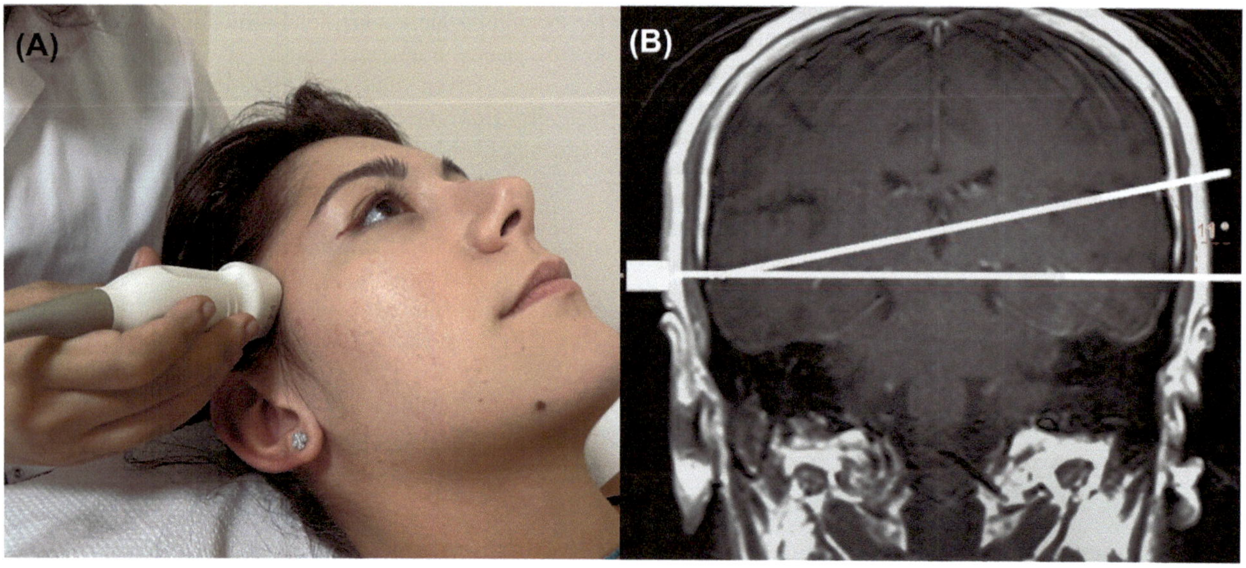

Fig.18.1. Image depicting the positioning of the examiner and the transducer, as well as the examination planes. (A) The patient lies in a supine position with the examiner seated at the head of the patient. The transducer should be positioned at the orbitomeatal line in front of the ear as shown. The transducer should then be moved back and forth to locate the bone window in the temporal bone, which allows optimal visualization of the structures. (B) After finding the bone window, the structures in the brain are examined in two planes as shown. The horizontal mesencephalic plane enables visualization of the mesencephalic brainstem. Examination of the third ventricle plane can be achieved by tilting the probe 10–15 degrees upward, as depicted.

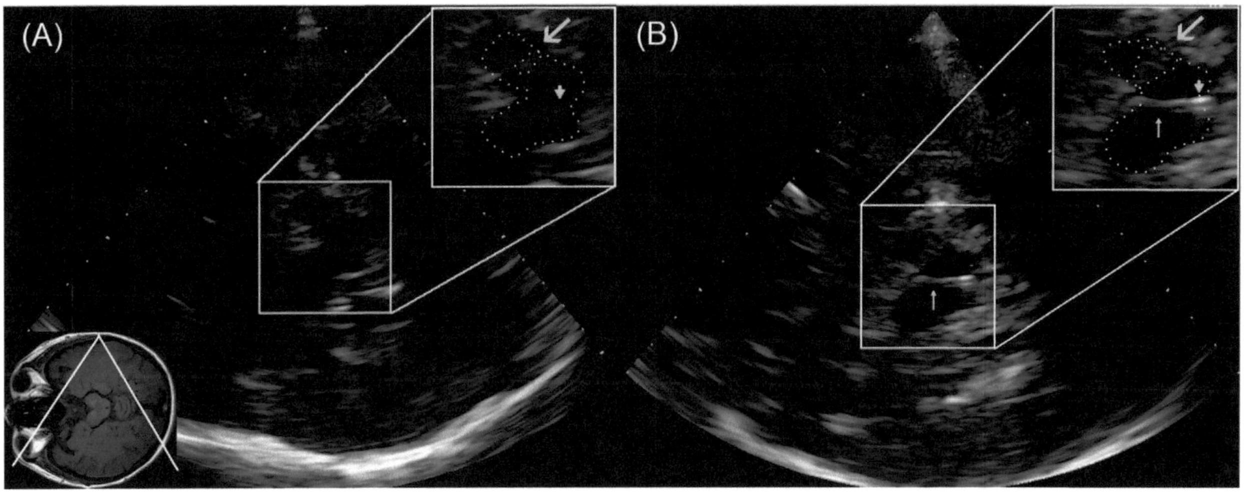

Fig.18.2. Images from the mesencephalic plane. The corresponding image from magnetic resonance imaging is shown at the bottom left. In both images, the butterfly-shaped hypoechogenic mesencephalon and the reflection of the contralateral hyperechogenic skull can be observed. On the ipsilateral side of the transducer, the substantia nigra is visible. The borders of the mesencephalon and the substantia nigra are delineated in the zoomed version at the top right. (A) Image of an individual with depression. Note the absence of the echogenic mesencephalic raphe in the middle of the mesencephalon compared to image B (vertical arrow). The substantia nigra appears normoechogenic (not enlarged, oblique arrow). (B) Image of an individual with Parkinson disease. The substantia nigra is enlarged, as depicted in the zoomed image at the top right, indicating substantia nigra hyperechogenicity (oblique arrow). The nucleus raphe is visible as an uninterrupted echogenic line (vertical arrow). Additionally, the aqueductus cerebri of the mesencephalon is shown in both images (arrowhead).

or the absence of the hyperechogenic line is abnormal and has been related to depression (Fig. 18.2A) (Drepper et al., 2018). Since the mesencephalic raphe is located at the midsection, assessment from one side is adequate if it looks normal (uninterrupted hyperechogenic line). To report an abnormal raphe, interruptions from both sides should be seen (Walter and Školoudík, 2014).

Within the mesencephalic plane, the SN is the second structure that should be evaluated. It becomes visible as a slightly hyperechogenic structure situated within the hypoechogenic mesencephalon inferior to the pedunculus cerebri. This evaluation should be conducted on the same side of the transducer. Since the SN is rather a small structure, the transducer should be carefully moved back and forth in order to find the image in which the SN is displayed at its largest dimensions. In contrast to the BR, the area of the SN should be measured to evaluate abnormality. This is done by freezing the image at its largest section and magnifying it two to threefold. After that, the anatomic area of the SN should be encircled manually using the measurement tools of the ultrasound, which is called the planimetric measurement (Fig. 18.2A and B). The cut-off value to determine whether the size of the SN is (pathologically) increased or not, is calculated by taking the 90th percentile of SN measurements from a sample of a minimum of 50 individuals aged 50 years or above, who are considered healthy (Berg et al., 2001). The cut-off values may vary based on both the ultrasound used and the specific population under assessment. Therefore, the detected cut-off value can only be used for that ultrasound machine and the population. For a detailed examination of cut-off values that vary according to different ultrasound systems, see Walter and Školoudík (2014). If the size of the SN area exceeds the established cut-off value, it is classified as SN hyperechogenicity (SN+) (Fig. 18.2B). It should be noted that, as explained above, the hyperechogenicity of a structure indicates its brightness in a given setting. However, in most of the literature, SN+ has been used as the terminology for the increase in the area of the SN (planimetric measurement) rather than its brightness.

The size of the hyperechogenic SN in the healthy population has been extensively investigated. The conducted studies have demonstrated that even healthy elderly individuals can exhibit a larger hyperechogenic area in the SN (Behnke et al., 2007). Another study compared the SN+ and mesencephalon size in individuals without any symptoms, including both children and adults. A positive correlation was found between age and hyperechogenic area of SN, as well as between age and SN hyperechogenic area/mesencephalon ratio (Hagenah et al., 2010). This may indicate an ongoing silent pathology in healthy individuals and is discussed below.

Of course, there were also attempts to establish a method for (additionally) evaluating the brightness of the SN. In one study, the echogenicity of the SN was divided into five grades: grade I: the same as the brainstem; grade II: very low but clearly detectable area of SN; grade III: medium echogenicity, lower than perimesencephalic cisterns; grade IV: same echogenicity as perimesencephalic cisterns; grade V: higher echogenicity than perimesencephalic cisterns (Bartova et al., 2008). Also, digital technology has been used to detect the brightness of the SN. In this method, the analysis is conducted by a computer program. By comparing the brightness of the SN area with the brightness of the rest of the image, an index was calculated. Values higher than the 90th percentile of a healthy population were considered indicative of SN+ (Školoudík et al., 2014). Of note, the given cut-off quantitative value of a structure for brightness should also include the current ultrasound settings (especially the dynamic range and time-gain compensation settings). Comparing the echogenicity of the structures in different settings is not possible; any ultrasound study reporting such values should take this into account (Yilmaz et al., 2023). Currently, both planimetric and digital analysis methods are being used for measurement in clinical settings. However, if not explained otherwise, the term SN+ in a published study indicates that the measured area of the SN is larger than the cut-off.

Although rare, visualization of the SN is also possible in the coronal plane. This plane is achieved by tilting the transducer 90 degrees. In this case, the SN would be seen as a slightly oblique vertical echogenic structure located below and laterally to the third ventricle. Evaluation of the SN in both (axial and coronal) sections may also be informative (Richter et al., 2017), although the established protocol involves only an axial examination (Walter and Školoudík, 2014; Yilmaz and Berg, 2017).

Third ventricle plane

The third ventricle plane (basal ganglia plane, thalamus plane) becomes visible by angling the ultrasound transducer approximately 10–15 degrees in the upward direction (Fig. 18.1B). By doing so, this plane is displayed in a slightly oblique plane, which prevents the evaluation of the structures on both sides due to asymmetric visualization. Also, in contrast to the structures in the mesencephalic plane, the structures of interest in this plane are located rather lateral to the midline (apart from the third ventricle and the pineal gland). This is important because as the monitor display of a phased-array transducer is cone-shaped, structures closer to the transducer appear smaller and "tightly packed" than the contralateral ones due to the increased spatial distribution of the ultrasound beam with increased distance from the source (Fig. 18.3A and B). This necessitates the examination of the structures in this plane on both (right and left) sides, and only structures contralateral to the transducer should be evaluated or measured, contrary to the evaluation of the SN, which is done on the ipsilateral side.

The evaluation of structures at this level begins with the detection of the pineal gland, which serves as a perfect orientation point as it is located in the middle and almost always appears very hyperechogenic due to calcification. After identifying the pineal gland, the borders of the third ventricle can be seen, which manifest as two parallel hyperechoic lines positioned anteriorly to the pineal gland. At this point, the transverse diameter of the third ventricle can be measured from its inner borders. Lateral to the third ventricle, the egg-shaped thalami can partially be detected on both sides. In healthy individuals, the thalami may appear isoechogenic or darker (hypoechogenic) than the surrounding structures. Progressing toward the nasal region, the frontal horn of the contralateral lateral ventricle comes into view on the opposite side of the transducer. Assessment of the frontal horn involves drawing a perpendicular line connecting the borders of the ventricle

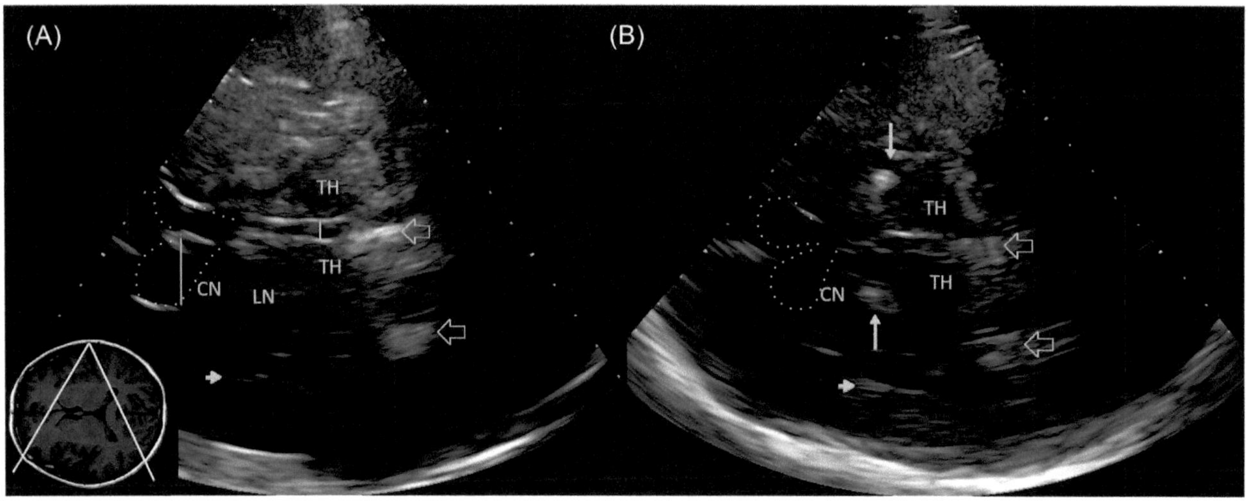

Fig.18.3. Images from the third ventricular plane. The corresponding image from magnetic resonance imaging is displayed at the bottom left. In both images, the reflection of the contralateral hyperechogenic skull is evident. (A) In the third ventricle plane, the centrally located pineal gland and the choroid plexus of the contralateral lateral ventricle are visible (open arrows). Both structures appear hyperechogenic due to calcification. In front of the pineal gland, the borders of the third ventricle are seen as two echogenic, parallel lines. The thalami are located on either side of the third ventricle and appear iso- or hypoechogenic. On the side opposite the transducer, the frontal horn of the lateral ventricle can be seen. The width of the third ventricle and the frontal horn can be determined by drawing a vertical line using the measurement tool. Adjacent to the frontal horn is the caudate nucleus. The lentiform nucleus is positioned between the caudate nucleus and the less echogenic thalamus. The insular cortex can also be observed on the far opposite side (arrowhead). (B) Image from the third ventricle plane of an individual with Wilson disease. Hyperechogenicity, due to copper accumulation in the LN, is clearly visible on both sides (vertical arrows). *CN*, caudate nucleus; *LN*, lentiform nucleus; *TH*, thalamus.

(Fig. 18.3A). The cut-off values for the third ventricle and the frontal horn are < 7/10 and < 17/20 mm for individuals younger or older than 60 years, respectively.

On this plane, the rest of the basal ganglia (excluding the subthalamic nucleus) can also be evaluated at the contralateral side of the insonating transducer. Adjacent to the frontal horn, the contralateral caudate nucleus (CN) is located. Positioned between the CN and thalamus is the lentiform nucleus (LN), which is not very clearly distinguishable and appears normoechogenic in healthy individuals due to its similar echogenicity with neighboring nuclei. The evaluation of the basal ganglia involves analyzing their echogenicity, specifically in relation to the echogenicity of the surrounding brain tissue or calcified structures like the pineal gland, choroid plexus, or even the skull. This kind of assessment is generally carried out in a semiquantitative manner, offering an approximate judgment rather than a quantitative measurement, such as indicating slight or moderately increased brightness. Notably, the CN may appear slightly more echogenic compared to the surrounding brain tissue, which aids in its identification (Walter and Školoudík, 2014; Yilmaz and Berg, 2018). On this plane, the insular cortex may also be visible as a horizontal hyperechogenic line as the farthest structure on the contralateral side, which has also been investigated in relation to disease (Fig. 18.3A and B) (Školoudík et al., 2016).

Other planes

In standard clinical TCBS examinations, the brain is visualized using mesencephalic and third ventricle planes. However, experts also use other planes and directions to evaluate deep brain stimulation (DBS) electrodes, dystonia, tremor patients, or evaluation of the medial temporal lobe in patients with Alzheimer disease or cerebellar pathologies (Walter and Školoudík, 2014; Yilmaz et al., 2021).

TRANSCRANIAL B-MODE SONOGRAPHY IN DISEASE

Movement disorders

Parkinson disease

Parkinson disease (PD) is the second most common neurodegenerative disorder (following Alzheimer disease) diagnosed around the ages of 55–60, when its core motor symptoms are detectable (bradykinesia accompanied by rigidity or 4–6 Hz tremor occurring when resting or walking; Shahed and Jankovic, 2007). The diagnostic criteria of PD have always been based on clinical examination; however, there are some imaging modalities that can support the diagnosis of parkinsonism, although not specific to PD. These include single-photon emission computerized tomography (SPECT), or positron emission tomography (PET) methods with several radiotracers, the most popular being ^{123}I-ioflupane SPECT (DATScan) (Scherfler et al., 2007). Detection of postganglionic cardiac autonomic denervation using I-123 metaiodobenzylguanidine scintigraphy is also another imaging method used in PD (Treglia et al., 2012).

Apart from those, in individuals diagnosed with idiopathic PD, a prominent observation through TCBS is the increased size of the echogenicity detected in the mesencephalic SN, which is called SN hyperechogenicity (SN+), as described above (Fig. 18.2B). The initial application of TCBS was carried out by Becker in 1995 to examine the brain parenchyma of PD. In this study, SN+ was demonstrated for the first time (Becker et al., 1995). Subsequent studies confirmed the presence of SN+ in patients with PD with a sensitivity of 85%–90%. In two meta-analyses involving more than 4000 PD patients each, the prevalence of SN+ was found to be 83% and 84% (Li et al., 2016; Shafieesabet et al., 2017), indicating that TCBS can be used for the diagnosis of PD.

Considering that detecting and appropriately measuring the SN requires expertise in TCBS, the reproducibility of this technique has also been studied. In one interesting study, sonographers examined patients who were covered with a blanket to conceal their symptoms (such as resting tremors) to secure the blindness of the sonographers and eliminate examiner bias in a darkened room. In that study, substantial intra and interrater reliability of the planimetric measurements was demonstrated (van de Loo et al., 2010). The following studies confirmed these findings with substantial to perfect intra- and interrater reliability values (Walter and Školoudík, 2014; Xu et al., 2020).

The pathologic meaning of the SN+ should also be mentioned. After the detection of SN+ in patients with PD, early studies focused on finding the correlates of this marker. In a study in which iron and 6-hydroxydopamine (a toxic substance for nigral dopaminergic neurons) were injected into the brains of rats, a dose-dependent increase in the echogenicity of the SN was detected. Interestingly, by adding chelators of iron to these substances, no increased echogenicity was found (Berg et al., 1999). This study was followed by three postmortem studies, in one of which the echogenicity of the SN was correlated with the iron content (but not calcium or copper) (Berg et al., 2002). The second study showed an increase in hyperechogenicity with decreased neuromelanin, which is a molecule binding iron and toxins, and decreases oxidative stress (Zecca et al., 2005). The third study additionally showed a correlation between SN echogenicity and microglia activation (Berg et al., 2010). These studies reported iron as the potential culprit in the pathophysiology behind increased echogenicity. This is reasonable given the fact that iron causes increased echogenicity like other metals in the brain. Also, it has been established that nonbound free iron increases nigral toxicity related to the pathology of PD. However, these studies still do not determine exactly whether the increased echogenicity is caused by an increase in iron, microglia, and gliosis, decrease in neuromelanin, or a combination of these. Probably, it is not related to only one cause as suggested in a well-designed recent study (Zhu et al., 2017).

In the field of PD, studies have also been conducted to investigate TCBS findings related to gene mutations that play a role in the pathogenesis of PD. It has been observed that PD patients with mutations in leucine-rich repeat kinase 2 (LRRK2), parkin (PRKN), PTEN-induced putative kinase 1 (PINK1), and DJ-1 have a smaller SN hyperechogenic area compared to (non-genetic) idiopathic PD patients, but a larger area compared to healthy controls (Schweitzer et al., 2007; Brockmann and Hagenah, 2010). When PD patients carrying LRRK2 G2019S mutation were compared with idiopathic PD patients, asymptomatic carriers of LRRK2 mutation, and healthy controls, no significant difference was observed in the echogenic area of SN between PD patients with and without LRRK2. Moreover, the area of SN hyperechogenicity in PD patients with LRRK2 mutation was similar compared to asymptomatic LRRK2 carriers (Brüggemann et al., 2011). The increased SN echogenicity in LRRK2 mutation carriers without any signs of parkinsonism has shed light on understanding the morphologic changes in the asymptomatic stage of the disease (Brockmann et al., 2011; Vilas et al., 2015). When individuals with homozygous and heterozygous PINK1 mutations within the same family were compared to healthy controls both with and without a family history of PD, it was observed that individuals with homozygous PINK1 mutations had a larger SN+ area compared to the control group, while those with heterozygous mutations showed no such effect (Hagenah et al., 2008). A relationship between the number of mutated alleles in the Parkin gene and the SN hyperechogenic area has also been demonstrated (Hagenah et al., 2007).

Beyond monogenic causes of PD, the effect of GBA1 mutation was also investigated, which is known to be a genetic risk factor for PD (Sidransky et al., 2009). The results showed that the area of hyperechogenicity was similarly enlarged in both GBA1 mutation carriers and patients with Gaucher disease relative to healthy controls (Arkadir et al., 2019). In another study, the mean SN hyperechogenic area was similar in sporadic PD and PD patients carrying GBA1 mutations. However, in healthy individuals with a GBA1 mutation, the SN size was not significantly different from controls (Kresojević et al., 2013).

When comparing the maximum area of nigral echogenicity between PD patients carrying GBA1 mutations and those carrying LRRK2 mutations, no significant difference was observed. In the same study, no relationship was found between the number of mutant alleles in the GBA1 gene and the size of the echogenic area (Barrett et al., 2013). In another study, a total of 34 individuals with and without GBA1 mutation patients, including 7 with parkinsonism, underwent examination with TCBS and [18F]-FDOPA PET. Larger SN hyperechogenic area and reduced striatal tracer-specific uptake were observed exclusively in patients with parkinsonism. In other words, there was a correspondence detected only in GBA1 mutation carriers with parkinsonism between nigral echogenicity and striatal presynaptic dopamine synthesis capacity. These findings are valuable for predicting future Parkinsonism in individuals with GBA1 mutations (Eisenberg et al., 2022).

The association of SN+ with several clinical features of PD has been extensively studied. These include parkinsonian motor and nonmotor symptoms or motor complications. A study with 300 PD patients reported that SN+ was more prevalent on the opposite side of where the motor symptoms began (Walter and Školoudík, 2014; Toomsoo et al., 2019). In another study involving a 5-year follow-up of 234 PD patients, asymmetry in SN echogenicity was associated with the asymmetry of symptoms (Fu et al., 2022). Furthermore, an association of SN+ with orthostatic hypotension was also studied, revealing a significantly larger SN in PD patients with this nonmotor symptom (Wu et al., 2021). With regard to motor complications, one frequent complication is levodopa-induced dyskinesia (LID), which can occur after prolonged levodopa use (Chen et al., 2020). In a study, the performance of TCBS in predicting the likelihood of developing LID was evaluated after measuring the SN quantitatively. While factors such as longer disease duration, a history of higher-dose levodopa usage, and greater disease severity are known to contribute to the development of LID, it was also observed that a larger hyperechogenic area was additionally associated with a higher likelihood of developing LID (Yan et al., 2023). In that regard, the question of whether the area of SN+ increases with disease duration is another matter of debate. This perspective has significant implications for the clinical application of SN assessments. In the case that SN+ does not change over time, it can be regarded as a risk factor for PD; however, demonstrating a change in the area with disease duration or severity would be more indicative of disease progression. Currently, it is believed that the SN+ does not change in the disease course, considering the majority of the studies have found no such change (Yilmaz and Berg, 2018).

Even though SN+ does not seem to change in the disease course, studies show that it is present before the diagnosis. Over the last decade, there has been a growing interest in the use of TCBS to diagnose prediagnostic stages of PD. These studies were performed in the context of prodromal PD, that is, to better identify individuals supposed to be in the prodromal stage of PD. Clinically, the prodromal stage is the phase of PD where certain symptoms or signs linked to PD appear, yet are not enough for a definitive diagnosis (Hustad and Aasly, 2020). Symptoms such as idiopathic rapid eye movement sleep behavior disorder (iRBD), depression, hyposmia, or autonomic disturbance may manifest, with some occurring as much as 15–20 years prior to a diagnosis (Heinzel et al., 2019). Identifying this prodromal phase is crucial, as it offers insights into nigral degeneration and presents an opportunity for potential disease-modifying treatments that might delay or stop the progression of the disease (Hung and Schwarzschild, 2020). In this context, one important multicenter longitudinal study on 1847 healthy individuals has shown that those with SN+ had a 17 times higher risk of developing PD within a three-year follow-up (Berg, 2011) and a 20 times higher risk after a five-year follow-up (Berg et al., 2013). These findings suggested the role of SN+ at the prediagnostic stages of PD and were followed by studies confirming these results, such as another study investigating healthy carriers of mutations of monogenetic PD which showed that these individuals exhibit the same echogenicity feature as their relatives who are already affected by the disease (Berg, 2006). A higher occurrence of SN+ in those with a family history of PD was also detected (Schweitzer et al., 2007).

Individuals with iRBD showed increased SN+ (Stockner et al., 2009; Iranzo et al., 2010). Moreover, in studies with DATScan, it was observed that healthy people with SN+ might exhibit reduced striatal uptake compared to those with standard echogenicity (Berg et al., 1999, 2002). Furthermore, individuals with no neurodegenerative diseases but with SN+ might perform poorer in motor activities than those with normoechogenicity (Berg et al., 2001; Walter et al., 2007c; Rupprecht et al., 2013). A larger area of SN in individuals with a high risk for future PD was also reported (Noyce et al., 2018). Moreover, subtle cognitive impairment has been shown in healthy individuals with SN+ (Liepelt et al., 2008; Yilmaz et al., 2016). Collectively, these findings suggest that increased SN echogenicity is present at the prodromal stage and may be used as a risk marker suggesting a predisposition to PD in the absence of a change in disease progression. This perspective has also been acknowledged in the Parkinson and Movement Disorders Society research criteria for prodromal PD, where SN+ has been named as a risk marker along with other risk or prodromal markers (Heinzel et al., 2019). In alignment with the published criteria for prodromal PD, studies in which the marker of SN+ has also been coupled with other risk or prodromal markers in PD patients were conducted, to find the best combination to better detect individuals with an ongoing PD-related neurodegeneration. In these studies, when TCBS, olfactory test, and motor assessments were combined, an increase in the accuracy rate for correct diagnosis was observed in early PD (Busse et al., 2012). Given that the clinical diagnosis of PD is still not that accurate, especially at the early stages (Rizzo et al., 2016), a combination of such markers may help to increase accuracy at the diagnostic stage. One study reported a sensitivity of 100% in the prediction of synucleinopathy in individuals with iRBD in 2.5 years when SN+ was combined with DATScan (Iranzo et al., 2010). These studies support that TCBS markers should be used in combination for the detection of people at the prodromal stage.

The strategy of combining markers for the detection of PD is the obvious consequence of the fact that no marker is robust enough to solely indicate future PD. As previously mentioned, this also applies to SN+. By definition (taking the 90th percentile of the measurements of healthy individuals older than 50 years old as a cutoff), 10% of healthy elderly individuals have SN+. However, most of them will never be diagnosed with PD given the prevalence of 1%–2% in this age group. Thus, an incidental finding of an SN+ during TCBS examination (such as intended for vascular imaging or detection of hydrocephalus) should not prompt disclosure of a potential underlying synucleinopathy, without being confirmed by additional markers including iRBD. However, in symptomatic individuals with suspected PD, SN+ detected by TCBS is a reliable imaging marker with high intra- and interrater reliability, as mentioned as a level A recommendation in the guidelines for the diagnosis of PD (Berardelli et al., 2013).

Atypical parkinsonism

Dementia with Lewy bodies

Dementia with Lewy bodies (DLB) is the second most common cause of neurodegenerative dementia after Alzheimer disease (AD). It is characterized by fluctuating cognition, visual hallucinations, and parkinsonism with cognitive impairment emerging before or within one year after the onset of parkinsonism (Yamada et al., 2020). The finding of bilaterally marked hyperechogenicity of the SN has been identified as a characteristic feature of DLB according to one German study (Favaretto et al., 2017). In that study, TCBS findings of 22 patients with DLB were compared with those of 28 patients with AD. The report addressed not only the size of the SN but also the brightness, which was graded 0–4. While the mean size of the SN was significantly larger in the DLB group, none in the DLB group had an SN brightness grade of 0. Similarly, no AD patient and control had an SN grade of 4. These findings indicate that the SN assessments of patients with AD should not be different than healthy individuals, supporting the notion that AD is not associated with SN pathology. Moreover, the findings underline a high level of echogenicity in patients with DLB with increased echo on both sides more than PD. This was previously suggested in a study comparing DLB patients with PD dementia and PD. While all groups exhibited SN+, the PD groups were associated with an asymmetry in SN+ contrary to DLB patients who showed marked bilateral SN+ with a rate of 80% (Walter et al., 2006). Using the proposed asymmetry index, patients with DLB could be differentiated from PD groups with 96% sensitivity and 80% specificity (Walter et al., 2006).

While comprehensive studies are lacking in the field of DLB, one study has hinted at the association of TCBS and prodromal DLB. In a 10-year longitudinal study, patients with iRBD were examined using TCBS to observe the risk of developing PD or DLB. Among the patients who exhibited SN+, a higher risk of developing both of these conditions was identified. This study indicates that SN+ may be regarded as a risk marker not only for PD but also for DLB (Miyamoto and Miyamoto, 2020).

Progressive supranuclear palsy

Progressive supranuclear palsy (PSP) is one of the most prevalent atypical parkinsonian syndromes with vertical gaze palsy, early gait difficulties, and parkinsonism (Giagkou et al., 2019). Recognizing the distinguishing features that can be obtained through TCBS from other parkinsonian syndromes is crucial for accurate diagnosis and effective treatment. In studies involving PSP patients using TCBS, a high prevalence of LN hyperechogenicity (LN+) was observed (Fig. 18.3B) (Behnke, 2005; Walter et al., 2007c; Kostić et al., 2013; Sadowski et al., 2015; Walter et al., 2003). According to these studies, LN+ was found to be approximately 73% in patients with PSP, whereas it was found to be 20% in PD.

An enlarged diameter of the third ventricle is another finding for PSP, corresponding to midbrain atrophy (Fig. 18.4B). In one study comparing corticobasal degeneration (CBD) and PSP, a

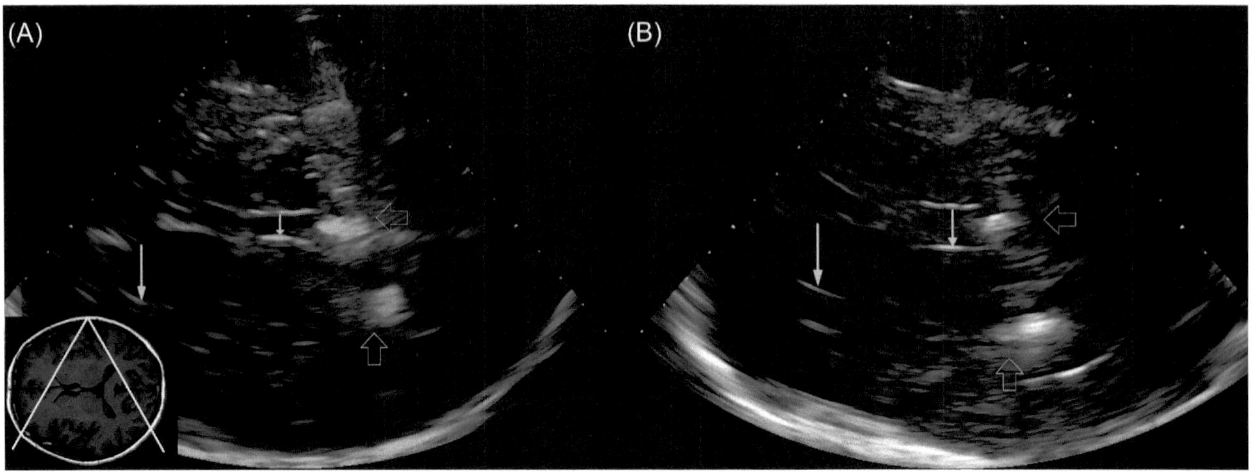

Fig.18.4. Images from the third ventricular plane. The corresponding image from magnetic resonance imaging is displayed at the bottom left. (A) Image from the third ventricle plane of a healthy individual, showing isoechogenic basal ganglia and no enlargement of the ventricles. (B) Image of an individual with normal pressure hydrocephalus. Note the enlarged third ventricle and the frontal horn of the lateral ventricle, which can be measured as depicted. Calcified pineal gland and choroid plexus of the lateral ventricle can also be observed as hyperechogenic structures (open arrows).

third ventricle width greater than 10 mm was observed solely in patients with PSP (Walter et al., 2004). This is also consistent with MRI studies, where enlargement of the third ventricle is also regarded as an imaging marker for PSP (Quattrone et al., 2021). These studies constitute an example that the findings of TCBS should be evaluated in combination rather than focusing on only SN+. In line with this, another TCBS study focused on atrophy rather than increased echogenicity in TCBS in patients with PSP and PD and demonstrated that the mesencephalic area was significantly smaller in PSP patients compared to PD patients (3.58 cm^2 vs 5.28 cm^2). In the same study, the third ventricle was also significantly enlarged in favor of PSP patients, as expected (8.84 mm vs 5.11 mm) (Sastre-Bataller et al., 2013). These findings were also confirmed by a recent study which showed decreased midbrain area and increased third ventricle in patients with PSP (Ghourchian et al., 2019).

As is well known, PSP has been clinically classified into subtypes according to the diagnostic criteria (Höglinger et al., 2017). One TCBS study compared the two most common forms of PSP: (1) PSP-parkinsonism (PSP-P), which is more associated with a better levodopa response, and (2) Richardson's syndrome (PSP-RS), which is the most frequent and the representative form of a classical PSP phenotype. Among the patients with PSP-RS and PSP-P, 14% and 73% of patients had SN+, respectively, whereas LN+ was observed in 67% and 36% of patients with PSP-RS and PSP-P, respectively. In addition, the third ventricle was significantly wider in patients with PSP-RS (11.2 ± 2.3 mm) when compared with patients with PSP-P (7.5 ± 1.4 mm) (Kostić et al., 2013). These results suggest that the two clinical forms might arise from distinct pathophysiologic processes. However, further studies are needed to validate this hypothesis. Nevertheless, these findings indicate a spectrum of PSP, where PSP-P is associated with more PD-like features both clinically (with levodopa response) and sonographically (with more SN+), whereas classical PSP is associated with no SN+ but with enlarged third ventricle and LN+.

Multiple system atrophy

Multiple system atrophy (MSA) is a form of atypical parkinsonism characterized by autonomic dysfunction, cerebellar syndrome, and parkinsonism with no family history (Fanciulli and Wenning, 2015). While cerebellar signs are absent in PD, differentiating PD and MSA may be challenging, especially in the early stages and particularly in patients with MSA-parkinsonism (MSA-P), where favorable response to levodopa or LID may be present (Poewe and Wenning, 2002). Therefore, TCBS has been investigated regarding its value in differentiating these two synucleinopathies.

In nontremor-dominant PD patients and MSA-P patients, SN+ was found to be 74.1% and 38.4%, respectively (Zhou et al., 2018). In another study, the specificity of SN+ was not high as a method to distinguish these two entities (Behnke, 2005). In a study that examined the value of TCBS, it was a more widespread marker for MSA compared to PD (sensitivity, 90%; specificity, 98%; positive predictive value, 86%) (Walter et al., 2007a). Also, one study, which included 101 patients with PD and 38 patients with atypical parkinsonism (MSA and PSP), revealed that the sensitivity and specificity of TCBS in making an accurate diagnosis were 53.1% and 91.7%, respectively (Fujita et al., 2016).

Corticobasal degeneration

CBD is the least frequent atypical parkinsonism subtype associated with parkinsonism, hyperkinesias (myoclonus or dystonia), higher cortical deficits (aphasia and/or apraxia), cognitive impairment, or alien limb phenomena (Armstrong et al., 2013). It has been reported that approximately 90% of CBD patients exhibit prominent bilateral SN+, which may help differentiate them from PSP. Apart from the SN+, another distinguishing feature between PSP and CBD is the width of the third ventricle (Walter et al., 2004), which has been reported to be larger in the PSP group. This is important because PSP may also present similar clinical features like CBD. Of note, PSP has a CBD subtype (PSP-CBD) (Giagkou et al., 2019). In another study comparing PSP and CBD patients, it was shown that the presence of parkinsonism symptoms accompanying SN+ indicates a diagnosis of CBD. However, LN+, especially when SN echogenicity is normal, tends to steer away from the diagnosis of CBD according to this study (Sadowski et al., 2015). In the largest-scale TCBS study conducted with CBD patients to date, the results showed that the sensitivity of SN and LN hyperechogenicity was remarkably better than previously reported, and the specificity was high, around 90%. Additionally, in this study, the enlargement of the third ventricle was found to be more frequent (43%) than expected, which can be attributed to the clinical overlap between PSP and CBD (Alonso-Canovas et al., 2019).

Most TCBS studies on atypical parkinsonism included several atypical groups and provided data on comparisons for atypical parkinsonism as a group, rather than one-to-one comparisons with PD. For instance, an early study on differential diagnosis between PD and atypical parkinsonian syndromes reported that the SN+ had a sensitivity of 94.8%, a specificity of 90%, and a diagnostic accuracy of 93.9% when atypical parkinsonism cases were taken as one group (Gaenslen et al., 2008). These early studies need confirmation given the possibility of inadequate image resolution compared to contemporary systems. It is also important to emphasize that LN should always be included in the TCBS examination of an individual. Results obtained from a meta-analysis involving 1091 patients with PD and 239 patients with atypical parkinsonism (MSA and PSP) confirmed that atypical parkinsonian syndromes like MSA-P and PSP display LN+, more frequently (Richter et al., 2019). Another study reported that the combination of marked SN+ along with normal LN echogenicity serves as a discriminating factor between PD, PSP, and MSA, with a positive predictive value exceeding 90% (Walter et al., 2007b).

In the differential diagnosis of parkinsonism using TCBS, the best accuracy is probably obtained by combining several TCBS markers (SN, LN, and third ventricle width), not the SN+ alone. This was also underlined in a recently published study, which reported values on the stratification of several TCBS markers. The study demonstrated that the combination of SN+ and LN+ had a higher specificity for atypical parkinsonism. Combining SN+ and third ventricle width also yielded high specificity for PSP. However, the sensitivity values were lower. And finally, combining the three markers resulted in 99% specificity for PSP and 91% specificity for CBD (Alonso-Canovas et al., 2019).

In summary, the literature supports the notion that the markers assessed in TCBS can be of use according to the suspected diagnosis of the patient, rather than a standard printout like a laboratory test. For instance, while the sensitivity of PD in a patient with SN+ may not be high enough for diagnosis only based on this echo feature, a normal SN echogenicity, LN+, and enlarged third ventricle rule out PD with an accuracy of 99% (Alonso-Cánovas et al., 2014). Therefore, in utilizing TCBS for differentiating atypical parkinsonism from PD, different sensitivity and specificity values of marker combinations for a specific comparison should be considered.

Secondary parkinsonism

Drug-induced parkinsonism

Drug-induced parkinsonism (DIP) is the second most common cause of secondary parkinsonism (Micheli and Cersosimo, 2007). It often develops as a result of dopamine receptor-blocking agents, mostly neuroleptics. There are studies in the literature where patients diagnosed with DIP have been examined using TCBS. The initial study was conducted on this topic in 2001 with 93 patients receiving neuroleptic treatment with and without DIP. The results showed that patients with DIP had significantly larger SN sizes compared to those with no DIP. Interestingly, this study also reported that the SN size before the initiation of a neuroleptic treatment might be correlated with future extrapyramidal symptoms (Berg et al., 2001). This was reported as an indication of the vulnerability of the SN, as in prodromal PD. Another study with 100 patients using neuroleptics showed an association between age and the size of the SN and parkinsonism in DIP (Jabs et al., 2003). It has been concluded that increased SN size is associated with increased sensitivity to extrapyramidal side-effects of neuroleptic drugs.

Whether SN+ represents future PD or nonresolution of the symptoms in patients with DIP was also investigated. In a prospective study, patients with DIP were followed for 6 months after discontinuation of the drug and were divided into two groups: the resolution of the parkinsonism group (DIP group) and the ongoing (subclinical) parkinsonism group. The blinded TCBS in these groups showed no significant difference in SN size. A similar study showed that patients with DIP have normal SN, which helps to distinguish them from PD (Mahlknecht et al., 2012). Another 2013 study also found no significant difference between DIP and PD. (Olivares Romero et al., 2013). However, these results were challenged in other studies. It has been proposed that the use of neuroleptics may trigger subclinic parkinsonism, which will not be resolved after drug discontinuation as indicated by the increased SN size before the start of the neuroleptic. To test this, 69 patients with DIP under dopamine receptor blocking agents were compared to PD patients and controls. While the DIP group did not have a significantly different SN size compared to controls, subgroup analyses showed more. The study reported normal SN (as in controls) in those with resolved parkinsonism after discontinuation. However, the baseline SN size was larger in those with permanent symptoms despite drug withdrawal (Oh et al., 2018).

In conclusion, it can be said that TCBS can be used in distinguishing between PD and DIP patients, considering that DIP is not associated with SN+. However, it is unclear whether the SN+ in some patients with DIP predicts future PD. More prospective studies are needed to clarify this issue.

Vascular parkinsonism

Vascular parkinsonism is another cause of secondary parkinsonism that predominantly impacts the elderly population, among whom vascular occlusive diseases are prevalent. The clinical manifestations of PD and vascular parkinsonism exhibit a significant degree of symptom overlap. However, the presentation of additional neurologic symptoms and signs, imaging findings of cerebrovascular disease, and a poor/variable levodopa response indicate rather vascular parkinsonism (Rektor et al., 2006).

Studies on vascular parkinsonism and TCBS are limited in the literature. Unlike PD patients, individuals with vascular parkinsonism typically do not display SN+. Also, in contrast to atypical parkinsonian syndromes, the basal ganglia are commonly observed as normoechogenic on TCBS in patients with vascular parkinsonism (Venegas-Francke, 2010). In one study, 80 PD, 30 vascular parkinsonism patients, and 60 controls were investigated using TCBS and Doppler ultrasound. The findings showed that 20% of patients with vascular parkinsonism displayed SN+ compared to 80% PD and 5% controls. Moreover, in this study, while flow velocities were not significantly different between the groups, the pulsatility index was significantly increased in patients with vascular parkinsonism (Tsai et al., 2007). These findings indicated that both modalities of ultrasound may be used to differentiate vascular parkinsonism from PD. However, additional studies are needed to confirm these results.

Hydrocephalus

Examination of a patient with suspected hydrocephalus using TCBS is mainly used in pediatrics through transfontanel ultrasound (El Adioui et al., 2023). However, in adults, this is possible through the temporal bone window as explained above. The TCBS examination may demonstrate hydrocephalus or ventricle enlargement, considering that the frontal horn of the lateral ventricle and the third ventricle can be measured using TCBS (Fig. 18.4B). The cut-off values for the ventricle sizes, which differ with regard to age, were also reported. The values for the third ventricle are < 7/10 mm and for the frontal horns are < 17/20 mm for individuals younger or older then 60 years, respectively (Go et al., 2012). These values may be of help in diagnosing normal pressure hydrocephalus. However, more data are needed on this subject. Having said that, assessment of ventricular enlargements may also be used in the intensive care units in patients with an acute increase in intracranial pressure. In a recent 2022 study with 177 patients hospitalized in neurologic intensive care units, examination of the third ventricle width and midline shift was compared between CT and TCBS, which showed a perfect correlation (r=0.96) indicating that TCBS can be used bedside to evaluate ventricle enlargement and midline shift without the need for transporting the patient to CT. The study also reported that the observed optimal threshold value for detecting hydrocephalus based on the width of the third ventricle was 5.72 mm (sensitivity=92%, specificity=92.1%) (Lasselin et al., 2022).

Other movement disorders

Essential tremor

Essential tremor (ET) is the most common movement disorder characterized by postural and kinetic tremor (action tremor) in contrast to resting tremor seen in PD. The prevalence of ET ranges from 2.8% to 4% in individuals older than 40 years (Dogu et al., 2003) and a prevalence of 14% in people older than 65 years (Wenning et al., 2005). It is necessary to distinguish between tremors of tremor-dominant PD and ET, especially in early disease stages. However, discrimination between ET and PD can be challenging.

Across all TCBS studies examining patients with ET, the low prevalence of SN+ was consistently reported, ranging from 0% to 18% (Kim et al., 2012; Alonso-Cánovas et al., 2014; Bártová et al., 2014; Richter et al., 2017). A meta-analysis that complied data from 1264 PD patients and 824 patients with ET revealed that TCBS demonstrated a combined sensitivity and specificity of 84.6% (95% confidence interval (CI), 79.4–88.6) and 83.9% (95% CI, 78.4–88.2), respectively, for the differential diagnosis between PD and ET (Heim et al., 2022). These findings focused more on the echogenicity of the SN rather than other TCBS features (such as LN echogenicity or ventricle enlargement), which play a more important role in atypical parkinsonism as described above. While the prevalence of individuals with ET having SN+ is much less frequent than PD, the percentage is higher than healthy controls. This is interesting, given that individuals with ET are approximately four times more likely to be diagnosed with PD compared to ones with no ET (Benito-Leon et al., 2008). Indeed, the higher prevalence of SN+ in the ET group may indicate an elevated risk of PD among the subset of ET patients displaying hyperechogenicity (Stockner and Wurster, 2010). One study tested this and showed that in a 3-year follow-up period, the risk of developing PD was shown in ET patients with SN+ (Cardaioli et al., 2019). In another study involving patients with ET, the relative risk for developing PD between those with baseline SN+ and those without was found to be 7.00 (95% CI, 1.62–30.34; sensitivity, 77.8%; specificity, 75.6%) (Sprenger et al., 2016). These studies indicate that in the differential diagnosis of ET and PD, SN normoechogenicity indicates ET. However, while SN+ is predominantly found in PD, it may also be detected in an individual with ET, in whom the likelihood of future PD diagnosis may be increased.

Huntington disease

Huntington disease (HD) is a hereditary neurodegenerative disorder characterized by an autosomal-dominant inheritance pattern with progressive motor, cognitive, and psychiatric deterioration mainly in adulthood (Ghosh and Tabrizi, 2018). In one early study conducted in 1999, 27% and 6% of the included 45 patients with HD displayed SN+ and LN+, respectively (Postert et al., 1999). This study also reported the increased echogenicity of the CN, which makes sense considering the established caudate pathology in HD patients (Rüb et al., 2015). Similarly, the CN echogenicity found in HD patients was correlated with higher signal intensities of CN in MRI (Postert et al., 1999). Studies also suggest that the increased number of CAG repeats in the Huntingtin gene is positively correlated with the presence of SN+ (Postert et al., 1999; Krogias et al., 2010).

One of the most comprehensive assessments was reported recently in a study of 21 individuals with HD and 23 healthy controls, in which echogenicity of the CN, SN, LN, insula, and BR was examined with TCBS-MRI fusion technique. Here, it should be noted that the assessment of the insular cortex may be regarded as a novelty, which is easily visible in TCBS-MRI fusion, a technique that enables the assessment of the structures that are not part of a standard TCBS assessment (Yilmaz and Berg, 2018). In this study, the authors reported that the echogenicity of CN, LN, and insula was significantly increased compared to controls. Moreover, the BR was significantly hypoechogenic. The study also reported very high sensitivity and specificity values for hyperechogenicity of SN and LN differentiating HD from controls, indicating that TCBS findings may potentially serve as diagnostic markers for HD.

The association of the symptoms of HD with several TCBS markers was also studied. For example, HD patients with hypokinesia exhibited a significantly higher prevalence of SN+ (93.3%), LN+ (53.3%), and CN+ (80%), surpassing the levels observed in the control group (Lambeck et al., 2015). Here, the very high prevalence of SN+ resembling PD is striking in HD patients with parkinsonism. A recent study in 2020, reported a lower prevalence of SN+ (66%). However, the degree of SN+ was correlated with motor scores on the Unified Huntington Disease Rating Scale (Witkowski et al., 2020). Moreover, they found a significant association with decreased hypoechogenicity of the BR and depression scores, in parallel with other reports on depression and raphe.

In light of the existing literature on HD and TCBS, it can be concluded that TCBS may be of help as a highly accurate diagnostic tool in differentiating patients with HD from controls. However, more studies are needed to differentiate HD from other forms of chorea or other neurodegenerative diseases that are listed in the differential diagnosis of HD.

Dystonia

Dystonia is a type of hyperkinetic movement characterized by sustained or intermittent twisting or tremulous muscle contractions causing abnormal movements or postures (Albanese et al., 2013). Studies on dystonia and TCBS date back to 1996, which reported increased LN echogenicity in patients with idiopathic dystonia. The authors stated that 75% of the patients with cervical dystonia, 83% of those with upper limb dystonia, and one-third of patients with facial dystonia display LN+ on the contralateral side of their symptoms. This finding was observed less frequently in the control group (12%) or in patients with secondary dystonia (0%) (Naumann et al., 1996). The finding of LN+ was supported by future studies with lower prevalence values (Becker et al., 1997; Defazio et al., 2007; Walter et al., 2014; Svetel et al., 2017). In addition, no such difference was reported in patients with oromandibular or facial dystonia (blepharospasm) (Zhang et al., 2016).

SN+ has been demonstrated in DYT6 dystonias (Saunders-Pullman et al., 2010). In another study, SN+ was identified in 63% of patients with dopa-responsive dystonia, a percentage more similar to those with PD (87%) than to individuals with other types of dystonia (20%) (Svetel et al., 2017). The literature also presents a case of a hairdresser's dystonia with contralateral SN+ accompanying the symptoms (Yoo et al., 2015).

Considering all of these studies, it becomes evident that TCBS findings in dystonia patients are variable. This phenomenon could potentially stem from the distinct subtypes of dystonia emerging through different mechanisms. Postmortem investigations have revealed copper accumulation in the LN of dystonia patients (Becker et al., 1999). To shed further light on the pathologic findings, more research is required.

Restless legs syndrome

The first study on TCBS findings in restless legs syndrome (RLS) was conducted with a small sample size. In this study, a reduced echogenicity of the SN was observed both compared to the healthy group and PD patients (Schmidauer et al., 2005). A decrease or loss of SN echogenicity more than controls was an interesting finding, considering that RLS may be accompanied by decreased ferritin, whereas the SN+ in PD may be associated with increased (nonbound) iron concentration in that region.

The echogenicity of the BR was also investigated in patients with RLS. In one Chinese study, 349 PD patients were divided into two groups, those with and without RLS. It was observed that patients with RLS had a significantly higher rate of hypoechogenicity in the BR (21.5% vs 7.8%). The patients with RLS also exhibited a significant increase in the size of the third ventricle compared to the group without these symptoms. However, interestingly, no significant differences in SN parameters were observed between the two groups (Zhuang et al., 2021). This is in contrast to previous studies in which SN hypoechogenicity in patients with RLS was reported. Only a 2010 study reported no change in SN echogenicity in patients with and without RLS, who were all diagnosed with PD. That study also reported SN hypoechogenicity in patients with idiopathic RLS (Kwon et al., 2010). To summarize, although the majority of the literature favors decreased echogenicity of SN in patients with RLS, more studies with larger sample sizes and more defined groups of RLS (secondary, idiopathic) are needed to determine whether TCBS can be of use in the daily routine for the diagnosis of RLS.

Transcranial B-mode sonography in deep brain stimulation

DBS has gained popularity, especially in the last decades for the symptomatic treatment of PD, dystonia, and tremor (Krack et al., 2019). For the success of the surgical process, the position of the DBS electrodes is of the utmost importance. Given that

DBS electrodes can be visualized using TCBS, and considering the difficulties in performing postsurgical MRI in these patients, the value of TCBS has been investigated for the detection of electrode localization. In one study, the position of electrodes in the subthalamic nucleus could be detected in eight patients using TCBS (Moringlane et al., 2005). Later in 2011, a study with 34 DBS patients showed that TCBS detected the electrode localization and displacement successfully. This study also reported that the suboptimal positioning detected by TCBS during the postoperative period predicted motor response in the long term (Walter et al., 2011). These findings were recently confirmed in an independent study conducted in 2018 with 19 PD patients, which investigated the distance between implanted electrodes and the third ventricle with TCBS, CT, and MRI. The study reported a significant correlation was found between distances from the electrode to the third ventricle by TCBS and CT/MRI. A cut-off value of 8.85 mm in these distances was the most sensitive (100%) and specific (90.5%) to predict if electrodes were positioned inside the SN (Ispierto et al., 2018). The image fusion technology was also used to evaluate electrode localization. In one study, the patient's MRI was overlaid in real time during TCBS. Through MRI-TCBS fusion, electrode displacements were successfully detected by integrating preoperative MRI images with postoperative ultrasound images in patients undergoing DBS (Walter et al., 2016). Apart from the studies focusing on postoperative electrode localization, the use of TCBS was also demonstrated in the perioperative setting (Walter et al., 2009).

In summary, considering the existing evidence, TCBS can be of use in expert hands to detect electrode localization in patients with DBS.

Other diseases/disorders

Accumulation syndromes

It is known that the primary location of the pathologic accumulation of elements such as calcium, iron, and other metals within the brain is the basal ganglia. Given that the echogenicity of the basal ganglia (especially the LN) can be evaluated using TCBS, and considering the detection of increased echogenicity of these elements in various tissues, TCBS emerges as a valuable noninvasive tool for detecting abnormal metal deposition in the basal ganglia.

Wilson's disease

Wilson's disease is caused by an autosomal recessive defect on chromosome 13, leading to defective biliary excretion and accumulation of copper in various tissues, primarily in the liver, cornea, and brain (European Association for the Study of the Liver, 2012). In the first TCBS study conducted on Wilson's disease, 89% of the 18 neuro-Wilson patients had LN+, with bilateral involvement. There was also a significant correlation between the extent of the hyperechogenic area and the severity of the disease. SN+ was observed in 25% of the patients (Walter et al., 2005). In another study, LN+ was detected in 82% of patients with neuro-Wilson and in 25% with hepatic-Wilson, compared to 7% of controls. SN+ was identified in 42% of neuro-Wilson patients, 8% of hepatic-Wilson patients, and 7% of healthy individuals (Svetel et al., 2012). A study regarding the prevalence of SN+ in neuro-Wilson found no significant difference between patients with Wilson's disease and controls (Mašková et al., 2016).

Another study in 41 treatment-naïve patients consisting of 20 hepatic-Wilson and 21 neuro-Wilson showed that 86% (n=18) and 45% (n=9) had LN+, respectively (Skowrońska et al., 2020). One study used quantitative assessments in 40 patients with neuro-Wilson and 40 controls (Tribl et al., 2018). They reported that Wilson's disease patients not only displayed significant LN or SN hyperechogenicity but also had increased echogenicity in the thalami and midbrain. Of note, the prevalence of LN+ in neuro-Wilson patients was 92.5% (in controls 20.4%), which validates the findings of previous studies. Interestingly, one prospective study on the effect of treatment on TCBS findings reported that the hyperechogenicity present in the LN did not show any changes, but the hyperechogenicity in the SN decreased with the treatment (Skowrońska et al., 2020). Carriers of the variants in the ATP7B gene were also investigated. Bilateral LN+ was found to be significantly more prevalent in heterozygous carriers of the ATP7B gene (73%) compared to healthy controls (0%). SN + was also observed to be more frequent in heterozygous carriers compared to healthy individuals (Skowronska et al., 2020).

These studies confirm the high prevalence of LN+ in patients with Wilson's disease (around 80%–90%), particularly in those with neurologic symptoms. However, SN+ seems to be similar or at least not strikingly more frequent than controls. The correlation of the LN+ with MRI findings and TCBS studies on postmortem brains also confirmed that the increased echogenicity is caused by copper accumulation (Walter and Školoudík, 2014). Thus, in patients with suspected neuro-Wilson disease, TCBS examination may be of help to support a diagnosis of Wilson's disease when bilateral LN+ is detected. However, it should also be noted that most of the listed studies included healthy controls to compare patients with Wilson's disease. To reach a conclusion on the value of TCBS in Wilson's disease, more studies should be conducted, including spectrum patients with accumulation syndromes or with diseases listed in the differential diagnoses of Wilson's disease.

Neurodegeneration with brain iron accumulation

Neurodegeneration with brain iron accumulation (NBIA) is a syndrome characterized by abnormal iron accumulation in specific regions of the brain, especially in the globus pallidus and SN pars reticulata (Koeppen and Dickson, 2001). The first findings of NBIA in TCBS were reported as a case report with LN+ and normal SN echogenicity (Brüggemann et al., 2011). In the following year, seven genetically confirmed NBIA patients were compared with 13 controls using TCBS. A significant increase in SN echogenicity and size was detected in NBIA patients (Liman et al., 2012). Among a group of 13 patients with "mitochondrial membrane protein-associated neurodegeneration," 12 individuals were observed to exhibit LN+, while none showed SN+ (Skowronska et al., 2018).

While MRI and genetic testing remain the gold standard for diagnosis, the use of TCBS may enable early and easier diagnosis of NBIA. More studies are needed to fully understand the role of TCBS in the differential diagnosis of NBIA or accumulation disorders.

Limitations of transcranial B-mode sonography

The primary limitation of TCBS is the absence of an adequate bone window in the preauricular area that permits examination of the structures. The bone window necessary for TCBS examination may narrow with increasing age (Školoudík and Walter, 2010). This is an important shortcoming considering the fact that TCBS is primarily used in adults, especially in the elderly population (mainly for PD and ventricle width). This may be attributed to the deterioration of the trabecular structure of the temporal bone, leading to the scattering of the ultrasound waves. Of note, the lack of a bone window appears to be more frequent in the female population (Školoudík and Walter, 2010). Studies also demonstrate that the bone window may differ according to population. While the absence of an adequate bone window was reported to be 10%–15% in Caucasians, data from the Chinese population were much higher, reaching 30% (Zhou et al., 2016). It should also be noted that in TCBS studies, the type of ultrasound, transducer type, brand, frequency, and settings are study-specific and might not allow the direct comparison of the quantitative findings between studies. Another limitation is the need for expertise in performing TCBS and evaluating structures. While some studies report on the blinded evaluations of the already-saved images, obtaining and saving the best image is heavily dependent on the examiner, which may be counted as bias in TCBS studies. On the other hand, widely organized hands-on TCBS courses worldwide and the growing culture of utilizing USG in neurology clinics may help to overcome this limitation.

CONCLUSION

TCBS, owing to its ease of application, cost-effectiveness, radiation-free nature, and patient-friendly approach, is a frequently utilized technique today and is expected to continue being used in the future. It is an imaging method that offers the opportunity for bedside examination of the brainstem, certain cortical structures, and the ventricular system, making it particularly valuable in the diagnosis, and differential diagnosis of PD and movement disorders. Challenges arising from interindividual differences in the bone window, the need for expertise, and issues related to device standardization are limitations that are expected to be addressed in the future.

References

Albanese A, Bhatia K, Bressman SB et al. (2013). Phenomenology and classification of dystonia: a consensus update. Mov Disord 28: 863–873.

Alonso-Cánovas A, López-Sendón JL, Buisán J et al. (2014). Sonography for diagnosis of Parkinson disease—from theory to practice. J Ultrasound Med 33: 2069–2074.

Alonso-Canovas A, Tembl Ferrairó JI, Martínez-Torres I et al. (2019). Transcranial sonography in atypical parkinsonism: how reliable is it in real clinical practice? A multicentre comprehensive study. Parkinsonism Relat Disord 68: 40–45.

Arkadir D, Dinur T, Becker Cohen M et al. (2019). Prodromal substantia nigra sonography undermines suggested association between substrate accumulation and the risk for *GBA*-related Parkinson's disease. Eur J Neurol 26: 1013–1018.

Armstrong MJ, Litvan I, Lang AE et al. (2013). Criteria for the diagnosis of corticobasal degeneration. Neurology 80: 496–503.

Barrett MJ, Hagenah J, Dhawan V et al. (2013). Transcranial sonography and functional imaging in glucocerebrosidase mutation Parkinson disease. Parkinsonism Relat Disord 19: 186–191.

Bártová P, Kraft O, Bernátek J et al. (2014). Transcranial sonography and ^{123}I-FP-CIT single photon emission computed tomography in movement disorders. Ultrasound Med Biol 40: 2365–2371.

Bartova P, Skoloudik D, Bar M et al. (2008). Transcranial sonography in movement disorders. Biomed Pap Med Fac Univ Palacky Olomouc Czech Repub 152: 251–258.

Becker G, Berg D, Rausch W-D et al. (1999). Increased tissue copper and manganese content in the lentiform nucleus in primary adult-onset dystonia. Ann Neurol 46: 260–263.

Becker G, Naumann M, Scheubeck M et al. (1997). Comparison of transcranial sonography, magnetic resonance imaging, and single photon emission computed tomography findings in idiopathic spasmodic torticollis. Mov Disord 12: 79–88.

Becker G, Seufert J, Bogdahn U et al. (1995). Degeneration of substantia nigra in chronic Parkinson's disease visualized by transcranial color-coded real-time sonography. Neurology 45: 182–184.

Behnke S (2005). Differentiation of Parkinson's disease and atypical parkinsonian syndromes by transcranial ultrasound. J Neurol Neurosurg Psychiatry 76: 423–425.

Behnke S, Double KL, Duma S et al. (2007). Substantia nigra echomorphology in the healthy very old: correlation with motor slowing. Neuroimage 34: 1054–1059.

Benito-Leon J, Louis ED, Bermejo-Pareja F (2008). Risk of incident Parkinson's disease and parkinsonism in essential tremor: a population based study. J Neurol Neurosurg Psychiatry 80: 423–425.

Berardelli A, Wenning GK, Antonini A et al. (2013). EFNS/MDS-ES recommendations for the diagnosis of Parkinson's disease. Eur J Neurol 20: 16–34.

Berg D (2006). Transcranial sonography in the early and differential diagnosis of Parkinson's disease. Parkinson's disease and related disorders. Vienna: Springer, pp. 249–254.

Berg D (2011). Enlarged substantia nigra hyperechogenicity and risk for Parkinson disease. Arch Neurol 68: 932.

Berg D, Becker G, Zeiler B et al. (1999). Vulnerability of the nigrostriatal system as detected by transcranial ultrasound. Neurology 53: 1026.

Berg D, Behnke S, Seppi K et al. (2013). Enlarged hyperechogenic substantia nigra as a risk marker for Parkinson's disease. Mov Disord 28: 216–219.

Berg D, Godau J, Riederer P et al. (2010). Microglia activation is related to substantia nigra echogenicity. J Neural Transm 117: 1287–1292.

Berg D, Grote C, Rausch W-D et al. (1999). Iron accumulation in the substantia nigra in rats visualized by ultrasound. Ultrasound Med Biol 25: 901–904.

Berg D, Jabs B, Merschdorf U et al. (2001). Echogenicity of substantia nigra determined by transcranial ultrasound correlates with severity of parkinsonian symptoms induced by neuroleptic therapy. Biol Psychiatry 50: 463–467.

Berg D, Roggendorf W, Schröder U et al. (2002). Echogenicity of the substantia nigra. Arch Neurol 59: 999.

Berg D, Siefker C, Ruprecht-Dörfler P et al. (2001). Relationship of substantia nigra echogenicity and motor function in elderly subjects. Neurology 56: 13–17.

Brockmann K, Gröger A, Di Santo A et al. (2011). Clinical and brain imaging characteristics in leucine-rich repeat kinase 2-associated PD and asymptomatic mutation carriers. Mov Disord 26: 2335–2342.

Brockmann K, Hagenah J (2010). TCS in monogenic forms of Parkinson's disease. Int Rev Neurobiol 90: 157–164.

Brüggemann N, Hagenah J, Stanley K et al. (2011). Substantia nigra hyperechogenicity with *LRRK2* G2019S mutations. Mov Disord 26: 885–888.

Brüggemann N, Wuerfel J, Petersen D et al. (2011). Idiopathic NBIA – clinical spectrum and transcranial sonography findings. Eur J Neurol 18: e58–e59.

Busse K, Heilmann R, Kleinschmidt S et al. (2012). Value of combined midbrain sonography, olfactory and motor function assessment in the differential diagnosis of early Parkinson's disease. J Neurol Neurosurg Psychiatry 83: 441–447.

Cardaioli G, Ripandelli F, Paolini Paoletti F et al. (2019). Substantia nigra hyperechogenicity in essential tremor and Parkinson's disease: a longitudinal study. Eur J Neurol 26: 1370–1376.

Chen X, Wang Y, Wu H et al. (2020). Research advances on L-DOPA-induced dyskinesia: from animal models to human disease. Neurol Sci 41: 2055–2065.

Defazio G, Berardelli A, Hallett M (2007). Do primary adult-onset focal dystonias share aetiological factors? Brain 130: 1183–1193.

Dogu O, Sevim S, Camdeviren H et al. (2003). Prevalence of essential tremor: door-to-door neurologic exams in Mersin Province, Turkey. Neurology 61: 1804–1806.

Drepper C, Geißler J, Pastura G et al. (2018). Transcranial sonography in psychiatry as a potential tool in diagnosis and research. World J Biol Psychiatry 19: 484–496.

Eisenberg DP, Lopez G, Gregory MD et al. (2022). Comparison of transcranial sonography and [^{18}F]-fluorodopa PET imaging in GBA1 mutation carriers. Mov Disord 37: 629–634.

El Adioui G, Ettouhami B, Haimeur Y et al. (2023). Transcranial sonographic measurement of the third ventricle to detect hydrocephalus in children population: correlation to computed tomography. J Clin Ultrasound 51: 995–1000.

European Association for the Study of the Liver (2012). EASL clinical practice guidelines: Wilson's disease. J Hepatol 56: 671–685.

Fanciulli A, Wenning GK (2015). Multiple-system atrophy. N Engl J Med 372: 249–263.

Favaretto S, Walter U, Baracchini C et al. (2017). Transcranial sonography in neurodegenerative diseases with cognitive decline. JAD 61: 29–40.

Fu XY, Zhang YC, Ding CW et al. (2022). Association between asymmetry of substantia nigra hyperechogenicity and clinical characteristics in different Parkinson disease subtypes: a 5-year follow-up study. Ultrasound Med Biol 48: 2139–2145.

Fujita H, Suzuki K, Numao A et al. (2016). Usefulness of cardiac MIBG scintigraphy, olfactory testing and substantia nigra hyperechogenicity as additional diagnostic markers for distinguishing between Parkinson's disease and atypical Parkinsonian syndromes. PLoS One 11: e0165869.

Gaenslen A, Unmuth B, Godau J et al. (2008). The specificity and sensitivity of transcranial ultrasound in the differential diagnosis of Parkinson's disease: a prospective blinded study. Lancet Neurol 7: 417–424.

Ghosh R, Tabrizi SJ Clinical features of Huntington's disease, pp.1–28.

Ghourchian S, Mousavi A, Zamani B et al. (2019). Midbrain area for differentiating Parkinson's disease from progressive supranuclear palsy. Clin Neurol Neurosurg 183: 105383.

Giagkou N, Höglinger GU, Stamelou M (2019). Progressive supranuclear palsy. Int Rev Neurobiol 49–86.

Go CL, Frenzel A, Rosales RL et al. (2012). Assessment of substantia nigra echogenicity in German and Filipino populations using a portable ultrasound system. J Ultrasound Med 31: 191–196.

Hagenah JM, Becker B, Bruggemann N et al. (2008). Transcranial sonography findings in a large family with homozygous and heterozygous PINK1 mutations. J Neurol Neurosurg Psychiatry 79: 1071–1074.

Hagenah JM, König IR, Becker B et al. (2007). Substantia nigra hyperechogenicity correlates with clinical status and number of Parkin mutated alleles. J Neurol 254: 1407–1413.

Hagenah J, König IR, Sperner J et al. (2010). Life-long increase of substantia nigra hyperechogenicity in transcranial sonography. Neuroimage 51: 28–32.

Heim B, Peball M, Hammermeister J et al. (2022). Differentiating Parkinson's disease from essential tremor using transcranial sonography: a systematic review and meta-analysis. J Parkinsons Dis 12: 1115–1123.

Heinzel S, Berg D, Gasser T et al. (2019). Update of the MDS research criteria for prodromal Parkinson's disease. Mov Disord 34: 1464–1470.

Höglinger GU, Respondek G, Stamelou M et al. (2017). Clinical diagnosis of progressive supranuclear palsy: the movement disorder society criteria. Mov Disord 32: 853–864.

Hung AY, Schwarzschild MA (2020). Approaches to disease modification for Parkinson's disease: clinical trials and lessons learned. Neurotherapeutics 17: 1393–1405.

Hustad E, Aasly JO (2020). Clinical and imaging markers of prodromal Parkinson's disease. Front Neurol 11: 395.

Iranzo A, Lomeña F, Stockner H et al. (2010). Decreased striatal dopamine transporter uptake and substantia nigra hyperechogenicity as risk markers of synucleinopathy in patients with idiopathic rapid-eye-movement sleep behaviour disorder: a prospective study. Lancet Neurol 9: 1070–1077.

Ispierto L, Muñoz J, Cladellas JM et al. (2018). Post-operative localization of deep brain stimulation electrodes in the subthalamus using transcranial sonography. Neuromodulation 21: 574–581.

Jabs BE, Bartsch AJ, Pfuhlmann B (2003). Susceptibility to neuroleptic-induced parkinsonism—age and increased substantia nigra echogenicity as putative risk factors. Eur Psychiatry 18: 177–181.

Kim J-S, Oh Y-S, Kim Y-I et al. (2012). Transcranial sonography (TCS) in Parkinson's disease (PD) and essential tremor (ET) in relation with putative premotor symptoms of PD. Arch Gerontol Geriatr 54: e436–e439.

Koeppen AH, Dickson AC (2001). Iron in the Hallervorden-Spatz syndrome. Pediatr Neurol 25: 148–155.

Kostić VS, Mijajlović M, Smajlović D et al. (2013). Transcranial brain sonography findings in two main variants of progressive supranuclear palsy. Eur J Neurol 20: 552–557.

Krack P, Volkmann J, Tinkhauser G et al. (2019). Deep brain stimulation in movement disorders: from experimental surgery to evidence-based therapy. Mov Disord 34: 1795–1810.

Kresojević N, Mijajlović M, Perić S et al. (2013). Transcranial sonography in patients with Parkinson's disease with glucocerebrosidase mutations. Parkinsonism Relat Disord 19: 431–435.

Krogias C, Eyding J, Postert T (2010). Transcranial sonography in Huntington's disease. Int Rev Neurobiol 90: 237–257.

Kwon D-Y, Seo W-K, Yoon H-K et al. (2010). Transcranial brain sonography in Parkinson's disease with restless legs syndrome. Mov Disord 25: 1373–1378.

Lambeck J, Niesen W-D, Matthias R et al. (2015). Substantia nigra hyperechogenicity in hypokinetic Huntington's disease patients. J Neurol 262: 711–717.

Lasselin P, Grousson S, Souza Netto EP et al. (2022). Accuracy of bedside bidimensional transcranial ultrasound versus tomodensitometric measurement of the third ventricle. J Neuroimaging 32: 629–637.

Li D-H, He Y-C, Liu J et al. (2016). Diagnostic accuracy of transcranial sonography of the substantia nigra in Parkinson's disease: a systematic review and meta-analysis. Sci Rep 6: 20863.

Liepelt I, Wendt A, Schweitzer KJ et al. (2008). Substantia nigra hyperechogenicity assessed by transcranial sonography is related to neuropsychological impairment in the elderly population. J Neural Transm 115: 993–999.

Liman J, Wellmer A, Rostasy K et al. (2012). Transcranial ultrasound in neurodegeneration with brain iron accumulation (NBIA). Eur J Paediatr Neurol 16: 175–178.

Mahlknecht P, Stockner H, Kiechl S et al. (2012). Is transcranial sonography useful to distinguish drug-induced parkinsonism from Parkinson's disease? Mov Disord 27: 1194–1196.

Mašková J, Školoudík D, Burgetová A et al. (2016). Comparison of transcranial sonography-magnetic resonance fusion imaging in Wilson's and early-onset Parkinson's diseases. Parkinsonism Relat Disord 28: 87–93.

Micheli F, Cersosimo M (2007). Drug-induced parkinsonism. Parkinson's disease and related disorders, Part II. Elsevier, pp. 399–416.

Miyamoto M, Miyamoto T (2020). Relationship of substantia nigra hyperechogenicity to risk of Lewy body disease in idiopathic REM sleep behavior disorder patients: a longitudinal study. Sleep Med 68: 31–34.

Moringlane JR, Fuss G, Becker G (2005). Peroperative transcranial sonography for electrode placement into the targeted subthalamic nucleus of patients with Parkinson disease: technical note. Surg Neurol 63: 66–69.

Naumann M, Becker G, Toyka KV et al. (1996). Lenticular nucleus lesion in idiopathic dystonia detected by transcranial sonography. Neurology 47: 1284–1290.

Noyce AJ, Dickson J, Rees RN et al. (2018). Dopamine reuptake transporter-single-photon emission computed tomography and transcranial sonography as imaging markers of prediagnostic Parkinson's disease. Mov Disord 33: 478–482.

Oh Y-S, Kwon D-Y, Kim J-S et al. (2018). Transcranial sonographic findings may predict prognosis of gastroprokinetic drug-induced parkinsonism. Parkinsonism Relat Disord 46: 36–40.

Olivares Romero J, Arjona Padillo A, Barrero Hernández FJ et al. (2013). Utility of transcranial sonography in the diagnosis of drug-induced parkinsonism: a prospective study. Eur J Neurol 20: 1451–1458.

Poewe W, Wenning G (2002). The differential diagnosis of Parkinson's disease. Eur J Neurol 9: 23–30.

Postert T, Lack B, Kuhn W et al. (1999). Basal ganglia alterations and brain atrophy in Huntington's disease depicted by transcranial real time sonography. J Neurol Neurosurg Psychiatry 67: 457–462.

Quattrone A, Antonini A, Vaillancourt DE et al. (2021). A new MRI measure to early differentiate progressive supranuclear palsy from de novo Parkinson's disease in clinical practice: an international study. Mov Disord 36: 681–689.

Rektor I, Rektorová I, Kubová D (2006). Vascular parkinsonism—an update. J Neurol Sci 248: 185–191.

Richter D, Katsanos AH, Schroeder C et al. (2019). Lentiform nucleus hyperechogenicity in Parkinsonian syndromes: a systematic review and meta-analysis with consideration of molecular pathology. Cells 9: 2.

Richter D, Woitalla D, Muhlack S et al. (2017). Coronal transcranial sonography and M-mode tremor frequency determination in Parkinson's disease and essential tremor. J Neuroimaging 27: 524–530.

Rizzo G, Copetti M, Arcuti S et al. (2016). Accuracy of clinical diagnosis of Parkinson disease. Neurology 86: 566–576.

Rüb U, Vonsattel JPG, Heinsen H et al. (2015). The neuropathology of Huntington's disease: classical findings. Recent developments and correlation to functional neuroanatomy. Cham: Springer International Publishing.

Rupprecht S, Walther B, Gudziol H et al. (2013). Clinical markers of early nigrostriatal neurodegeneration in idiopathic rapid eye movement sleep behavior disorder. Sleep Med 14: 1064–1070.

Sadowski K, Serafin-Król M, Szlachta K et al. (2015). Basal ganglia echogenicity in tauopathies. J Neural Transm 122: 863–865.

Sastre-Bataller I, Vázquez JF, Martínez-Torres I et al. (2013). Mesencephalic area measured by transcranial sonography in the differential diagnosis of parkinsonism. Parkinsonism Relat Disord 19: 732–736.

Saunders-Pullman R, Stanley K, Brüggemann N et al. (2010). Substantia nigra hyperechogenicity in DYT6 dystonia: a pilot study. Parkinsonism Relat Disord 16: 420–422.

Scherfler C, Schwarz J, Antonini A et al. (2007). Role of DAT-SPECT in the diagnostic work up of Parkinsonism. Mov Disord 22: 1229–1238.

Schmidauer C, Sojer M, Seppi K et al. (2005). Transcranial ultrasound shows nigral hypoechogenicity in restless legs syndrome. Ann Neurol 58: 630–634.

Schweitzer KJ, Behnke S, Liepelt I et al. (2007). Cross-sectional study discloses a positive family history for Parkinson's disease and male gender as epidemiological risk factors for substantia nigra hyperechogenicity. J Neural Transm 114: 1167–1171.

Schweitzer KJ, Brüssel T, Leitner P et al. (2007). Transcranial ultrasound in different monogenetic subtypes of Parkinson's disease. J Neurol 254: 613–616.

Shafieesabet A, Fereshtehnejad S-M, Shafieesabet A et al. (2017). Hyperechogenicity of substantia nigra for differential diagnosis of Parkinson's disease: a meta-analysis. Parkinsonism Relat Disord 42: 1–11.

Shahed J, Jankovic J (2007). Motor symptoms in Parkinson's disease. Handb Clin Neurol 83: 329–342.

Sidransky E, Nalls MA, Aasly JO et al. (2009). Multicenter analysis of glucocerebrosidase mutations in Parkinson's disease. N Engl J Med 361: 1651–1661.

Školoudík D, Bártová P, Mašková J et al. (2016). Transcranial sonography of the insula: digitized image analysis of fusion images with magnetic resonance. Ultraschall Med 37: 604–608.

Školoudík D, Jelínková M, Blahuta J et al. (2014). Transcranial sonography of the substantia nigra: digital image analysis. Am J Neuroradiol 35: 2273–2278.

Školoudík D, Walter U (2010). Method and validity of transcranial sonography in movement disorders. Int Rev Neurobiol 7–34.

Skowronska M, Kmiec T, Czlonkowska A et al. (2018). Transcranial sonography in mitochondrial membrane protein-associated neurodegeneration. Clin Neuroradiol 28: 385–392.

Skowrońska M, Litwin T, Kurkowska-Jastrzębska I et al. (2020). Transcranial sonography changes in patients with Wilson's disease during de-coppering therapy. Neurol Neurochir Pol 54: 185–192.

Skowronska M, Litwin T, Kurkowska-Jastrzębska I et al. (2020). Transcranial sonography changes in heterozygotic carriers of the ATP7B gene. Neurol Sci 41: 2605–2612.

Sprenger FS, Wurster I, Seppi K et al. (2016). Substantia nigra hyperechogenicity and Parkinson's disease risk in patients with essential tremor. Mov Disord 31: 579–583.

Stockner H, Iranzo A, Seppi K et al. (2009). Midbrain hyperechogenicity in idiopathic REM sleep behavior disorder. Mov Disord 24: 1906–1909.

Stockner H, Wurster I (2010). Transcranial sonography in essential tremor. Int Rev Neurobiol 90: 189–197.

Svetel M, Mijajlović M, Tomić A et al. (2012). Transcranial sonography in Wilson's disease. Parkinsonism Relat Disord 18: 234–238.

Svetel M, Tomić A, Mijajlović M et al. (2017). Transcranial sonography in dopa-responsive dystonia. Eur J Neurol 24: 161–166.

Toomsoo T, Liepelt-Scarfone I, Berg D et al. (2019). Effect of age on substantia nigra hyper-echogenicity in Parkinson's disease patients and healthy controls. Ultrasound Med Biol 45: 122–128.

Treglia G, Cason E, Stefanelli A et al. (2012). MIBG scintigraphy in differential diagnosis of Parkinsonism: a meta-analysis. Clin Auton Res 22: 43–55.

Tribl GG, Trindade MC, Almeida KJ et al. (2018). Quantitative transcranial sonography in Wilson's disease and healthy controls: cut-off values and functional correlates. J Neurol Sci 385: 69–74.

Tsai C-F, Wu R-M, Huang Y-W et al. (2007). Transcranial color-coded sonography helps differentiation between idiopathic Parkinson's disease and vascular parkinsonism. J Neurol 254: 501–507.

van de Loo S, Walter U, Behnke S et al. (2010). Reproducibility and diagnostic accuracy of substantia nigra sonography for the diagnosis of Parkinson's disease. J Neurol Neurosurg Psychiatry 81: 1087–1092.

Venegas-Francke P (2010). Transcranial sonography in the discrimination of Parkinson's disease versus vascular parkinsonism. Int Rev Neurobiol 90: 147–156.

Vilas D, Ispierto L, Álvarez R et al. (2015). Clinical and imaging markers in premotor LRRK2 G2019S mutation carriers. Parkinsonism Relat Disord 21: 1170–1176.

Walter U, Blitzer A, Benecke R et al. (2014). Sonographic detection of basal ganglia abnormalities in spasmodic dysphonia. Eur J Neurol 21: 349–352.

Walter U, Dressler D, Wolters A et al. (2004). Sonographic discrimination of corticobasal degeneration vs progressive supranuclear palsy. Neurology 63: 504–509.

Walter U, Dressler D, Wolters A et al. (2006). Sonographic discrimination of dementia with Lewy bodies and Parkinson's disease with dementia. J Neurol 253: 448–454.

Walter U, Behnke S, Eyding J et al. (2007a). Transcranial brain parenchyma sonography in movement disorders: state of the art. Ultrasound Med Biol 33: 15–25.

Walter U, Dressler D, Probst T et al. (2007b). transcranial brain sonography findings in discriminating between Parkinsonism and idiopathic Parkinson disease. Arch Neurol 64: 1635.

Walter U, Dressler D, Wolters A et al. (2007c). Transcranial brain sonography findings in clinical subgroups of idiopathic Parkinson's disease. Mov Disord 22: 48–54.

Walter U, Kirsch M, Wittstock M et al. (2011). Transcranial sonographic localization of deep brain stimulation electrodes is safe, reliable and predicts clinical outcome. Ultrasound Med Biol 37: 1382–1391.

Walter U, Krolikowski K, Tarnacka B et al. (2005). Sonographic detection of basal ganglia lesions in asymptomatic and symptomatic Wilson disease. Neurology 64: 1726–1732.

Walter U, Müller J-U, Rösche J et al. (2016). Magnetic resonance-transcranial ultrasound fusion imaging: a novel tool for brain electrode location. Mov Disord 31: 302–309.

Walter U, Wolters A, Wittstock M et al. (2009). Deep brain stimulation in dystonia: sonographic monitoring of electrode placement into the globus pallidus internus. Mov Disord 24: 1538–1541.

Walter U, Niehaus L, Probst T et al. (2003). Brain parenchyma sonography discriminates Parkinson's disease and atypical parkinsonian syndromes. Neurology 60: 74–77.

Walter U, Školoudík D (2014). Transcranial sonography (TCS) of brain parenchyma in movement disorders: quality standards, diagnostic applications and novel technologies. Ultraschall Med 35: 322–331.

Wenning GK, Kiechl S, Seppi K et al. (2005). Prevalence of movement disorders in men and women aged 50–89 years (Bruneck Study cohort): a population-based study. Lancet Neurol 4: 815–820.

Witkowski G, Jachinska K, Stepniak I, Ziora-Jakutowicz K, Sienkiewicz-Jarosz H (2020). Alterations in transcranial sonography among Huntington's disease patients with psychiatric symptoms. Neural Transm 127: 1047–1055.

Wu J, Jin H, Shao Y et al. (2021). Cognition and transcranial sonography in Parkinson's disease patients with or without orthostatic hypotension. Brain Behav 11: e2252.

Xu R, Chen G, Mao Z et al. (2020). Diagnostic performance of transcranial sonography for evaluating substantia nigra hyperechogenicity in patients with Parkinson's disease. Ultrasound Med Biol 46: 1208–1215.

Yamada M, Komatsu J, Nakamura K et al. (2020). Diagnostic criteria for dementia with Lewy bodies: updates and future directions. J Mov Disord 13: 1–10.

Yan J-H, Li K, Ge Y-L et al. (2023). Quantitative transcranial sonography evaluation of substantia nigra hyperechogenicity is useful for predicting levodopa-induced dyskinesia in Parkinson disease. Ultrasound Med Biol 49: 607–615.

Yilmaz R, Behnke S, Liepelt-Scarfone I et al. (2016). Substantia nigra hyperechogenicity is related to decline in verbal memory in healthy elderly adults. Eur J Neurol 23: 973–978.

Yilmaz R, Berg D (2017). Evaluating a patient with transcranial sonography to look for echogenicity. Mov Disord Clin Pract 4: 907.

Yilmaz R, Berg D (2018). Transcranial B-mode sonography in movement disorders. Int Rev Nuerobiol 143: 179–212.

Yilmaz R, Granert O, Schäffer E et al. (2021). Transcranial sonography findings in Alzheimer's disease: a new imaging biomarker. Ultraschall Med 42: 623–633.

Yilmaz R, Wolke R, Puls N et al. (2023). Characterizing camptocormia in Parkinson's disease using muscle ultrasonography. J Parkinsons Dis 13: 819–827.

Yoo S-W, Park H-E, Lee K-S et al. (2015). Hairdresser dystonia: an unusual substantia nigra hyperechogenicity. J Neurol Sci 357: 314–316.

Zecca L, Berg D, Arzberger T et al. (2005). In vivo detection of iron and neuromelanin by transcranial sonography: a new approach for early detection of substantia nigra damage. Mov Disord 20: 1278–1285.

Zhang Y, Zhang Y-C, Sheng Y-J et al. (2016). Sonographic alteration of basal ganglia in different forms of primary focal dystonia. Chin Med J (Engl) 129: 942–945.

Zhou H-Y, Huang P, Sun Q et al. (2018). The role of substantia nigra sonography in the differentiation of Parkinson's disease and multiple system atrophy. Transl Neurodegener 7: 15.

Zhou H-Y, Sun Q, Tan Y-Y et al. (2016). Substantia nigra echogenicity correlated with clinical features of Parkinson's disease. Parkinsonism Relat Disord 24: 28–33.

Zhu Y, Wang B, Tao K et al. (2017). Iron accumulation and microglia activation contribute to substantia nigra hyperechogenicity in the 6-OHDA-induced rat model of Parkinson's disease. Parkinsonism Relat Disord 36: 76–82.

Zhuang S, Gu X, Chai B et al. (2021). Transcranial sonographic characteristics of Parkinson's disease with symptoms of restless legs syndrome. Zhonghua Yi Xue Za Zhi 101: 1566–1571.

Handbook of Clinical Neurology, Vol. 216 (3rd series)
The Brainstem
L.T. Grinberg and A.D. Korczyn
https://doi.org/10.1016/B978-0-443-15736-3.00023-8

Chapter 19

Anatomic imaging: Angiography

MADHAVI DUVVURI[1], EYTAN RAZ[2], AND MATTHEW AMANS[3,*]

[1]*Department of Neuroimaging and Neurointervention, Stanford University, Stanford, CA, United States*

[2]*Department of Radiology and Neurosurgery, New York University Grossman School of Medicine, New York, NY, United States*

[3]*Department of Neuroendovascular Surgery, University of California San Francisco, San Francisco, CA, United States*

Abstract

This chapter reviews the vascular anatomy of the brainstem and posterior circulation with emphasis on clinical neurology and angiographic interpretation. Using a caudal-to-cranial approach, we describe the vertebral artery at the craniocervical junction and its key branches, including meningeal, medullary, spinal, and posterior inferior cerebellar arteries (PICA), highlighting common variants and clinically important anastamoses. We then outline basilar artery segments, pontine perforator patterns, and vertebrobasilar perforator groups, linking lesion location to characteristic neurologic syndromes. The cerebellar arteries (PICA, AICA, and superior cerebellar artery) are discussed with attention to variant origins, shared trunks, and collateral pathways that influence infarct topography. Finally, we summarize basilar tip and posterior cerebral artery (PCA) anatomy, including fetal PCA configurations, and thalamoperforator variants, such as the artery of Percheron.

INTRODUCTION

This chapter explores the intricate vascular anatomy crucial for sustaining key regions of the brainstem. We address the complexity of the vascular network, emphasizing redundancy in the supply. We specifically focus on challenges in angiographic imaging of the posterior circulation, navigating through small branching arteries and anastomoses as well as variant anatomy, requiring careful review prior to intervention to prevent inadvertent damage to nontargeted areas.

We start by examining the craniocervical junction anatomy, which includes the vertebral artery and its branches. Moving in a caudal-to-cranial approach, we cover the branches of the basilar artery below the superior cerebellar arteries (SCAs), concluding with a discussion of basilar tip anatomy. Each major vessel of the posterior circulation and its branches are discussed in detail, providing insights into normal and variant anatomic presentations, relevant anastomoses, and perforator vasculature. We highlight the brain regions these vessels supply and their clinical implications, offering a useful guide for neurologists and neurointerventionalists in their daily practice.

CRANIOCERVICAL JUNCTION

Vertebral artery

The vertebral artery (VA), the first vessel branching from the subclavian artery, displays a range of anatomic variations. Among these, the most common is the direct aortic origin of the left vertebral artery. The left vertebral artery is typically dominant in 50%–60% of individuals, characterized by a larger lumen (Polguj et al., 2013). The VA is divided into four segments: the first extends from its origin to the entry into the foramen transversarium of the C6 vertebra; the second part runs from the foramina transversarium of C6 to C2 vertebrae; the third segment extends from the foramen transversarium of the C2 vertebra to its entry into the subarachnoid space; and the fourth spans from the subarachnoid space to its junction with the contralateral VA, forming the basilar artery at the lower border of the pons.

There are several anatomic variations of the vertebral artery, which can have clinical consequences. Hypoplasia of the VA, characterized by a lumen smaller than 3 mm, can predispose patients to posterior circulation ischemia and occlusion due to increased susceptibility to thrombosis and atherosclerosis

*Correspondence to: Matthew Amans, MD, Department of Neuroendovascular Surgery, University of California, San Francisco, CA, United States. Tel: +1415-353-1863; E-mail: matthew.amans@ucsf.edu.

(Mitsumura et al., 2016). Duplication of the VA, where the VA has two origins that fuse at the neck, is also observed. An additional variation occurs when one of the vertebral arteries exits the transversarium to enter the spinal canal, following a subarachnoid course and ending at a cervical root (Lasjaunias and Berenstein, 1983). This can predispose patients to vertebral dissection (Dare et al., 1997), and patients with this duplication may present with symptoms like vertigo or dizziness, possibly due to an increased predisposition to kinking from a decreased lumen size (Ionete and Omojola, 2006). Fenestration of the vertebral artery, a rare condition, can increase the chance of arterial dissection in 65% of patients (Schievink, 2001). Finally, termination of the VA in the posterior inferior cerebellar artery (PICA) occurs in about 1% of cases (Polguj et al., 2013).

Clinically, the VA is particularly vulnerable to obstruction in patients with atlantoaxial dislocation. This is due to the susceptibility of the VA to mechanical compression during head and neck rotation between C1 and C2. The long course of the third segment normally accommodates stretching of the VA, but in patients with atlantoaxial dislocation, this segment may be shorter with supernumerary coiling (Sawlani et al., 2006). Additionally, dolichoectasia, focal narrowing, elongation, and arterial looping are also observed with the cervical vertebral artery. Cervicovertebral junction anomalies can also be associated with dissection or occlusion of the VA, leading to conditions like rotational C1 through C2 VA occlusion causing vertebrobasilar insufficiency (Bow Hunter's syndrome).

The vertebral artery gives rise to the anterior meningeal artery, the posterior meningeal artery, the medullary arteries, the anterior spinal artery (ASA), the posterior spinal arteries (PSAs), and the PICA, which will be discussed in the following sections.

Anterior meningeal artery

The anterior meningeal artery arises from the vertebral artery between the C2 and C3 level. It courses medially through the intervertebral foramen and then takes a dorsal path along the posterior longitudinal ligament (PLL), closely aligned with the anterior internal vertebral venous plexus. This artery is known for branching off multiple small tributaries that supply the PLL, ligaments, and soft tissues above the odontoid process, as well as even smaller branches that extend to the dura at the level of the foramen magnum (Shimizu et al., 2004). Its primary role includes feeding the ventral dura of the foramen magnum and the bony and ligamentous structures of the craniovertebral junction. Additionally, it forms a significant anastomosis with the meningeal branch of the ascending pharyngeal artery. In radiographic imaging, especially in the anterior projection, the anterior meningeal artery is most easily identifiable. Notably, in cases of tumors adherent to the meninges, this artery can be observed to be enlarged (Newton, 1968).

Posterior meningeal artery

The posterior meningeal artery originates from the extracranial vertebral artery, specifically above the arch of the atlas and just below the foramen magnum. Occasionally, the posterior meningeal artery can be seen arising from the PICA near its origin. This artery, situated within the dura, enters the skull via the lateral portion of the foramen magnum. Its primary function is to supply the occipital portion of the posterior fossa dura, as well as the posterior tentorium and falx. The artery arises from the vertebral artery between the "second bend" of the vertebral artery and the foramen magnum (Greitz and Lauren, 1968). It extends lateral branches along the posteromedial aspects of the posterior fossa dura. Radiographically, the posterior meningeal artery is best identified on the lateral projection. In clinical pathology, this artery can be enlarged in the setting of meningiomas, dural arteriovenous fistulas (DAVFs), and hemangioblastomas (Newton, 1968).

Medullary arteries

The vascular supply to the medulla is intricately organized into four groups of medullary arteries: the perforating arteries, short anterolateral, long lateral, and dorsal medullary arteries. These groups emanate predominantly from the vertebral artery, ASA, PICA, and the PSA. The perforating arteries arise from the vertebral arteries and the ASA, penetrate the foramen cecum, and supply the pyramids and the medial region of the medulla (Tatu et al., 1996; Djukić et al., 2023). The short anteromedial perforating branches, arising from the vertebral arteries and the ASA, specifically nourish the medial region of the medulla, including the medial part of the pyramidal tract, corticobulbar fascicles, and the arcuate nucleus (Tatu et al., 1996). The paramedian branches supply critical structures such as the medial lemniscus, tectospinal tract, medial longitudinal fasciculus, and the hypoglossal nucleus. The short anterolateral arteries, originating from the perforating arteries and the ASA, are responsible for the vascularization of the lateral part of the pyramidal and corticobulbar tracts, the hypoglossal nerve, and the inferior olivary complex. The lateral arteries, often originating from the vertebral artery and PICA, supply the lateral medullary territory. This includes somatosensory and autonomic pathways such as the spinothalamic, spinal trigeminal, trigeminothalamic, and solitary tracts, as well as motor pathways including the nucleus ambiguous, and various spinal tracts. Lateral medullary infarcts can manifest with symptoms such as facial hemihypesthesia, dysarthria, dysphagia, ataxia, nausea, vertigo, unsteadiness, and Horner's syndrome. Lastly, the dorsal arteries, which branch from the PICA and the PSA, supply the posterior and dorsal aspects of the medulla, as well as the inferior medullary velum, the lateral rhomboid fossa, and the tela choroidea and choroid plexus. Isolated dorsal medullary infarctions are rare due to the presence of anastomoses (Djukić et al., 2023).

Anterior spinal artery

The ASA arises from the anteromedial aspect of the V4 portion of the vertebral artery (Er et al., 2008). It originates as left and right rami, which subsequently fuse to form the ASA. This fusion in addition to the vertebral arteries from which they arose and subsequently their fusion into the basilar artery often forms a diamond shape and is thus referred to as the "spinal diamond." The anatomic variations of the ASA are classified into three types. Type I features two rami from bilateral VAs that fuse to form the ASA. Type II is characterized either by the ASA arising from either the right or left VA, or by two rami arising from both VAs, with one ramus being shorter and the other forming the primary ASA. Type III involves a left ASA arising from the left VA and a right ASA from the right VA.

Clinical implications of these variations are significant. Occlusion of the ASA can lead to bilateral pyramidal damage, resulting in quadriplegia that may even be ventilator-dependent quadriplegia depending on the extent of medulla and upper cord supply, as reported by Kim et al. Understanding these anatomic variations is critical, especially for endovascular or surgical procedures involving the CV junction. Inadvertent iatrogenic occlusion of the distal VA, particularly in procedures targeting aneurysms of the vertebrobasilar system, can result in the occlusion of the ASA at its origin. For instance, in a Type II ASA, where blood flow to the ASA is asymmetrically provided by one of the vertebral arteries, occlusion at the origin of the ASA trunk can cause bilateral medial and lateral medullary syndrome (Er et al., 2008). Conversely, in a Type III ASA, occlusion of the ASA trunk may only result in ipsilateral symptoms (Mahmood et al., 1991).

Posterior spinal artery

The PSA most commonly originates from the V3 segment of the vertebral artery and, to a lesser extent, from the PICA when it arises extradurally. Intracranially, it often arises from the V4 segment of the vertebral artery, subsequently dividing into ascending or descending branches. The PSA is a paired artery, with each branch having a distinct trajectory. The descending branch courses caudally with the posterior longitudinal spinal trunk and the ascending artery courses along the inferior cerebellar peduncle, and eventually terminates at the obex and foramen of Magendie (Djukić et al., 2023). In cases where there is early branching of the PICA from the vertebral artery, it is advisable to consider the PSA's potential origin from the PICA (Seçkin et al., 2009). There also exist multiple anastomoses with the contralateral PSA and the ASA to form a plexus that wraps around the spinal cord. The PSA is often difficult to visualize on angiography and in surgical interventions, underscoring the challenges in diagnosing conditions affecting this artery. Damage to the PSA may result in sensory and motor weaknesses, affecting both the bilateral upper and lower extremities (Seçkin et al., 2009).

Posterior inferior cerebellar artery

The PICA commonly arises from the vertebral artery approximately 16 mm proximal to the vertebrobasilar junction, traversing the lateral medulla (Lister et al., 1982). Notably, PICA is the largest branch of the distal segment of the vertebral artery, encircling the lateral medullary surface and entering the tonsillomedullary fissure. It then courses along the dorsal surface of the medulla and the roof of the fourth ventricle, where it divides into its terminal trunks. Variations in the origin of PICA are important to note. A common variant is the extradural vertebral origin of PICA, where dorsal-perforating arteries arise from the distal vertebral artery instead of the proximal PICA and supply the lateral medulla (Lasjaunias and Berenstein, 1983). This variant is suspected when an artery rises below the foramen magnum and penetrates the dura between the atlas and foramen magnum; extradural PICA origins can also penetrate the dura between the levels of C1 and C2. Clinically, in the setting of vertebral artery dissection, if it arises at the level of V3, the origin of an extradural PICA may place the patient at risk for medullary infarction (Fine et al., 1999). When there is a low origin of the PICA from the vertebral artery at C1, the low PICA may demonstrate features of the PSA (Siclari et al., 2007). Other variations include PICA arising from the occipital artery, ascending pharyngeal artery, or less commonly from the ASA, or as duplicated origin from the vertebral artery. An additional variation includes the PICA rising from the AICA (Robert et al., 2023).

The anatomy of PICA is segmented into five distinct parts (Robert et al., 2023): the first segment or the anterior medullary segment, courses below the hypoglossal nerve and ends at the olive; the second segment, or the lateral medullary segment, extends to the origin of the vagal, glossopharyngeal, and accessory nerves; the third or the tonsillomedullary segment, runs from the origin of these nerves to the medial surface of the tonsil; the fourth or the telovelotonsillar segment, begins at the medial surface of the tonsil and ends at the exit from the fissure between the vermis and cerebellar hemisphere; and the fifth segment, or cortical segment, consists of the terminal branches which supply the cerebellar surface.

The terminal trunks of the PICA include the tonsillohemispheric and vermian trunks. The tonsillohemispheric trunk descends over the posterior surface of the cerebellar tonsil, bifurcating into a tonsillar branch that courses downward and anteriorly around the tonsil, and a hemispheric branch that extends over the cerebellar hemisphere. The vermian trunk travels along the lower part of the inferior vermis, between the vermis and the cerebellar hemisphere (Fujii et al., 1980). Branches of the vermian trunk anastomose with the superior vermian branches of the SCA at the posterior vermis. Furthermore, the PICA forms cranial and caudal loops, which bifurcate into medial and lateral trunks, each serving specific cerebellar regions.

The perforating branches from the PICA, which consist of direct and circumflex branches form an arterial plexus that supplies the lateral part of the medulla. In cases of ectopic or low-lying PICA, these branches may emerge directly from the vertebral artery. The PICA also gives rise to choroidal arteries, which originate from the distal segment of the PICA, and the posterior medullary and supratonsillar segments (Fujii et al., 1980) can also arise, albeit less commonly, from the vermian trunk. These arteries are pivotal in supplying the choroid plexus and are known to enlarge in the presence of fourth ventricle tumors, such as meningiomas, papillomas, and intraventricular ependymomas. Lastly, the cortical branches of the PICA, arising from its distal segment, are divided into vermian, hemispheric, and tonsillar branches. The vermian branches supply the inferior vermis, whereas the hemispheric branches supply the suboccipital cerebellar hemisphere lateral to the vermis.

BASILAR ARTERY BELOW THE SUPERIOR CEREBELLAR ARTERIES

Basilar artery

Origins

The basilar artery, formed by the union of the two vertebral arteries, is a crucial component in the brain's vascular system. In addition to giving rise to the paired anterior inferior cerebellar arteries, the labyrinthine arteries, and the SCAs, it also gives rise to the pontine arteries and occasionally the PICA. Moreover, the posterior and lateral surfaces of the basilar artery are a rich source of perforating arteries that supply the brainstem (Saeki and Rhoton, 1977) (Fig. 19.1).

Segments

The basilar artery is primarily divided into three anatomic sections: 1) the proximal segment, which extends from the vertebral artery junction to the origin of the anterior inferior cerebellar artery; 2) the middle segment, which extends from the AICA to the SCA; and 3) the distal segment, which extends from the origin of the SCAs to the basilar tip just prior to the bifurcation into the posterior cerebral arteries. There are several anatomic variants of the basilar artery which can have clinical consequences. The fenestration of the basilar artery is a rare small duplication of the basilar artery, which is most commonly seen in the lower third of the basilar artery (Gao et al., 2013). Although fenestrations by themselves are not clinically symptomatic, it is hypothesized that the thinned vascular intima of the fenestrated portion can predispose the area to aneurysms and possible subsequent subarachnoid hemorrhage (Campos et al., 1987).

Vascular supply

Through its many major and minor branch arteries, which will be discussed in the following sections, the basilar artery supplies the pons, the mesencephalon, the midbrain, and the thalamus. Occlusions of the basilar artery at different segments can result in varying burdens of ischemia. For example, occlusion of the proximal or middle segments can result in large pontine strokes resulting in quadriplegia or hemiplegia with loss of consciousness, dysarthria, dysphagia, and several cranial nerve palsies. Occlusions of the distal segment of the basilar artery can cause ischemia of the mesencephalon and the

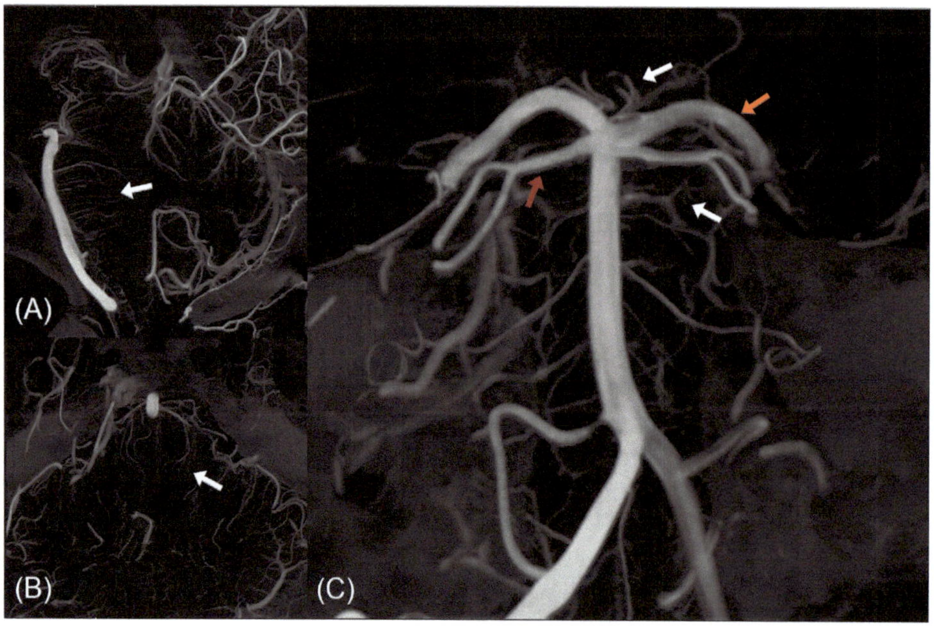

Fig. 19.1. Cone beam CT (CBCT) reconstructions demonstrating vertebrobasilar perforators. Panel A, B and C demonstrate sagittal, axial and coronal maximum intensity projection (MIP) reformats of the brainstem and posterior fossa. Panel A and B showcase the basilar artery and its many perforating arteries, such as the pontine perforating arteries (white arrow). In Panel C, again the perforating arteries of the basilar artery are highlighted (white arrows). The posterior cerebral arteries (orange arrow), and the superior cerebellar arteries (red arrow) are also noted.

thalamus, resulting in loss of consciousness, quadriparesis, confusion, amnesia, and oculomotor dysfunction (Mattle et al., 2011). Basilar artery thrombi often lead to large pontine infarcts, whereas basilar artery emboli lead to midbrain infarcts (Kubik and Adams, 1946). Occlusion of smaller branch arteries of the basilar artery can lead to pontine warning syndrome, which is characterized by motor and speech disturbances that wax and wane. Transient occlusion of the vertebrobasilar system can also occur, resulting in reversible ischemia of the cerebellum and/or brainstem, which usually present as drop attacks.

Carotid-vertebrobasilar anastomoses

Carotid-vertebrobasilar anastomoses are present at the embryonic stage, and serve to provide arterial supply from the internal carotid artery (ICA) to the vertebrobasilar artery in the hindbrain. Persistence of these anastomoses in the adult can have clinical significance. The four carotid-vertebrobasilar anastomoses include the persistent trigeminal artery (PTA), the persistent hypoglossal artery, the persistent proatlantal artery, and the often-debated persistent otic artery.

The PTA (Fig. 19.2) is the most common of these anastomoses with an incidence of 0.1%–0.6% (Wise, 1957). Originating from the ICA at the carotid canal and entering the cavernous sinus, the PTA takes a posterior course to reach the subarachnoid space of the posterior fossa. It often runs close to or within the sella, coursing medially to the root of the trigeminal nerve, and communicates with the basilar artery between the origins of the AICA and SCA (Meckel et al., 2013). The PTA has multiple variants, known as the Saltzman variants. Type 1, or fetal PTA, involves a hypoplastic basilar artery proximal to the PTA with an absent posterior communicating artery (PCOM), making the PTA the sole supply to the distal basilar artery and cerebellum (Boyko et al., 1996; Meckel et al., 2013). Type 2 PTA joins the basilar artery below the SCA, with the posterior cerebral arteries receiving blood supply via the PCOMs. Type 3 PTA is characterized by anomalous cerebellar arteries originating from the ICA and coursing similarly to the SCA, AICA, and PICA, with AICA being more common in these variants. The basilar origin of the ophthalmic artery is also on the spectrum of PTA variants. Clinically, the PTA is associated with vertebrobasilar embolic phenomena and trigeminal neuralgia due to its proximity to the trigeminal nerve (Morita et al., 1989). Trigeminal-cavernous sinus fistulas, usually seen with Saltzman Type II variants, present with symptoms similar to carotid cavernous fistulas. These require careful angiographic evaluation prior to treatment, as ligation of a Type 1 PTA can significantly reduce blood flow to the posterior circulation.

The persistent hypoglossal artery (PHA) is a rare anastomosis (Fig. 19.3), with an incidence of 0.27%–0.26% in the population (Brismar, 1976). It is the second most common carotid-vertebrobasilar anastomosis, typically originating from the ICA near C1 or C1/2 as a large extracranial branch, and passing through the hypoglossal canal to form a basilar trunk, often accompanying a hypoplastic ipsilateral vertebral artery (Uchino et al., 2013). Occasionally, it may arise from the external carotid artery or even from the ascending pharyngeal artery, functioning as a collateral pathway. A very rare variant includes the PICA emerging from the ICA and passing through the hypoglossal canal without connecting to the basilar artery (Andoh et al., 2001). PHAs are also rarely associated with basilar artery fenestrations. The PHA's clinical significance lies in its potential role in cerebrovascular pathologies, particularly in patients with atherosclerotic disease of the ICA, where emboli may travel through the PHA into the posterior circulation, causing ischemic events in both the anterior and posterior circulation.

The persistent proatlantal intersegmental artery (Fig. 19.4) is a very rare anastomosis and can present as two distinct types (Bahşi et al., 1993). Type I originates from the caudal ICA or external carotid artery (ECA) at C2/3, ascending to the occipitoatlantal space and merging with the vertebral artery's fourth segment, before passing above the atlas' transverse process into the foramen magnum. Type II, which arises from

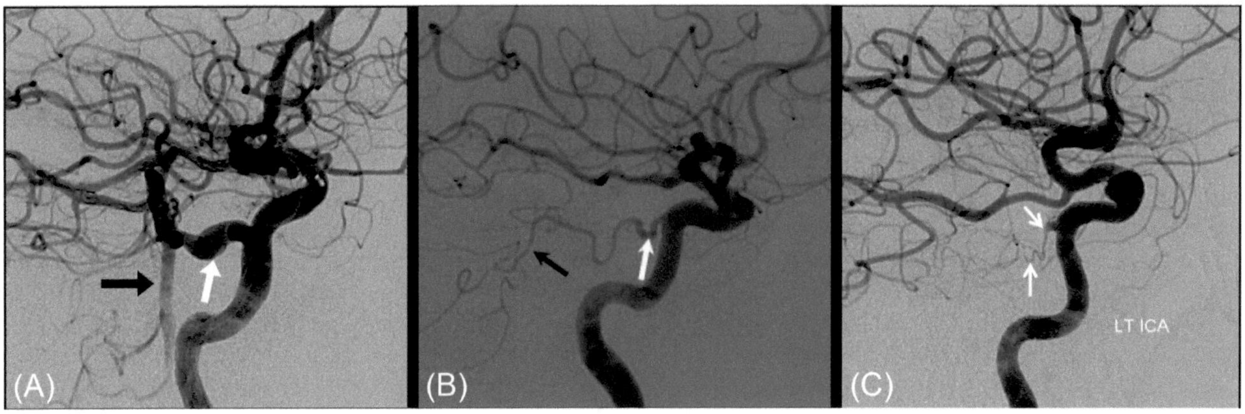

Fig. 19.2. Visualization of the spectrum of presentations of the persistent trigeminal artery. Panel A, B, and C show digital subtraction angiography of the left internal carotid artery in lateral projections. In Panel A, a large persistent trigeminal artery (white arrow), arises from the caudal aspect of the left internal carotid artery and communicates with the basilar artery (black arrow). In Panel B, a smaller persistent trigeminal artery (white arrow) arises from the left internal carotid artery and communicates with the anterior inferior cerebellar artery (black arrow). This is consistent with a Type IIIB Saltzman classification. In Panel C, an infundibulum of a small trigeminal artery is seen, also communicating with a small anterior inferior cerebellar artery (smaller white arrow).

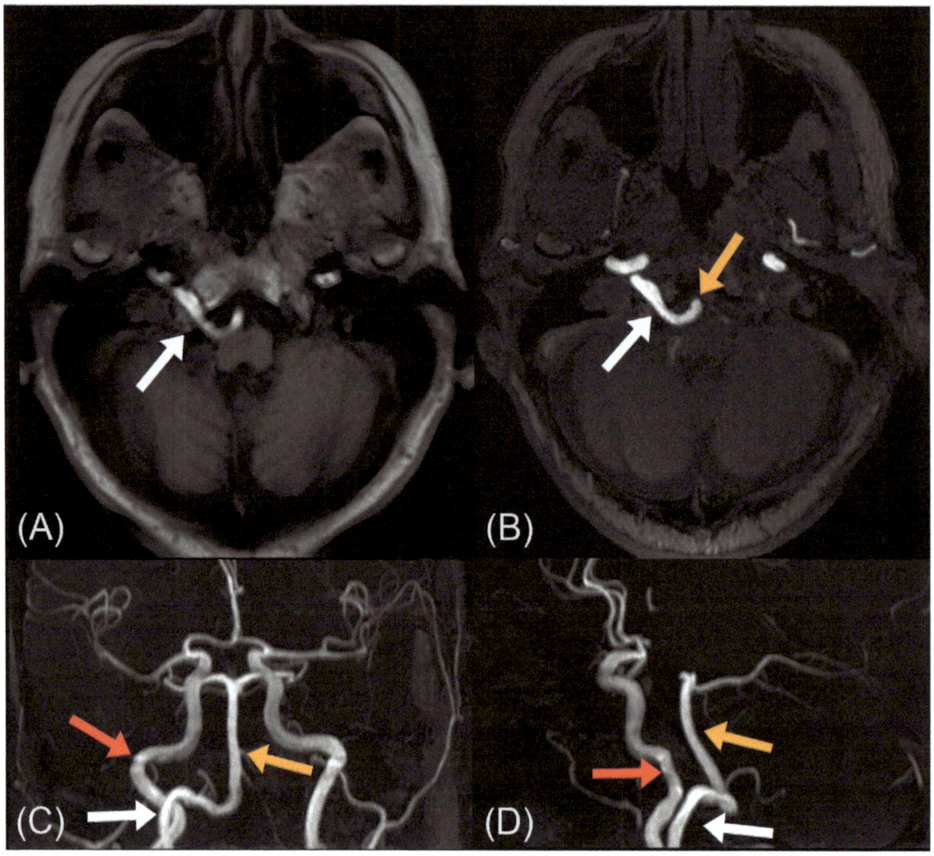

Fig. 19.3. Visualization of a persistent right hypoglossal artery. Panel A, a post-contrast non-fat-suppressed T1-weighted axial image of the skull base and Panel B, a time of flight (TOF) magnetic resonance angiography axial image of the skull base demonstrate a prominent vessel (white arrow) in location of the right hypoglossal canal, coursing medially to form the basilar artery (orange arrow). Coronal and sagittal maximum intensity projections (MIPs) from TOF MRA are shown in Panel C and Panel D, demonstrate a vessel (white arrow) arising from the right internal carotid artery (red arrow), and coursing medially to meet the basilar artery (orange arrow). This vessel represents the persistent hypoglossal artery.

the ECA, joins the V3 segment of the vertebral artery, traversing the atlas to enter the skull. Usually, in the setting of a proatlantal intersegmental artery, the ipsilateral vertebral artery is aplastic, with the anastomosis providing vascular supply to the vertebral artery territories. Generally, the presence of the persistent proatlantal intersegmental artery is asymptomatic, but its occlusion can precipitate ischemia in posterior circulation territories. In cases of ICA atherosclerosis, it can become a conduit for emboli to the basilar system, potentially leading to basilar occlusion and symptoms akin to "top of the basilar syndrome," including cognitive dysfunction, vertical gaze palsy, and somnolence (Bahşi et al., 1993).

The persistent otic artery, a very rare anastomosis, originates from the carotid artery within the carotid canal, traverses the internal auditory canal (IAC), and joins the basilar artery. Its identification is challenging; it is often not visible on standard angiography, and only two case reports have documented its existence on angiography (Patel et al., 2003).

Labyrinthine artery

The labyrinthine artery predominantly originates from the AICA, while in a minority of cases, it arises from the basilar artery. This artery supplies the vestibulocochlear and facial nerves, as well as the labyrinth and cochlea of the inner ear. Anatomically, the labyrinthine artery emerges from the AICA below the level of the IAC, specifically inferomedial to the facial and vestibulocochlear nerves (Salgado-Lopez et al., 2020). Postemergence, it closely follows the vestibulocochlear nerve into the IAC, culminating in the inferior part of the canal. The location and origin of the labyrinthine artery are important, particularly in the context of skull base surgery. Pathologies such as thrombosis of this artery can lead to a constellation of symptoms, including hearing loss, vertigo, tinnitus, and Meniere disease (Kim et al., 1999). Furthermore, aneurysms of the labyrinthine artery manifest with clinical presentations akin to its thrombosis or ischemia or may mimic an IAC tumor, thus necessitating careful consideration in diagnostic evaluations. This is also important in that it supplies the facial nerve and has an anastomosis at its apex with the posterior division of the middle meningeal artery. Therefore, when embolizing the middle meningeal artery (MMA) with liquid embolic materials, it is paramount to stop the infusion prior to reaching the IAC where this anastomosis *always* exists. Otherwise, not only will one cause a facial nerve palsy, but also cause potential nontarget embolization into the basilar artery (via the labyrinthine artery to the AICA to the basilar artery).

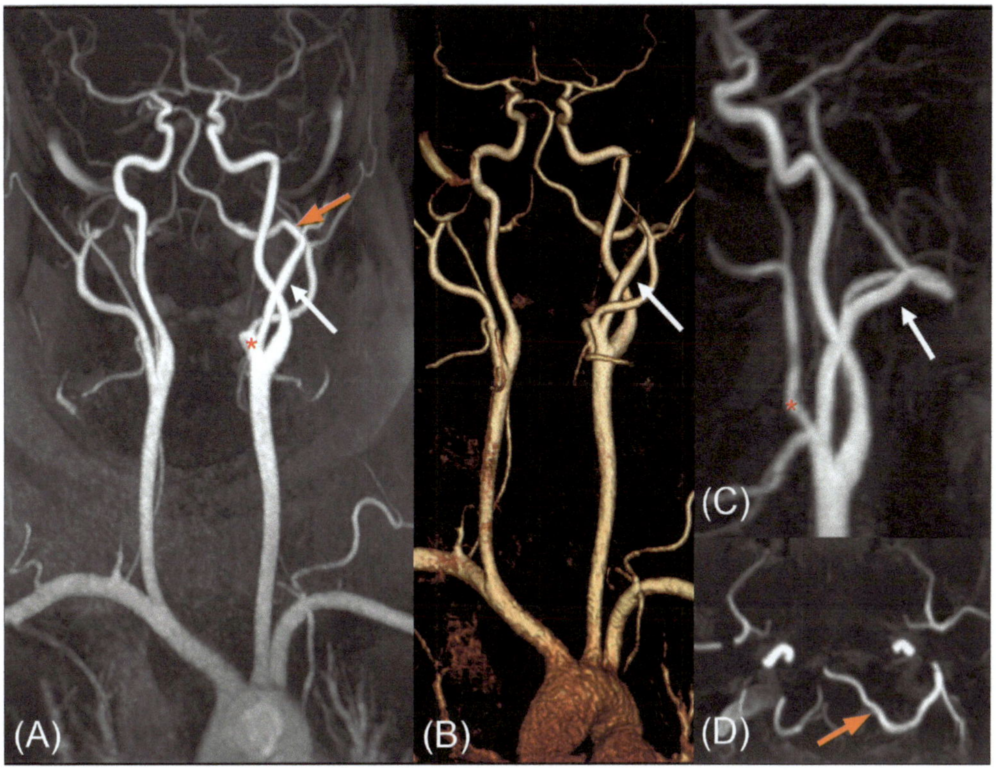

Fig. 19.4. Visualization of a persistent left Type II left proatlantal intersegmental artery on post-contrast magnetic resonance angiography (MRA) of the neck with coronal, sagittal and axial maximum intensity projection images (Panels A, C, D) and 3-D reconstruction images (Panel B). Panel A Panel B, and Panel C shows a vessel (white arrow) arising from the caudal aspect of the left external carotid artery (red star). This vessel anastomoses with the V3 segment of the left vertebral artery (orange arrow) and enters the skull base (Panel D). Note that the proximal left vertebral artery is aplastic.

Pontine arteries/perforators

The pontine arteries, which are central to the vascular supply of the brainstem, typically number seven on each side, all branching from the basilar artery. These arteries are diverse, each type playing a distinct role in brainstem vascularization. The paramedian branches originate from the pontine quadrant of the basilar artery and have a relatively short course before they divide into terminal perforators (Kwiatkowska et al., 2023). These perforators enter the pons paramedically, supplying vital areas of the brainstem. The short circumferential branches also arise from the pontine quadrant of the basilar artery. Unlike the paramedian branches, they encompass the pons and give off terminal perforators (Kwiatkowska et al., 2023). Within this group, the superolateral pontine artery stands out as a noteworthy short circumferential branch. Moreover, there exists a combination of paramedian and short circumferential branches, which emerge from the same pontine quadrant of the basilar artery and are characterized by a higher count of terminal perforators compared to either the paramedian or short circumferential branches alone. Long circumferential branches are differentiated by their extensive course around the pons and a lesser number of terminal perforating branches. Key examples of long circumferential branches include the posterolateral pontine artery and the pontomedullary artery. The median branches, distinct from paramedian branches, are notable for their entry into the pons at the midline and subsequent branching into a small number of terminal perforators.

Each of the seven pontine arteries has a unique clinical significance and area of supply. The posterolateral pontine artery, arising 1 mm below the SCA, is bilaterally present and supplies areas associated with cranial nerves III and V, the ventrolateral area, and the ventrointermedial pons. The superolateral pontine artery, a bilaterally present short circumflex branch arising at the distal portion of the basilar artery, supplies regions related to cranial nerves II and V, and parts of the pons. The third, fourth, and fifth pontine arteries, typically present bilaterally, have perforators that supply the entire pons. The sixth pontine artery, the anterolateral pontine artery (a short circumflex branch type), emerges from the basilar artery above the AICA and supplies the ventromedial and ventrointermedial pons, as well as areas of cranial nerves VI, VII/VIII. The seventh pontine artery, a long circumflex type, supplies the inferolateral pontine surface.

The proximity of pontine arteries to various cranial nerves, such as cranial nerve (CN) II, CN V, CN VI, CN VII, and CN VIII, underscores their clinical importance. The paramedian branches, in particular, may contact these nerves at the midline pons, including CN VI (Kwiatkowska et al., 2023). Additionally, pontine perforators can originate from the SCA or AICA, with those from the SCA supplying the upper ventrolateral area of the pons and those from the AICA catering to the ventrointermedial area and the middle cerebellar peduncle.

The clinical relevance of these arteries is underscored by their susceptibility to various pathologies. Conditions such as atherosclerosis, aneurysms, and arteriovenous malformations involving the pontine arteries can lead to strokes, lacunar infarcts, and hemorrhages. Isolated pontine strokes are a frequent occurrence, with ventral infarcts being the most common type of isolated pontine infarction. More specific occlusions of these arteries can lead to a range of syndromes. For instance, occlusion of the anteromedial or anterolateral groups can lead to ventral infarcts, while occlusion of the posterolateral artery can result in ipsilateral loss of facial sensation, paralysis of muscles of mastication, and contralateral impairment of sensory modalities, affecting the trigeminal nuclei and the motor nucleus of CN V. Occlusion of the paramedian and short circumflex arteries can lead to contralateral hemiparesis, facial sensory loss, and paralysis of muscles of mastication, impacting the corticospinal tract and root fibers of CN V and the superior cerebellar peduncle pontine nuclei. Occlusion of the anterolateral pontine artery can affect the pyramidal tract, CN VI, CN VII, lateral spinothalamic tracts, and the medial lemniscus, leading to Millard–Gubler or Foville's syndrome, characterized by contralateral spastic hemiplegia and other sensory modalities, except for ipsilateral CN VI palsy and facial palsy. Occlusion of the long circumflex branch can cause ipsilateral CN VI and VII palsy and contralateral impairment of all sensory modalities, affecting the CN VI, CN VII, lateral spinothalamic tract, medial lemniscus, central sympathetic tract, middle cerebellar peduncle, and vestibular nuclei. Furthermore, occlusion of the proximal basilar artery paramedian and median branches can lead to medial medullary syndrome (Dejerine–Roussy syndrome), characterized by contralateral hemiparesis, notable flaccid ipsilateral hypoglossal nerve palsy, and contralateral loss of vibration and position sense, impacting the pyramidal tract and fibers of CN XII (de Mendivil et al., 2013). Additional syndromes include Foix–Chavany–Marie syndrome, a dorsal infarct syndrome featuring ipsilateral ataxia, contralateral sensory loss to pinprick sensation and temperature, and contralateral hemiparesis; Landry syndrome, characterized by ipsilateral CN VI palsy and contralateral hemiparesis; and Raymond–Céstan syndrome, presenting with ipsilateral conjugate gaze palsy and contralateral sensory and motor deficits.

Superior cerebellar artery

The SCA is typically the first main artery to branch off from the basilar artery and is situated below the posterior cerebral artery (PCA). It arises rostrally at the apex of the basilar artery, extending to the pontomesencephalic junction, and navigates laterally through the cerebellomesencephalic fissure, encircling the brainstem (Malicki et al., 2023). The SCA predominantly supplies the pons, cerebral peduncle, the tentorial, and the superior petrosal surfaces of the cerebellum, and the cerebellar peduncles.

The SCA most commonly emerges as a single trunk from the distal basilar artery, while duplicated SCAs function similarly to solitary SCAs in their distribution (Hardy et al., 1980). There are rare variations in its origin, including arising from the PCA (Uchino et al., 2003), or as a common trunk with the PCA. Additional rare variants include bifurcation and origin from the ICA (Malicki et al., 2023).

The main trunk of the SCA divides into rostral and caudal branches lateral to the brainstem at the level of the ambient cistern. The rostral trunk supplies the vermis and the adjacent hemisphere, giving off hemispheric, vermian, and perforating arteries. The superior vermian artery, a branch of the rostral trunk, specifically supplies the vermis. Conversely, the caudal trunk is responsible for supplying the hemispheric surfaces lateral to the territory of the rostral trunks and emits hemispheric, marginal, and perforating arteries. The SCA can be separated into four segments visible in the upper cerebellopontine (CP) angle (Hardy et al., 1980); the anterior pontomesencephalic segment, which passes inferior to the oculomotor nerve; the lateral pontomesencephalic segment, situated below the trochlear and above the trigeminal nerve; the cerebellomesencephalic segment, coursing in the groove between the cerebellum and upper brainstem; and the cortical segment, distributed to the cerebellar surface. The cortical branches of the SCA supply the upper part of the suboccipital surface, the superior vermis, and the superior part of the petrosal surface. Of note, superior vermian branches of the SCA anastomose with the PICA's vermian branches.

The SCA's perforators, typically branching off prior to the bifurcation into rostral and caudal trunks, include the long circumflex, which terminate in the junction between the superior and middle cerebellar peduncles, the inferior colliculus, the cerebral peduncle, and the interpeduncular fossa. Additionally, direct perforators from the proximal SCA supply critical areas including the oculomotor nerves, posterior thalamus, paramedian and paraventricular nuclei of the thalamus, substantia nigra, red nucleus, midline midbrain structures, and the rostromedian floor of the fourth ventricle (Garcia-Gonzalez et al., 2012).

The SCA also provides supply to the tenorium. As an aside, the artery of Davidoff and Schecter (a branch of the PCA) (ADS) courses inferior to the SCA in the ambient cistern and also suppplies the tentorium. Dural arteriovenous fistulas that arise at the tentorium often involve the ADS. A notable corollary tentorial branch of the SCA is the artery of Wollschlaeger and Wollschlaeger. This artery is very difficult to see on angiography and can be particularly prominent in the setting of a tentorial DAVF. A case report notes that this branch of the SCA originates near the edge of the tentorium and courses posteriorly to the interior surface of the tentorium (Weil et al., 2011).

The clinical significance of the SCA is underscored by its proximity to various cranial nerves. Its anatomic variations, such as early bifurcation from the ICA and the caudal course of a duplicated SCA trunk, can predispose patients to neurovascular compression, contributing to conditions like trigeminal neuralgia. Additional complications may include oculomotor nerve palsy (Joshi et al., 2017) and ocular neuromyotonia (Inoue et al., 2012). When addressing aneurysms of the basilar artery or basilar tip, it is crucial to consider potential damage to the proximal perforators emanating from the SCA. Strokes involving the proximal SCA can manifest as ipsilateral cerebellar and brainstem deficits, along with contralateral sensory deficits (Garcia-Gonzalez et al., 2012). If patients exhibit only cerebellar involvement and dysarthria, it is likely that the lateral or medial branches of the SCA are affected rather than the origin, often due to an embolic source (Terao et al., 1996). Compromise of the SCA's direct perforators can lead to oculomotor disturbances and ataxia, primarily affecting the anteromedial midbrain.

Anterior inferior cerebellar artery

The anterior inferior cerebellar artery supplies the lower pons, medulla, and the anterior portion of the cerebellum. It originates from the basilar artery approximately 1 cm from the junction of the vertebral arteries and courses laterally (Takahashi et al., 1968). It notably crosses the vestibulocochlear nerve in the CP angle cistern. The AICA gives off the labyrinthine or internal auditory artery before dividing into two major branches within the CP cistern.

The lateral branch of the AICA courses laterally toward the superior and inferior semilunar lobules of the cerebellum. It anastomoses with the branches of the SCA and the PICA, curving around the flocculus. Small branches from the lateral branch supply the middle cerebellar peduncle. Conversely, the medial branch moves downward toward the biventral lobule, anastomosing with branches of the PICA. It also supplies small branches to the restiform body. At the facial–vestibulocochlear nerve complex, the AICA bifurcates into rostral and caudal trunks (Rodríguez-Hernández et al., 2011). The rostral trunk sends lateral branches to the middle cerebellar peduncle and superior CP fissure, while the caudal trunk supplies the inferior petrosal surface, including the flocculus and choroid plexus.

The AICA's perforating arteries include perforating branches to the brainstem and choroidal branches to the choroid plexus (Rodríguez-Hernández et al., 2011). The AICA can also give rise to the labyrinthine artery, recurrent perforating arteries, and subarcuate arteries.

The segments of the AICA are distinct and clinically relevant. The anterior pontine segment begins at the clivus, extends upward to the belly of the pons, and contacts the root of the abducens nerve. The lateral pontine segment starts at the anterolateral margin of the pons, proceeds to the CP angle, courses between cranial nerves VII and VIII, and is closely associated with the internal auditory meatus. This segment gives off the labyrinthine artery, recurrent perforating, and subarcuate arteries, and is subdivided into premeatal, meatal, and postmeatal portions. The meatal segment is the most common site of AICA aneurysms, which are challenging to treat due to the proximity of cranial nerves, making endovascular treatment a preferred option (Lv et al., 2016). The flocculopeduncular segment of the AICA passes rostral or caudal to the flocculus to reach the middle cerebellar peduncle, and the cortical segment branches to the petrosal surface of the cerebellum.

Variants of the AICA include the medial branch supplying areas typically served by a hypoplastic PICA (and vice versa), an origin from the vertebral artery, and a common origin with the PICA from a common trunk from the basilar artery. A particularly rare variant, as noted by Hou et al. (2020) is the AICA-PICA common trunk, where a single artery from either the vertebral or basilar artery bifurcates to supply both the cerebellum and brainstem. These branching patterns can occur at various points along the basilar or vertebral arteries. Clinically, unrecognized injury to the common AICA-PICA trunk can lead to ischemia affecting both the cerebellum and brainstem.

Clinically, occlusion of the AICA can lead to vertigo with nausea and vomiting (Takahashi et al., 1968). AICA territory infarcts, particularly common in individuals with high vascular risks such as diabetes and hypertension, often present with vertigo (Amarenco et al., 1993). Pure AICA infarcts, or branch infarcts, typically lead to vertigo, especially in diabetic patients. More extensive arterial occlusions involving the vertebral artery or the basilar artery can present with additional symptoms like diplopia, blurred vision, and numbness, and are usually seen in patients with hypertensive disease. In cases where the PICA is unusually small or hypoplastic, it is often observed that the AICA supplies more blood flow, a factor crucial in angiographic procedures. Furthermore, displacement of the AICA in a downward direction can be indicative of intra-axial tumors.

BASILAR ARTERY ABOVE THE SUPERIOR CEREBELLAR ARTERIES

Posterior cerebral artery

The PCA typically arises as a branch of the basilar artery. Variant origins include the complete and partial fetal PCA. In the complete fetal PCA, the artery originates solely from the ICA with no basilar connection. The partial fetal PCA maintains a diminutive connection to the basilar artery. Clinically, these variations imply greater posterior circulation reliance on the ICA, which can be concerning in ICA steno-occlusive disease. However, large-scale studies have not conclusively linked fetal PCAs to increased stroke risk (Shaban et al., 2013).

The course of the PCA is divided into four segments. The P1 segment, also known as the peduncular segment, includes the PCA prior to the PCOM. This segment extends from its origin at the basilar artery to where the PCOM arises. The P2 segment, or ambient segment, starts at the PCOM and courses around the cerebral peduncle, inferior to the optic tract and basal vein and has two divisions, the anterior P2 segment (P2A) and the posterior P2 segment (P2P). It runs medially to the hippocampus into the ambient cistern, then along the midbrain, superomedial to the trochlear nerve and tentorial edge. The P3 segment, the quadrigeminal segment, courses along the

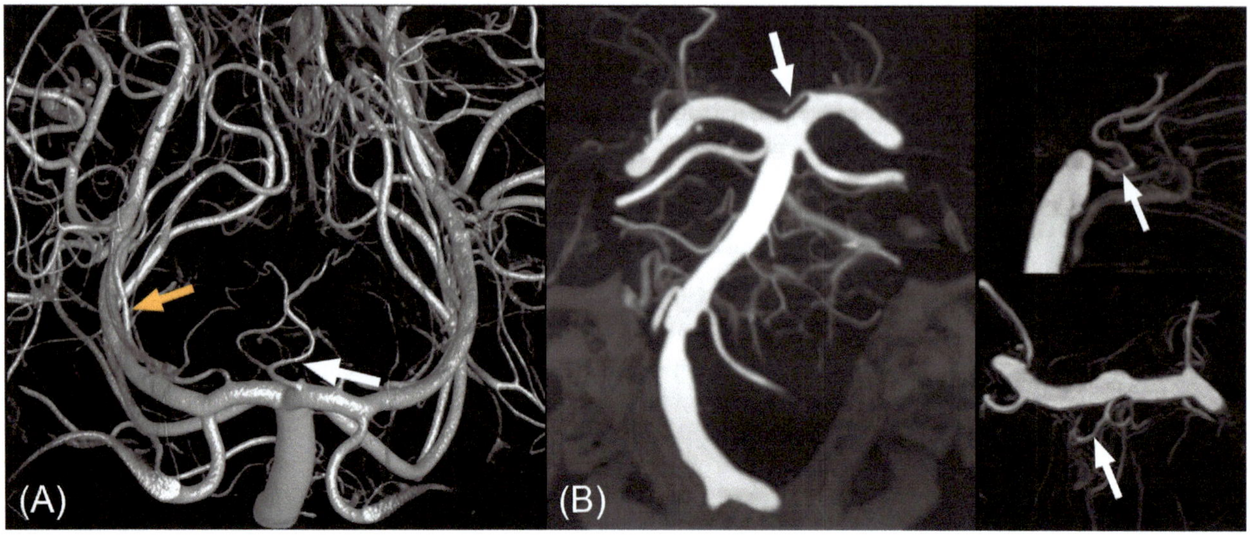

Fig. 19.5. Visualization of the midbrain and thalamic perforators by high resolution cone beam CT (CBCT). Three-dimensional volume rendering in Panel A and coronal, sagittal and axial maximum intensity projections in Panel B of the distal basilar artery and bilateral PCAs demonstrate a dominant thalamoperforator, also known as the artery of Percheron (white arrow). Note also the collicular arteries (orange arrow) in Panel A which follow the course of the bilateral posterior cerebral arteries.

quadrigeminal cistern, posteriorly from the pulvinar to the anterior calcarine fissure, dividing into calcarine and parieto-occipital arteries.

There are five different sets of perforating arteries arising from the various segments of the PCA. The long and short circumflex arteries, arising from the P1 segment of the PCA, have distinct destinations and functions. The short circumflex arteries target the geniculate bodies, while the long circumflex arteries extend further to supply the colliculi and the quadrigeminal bodies, encircling the midbrain medial to the PCA. These terminal branches anastomose with branches from the SCA. Clinically, occlusion of these arteries can lead to vertical gaze defects.

The direct perforating arteries of the PCA are the thalamoperforating and thalamogeniculate arteries. The thalamoperforating arteries (Fig. 19.5) arise from the P1 segment and supply critical areas including the anterior and posterior thalamus, hypothalamus, substantia nigra, red nucleus, and the nuclei and nerve of cranial nerves III and IV (Zeal and Rhoton, 1978). The anterior thalamoperforating arteries originate from the posterior communicating artery, with the premammillary artery being the largest. The posterior thalamoperforating arteries emerge from the proximal P1 segment on the superior or posterior aspect (Saeki and Rhoton, 1977). Occlusion of the thalamoperforating arteries can lead to contralateral hemiplegia, cerebellar ataxia, and contralateral hemiballismus. The thalamogeniculate arteries, of which there can be up to seven, arise from the posterior aspect of the P2 segment part and supply the geniculate bodies, posterior limb of the internal capsule, and optic tract. Occlusion of the proximal PCA can lead to contralateral loss of superficial and deep sensation, hypersensitivity to pain and temperature stimuli, and contralateral hemiplegia. This is because some parts of the thalamus are supplied by the thalamogeniculate arteries, but not all, as the circumflex and choroidal branches of the PCA also supply the thalamus. An isolated thalamogeniculate artery occlusion is unlikely to cause the aforementioned symptoms.

One prominent anatomic variant of the thalamoperforating arteries arising from the P1 segment is a single or dominant thalamoperforating artery arising from either P1 segment (Fig. 19.5), that supplies both of the thalamus and occasionally, the midbrain. This single or dominant thalamoperforating artery is named the artery of Percheron. Occlusion of this single artery can present with bilateral thalamic infarcts, and occasionally rostral midbrain infarcts.

The long and short circumflex arteries, originating from the PCA's P1 segment, have distinct roles. The short circumflex arteries supply the geniculate bodies, while long circumflex arteries extend to the colliculi (hence called the collicular arteries), supplying the quadrigeminal bodies and encircling the midbrain medial to the PCA. Their terminal branches anastomose with branches from the SCA.

The PCA's anterior P2 segment gives rise to several arteries: hippocampal, anterior temporal, peduncular perforating, posterior temporal, and medial posterior choroidal arteries. The hippocampal artery, usually the first cortical PCA branch, supplies the uncus, hippocampal gyrus, and the dentate gyrus. When there is bilateral occlusion of the hippocampal arteries, patients can present with Korsakoff's syndrome. The anterior temporal artery, typically the second cortical branch from the PCA's P2 segment, supplies the anterior inferior temporal lobe. The posterior temporal artery, arising from the PCA's inferior and lateral aspect and courses posteriorly to the occipital pole, supplies the inferior temporal and occipital poles and the lingual gyrus. Its occlusion can cause mild dysphagia and right homonymous hemianopsia with temporary amnesia (Zeal and Rhoton, 1978). The medial posterior choroidal artery, potentially single or multiple, course around the midbrain, and past the pineal gland to and through the foramen of Monro to reach the choroid plexus while sending branches to the cerebral peduncle, the tegmentum, pulvinar, pineal gland, and thalamus. Rarely, the medial posterior choroidal artery can arise from the parieto-occipital or calcarine arteries. It can be displaced by pineal or callosal tumors.

The posterior P2 segment of the PCA gives rise to the middle temporal, common temporal, and lateral posterior choroidal arteries (Zeal and Rhoton, 1978). The middle temporal artery is the smallest cortical branch and may be absent. The common temporal artery supplies the left temporal lobe. The lateral posterior choroidal artery extends over the pulvinar into the trigone of the lateral ventricle, and gives off branches to the cerebral peduncle, posterior commissure, fornix, lateral geniculate body, and caudate nucleus.

The P3 segment of the PCA terminates in two primary arteries: the calcarine and parieto-occipital arteries. The calcarine artery courses through the calcarine fissure, and supplies branches to the occipital pole, lingual gyrus, and inferior cuneus, primarily supplying the visual cortex. Its occlusion typically results in homonymous visual field defects, often sparing the macula. The parieto-occipital artery courses through the parieto-occipital fissure, and supplies the posterior parasagittal region, cuneus, precuneus, lateral occipital gyrus, precentral lobule, and parietal lobules. It also gives rise to the splenial artery, which supplies the dominant occipital pole and splenium of the corpus callosum. There can also be an anastomosis between the pericallosal artery and the splenial artery. If there is retrograde filling of the splenial artery through the pericallosal artery, there may be a PCA occlusion (Hoyt and Newton, 1970).

Vertebrobasilar perforators

The vertebrobasilar perforators comprise of the caudal, middle, and rostral arteries. The caudal perforating arteries originate from the proximal dorsal aspect of the basilar artery, situated between the vertebral artery junction and the AICA origin sites. They descend along the basilar sulcus and enter the foramen cecum (Marinkovic and Gibo, 1993). Typically, there are 2–5 of these arteries, with an average of two on each side of the basilar artery. In cases where perforators are absent on one side, a larger trunk from one side may supply both sides. These arteries supply the pyramids, medulla, and hypoglossal nerve, and form anastomoses with perforating arteries from the vertebral or anterior spinal arteries.

The middle perforating arteries originate from the dorsal aspect of the basilar artery between the AICA and the posterolateral artery. These arteries course both caudally and rostrally, giving rise to collateral and terminal branches such as the pontomedullary artery, long pontine arteries, and anterolateral branches. They also contribute to complex arteries comprising perforating, anterolateral vessels, and a branch to the abducens nerve. Generally, these arteries have anastomoses only with the caudal perforators. The paramedian, short circumferential, and long circumferential arteries are the main axial perforators of the hindbrain. The short and long circumferential arteries correspond to the vasa corona, entering at the anteromedial and posterolateral portions, respectively (Ota and Komiyama, 2022). The vasa corona, which is difficult to appreciate on angiography or other types of imaging, serves as anastomotic connections *between* perforators that have persisted since the embryologic stage. Occlusion of the vasa corona can lead to infarctions of the medulla. Rarely, aneurysms of the vasa corona can also arise (Shimizu et al., 2004).

The rostral perforating arteries emerge from the terminal part of the basilar artery and proceed into the interpeduncular fossa. These arteries are primarily responsible for giving rise to anterolateral branches and form anastomoses with other rostral perforators. This network is crucial for supplying blood to the upper brainstem regions.

CONCLUSION

The vascular anatomy of the brainstem and posterior circulation demonstrates dense redundancy, anatomic variation, and a network of perforating arteries supplying critical and eloquent brain structures with limited physiologic reserve. This anatomy illustrates why posterior circulation syndromes are often mixed, or deceptively nonlocalizing as overlapping territories (e.g., AICA-PICA-SCA), variable dominance patterns (vertebral, fetal PCA), and small but critical perforators (pontine, thalamoperforating, circumflex branches) can produce disproportionate deficits from lesions that appear subtle on initial imaging. The structured caudal-to-cranial understanding discussed in this chapter provides a practical framework to anticipate which brainstem nuclei and long tracts are at risk, to recognize classic and atypical stroke presentation and to interpret imaging findings with greater anatomic specificity. The safe diagnosis and treatment of pathologies in the posterior circulation depends on recognizing the full network in which the culprit vessel resides. Careful review of vascular and angiographic imaging is essential to avoid nontarget embolization, perforator infarction or compromise of shared arterial trunks, or embryologic carotid-vertebrobasilar connections in order to improve outcomes for patients.

References

Amarenco P, Rosengart A, DeWitt LD et al. (1993). Anterior inferior cerebellar artery territory infarcts: mechanisms and clinical features. Arch Neurol 50: 154–161.

Andoh K, Tanohata K, Moriya N et al. (2001). The posterior inferior cerebellar artery arising from the extracranial segment of the internal carotid artery via the hypoglossal canal without an interposed segment of the basilar artery: a persistent primitive hypoglossal artery variant. Clin Imaging 25: 86–89.

Bahşi YZ, Uysal H, Peker S et al. (1993). Persistent primitive proatlantal intersegmental artery (proatlantal artery I) results in "top of the basilar" syndrome. Stroke 24: 2114–2117.

Boyko OB, Curnes JT, Blatter DD et al. (1996). MRI of basilar artery hypoplasia associated with persistent primitive trigeminal artery. Neuroradiology 38: 11–14.

Brismar J (1976). Persistent hypoglossal artery, diagnostic criteria: report of a case. Acta Radiol Diagn (Stockh) 17: 160–166.

Campos J, Fox AJ, Viñuela F et al. (1987). Saccular aneurysms in basilar artery fenestration. Am J Neuroradiol 8: 233–236.

Dare AO, Chaloupka JC, Putman CM et al. (1997). Vertebrobasilar dissection in a duplicated cervical vertebral artery: a possible pathoetiologic association?: a case report. Vasc Surg 31: 103–109.

Djukić B, Djukić-Macut N, Djulejić V et al. (2023). Medullary branches of the vertebral artery: microsurgical anatomy and clinical significance. Acta Neurochir 165: 1807–1819.

Er U, Fraser K, Lanzino G (2008). The anterior spinal artery origin: a microanatomical study. Spinal Cord 46: 45–49.

Fine AD, Cardoso A, Rhoton AL (1999). Microsurgical anatomy of the extracranial—extradural origin of the posterior inferior cerebellar artery. J Neurosurg 91: 645–652.

Fujii K, Lenkey C, Rhoton AL (1980). Microsurgical anatomy of the choroidal arteries. Fourth ventricle and cerebellopontine angles. J Neurosurg 52: 504–524.

Gao L, Guo X, Zhou J et al. (2013). Basilar artery fenestration detected with CT angiography. Eur Radiol 23: 2861–2867.

Garcia-Gonzalez U, Cavalcanti DD, Agrawal A et al. (2012). Anatomical study on the "Perforator-free Zone": reconsidering the proximal superior cerebellar artery and basilar artery perforators. Neurosurgery 70: 764.

Greitz T, Lauren T (1968). Anterior meningeal branch of the vertebral artery. Acta Radiol Diagn 7: 219–224.

Hardy DG, Peace DA, Rhoton ALJ (1980). Microsurgical anatomy of the superior cerebellar artery. Neurosurgery 6: 10.

Hou K, Li G, Luan T, Xu K, Xu B, Yu J (2020). Anatomical study of anterior inferior cerebellar artery and its reciprocal relationship with posterior inferior cerebellar artery based on angiographic data. World Neurosurgery 133: e459–e472.

Hoyt W, Newton T (1970). Angiographic changes with occlusion of arteries that supply the visual cortex. NZ Med J 72: 310–317.

Inoue T, Shimizu T, Hirai et al. (2012). Ocular neuromyotonia treated by microvascular decompression: usefulness of preoperative 3D imaging. J Neurosurg 117: 1166–1169.

Ionete C, Omojola MF (2006). MR angiographic demonstration of bilateral duplication of the extracranial vertebral artery: unusual course and review of the literature. AJNR Am J Neuroradiol 27: 1304–1306.

Joshi S, Tee WWH, Franconi C et al. (2017). Transient oculomotor nerve palsy due to non-aneurysmal neurovascular compression. J Clin Neurosci 45: 136–137.

Kim JS, Lopez I, DiPatre PL et al. (1999). Internal auditory artery infarction. Neurology 52: 40.

Kubik CS, Adams RD (1946). Occlusion of the basilar artery—a clinical and pathological study. Brain 69: 73–121.

Kwiatkowska M, Rzepliński R, Ciszek B (2023). Anatomy of the pontine arteries and perforators of the basilar artery in humans. J Anat 243: 997–1006.

Lasjaunias PL, Berenstein A (1983). Collateral circulation and angiographic protocols. Craniofacial and upper cervical arteries. Williams & Wilkins.

Lister JR, Rhoton AL, Matsushima T et al. Microsurgical anatomy of the posterior inferior cerebellar arteryNeurosurgery, 10, pp. 170–199.

Lv X, Ge H, He H et al. (2016). Anterior inferior cerebellar artery aneurysms: segments and results of surgical and endovascular managements. Interv Neuroradiol 22: 643–648.

Mahmood A, Dujovny M, Torch M et al. (1991). Microvascular anatomy of foramen caecum medullae oblongatae. J Neurosurg 75: 299–304.

Malicki M, Szmyd BM, Bobeff EJ et al. (2023). The superior cerebellar artery: variability and clinical significance. Biomedicines 11: 2009.

Marinkovic SV, Gibo H (1993). The surgical anatomy of the perforating branches of the basilar artery. Neurosurgery 33: 80.

Mattle HP, Arnold M, Lindsberg PJ et al. (2011). Basilar artery occlusion. Lancet Neurol 10: 1002–1014.

Meckel S, Spittau B, McAuliffe W (2013). The persistent trigeminal artery: development, imaging anatomy, variants, and associated vascular pathologies. Neuroradiology 55: 5–16.

de Mendivil AO, Alcalá-Galiano A, Ochoa M et al. (2013). Brainstem stroke: anatomy, clinical and radiological findings. Semin Ultrasound CT MR 34: 131–141.

Mitsumura H, Miyagawa S, Komatsu T et al. (2016). Relationship between vertebral artery hypoplasia and posterior circulation ischemia. J Stroke Cerebrovasc Dis 25: 266–269.

Morita A, Fukushima T, Miyazaki S et al. (1989). Tic douloureux caused by primitive trigeminal artery or its variant. J Neurosurg 70: 415–419.

Newton T (1968). The anterior and posterior meningeal branches of the vertebral artery. Radiology 91.

Ota T, Komiyama M (2022). Vascular supply of the hindbrain: basic longitudinal and axial angioarchitecture. Interv Neuroradiol 28: 756–764.

Patel AB, Gandhi CD, Bederson JB (2003). Angiographic documentation of a persistent otic artery. AJNR Am J Neuroradiol 24: 124–126.

Polguj M, Podgórski M, Jędrzejewski K et al. (2013). Fenestration and duplication of the vertebral artery. Clin Anat 26: 933–943.

Robert T, Cicciò G, Bonasia S (2023). Anatomy and variations of the posterior inferior cerebellar artery. In: Robert T, Bonasia S, Bojanowski MW. (Eds.). Anatomy of cranial arteries, embryology and variants. Cham: Springer International Publishing, pp. 191–202.

Rodríguez-Hernández A, Rhoton AL, Lawton MT (2011). Segmental anatomy of cerebellar arteries: a proposed nomenclature: laboratory investigation. J Neurosurg 115: 387–397.

Saeki N, Rhoton AL (1977). Microsurgical anatomy of the upper basilar artery and the posterior circle of Willis. J Neurosurg 46: 563–578.

Salgado-Lopez L, Leonel LCP, Aydin SO et al. (2020). Surgical anatomy of the labyrinthine and subarcuate arteries and clinical implications. World Neurosurg 141: e880–e887.

Sawlani V, Behari S, Salunke P et al. (2006). Stretched loop sign" of the vertebral artery: a predictor of vertebrobasilar insufficiency in atlantoaxial dislocation. Surg Neurol 66: 298–304.

Schievink WI (2001). Spontaneous dissection of the carotid and vertebral arteries. N Engl J Med 344: 898–906.

Seçkin H, Ateş Ö, Bauer AM et al. (2009). Microsurgical anatomy of the posterior spinal artery via a far-lateral transcondylar approach: laboratory investigation. J Neurosurg Spine 10: 228–233.

Shaban A, Albright KC, Boehme AK et al. (2013). Circle of Willis variants: fetal PCA. Stroke Res Treat 2013: 105937.

Shimizu S, Garcia AS, Tanriover N et al. (2004). The so-called anterior meningeal artery: an anatomic study for treatment modalities. Interv Neuroradiol 10: 293–299.

Siclari F, Burger IM, Fasel JHD et al. (2007). Developmental anatomy of the distal vertebral artery in relationship to variants of the posterior and lateral spinal arterial systems. Am J Neuroradiol 28: 1185–1190.

Takahashi M, Wilson G, Hanafee W (1968). The anterior inferior cerebellar artery: its radiographic anatomy and significance in the diagnosis of extra-axial tumors of the posterior fossa. Radiology 90.

Tatu L, Moulin T, Bogousslavsky J et al. (1996). Arterial territories of human brain. Neurology 47: 1125–1135.

Terao S, Sobue G, Izumi M et al. (1996). Infarction of superior cerebellar artery presenting as cerebellar symptoms. Stroke 27: 1679–1681.

Uchino A, Saito N, Okada Y et al. (2013). Persistent hypoglossal artery and its variants diagnosed by CT and MR angiography. Neuroradiology 55: 17–23.

Uchino A, Sawada A, Takase Y et al. (2003). Variations of the superior cerebellar artery: MR angiographic demonstration. Radiat Med 21: 235–238.

Weil AG, McLaughlin N, Denis D et al. (2011). Tentorial branch of the superior cerebellar artery. Surg Neurol Int 2: 71.

Wise BL (1957). Angiographic demonstration of the primitive trigeminal artery (carotid-basilar anastomosis). Radiology 68: 731–732.

Zeal AA, Rhoton AL (1978). Microsurgical anatomy of the posterior cerebral artery. J Neurosurg 48: 534–559.

Handbook of Clinical Neurology, Vol. 216 (3rd series)
The Brainstem
L.T. Grinberg and A.D. Korczyn
https://doi.org/10.1016/B978-0-443-15736-3.00007-X

Chapter 20

Brainstem evoked responses

GABRIEL BYCZYNSKI[1] AND SVEN VANNESTE[1,2,*]

[1]*Trinity College Institute of Neuroscience, School of Psychology, Trinity College Dublin, Dublin, Ireland*

[2]*Global Brain Health Institute, School of Psychology, Trinity College Dublin, Dublin, Ireland*

Abstract

The brainstem's activity and control over both central and peripheral functioning is notably widespread, yet difficult to quantify due to its location and size. Methods have been developed to improve our understanding of the brainstem's activity, particularly through the use of proxy-measures, via physiologic imaging. Heart rate variation, pupillometry, skin conduction, P300 alterations, and salivary α-amylase are popular measures which allow for a determination of brainstem activity. This chapter discusses the use and findings of physiologic imaging for brainstem activity. It also discusses how researchers may attempt to alter brainstem activity with stimulation. Vagus nerve, occipital nerve, and auditory nerve stimulation are discussed, with their therapeutic contributions and functional effects for context. We discuss evoked brainstem responses including brainstem evoked audiometry, which additionally provide valuable information as a direct representation of brainstem activation and pathology. Lastly, we discuss the application of brainstem responses and nerve stimulation in a surgical setting via intraoperative recording. We conclude by discussing the value of these measures and methods in brainstem research.

INTRODUCTION

There has been significant advancement through the past few decades in the understanding of the brainstem's contributions to aspects of cognitive and somatic function. Innervations of the heart, lungs, and muscles, alongside projections to the cortex, mean that the brainstem has an exceptional amount of influence over the body and brain. There is therefore considerable interest in methodologies of measurement, clinical uses for the reported results, and means to augment and modulate its activity. Brainstem activity can provide valuable information about the state of the body and brain, with such exceptional specificity and validity that its measurement has become commonplace in medicine and research. This chapter will focus on the current methodologies used to physiologically decipher the activity of the brainstem, including reference to essential anatomy and connectivity. Broadly, these methods range from classical methods such as skin conduction, formerly the galvanic skin response, to more electrophysiologic measures such as P300-correlates of locus coeruleus (LC) activity. Through development, these measures have become popular methods of estimating brainstem activity. In research, physiologic measures can be used to quantify the effect of a treatment on the brainstem indirectly, and in medicine, some measures are already commonplace as a diagnostic or screening measure (i.e., audiometry). Due to its connectivity and influence, the brainstem and its nuclei are attractive targets for interventive therapies. Epilepsy, migraines, and even alterations in memory performance, are examples of the domains which neural modulation via stimulation of the brainstem through cranial nerves have been accomplished. Accordingly, we discuss the methodology of vagus, occipital, trigeminal, and auditory nerve stimulations. We approach these methods with both implanted stimulation, and transcutaneous nerve stimulation, and discuss the surgical approaches, effectiveness, and outcomes. Finally, we discuss key responses of the brainstem which are used in diagnostic and research areas. This includes brainstem evoked response audiometry (BERA) and intraoperative recording. We aim to provide a clear account of the current physiologic imaging methods, stimulation approaches, responses produced, and utility of the brainstem, specifically relating to brainstem activity and connectivity.

PHYSIOLOGIC IMAGING OF THE BRAINSTEM

Typical imaging of the brain and body is generally very effective and provides valuable information about the size, shape, and functionality of the organ in question. There are, however, regions which are difficult to image for several

*Correspondence to: Sven Vanneste, Global Brain Health Institute, Trinity College Dublin, Lloyd Building, Dublin 2, D02 X9W9, Ireland. Tel: +353877211893; E-mail: sven.vanneste@tcd.ie.

reasons. Due to its location and its anatomic and functional complexity, and with its added susceptibility to physiologic noise in imaging, methods of directly measuring brainstem activity can often be trivial (Sclocco et al., 2018). Accordingly, numerous methods have been established that allow researchers to probe the activity and influence of the brainstem in the body and brain. Since there is difficulty in imaging it directly, proxy-measures can be used based on the functional aspects of the brainstem, and the systems it is involved in. One such system includes the autonomic nervous system (ANS), for which there are numerous methodologic approaches to estimating its activity. Such methods that are commonplace in ANS investigation include heart rate variation, pupillometry, skin conduction, and electrophysiologic imaging (i.e., electroencephalography (EEG) recording). We will discuss the origins and current uses for these physiologic markers and how they relate to brainstem activity, in reference to ANS activation.

The brainstem has numerous innervations throughout the body; however, this section will restrict discussion to those which are relevant for so-called physiologic imaging. Beginning with the influence of the cardiac muscles, brainstem regions including the medulla oblongata (MO), nucleus of the solitary tract (NTS), the nucleus ambiguus, and the dorsal motor nucleus of the vagus are the source of vagal efferent outflow. That is, they possess some aspect of control over the cardiac muscles. These regions are thus considered to be modulators of cardiac activity via glutamatergic, cholinergic, and Gamma-aminobutyric acid (GABA) inputs (Wang et al., 2001; Park et al., 2020). Vagal postganglionic nerves modulate cardiac activity through the ventricular ganglionated plexuses, within which numerous neuronal subtypes modulate the heart's activity. Vagal fibers create cholinergic activation of, primarily, the Gi-protein-coupled M2 receptor in cardiomyocytes, which in turn elicits cardiac regulation (Brack et al., 2013). To control this innervation and to maintain optimal cardiac function, there is the baroreceptor (BR) reflex. The BR reflex is a control system that regulates sympathetic and parasympathetic nervous system (SNS and PNS, respectively) activity via brainstem nuclei (Park et al., 2020). It does so by controlling blood pressure (BP), heart rate, and thus cardiac output, in an attempt to maintain cardiovascular homeostasis. BRs function by detecting arterial BP with mechanoreceptors, and elicit their response in three steps. First, there is vascular distention of the BR nerve endings that results from alternations in BP. This distention leads to depolarization of the nerve endings, which finally produces an action potential discharge. BR inputs travel to the NTS and complete the BR reflex arc with connections to the caudal ventrolateral medulla and rostral ventrolateral medulla (RVLM), which concludes with control of the sympathetic preganglionic neurons (McCall et al., 2017). It is here that the activity of the heart becomes a marker of brainstem activity through arousal. Due to the relationship between the brainstem and cardiac behavior, heart-rate variability (HRV) has been considered an effective representation of sympathetic and parasympathetic brainstem activity. HRV describes the changes of interbeat intervals of the heart rate, reflecting changes in SNS and PNS recruitment by the ANS. Increased HRV often reflects adaptation to the environment, and the individual SNS and PNS contributions to HRV can be estimated with cardiac signal information (Thome et al., 2022).

Given that the brainstem has close association with the SNS and PNS, reducing activity in the brainstem has been shown to result in reduced ANS control (Thome et al., 2022). In further discussion of the vagal-cardiac connection, there are studies that have shown that vagal fiber activity relates to coronary artery occlusion, and can predict increased risk of sudden cardiac death (Brack et al., 2013). Indeed, it has been shown that lesions of the brainstem, (medullar) infarcts, and reduced gray matter connectivity from the brainstem associate with reductions or abnormalities in HRV (Korpelainen et al., 1996; Monge Argilés et al., 2000; Mueller et al., 2019). Monge Argilés et al. (2000) further reported that individuals with multiple sclerosis (MS) possess autonomic dysfunction as reflected in reduced HRV and that this dysfunction was exaggerated in those with medullary infarction. These findings echo those of Korpelainen et al. (1996), who also reported that HRV was lower in those with brainstem infarction compared to controls and further that the same effect was not seen in those with pontine infarcts, positing subregional specificity. Altogether, literature supports the use of HRV as a proxy-measure of brainstem activity, through its connection with ANS functioning. There are therefore several other autonomic-related physiologic events which are considered to be effective markers of activity, and thus may be of use in estimating brainstem activity.

Another well-established method is pupillometry, which studies the effect of cognitive or physiologic changes via pupil size (Sirois and Brisson, 2014). There is a noted close connection between the brainstem, particularly the noradrenergic LC, and the pupillary response. LC activation leads to dilation of the pupils (Joshi et al., 2016), this occurring during phasic firing of neurons in the LC (Gilzenrat et al., 2010). In one prominent study by Joshi et al. (2016), it was shown that in macaque monkeys, spiking activity of the LC was positively correlated with changes in pupil size, and furthermore that there was increased spiking activity in the LC, inferior colliculus (IC), and superior colliculus (SC) preceding pupil dilation compared to constriction. In order to confirm a more causal relationship, Joshi et al. (2016) microstimulated the LC, IC, and SC, finding that stimulation of these regions evoked pupil size changes (with this effect being strongest in the LC). Lastly, when presented with a startling auditory stimulus, there was also LC-correlated pupil responses, supporting that the LC-pupil dilation relationship is also task-related and not strictly a transient phenomenon. Thus, at least in animal models, it appeared that brainstem activity, specifically the LC, might be appropriately (and sensitively) represented using pupillometry.

In humans, since there is more barrier in directly recording and stimulating the LC, research can be carried out instead using functional magnetic resonance imaging (fMRI). Combined fMRI and pupillometry has provided further evidence that activity of the brainstem correlates with pupil diameter in humans. The mechanism of LC-modulated pupil dilation is proposed to be via the Edinger–Westphal (E–W) nucleus, where the LC connects via inhibitory adrenergic receptors (Nobukawa et al., 2021). When the LC inhibits the E–W nucleus, parasympathetic activity is reduced, leading to an autonomic imbalance. This imbalance triggers pupil dilation through the sympathetic pathway (Carrick et al., 2021). Important, however, is the observation that due to the complexity of the LC's contribution to arousal and cognitive function, the modern understanding of LC activation is not linear, but rather follows the Yerkes–Dodson law (i.e., inverted U-shape), with an optimal activation occurring during phasic firing, low-to-no tonic firing associated with inattention or drowsiness, and higher tonic-mode LC firing correlated with poor performance (Gilzenrat et al., 2010). The use of pupillometry to measure brainstem activity found its way into the cognitive domain, with research showing correlations between pupil dilation and memory load (Kahneman and Beatty, 1966), task engagement (Gilzenrat et al., 2010), and emotional response (Bradley et al., 2008). The pathways outlining the brainstem's influence on measurable physiologic changes that are discussed in this section, are shown in Fig. 20.1.

Despite the difficulties surrounding direct imaging of the brainstem, justifying the use of physiologic measures, there is constant development of methods to mitigate the physiologic noise that complicates the use fMRI for brainstem research. Methods including neuromelanin weighted T1 (nmT1) and multiecho functional magnetic resonance imaging (ME-fMRI) are two possibilities. One substantial issue with the direct imaging and measuring of the LC region of the brainstem is its homogeneity with the surrounding regions, its location variability between individuals, and its comparably small size. It is proposed that an alternative to single-echo MRI is ME-fMRI, which involves multiple acquisitions of each voxel per scan, reducing noise and maximizing contrast (Turker et al., 2021). This enhancement improves the ability to localize the LC within the brainstem. Alternatively, nmT1 MRI is also employed to reduce the problem of LC localization. nmT1 involves developing imaging sequences that increases contrast for regions containing neuromelanin, which includes the LC (Turker et al., 2021). Combining fMRI and physiologically derived measures has confirmed correlation between LC blood oxygen level–dependent (BOLD) activity and pupil dilation, providing internal validity for both measures (Murphy et al., 2014). In memory work, a similar nonlinear trend of LC activation and memory performance is seen, as increased LC BOLD signal variability was correlated with reduced parasympathetic activity, and thus greater sympathetic modulation, leading to improved memory (Jacobs et al., 2020). In their study, Jacobs and colleagues also employed HRV measurements and found indeed that emotional memory performance relationships between HRV and LC BOLD signals. They also found that emotional memory performance was similarly linked with salivary α-amylase (sAA) levels. sAA, an enzyme whose secretion is largely controlled by the SNS, is also considered to be a measure of SNS activity, originally through its relationship with exercise and stress (Rohleder and Nater, 2009; Liu et al., 2020). sAA levels correlate with LC activation and connectivity changes, placing it as a plausible and feasible physiologic measure of sympathetic tonal increases (Jacobs et al., 2020). In their study, Jacobs and colleagues also found that activation of the LC corresponded to greater sympathetic modulation, indexed by changing sAA levels, and ultimately improved memory performance.

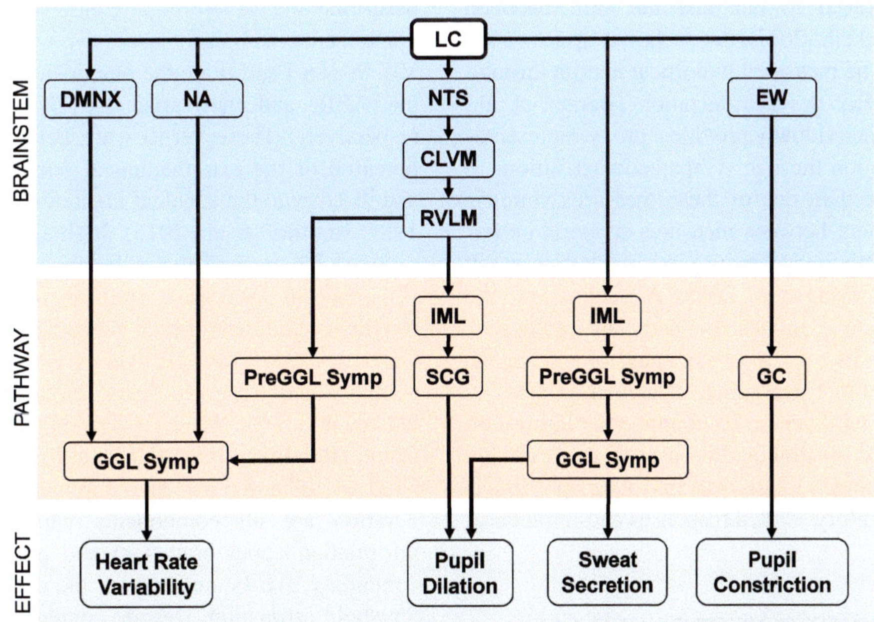

Fig. 20.1. Schematic illustration of the pathway of brainstem efferents to the body involved in heart rate variability, pupil dilation, sweat secretion, and pupil constriction. *CVLM*, caudal ventrolateral medulla; *DMNX*, dorsal motor nucleus of the vagus nerve; *EW*, Edinger–Westphal nucleus; *GC*, ganglion ciliare; *GGL Symp*, sympathetic ganglion; *NA*, nucleus ambiguus; *NTS*, nucleus of the solitary tract; *IML*, intermediolateral column; *PreGGL Symp*, sympathetic preganglionic neurons; *RVLM*, rostral ventrolateral medulla; *SCG*, superior cervical ganglion.

Another measure that reflects brainstem activity and cognitive load, although not as sensitive as the pupillary response, is measurement of conductivity of the skin (Hogervorst et al., 2014). Electrodermal activity (EDA) or the skin conductance response (SCR) is used to measure sweat gland responses, which are innervated by the sympathetic branch of the ANS (Laine et al., 2009). SCR measures changes in skin conductance, which fluctuates with sweat secretion (Schumann et al., 2018). An early study found that via transection through the spinal cord and pons, induced an increase in EDA (Glasser et al., 1964). Later, a study demonstrated that EDA in cats was silenced following chemical lesions in the ascending noradrenergic bundle of the brainstem (although no such effect was seen after lesioning the ventral tegmental area) (Yamamoto et al., 1990). Indeed, EDA/SCR provide markers of SNS arousal, with the LC often being considered a probable influencer of SCR and their activity being correlated (Eisenbarth et al., 2016; Le et al., 2019). Although the ANS measures of brainstem activity are strongly endorsed as a representation of its function, the robustness of imaging (fMRI) and direct measuring (electrophysiologic methods) may still be considered of a higher standard.

Although fMRI has progressed in its ability to capture brainstem activity, debate persists on the reliability of measuring brainstem activity with other imaging methods such as EEG or magnetoencephalographic, beyond the well-studied BERA. Given its medial depth, methods of recording the brainstem cannot be achieved directly; however, measurement of thalamocortical oscillations can be used to derive information on activity of neuromodulation originating at the brainstem (Scheib, 2017). Commonly discussed in reference to sleep and consciousness, brainstem-induced changes to the thalamus and cortex are widely believed to facilitate the shift between awareness and sleep (Scheib, 2017). As such, the effect of the brainstem's arousal can be measured in cortical neuron through phenomena such as EEG desynchronization (Berger et al., 2021). These measures individually provide a proxy representation of brainstem activation through sympathetic activation. In studies employing more than one of these measures simultaneously, correlated activity between measures supports individual robustness and corroborates their use in brainstem research (Eisenbarth et al., 2016; Wang et al., 2018). As such, due to the extent of the brainstem's influence, there are numerous approaches to measuring its activity through physiologic means. Combining this with direct recording creates a picture of brainstem activity and its influence. To continue the discussion of brainstem physiologic imaging and its uses in research and medicine, the next section will discuss two uses of (direct) brainstem recording: auditory evoked responses and intraoperative recording.

AUDITORY EVOKED RESPONSES AND INTRAOPERATIVE RECORDING

Similar to the well-known event-related potentials (ERPs), which are prevalent in EEG research, recordings from the brainstem provide a similar evoked response that can be used, among other indices, to detect hearing loss, lesions, and tumors (Young et al., 2022). Originating from the 1970s, the auditory brainstem response (ABR) is a key component of BERA and a well-studied physiologic measure that consists of seven positive peaks (I–VII). It is often employed in infant auditory screening due to its sensitivity to detect abnormalities, including those listed above, but extending even to hyperbilirubinemia (Soni et al., 2016), MS (Chiappa, 1980; Robinson and Rudge, 1980), and mild traumatic brain injury (Washnik et al., 2019). The encompassing factor in the diagnostic ability provided by BERA is the brainstem's contribution in generating consistent and accurate waveforms. In addition, its relative simplicity (and even automatability) places it as a popular measure in healthcare.

To perform ABR testing, an EEG recording system is used, typically involving electrode placement on the vertex, right and left earlobes, and a ground electrode. Headphones are used to deliver acoustic stimuli. The intensity of the stimulus, usually a click sound, is then increased in 10 dB increments to produce analyzable waveforms (Norton et al., 2005). Each of the seven waves of the ABR have specific anatomic origins (although sometimes disputed) and functional roles which can be used to provide diagnostic information about the state of the brainstem and its associate regions. Beginning when an auditory click stimulus is presented, recording of electrophysiologic activity produces several waveforms (usually seven). As the signal travels through established brainstem-auditory pathways, measurable potentials can be recorded and produce a particularly sensitive representation of neural functioning of the brainstem and auditory system. A visualization of this response with their corresponding anatomic origins is shown in Fig. 20.2. In describing the waveform, attention is often given to the positive components, with each successive wave being named from I to VII. Waves I and II are the first positive vertices to occur during the ABR, and they arise at 1.5 and 2.5 ms poststimulus, respectively. These vertices are believed to arise from action potential of the extrameningeal portion of the auditory nerve and its entry to the cochlear nucleus (CN) (Rowe, 1981; Moore, 1987; Rouillon et al., 2016). It should be noted, however, that research has posited that subcomponents of wave II (i.e., IIa, IIb) may in fact arise from different locations along the proximal auditory nerve and CN (Ananthanarayan and Durrant, 1991). Wave III likely arises from the superior olivary complex and trapezoid body, bilaterally, and occurs with a latency of 3.5 ms. Wave IV is said to originate from the nucleus of the lateral lemniscus (NLL) at 4.5 ms and wave V from the NLL and IC at 5.5 ms. The amplitude, latency, and intervertex duration are all components which are used to provide information about brainstem and auditory functioning. Most commonly, ABRs are used for the detection of hearing loss or threshold estimation. Despite contention surrounding the exact localization of the source of each peak, the anatomic representations presented above are considered reliable, representations, further validated by their success in diagnostic and pathologic associations. For example, in children, the time between waves

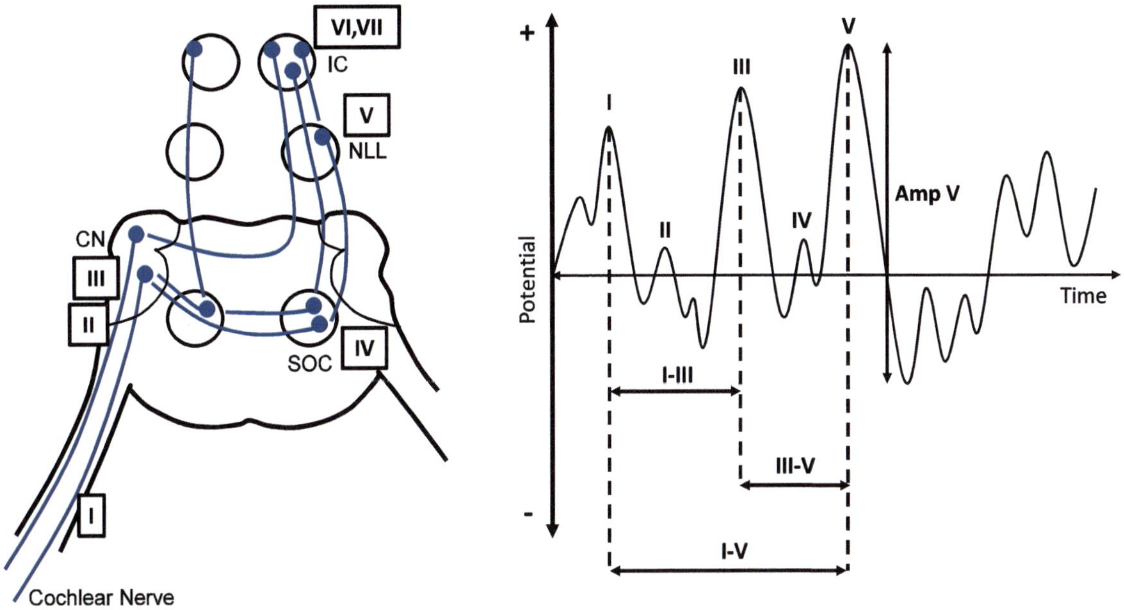

Fig. 20.2. Schematic illustration of the brainstem beginning at cochlear nerve, with indicators of region (I–VII) and its corresponding auditory evoked potential latency. Latency calculations are also provided (e.g., I–III, III–V, and I–V). *CN*, cochlear nucleus; *IC*, inferior colliculus; *NLL*, nucleus of the lateral lemniscus; *SOC*, superior olivary complex.

I and III or III and V can be used to determine the possibility of lesions, that is, when the intervals exceed an interaural difference of 0.35 ms (Rouillon et al., 2016).

There is also extensive diagnostic use for ABRs beyond newborn auditory screening. ABR measurements have a role in detecting both psychiatric and physical pathologies. In one paper, it was shown that there was a relationship between ABR pathology and auditory hallucinations in individuals with schizophrenia (Lindstrom et al., 1987), and an ABR profiling test was able to discriminate between individuals with ADHD or schizophrenia from controls with high sensitivity (>84.6%) using a range of traits that included sex-specific correlations with 10 ms norm curves, compiling an index that allocated control from pathology groups (Juselius Baghdassarian et al., 2018). Using variations of ABR including speech-evoked ABR, it has also been possible to investigate auditory processing disorder, as it has been shown in a study that abnormal (speech-evoked) ABR identified children with auditory processing disorder or language impairment (Rocha-Muniz et al., 2012). More broadly, it was shown that children with learning problems had both click and speech-evoked ABR abnormalities, specifically in regards to wave latencies and interpeak latencies (Malayeri et al., 2014). Considerable research also exist on the relationship between ABRs and individuals autism spectrum disorder (ASD). In a meta-analysis, it has been revealed that wave V is prolonged in infants and children with ASD and shortened in adults with ASD (Miron et al., 2018). The authors report that this trend may present a valuable biomarker for ASD in young children and may better inform referrals for more advanced and less-available testing.

While ABRs are often used in children for hearing loss detection or pathology surveillance, BERA is also significantly used for intraoperative recording to safeguard the auditory systems during surgical procedures. In a similar manner to investigational ABR procedures, short-latency auditory evoked potentials (SAEPs), which are produced by abrupt auditory stimuli, can be used to determine the functional state of the brainstem. SAEPs include electrocochleogram (ECochG) and brainstem auditory evoked potentials (BAEPs) (Harper, 2004; Zamel, 2010). ECochG is often used as an alternative to BAEP and is typically used for detecting acoustic neuromas. ECochG involves measuring the electrical potentials generated by hair cells and neurons, producing both the cochlear microphonic for the hair cells and the auditory nerve neurophonic for the neural response (Gendt et al., 2020). ECochG can inform hearing retention using its N1 potential, although due to the N1 origin (lateral on the auditory nerve), damage occurring medially may go unnoticed (Harper, 2004). Intraoperative BAEP monitoring has been shown to be beneficial not only in helping to ensure preservation of function, but also in predicting postoperative hearing condition of patient after surgeries involving the auditory pathway (Gu et al., 2022). Intracochlear ECochG can be used during surgery to monitor damage to the hair cells and optimize placement of cochlear implants, both as a form of intraoperative recording (O'Connell et al., 2017; Helmstaedter et al., 2018). Its efficiency in predicting postoperative hearing loss, however, remains limited (Yin et al., 2021).

ECochG is one of many forms of intraoperative monitoring, which involves essentially the monitoring of the nervous system function during surgery, ultimately beneficial for the preservation of function. There are several other types of evoked nerve responses used for intraoperative recording including compound muscle action potentials (CMAPs), nerve action potentials (NAPs), and motor unit potentials. Similar to ABRs, these evoked responses can be used intraoperatively for monitoring of the patient's functional retention, and to inform future surgical methods. Using an intramuscular recording method that involves the placement of needle electrodes in the muscle of interest, CMAPs record the activity of muscles. Alternatively, NAPs record directly from the nerves (Harper, 2004). Through the use of a stimulator, the nerve is activated and produce a flow of current that is subsequently measured. The magnitude of the CMAP correlates to the functionality of the axons located between the stimulation and recording electrodes (Harper, 2004). There is proven success with intraoperative monitoring, with reductions in adverse outcomes being shown as a result (Radtke et al., 1989; Nuwer et al., 2012). In effect, the methods used for intraoperative recording are consistent in that they utilize a stimulus and observe the response. In line with this approach, we will continue this chapter by discussing the use of stimulation of the cranial nerves.

NERVE STIMULATION

As researchers understood the large contribution of the brainstem to all aspects of cortical control, audiometry, arousal, attention, and disease, therapeutic interventions targeting the brainstem via cranial nerves became a topic of interest. This section will discuss two forms of cranial nerve stimulation, vagus nerve stimulation (VNS) and occipital nerve stimulation (ONS). Here, we discuss the background for each stimulation technique, including animal models, its transfer to human research, mechanisms of action, surgical approaches, and its current uses in medicine and research.

Vagus nerve stimulation

VNS involves using an electric current either directly or transcutaneous to modulate the activity of the vagus nerve and its innervations. The longest nerve in the body, the vagus nerve (cranial nerve X) is a key moderator of the ANS and possesses immune system regulatory aspects (Noller et al., 2019). It has afferent connectivity to the NTS through the nodose ganglion, which further relays to the cortex (Zhuo et al., 1997). It is this connectivity to the brain which has made VNS an intriguing target. Early work in cats has shown EEG synchronization via VNS, with later work unearthing more specific anatomic activations including the cingulate and limbic regions (George et al., 2003). In rodents, VNS can be delivered via several electrode configurations including a cuff or needle electrode. Using this technique, researchers have been able to investigate the impact of VNS on cognitive, motor, and immunologic disorders. For example, one study showed that rat corticosterone levels were increased following 30 Hz VNS, suggesting that VNS results in immunomodulation, as well as immediate effects on the hypothalamic-pituitary-adrenal axis (De Herdt et al., 2009). In another study which used cervical VNS to reduce the effect of induced lipopolysaccharide endotoxemia in mice, it was found that VNS was able to reduce brain inflammation, support cognitive improvement, and produce observable changes in microglial activation (Huffman et al., 2019). VNS, given its relationship with cardiac functioning, has also been suggested to play a supportive role as an antiarrhythmogenic, with studies in dogs and rats showing the VNS reverses protein expression associated with heart failure (Brack et al., 2013).

VNS in mice for improvement of motor rehabilitation is also a topic of interest and has led to findings that suggest VNS effects may be task-dependent as an interventive. In a study by Khodaparast et al. (2014), pairing VNS with rehabilitation after ischemic lesion to the motor cortex improved forelimb function to before-lesion levels, but VNS or rehabilitation alone did not elicit the same level of improvement. The same findings were also presented in another study, where VNS and rehabilitation combined yielded a 77% recovery of function, while rehabilitation alone only showed 29% recovery (Hays et al., 2014). Further research revealed that by lesioning the nucleus basalis of Meynert, VNS-dependent plasticity changes in the motor cortex were prevented, positing that VNS-motor improvement may function by cholinergic innervation (Hulsey et al., 2016). These studies suggest that VNS may act as a promising therapy for neurologic injury, where insufficient plasticity is preventing recovery. VNS-driven plasticity is not restricted to the presented motor domain. When paired with tones, VNS has been shown to induce map reorganization and interestingly, in an event-specific manner as opposed to a general excitability change (Engineer et al., 2011). In cognition, it has been observed that pairing VNS with behavioral experiences can lead to improved memory retention, as one study found that rats receiving VNS directly after an avoidance task had increased retention 24 hours late (Clark et al., 1998). This finding was then additionally shown transfer well to humans (Clark et al., 1999).

In humans, VNS is found to be effective when delivered both directly via surgical implantation and transcutaneous through external electrodes (Ben-Menachem et al., 2015). Implanted VNS is often used in the treatment of epilepsy, first discussed in the 1992 publication by Jacob Zabara (1992). It has since become widely accepted as an effective treatment, demonstrating 50% seizure reduction in between 25% and 50% of patients, being approved in both Europe and the United States (Penry and Dean, 1990; Vonck et al., 2004; O'Reardon et al., 2006; Ben-Menachem et al., 2015). Generally, VNS implantation involves a small device with a cuff or helical winding that attaches to the vagus nerve, with power device implanted into the chest. For implanted VNS, the implantation technique is relatively unvaried and is reproduced here from *Vagus nerve stimulation: Surgical technique of implantation and revision and related morbidity* (Giordano et al., 2017). Due to potential complications involving the right vagus nerve that could lead to cardiac side-effects, the stimulator is implanted on the left side. Patients are positioned supine, with the head extended and turned slightly to the right. A skin incision is made in the left neck, carried out between the mastoid and clavicle and extending to the medial border of the sternocleidomastoid. The vagus nerve, which is most often located between the carotid artery and internal jugular, is then bluntly dissected for placement of the stimulator. For a coiled stimulator, the process of stimulator placement is shown in Fig. 20.3. A tethering anchor is secured to the vagus nerve, and the positive and negative electrodes are then positioned, being placed inferior to the cardiac branches to prevent cardiac effect. A subcutaneous pocket is created typically 5 cm below the clavicle, and the stimulator lead is positioned using a tunnelizer from the thoracic to cervical incisions. The electrodes and leads are anchored with suturing, with a strain relief bend for head movement freedom.

Vagal afferents in the NTS, LC, and MO are the likely pathways for VNS as an antiepileptic. Through this pathway, it is understood that VNS results in reductions of epinephrine and increases in GABA. Beyond epilepsy, VNS has also been considered a possible treatment for major depression, heart failure, migraine, and many other illnesses (De Ferrari et al., 2011; Ben-Menachem et al., 2015). As a less-invasive and safer alternative to implantation, is employing VNS via transcutaneous methods (tVNS). In tVNS, the vagus nerve is stimulated with electrodes placed on the skin on the back of the head. The stimulation can be delivered at either a constant or oscillating current, which further allows for frequency-specific effects to be investigated. Indeed, for migraines it has been shown that 1 and 25 Hz transcutaneous auricular VNS (taVNS) were both effective in reducing migraine frequency (Straube et al., 2015). One possible explanation for this effect is that VNS reduces nociception by activating descending pain inhibitory systems. It is additionally believed that VNS activates these descending pathways by targeting projections from the brainstem (i.e., NTS and LC) (Straube et al., 2015). Another explanation is the alteration of cortical excitability as a function of LC and NTS modulation, which alters activity of the serotonergic and noradrenergic innervations of cortical regions (Giraudier et al., 2022). This is supported by the observed reductions in SNS activity from taVNS, which may be a contributor to altered cortical excitability (Clancy et al., 2014; Straube et al., 2015). Drawing from the above-discussed LC proxy-measures such as sAA, there is supporting evidence that taVNS activates the LC, as taVNS correlated with increasing sAA levels (Warren et al., 2019; Giraudier et al., 2022). It has also been observed that vagus afferents project to the adrenal gland and onward to the NTS, which in turn connect to the LC through the paragigantocellularis nucleus, corroborated by evidence that VNS induces current-correlating pattern changes in LC phasic activity (Giraudier et al., 2022). The modulation produced by taVNS is not restricted to mere upregulation at any frequency, as taVNS has also been shown to be frequency-dependent. In one study, 100 Hz stimulus frequency produced a comparatively larger brainstem response to 2 Hz and correlated with increase fMRI responses for LC nuclei (Sclocco et al., 2020). Here, we explore how combining the identified physiologic markers of brainstem activity (notably from the LC) with the observed effects of VNS can deepen our understanding of VNS's potential to modify brainstem function. Chronic VNS has been shown to alter resting pupillary diameter, mimicking the same effect seen in medications targeting serotonin and noradrenaline reuptake. This finding was interpreted as autonomic imbalance via sympathetic activation and was attributed to LC-autonomic regulation alteration (Desbeaumes Jodoin et al., 2015). In a pilot study that combined the P300 measure (a positive potential occurring approximately 300 ms after stimulus onset) of LC activity and sAA levels, it was found that tVNS elicited higher sAA following stimulation, and P300(b) amplitude increases, reflecting tVNS-induced changes in LC activity via proxy-measure (Ventura-Bort et al., 2018). In another study, it was shown that tVNS resulted in sAA increases and alters salivary cortisol release, both representing markers of norepinpherine (NE) activity (Warren et al., 2019). This study, however, reported no change in ERP markers nor pupillometry, although the authors suggest this could represent underpowering in the study. This underscores that despite a robust scientific foundation supporting the outcomes and theories related to peripheral nerve stimulation, the field continues to expand and evolve. Research has also investigated the plausibility of VNS for improvement of cognitive functions (Clark et al., 1999). Since memory storage is modulated by sympathetic activity, there is a logical connection between the alteration of autonomic balance and memory (Cahill and McGaugh, 1996). VNS in rats improved memory and cognitive flexibility, and in humans was shown to improve memory consolidation and retention, possibly as a result of increased NTS and LC activity which led to increases in hippocampal activity (Ghacibeh et al., 2006; Driskill et al., 2022). Conclusively, it is clear that VNS

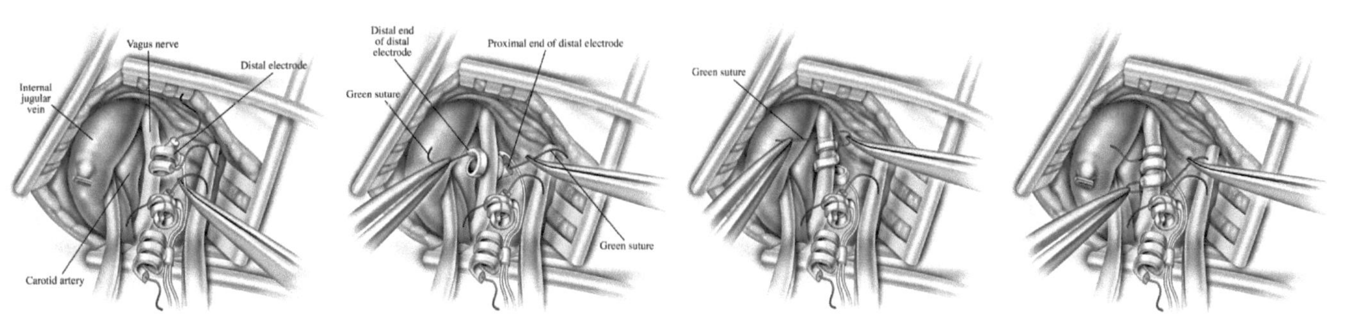

Fig. 20.3. Surgical technique for implantation of vagus nerve stimulator, illustrating (from left to right) vagus nerve identification and helical electrode positioning parallel to the nerve, distal spiral placement under the vagus nerve, and subsequent suturing (Santos, 2004). Reproduced with permission from Santos PM (2004) Surgical placement of the vagus nerve stimulator. Oper Tech Otolaryngol Head Neck Surg 15: 201–209. https://doi.org/10.1016/j.otot.2004.09.002.

offers a well-studied and promising field of research, particularly in regards to brainstem modulation and subsequent cognitive and somatic changes. This establishes the exploration of other cranial nerves for similar modulatory impacts as a valid and intriguing direction for future research.

Occipital nerve stimulation

Alongside VNS, ONS is also a field of study. As with VNS, in this section we will discuss the history, surgical approach, and research uses for ONS. First used in 1977 for treatment of headache (Picaza et al., 1977), implanted ONS has been effective in reducing drug-resistant cluster headache, during which brainstem metabolism was noted to change, possibly indicating ONS influence of the trigeminocervical complex (Magis et al., 2011). As described in Trentman and Zimmerman (2008), ONS implantation involves a 2 cm midline incision at C1, or alternatively an incision medial and inferior to the mastoid process. A pocket is created in the subcutaneous tissue for lead looping and anchoring. It is specifically noted that a placement that is too close to the fascia may result in muscle stimulation, and not the intended ONS. The leads are placed and anchored to the fascia and tunneled to the internal pulse generator. The generators are commonly placed in the buttock, abdomen, midaxillary line, intrascapular, or infraclavicular regions. The decision in this case is largely made with the aim to reduce traction and take into account patient concerns, cosmetics, and patient preference. NS and ONS share common complications including lead migration, infection, and general discomfort being the most common (Trentman and Zimmerman, 2008). As new methods of nerve stimulation become developed, there is also a push for more transportable, lower cost, and lower risk alternative to permanent implantation. One alternative presented, like in VNS, is transcutaneous delivery. Transcutaneous delivery of occipital nerve stimulation (tONS) may act through inputs to the NTS and LC. Thus far, it has been shown to improve migraine in a randomized control study, whereby a significant reduction in headache intensity was observed in the treatment group compared to control (Liu et al., 2017). Not only a treatment method, tONS has also been used as a predictor of response to implanted nerve stimulators. It was shown in Nguyen et al. (2016) that the outcome of implanted ONS corresponded to the response to preoperative tONS, making it a potentially valuable tool for predicting treatment outcomes and informing surgical decisions. Recently, the use of tONS is expanding beyond migraine and headache treatment, as it has been posited that modulation of neural activity may be possible via tONS. In a recent study, it was shown that 40 Hz ONS enhanced learning and recall during a memory task, which the authors posit works in a similar pathway as VNS, but with ease of access via the C2 dermatome which the ON innervates (Luckey et al., 2022). It is suggested in this study that LC activation via the occipital nerves can improve learning through arousal-based mechanisms, in line with the previous research discussed here.

Trigeminal nerve stimulation

We will complete this section with discussion of another cranial nerve, which has recently become the target of stimulation. The trigeminal nerve is the fifth cranial nerve and innervates much of the face and cerebral vasculature. The trigeminal nerve spreads extensively, having overlaps with other nerves including the facial nerve. In fact, it is suggested that the brainstem is a probable intersection point between the facial and trigeminal nerves (White et al., 2021). Trigeminal nerve stimulation (TNS) results in activation of the RVLM, which itself has been shown to cause peripheral vasoconstriction, elevations in BP, and increased cerebral oxygen consumption (White et al., 2021). Approaches for TNS have included both transcutaneous and subcutaneous stimulation. Commonly, TNS has been applied to treat pain disorders including trigeminal neuralgia and epilepsy (DeGiorgio et al., 2006, 2013). TNS has also been implicated in treatment for depression, where in one study it was observed that delivering external TNS improved depression symptoms (Cook et al., 2013). TNS, like other nerve stimulation approaches, has also been considered as a potential means of improving cognitive functioning. In a study by McGough et al. (2015), individuals aged 7–14 with attention-deficit/hyperactivity disorder (ADHD) received TNS during sleep over 8 weeks, and significant improvements in attention and executive functions were observed. Then, in a larger study, the same effect was seen in children aged 8–12 years old, in a 4-week nightly TNS treatment paradigm (McGough et al., 2019). Lastly, we discuss a less-classical but interesting method of nerve stimulation, which is through the tongue. The approach, used to treat tinnitus, uses bimodal auditory-somatosensory stimulation at the tongue to stimulate the trigeminal nerve. In one study, adults with chronic tinnitus were stimulated using 32 electrodes placed on the antero-dorsal tongue surface, with headphones playing biphasic pulses of 17.5 μs duration and varying amplitude. After a period of 10 weeks, with daily stimulation of between 30 and 60 minutes, Tinnitus Handicap Inventory scores, tinnitus loudness matching, and minimum masking level were all reduced significantly in the group, which was determined to be compliant (Hamilton et al., 2016).

We can consider then that through stimulation of various nerves including the vagus, occipital, and trigeminal, there is a growing field of interventions for a host of disorders and diseases. As methodologies become more refined and mechanisms well understood, it is expected that the range of applications for peripheral nerve stimulation will similarly grow.

CONCLUSIONS

In this chapter, we have discussed numerous aspects of brainstem activity measurement and nerve stimulation. Brainstem activity can be reliably measured through various physiologic means, evoked responses, and is a key structure in the effects of nerve stimulation. The activity of brainstem nuclei including the LC and NTS has been accessibly measured using autonomic imbalance representations including pupillary responses, heart rate variation, sAA, skin conduction, and the P300 component. These measures are consistently related to LC activity, and thus, we find to be a sound proxy representations of brainstem arousal. Use of these measures has been an effective method of monitoring changes in brainstem activity that may result from nerve stimulation, including VNS, ONS, and TNS. In this chapter, we present vagal stimulation as an effective method of intervention for numerous disorders and diseases such as epilepsy and depression. ONS has similarly been successful as an intervention for headaches, as was also found for trigeminal stimulation, which is also showing promise as a treatment for tinnitus. Through peripheral methods, VNS, ONS, and TNS are on the rise as viable modulators of brainstem activity, likely through functional connectivity to the brainstem, and in some cases involving the LC in its role as the center for autonomic control. Another form of nerve stimulation, which produce brainstem evoked potentials (e.g., ABRs), have also been discussed here as providing not only a sensitive measure of brainstem activity, but also a valid and proven diagnostic approach for hearing loss and brainstem dysfunction. Similar to ABRs, nerve stimulation intraoperatively can be used to monitor functional retention and improve outcomes. In all, the stimulation of the brainstem through cranial nerves and the advancements made in measuring its activity are continually evolving fields that show promise for interventive therapies and a better understanding of brain connectivity.

References

Ananthanarayan AK, Durrant JD (1991). On the origin of wave II of the auditory brain stem evoked response. Ear Hear 12: 174–179.

Ben-Menachem E, Revesz D, Simon BJ et al. (2015). Surgically implanted and non-invasive vagus nerve stimulation: a review of efficacy, safety and tolerability. Eur J Neurol 22: 1260–1268.

Berger C, Dück A, Perin F et al. (2021). Brain arousal as measured by EEG-assessment differs between children and adolescents with attention-deficit/hyperactivity disorder (ADHD) and depression. Front Psychiatry 12: 633880.

Brack KE, Winter J, Ng GA (2013). Mechanisms underlying the autonomic modulation of ventricular fibrillation initiation—tentative prophylactic properties of vagus nerve stimulation on malignant arrhythmias in heart failure. Heart Fail Rev 18: 389–408.

Bradley MM, Miccoli L, Escrig MA et al. (2008). The pupil as a measure of emotional arousal and autonomic activation. Psychophysiology 45: 602–607.

Cahill L, McGaugh JL (1996). Modulation of memory storage. Curr Opin Neurobiol 6: 237–242.

Carrick FR, Azzolino SF, Hunfalvay M et al. (2021). The pupillary light reflex as a biomarker of concussion. Life 11: 1104.

Chiappa KH (1980). Pattern shift visual, brainstem auditory, and short-latency somatosensory evoked potentials in multiple sclerosis. Neurology 30: 110–123.

Clancy JA, Mary DA, Witte KK et al. (2014). Non-invasive vagus nerve stimulation in healthy humans reduces sympathetic nerve activity. Brain Stimul 7: 871–877.

Clark KB, Naritoku DK, Smith DC et al. (1999). Enhanced recognition memory following vagus nerve stimulation in human subjects. Nat Neurosci 2: 94–98.

Clark KB, Smith DC, Hassert DL et al. (1998). Posttraining electrical stimulation of vagal afferents with concomitant vagal efferent inactivation enhances memory storage processes in the rat. Neurobiol Learn Mem 70: 364–373.

Cook IA, Schrader LM, DeGiorgio CM et al. (2013). Trigeminal nerve stimulation in major depressive disorder: acute outcomes in an open pilot study. Epilepsy Behav 28: 221–226.

De Ferrari GM, Crijns HJ, Borggrefe M et al. (2011). Chronic vagus nerve stimulation: a new and promising therapeutic approach for chronic heart failure. Eur Heart J 32: 847–855.

De Herdt V, Puimege L, De Waele J et al. (2009). Increased rat serum corticosterone suggests immunomodulation by stimulation of the vagal nerve. J Neuroimmunol 212: 102–105.

DeGiorgio CM, Shewmon A, Murray D et al. (2006). Pilot study of trigeminal nerve stimulation (TNS) for epilepsy: a proof-of-concept trial. Epilepsia 47: 1213–1215.

DeGiorgio CM, Soss J, Cook IA et al. (2013). Randomized controlled trial of trigeminal nerve stimulation for drug-resistant epilepsy. Neurology 80: 786–791.

Desbeaumes Jodoin V, Lespérance P, Nguyen DK et al. (2015). Effects of vagus nerve stimulation on pupillary function. Int J Psychophysiol 98: 455–459.

Driskill CM, Childs JE, Itmer B et al. (2022). Acute vagus nerve stimulation facilitates short term memory and cognitive flexibility in rats. Brain Sci 12: 1137.

Eisenbarth H, Chang LJ, Wager TD (2016). Multivariate brain prediction of heart rate and skin conductance responses to social threat. J Neurosci 36: 11987–11998.

Engineer ND, Riley JR, Seale JD et al. (2011). Reversing pathological neural activity using targeted plasticity. Nature 470: 101–104.

Gendt MJV, Koka K, Kalkman RK et al. (2020). Simulating intracochlear electrocochleography with a combined model of acoustic hearing and electric current spread in the cochlea. J Acoust Soc Am 147: 2049–2060.

George MS, Rush AJ, Sackeim HA et al. (2003). Vagus nerve stimulation (VNS): utility in neuropsychiatric disorders. Int J Neuropsychopharmacol 6: 73–83.

Ghacibeh GA, Shenker JI, Shenal B et al. (2006). The influence of vagus nerve stimulation on memory. Cognit Behav Neurol 19: 119–122.

Gilzenrat MS, Nieuwenhuis S, Jepma M et al. (2010). Pupil diameter tracks changes in control state predicted by the adaptive gain theory of locus coeruleus function. Cogn Affect Behav Neurosci 10: 252–269.

Giordano F, Zicca A, Barba C et al. (2017). Vagus nerve stimulation: surgical technique of implantation and revision and related morbidity. Epilepsia 58: 85–90.

Giraudier M, Ventura-Bort C, Burger AM et al. (2022). Evidence for a modulating effect of transcutaneous auricular vagus nerve stimulation (taVNS) on salivary alpha-amylase as indirect noradrenergic marker: a pooled mega-analysis. Brain Stimul 15: 1378–1388.

Glasser RL, Perez-Reyes M, Tippett JW (1964). Brain-stem inhibition of electrodermal (galvanic skin) activity. Am J Physiol 207: 1133–1138.

Gu F, Yang X, Wang Z et al. (2022). Diagnostic accuracy of intraoperative brainstem auditory evoked potential for predicting hearing loss after vestibular schwannoma surgery. Front Neurol 13: 1018324.

Hamilton C, D'Arcy S, Pearlmutter BA et al. (2016). An investigation of feasibility and safety of bi-modal stimulation for the treatment of tinnitus: an open-label pilot study. Neuromodulation 19: 832–837.

Harper CM (2004). Intraoperative cranial nerve monitoring. Muscle Nerve 29: 339–351.

Hays SA, Khodaparast N, Hulsey DR et al. (2014). Vagus nerve stimulation during rehabilitative training improves functional recovery after intracerebral hemorrhage. Stroke 45: 3097–3100.

Helmstaedter V, Lenarz T, Erfurt P et al. (2018). The summating potential is a reliable marker of electrode position in electrocochleography: cochlear implant as a theragnostic probe. Ear Hear 39: 687–700.

Hogervorst MA, Brouwer A-M, van Erp JBF (2014). Combining and comparing EEG, peripheral physiology and eye-related measures for the assessment of mental workload. Front Neurosci 8: 322.

Huffman WJ, Subramaniyan S, Rodriguiz RM et al. (2019). Modulation of neuroinflammation and memory dysfunction using percutaneous vagus nerve stimulation in mice. Brain Stimul 12: 19–29.

Hulsey DR, Hays SA, Khodaparast N et al. (2016). Reorganization of motor cortex by vagus nerve stimulation requires cholinergic innervation. Brain Stimul 9: 174–181.

Jacobs HI, Priovoulos N, Poser BA et al. (2020). Dynamic behavior of the locus coeruleus during arousal-related memory processing in a multi-modal 7T fMRI paradigm. Elife 9: e52059.

Joshi S, Li Y, Kalwani RM et al. (2016). Relationships between pupil diameter and neuronal activity in the locus coeruleus, colliculi, and cingulate cortex. Neuron 89: 221–234.

Juselius Baghdassarian E, Nilsson Markhed M, Lindström E et al. (2018). Auditory brainstem response (ABR) profiling tests as diagnostic support for schizophrenia and adult attention-deficit hyperactivity disorder (ADHD). Acta Neuropsychiatr 30: 137–147.

Kahneman D, Beatty J (1966). Pupil diameter and load on memory. Science 154: 1583–1585.

Khodaparast N, Hays SA, Sloan AM et al. (2014). Vagus nerve stimulation delivered during motor rehabilitation improves recovery in a rat model of stroke. Neurorehabil Neural Repair 28: 698–706.

Korpelainen JT, Huikuri HV, Sotaniemi KA et al. (1996). Abnormal heart rate variability reflecting autonomic dysfunction in brainstem infarction. Acta Neurol Scand 94: 337–342.

Laine CM, Spitler KM, Mosher CP et al. (2009). Behavioral triggers of skin conductance responses and their neural correlates in the primate amygdala. J Neurophysiol 101: 1749–1754.

Le TM, Wang W, Zhornitsky S et al. (2019). Reward sensitivity and electrodermal responses to actions and outcomes in a go/no-go task. PLoS One 14: e0219147.

Lindstrom L, Klockhoff I, Svedberg A et al. (1987). Abnormal auditory brain-stem responses in hallucinating schizophrenic patients. Br J Psychiatry 151: 9–14.

Liu Y, Dong Z, Wang R et al. (2017). Migraine prevention using different frequencies of transcutaneous occipital nerve stimulation: a randomized controlled trial. J Pain 18: 1006–1015.

Liu Y, Yin H, Ma H et al. (2020). The salivary-α-amylase level after Stroop test in anxious patients can predict the severity of anxiety. Neurosci Lett 715: 134613.

Luckey AM, McLeod SL, Mohan A et al. (2022). Potential role for peripheral nerve stimulation on learning and long-term memory: a comparison of alternating and direct current stimulations. Brain Stimul 15: 536–545.

Magis D, Bruno MA, Fumal A et al. (2011). Central modulation in cluster headache patients treated with occipital nerve stimulation: an FDG-PET study. BMC Neurol 11: 25.

Malayeri S, Lotfi Y, Moossavi SA et al. (2014). Brainstem response to speech and non-speech stimuli in children with learning problems. Hear Res 313: 75–82.

McCall AA, Miller DM, Yates BJ (2017). Descending influences on vestibulospinal and vestibulosympathetic reflexes. Front Neurol 8.

McGough JJ, Loo SK, Sturm A et al. (2015). An eight-week, open-trial, pilot feasibility study of trigeminal nerve stimulation in youth with attention-deficit/hyperactivity disorder. Brain Stimul 8: 299–304.

McGough JJ, Sturm A, Cowen J et al. (2019). Double-blind, sham-controlled, pilot study of trigeminal nerve stimulation for attention-deficit/hyperactivity disorder. J Am Acad Child Adolesc Psychiatry 58: 403–411.e403.

Miron O, Beam AL, Kohane IS (2018). Auditory brainstem response in infants and children with autism spectrum disorder: a meta-analysis of wave V. Autism Res 11: 355–363.

Monge Argilés JA, Palacios Ortega F, Vila Sobrino JA et al. (2000). Brainstem lesions decrease heart rate variability. Neurologia 15: 158–163.

Moore JK (1987). The human auditory brain stem as a generator of auditory evoked potentials. Hear Res 29: 33–43.

Mueller SG, Bateman LM, Nei M et al. (2019). Brainstem atrophy in focal epilepsy destabilizes brainstem-brain interactions: preliminary findings. Neuroimage Clin 23: 101888.

Murphy PR, O'Connell RG, O'Sullivan M et al. (2014). Pupil diameter covaries with BOLD activity in human locus coeruleus. Hum Brain Mapp 35: 4140–4154.

Nguyen J-P, Nizard J, Kuhn E et al. (2016). A good preoperative response to transcutaneous electrical nerve stimulation predicts a better therapeutic effect of implanted occipital nerve stimulation in pharmacologically intractable headaches. Neurophysiol Clin 46: 69–75.

Nobukawa S, Shirama A, Takahashi T et al. (2021). Pupillometric complexity and symmetricity follow inverted-U curves against baseline diameter due to crossed locus coeruleus projections to the Edinger-Westphal nucleus. Front Physiol 12: 614479.

Noller CM, Levine YA, Urakov TM et al. (2019). Vagus nerve stimulation in rodent models: an overview of technical considerations. Front Neurosci 13: 911.

Norton S, Perkins J, Cummings C (2005). Early detection and diagnosis of infant hearing impairment. Cummings otolaryngology: head and neck surgery. 4th edition. New York: WB Saunders Company, pp. 4387–4398.

Nuwer MR, Emerson RG, Galloway G et al. (2012). Evidence-based guideline update: intraoperative spinal monitoring with somatosensory and transcranial electrical motor evoked potentials: report of the Therapeutics and Technology Assessment Subcommittee of the American Academy of Neurology and the American Clinical Neurophysiology Society. Neurology 78: 585–589.

O'Connell BP, Holder JT, Dwyer RT et al. (2017). Intra- and postoperative electrocochleography may be predictive of final electrode position and postoperative hearing preservation. Front Neurosci 11: 291.

O'Reardon JP, Cristancho P, Peshek AD (2006). Vagus nerve stimulation (VNS) and treatment of depression: to the brainstem and beyond. Psychiatry (Edgmont) 3: 54–63.

Park JH, Gorky J, Ogunnaike B et al. (2020). Investigating the effects of brainstem neuronal adaptation on cardiovascular homeostasis. Front Neurosci 14: 470.

Penry JK, Dean JC (1990). Prevention of intractable partial seizures by intermittent vagal stimulation in humans: preliminary results. Epilepsia 31: S40–S43.

Picaza JA, Hunter SE, Cannon BW (1977). Pain suppression by peripheral nerve stimulation. Chronic effects of implanted devices. Appl Neurophysiol 40: 223–234.

Radtke RA, Erwin CW, Wilkins RH (1989). Intraoperative brainstem auditory evoked potentials. significant decrease in postoperative morbidity. Neurology 39: 187.

Robinson K, Rudge P (1980). The use of the auditory evoked potential in the diagnosis of multiple sclerosis. J Neurol Sci 45: 235–244.

Rocha-Muniz CN, Befi-Lopes DM, Schochat E (2012). Investigation of auditory processing disorder and language impairment using the speech-evoked auditory brainstem response. Hear Res 294: 143–152.

Rohleder N, Nater UM (2009). Determinants of salivary α-amylase in humans and methodological considerations. Psychoneuroendocrinology 34: 469–485.

Rouillon I, Parodi M, Denoyelle F et al. (2016). How to perform ABR in young children. Eur Ann Otorhinolaryngol Head Neck Dis 133: 431–435.

Rowe MJI (1981). The brainstem auditory evoked response in neurological disease: a review. Ear Hear 2: 41–51.

Santos PM (2004). Surgical placement of the vagus nerve stimulator. Oper Tech Otolaryngol Head Neck Surg 15: 201–209.

Scheib CM (2017). Brainstem influence on thalamocortical oscillations during anesthesia emergence. Front Syst Neurosci 11: 66.

Schumann A, Köhler S, de la Cruz F et al. (2018). The use of physiological signals in brainstem/midbrain fMRI. Front Neurosci 12: 718.

Sclocco R, Beissner F, Bianciardi M et al. (2018). Challenges and opportunities for brainstem neuroimaging with ultrahigh field MRI. Neuroimage 168: 412–426.

Sclocco R, Garcia RG, Kettner NW et al. (2020). Stimulus frequency modulates brainstem response to respiratory-gated transcutaneous auricular vagus nerve stimulation. Brain Stimul 13: 970–978.

Sirois S, Brisson J (2014). Pupillometry. WIREs Cogn Sci 5: 679–692.

Soni A, Kanaujia SK, Kaushik S (2016). Brainstem evoked response audiometry (BERA) in neonates with hyperbillirubinemia. Indian J Otolaryngol Head Neck Surg 68: 334–338.

Straube A, Ellrich J, Eren O et al. (2015). Treatment of chronic migraine with transcutaneous stimulation of the auricular branch of the vagal nerve (auricular t-VNS): a randomized, monocentric clinical trial. J Headache Pain 16: 63.

Thome J, Densmore M, Terpou BA et al. (2022). Contrasting associations between heart rate variability and brainstem-limbic connectivity in posttraumatic stress disorder and its dissociative subtype: a pilot study. Front Behav Neurosci 16: 862192.

Trentman TL, Zimmerman RS (2008). Occipital nerve stimulation: technical and surgical aspects of implantation. Headache 48: 319–327.

Turker HB, Riley E, Luh W-M et al. (2021). Estimates of locus coeruleus function with functional magnetic resonance imaging are influenced by localization approaches and the use of multi-echo data. Neuroimage 236: 118047.

Ventura-Bort C, Wirkner J, Genheimer H et al. (2018). Effects of transcutaneous vagus nerve stimulation (tVNS) on the P300 and alpha-amylase level: a pilot study. Front Hum Neurosci 12: 202.

Vonck K, Thadani V, Gilbert K et al. (2004). Vagus nerve stimulation for refractory epilepsy: a transatlantic experience. J Clin Neurophysiol 21: 283–289.

Wang J, Irnaten M, Neff RA et al. (2001). Synaptic and neurotransmitter activation of cardiac vagal neurons in the nucleus ambiguus. Ann N Y Acad Sci 940: 237–246.

Warren CM, Tona KD, Ouwerkerk L et al. (2019). The neuromodulatory and hormonal effects of transcutaneous vagus nerve stimulation as evidenced by salivary alpha amylase, salivary cortisol, pupil diameter, and the P3 event-related potential. Brain Stimul 12: 635–642.

Washnik NJ, Anjum J, Lundgren K et al. (2019). A review of the role of auditory evoked potentials in mild traumatic brain injury assessment. Trends Hear 23: 2331216519840094.

White TG, Powell K, Shah KA et al. (2021). Trigeminal nerve control of cerebral blood flow: a brief review. Front Neurosci 15: 649910.

Yamamoto K-i, Arai H, Nakayama S (1990). Skin conductance response after 6-hydroxydopamine lesion of central noradrenaline system in cats. Biol Psychiatry 28: 151–160.

Yin LX, Barnes JH, Saoji AA et al. (2021). Clinical utility of intraoperative electrocochleography (ECochG) during cochlear implantation: a systematic review and quantitative analysis. Otol Neurotol 42: 363–371.

Young A, Cornejo J, Spinner A (2022). Auditory brainstem response. StatPearls [Internet]. StatPearls Publishing.

Zabara J (1992). Inhibition of experimental seizures in canines by repetitive vagal stimulation. Epilepsia 33: 1005–1012.

Zamel KM (2010). Brainstem auditory evoked potential monitoring. In: Galloway GM, Lopez JR, Zamel KM et al. (eds.) Intraoperative neurophysiologic monitoring. Cambridge University Press, pp. 90–100

Zhuo H, Ichikawa H, Helke CJ (1997). Neurochemistry of the nodose ganglion. Prog Neurobiol 52: 79–107.

Handbook of Clinical Neurology, Vol. 216 (3rd series)
The Brainstem
L.T. Grinberg and A.D. Korczyn
https://doi.org/10.1016/B978-0-443-15736-3.00009-3

Chapter 21

Functional and molecular brainstem imaging

HEIDI I.L. JACOBS[1,*], PROKOPIS C. PROKOPIOU[1], AND TARUN SINGHAL[2]

[1]*Athinoula A. Martinos Center for Biomedical Imaging, Department of Radiology, Massachusetts General Hospital, Harvard Medical School, Boston, MA, United States*

[2]*Ann Romney Center for Neurologic Diseases, Department of Neurology, Brigham and Women's Hospital, Harvard Medical School, Boston, MA, United States*

Abstract

The brainstem connects the cerebrum, cerebellum, and spinal cord and includes several nuclei that produce key neurotransmitters important for the modulation of behavior and cognition. Since the 16th century, humans have tried to understand the importance of the brainstem in health and disease. The development of modern neuroimaging methods, using functional magnetic resonance imaging or positron emission tomography, has ushered in an era of understanding function and molecular changes in the brainstem in many neurologic diseases. This chapter discusses the methodologic complexities of imaging the neuromodulatory subcortical nuclei in the brainstem. Furthermore, we discuss the current state-of-the-art clinical applications using these techniques for early detection, differential diagnostics, monitoring of disease and intervention, and drug development, highlighting future perspectives for the field.

INFERRING CLINICAL INSIGHTS FROM BRAINSTEM FUNCTION THROUGHOUT HISTORY

Ever since ancient history, people have explored the anatomy and function of the human brain. The first attempts to understand brain function can be traced back to trepanning performed by the Incas and to descriptions of treatments for several neurologic injuries on papyrus by the ancient Egyptians. The Greek Aristoteles (347 BCE) dissected the nervous system and documented the meninges, cerebrum, and cerebellum. Around the same time, in Rome, Galen performed animal studies and – wrongfully – concluded that the fourth ventricle was responsible for maintaining consciousness (Freemon, 1994). Nonetheless, the foundation of the current functional neurology examination of the brainstem via the cranial nerves was laid by Galen in 173 AD (Porras-Gallo et al., 2019). Throughout history, the study of the function of the brainstem has been overshadowed by an encephalocentric view, but the development of new imaging methods and insights into the importance of the brainstem for many neurologic disorders have shifted the attention to the brainstem, one of the phylogenetically oldest structures of the brain.

One of the first drawings of the brainstem can be found in Andreas Vesalius' *De humani corporis fabrica libri septum* (1543), later anatomically identified as hindbrain. At the time, Thomas Willis, rightly considered as the founder of modern neuroscience and an innovator in brain preservation methods, accurately described the gross anatomy of the brainstem, recognizing the midbrain, pons, and medulla (Cucu et al., 2021). Colored images of brainstem anatomy became available in 1786 (by Vicq d'Azyr), which included the first descriptions of the dark blue and black spots, respectively, the locus coeruleus and the substantia nigra (Parent, 2007). While many of these writings focused on the anatomy, there were ideas about localization, reflexes, and possible associations with neurologic symptoms, including epilepsy.

Linking morphology to clinical symptoms or functions, even when still crude, inspired many scientists. Early in the 19th century, attempts to localize mental functions in the brain through phrenology by Franz Joseph Gall were highly criticized, and motivated Jean-Pierre Flourens to test and disprove these assumptions by dissecting animals (Finger and Eling, 2020). Flourens discovered that destroying the medulla oblongata in animals caused death, leading to the inference that the medulla oblongata controls breathing and circulation. Subsequent animal experiments throughout the 19th and 20th centuries confirmed the importance of the medulla oblongata for respiration and eye movements and led to the identification of key brainstem nuclei, including the Edinger–Westphal and

*Correspondence to: Dr. Heidi I.L. Jacobs, PhD, The Athinoula A. Martinos Center for Biomedical Imaging, Department of Radiology, Massachusetts General Hospital, Harvard Medical School, 149 13th Street, Charlestown, MA 02129, United States. E-mail: hjacobs@mgh.harvard.edu.

Perlia nuclei. The ability to identify the neurochemistry in the brain through the development of the Falck–Hillarp method of fluorescence in the early 1960s made it possible to study the presence of monoamines. Groups of noradrenergic, dopaminergic, and serotonergic neurons were identified and provided support for Cajal's idea of communication between brain cells through neurotransmission and contributing to behavior.

In the 19th century, correlating symptoms to specific brain systems in vivo in humans was limited to the examination of the cranial nerves III–XII that nicely relay to brainstem nuclei or to the study of specific lesions cases – either following accidents, diseases, or surgical removal. But a systematic clinical examination of the brainstem did not occur till 1952, studying eye movements and the vestibulo-ocular reflex in comatose patients and patients with basilar artery occlusion or glioblastoma with brainstem hematoma (Cucu et al., 2021).

The ability to image the brain in vivo in the 1970s revolutionized our understanding of the brain. At first, computerized axial tomography (CAT) allowed to visualize brain anatomy, but a few years later, the development of positron emission tomography (PET) allowed researchers to examine brain function by using high-energy ionizing radiation. Around the same time, in 1970, Paul Lauterbur (State University of New York, Stony Brook) and Paul Mansfield (University of Nottingham) independently discovered magnetic resonance imaging (MRI), providing greater anatomic detail than the CAT images (Lauterbur, 1989). In the early 1990s, Seiji Ogawa (Bell Laboratories) and Kenneth Kwong (Massachusetts General Hospital) independently demonstrated how brain function can be measured with MRI using changes in deoxyhemoglobin (Ogawa and Lee, 1990; Kwong et al., 1992). The use of molecular and functional imaging with PET or MRI has led to exciting and growing insights into healthy brain function and the impact of neurologic diseases on the brain. Though the neuroimaging field was largely corticocentric, new advances in tracers, methods, hardware, and processing methods have spurred an interest in the exploration of brainstem function in neurology. In this chapter, we will discuss the specifics of brainstem functional and molecular imaging, its challenges, clinical applications and future perspectives.

THE NEED TO IMAGE THE BRAINSTEM IN NEUROLOGIC DISORDERS

Many clinical decisions, diagnostic or intervention-based, are based on a review of neuroimages, both at the anatomic and functional level. The brainstem is a unique part of the central nervous system, which has a very high concentration of extremely diverse types of neurons. The brainstem has several nuclei with unique neurochemical organization, which project to various parts of the central neuraxis, rostrally and caudally. Interestingly, the brainstem lies at the interaction of the cerebrum, cerebellum, and spinal cord and is also the seat of the regulation of critical homeostatic processes such as breathing, respiration, and circulation. Hence, the brainstem is a critical component of the central nervous system that has an integrative role in the function of the entire human body.

The brainstem has a rich array of nuclei that serve as hubs for specific neurotransmitter systems in the brain. These neurotransmitter nuclei have specific and widespread projections to subcortical and cortical brain regions and are critical for the normal homeostatic functioning of the nervous system itself. Some of these nuclei may be sites of earliest involvement in various neurodegenerative and neuropsychiatric diseases. For example, brainstem nuclei such as the locus coeruleus, midbrain and pontine raphe, ventral tegmental area, substantia nigra, and others are hubs for noradrenergic, serotonergic, and dopaminergic neurons and may be involved early in Alzheimer disease, Parkinsonian disorders, traumatic brain injury, depression, and many other neurologic and neuropsychiatric diseases.

Thus, there is a need for imaging of small brainstem nuclei. Intriguingly, functional and molecular changes often precede structural changes under disease conditions, and various brainstem nuclei undergo early and specific involvement, often before severe symptoms in several neurodegenerative conditions such as Alzheimer disease and Parkinson disease. Detection of these earliest changes in the brainstem can lead to early diagnosis and help with narrowing the differential diagnosis in such conditions. Moreover, imaging of the brainstem nuclei can lead to improved risk stratification and prognostication and can also potentially facilitate treatment selection, planning, and monitoring. Overall molecular imaging of the brainstem can improve the understanding of the brain function and structure and contribute to reducing human suffering.

FUNCTIONAL MAGNETIC RESONANCE IMAGING

Brief background

Over the last 30 years, functional MRI (fMRI) has emerged as the primary imaging tool to investigate brain function and its organization into functional networks in health and disease. The contrast mechanism in most fMRI imaging studies has been based on the level of blood oxygenation, which is known as the blood oxygenation level–dependent (BOLD) contrast (Kwong et al., 1992). The contrast agent of BOLD-fMRI is hemoglobin, a macromolecule exhibiting different magnetization properties depending on its oxygenation status: when fully oxygenated, hemoglobin is diamagnetic, but when deoxygenated, it becomes paramagnetic. This paramagnetism induces local permutations in magnetic fields generated in and around blood vessels, whose strength depends on the total concentration of deoxygenated hemoglobin.

When a region in the brain is activated, either spontaneously or in response to cognitive or sensory stimulation, it leads to an increase in glycolytic metabolic demands. This, in turn, results in increases in blood flow through the local vascular network, delivering an enriched supply of oxygen and essential nutrients to meet the increased metabolic demands of the activated neurons (Glover, 2011). Hence, BOLD-fMRI is an indirect measure of neuronal activity through a series of physiologic events that are collectively referred to as the hemodynamic response to neuronal activation. Despite this indirect link between neuronal activity and cerebral blood flow, BOLD-fMRI has gained increased popularity due to its good spatiotemporal resolution and non-invasiveness nature, particularly when compared to other imaging techniques (Engels-Dominguez et al., 2023), such as PET.

Challenges in imaging the brainstem with functional magnetic resonance imaging

Imaging the brainstem using BOLD-fMRI has thus far received little attention from the scientific and clinical community because it comes with a set of unique challenges (Beissner, 2015; Sclocco et al., 2018). These challenges stem partly from methodologic and technical aspects of fMRI data acquisition, as well as from contributions from anatomic and physiologic properties of the brainstem that can confound the BOLD signal. In the last decade, many advances have been made to improve BOLD-fMRI quantification in the brainstem, allowing for an in-depth investigation and understanding of brainstem function in clinically normal individuals as well as in neurologic populations.

The human brainstem consists of a complex mixture of neuronal groups forming nuclei, and so far, Paxinos, Furlong, and Watson have identified more than 60 nuclei using chemoarchitectonic comparisons. Many of these nuclei are tiny, and this represents one of the primary challenges in brainstem fMRI. Standard BOLD-fMRI sequences, the echoplanar imaging (EPI)-fMRI sequences, at commonly available field strengths (3.0 T) are limited in terms of spatial resolution and signal quality (Beissner, 2015). While the actual implementation of these sequences is often a trade-off between spatial and temporal resolution, the majority of fMRI studies operate around 2 mm^2 isotropic. Going to smaller voxel sizes at these systems can lower the signal-to-noise ratio. Working with this spatial resolution can be a problem for reliably capturing signal specific to nuclei at or smaller than this voxel size in width or length. In addition to the confluence of the voxel size and anatomic size of the nuclei of interest, other factors can also impact the specificity and accuracy of measuring signal from a particular nucleus. Spatial distortions, for example, due to magnetic field inhomogeneities, can reduce the quality of the image, deform the image, and degrade its accuracy. In addition, as EPI-fMR-images do not provide sufficient anatomic detail to identify structures in the brain, these images are often registered to high-resolution individual anatomic images or to anatomic atlases. These manipulations are associated with small errors, but as many of these functionally heterogeneous nuclei are close to each other, such small errors can confound the sensitivity and specificity of the measurement and will make it difficult to compare an individual to another individual.

The increasing availability of ultrahigh-field (>7 T) MRI systems over the last decade has led to the development of robust, accelerated, simultaneous multislice EPI fMRI sequences (also referred to as multiband EPI), which allow the acquisition of images with a submillimeter resolution in a relatively short acquisition time. Ultrahigh-field MRI systems have so far mainly been used in the research setting, and their application to the brainstem has just started. In order to obtain a reliable BOLD-fMRI signal from the brainstem at 7 T MRI and be able to translate this technique to the clinic, more research is warranted. An important area of investigation is the trade-off between achieving faster image acquisition through higher acceleration factors and the potential impact on image quality, such as increased crosstalk. Furthermore, in order to not push the acceleration factor too far but achieve small voxel sizes with faster scan times, researchers will often opt for partial field-of-view acquisitions. Even though these images have greater anatomic detail, there is still a need to register these functional images to higher-resolution structural images, and the lack of information on the boundaries of the entire brain imposes additional complexity on accurate registration.

The increasing interest in in vivo imaging of the brainstem in the last few decades, together with these new technologic advances, has resulted in several methodologic developments toward optimized brainstem-specific solutions for improving anatomic localization, allowing comparisons across individuals, and mitigating noise confounding. As mentioned earlier, registration is key for anatomic localization (Fig. 21.1B). Most registration methods work by calculating a transformation matrix based on the similarity or common geometric features (e.g., the general shape and lateral ventricles), but their accuracy is poor for specific brain subregions. These registration methods can be improved by providing the algorithm landmark-based information, that is, by providing it with a segmentation of the brainstem volume and introducing higher weight values around and within the brainstem. These methods have been shown to significantly improve brainstem registrations and contribute to more accurate anatomic localization (Napadow et al., 2006). Researchers perform spatial smoothing on the data to allow for comparisons between individuals and mitigate some of the noise inherent to fMRI data. This means that the signal of a voxel is replaced with a weighted average of that voxel's neighbors. The number of neighbors included in this manipulation can hold implications for the functional interpretation of the signal, as larger smoothing kernels can bring in contributions from other nuclei or tissue compartments (e.g., contributions from cerebrospinal fluid from the fourth ventricle). Balancing the need for smoothing to reduce noise and allow for group analyses while also not introducing other noise sources can be achieved by the utilization of properly shaped Gaussian kernels that account for the peculiar shapes of the regions of interest (e.g., the locus coeruleus) (Prokopiou et al., 2022a). While the ultrahigh magnetic field strength can enhance the detection of functional activation in small brainstem nuclei, EPI-based fMRI acquisitions are even more vulnerable to susceptibility-induced distortions at higher fields, especially in the brainstem and medial temporal lobe regions (Fig. 21.1A). These distortions result from the magnetic field inhomogeneity at the air–tissue boundary in the posterior oral cavity, which can lead to signal loss or deformations on images. Corrections based on field maps or other reference data have

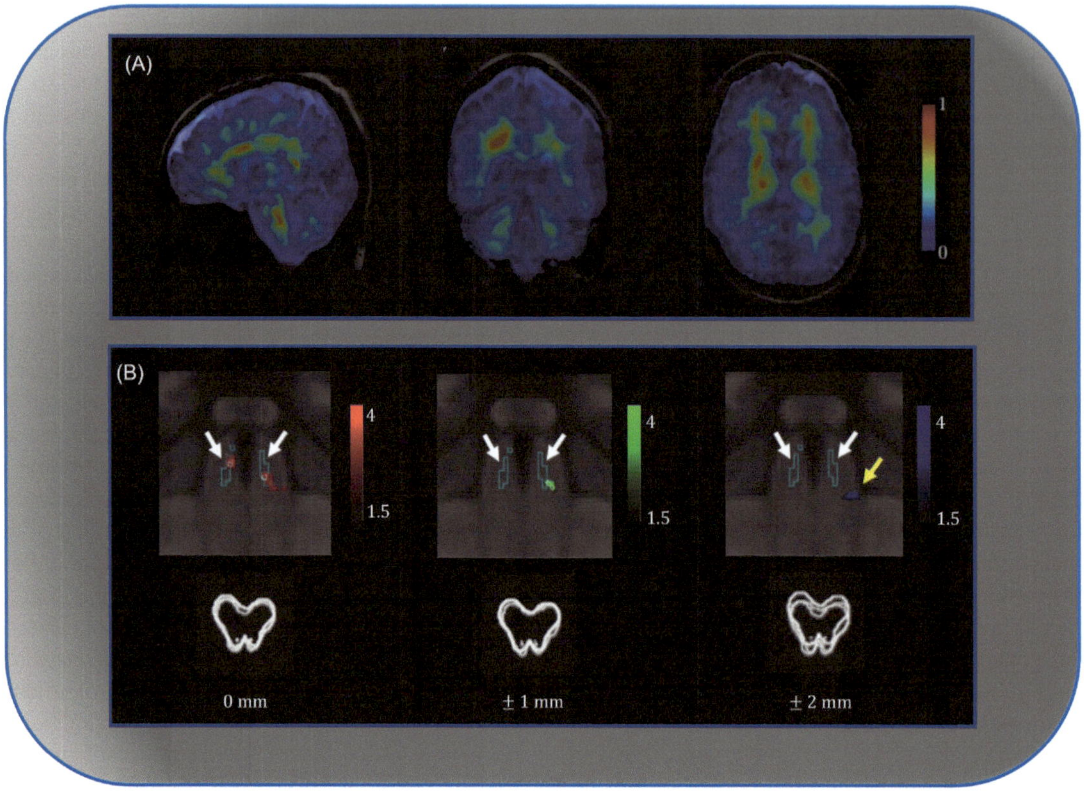

Fig. 21.1. Examples of methodologic challenges in imaging the brainstem with functional magnetic resonance imaging (fMRI). (**A**) Normalized amplitude of the brain's 3D displacement evaluated using 3D cine flow MRI, which is sensitive to the flow of the cerebrovascular fluid as it moves throughout the brain. The brain undergoes substantial deformation near the lateral ventricles and the pons (adapted from Supplemental Material Video S1 – Abderezaei et al., 2021). (**B**) Reductions in group-level functional activation in small brainstem structures such as the locus coeruleus at different levels of spatial deviations of subject-level (first level) contrast maps in a group-level (second level) fMRI analysis (Yi et al., 2023). The white arrows indicate the locus coeruleus area identified using an aggregated meta-locus coeruleus template. (A) Adapted from Abderezaei J, Pionteck A, Terem I, et al. (2021) Development, calibration, and testing of 3D amplified MRI (aMRI) for the quantification of intrinsic brain motion. Brain Multiphysics 2: 100022. (B) Adapted from Yi YJ, Lusebrink F, Ludwig M, et al. (2023) It is the locus coeruleus! Or… is it?: a proposition for analyses and reporting standards for structural and functional magnetic resonance imaging of the noradrenergic locus coeruleus. Neurobiol Aging 129: 137–148.

been proposed for mitigating these distortion effects. In practice, correcting susceptibility distortions is crucial, especially for ultrahigh-field image acquisitions, to improve the signal-to-noise ratio and thus increase the sensitivity of the data and facilitate coregistration of functional to high-resolution structural images.

So far, we have discussed the challenges of brainstem fMRI from a methodologic perspective, but there are also important biologic challenges, particularly those related to the anatomy and physiology of the brainstem. One such factor is the weak contrast of the brainstem gray/white matter. The brainstem gray matter parenchyma and white matter tracks are often intermingled, resulting in a seemingly homogeneous structure, particularly at lower magnetic field strengths. Brainstem tissue discrimination can be improved at higher magnetic fields (Sclocco et al., 2018), particularly when fMRI is used in tandem with advanced structural MRI techniques that have been used for imaging the structural properties of the brainstem, such as magnetization transfer MRI (Priovoulos et al., 2018), quantitative susceptibility mapping, and multi-parametric mapping.

The most significant biologic source of noise affecting brainstem fMRI is physiology. Physiologic noise in fMRI originates from autonomic activity, including local magnetic field distortions due to chest motion, blood flow pressure waves resulting from cardiac and respiratory-related activity, and cerebrovascular reactivity to spontaneous changes in arterial CO_2, affecting the brainstem surrounding vasculature, cerebrospinal fluid (CSF) flow, and parenchyma (Glover et al., 2000; Kassinopoulos and Mitsis, 2019). Respiratory-related chest motion changes the local oxygen (O_2) concentrations inside the lungs, leading to signal dropout artifacts in fMRI data. The respiratory-related chest motion artifact is more pronounced in the brainstem due to its proximity to the lungs than other brain regions. This artifact can be partially mitigated using retrospective physiologic noise correction strategies (Glover et al., 2000; Bianciardi et al., 2014) or prospective external shimming strategies. Cardiac pulsatility and respiratory-related effects generate significant distortions in the BOLD-fMRI signal, particularly in the brainstem regions adjacent to the aqueduct, the fourth ventricle, the circle of

Willis, and the vertebral/basilar arteries (Sclocco et al., 2018), and can be corrected using retrospective physiologic noise correction strategies (Glover et al., 2000; Prokopiou et al., 2019), or data-driven independent component analysis (ICA)-based techniques (Brooks et al., 2013; Jacobs et al., 2020). These methods try to identify and remove patterns related to motion or other noise sources based on previously trained classifiers.

Lastly, there are disease-related challenges in brainstem fMRI, with a notable concern being brainstem atrophy. This condition complicates the registration of the brainstem and subsequent analysis of fMRI data. In particular, structural atrophy in the brainstem has been observed in conditions such as progressive supranuclear palsy (Brown et al., 2017) and multiple system atrophy (Miyatake et al., 2010). These disease-related structural changes in the brainstem can significantly affect the use of anatomic landmarks during brainstem registration, posing difficulties in the identification and delineation of specific nuclei or regions. Moreover, atrophic brainstem tissue may exhibit increased susceptibility to distortions and signal dropout, further impeding the accuracy of fMRI data (Engels-Dominguez et al., 2023).

Clinical applications of brainstem functional magnetic resonance imaging

fMRI has been instrumental in advancing our understanding of brain function, and it has been proposed as an important tool in the clinical setting for the prognosis and diagnosis of neurodegenerative and neuropsychiatric diseases. To find broad applications in the clinic, fMRI must offer biomarkers that are robust, reproducible, and provide high accuracy in either early detection or disease monitoring. Thus far, two primary classes of fMRI measures have been suggested to have potential as biomarkers: quantification of functional activation in response to a cognitive or sensory task and functional connectivity.

Detection of functional activation is typically performed using a mass-univariate linear model analysis. In this context, the cognitive or sensory events included in a task are grouped together based on their type or condition and subsequently convolved with a hemodynamic response function (HRF) to account for the sluggishness of the hemodynamic response and the slow oscillations in the BOLD signal. Each condition is then used as an independent variable to predict the BOLD-fMRI signal changes at each voxel throughout the brain. This procedure produces a parameter estimate for each condition included in the model. The parameters obtained for the conditions whose effects are of interest are subtracted from each other, creating contrast images that enhance the signal intensity in regions activated in response to the task. Finally, the contrast images are thresholded based on statistical criteria to isolate the activated regions. In addition to quantifying the levels of activation in different brain regions, this process also serves as a valuable tool for exploring how these activation levels change in the context of various neurologic or neuropsychiatric conditions. However, it is crucial to note that the effectiveness of this process heavily relies on selecting the appropriate HRF. Research has demonstrated significant variations in this function between different brain regions and individuals. Failing to account for this variability may lead to biased results in studies focusing on functional activation (Handwerker et al., 2004).

Unlike the focus on detecting functional activation, which heavily depends on a model of the hemodynamic response to neuronal activation, functional connectivity analyses involve quantifying the extent of temporal covariation in the BOLD-fMRI signal between different brain regions (Rogers et al., 2007). In BOLD-fMRI functional connectivity, assessments are typically conducted through voxel-wise Pearson's correlations between a seed region and each voxel throughout the brain. This evaluation helps to understand the coactivation patterns of different brain regions with the chosen seed. Alternatively, blind source separation techniques, such as ICA, can also be used to detect multiple sources of coactivation that are typically referred to as functional connectivity networks.

In neurologic disorders, alterations in brainstem function can reflect the disease stage and hence be relevant for diagnostics, correspond to the symptomatology, and provide information on clinical heterogeneity. They can also be a downstream consequence and important for prognostics. Furthermore, when brainstem changes are inherent to the disease, either as part of the staging or clinical symptomatology, monitoring brainstem changes during interventions can provide information on the effectiveness of the treatment.

Even though neurodegenerative disorders were investigated through a corticocentric lens over the last two to three decades, the involvement of the brainstem in the earliest stages of the disease is now well established. In Alzheimer disease, the most common form of dementia, several brainstem nuclei – and in particular, the locus coeruleus – are considered among the first regions to accumulate hyperphosphorylated tau. fMRI imaging of the locus coeruleus has been used to identify early changes in the locus coeruleus function, starting from the preclinical stages of Alzheimer disease, which could be used to distinguish individuals on an Alzheimer disease–related trajectory even before the onset of cognitive symptoms (Kelberman and Weinshenker, 2022). During the resting state, whole-brain functional connectivity of the locus coeruleus was found to be lower in asymptomatic individuals at risk for familial Alzheimer disease (Del Cerro et al., 2020). Similarly, lower task-related functional connectivity of the locus coeruleus has been reported in individuals with an increased risk for Alzheimer disease. Work from our group demonstrated that lower novelty-related locus coeruleus activity was related to greater tau deposition in the entorhinal cortex and to steeper memory decline in a group of asymptomatic older adults (Prokopiou et al., 2023) (Fig. 21.2). Lower locus coeruleus activity and connectivity between the locus coeruleus, amygdala, and hippocampus were also associated with steeper Aβ-related cognitive decline, indicating an early involvement of the locus coeruleus in tau accumulation, but as the disease progresses and pathology accumulates, the effects on cognition are detectable (Prokopiou et al., 2022a).

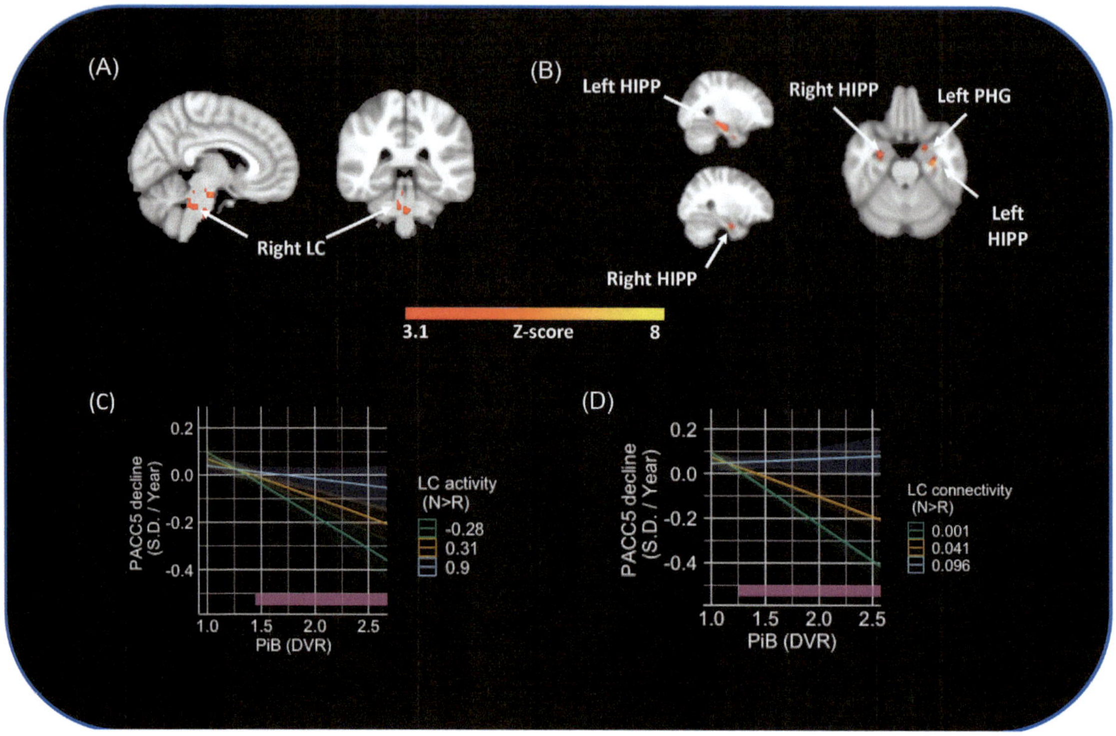

Fig. 21.2. Example of clinical applications of brainstem functional magnetic resonance imaging. Lower locus coeruleus (LC) function relates to steeper β-amyloid-related cognitive decline in cognitively healthy older adults. (**A**) Brain activation map showing significant activation during novel versus repeated (NvR) face-name associations in the LC. (**B**) Brain connectivity map showing significant activation in medial temporal regions showing significant connectivity with the LC during NvR. (**C**) Lower novelty-related LC activity was related to a steeper β-amyloid-related cognitive decline. The pink bar illustrates the range of [C-11] Pittsburgh Compound B (PiB) values at which NvR LC activity was associated with PACC5 decline (PiB > 1.44, distribution volume ratio – partial volume corrected). (**D**) Lower novelty-related connectivity between the bilateral LC and hippocampus was related to a steeper β-amyloid-related cognitive decline. The pink bar illustrates the range of PiB values at which NvR connectivity is associated with PACC5 decline (PiB > 1.32, distribution volume ratio – partial volume corrected). Adapted from Prokopiou PC, Engels-Domínguez N, Papp KV, et al. (2022a) Lower novelty-related locus coeruleus function is associated with Aβ-related cognitive decline in clinically healthy individuals. Nat Commun 13: 1571.

While these findings on the norepinephrine locus coeruleus system in Alzheimer disease hold implications for early detection of the disease, the role of other brainstem nuclei remains underexplored. fMRI studies of the dopaminergic ventral tegmental area in individuals across the entire Alzheimer disease spectrum demonstrated that lower resting-state functional connectivity between the ventral tegmental area and hippocampus was related to lower hippocampal volume and worse memory performance (De Marco and Venneri, 2018). Additionally, research on the serotonergic dorsal raphe nucleus revealed lower resting-state functional connectivity with the default mode network in a group of Alzheimer disease patients with depression, possibly reflecting the symptoms of these patients. Further characterization of the involvement of these nuclei and their specific correlation to symptoms can be key in understanding heterogeneity in Alzheimer disease.

Lewy body dementia, similar to Alzheimer dementia, encompasses a spectrum of related neurodegenerative disorders, including Parkinson disease and Lewy body disease. These disorders are often characterized by the deposition of α-synuclein inclusions (Lewy bodies) in the brainstem early in the disease process, affecting major brainstem neurotransmitter systems. In Parkinson disease, Lewy body pathology begins in the brainstem, the dorsal motor nucleus of the vagal and glossopharyngeal nerves of the medulla oblongata. It then ascends to the forebrain to ultimately affect the neocortex. In contrast, in Lewy body disease pathology is believed to spread from the olfactory bulb to the neocortex via either a brainstem- or limbic-predominant pathway (Braak et al., 2003; Beach et al., 2009; McKeith et al., 2017). Whereas Parkinson disease is primarily characterized by movement-related symptoms, Lewy body dementia includes a broader range of symptoms that can include cognitive impairment, visual hallucinations, fluctuating alertness, and sleep disturbances.

Studies using resting-state brainstem fMRI demonstrated that functional connectivity patterns could provide insights into the clinical phenotype and its prognosis. In Parkinsonian syndromes (Filippi et al., 2019), lower functional connectivity between the brainstem midbrain and anterior putamen was related to a worsening of Parkinson disease severity (Manza et al., 2016), and lower functional connectivity between midbrain regions and the striatum has been linked to advanced stages of Parkinson disease (Hacker et al., 2012). Lower functional connectivity between the midbrain, including the substantia nigra, and the anterior putamen was associated with greater declines in motor function after one year. Consistent with our understanding of brain–behavior relationships, in Parkinson disease patients with a faster decline in cognitive functions instead of motor functions, the functional connectivity patterns were predominantly constricted to the cortex and did not involve brainstem regions (Manza et al., 2016). Similarly, consistent with the hypothesized underlying neurobiologic mechanisms, Parkinson disease patients with freezing of gait exhibited lower functional connectivity between the dorsal raphe nucleus and cortical regions compared to controls and Parkinson disease patients with no freezing of gait (Lv et al., 2022). Furthermore, fMRI may also facilitate the differential diagnostics between atypical forms of Parkinson disease, as these patterns correlate well with the clinical phenotype and can contribute to the prediction of disease trajectories. Lower functional connectivity in the dorsal midbrain tegmentum network was associated with a lower Frontal Assessment Battery score (Piattella et al., 2015) and vertical gaze impairment (Rosskopf et al., 2017) in patients with progressive supranuclear palsy, as well as with cognitive decline in the more advanced stages of the disease (Gardner et al., 2013). Also, patients with progressive supranuclear palsy demonstrated lower functional connectivity between the rostral midbrain tegmentum, subcortical areas, and areas in the prefrontal cortex as compared to healthy controls (Brown et al., 2017), and at a six-month follow-up, this was associated with progressive atrophy in the midbrain, globus pallidus, and posterior frontal cortex, reflecting the progression of the disease.

Many neurologists are also board-certified in pain medicine. Pain regulation involves many noradrenergic and serotonergic pathways, emerging from the brainstem. Brainstem fMRI in patients with pain can have clinical implications, in particular when aiming to optimize the treatment to the pain severity. Patients with migraine without aura exhibited lower functional connectivity between the raphe nucleus and substantia nigra with several cortical regions involved in pain regulation, but higher functional connectivity between the raphe nucleus and the cerebellum was associated with a greater impact of the headache (Huang et al., 2019). Applying repeated transcutaneous vagus nerve stimulation, a noninvasive form of brainstem stimulation that has been associated with increased norepinephrine release, revealed that migraine patients receiving the treatment exhibited lower functional connectivity patterns as compared to the sham condition between the nucleus tractors solitarius, locus coeruleus, and raphe nucleus with cortical regions involved in pain processing (Garcia et al., 2017). Furthermore, transcutaneous vagus nerve stimulation has been related to stronger functional connectivity between the raphe nucleus and dopaminergic regions (putamen and caudate) compared to the sham condition. In addition, these functional connectivity patterns of the intervention were associated with fewer migraine attacks (Huang et al., 2023).

So far, our discussion has centered around neurologic diseases in which brainstem regions or nuclei play a key role in the etiology or the symptomatology. However, brainstem fMRI can also be valuable in disease populations where lesions are focal or involve predominantly the cortex but have widespread functional effects and can explain the emergence of symptoms or complications that may be brainstem-related. Patients with multiple sclerosis can have lesions in brain regions with projections to the monoaminergic nuclei of the brainstem and can contribute to axonal loss of these projections. As expected, patients with relapse-remitting multiple sclerosis showed lower functional connectivity between the ventral tegmental area and bilateral posterior cingulate cortex, between the locus coeruleus and prefrontal cortex, and between the raphe nucleus and the cerebellum as compared to healthy controls (Carandini et al., 2021). Furthermore, patients with multiple sclerosis with depression exhibited lower raphe nucleus–related resting-state functional connectivity compared to those patients with no depressive symptoms, reflective of potential damage to a specific disconnection of this brainstem monoaminergic pathway (Martino et al., 2021). Brainstem imaging can provide insight into the mechanisms underlying complications. Sudden, unexpected death in epilepsy often occurs in patients with pharmaco-resistant epilepsy and is associated with loss of breathing. Brainstem imaging of patients undergoing intracranial electroencephalography for intractable epilepsy revealed lower BOLD activity in the pons and medulla, including the parabrachial nucleus and the pre-Bötzinger complex, presumably indicating suppression of breathing and impaired chemosensation (Harmata et al., 2023).

Future directions for brainstem functional magnetic resonance imaging

Ever since the inception of fMRI in the 1990s, the application, sophistication, and clinical implementation have grown tremendously. In the beginning of the 21st century, the spatial resolution was being pushed along with the development of multislice echoplanar sequences, enabling for brainstem fMRI in the last decade. As discussed in this chapter, in vivo functional imaging of the brainstem holds great promise in the diagnostics and prognostics of neurodegenerative and other neurologic diseases. However, these efforts have been challenging due to the technical as well as biologic limitations. An important development in mitigating physiologic noise and signal dropout is the use of multiecho fMRI. Multiecho fMRI measures the MRI signal two or more times per volume acquisition and thus provides greater temporal signal-to-noise ratio compared to conventional single-echo EPI-fMRI techniques (Markello et al., 2018). This allows for the maximization of BOLD contrast through the optimal combination of echoes at each voxel during the data analysis (Markello et al., 2018). As the severity of head motion and physiology artifacts on the MR signal has been shown to depend on the echo time, multiecho data-driven ICA techniques have also been shown quite effective in removing contributions of non-BOLD signal components from the data, providing further improvements in the temporal signal-to-noise ratio, even in smaller brainstem structures that are particularly vulnerable to MR artifacts.

Another important development is in ultrahigh-field imaging, particularly with the recent U.S. Food and Drug Administration (FDA) approval for clinical use of 7 T MRI. Ultrahigh-field MRI provides increased sensitivity in blood oxygenation levels compared to lower magnetic field systems, which translates into a higher contrast-to-noise ratio (Sclocco et al., 2018). Importantly, this results in a general increase in statistical power, allowing for a reduction in cohort sizes compared to lower field strengths. Ultrahigh-field fMRI can also be used in tandem with other MRI modalities, such as high-resolution structural imaging, which take advantage of the specific macromolecular composition of specific tissues to provide more accurate localization of specific brainstem nuclei at the individual level.

Another future direction of brainstem fMRI is to understand the complexity of the BOLD signal, as this reflects an indirect measure of neuronal activity and is heuristically modeled. As discussed in a previous section, the relationship between local changes in the BOLD-fMRI signal and the underlying changes in neuronal activity is quantified in terms of the HRF. The HRF has been shown to vary considerably between subjects, brain regions, tasks, and stimulus intensity (Handwerker et al., 2004; Prokopiou et al., 2022b), particularly when imaging subcortical structures at higher magnetic fields. Understanding and modeling this hemodynamic variability is a perpetual challenge due to the inherent complexity of the cerebral hemodynamics (Polimeni and Lewis, 2021), and failure to account for this variability has been shown to result in erroneous fMRI data analyses. Similar challenges are also faced in the development of physiology response models, which are used to remove the nonneuronal-related artifacts in the BOLD-fMRI signal. Cardiac and respiratory activity affect the BOLD-fMRI signal via cerebral blood circulation and respiratory-related motion effects due to its proximity to the lung expansions and contractions, ascending arteries, and pulsatile CSF flow. Hence, brainstem-specific physiology models could be more effective in removing BOLD-fMRI artifacts of physiology origins (Kassinopoulos and Mitsis, 2019; Prokopiou et al., 2019). It should be noted that as physiology activity is related to brainstem structures involved in autonomic control, physiologic noise correction might result in the removal of signal of interest. Finding an optimal BOLD-fMRI data preprocessing pipeline specific to the brainstem remains an open question. While general suggestions and recommendations for processing brainstem BOLD-fMRI have been proposed (Brooks et al., 2013), this process is often a trial-and-error procedure, depending on the specific research question or clinical application.

Lastly, brainstem fMRI can also be used in tandem with other data domains, such as high-resolution structural imaging, molecular (PET) imaging, and behavioral or cognitive testing, for a better understanding of the early stages of neurodegenerative diseases and the development of early biomarkers (Engels-Dominguez et al., 2023). Recent advances in structural MR imaging have led to the development of dedicated MR imaging sequences that take advantage of the cellular and macromolecular composition of specific regions, such as the locus coeruleus (Priovoulos et al., 2018, 2020) and substantia nigra, to achieve more precise localization. Dedicated structural imaging of the brainstem can be used as an anatomic prior to isolating the BOLD-fMRI signal from specific brainstem nuclei at the subject level, facilitating the investigation of their isolated contributions to disease progression. Multimodal data fusion methods are rapidly evolving and can further our understanding of brainstem-related processes or symptoms and their relevance in the etiology of neurologic disorders, facilitating differential diagnostics or understanding disease heterogeneity, supporting or monitoring intervention outcomes, and contributing to the development of new targeted therapeutic approaches.

Given that brainstem dysfunction has been described in almost all neurodegenerative diseases, either correlating with its symptoms or constituting part of the initial neuropathology of the disease, it will be critical to disentangle further unique versus shared functional brainstem changes representing their distinct phenotypes. Relating these differences to clinical phenotypes can be performed at the location level (voxel/nuclei/region), at the neuronal firing pattern level, or may only be evident in relation to other brainstem-specific properties, such a transcriptional profile of cells, neurotransmission, or receptor density changes. Such an endeavor will require large-scale multimodal studies, including the various disorders, ideally at different sampling moments across its progression. One should remain cautious not to consider brainstem function or dysfunction in an isolated manner, as the brainstem interacts in nonlinear and dynamic ways with other brain regions, and a comprehensive account of the brain in its entire complexity integrated with clinical information is warranted to understand disease-specific perturbations, relationships to symptomatology, and intervention responses.

MOLECULAR IMAGING

Background

PET is a molecular imaging technique that brings biology, physics, and chemistry together in real time to perform cellular imaging through molecular targeting of biomolecular processes in vivo (Phelps, 2012). Positrons are antimatter particles that have the same mass as electrons but are positively charged, as opposed to electrons that are negatively charged. From an astrophysics perspective, antimatter was present in the first 10^{-36} seconds after the Big Bang, but due to a phenomenon called "baryonic asymmetry," the immanent universe was created out of matter particles, such as protons, neutrons, and electrons. The fate of the initial antimatter particles remains uncertain and is an active area of astrophysics research. Hence, these positrons do not exist spontaneously in nature except under extraneous circumstances such as large storms where energy from high-voltage lightning leads to subatomic production of positron antimatter particles. However, in the clinical setting, these positrons can be created in an artificial environment such as that created by a medical cyclotron. In a medical cyclotron, a high-energy beam of charged particles such as a proton is directed toward an atomic target, which, when bombarded by this high-energy beam, undergoes a nuclear reaction, leading to the production of positron-emitting isotopes. These positron-emitting isotopes are then used for chemical reactions that lead to the synthesis of radiopharmaceutics. These radiopharmaceutics often have positron-emitting isotopes substituting for their stable (nonradioactive) counterparts in a biologic relevant molecule of interest. Therefore, the radiopharmaceutic maintains its chemical properties that may be identical or very similar to the parent molecule and can be used for tracing the kinetics of various biologic and physiologic processes. Alternatively, the positron-emitting isotope is chelated to a molecule of biologic interest, and the kinetics of the radiopharmaceutics are hence traced to provide biologic and clinically relevant information.(Phelps, 2012; Singhal, 2012)

When the positron is emitted by a positron-emitting radioisotope, it interacts with an electron in its immediate environment after following a curvilinear path of various lengths, depending on the energy of the positron emitted, which in turn depends on the parent positron-emitting isotope. This interaction between the "matter," electron particle, and the "antimatter," positron particle, leads to the annihilation of their mass into energy. This energy that is then emitted has two photons of 511 keV in two opposite directions. In reality, these gamma rays are emitted at an angle to each other that is not completely 180 degrees, to fulfill the laws of conservation of momentum, as at the time of annihilation, the positron and the electron have not come to complete rest, and therefore the net momentum cannot be zero, which would be the case if the two gamma photons were emitted at exactly 180 degrees to each other. These gamma photons are then detected by an array of detectors in the PET scanner. These detectors are often made of semiconductor material like bismuth germanate (BGO), (lutetium oxyorthosilicate) LSO, or (lutetium yttrium oxyorthosilicate) LYSO, and the energy from these photons is converted into a signal, which is then detected by the electrical circuitry in the scanner, and the data obtained are converted into a three-dimensional image. This image provides a map of the spatial distribution of the magnitude of radioactivity concentration based on the amount of radiopharmaceutics present in the target of interest, at a given point in time. The images can then be visualized as sections of the target tissue; hence, the term tomography, as "tomos," in Greek, means sections. PET imaging is different from conventional computer tomography, which utilizes the principle of attenuation of high-energy photons that are *transmitted* through the target tissue, whereas in PET imaging, the high-energy photons are *emitted* from the tissue and detected by the PET scanner. Hence, PET imaging is an *emission* tomography as opposed to conventional CT scans, which are a result of tomographic imaging based on the *transmission* of photons through the target tissue.

PET imaging is different from single photon emission computed tomography (SPECT) as well. In SPECT imaging, the radioisotopes that are used produce only one gamma photon because of their radioactive decay, as opposed to the production of two gamma photons as explained above, by radioisotopes used for PET imaging. The principles of signal localization used for SPECT imaging are also different from PET, as imaging in the latter is eventually based on coincidence detection of the photons emitted in opposite directions by radioisotopes, while SPECT localization is based on the detection of single photons representing each radioactive decay event and requires collimation of the emitted photons.

The most commonly used PET isotopes are isotopes of the naturally occurring building blocks of biologic systems, that is, carbon, oxygen, nitrogen, hydrogen, or analogs of the hydroxyl group. To a large extent, PET imaging derives its versatility from the fact that the PET isotopes can mimic the chemical behavior of the parent biomolecules and provide unique tracer information about the behavior of the molecular target. Overall, PET imaging is based on the "tracer principle," which was first elucidated by De Hevesy, and which states that "tracers" undergo chemical processes that are similar to their naturally occurring counterpart molecules. Further, it states that these tracers are introduced in such low quantities that they do not interfere with the underlying biologic processes. Specifically, in PET imaging, radiopharmaceutics are often introduced in picomolar concentrations and yet are able to provide enough signal for detection and quantitation without interfering with the biologic process of interest.

From a nuclear physics perspective, positron-emitting isotopes have an excess of protons as compared to neutrons in their atomic nucleus and deviate from the "line of stability" of the proton-to-neutron ratio, which makes these isotopes unstable due to excessive repulsion amongst the positively charged protons as compared to the nuclear strong forces that keep the nucleons together. Therefore, these isotopes attain nuclear stability by emission of a positively charged particle, that is, the positron, and decreasing the burden of repulsive forces in the atomic nucleus. PET isotopes have short half-lives. For example, carbon 11 has a half-life of approximately 20 minutes, oxygen 15 has a half-life of two minutes, nitrogen 13 has a half-life of 10 minutes, and fluorine 18, which often serves as an isotope substituting for hydroxyl groups (due to the single negative charge in the fluoride ion), has a half-life of approximately 110 minutes. The short half-lives of positron-emitting isotopes can make their use logistically challenging, as they need to be produced in real-time and utilized within a short period of time after their production. In this regard, fluorine 18 labeled isotopes have several advantages over the other PET isotopes. Given its 110-minute half-life, it can be produced in a central radiochemistry facility, labeled to synthesize the appropriate radiopharmaceutical, and then transported to individual imaging facilities where the radiopharmaceutical can be used for imaging. Therefore, in this case, individual facilities do not need to each have an onsite cyclotron to perform PET imaging.

The spatial resolution of PET can range from 2.5 to 5 mm, depending upon the type of PET scanner used. Conventionally, a brain-dedicated scanner such as the "HR+" has a full-width half maximum of 5 mm as opposed to the subsequently developed high-resolution research tomography (HRRT) scanner, which is also a dedicated brain and small animal scanner with a spatial resolution of 2.3–3.4 mm. Ultrahigh-resolution PET scanners are currently being developed for dedicated brain imaging. Positrons emitted by each type of PET isotope have a different length of the path that they travel before losing enough energy, and interacting with the corresponding electrons leads to their annihilation and production of the two gamma photons that were described before. These differences in the path length among PET isotopes also contribute to the slight variability in the best spatial resolution that can be theoretically achieved in PET imaging for each given pet isotope.

Clinical applications of brainstem molecular imaging

Various PET radiopharmaceutics have been employed for molecular imaging of the brainstem. We will cover here the clinical applications of these tracers per neurotransmitter system (Table 21.1).

Molecular imaging of locus coeruleus and noradrenergic system

Using a novel PET ligand, (S,S)[C-11]Methylreboxetine ([C-11]MRB, a ligand for the norepinephrine transporter (NET)), we reported on one of the earliest descriptions of the noradrenergic system in the brainstem and beyond using PET (Ding et al., 2010). Using [C-11]MRB and HRRT, higher levels of concentrations of the tracer were observed in the locus coeruleus, midbrain and pontine raphe, red nucleus, hypothalamus, and the thalamus, including the thalamic subnuclei. Tracer concentration in the pontine raphe and locus coeruleus is manifested in the so-called "wing-shaped" appearance on transaxial views of averaged [C-11]MRB-PET scans (Fig. 21.3). In clinical populations, upregulation of the NET was described in cocaine abusers in various regions of the brainstem, including locus coeruleus, whereas the NET signal in the locus coeruleus and other structures in the brain declines with aging. These results were consistent with the results obtained in in vitro studies in humans and animal models (Ding et al., 2010).

Beyond cocaine abusers, NET sites have been shown to be decreased in the locus coeruleus in Alzheimer disease postmortem studies. Using [F-18]MeNer (another PET ligand for NET), a significant reduction in NET density in Alzheimer disease was shown in a postmortem autoradiographic study in humans (Tejani-Butt et al., 1993; Gulyas et al., 2010), Molecular imaging of the locus coeruleus in Alzheimer disease has therapeutic and pathogenetic implications. Norepinephrine stimulation of microglia suppressed amyloid-β-induced cytokine and chemokine production, and increased microglial migration and phagocytosis of amyloid β, thereby contributing to attenuating the Alzheimer disease pathology (Heneka et al., 2010). Norepinephrine-depleted transgenic mice showed reduced recruitment of microglia to amyloid-β plaque sites and impaired microglial amyloid-β phagocytosis (Heneka et al., 2010). Future in vivo studies of brainstem imaging with various PET ligands targeting the NET, amyloid, tau, and neuroinflammation are needed in Alzheimer disease.

In Parkinson disease, locus coeruleus degeneration has also been demonstrated in vivo using [C-11]MeNer-PET (NET ligand) imaging (Sommerauer et al., 2018). This reduction

Table 21.1

Brainstem molecular imaging targets and examples for corresponding molecular imaging agents (PET or SPECT radiopharmaceutics) and clinical/research applications.

Brainstem target	Molecular imaging agent	Clinical/research applications
Locus coeruleus and noradrenergic system	[C-11]MRB or [C-11]MeNer, [F-18]MeNer	Substance abuse, Alzheimer disease, multiple sclerosis, Parkinson disease, post-traumatic stress disorder
Serotonergic system	[I-123]-FP-CIT [C-11]-WAY-100635 [C-11]DASB	Parkinson disease, tremors
Dopaminergic system	[I-123]-FP-CIT [C-11]PHNO (D2/D3 receptor) [F-18]FDOPA	Parkinson disease, atypical Parkinsonian disorders
Neuroinflammation	[F-18]PBR06 [C-11]PBR28 [C-11]PK11195	Multiple sclerosis, multiple system atrophy, chronic fatigue syndrome
Glucose metabolism	[F-18]FDG	Atypical Parkinsonian disorders including multiple system atrophy, progressive supranuclear policy
Synaptic vesicle density	[C-11]UCB-J [F-18]SynVesT-1	Parkinson disease, corticobasal degeneration
White matter/demyelination	[C-11]PiB [C-11]Florbetaben [F-18]3F4AP	Multiple system atrophy, multiple sclerosis

PET, positron emission tomography; *SPECT*, single photon emission computed tomography.

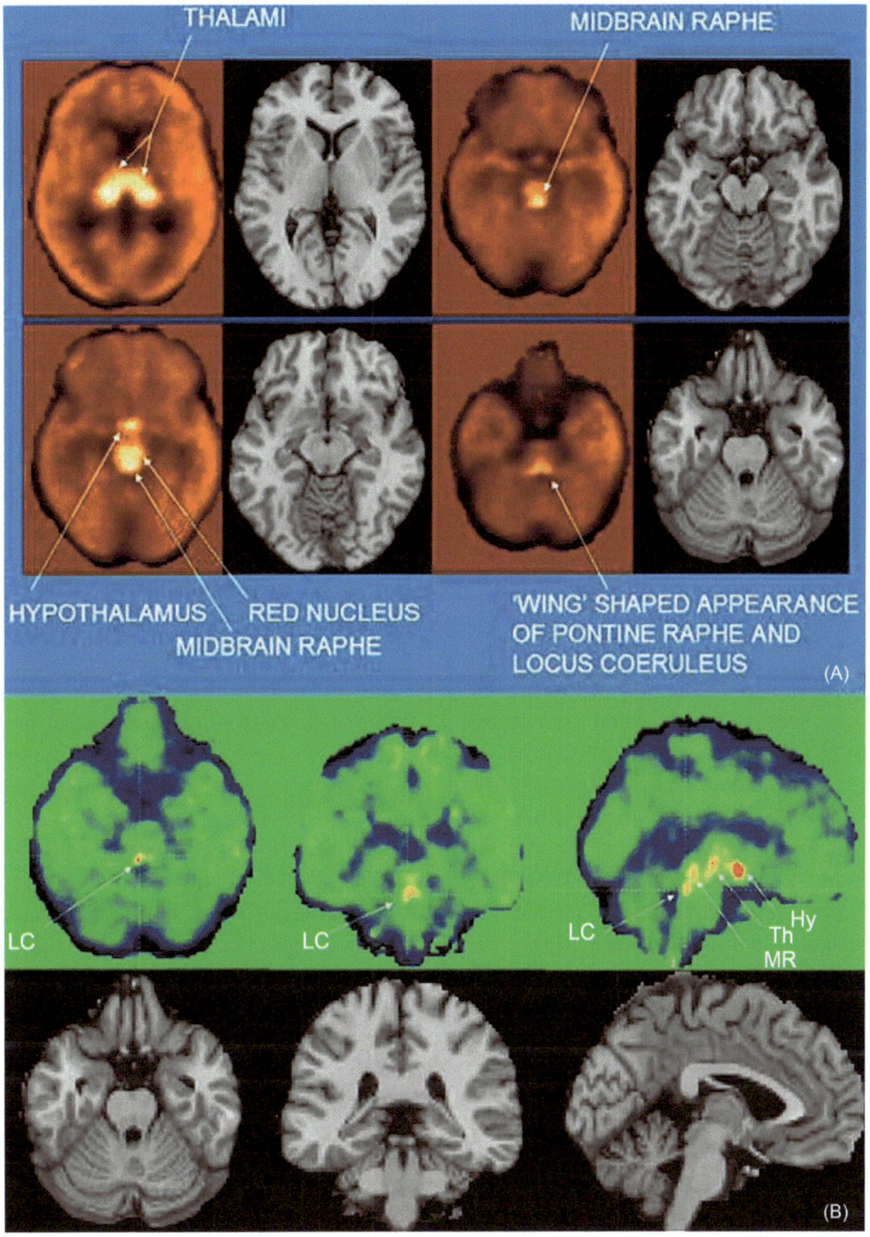

Fig. 21.3. (**A**) Averaged [C-11]MRB-PET images from healthy controls depicting the normal distribution of radiotracer in thalamus, midbrain raphe, hypothalamus, red nucleus, pontine raphe, and locus coeruleus. (**B**) Individual [C-11]MRB-PET images showing high radiotracer uptake in locus coeruleus in a cocaine subject. *LC*, locus coeruleus; *Hy*, hypothalamus; *MR*, midbrain Raphe; *PET*, positron emission tomography; *Th*, thalamus. Reproduced with permission from Ding YS, Singhal T, Planeta-Wilson B, et al. (2010) PET imaging of the effects of age and cocaine on the norepinephrine transporter in the human brain using (S,S)-[(11)C]O-methylreboxetine and HRRT. Synapse 64: 30–38.

was more prominent in Parkinson disease patients with rapid eye movement (REM) sleep behavioral disorder. A comparison with locus coeruleus-specific anatomic imaging demonstrated a correlation between PET signal and neuronal loss. Tracer uptake of [C-11]MeNer-PET has been reported to correlate with symptoms. In Parkinson disease patients with REM sleep behavioral disorders, locus coeruleus NET concentration was associated with global cognitive performance (Sommerauer et al., 2018), whereas tremulous Parkinson disease patients demonstrated reduced tracer uptake in the raphe nucleus (Kinnerup et al., 2021; Pasquini and Ceravolo, 2021). Beyond the locus coeruleus, significant reductions in tracer uptake of [C-11]MeNer-PET in Parkinson disease patients relative to controls were also reported in the red nucleus and thalamus (Nahimi et al., 2018; Pasquini and Ceravolo, 2021).

Similar to Alzheimer disease and Parkinson disease, there is evidence for locus coeruleus injury in multiple sclerosis (Polak et al., 2011). Both experimental animal and postmortem human studies have demonstrated injury to the locus coeruleus and reduced norepinephrine levels in the brain in the experimental allergic encephalomyelitis model and in multiple sclerosis patient autopsy specimens. Concomitant reduced sympathetic activity is seen in the early stages of multiple sclerosis and may contribute to a higher incidence of depression and fatigue in these patients. Moreover, neuronal injury concomitant with neuroinflammation has been demonstrated in the locus coeruleus in postmortem human specimens (Polak et al., 2011). We have observed increased microglial activation in locus coeruleus in multiple sclerosis patients with fatigue, using [F-18] PBR06 (an 18-kDa translocator protein (TSPO) ligand targeting microglia and astrocytes) PET, as compared to healthy controls. These studies provide the basis for pursuing brainstem PET imaging studies targeting NET in combination with neuroinflammation imaging in multiple sclerosis. The interaction between norepinephrine and the central and peripheral immune system cannot be overemphasized (Feinstein et al., 2016) and underscores the importance of brainstem imaging of the noradrenergic system in tandem with the evaluation of the inflammatory pathways.

Molecular imaging of serotonergic and dopaminergic system

Dopamine transporter imaging is a well-established clinical tool for the evaluation of neurodegenerative Parkinsonian syndromes, such as multiple-system atrophy, progressive supranuclear palsy, corticobasal degeneration, Parkinson, and Lewy body disease (Wallert et al., 2022). [I-123]-FP-CIT is a clinically approved SPECT tracer for distinguishing Parkinsonian tremors from essential tremors and from drug-induced tremors. Normal appearance of the basal ganglia using this tracer (on the so-called "DaT scans") reveals a "comma-shaped appearance," while degeneration of the putamen leads to a "period sign" where there is loss of tracer uptake in the putamen due to degeneration of presynaptic dopamine transporter in patients with Parkinson disease and other forms of neurodegenerative Parkinsonian disorders, and relative preservation of the tracer uptake in the caudate. More advanced cases may demonstrate reduced tracer concentration in the caudate as well. Often, the reduced tracer uptake seen on DaT scans is asymmetric and matches with the contralateral presence of tremors consistent with known neuroanatomic pathways. [F-18]Fluorodopa is a PET ligand that has been used in the evaluation of Parkinson disease but may be normal in the initial stages of Parkinson disease due to compensatory upregulation of the dopa decarboxylase enzyme (Singhal, 2012; Morbelli et al., 2020).

In addition to dopaminergic abnormalities, brainstem serotonergic abnormalities have also been reported in Parkinson disease using various PET and SPECT ligands. In a study of 23 Parkinson disease subjects and eight healthy controls, Unified Parkinson Disease Rating Scale total and rest tremor scores were inversely correlated with tracer binding of [C-11]-WAY-100635 (a $5HT_{1A}$ receptor ligand) in midbrain raphe (Doder et al., 2003; Pasquini and Ceravolo, 2021). Another study showed that the uptake of [C-11]DASB (a serotonin transporter (SERT) ligand) in raphe nuclei (along with caudate and putamen) was inversely correlated with the severity of action tremor but not rest tremor in tremor-predominant Parkinson disease patients (Loane et al., 2013; Pasquini and Ceravolo, 2021). Further, as mentioned above, [I-123]-FP-CIT is a SPECT ligand that is traditionally used for dopamine transporter binding in basal ganglia, but it has also been shown to bind to SERT and NET in the brainstem (Wallert et al., 2022). Interestingly, in a study comprising 345 de Novo Parkinson disease patients and 145 healthy controls, [I-123]-FP-CIT uptake in the raphe was studied as a marker of SERT availability, and was shown to be inversely correlated with rest tremor severity in the whole cohort and in a subgroup of tremulous patients (Qamhawi et al., 2015; Pasquini and Ceravolo, 2021). In a subsequent longitudinal study comprising 378 de novo Parkinson disease patients and 145 healthy controls, rest tremor severity was inversely correlated with raphe SERT availability at baseline and after two years (as measured by raphe [I-123]-FP-CIT SPECT). This study provided information about the role of brainstem serotonergic pathways in the pathogenesis of Parkinson disease tremor, as more severe raphe serotonergic dysfunction in association with less severe putaminal dopaminergic loss (measured by [I-123]-FP-CIT SPECT uptake) was predictive of higher rest tremors scores and poorer response of the rest tremor to dopaminergic therapy (Pasquini et al., 2018; Pasquini and Ceravolo, 2021).

Molecular imaging of brainstem neuroinflammation

The 18-kDa TSPO is a transmembrane protein present in the outer layer of mitochondria that has a role in the translocation of cholesterol from the cytosol to mitochondria for lipid synthesis (Singhal et al., 2017). Under disease conditions, there is overexpression of TSPO at the tissue level in the central nervous system, which has been localized to microglia and astrocytes, and, hence, studied as a marker of neuroinflammation. Several PET ligands have been developed that bind to TSPO, and a range of TSPO-PET studies have been performed to evaluate innate immune activation in neurologic and neuropsychiatric diseases (Singhal et al., 2017; Kreisl et al., 2020). In the context of brainstem imaging, we demonstrated that increased neuroinflammation (as measured by [F-18]PBR06, a second-generation TSPO-PET ligand) in substantial nigra is closely linked with fatigue in multiple sclerosis (Singhal et al., 2020). Multiple sclerosis patients with fatigue had higher levels of inflammation as measured by [F-18]PBR06 uptake in substantial nigra as compared to those multiple sclerosis patients who did not have fatigue, and as compared to healthy controls. Further, we have reported our initial observations of increased pontine and middle cerebellar peduncle [F-18]PBR06 uptake in multiple system atrophy, which has been verified on histopathologic evaluation (Ndayisaba et al., 2022). Another study assessing neuroinflammation using [C-11]PK11195, a conventional TSPO-PET ligand, demonstrated increased microglial activation in the brainstem in patients with chronic fatigue syndrome (Nakatomi et al., 2014). Further studies that combine PET imaging of neuroinflammation with other targets, such as metabolism, neurochemistry, and metabolism, are needed.

Molecular imaging of brainstem metabolism

[F-18]Fluorodeoxyglucose (FDG) has been used as a marker of glucose metabolism for more than 45 years since its initial synthesis in 1976 (Jones et al., 2012). FDG has been a workhorse radiopharmaceutical for systemic PET imaging but has also played a critical role in neuroimaging. FDG-PET is currently approved for the evaluation of dementia patients and for distinguishing Alzheimer disease from frontotemporal dementia. FDG-PET is also used for the evaluation of refractory epilepsy patients and the identification of potential seizure focus. Interestingly, brainstem metabolic changes evaluated by FDG PET have been shown to contribute to the differential diagnosis of Parkinsonian syndromes. Specifically, midbrain hypometabolism is seen in progressive supranuclear palsy, which can be used to distinguish this condition from Parkinson disease and multiple system atrophy. Multiple system atrophy is associated with pontine hypometabolism, and Parkinson disease may initially show normal-appearing glucose metabolism, and advanced cases with Parkinson disease dementia may demonstrate a hypometabolic pattern similar to Alzheimer disease (Singhal, 2012).

Molecular imaging of synaptic density in the brainstem

In the last few years, novel pet ligands have been developed that target the glycoprotein SV2a that serves as a measure of synaptic density (Carson et al., 2022). Both F-18 labeled and C-11 labeled synaptic density PET tracers have been developed. Molecular brainstem pathology is beginning to be elucidated using this set of PET tracers. Specifically, in Parkinson disease, a loss of synaptic density in substantial nigra, red nucleus, dorsal raphe, and locus coeruleus has recently been demonstrated using [C-11]UCB-J, a SV2a PET ligand (Matuskey et al., 2020; Carson et al., 2022). Reduced synaptic density in the medulla, pons, and midbrain has been shown in progressive nuclear palsy. In the same study, corticobasal degeneration showed reduced synaptic density in the medulla as compared to healthy controls (Holland et al., 2020; Carson et al., 2022).

Molecular imaging of brainstem white matter

The brainstem is unique in the central neuraxis, given its intermingled organization of gray and white matter tissue in close proximity. Imaging of brainstem white matter can be critical for the evaluation of diseases such as multiple sclerosis, multiple system atrophy, and other disorders. Novel applications of PET tracers targeting white matter pathology, including repurposed use of some of the conventional amyloid tracers (de Paula Faria, 2020), can demonstrate demyelination in the brainstem, which may be more sensitive than conventional MRI scans. Moreover, radiolabeled 4-aminopyridine has been validated as a marker of demyelination (Guehl et al., 2021) and can be an interesting PET ligand for imaging demyelination in the brainstem.

Drug development studies

There is a high failure rate of novel CNS drugs development that has been at least partially attributed to a lack of availability of adequate pharmacokinetic and pharmacodynamic data in humans prior to clinical trials (Zimmer, 2023). Molecular PET and SPECT imaging provides a unique set of tools that allow the evaluation of the ability of a candidate drug to cross the blood–brain barrier, its target binding, and its pharmacodynamic effects. Drug candidates can be radio-labeled with carbon 11 or fluorine 18 or, in some cases, zirconium 89 (half-life 3.3 days) or iodine 124 (half-life 4.2 days), and their distribution can be assessed in terms of their ability to cross the blood–brain barrier. These studies can be performed at tracer doses or the so-called microdosing approach (Zimmer, 2023). According to some criteria, in the microdosing approach, the injected dose of a drug candidate is supposed to be less than 1% of the expected pharmacodynamic active dose, less than 100 μg, administered to fewer than 10 healthy volunteers and for a period of less than 7 days (Zimmer, 2023). Microdosing has been applied to evaluate the extent of brain uptake for a variety of psychotropic drugs, and a huge variability was observed (Schou et al., 2015; Zimmer, 2023). This approach is used when appropriate radio ligands are not available for estimating target occupancy for candidate drug molecules (Varrone et al., 2022; Zimmer, 2023). Future studies using microdosing can target specific evaluation of brainstem exposure to emerging drug molecules. Beyond microdosing, qualitative and quantitative PET imaging studies can be performed to demonstrate the occupancy of a particular target by emerging molecules (Varrone et al., 2022; Zimmer, 2023). Multiple PET imaging studies have shown occupancy of the NET by a variety of psychotropic drugs such as quetiapine, nortriptyline, and venlafaxine (Nyberg et al., 2013; Takano et al., 2014; Arakawa et al., 2019). Moreover, PET imaging allows quantification of the percent receptor occupancy, and the extent of receptor occupancy has been shown to be associated with a meaningful therapeutic effect versus risk of adverse effects, depending on the extent of receptor occupancy (Madsen et al., 2019). Further, PET imaging can be used to evaluate the downstream pharmacodynamic effects of emerging drugs. Antiamyloid therapies have now been approved by FDA based on their ability to demonstrate a significant reduction in brain amyloid deposition (Zimmer, 2023). It is conceivable that in the future, emerging drugs will target brainstem pathology and that PET imaging will be used to demonstrate the effect of emerging drugs on brainstem pathophysiology and enhance novel drug development.

Future developments

Given the somewhat limited spatial resolution of PET and SPECT, these techniques may afford an underestimation of molecular signal due to the small size of the target nuclei in the brainstem. However, despite the relatively poor spatial resolution of PET, the unique strength of the technique is its ability to detect this signal based on contrast-to-noise ratio despite the small size of the target structures in the brainstem.

Detectability on PET is determined by the so-called rose criteria, which states that the contrast-to-noise ratio should be greater than or equal to 4 to allow adequate detection and is summarized by the formula $|C_l| * d_t * \sqrt{IDo} \geq 4$, where C_l is lesional contrast, d_t is diameter, and IDo is background information density (Cherry et al., n.d.). Conventional MRI measures are unable to provide this level of contrast and, therefore, may not detect molecular changes despite high resolution. Novel PET radiopharmaceutics with higher specificity can improve the contrast-to-noise ratio and detectability by PET imaging. Radiotracers targeting early, disease-specific molecular formations are highly desired, as they could aid in the differential diagnosis of neurodegenerative disorders with common pathologic aggregates, such as tauopathies and synucleinopathies (Groot et al., 2022; Xiang et al., 2023). Moreover, PET imaging has great versatility in terms of the variety of targets that can be exploited to provide unique molecular information about the brainstem. Future studies need to concentrate on improved quantification approaches that are preferably noninvasive and do not require arterial sampling. Improved scanner resolution and the development of ultrahigh-resolution and wider field-of-view PET scanners (such as NeuroEXPLORER) for brain imaging are much-anticipated developments that will lead to reduced partial volume effects and improved quantification of brainstem molecular pathophysiology and may even eliminate the need for arterial blood sampling. Further, there is clearly a need for multimodal imaging that combines PET, MRI, SPECT, EEG, and optical imaging. In particular, combining brainstem PET imaging with functional imaging of the brainstem and interregional associations of brainstem structures with other subcortical and cortical regions are likely to be particularly high yield in terms of providing novel insights. This would require novel and improved methods of coregistration across modalities, including nonlinear approaches. Further, clearly, there is a need for improved anatomic delineation in which cross-modality coregistration and improved atlases can clearly play a role. From a biologic standpoint, further studies identifying relationships between genetic and epigenetic markers and brainstem molecular imaging are needed. In terms of patient selection, future studies should also include subjects with prodromal disease states and subjects across the age spectrum to provide disease-specific information without confounding effects of concomitant treatment, downstream effects of the disease state, and comorbidities.

CONCLUSION

The brainstem is the source of monoamine and serotonergic neuromodulators. Through its many ascending and descending pathways, it plays a key role in modulating and integrating behavior, mood, and cognitive functions, and in the pathophysiology or symptomatology of neurologic disorders. The development of functional BOLD-imaging of the brainstem and radioactive ligands binding to neurotransmitters and other molecular targets in the brainstem and beyond revealed new insights into the clinical applications of the brainstem in neurodegenerative and other neurologic disorders and unveiled its potential transformative role in the management of patients with diseases affecting the brainstem.

References

Abderezaei J, Pionteck A, Terem I et al. (2021). Development, calibration, and testing of 3D amplified MRI (aMRI) for the quantification of intrinsic brain motion. Brain Multiphys 2: 100022.

Arakawa R, Stenkrona P, Takano A et al. (2019). Venlafaxine ER blocks the norepinephrine transporter in the brain of patients with major depressive disorder: a PET study using [18F]FMeNER-D2. Int J Neuropsychopharmacol 22: 278–285.

Beach TG, Adler CH, Lue L et al. (2009). Unified staging system for Lewy body disorders: correlation with nigrostriatal degeneration, cognitive impairment and motor dysfunction. Acta Neuropathol 117: 613–634.

Beissner F (2015). Functional MRI of the brainstem: common problems and their solutions. Clin Neuroradiol 25: 251–257.

Bianciardi M, van Gelderen P, Duyn JH (2014). Investigation of BOLD fMRI resonance frequency shifts and quantitative susceptibility changes at 7 T. Hum Brain Mapp 35: 2191–2205.

Braak H, Del Tredici K, Rub U et al. (2003). Staging of brain pathology related to sporadic Parkinson's disease. Neurobiol Aging 24: 197–211.

Brooks JC, Faull OK, Pattinson KT et al. (2013). Physiological noise in brainstem FMRI. Front Hum Neurosci 7: 623.

Brown JA, Hua AY, Trujllo A et al. (2017). Advancing functional dysconnectivity and atrophy in progressive supranuclear palsy. Neuroimage Clin 16: 564–574.

Carandini T, Mancini M, Bogdan I et al. (2021). In vivo evidence of functional disconnection between brainstem monoaminergic nuclei and brain networks in multiple sclerosis. Mult Scler Relat Disord 56: 103224.

Carson RE, Naganawa M, Toyonaga T et al. (2022). Imaging of synaptic density in neurodegenerative disorders. J Nucl Med 63: 60S–67S.

Cherry S, Sorensen J, Phelps M (n.d.). Physics in nuclear medicine. Elsevier.

Cucu AI, Turliuc S, Costea CF et al. (2021). The brainstem and its neurosurgical history. Neurosurg Rev 44: 3001–3022.

Del Cerro I, Villarreal MF, Abulafia C et al. (2020). Disrupted functional connectivity of the locus coeruleus in healthy adults with parental history of Alzheimer's disease. J Psychiatr Res 123: 81–88.

De Marco M, Venneri A (2018). Volume and connectivity of the ventral tegmental area are linked to neurocognitive signatures of Alzheimer's disease in humans. J Alzheimers Dis 63: 167–180.

de Paula Faria D (2020). Myelin positron emission tomography (PET) imaging in multiple sclerosis. Neural Regen Res 15: 1842–1843.

Ding YS, Singhal T, Planeta-Wilson B et al. (2010). PET imaging of the effects of age and cocaine on the norepinephrine transporter in the human brain using (S,S)-[(11)C]O-methylreboxetine and HRRT. Synapse 64: 30–38.

Doder M, Rabiner EA, Turjanski N et al. (2003). Tremor in Parkinson's disease and serotonergic dysfunction: an 11C-WAY 100635 PET study. Neurology 60: 601–605.

Engels-Dominguez N, Koops EA, Prokopiou PC et al. (2023). State-of-the-art imaging of neuromodulatory subcortical systems in aging and Alzheimer's disease: challenges and opportunities. Neurosci Biobehav Rev 144: 104998.

Feinstein DL, Kalinin S, Braun D (2016). Causes, consequences, and cures for neuroinflammation mediated via the locus coeruleus: noradrenergic signaling system. J Neurochem 139: 154–178.

Filippi M, Sarasso E, Agosta F (2019). Resting-state functional MRI in Parkinsonian syndromes. Mov Disord Clin Pract 6: 104–117.

Finger S, Eling P (2020). Franz Joseph Gall: naturalist of the mind, visionary of the brain. New York: Oxford University Press.

Freemon FR (1994). Galen's ideas on neurological function. J Hist Neurosci 3: 263–271.

Garcia RG, Lin RL, Lee J et al. (2017). Modulation of brainstem activity and connectivity by respiratory-gated auricular vagal afferent nerve stimulation in migraine patients. Pain 158: 1461–1472.

Gardner RC, Boxer AL, Trujillo A et al. (2013). Intrinsic connectivity network disruption in progressive supranuclear palsy. Ann Neurol 73: 603–616.

Glover GH (2011). Overview of functional magnetic resonance imaging. Neurosurg Clin N Am 22: 133–139.

Glover GH, Li TQ, Ress D (2000). Image-based method for retrospective correction of physiological motion effects in fMRI: RETROICOR. Magn Reson Med 44: 162–167.

Groot C, Villeneuve S, Smith R et al. (2022). Tau PET imaging in neurodegenerative disorders. J Nucl Med 63: 20S–26S.

Guehl NJ, Ramos-Torres KM, Linnman C et al. (2021). Evaluation of the potassium channel tracer [(18)F]3F4AP in rhesus macaques. J Cereb Blood Flow Metab 41: 1721–1733.

Gulyas B, Brockschnieder D, Nag S et al. (2010). The norepinephrine transporter (NET) radioligand (S,S)-[18F]FMeNER-D2 shows significant decreases in NET density in the human brain in Alzheimer's disease: a post-mortem autoradiographic study. Neurochem Int 56: 789–798.

Hacker CD, Perlmutter JS, Criswell SR et al. (2012). Resting state functional connectivity of the striatum in Parkinson's disease. Brain 135: 3699–3711.

Handwerker DA, Ollinger JM, D'Esposito M (2004). Variation of BOLD hemodynamic responses across subjects and brain regions and their effects on statistical analyses. Neuroimage 21: 1639–1651.

Harmata GI, Rhone AE, Kovach CK et al. (2023). Failure to breathe persists without air hunger or alarm following amygdala seizures. JCI Insight 8: e172423.

Heneka MT, Nadrigny F, Regen T et al. (2010). Locus coeruleus controls Alzheimer's disease pathology by modulating microglial functions through norepinephrine. Proc Natl Acad Sci U S A 107: 6058–6063.

Holland N, Jones PS, Savulich G et al. (2020). Synaptic loss in primary tauopathies revealed by [(11) C]UCB-J positron emission tomography. Mov Disord 35: 1834–1842.

Huang X, Zhang D, Chen Y et al. (2019). Altered functional connectivity of the red nucleus and substantia nigra in migraine without aura. J Headache Pain 20: 104.

Huang Y, Zhang Y, Hodges S et al. (2023). The modulation effects of repeated transcutaneous auricular vagus nerve stimulation on the functional connectivity of key brainstem regions along the vagus nerve pathway in migraine patients. Front Mol Neurosci 16: 1160006.

Jacobs HI, Priovoulos N, Poser BA et al. (2020). Dynamic behavior of the locus coeruleus during arousal-related memory processing in a multi-modal 7T fMRI paradigm. Elife 9: e52059.

Jones T, Rabiner EA, PET Research Advisory Company (2012). The development, past achievements, and future directions of brain PET. J Cereb Blood Flow Metab 32: 1426–1454.

Kassinopoulos M, Mitsis GD (2019). Identification of physiological response functions to correct for fluctuations in resting-state fMRI related to heart rate and respiration. Neuroimage 202: 116150.

Kelberman MA, Weinshenker D (2022). A novel link between locus coeruleus activity and amyloid-related cognitive decline. Trends Neurosci 45: 651–653.

Kinnerup MB, Sommerauer M, Damholdt MF et al. (2021). Preserved noradrenergic function in Parkinson's disease patients with rest tremor. Neurobiol Dis 152: 105295.

Kreisl WC, Kim MJ, Coughlin JM et al. (2020). PET imaging of neuroinflammation in neurological disorders. Lancet Neurol 19: 940–950.

Kwong KK, Belliveau JW, Chesler DA et al. (1992). Dynamic magnetic resonance imaging of human brain activity during primary sensory stimulation. Proc Natl Acad Sci U S A 89: 5675–5679.

Lauterbur PC (1989). Image formation by induced local interactions. Examples employing nuclear magnetic resonance. 1973. Clin Orthop Relat Res 244: 3–6.

Loane C, Wu K, Bain P et al. (2013). Serotonergic loss in motor circuitries correlates with severity of action-postural tremor in PD. Neurology 80: 1850–1855.

Lv L, Zhang H, Tan X et al. (2022). Associated factors and abnormal dorsal raphe nucleus connectivity patterns of freezing of gait in Parkinson's disease. J Neurol 269: 6452–6466.

Madsen MK, Fisher PM, Burmester D et al. (2019). Psychedelic effects of psilocybin correlate with serotonin 2A receptor occupancy and plasma psilocin levels. Neuropsychopharmacology 44: 1328–1334.

Manza P, Zhang S, Li CS et al. (2016). Resting-state functional connectivity of the striatum in early-stage Parkinson's disease: cognitive decline and motor symptomatology. Hum Brain Mapp 37: 648–662.

Markello RD, Spreng RN, Luh WM et al. (2018). Segregation of the human basal forebrain using resting state functional MRI. Neuroimage 173: 287–297.

Martino M, Magioncalda P, El Mendili MM et al. (2021). Depression is associated with disconnection of neurotransmitter-related nuclei in multiple sclerosis. Mult Scler 27: 1102–1111.

Matuskey D, Tinaz S, Wilcox KC et al. (2020). Synaptic changes in Parkinson disease assessed with in vivo imaging. Ann Neurol 87: 329–338.

McKeith IG, Boeve BF, Dickson DW et al. (2017). Diagnosis and management of dementia with Lewy bodies: fourth consensus report of the DLB Consortium. Neurology 89: 88–100.

Miyatake S, Mochizuki H, Naka T et al. (2010). Brain volume analyses and somatosensory evoked potentials in multiple system atrophy. J Neurol 257: 419–425.

Morbelli S, Esposito G, Arbizu J et al. (2020). EANM practice guideline/SNMMI procedure standard for dopaminergic imaging in Parkinsonian syndromes 1.0. Eur J Nucl Med Mol Imaging 47: 1885–1912.

Nahimi A, Sommerauer M, Kinnerup MB et al. (2018). Noradrenergic deficits in Parkinson disease imaged with (11)C-MeNER. J Nucl Med 59: 659–664.

Nakatomi Y, Mizuno K, Ishii A et al. (2014). Neuroinflammation in patients with chronic fatigue syndrome/myalgic encephalomyelitis: an ^{11}C-(R)-PK11195 PET study. J Nucl Med 55: 945–950.

Napadow V, Dhond R, Kennedy D et al. (2006). Automated brainstem co-registration (ABC) for MRI. Neuroimage 32: 1113–1119.

Ndayisaba A, Pitaro AT, Willett AS et al. (2022). Clinical trial-ready patient cohorts for multiple system atrophy: coupling biospecimen and iPSC banking to longitudinal deep-phenotyping. Cerebellum 23: 31–51.

Nyberg S, Jucaite A, Takano A et al. (2013). Norepinephrine transporter occupancy in the human brain after oral administration of quetiapine XR. Int J Neuropsychopharmacol 16: 2235–2244.

Ogawa S, Lee TM (1990). Magnetic resonance imaging of blood vessels at high fields: in vivo and in vitro measurements and image simulation. Magn Reson Med 16: 9–18.

Parent A (2007). Felix Vicq d'Azyr: anatomy, medicine and revolution. Can J Neurol Sci 34: 30–37.

Pasquini J, Ceravolo R (2021). The molecular neuroimaging of tremor. Curr Neurol Neurosci Rep 21: 74.

Pasquini J, Ceravolo R, Qamhawi Z et al. (2018). Progression of tremor in early stages of Parkinson's disease: a clinical and neuroimaging study. Brain 141: 811–821.

Phelps ME (2012). PET: molecular imaging and its biological applications. New York: Springer-Verlag.

Piattella MC, Tona F, Bologna M et al. (2015). Disrupted resting-state functional connectivity in progressive supranuclear palsy. AJNR Am J Neuroradiol 36: 915–921.

Polak PE, Kalinin S, Feinstein DL (2011). Locus coeruleus damage and noradrenaline reductions in multiple sclerosis and experimental autoimmune encephalomyelitis. Brain 134: 665–677.

Polimeni JR, Lewis LD (2021). Imaging faster neural dynamics with fast fMRI: a need for updated models of the hemodynamic response. Prog Neurobiol 207: 102174.

Porras-Gallo MI, Pena-Meliaan A, Viejo F et al. (2019). Overview of the history of the cranial nerves: from Galen to the 21st century. Anat Rec (Hoboken) 302: 381–393.

Priovoulos N, Jacobs HIL, Ivanov D et al. (2018). High-resolution in vivo imaging of human locus coeruleus by magnetization transfer MRI at 3T and 7T. Neuroimage 168: 427–436.

Priovoulos N, van Boxel SCJ, Jacobs HIL et al. (2020). Unraveling the contributions to the neuromelanin-MRI contrast. Brain Struct Funct 225: 2757–2774.

Prokopiou PC, Engels-Domínguez N, Papp KV et al. (2022a). Lower novelty-related locus coeruleus function is associated with Aβ-related cognitive decline in clinically healthy individuals. Nat Commun 13: 1571.

Prokopiou PC, Engels-Domínguez N, Schultz AP et al. (2023). Association of novelty-related locus coeruleus function with entorhinal tau deposition and memory decline in preclinical Alzheimer disease. Neurology 101: e1206–e1217.

Prokopiou PC, Pattinson KTS, Wise RG et al. (2019). Modeling of dynamic cerebrovascular reactivity to spontaneous and externally induced CO(2) fluctuations in the human brain using BOLD-fMRI. Neuroimage 186: 533–548.

Prokopiou PC, Xifra-Porxas A, Kassinopoulos M et al. (2022b). Modeling the hemodynamic response function using EEG-fMRI data during eyes-open resting-state conditions and motor task execution. Brain Topogr 35: 302–321.

Qamhawi Z, Towey D, Shah B et al. (2015). Clinical correlates of raphe serotonergic dysfunction in early Parkinson's disease. Brain 138: 2964–2973.

Rogers BP, Morgan VL, Newton AT et al. (2007). Assessing functional connectivity in the human brain by fMRI. Magn Reson Imaging 25: 1347–1357.

Rosskopf J, Gorges M, Müller HP et al. (2017). Intrinsic functional connectivity alterations in progressive supranuclear palsy: differential effects in frontal cortex, motor, and midbrain networks. Mov Disord 32: 1006–1015.

Schou M, Varnas K, Lundquist S et al. (2015). Large variation in brain exposure of reference CNS drugs: a PET study in nonhuman primates. Int J Neuropsychopharmacol 18: pyv036.

Sclocco R, Beissner F, Bianciardi M et al. (2018). Challenges and opportunities for brainstem neuroimaging with ultrahigh field MRI. Neuroimage 168: 412–426.

Singhal T (2012). Positron emission tomography applications in clinical neurology. Semin Neurol 32: 421–431.

Singhal T, Cicero S, Pan H et al. (2020). Regional microglial activation in the substantia nigra is linked with fatigue in MS. Neurol Neuroimmunol Neuroinflamm 7: e854.

Singhal T, Weiner HL, Bakshi R (2017). TSPO-PET imaging to assess cerebral microglial activation in multiple sclerosis. Semin Neurol 37: 546–557.

Sommerauer M, Fedorova TD, Hansen AK et al. (2018). Evaluation of the noradrenergic system in Parkinson's disease: an 11C-MeNER PET and neuromelanin MRI study. Brain 141: 496–504.

Takano H, Arakawa R, Nogami T et al. (2014). Norepinephrine transporter occupancy by nortriptyline in patients with depression: a positron emission tomography study with (S,S)-[^{18}F]FMeNER-D (2). Int J Neuropsychopharmacol 17: 553–560.

Tejani-Butt SM, Yang J, Zaffar H (1993). Norepinephrine transporter sites are decreased in the locus coeruleus in Alzheimer's disease. Brain Res 631: 147–150.

Varrone A, Bundgaard C, Bang-Andersen B (2022). PET as a translational tool in drug development for neuroscience compounds. Clin Pharmacol Ther 111: 774–785.

Wallert ED, van de Giessen E, Knol RJJ et al. (2022). Imaging dopaminergic neurotransmission in neurodegenerative disorders. J Nucl Med 63: 27S–32S.

Xiang J, Tao Y, Xia Y et al. (2023). Development of an alpha-synuclein positron emission tomography tracer for imaging synucleinopathies. Cell 186: 3350–3367. e19.

Yi YJ, Lusebrink F, Ludwig M et al. (2023). It is the locus coeruleus! Or… is it?: a proposition for analyses and reporting standards for structural and functional magnetic resonance imaging of the noradrenergic locus coeruleus. Neurobiol Aging 129: 137–148.

Zimmer L (2023). Positron emission tomography for the discovery of new drugs in psychiatry. ACS Chem Neurosci 14: 524–526.

Handbook of Clinical Neurology, Vol. 216 (3rd series)
The Brainstem
L.T. Grinberg and A.D. Korczyn
https://doi.org/10.1016/B978-0-443-15736-3.00013-5

Chapter 22

Brain death and the brainstem

CALIXTO MACHADO*

Department of Clinical Neurophysiology, Institute of Neurology and Neurosurgery, Havana, Cuba

Abstract

Brain death/death by neurologic criteria (BD/DNC) represents a distinct state after a catastrophic brain injury. In the 19th century, anatomists and physiologists progressively began to sustain the conception of the brainstem as a mediator of vital physiologic functions. A primary supratentorial brain lesion usually provokes a rostrocaudal transtentorial brain herniation syndrome, resulting in brain function impairment in both cerebral hemispheres and the brainstem. When a secondary brain lesion (i.e., cerebral hypoxia) is present, the brainstem, like the forebrain, is also affected. Nevertheless, some patients with a primary infratentorial brain lesion may retain cerebral blood flow and electroencephalographic (EEG) activity without a significant increment of intracranial pressure. Jahi McMath was declared brain-dead, leading to a legal controversy about keeping life-support despite the BD/DNC diagnosis. She presented a huge lesion in the brainstem, but ancillary tests, performed nine months after the initial brain insult, showed conservation of intracranial structures, EEG activity, and autonomic reactivity to the "Mother Talks" stimulus, rejecting the BD/DNC diagnosis. Jahi represented a new state of disorder of consciousness, nonpreviously described, termed "responsive unawakefulness syndrome."

BRAIN DEATH

Brain death, also known as death by neurologic criteria (BD/DNC), represents a distinct state that occurs when a person experiences a catastrophic brain injury, leading to the complete and irreversible loss of all brain function. This topic has a fascinating history that has significantly shaped our understanding of life and death (Pernick, 1988; Machado, 2007).

Before modern technology, there was widespread agreement that death occurred when the heartbeat and breathing ceased and the soul left the body. Methods such as observing the absence of fog on a glass or mirror placed under the nostrils or waiting for a person to respond after being called three times by name were commonly used to document death (Pernick, 1988; Truog et al., 2018c; Grasset et al., 2023; Gambardella et al., 2024; Larson et al., 2024; Wijdicks, 2024).

However, the last century witnessed remarkable technical and scientific advances, including developing effective mechanical ventilators and cardiopulmonary resuscitation. The dramatic first cardiac defibrillation in 1947 and the introduction of mechanical ventilation in 1950 created a previously unimaginable condition: a brain that is massively damaged and nonfunctional, while other organs continue to function.

In 1954, Schwab reported a pivotal case: withdrawing treatment from a patient without reflexes and spontaneous breathing in electroencephalographic (EEG) silence. This moment marked a fundamental turning point for future developments in this field (Grasset et al., 2023).

Wertheimer and Jouvet (Wertheimer et al., 1959) aligned with Schwab on the diagnosis and treatment withdrawal for what they defined as "death of the nervous system" (Schwab et al., 1954; Alderete et al., 1968; Sament et al., 1969). They further proposed stopping the ventilator if the death of the nervous system was clinically diagnosed and confirmed by "repeatedly verified absence of electroencephalographic (EEG) activity both in the cortex and in the diencephalon," provided sufficient time for resuscitative efforts had elapsed (typically 18–24 hours).

In 1959, during the 23rd International Neurological Meeting held in Paris, two notable neurologists – Pierre Mollaret and Maurice Goulon – significantly contributed by describing a clinical condition known as "coma dépassé." This condition was based on a series of 23 patients with severe neurologic impairment who were hospitalized at the Claude Bernard Hospital in Paris. Coma dépassé manifested as a deep coma with no spontaneous respiration, absence of reflexes, polyuria,

*Correspondence to: Calixto Machado, MD, PhD, FAAN, Department of Clinical Neurophysiology, Institute of Neurology and Neurosurgery, 2t9 y D, Vedado, La Habana 10400, Cuba. Email: cmachado180652@gmail.com.

and hypotension (unless norepinephrine was continuously administered). Additionally, there was a complete absence of EEG activity. Mollaret and Goulon observed that if ventilation or norepinephrine infusion were stopped, patients would rapidly die, typically experiencing cardiac arrest within days after the establishment of the deep coma. This term describes a state beyond the deepest form of coma, characterized by a total and definitive abolition of neurologic function (Mollaret and Goulon, 1959; Goulon et al., 1983).

While Mollaret and Goulon's article represented a breakthrough in describing brain death, they did not share Wertheimer et al.'s perspective regarding the death of the nervous system. Instead, they defended the concept of cardiac death as the defining phenomenon, leading to a major controversy (Mollaret and Goulon, 1959; Wertheimer et al., 1959; Goulon et al., 1983).

Robert S. Schwab and colleagues proposed that "the total absence of EEG activity after 30 minutes of recording is the most important evidence of death of the nervous system." They were among the first to accept the diagnosis of human death based on neurologic grounds, even in the presence of a beating heart. Schwab and his team called these cases "human heart-lung preparation." Notably, they went beyond recommending EEG as an ancillary test; they incorporated EEG findings into their diagnostic criteria, similar to those established by the Harvard Committee (Schwab et al., 1963). Interestingly, the Harvard Committee (1968) participants appeared unaware of the French contribution in this area (Mollaret and Goulon, 1959; Wertheimer et al., 1959), resulting in parallel approaches to brain death.

In 1965, Swedish neurosurgeon Frykholm circulated a memorandum during a meeting arranged by the Swedish Medical Authorities to discuss transplantation rules. He suggested that the ventilator should be stopped in patients with no reflexes from the cranial nerves, an isoelectric EEG, and no blood flow through the brain, as observed in angiography. Frykholm aimed to implement this definition of death in organ transplantation, arguing that if the public accepted it, it should be included in the law. He believed that if these deceased patients were suitable as organ donors, continuing ventilation could be justified. However, similar to Wertheimer, Frykholm faced criticism for his opinion (Frykholm, 1965).

Two major medical events occurred in 1968, a crucial year for the development of the concept of BD/DNC. On August 5, the 22nd World Medical Assembly met in Sydney (Australia) and issued an essential statement on human death (Machado et al., 2007). Additionally, the Journal of the American Medical Association (JAMA) published a landmark article titled "A Definition of Irreversible Coma," which was the Ad Hoc Committee of the Harvard Medical School report tasked with examining the definition of brain death (1968).

The Sydney Declaration touched on philosophic issues related to death, emphasizing that death is a gradual process at the cellular level, with tissues varying in their ability to withstand oxygen deprivation. It stated that death lies in preserving isolated cells and the person's fate.

The Harvard Report had momentous repercussions and established a paradigm for defining death based on neurologic criteria. Following the Harvard Committee's Report, the concept of brain death gained widespread acceptance.

Interestingly, before the Harvard criteria emerged in 1968, Belgian surgeon Guy Alexandre, from the Catholic University of Louvain, introduced a set of brain death criteria based on the description of "coma dépassé." On June 3, 1963, Alexandre and his team performed the first transplant in Belgium. Notably, they did not discontinue mechanical ventilation and waited for the donor's heart to stop beating. Their transplantation marked the first use of a heart-beating, brain-dead donor.

The idea of BD/DNC was gradually conceptualized over the following years on both sides of the Atlantic Ocean and later codified by the Harvard Ad Hoc Committee in 1968 and endorsed by the World Medical Association. In 1976, the U.K. Conference of Medical Royal Colleges and their Faculties published guidance on diagnosing BD/DNC, indicating that the loss of brainstem function is pivotal to the state of BD/DNC, as described in the Harvard expert opinion (1976). Three years later, a follow-up memorandum proposed brainstem death as equivalent to the death of the whole person (1979). In the United States, the National Institutes of Neurological Diseases and Stroke suggested adding an assessment of cerebral blood flow (CBF) to the Harvard Criteria, giving additional validation to BD/DNC as a clinical entity (1977). National legal validation was provided in 1981 by the President's Commission for the Study of Ethical Problems in Medicine and Biomedical and Behavioral Research (1981). The Commission adopted the conceptualization of BD/DNC as proposed by the senior author James Bernat (Bernat et al., 1981a, 1981b) and legally sanctioned it with the Uniform Determination of Death Act (UDDA) is a model state law that was approved for the United States in 1981 by the National Conference of Commissioners on Uniform State Laws, in cooperation with the American Medical Association, the American Bar Association, and the President's Commission for the Study of Ethical Problems in Medicine and Biomedical and Behavioral Research (1981). The American Academy of Neurology (AAN) standardized the diagnostic criteria for BD in 1995 (Wijdicks, 1995), in 2010 (Wijdicks et al., 2010), and again in 2023 (Greer et al., 2023).

Were these patients alive or dead? This radically changed the course of the debates about human death, marking a turning point when brain-oriented definitions of death started to be formulated, and BD/DNC was gradually accepted as the death of the individual (Machado, 2007).

Historically, there are three conceptual formulations of BD/DNC: the "whole brain" formulation, the "brainstem" formulation, and the "higher brain" formulation (Machado, 1994). Any full account of death should include three distinct elements: the definition of death, the criterion (anatomic-functional substratum) of BD/DNC, and the tests to prove that the criterion has been satisfied. Undoubtedly, the term "criterion" for referring to the anatomic-functional substratum introduces confusion in this discussion because protocols of tests (clinical and instrumental) for BD/DNC diagnosis are called "diagnostic criteria" or "sets of diagnostic criteria."

Therefore, I will use the term "anatomic-functional substratum" instead of criterion (Machado, 1994, 1999, 2003).

The "higher brain" formulation, which focuses on consciousness and cognition (Veatch, 1992, 1993), is related to the function of the cortex rather than brainstem function (Gambardella et al., 2024). However, it is not medically or legally recognized in any country worldwide. On the other hand, the "brainstem" formulation concentrates solely on brainstem function and was initially defended by Mohandas and Chou (1971) with further development by Christopher Pallis (Pallis and Macgillivary, 1981; Pallis, 1982, 1987).

James Bernat and his collaborators have presented the most comprehensive defense of the "whole brain" *formulation*, considering both supratentorial and infratentorial structures (Bernat, 1987, 2004, 2006, 2013, 2014, 2023a, 2023b, 2024). The ongoing debate between the "whole brain" and "brainstem" formulations has spanned over 40 years and is often referred to as the "transatlantic divide." The United Kingdom primarily supports the "brainstem" formulation, while the United States advocates for the "whole brain" formulation (Wijdicks, 2012; Machado, 2019). However, this distinction becomes relevant mainly when evaluating patients with posterior fossa lesions (Machado, 2022c).

The brainstem standard was adopted in several Commonwealth countries. Pallis emphasized that the capacity for consciousness and respiration are two hallmarks of human life, and that brainstem death predicts an inescapable asystole (Pallis, 1990, 1995; Machado, 2023a).

1. Definition
   There is only one kind of human death: the irreversible loss of the capacity for consciousness, combined with the irreversible loss of the capacity to breathe (and hence to sustain a spontaneous heartbeat).
2. Anatomic-physiologic substratum
   The permanent cessation of the functioning of the brainstem.
   Pallis postulated that the capacity for consciousness and respiration are two hallmarks of life for human beings and that brainstem death predicts an inescapable asystole. This author emphasized that the ascending reticular formation gives rise to a generalized activation of the cortex, producing the necessary arousal to endow the functioning of the "brain as a whole." He considered that the physiologic and anatomic mechanism to abolish the capacity for consciousness is the irreversible damage of the paramedial tegmental areas of the mesencephalon and rostral pons (Pallis, 1990).
   For centuries, respiration was the crucial function defining the frontiers between life and death. In several ancient cultures, death was considered "the departure of the soul from the body"; hence, the words that stand for the soul are, in many idioms, the same as those standing for breath (Pernick, 1988). Pallis also documented that the "loss of breath" or apnea relies on irreversible damage of the lower brainstem, where "crucial mechanisms concerned with breathing are located" (Pallis, 1987).
3. Tests
   According to Pallis, "brainstem death is a clinical concept," and therefore, "a dead brainstem can be diagnosed at the bedside." The procedure is to diagnose an unconscious patient with irreversible apnea and irreversible loss of brainstem reflexes, provided that all reversible causes
   However, a physiopathologic review of consciousness generation will provide a basis for not accepting Pallis' definition of death (Machado, 1999). Moreover, recent clinical cases have shown that brain death does not always predict an "inevitable asystole within a short while" (Shewmon, 1998).
   The World Brain Death Project (WBDP) published an international consensus statement that provides historical context and recommendations for BD/DNC determination (Greer et al., 2020). In 2023, the ANN collaborated with the American Academy of Pediatrics, Child Neurology Society, and Society for Critical Care Medicine to formulate an updated, evidence-informed, consensus-based guideline for pediatric and adult brain BD/DNC determination (Greer et al., 2023). This guideline aims to (1) ensure that recommendations are conservative yet practical, emphasizing circumstances where BD/DNC determination should be delayed or deferred to reduce the risk of false-positive determinations, and (2) offer guidance on aspects of BD/DNC determination that clinicians find thought-provoking or contentious.
   The UDDA revision series in *Neurology* originated in response to the plan of the Uniform Law Commission to create a revised Uniform Determination of Death Act (rUDDA) to address contemporary controversies associated with BD/DNC determination (Lewis, 2022). Although other articles in this series use the term "brain death," the term "brain death/death by neurologic criteria" (or "BD/DNC") is used in this article, which concludes the series, to both embrace the colloquial term and reinforce its equivalency to death by circulatory-respiratory criteria, consistent with the approach taken by the WBDP, a 2020 international consensus statement on BD/DNC determination (Greer et al., 2020). Although it is unknown whether and how the rUDDA will address each of these controversies, this article contextualizes these and other controversies. It reviews the extent to which they represent potential threats and impediments to the clinical practice of BD/DNC determination. It also explains why our rapidly evolving understanding of the brain's ability to recover from injury should not affect the clinical practice of BD/DNC determination. Finally, it explores the myriad ways in which the AAN has addressed potential threats and impediments to the clinical practice of BD/DNC determination, as well as the implications that potential changes to the UDDA may have for the future of this practice.
   In summary, while revising the UDDA is necessary, the rUDDA's proposed changes raise valid concerns. The ongoing debate underscores the importance of clarity, accuracy, and respect for individual perspectives in defining death (Bernat, 2023a, 2023b; Machado, 2023a, 2023b, 2023c, 2023d; Bernat, 2024; Machado, 2024).

Fig. 22.1. The term "consciousness" was defined by William James in 1890 "as awareness of the self and the environment." Photo of William James: Courtesy of the National Library of Medicine.

THE BRAINSTEM AND CONSCIOUSNESS GENERATION

The term "consciousness" was defined by William James in 1890 as awareness of the self and the environment: "Normal consciousness entails a state of being awake and aware of self and environment" (James, 1950) (Fig. 22.1).

Plum and Posner also provided a similar definition of consciousness: "the state of awareness of self and the environment" (Plum and Posner, 1972, 2015).

Two physiologic components govern conscious behavior: Arousal and awareness (Zeman, 2001; Korein and Machado, 2004; Machado, 2007; Bostanciklioglu, 2020, 2021; Leisman and Machado, 2021; Yu et al., 2023; Zheng et al., 2023). Arousal represents several behavioral changes occurring when a person awakens from sleep or transits to a state of alertness (Fig. 22.2).

The ascending reticular activating system is closely linked to arousal, one component of consciousness (Kwak and Chang, 2020; Jang and Kwon, 2022; Taran et al., 2023) (Fig. 22.3).

Awareness relies mainly on the functional integrity of the cerebral cortex. Nonetheless, both components of consciousness are based on networks extensively located throughout the brain (Mountcastle, 1957, 1978, 1997; Mountcastle et al., 1957) (Fig. 22.4).

NEW STANDARD OF HUMAN DEATH

Based on the pathophysiologic mechanisms of consciousness generation, I proposed a new standard of death on neurologic grounds.

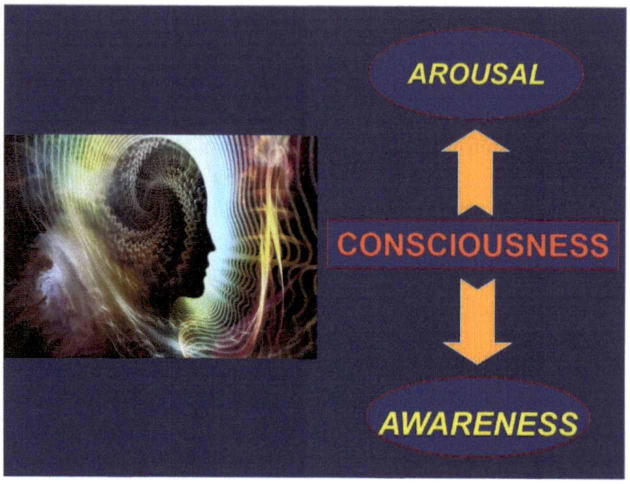

Fig. 22.2. Components of consciousness. Figure **designed by Dr. Calixto Machado**.

1. Definition
   Irreversible loss of the consciousness which provides the key human attributes and the highest level of control in the hierarchy of integrating functions within the human organism.
2. Anatomic-physiologic substratum
   Irreversible destruction of the anatomic and functional substratum of consciousness throughout the whole brain.
   Substantial interconnections between the brainstem, subcortical structures, and the neocortex are essential for subserving and integrating both components of human consciousness. Consequently, consciousness generation is based on anatomy and physiology throughout the brain (Machado, 1994, 1999; Machado et al., 2009; Machado, 2022b).
3. Tests
   Unresponsiveness, no arousal to stimuli, and no cognitive and affective functions.

BRAINSTEM LESIONS AND THE DETERMINATION OF BRAIN DEATH/DEATH BY NEUROLOGIC CRITERIA

A primary lesion localized to the posterior fossa sets a controversial debate on BD/DNC determination. A primary supratentorial brain lesion usually provokes a rostrocaudal transtentorial brain herniation syndrome, resulting in brain function impairment in both cerebral hemispheres and the brainstem. When a secondary brain lesion (i.e., cerebral hypoxia) is present, the brainstem, like the forebrain, is also affected. Nevertheless, some patients with a primary infratentorial brain lesion (e.g., in basilar artery thrombosis or large brainstem or cerebellar bleeds), without a significant increment of intracranial pressure, may retain CBF and EEG activity (Varelas and Lewis, 2016; Varelas, 2016, Varelas et al., 2017; Machado and Estevez, 2018; Machado et al., 2018; Walter et al., 2018; Machado, 2019, 2020, 2022c; Manara et al., 2019; Roth et al., 2019).

Fig. 22.3. Giussepe Moruzzi and Horace Winchell Magoun discovered the ascending reticular activating system. From National Library of Medicine, with permission.

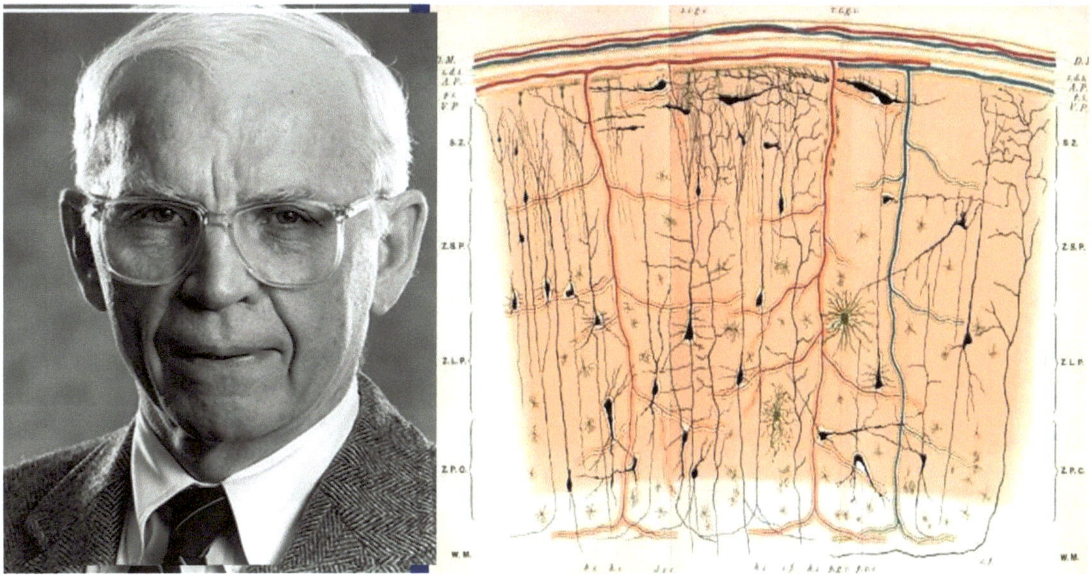

Fig. 22.4. Cortical columns of the cerebral cortex. Figure designed by Dr. Calixto Machado. Photo courtesy: National Library of Medicine, with permission.

Grigg et al. studied 56 patients fulfilling BD/DNC clinical criteria, and 20% preserved EEG activity, suggesting that EEG is not required as an ancillary test to confirm BD/DNC. Nonetheless, the apnea test was not performed in one-third or more of the cases, and these authors did not describe which PCO_2 thresholds they took to confirm BD/DNC in the remaining patients. Pathologic studies were realized in only two cases demonstrating ischemic brainstem lesions, relatively sparing the cerebral cortex. The pathologic examination of the rest of the cases was not described (Grigg et al., 1987).

Ferbert reported preservation of EEG and visual evoked potentials in a patient with a hematoma in the cerebellum and the pons who fulfilled brainstem clinical BD/DNC criteria. Hence, this author stated that EEG should be mandatory for BD/DNC diagnosis in the presence of posterior fossa lesions (Ferbert et al., 1986). Esteban et al. also reported five patients who fulfilled clinical BD/DNC criteria and showed long-lasting EEG activity. Three of those patients had primary posterior fossa lesions (Esteban et al., 1995).

From a series of 161 brain-dead cases, Varela et al. studied three cases (represented 1.9%). They added another patient from a different hospital (Varelas et al., 2017). These four patients presented posterior fossa injuries fulfilling the UK-BD/DNC clinical criteria, including the apnea test. All four patients suffered from catastrophic posterior fossa injuries and therefore fulfilled UK-BD/DNC criteria, including the apnea test. These four patients showed preservation of supratentorial CBF, which vanished after a period of between 2 and 6 days, then permitting BD/DNC diagnosis, according to the US accepted standard of whole brain criteria. These authors concluded that fulfilling US-BD/DNC clinical criteria would characteristically progress from conserving in patients complaining of primary posterior fossa catastrophic lesions. Therefore, the authors stated that if CBF assessment is used as a confirmatory test, those patients are not dissimilar from those brain-dead patients with supratentorial lesions. Nonetheless, the challenge of the cases mentioned above focuses on determining when the patients were brain-dead according to the US or UK-BD/DNC criteria.

The main difference in applying "whole brain" or "brainstem" standard BD/DNC criteria is precisely when a primary posterior fossa lesion is present (Machado et al., 2018).

Wijdicks commented about a study of 56 cases fulfilling clinical BD/DNC criteria, that 20% retained EEG activity, lasting up to 168 hours. This author also stated that the EEG sensitivity and specificity in BD/DNC determination is about 90% (Wijdicks, 2015). This author remarked there is not one patient fulfilling brainstem death criteria who has survived (Wijdicks, 2012), but this statement is related to prognosis and not to diagnosis (Machado, 2007).

Varela et al. commented that in the case of isolated brainstem lesions, sparing the mesopontine tegmental reticular formation, this condition would theoretically lead to a fully apneic locked-in syndrome (LIS), which imitates brainstem death (Varelas, 2016; Manara et al., 2019). Walter et al. affirmed that a patient with a primary posterior fossa injury could retain some degree of consciousness, lasting some time (Walter et al., 2018).

Varela et al. also stated that there are no reports of brain-dead patients diagnosed correctly in either the whole brain or brainstem views who have shown consciousness recovery or breathing recommence. Then, they erroneously stated that persistent supratentorial CBF or EEG activity does not indicate remaining brain function (Varelas et al., 2017). This was not the case in Jahi McMath (Machado et al., 2017, 2018).

Moreover, experimental studies in animals with critical neuronal damage to the mesopontine tegmental reticular formation have shown that a quasi-normal EEG, exhibiting significant alpha, beta, or theta activity, may still occur. Therefore, these authors remarked that confirmatory tests should be mandatory for diagnosing BD/DNC to find no EEG activity or, even better, no CBF. Thus, BD/DNC diagnosis has to be doubted in this condition, and as long as EEG activity is present, it cannot be excluded that patients might be at least partially conscious (Laureys, 2005; Walter et al., 2015, 2018). Arguing about brainstem death formulation, Bernat points out the possibility of a "profound locked-in syndrome" in which awareness might be retained in the absence of brainstem activity (Bernat, 2002).

These reports of near-normal EEG preserved supratentorial CBF and cortical visual evoked potentials in cases with posterior fossa injury, leading to harsher BD/DNC codes in Europe. When a primary posterior fossa lesion is present, stricter BD/DNC protocols mandate the demonstration of electrocortical silence and the confirmation of absent cerebral blood flow, assessed by cerebral angiography, perfusion scintigraphy, or Doppler sonography (Frowein et al., 1987; Haupt, 1987, 1996; Rudolf et al., 1998; Sawicki et al., 2015, 2018, 2019; Forderreuther and Angstwurm, 2003).

THE JAHI MCMATH CASE AND A LESION IN THE BRAINSTEM

Jahi McMath is a relevant case to discuss the need to update the classification of disorders of consciousness (DoCs) (Paola, 2014; Lewisg, 2015; Luce, 2015; Yanke et al., 2016; Lewis, 2018a, 2018b; Machado et al., 2018; Truog, 2018a, 2018b; Machado, 2020c; Scott, 2020; Shewmon and Salamon, 2021; Machado, 2021a). She underwent pharyngeal surgery for obstructive sleep apnea and suffered a massive hemorrhage inside her respiratory ways, leading to a cardiorespiratory arrest. Jahi McMath was diagnosed as being in a BD/DNC. This case promoted a substantial legal debate between Jahi McMath's relatives and the hospital.

Dr. Calixto Machado was invited as an expert advisor. Jahi McMath was studied at Rutgers Hospital. Dr. Charles Prestigiacomo, a US-licensed neurologist, examined her and prescribed the ancillary tests Dr. Machado suggested. EEGs were also recorded at the apartment where Jahi stayed in New Jersey.

After nine and a half months of her initial diagnosis, these were the findings in Jahi McMath (Machado et al., 2017, 2018).

She presented a massive lesion in the brainstem, but ancillary tests, performed nine months after the initial brain insult, showed conservation of intracranial structures, EEG activity, and autonomic reactivity to the "Mother Talks" stimulus, rejecting the BD/DNC diagnosis. Therefore, Jahi represented a new state of DoC, nonpreviously described, termed "responsive unawakefulness syndrome" (RUS).

The main findings in Jahi McMath were: Her magnetic resonance imaging (MRI) showed preservation of intracranial structures despite enormous abnormalities; EEG activity over 2 μV of amplitude was recorded. Moreover, the delta-theta range's predominant activity was found. Heart rate variability (HRV) methodology showed preservation of all HRV bands; "Mother Talks" stimulation demonstrated autonomic reactivity.

Preservation of intracranial structures

A striking finding was preserving intracranial structures in the brainstem and cerebral hemispheres (Figs. 22.5 and 22.6). This was observed nine months after cardiac arrest, resulting in severe brain injury. It has been extensively described that a brain-dead patient has a complete absence of CBF. Moreover, by using MRI diffusion tensor imaging, (Fig. 22.7), which is a MRI technique that uses anisotropic diffusion to estimate the axonal (white matter) organization of the brain, it was possible to demonstrate showed preservation of tracts from the brainstem to the cerebral cortex

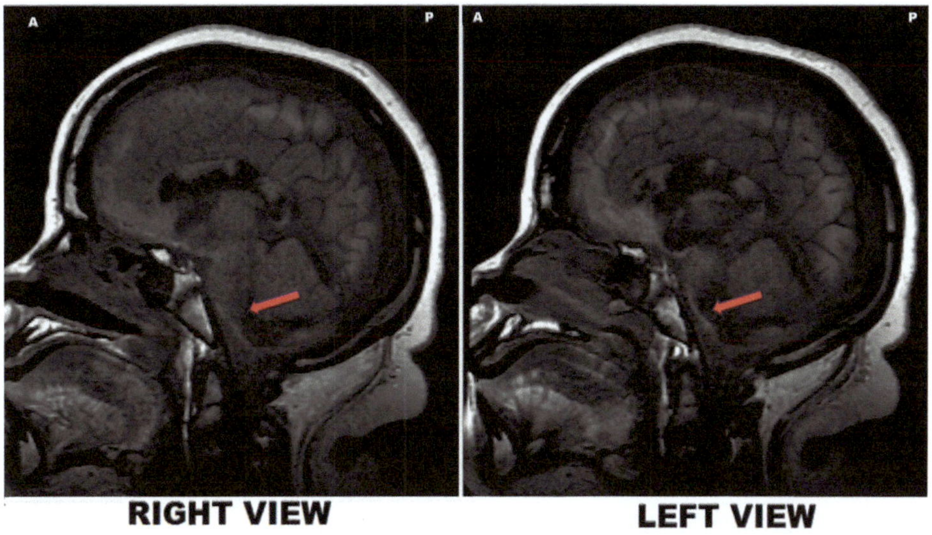

Fig. 22.5. MRI T1 axial planes (right and left views). Red arrows reveal a midline cleft in the medulla and pons, more lateralized to the left side. From Machado C, Defina P, Estevez M, et al. (2017). A reason for care in the clinical evaluation of function on the spectrum of consciousness. J Funct Neurol Rehabil Ergon Rehabil 4: 542–556, with permission.

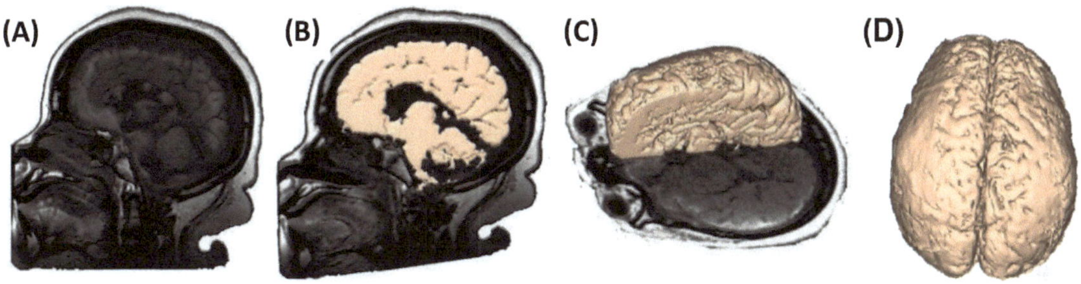

Fig. 22.6. MRI of Jahi McMath. (**A**) T1 MRI sagittal slice, (**B** and **C**) 3D reconstruction of the left hemisphere plotted over axial and sagittal slices, and (**D**) 3D reconstruction, upper view. From Machado C, Defina P, Estevez M, et al. (2017) A reason for care in the clinical evaluation of function on the spectrum of consciousness. J Funct Neurol Rehabil Ergon Rehabil 4: 542–556, with permission.

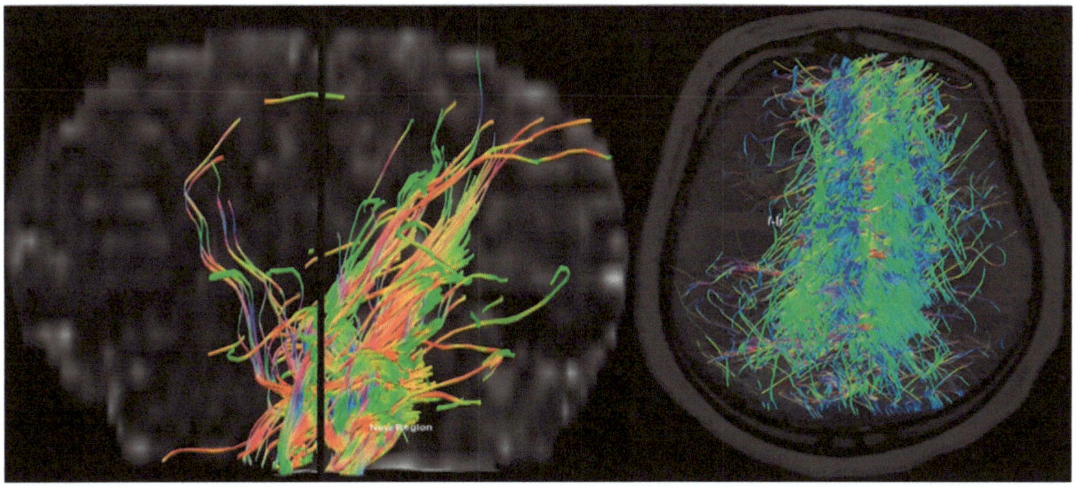

Fig. 22.7. MRI diffusion tensor imaging shows preservation of tracts from the brainstem to the cerebral cortex in a frontal view (left panel) and in the cerebral cortex (axial view – right panel). From Machado C, Defina P, Estevez M, et al. (2017) A reason for care in the clinical evaluation of function on the spectrum of consciousness. J Funct Neurol Rehabil Ergon Rehabil 4: 542–556, with permission.

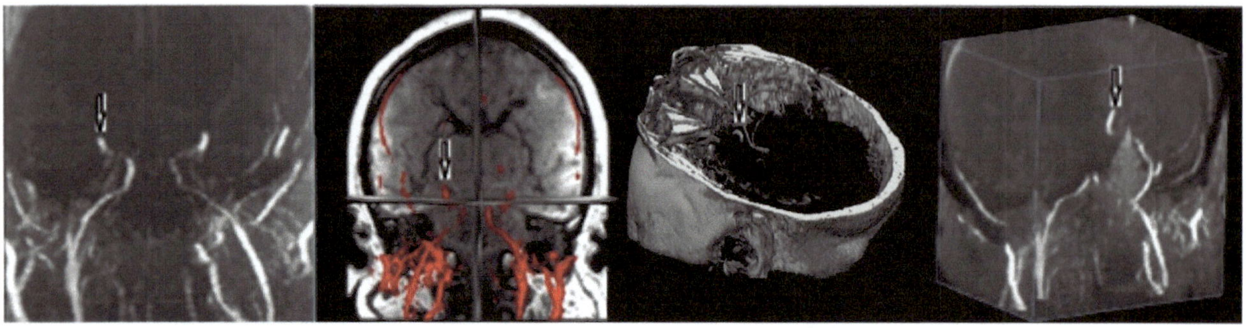

Fig. 22.8. Magnetic resonance angiography (MRA) is shown in Panel **A**. Panels **B**, **C**, and **D** show MRA-3D reconstructions. The arrows show the persistence of flow in the supraclinoid segment of the internal carotid artery. From Machado C, Defina P, Estevez M, et al. (2017) A reason for care in the clinical evaluation of function on the spectrum of consciousness. J Funct Neurol Rehabil Ergon Rehabil 4: 542–556, with permission.

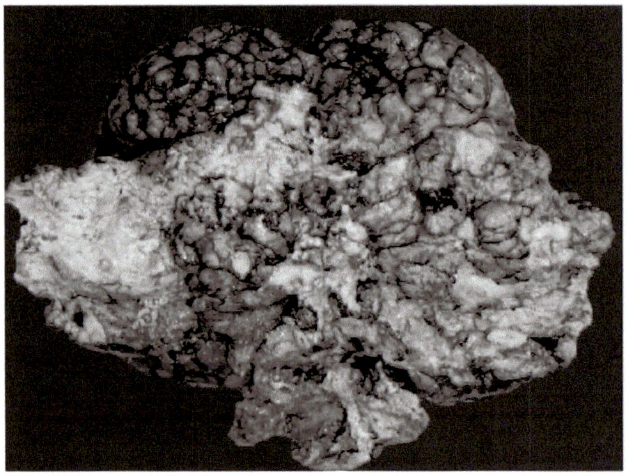

Fig. 22.9. Respirator brain from a brain-dead patient. Autopsy was performed after 10 days from the brain death/death by neurologic criteria diagnosis. The brain is fragmented, dusky, congested, and discolored, containing liquid portions. There are extensive destructive lesions with cerebral edema, brain herniation, and mass effect. Case from Dr. Calixto Machado's database.

and inside the cortex (Fig. 22.7) (Machado et al., 2017; Shewmon and Salamon, 2021; Machado et al., 2022c).

The MRA showed patency of external carotid artery branches, including the middle meningeal arteries (Fig. 22.8). Still, the only visualizable intracranial flow was in the supraclinoid segment of the internal carotid and ophthalmic arteries. No blood flow in the dural sinuses or the deep venous system was evident on MRA.

After nine and a half months from her initial BD/DNC diagnosis, we would have expected a "respirator brain" finding (Fig. 22.9).

Preservation of electroencephalographic activity

EEG bioelectrical activity was found in this case over 2 μV of amplitude. The power spectral analysis showed a predominant activity within the delta-theta range, demonstrating that this patient did not show a pattern of "electrocerebral inactivity" (Machado et al., 2018; Machado, 2021a, 2021b; Machado, 2021c; Machado, 2022a, 2022b) (Figs. 22.10 and 22.11).

Preservation of the central nervous autonomic system function

Using HRV methodology, it was possible to reveal the conservation of all HRV bands in this patient. The loss of all HRV power has characterized BD/DNC (Machado, 2010, Machado et al., 2011; Del Giudice et al., 2016) (Fig. 22.12).

Another key finding, in this case, was the autonomic reactivity to "Mother Talks" stimulation compared to the aged-matched voices of other women. These findings in Jahi McMath demonstrated the remaining function at different levels of the central autonomic system and suggested recognition of the mother's voice based on residual linguistic processing (Machado, 2020).

Therefore, our findings on autonomic reactivity to the "Mother Talks" stimulus might explain the video findings reported by Dr. Shewmon, who observed Jahi's movements, which he interpreted as volitional responses to commands and might be consistent with cognitive awareness (Machado, 2020).

Jahi McMath with a massive lesion in the brainstem

As previously noted, Jahi McMath's MRI demonstrated a massive lesion in the pons extending to the medulla. Hence, several reasons might explain the intermittent conscious responses in this patient: relative integrity of the upper part of the brainstem, paramedian thalamus, and cortex, as well as the partial sparing of the mesopontine tegmental reticular formation.

Therefore, after analyzing Jahi McMath's results, she did not fulfill any state in describing the DoCs. Jahi McMath was diagnosed as being brain-dead because her clinical examination demonstrated a sleep-like state of unarousable

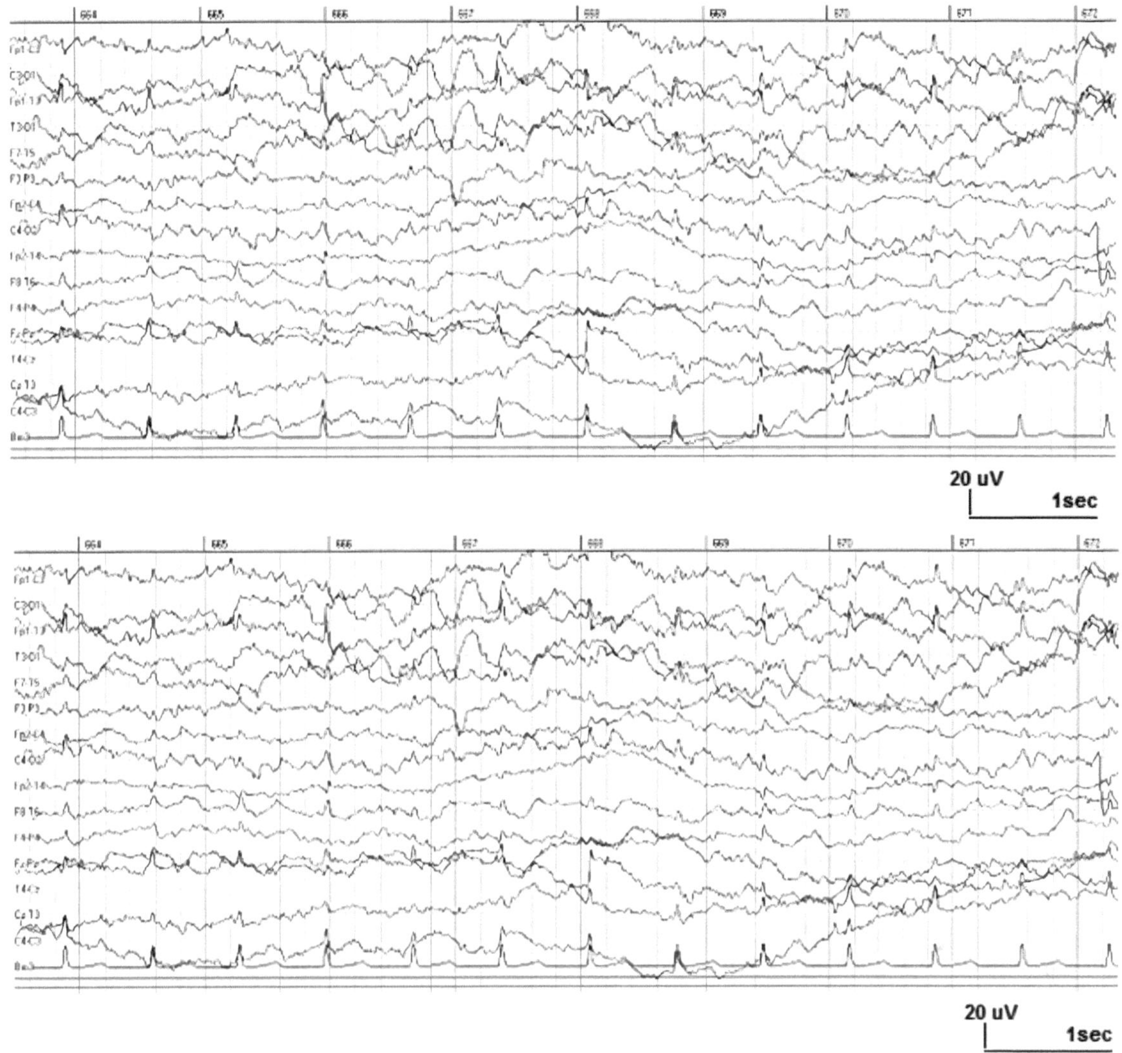

Fig. 22.10. QEEG sample from the home recording on 09/01/2014 (fragments 7 from "Exhibit B" of the court declaration of Elena Labkovsky, PhD). Recording equipment was a Mitsar 21-Channel EEG Amplifier with WinEEG software, with bandpass of 0.53–30 Hz and electrode impedances between 100 and 10,000 Ω.

unresponsiveness without evidence of awareness of self or environment, and showed the absence of brainstem reflexes, no spontaneous respiratory drive (apnea), and required permanent mechanical ventilation. Nonetheless, ancillary tests performed nine months after the initial brain insult showed conservation of intracranial structures, EEG activity, and autonomic reactivity to the "Mother Talks" stimulus, demonstrating preserved function in both the brainstem and the cerebral hemispheres, all of which speak against a diagnosis of BD/DNC (Machado et al., 2017).

She was not comatose because although she was clinically in a sleep-like state of unarousable, unresponsiveness without evidence of awareness of self or environment, and an entire absence of brainstem reflexes rejected the possibility of being in a coma (Arseni et al., 1976; Demertzi et al., 2008, 2013a, 2013b; Di Perri et al., 2016; Alnagger et al., 2023; Bauer et al., 2023).

Jahi McMath was not an unresponsive wakefulness syndrome (UWS) patient because Jahi was not in a wakefulness state and showed partial responsiveness. Moreover, UWS patients preserve full or partial brainstem reflexes and usually breathe independently without mechanical ventilation (Machado, 2020, 2022a, 2022b, 2022c; Comanducci et al., 2023; Goldstein and Poduri, 2024; Golden et al., 2024).

Jahi McMath did not have a LIS because she was not awake and only showed partial responsiveness (Machado et al., 2018; Machado, 2020, 2021b, 2022a, 2022b, 2022c). Moreover, LIS patients, although quadriplegic, usually preserve full or partial

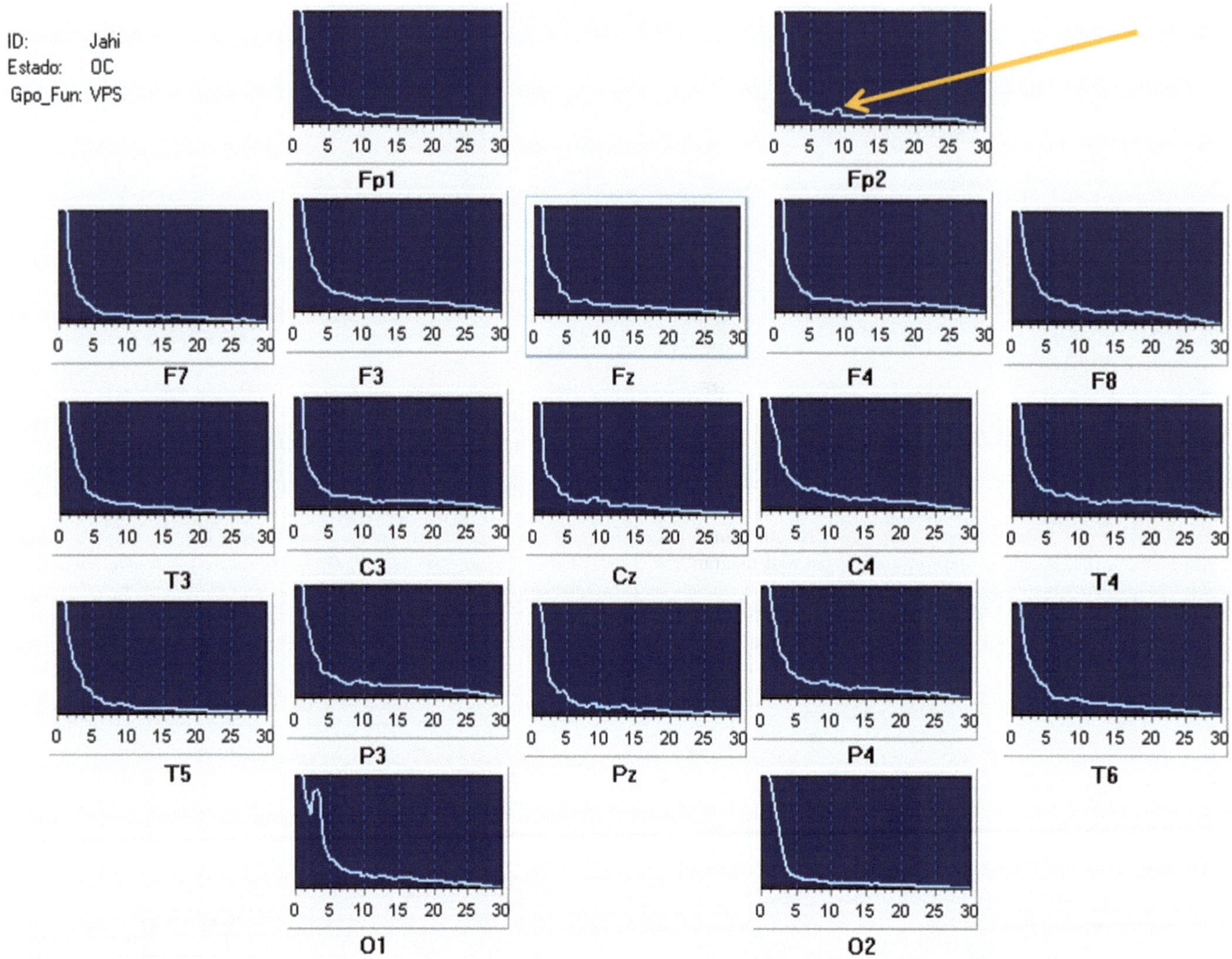

Fig. 22.11. Machado et al.'s QEEG processing was done with permission (Machado et al., 2017). The QEEG spectrum predominates activity within the delta and theta bands. The Fp2 derivation peaks at 9 Hz, and the O1 lead peaks at 3 Hz. From Machado C, Defina P, Estevez M, et al. (2017) A reason for care in the clinical evaluation of function on the spectrum of consciousness. J Funct Neurol Rehabil Ergon Rehabil 4: 542–556, with permission.

brainstem reflexes, vertical eye movements, and/or blinking, and most patients respire on their own, not requiring mechanical ventilation (Laureys et al., 2004, 2005; Bruno et al., 2009, 2013; Bodart et al., 2018). However, LIS patients may also be apneic with permanent or transient failure of triggering the ventilator (Wijdicks, 2019).

Jahi McMath was not a minimally conscious state (MCS) patient, but this requires a more detailed discussion. MCS is a condition of severely altered consciousness characterized by minimal, inconstant, yet definite behavioral signs of awareness of self and surroundings. Based on their purposeful behavioral signs, MCS patients were recently subcategorized as MCS plus (showing command following, intelligible verbalizations, or nonfunctional communication). Hence, Jahi McMath was not either an MCS or an MCS plus (Machado et al., 2017).

Therefore, Jahi McMath was not brain-dead or in an unresponsive wakefulness syndrome (UWS), previously termed persistent vegetative state, in a minimally conscious state (MCS), or a LIS (Machado et al., 2017; Machado, 2020, 2022c).

In conclusion, Jahi McMath represented a new state of DoC, nonpreviously described, termed "responsive unawakefulness syndrome" (RUS) (Machado, 2023a, 2023b, 2023c, 2023d, 2024) (Figs. 22.13 and 22.14).

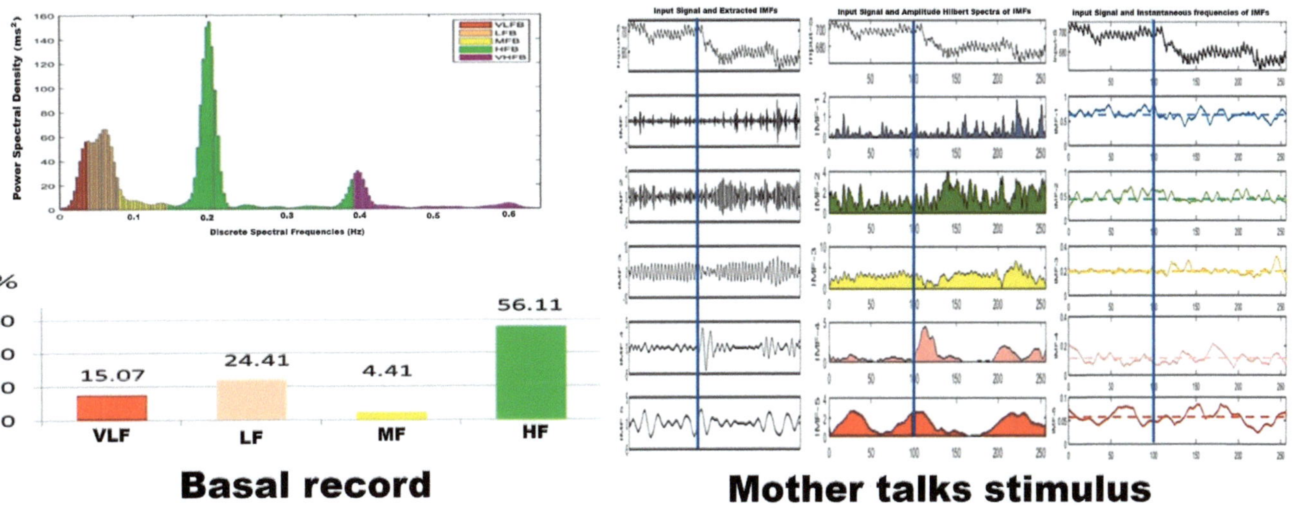

Fig. 22.12. The heart rate variability (HRV) power spectral density spectrum. The left panel shows spectral frequencies within the very low-frequency (VLF), low-frequency (LF), middle-frequency (MF), and high-frequency (HF) bands. The right panel shows the valid intrinsic mode functions (*IMFs*) (left column), the instantaneous spectral amplitudes (middle column), and the instantaneous frequency (right column) values of the respective IMFs. Tacograms are present in the first diagrams of the three columns. A thick blue vertical line indicates the beginning of the "Mother Talks" stimulus vs the "Basal Record." Comparing both experimental conditions, a seeming dynamic in the different HRV frequencies is present. From Machado C, Defina P, Estevez M, et al. (2017). A reason for care in the clinical evaluation of function on the spectrum of consciousness. J Funct Neurol Rehabil Ergon Rehabil, 4: 542–556, with permission.

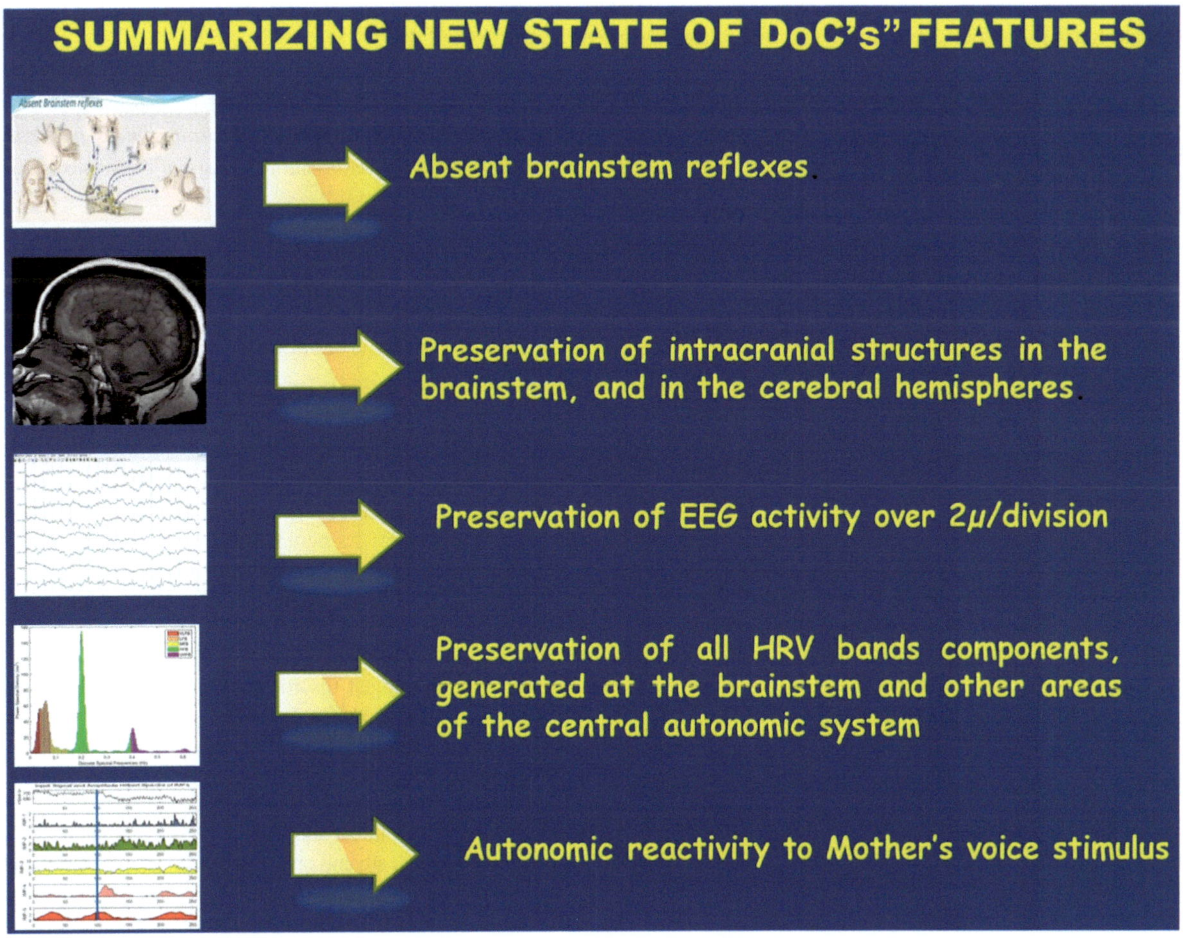

Fig. 22.13. Summary of the new state of disorder of consciousness (DoC) features. Figure designed by Dr. Calixto Machado.

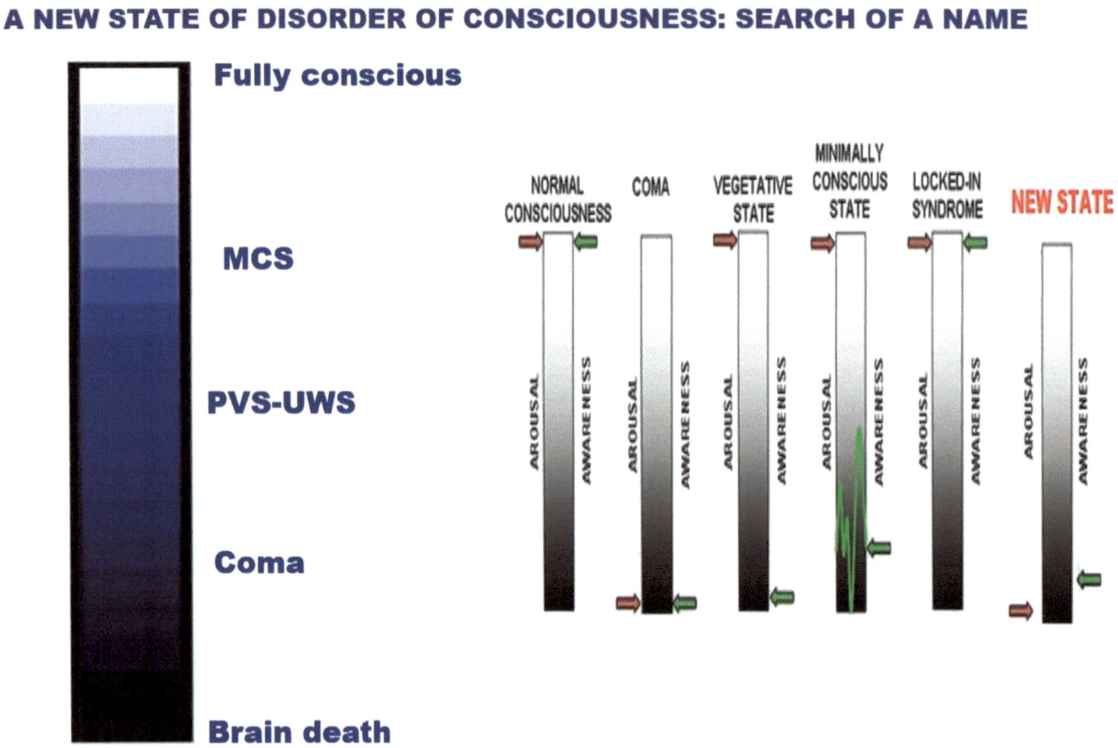

Fig. 22.14. Comparison of the known disorders of consciousness (DoCs) with the new-state DoCs. Modified from Laureys S, Owen AM, Schiff ND (2004) Brain function in coma, vegetative state, and related disorders. Lancet Neurol 3: 537–546.

CONCLUSION

The Jahi McMath case demonstrated the importance of the brainstem for diagnosing BD/DNC and DoCs. The symptoms associated with brainstem lesions are often severe due to the critical functions housed within this area, including long tract signs, cranial nerve deficits, and cerebellar signs.

Understanding the anatomy and physiology of the brainstem is crucial for accurately diagnosing BD/DNC and differentiating various DoCs. When Jahi McMath was studied, she did not fit into any classification of BD/DNC or DoCs.

The assertion that Jahi McMath was in a new state of DoC underscored the complexities and limitations of current medical definitions of BD/DNC. This perspective suggested that Jahi's neurologic condition did not fit neatly into existing categories, prompting a call for further research and potentially reexamining how such conditions are classified and understood. This case catalyzes ongoing discussion about the role of studying the brainstem in neurology.

This chapter demonstrates that studying the anatomy and physiology of the brainstem is indispensable in neurology. It is fundamental for diagnosing and managing neurologic disorders, understanding and treating DoCs, determining BD/DNC, and advancing research and treatment methodologies. Comprehensive knowledge of the brainstem enhances clinical practice and improves patient outcomes, making it a cornerstone of neurologic science.

Importance:

I review the importance of the brainstem to understand the concept of brain death and disorders of consciousness.

DISCLOSURE

The author reports no conflict of interest.

References

Alderete JF, Jeri FR, Richardson EP Jr et al. (1968). Irreversible coma: a clinical, electroencephalographic and neuropathological study. Trans Am Neurol Assoc 93: 16–20.

Alnagger N, Cardone P, Martial C et al. (2023). The current and future contribution of neuroimaging to the understanding of disorders of consciousness. Presse Med 52: 104163.

Arseni C, Golu M, Danaila L et al. (1976). Consciousness, awareness and their prolonged disorders. A philosophical study and terminology. Rev Med Interna Neurol Psihiatr Neurochir Dermatovenerol Neurol Psihiatr Neurochir 21: 241–252.

Bauer ZA, De Jesus O, Bunin JL (2023). Unconscious patient. StatPearls [Internet]. Treasure Island, FL: StatPearls Publishing.

Bernat JL (1987). Ethical and legal aspects of the emergency management of brain death and organ retrieval. Emerg Med Clin North Am 5: 661–676.

Bernat JL (2002). The biophilosophical basis of whole-brain death. Soc Philos Policy 19: 324–342.

Bernat JL (2004). On irreversibility as a prerequisite for brain death determination. In: Machado C, Shewmon DL (eds.) Brain death and disorders of consciousness. New York: Kluwer Academic/ Plenum Publishers.
Bernat JL (2006). The whole-brain concept of death remains optimum public policy. J Law Med Ethics 34: 35–433.
Bernat JL (2013). On noncongruence between the concept and determination of death. Hastings Cent Rep 43: 25–33.
Bernat JL (2014). Whither brain death? Am J Bioeth 14: 3–8.
Bernat JL (2023a). The brain-as-a-whole criterion and the Uniform Determination of Death Act. AJOB Neurosci 14: 271–274.
Bernat JL (2023b). Challenges to brain death in revising the Uniform Determination of Death Act: The UDDA Revision Series, 101, 30-37.
Bernat JL (2024). Author response: challenges to brain death in revising the Uniform Determination of Death Act: the UDDA Revision Series. Neurology 102: e208045.
Bernat JL, Culver CM, Gert B (1981a). Definition of death. Ann Intern Med 95: 652.
Bernat JL, Culver CM, Gert B (1981b). On the definition and criterion of death. Ann Intern Med 94: 389–394.
Bodart O, Amico E, Gomez F et al. (2018). Global structural integrity and effective connectivity in patients with disorders of consciousness. Brain Stimul 11: 358–365.
Bostanciklioglu M (2020). Temporal correlation between neurological and gastrointestinal symptoms of SARS-CoV-2. Inflamm Bowel Dis 26: e89–e91.
Bostanciklioglu M (2021). Unexpected awakenings in severe dementia from case reports to laboratory. Alzheimers Dement 17: 125–136.
Bruno MA, Laureys S, Demertzi A (2013). Coma and disorders of consciousness. Handb Clin Neurol 118: 205–213.
Bruno MA, Schnakers C, Damas F et al. (2009). Locked-in syndrome in children: report of five cases and review of the literature. Pediatr Neurol 41: 237–246.
Comanducci A, Casarotto S, Rosanova M et al. (2023). Unconsciousness or unresponsiveness in akinetic mutism? Insights from a multimodal longitudinal exploration. Eur J Neurosci 59: 860–873.
Del Giudice R, Blume C, Wislowska M et al. (2016). Can self-relevant stimuli help assessing patients with disorders of consciousness? Conscious Cogn 44: 51–60.
Demertzi A, Soddu A, Laureys S (2013a). Consciousness supporting networks. Curr Opin Neurobiol 23: 239–244.
Demertzi A, Vanhaudenhuyse A, Bredart S et al. (2013b). Looking for the self in pathological unconsciousness. Front Hum Neurosci 7: 538.
Demertzi A, Vanhaudenhuyse A, Bruno MA et al. (2008). Is there anybody in there? Detecting awareness in disorders of consciousness. Expert Rev Neurother 8: 1719–1730.
Di Perri C, Bahri MA, Amico E et al. (2016). Neural correlates of consciousness in patients who have emerged from a minimally conscious state: a cross-sectional multimodal imaging study. Lancet Neurol 15: 830–842.
Esteban A, Traba J, Prieto J et al. Prolonged EEG activity in brainstem death. In: Machado C (ed.) Brain death (proceedings of the second international symposium on brain death). Amsterdam: Elsevier Science, BV.
Ferbert A, Buchner H, Ringelstein EB et al. (1986). Isolated brain-stem death. Case report with demonstration of preserved visual evoked potentials (VEPs). Electroencephalogr Clin Neurophysiol 65: 157–160.
Forderreuther S, Angstwurm H (2003). Re: W.F. Haupt, W. Holling: Diagnosis of brain death: medical and legal aspects under the transplantation law of the German Republic Fortschr Neurol Psychiat 2002; 70: 583-590. Fortschr Neurol Psychiatr 71: 221 author reply 221.
Frowein RA, Ganshirt H, Hamel E et al. (1987). Diagnosis of brain death in primary infratentorial brain damage. Nervenarzt 58: 165–170.
Frykholm RPM. Beträffande begreppet 'cerebral död', Stockholm 1965 [Memorandum about the concept of 'brain death', Stockholm 1965]. Frykholm K, Hjärndödsbegreppet ett svenskt dilemma [The concept of brain death, a Swedish dilemma]. Stockholm: Elanders Svenskt Tryck.
Gambardella I, Nappi F, Worku B et al. (2024). Taking the pulse of brain death: a meta-analysis of the natural history of brain death with somatic support. Eur J Neurol 31: e16243.
Golden K, Bodien YG, Giacino JT (2024). Disorders of consciousness: classification and taxonomy. Phys Med Rehabil Clin N Am 35: 15–33.
Goldstein RD, Poduri A (2024). Seizures and sudden death beyond SUDEP. Neurology 102: e208119.
Goulon M, Babinet P, Simon N (1983). Brain death or coma dépasséCare of the critically ill patient. Springer.
Grasset L, Power MC, Crivello F et al. (2023). How traumatic brain injury history relates to brain health MRI markers and dementia risk: findings from the 3C Dijon Cohort. J Alzheimers Dis 92: 183–193.
Greer DM, Kirschen MP, Lewis A et al. (2023). Pediatric and adult brain death/death by neurologic criteria consensus guideline. Report of the AAN Guidelines Subcommittee, AAP, CNS, and SCCM. Neurology 101: 1112–1132.
Greer DM, Shemie SD, Lewis A et al. (2020). Determination of brain death/death by neurologic criteria: The World Brain Death Project. JAMA 324: 1078–1097.
Grigg MM, Kelly MA, Celesia GG et al. (1987). Electroencephalographic activity after brain death. Arch Neurol 44: 948–954.
Haupt WF (1987). Multimodality evoked potentials and brain death. Prerequisites, findings and problems. Nervenarzt 58: 653–657.
Haupt WF (1996). Brain death. Med Klin (Munich) 91: 46.
James W (1950). The principles of psychology. New York: Dover.
Jang SH, Kwon YH (2022). Recovery of the ascending reticular activating system and consciousness following comprehensive management in a patient with traumatic brain injury: a case report. J Yeungnam Med Sci 39: 332–335.
Korein J, Machado C (2004). Brain death: updating a valid concept for 2004. Adv Exp Med Biol 550: 1–14.
Kwak S, Chang MC (2020). Impaired consciousness due to injury of the ascending reticular activating system in a patient with bilateral pontine infarction: a case report. Transl Neurosci 11: 264–268.
Larson NJ, Dries DJ, Blondeau B et al. (2024). Brain death/death by neurologic criteria: what you need to know. J Trauma Acute Care Surg 97: 165–174.
Laureys S (2005). The neural correlate of (un)awareness: lessons from the vegetative state. Trends Cogn Sci 9: 556–559.
Laureys S, Owen AM, Schiff ND (2004). Brain function in coma, vegetative state, and related disorders. Lancet Neurol 3: 537–546.
Laureys S, Pellas F, Eeckhout VAN et al. (2005). The locked-in syndrome: what is it like to be conscious but paralyzed and voiceless? Prog Brain Res 150: 495–511.

Leisman G, Machado C (2021). Many paths to consciousness or just one? Life in a bounded continuum. J Conscious Stud 28: 83–96.
Lewis A (2018a). The legacy of Jahi McMath. Neurocrit Care 29: 519–520.
Lewis A (2018b). Reconciling the case of Jahi McMath. Neurocrit Care 29: 20–22.
Lewis A (2022). The Uniform Determination of Death Act is being revised. Neurocrit Care 36: 335–338.
Lewisg A (2015). The history of brain death and how the Jahi McMath case will impact future brain dead patients. Med Law 34: 497–507.
Luce JM (2015). The uncommon case of Jahi McMath. Chest 147: 1144–1151.
Machado C (1994). Death on neurological grounds. J Neurosurg Sci 38: 209–222.
Machado C (1999). Consciousness as a definition of death: its appeal and complexity. Clin Electroencephalogr 30: 156–164.
Machado C (2003). A definition of human death should not be related to organ transplants. J Med Ethics 29: 201–202 author reply 202.
Machado C (2007). Brain death: a reappraisal. New York: Spinger Science+Bussiness Media, LLC.
Machado C (2010). Diagnosis of brain death. Neurol Int 2: e2.
Machado C (2019). Further thoughts about the "transatlantic divide" in brain death determination. Anaethesia 74. http://www.respond2articles.com/ANA/forums/thread/2778.asp.
Machado C (2020). Reader response: variability in reported physician practices for brain death determination. Neurology 94: 96–97.
Machado C (2021a). Jahi McMath: a new state of disorder of consciousness. J Neurosurg Sci 65: 211–213.
Machado C (2021b). Jahi McMath, a new disorder of consciousness. Rev Latinoam Bioét 21: 137–154.
Machado C (2022a). Brain death diagnosis in primary posterior fossa lesions. Neurol India 70: 670–675.
Machado C (2022b). The Jahi McMath case: first detailed study of her brain. Neurol India 70: 2235–2236.
Machado C (2022c) Brain death diagnosis in primary posterior fossa lesions. Neurol India 70: 670-675
Machado C (2023a). Christopher Agamemnon pallis and Julius m. Korein: Two main inspiring pioneers leading the U.K.-U.S. Divide on brain death determination. Interdiscip Neurosurg 31: 101672.
Machado C (2023b). Comment on: Canadian clinical practice guideline on brain death. Can J Anaesth 70: 2013–2014.
Machado, C. (2023c). Reader response: infratentorial brain injury among patients suspected of death by neurologic criteria: a systematic review and meta-analysis. 100: 494-495.
Machado C (2023d). The Uniform Determination of Death Act should be revised. Neurocrit Care 38: 212–213.
Machado C (2024). Reader response: challenges to brain death in revising the Uniform Determination of Death Act: The UDDA Revision Series. Neurology 102: e208044.
Machado C, Defina P, Estevez M et al. (2017). A reason for care in the clinical evaluation of function on the spectrum of consciousness. J Funct Neurol Rehabil Ergon Rehabil 4: 542–556.
Machado C, Estevez M (2018). Reader response: practice current: when do you order ancillary tests to determine brain death? Neurol Clin Pract 8: 364.
Machado C, Estevez M, Defina PA et al. (2018). Response to Lewis A: reconciling the case of Jahi McMath. Neurocrit Care 29: 521–522.
Machado C, Estevez M, Gutierrez J et al. (2011). Recognition of the mom's voice with an emotional content in a PVS patient. Clin Neurophysiol 122: 1059–1060 author reply 1061–2.
Machado C, Korein J, Ferrer Y et al. (2007). The Declaration of Sydney on human death. J Med Ethics 33: 699–703.
Machado C, Rodriguez R, Carballo M et al. (2009). Brain anatomy, cerebral blood flow, and connectivity in the transition from PVS to MCS. Rev Neurosci 20: 177–180.
Manara A, Varelas P, Wijdicks EF (2019). Brain death in patients with "Isolated" brainstem lesions: a case against controversy. J Neurosurg Anesthesiol 31: 171–173.
Mohandas A, Chou SN (1971). Brain death. A clinical and pathological study. J Neurosurg 35: 211–218.
Mollaret P, Goulon M (1959). Coma dépassé (preliminary memoir). Rev Neurol (Paris) 101: 3–15.
Mountcastle VB (1957). Modality and topographic properties of single neurons of cat's somatic sensory cortex. J Neurophysiol 20: 408–434.
Mountcastle V (1978). An organization principle for cerebral function: the unit module and the distributed system. In: Edelman GE, Mouncastle V (eds.) The mindful brain. Cambridge: The MIT Press.
Mountcastle VB (1997). The columnar organization of the neocortex. Brain 120: 701–722.
Mountcastle VB, Davies PW, Berman AL (1957). Response properties of neurons of cat's somatic sensory cortex to peripheral stimuli. J Neurophysiol 20: 374–407.
Pallis C (1982). ABC of brain stem death. Diagnosis of brain stem death – I. Br Med J (Clin Res Ed) 285: 1558–1560.
Pallis C (1987). Brain stem death – the evolution of a concept. Med Leg J 55: 84–107.
Pallis C (1990). Brainstem death: the evolution of a concept. Semin Thorac Cardiovasc Surg 2: 135–152.
Pallis C (1995). Further thoughts on brainstem death. Anaesth Intensive Care 23: 20–23.
Pallis C, Macgillivary B (1981). Brain death. Lancet 1: 223.
Paola FA (2014). The case of Jahi McMath: professionalism and knowing one's limitations. J Physician Assist Educ 25: 59–61.
Pernick M (1988). Back from the grave: recurring controversies over defining and diagnosing death in history. In: Zaner R (ed.) Death: beyond the whole- brain criteria. New York: Kluwer Academic.
Plum F, Posner JB (1972). The diagnosis of stupor and coma. Contemp Neurol Ser 10: 1–286.
Plum F, Posner JB (2015). The diagnosis of stupor and coma. Brain Nerve 67: 344–345.
President's Commission for the Study of Ethical Problems in Medicine and Biomedical and Behavioral Research (1981). Defining death: medical, legal and ethical issues in the determination of death. Washington, DC: US Government Printing Office.
Roth C, Ferbert A, Matthaei J et al. (2019). Progress of intracranial pressure and cerebral perfusion pressure in patients during the development of brain death. J Neurol Sci 398: 171–175.
Rudolf J, Haupt WF, Neveling M et al. (1998). Potential pitfalls in apnea testing. Acta Neurochir (Wien) 140: 659–663.
Sament S, Alderete JF, Schwab RS (1969). The persistence of the electroretinogram in patients with flat isoelectric EEGs. Electroencephalogr Clin Neurophysiol 26: 121.
Sawicki M, Solek-Pastuszka J, Chamier-Cieminska K et al. (2018). Accuracy of computed tomographic perfusion in diagnosis of brain death: a prospective cohort study. Med Sci Monit 24: 2777–2785.
Sawicki M, Solek-Pastuszka J, Chamier-Cieminska K et al. (2019). Computed tomography perfusion is a useful adjunct to computed

tomography angiography in the diagnosis of brain death. Clin Neuroradiol 29: 101–108.

Sawicki M, Solek-Pastuszka J, Jurczyk K et al. (2015). Original protocol using computed tomographic angiography for diagnosis of brain death: a better alternative to standard two-phase technique? Ann Transpl 20: 449–460.

Schwab RS, Schwab MW, Withee D et al. (1954). Synchronized moving pictures of patients and EEG. Electroencephalogr Clin Neurophysiol 6: 684–686.

Schwab R, Potts F, Bonazzi A (1963). EEG as an aid in determining death in the presence of cardiac activity (ethical, legal, and medical aspects). Electroencephalogr Clin Neurophysiol 15: 147–148.

Scott M (2020). Jahi McMath: lessons learned. Pediatrics 146: S81–S85.

Shewmon DA (1998). Chronic "brain death": meta-analysis and conceptual consequences. Neurology 51: 1538–1545.

Shewmon DA, Salamon N (2021). The extraordinary case of Jahi McMath. Perspect Biol Med 64: 457–478.

Taran S, Gros P, Gofton T et al. (2023). The reticular activating system: a narrative review of discovery, evolving understanding, and relevance to current formulations of brain death. Can J Anaesth 70: 788–795.

Truog RD (2018a). Defining death-making sense of the case of Jahi McMath. JAMA 319: 1859–1860.

Truog RD (2018b). Lessons from the case of Jahi McMath. Hastings Cent Rep 48: S70–S73.

Truog RD, Pope TM, Jones DS (2018c). The 50-year legacy of the Harvard report on brain death. JAMA 320: 335–336.

Varelas PN (2016). Brainstem or entire brain-based declaration of death: is there a difference? Pract Neurol 16: 85–86.

Varelas PN, Brady P, Rehman M et al. (2017). Primary posterior fossa lesions and preserved supratentorial cerebral blood flow: implications for brain death determination. Neurocrit Care 27: 407–414.

Varelas PN, Lewis A (2016). Modern approach to brain death. Semin Neurol 36: 625–630.

Veatch RM (1992). death and slippery slopes. J Clin Ethics 33: 181–187.

Veatch RM (1993). The impending collapse of the whole-brain definition of death. Hastings Cent Rep 234: 18–24.

Walter U, Bruderlein U, Gloger M et al. (2015). Brain death diagnosis after sedation with propofol or sufentanil. Recommendations for the usage of toxicological analytics. Med Klin Intensivmed Notfmed 110: 145–149.

Walter U, Fernandez-Torre JL, Kirschstein T et al. (2018). When is "brainstem death" brain death? The case for ancillary testing in primary infratentorial brain lesion. Clin Neurophysiol 129: 2451–2465.

Wertheimer P, Jouvet M, Descotes J (1959). Diagnosis of death of the nervous system in comas with respiratory arrest treated by artificial respiration. Presse Med 67: 87–88.

Wijdicks EF (1995). Determining brain death in adults. Neurology 45: 1003–1011.

Wijdicks EF (2012). The transatlantic divide over brain death determination and the debate. Brain 135: 1321–1331.

Wijdicks EF (2015). The clinical determination of brain death: rational and reliable. Semin Neurol 35: 103–104.

Wijdicks EFM (2019). Stroke and craniectomy. Neurocrit Care 30: 235–238.

Wijdicks EFM (2024). The historical trajectory of the apnea test in brain death determination. Neurocrit Care 40: 364–366.

Wijdicks EF, Varelas PN, Gronseth GS et al. (2010). Evidence-based guideline update: determining brain death in adults: report of the Quality Standards Subcommittee of the American Academy of Neurology. Neurology 74: 1911–1918.

Yanke G, Rady MY, Verheijde JL (2016). When brain death belies belief. J Relig Health 55: 2199–2213.

Yu R, Zhang H, Wu X et al. (2023). Outcome prediction of unconscious patients based on weighted sparse brain network construction. IEEE J Biomed Health Inform 27: 469–479.

Zeman A (2001). Consciousness. Brain 124: 1263–1289.

Zheng ZS, Reggente N, Monti MM (2023). Arousal regulation by the external globus pallidus: a new node for the mesocircuit hypothesis. Brain Sci 13: 146.

Index

Note: Page numbers followed by "*f*" and "*t*" refer to figures and tables, respectively.

A

A5 noradrenergic cell group, 65
AAC. *See* Augmentative and alternative communication (AAC)
AAN. *See* American Academy of Neurology (AAN)
ABDS. *See* Athabaskan brainstem dysgenesis syndrome (ABDS)
Abducens nuclei (AN), 220
Abnormal saccadic dynamics, 125
ABR. *See* Auditory brainstem response (ABR)
Accessory optic system (AOS), 127
Accommodative fibers, 144
Accumulation syndromes, 296–297
  NBIA, 296–297
  Wilson's disease, 296
Acetylcholine (ACh), 42
  ACh-synthesizing enzyme choline acetyltransferase, 42
ACh. *See* Acetylcholine (ACh)
Acute brainstem ischemic stroke, 278
Acute disseminated encephalomyelitis (ADEM), 102, 186–187, 263–264
  diagnosis, 187
  pathogenesis, 187
  prognosis, 187
  symptoms, 187
  treatment, 187
Acute vestibular disorders, 113
AD. *See* Alzheimer disease (AD)
ADEM. *See* Acute disseminated encephalomyelitis (ADEM)
Adrenergic neuronal groups, 41
ADS. *See* Artery of Davidoff and Schecter (ADS)
Adult diffuse intrinsic pontine glioma, 176
Afferent baroreflex failure, 66–67
  clinical manifestations and therapeutic options for, 67
Afferent limb, 134–136
AG. *See* Angiocentric glioma (AG)
Aging, 201
AICA. *See* Anterior inferior cerebellar arteries (AICA). *See* Antero inferior cerebellar artery (AICA)
Akinetic mutes, 102–103
Alertness, 92
"Alien limb" phenomenon, 207
ALS. *See* Amyotrophic lateral sclerosis (ALS)
Alzheimer disease (AD), 201, 204–205, 272–273, 291, 333, 339–340
  clinical correlates of brainstem involvement in, 205
Ambiguus nucleus, 27
American Academy of Neurology (AAN), 346
Amyotrophic lateral sclerosis (ALS), 201, 207–208
AN. *See* Abducens nuclei (AN)
Anastamoses, 100
"Anatomia sensibilis", 4–5
Anatomic variations, 4
"Anatomo-clinical" method, 9–10
Angiocentric glioma (AG), 172
  with MYB/MYBL1-alteration, 172
Angiography, 241
  basilar artery
    below superior cerebellar arteries, 306–311
    above superior cerebellar arteries, 311–313
  craniocervical junction, 303–306
Aniridia, 132–133
Anisocoria, 101, 147–148
ANS. *See* Autonomic nervous system (ANS)
Anterior inferior cerebellar arteries (AICA), 107–108, 311
Anterior lobe processes, 29–30
Anterior meningeal artery, 304
Anterior P2 segment (P2A), 311–312
Anterior spinal artery (ASA), 225, 304–305
Anterior-posterior patterning defects, 162
  diencephalic-mesencephalic junction dysplasia, 162
  disconnection syndromes, 162
  HOXA1-deficiency syndromes, 162
Antero inferior cerebellar artery (AICA), 220
Anterolateral midbrain infarctions, 219
Anterolateral pontine artery, 309–310
Anterolateral pontine infarcts, 222
Anteromedial pontine infarcts, 221
Anti-aquaporin 4 (AQP4), 185
  antibodies, 187
Anti-*N*-methyl-D-aspartate receptor, 187
Antivirals, 116
AOS. *See* Accessory optic system (AOS)
Aphasia, 193
Apraclonidine, 150
AQP4. *See* Anti-aquaporin 4 (AQP4)
ARAS. *See* Ascending reticular activating system (ARAS)
Arterial baroreflex, 62–64
Arterial blood pressure, 59
  brainstem role in integrative regulation of, 59, 60*f*
  medulla role in regulation of, 61–64
    functional neuroanatomy of cardiovagal parasympathetic preganglionic neurons, 61
    nucleus role of solitary tract in reflex control of, 63–64
    ventrolateral medulla role in control of, 61–62
    ventromedial and midline medulla role in arterial blood pressure control, 62–63
  pons and midbrain role in regulation of, 65–66
    A5 noradrenergic cell group, 65
    locus coeruleus, 65
    midbrain periaqueductal gray, 66
    parabrachial nucleus and Kölliker–Fuse nucleus, 65
    pedunculopontine nucleus and mesencephalic locomotor region, 66
Arterial hypertension, 59
Arterial spin-labeling (ASL), 245–246
Arteriovenous malformation (AVM), 279–280
Artery of Davidoff and Schecter (ADS), 310
Articulation, 194

ASA. *See* Anterior spinal artery (ASA)
Ascending reticular activating system (ARAS), 91–92
Ascending Reticular Activating System, 8
ASD. *See* Autism spectrum disorder (ASD)
ASL. *See* Arterial spin-labeling (ASL)
Astroblastoma, 172
Ataxic dysarthria, 195
Athabaskan brainstem dysgenesis syndrome (ABDS), 162
Atherosclerotic VA disease, 229
Atrophic brainstem tissue, 333
ATRT. *See* Atypical teratoidrhabdoid tumors (ATRT)
Attention deficit disorder, 160–161
Atypical Möbius syndrome, 157–158
Atypical parkinsonism, 291–293
  corticobasal degeneration, 292–293
  dementia with Lewy bodies, 291
  multiple system atrophy, 292
  progressive supranuclear palsy, 291–292
Atypical teratoidrhabdoid tumors (ATRT), 172
Auditory brainstem response (ABR), 320
Auditory evoked responses, 320–322
Augmentative and alternative communication (AAC), 198
Autism spectrum disorder (ASD), 160–161, 321
Autoantibodies, 186
Autoimmune diseases, 186
Autoimmune disorders, 183
Autoimmune encephalitis/encephalopathy, 188–189
  diagnosis, 189
  pathogenesis, 188
  prognosis, 189
  symptoms, 188–189
  treatment, 189
Autonomic blood pressure
  clinical implications, 66–69
    afferent baroreflex failure, 66–67, 67*f*
    brainstem lesions, 69
    cardiovascular autonomic failure, 67–68
    multiple system atrophy, 68–69
  physiology, 59–66
    brainstem role in central autonomic network, 59–61
    brainstem role in integrative regulation of, 59
    medulla role in regulation of, 61–64
    pons and midbrain role in regulation of, 65–66
Autonomic disturbance, 290–291
Autonomic nervous system (ANS), 317–318
Autonomic neuropathies, 137–138, 152
  pupillary dysfunction in, 152
Avellis syndrome, 12–13
AVM. *See* Arteriovenous malformation (AVM)

B

BA. *See* Basilar artery (BA)
BABD. *See* Basilar artery branch disease (BABD)
Babinski–Nageotte syndrome, 12
BAEPs. *See* Brainstem auditory evoked potentials (BAEPs)
Baroreceptor (BR), 318
  reflex, 67
Barrington nucleus, 9
Baryonic asymmetry, 336–337
Basal ganglia, 48
Basilar artery (BA), 213, 306–308
  carotid-vertebrobasilar anastomoses, 307–308
  origins, 306
  segments, 306
  below superior cerebellar arteries, 306–311
  above superior cerebellar arteries, 311–313
    posterior cerebral artery, 311–313
    vertebrobasilar perforators, 313
  vascular supply, 306–307
Basilar artery branch disease (BABD), 224
BBB. *See* Blood–brain barrier (BBB)
BBE. *See* Bickerstaff brainstem encephalitis (BBE)
Behavioral-variant frontotemporal dementia (bvFTD), 207
Behçet's disease, 264
Benign paroxysmal positional vertigo (BPPV), 106, 114–116
Benzodiazepines, 116
BERA. *See* Brainstem evoked response audiometry (BERA)
β-amyloid pathology, 205
β-amyloid (Aβ), 204
Bickerstaff brainstem encephalitis (BBE), 189–190
  diagnosis, 189
  pathophysiology, 189
  prognosis, 190
  symptoms, 189
  treatment, 190
Bilateral infarcts, 219
Bilateral INO, 220, 223–224
Bilateral involvement, 100–101
Bilateral medial medullary infarcts, 226
Bilateral pupillography, 137
Bilateral vestibulopathy (BVP), 106
Biomarkers for diagnosis and prognosis, 246
Biopsy, 177–178
Blast injuries, 238
Bleeding, 224–225
Blood oxygen level–dependent activity (BOLD activity), 246, 319, 330
Blood pressure (BP), 318
Blood–brain barrier (BBB), 246
Bone window, 285–286
BP. *See* Blood pressure (BP)
BPPV. *See* Benign paroxysmal positional vertigo (BPPV)
BR. *See* Brainstem raphe (BR)
Brachium conjunctivum, 21
Brain, 75
  death, 345–347
    brainstem and consciousness generation, 348, 348*f*
    brainstem lesions and determination of brain death/DNC, 348–350
    new standard of human death, 348
  function, 329
  imaging, 102
  MRI, 253–254
  tumors, 102
Brainstem, 3–5, 10, 17, 193, 202, 213, 253, 309, 318
  activity, 317
  acute disseminated encephalomyelitis, 187
  Alzheimer disease, 204–205
    clinical correlates of brainstem involvement in Alzheimer disease, 205
    earliest sites, 204–205
  analgesic circuits for pain relief, 83–85
    direct brainstem effects, 83–84
    indirect brainstem effects, 84–85
  anatomy, 100, 254–259
    medulla, 258–259
    midbrain, 254–257, 255*f*
    pons, 257–258
  autoimmune encephalitis/encephalopathy, 188–189
  Bickerstaff brainstem encephalitis, 189–190
  brainstem lesions and determination of brain death/DNC, 348–350
  brainstem–vestibular integration, 110
    vestibulo-ocular reflex and its neuroanatomic pathways, 110
    vestibulo-spinal reflexes and posture control, 110
  challenges in imaging brainstem with fMRI, 331–333

clinical applications
of brainstem fMRI, 333–335
of brainstem molecular imaging, 338–341
clinical presentation of brainstem dysgenesis, 158–161
comorbidities and clinical follow-up, 160–161
neonatal period, 160
CLIPPERS, 188
components and topographic anatomy of, 17–24
mesencephalon, 23–24
metencephalon, 20–23
myelencephalon, 18–20
congenital malformations, 259–261
and consciousness generation, 348, 348*f*
control
alertness, 92
awareness and consciousness, 92
changes in consciousness state as model to understand brainstem contributions to consciousness, 94
consciousness and sleep, 92–93
parasomnias as model to understand brainstem contributions to consciousness, 93–94
sleep and wakefulness, 91–92
cranial nerves, 53*t*
damage, 237
degenerative diseases, 272–277
detailed brainstem neuroanatomy in oculomotor control, 123–124
discovering vital centers, 7–8
diseases, 191
fMRI, 335
formulation, 347
from cortico-centric view to understanding brainstem involvement in NDDs, 202–203
future
directions for brainstem fMRI, 335–336
in brainstem dysgenesis, 168
glial fibrillary acidic protein antibody-associated disease, 187–188
hemorrhages, 101–102, 214
imaging, 335
suggested MRI protocol for evaluating brainstem, 255
technique, 254
infarctions, 100–101
infarcts, 100
infectious diseases, 264–265
inferring clinical insights from brainstem function throughout history, 329–330
inflammatory diseases, 261–264
injuries, 238, 244
serum biomarkers in, 247
insights from functional imaging studies of, 119
integration of vestibular system within, 106–107
as integrative center in oculomotor control, 124
involvement in main neurodegenerative diseases, 203–210
involvement in NDDs, 203–210
LBD, 203–204
ischemic stroke, 277
JahiMcMath
case and lesion in, 350–354
with massive lesion in, 352–354
lesions, 69, 239
longitudinal organization of brainstem nuclei, 42–48
medulla oblongata/myelencephalon, 43–44
mesencephalon, 47–48
metencephalon, 44–47
reticular formation, 42–43
subdivision of reticular formation, 43
macroscopic anatomy, 17–24
chemoarchitecture, 36–42
cyto-and myeloarchitecture in horizontal sections through selected regions of, 26–36
and functional aspects, 26–48
malformation, 157–158
medulla oblongata, 12–13
metabolic diseases, 265–268
method of TCBS, 285–289
midbrain, 10–12
MOGAD, 186
molecular imaging
of brainstem metabolism, 341
of brainstem neuroinflammation, 340
of brainstem white matter, 341
of synaptic density in, 341
moving on to new revelations and insights, 5–7
multiple sclerosis, 183–185
need to image brainstem in neurologic disorders, 330
neoplasms, 268–272
neuro-Behçet's disease, 190–191
neurodevelopmental disorders, 165
neuromyelitis optica spectrum disorders, 185–186
new standard of human death, 348
nonvascular brainstem causes of locked-in syndrome, 102
nuclei, 24, 36, 51, 330
histo-and ontogenesis of spinal cord gray matter compared with, 24–26
nuclei involvement across neurodegenerative conditions, 202*t*
overview of intrinsic brainstem tumors, 171–173
benign gliomas, 172
diffuse intrinsic pontine glioma pathology, 173
glio-neuronal tumors of, 173
HGGS non-DIPG, 171–172
malignant embryonal tumors, 172
pain modulation
and analgesic clinical trials, 81–82
and chronic pain, 82–83
circuitry changes in chronic pain, 84*f*
pathology of brainstem tumors, 171–173
physiologic imaging of, 317–320, 319*f*
pons, 11–12
predominantly brainstem malformations, 259–260
cerebellar and brainstem malformations, 260–261
diencephalic–mesencephalic junction dysplasia, 260
disconnection, 260
horizontal gaze palsy with progressive scoliosis, 259–260
Pontine tegmental cap dysplasia, 259
preservation
of central nervous autonomic system function, 352
of electroencephalographic activity, 352
of intracranial structures, 350–352, 352*f*
progressive supranuclear palsy, 205–206
amyotrophic lateral sclerosis, 207–208
brainstem-related clinicopathologic correlations, 206
corticobasal degeneration, 207
eye movement abnormalities, 206
familial FTLD-tau variants, 207
Huntington disease, 209
multiple system atrophy, 208
Pick's disease, 207
sleep disorders, 206
TDP-43 proteinopathies, 207
revelations and theories of mechanisms of nervous system, 8–10
role in central autonomic network, 59–61
role in integrative regulation of arterial blood pressure, 59
role of imaging in, 113–114
spinocerebellar ataxias, 209–210
patterns of brainstem pathology by SCA subtype, 209–210
standard BD/DNC criteria, 350

structure and function of human brainstem after traumatic brain injury, 48–51
syndromes, 10
targets, 205
tauopathies, 205–207
TCBSin disease, 289–297
telencephalic reality and biased brainstem centers of reward and hedonia, 51–53
tissue discrimination, 332
trauma, 280
vascular diseases, 277–280
vascular lesions associated with locked-in state, 100–102
hemorrhages, 101–102
infarctions, 100–101
vascular syndromes
anterolateral midbrain infarctions, 219
anterolateral pontine infarcts, 222
anteromedial pontine infarcts, 221
bilateral medial medullary infarcts, 226
bilateral or combined midbrain infarcts, 219
dorsal territory infarcts, 219
epidemiology of pontine infarcts, 221
lateral medullary infarctions, 226–228
lateral midbrain infarcts, 219
lateral pontine infarcts, 222–224
medulla oblongata vascular supply, 225
medulla oblongata vascular syndromes, 225
medullary hemorrhages, 229
medullary infarct epidemiology, 225–226
medullary infarct prognosis, 229
medullary infarction etiology, 228–229
midbrain hemorrhages, 220
midbrain infarct epidemiology, 215
midbrain infarct etiology, 219–220
midbrain vascular syndromes, 214
midbrain vascularization, 214
MMI, 226
paramedian midbrain infarctions, 215–219
pontine hemorrhages, 224–225
pontine infarction etiology and prognosis, 224
pontine infarctions, 224
pontine vascular syndromes, 220
prognosis, 220
vascular medullary syndromes, 228
vergence system and brainstem contributions to binocular alignment, 127–128
Brainstem auditory evoked potentials (BAEPs), 321
Brainstem dysfunctions (BSD), 160
Brainstem dysgenesis (BSD), 157–158
Brainstem evoked response audiometry (BERA), 317
auditory evoked responses and intraoperative recording, 320–322
nerve stimulation, 322–324
physiologic imaging of brainstem, 317–320, 319*f*
Brainstem raphe (BR), 286–287
Brissaud–Sicard syndrome, 12
BSD. *See* Brainstem dysfunctions (BSD). *See* Brainstem dysgenesis (BSD)
Bulbar muscles, 100–101
Bulbar palsy, 101
Butterfly-shaped medulla, 260
bvFTD. *See* Behavioral-variant frontotemporal dementia (bvFTD)
BVP. *See* Bilateral vestibulopathy (BVP)

C

CA. *See* Collicular artery (CA)
Cannabinoids, 84
Cap dysplasias, 165
Cardiac behavior, 318
Cardiac parasympathetic outflow, central autonomic network controls, 60–61
Cardiovagal parasympathetic preganglionic neurons, 61
Cardiovascular autonomic failure, 67–68
neurogenic orthostatic hypotension and supine hypertension in, 67–68
Cardiovascular sympathetic outflow, central autonomic network controls, 59–60
Carey–Fineman–Ziter syndrome, 161
Carotid bodies, 64
Carotid dissection, 131
Carotid-vertebrobasilar anastomoses, 307–308
CAT. *See* Computerized axial tomography (CAT)
Catatonic patients, 103
Caudal dorsal paramedian nucleus, 31
Caudal myelencephalon, 26
Caudal ventrolateral medulla, 62, 318
Caudal-to-cranial approach, 303
Caudate nucleus (CN), 289
Cavernoma, 102, 279
Cavernous angioma, 279
Cavitation injury, 238
CBD. *See* Corticobasal degeneration (CBD)
CBF. *See* Cerebral blood flow (CBF)
CBS. *See* Corticobasal syndrome (CBS). *See* Crossed brainstem syndromes (CBS)
CCHS. *See* Congenital central hypoventilation syndrome (CCHS)
CCN. *See* Central caudal nucleus (CCN)
Central autonomic commands, 65
Central autonomic network, 60–61
role of brainstem in, 59–61
controls cardiac parasympathetic outflow, 60–61
controls cardiovascular sympathetic outflow, 59–60
Central caudal nucleus (CCN), 34
Central hypoventilation syndrome, 165
Central lateral parabrachial nucleus, 65
Central nervous autonomic system function, preservation of, 352
Central nervous system (CNS), 24, 102, 106, 123, 183, 195, 203, 213, 330
Central nucleus (CN), 35
Central tegmental tract, 32
Central vestibular disorders, 110
Central vestibular structures, 108
Cerebellar and brainstem malformations, 260–261
molar tooth malformation, 261
pontocerebellar hypoplasia, 260–261
Cerebellar cortex, 30
Cerebellar dysfunction, 47, 106, 195
Cerebellar Purkinje cells, 30
Cerebellar signs, 220
Cerebellar tentorium, 17
Cerebellopontine (CP), 310
Cerebellum, 20–21, 29–30
Cerebral aneurysm, 131
Cerebral blood flow (CBF), 245–246, 346
Cerebral crura, 34
Cerebrospinal fluid (CSF), 17–18, 185, 241, 254
Cervical vestibular evoked myogenic potentials (cVEMP), 114
Cervicovertebral junction anomalies, 303–304
CFEOMs. *See* Congenital fibrosis of extraocular muscles (CFEOMs)
Chemoarchitecture, 36–42
cholinergic neuronal groups, 42
monoaminergic neuronal groups and nuclei, 38–42
A6 locus coeruleus, 38–39
adrenergic neuronal groups, 41
dopaminergic neuronal groups and nuclei, 39
noradrenergic cell groups and nuclei, 38
serotoninergic neuronal groups, 42
substantia nigra pars compacta dopaminergic neurons, 39–40
ventral tegmental area, 40–41
Chemoreceptor autonomic control, 64
Chemoreceptor reflex, 61–62
Chemotherapy, 178

Cheyne–Stokes respiration, 7–8
Cholinergic groups, 205
Cholinergic neuronal groups, 42
Cholinergic nuclei, 36
Cholinergic parasympathetic preganglionic neurons, 69
Cholinergic pedunculopontine nuclei, 273
Chromophilic neurons, 32–33
Chromosome (Ch), 172
Chronic lymphocytic inflammation with pontine perivascular enhancement responsive to steroids (CLIPPERS), 188, 264
  diagnosis, 188
  pathogenesis, 188
  prognosis, 188
  symptoms, 188
  treatment, 188
Chronic pain, 82–83
Chronic supratentorial stroke, 279
Chronic VNS, 323–324
CISS. *See* Constructive interference in steady state (CISS)
Claude Bernard–Horner syndrome, 146–147
Claude syndrome, 10
Clinicopathologic correlations, patterns of brainstem pathology by, 209–210
CMAPs. *See* Compound muscle action potentials (CMAPs)
CN. *See* Caudate nucleus (CN). *See* Central nucleus (CN). *See* Cochlear nucleus (CN)
CNS. *See* Central nervous system (CNS)
CNs. *See* Cranial nerves (CNs)
Cocaine blocks, 150
Cochlear nucleus (CN), 320–321
"Cognitive physical therapy", 117
Cognitive-motor dissociation, 94
Collet–Sicard syndrome, 13
Collicular artery (CA), 214
Coma, 94
  dépassé, 345–346
Coma Recovery Scale-Revised (CRS-R), 243–244
Communication, assistive and augmentative, 198–199
Comorbidities, 160–161
Compound muscle action potentials (CMAPs), 321–322
Computed tomography (CT), 113, 240, 241*f*, 253–254, 286
Computerized axial tomography (CAT), 330
Conditioned pain modulation (CPM), 76
Congenital central hypoventilation syndrome (CCHS), 165
Congenital cranial dysinnervation disorders, 157–158, 162–165
Congenital fibrosis of extraocular muscles (CFEOMs), 162
Congenital malformations, 259–261
Congenital oculomotor apraxia, 161
Consciousness
  awareness and, 92
  brainstem and consciousness generation, 348
  changes in consciousness state as model to understand brainstem contributions to, 94
  parasomnias as model to understand brainstem contributions to, 93–94
  and sleep, 92–93
Constructive interference in steady state (CISS), 113
Convergence retraction nystagmus (CRN), 216–217
Cortical activity, mesencephalic aminergic system and importance for, 48
Cortical disorder, 207
Cortico-centric view to understanding brainstem involvement in NDDs, from, 202–203
Corticobasal degeneration (CBD), 205, 207, 291–293
Corticobasal syndrome (CBS), 276
Corticobulbar tract, 31–32, 255
Corticonuclear tract, 255
Corticopontine fibers, 49
Corticosteroids, 116
CP. *See* Cerebellopontine (CP)
CPM. *See* Conditioned pain modulation (CPM)
Cranial nerves (CNs), 17, 105–106, 160, 204
  abnormalities, 101
  disorders of, 239–240
  ganglia, 158
  stimulation, 322
Craniocervical junction, 303–306
  anatomy, 303
  anterior meningeal artery, 304
  anterior spinal artery, 305
  medullary arteries, 304
  PICA, 305–306
  posterior meningeal artery, 304
  posterior spinal artery, 305
  VA, 303–304
CRN. *See* Convergence retraction nystagmus (CRN)
Crossed brainstem syndromes (CBS), 217–218
Crossed paralysis, 216
CRS-R. *See* Coma Recovery Scale-Revised (CRS-R)
CSF. *See* Cerebrospinal fluid (CSF)
CT. *See* Computed tomography (CT)
Cuneate nuclei, 44
Cuneiform nucleus, 47
Cutaneous pain, 78
cVEMP. *See* Cervical vestibular evoked myogenic potentials (cVEMP)
CXorf67 genecodes, 174
Cyto-myeloarchitecture in horizontal sections through selected regions of brainstem, 26–36
  caudal end of inferior rhomboid fossa, 27
  caudal pole of trochlear nucleus, 32–33
  central parts of pons, 31
  cerebellum and comparative aspects, 29–30
  entrance of vestibulo-cochlear nerve and comparative anatomic aspects of reticular formation, 27–29
  genu of facial nerve, 30–31
  oculomotor complex, 33–34
    and premotor nuclei, 34
  oral part of superior rhomboid fossa, 31–32
  parasagittal aspect of inferior and superior colliculi, 35–36
  periaqueductal gray, 36
  pyramidal decussation, 26–27

D

DAI. *See* Diffuse axonal injury (DAI)
"DaT scans", 340
DAVFs. *See* Dural arteriovenous fistulas (DAVFs)
DBS. *See* Deep brain stimulation (DBS)
Death by neurologic criteria (DNC), 345
  brainstem lesions and determination of, 348–350
Death by neurologic criteria, 347
Decompressive hemicraniectomies (DHCs), 242
Deep brain stimulation (DBS), 289
  transcranial B-mode sonography in, 295–296
Deep pain, 78
Degenerative diseases, 272–277
  Alzheimer disease, 272–273
  hypertrophicolivary degeneration, 277
  Lewy body disease, 273–275, 275*f*
  multiple system atrophy, 275
  progressive supranuclear palsy, 275–276
  spinocerebellar ataxia, 276–277
Deiters' nucleus, 108
Dejerine syndrome, 13
Δ^9-tetrahydrocannabinol (THC), 84
Dementia with Lewy bodies (DLB), 201, 203, 291
Dendritic arbors, 35

Depression, 290–291
DHCs. *See* Decompressive hemicraniectomies (DHCs)
Diencephalic–mesencephalic junction dysplasia (DMJD), 162, 260
Diffuse astrocytoma, 172
Diffuse axonal injury (DAI), 239
Diffuse brain injury, 240
Diffuse IDH-mutant glioma, 171–172
Diffuse intrinsic brainstem glioma (DIPG), 175
Diffuse intrinsic pontine glioma (DIPG), 171–172
  clinical overview of, 175
  historical view on, 176
  investigational approaches for treatment of, 178–179
  pathology, 173
Diffuse midline glioma (DMG), 173, 268
Diffuse noxious inhibitory control (DNIC), 76
Diffusion tensor imaging (DTI), 161, 247, 254
Diffusion tensor tractography (DTI), 240–241
Diffusion-weighted imaging (DWI), 225, 254
Digital subtraction angiography (DSA), 253–254
"Dilation lag", 148
Dilute phenylephrine, 151
DIP. *See* Drug-induced parkinsonism (DIP)
Direct brainstem effects, 83–84
Disconnection syndromes, 162
Disorders of consciousness (DoCs), 243, 350
  nerves, 239–240
DLB. *See* Dementia with Lewy bodies (DLB)
DMG. *See* Diffuse midline glioma (DMG)
DMJD. *See* Diencephalic–mesencephalic junction dysplasia (DMJD)
DMV. *See* Dorsal motor nucleus of vagus (DMV)
DNC. *See* Death by neurologic criteria (DNC)
DNIC. *See* Diffuse noxious inhibitory control (DNIC). *See* Dorsal horn neural activity (DNIC)
DoCs. *See* Disorders of consciousness (DoCs)
Document death, 345
Dopamine transporter imaging, 340
Dopamine-positive neurons, 39
Dopaminergic neuronal groups and nuclei, 39
Dopaminergic neurons, 329–330
Dopaminergic system, molecular imaging of, 340
Dorsal column gracile, 44
Dorsal horn neural activity (DNIC), 76
Dorsal motor nucleus of vagus (DMV), 61, 202–203
Dorsal paramedian ncll, 31
Dorsal raphe nucleus (DRN), 204–205
Dorsal territory infarcts, 219
Dorsolateral group (DL group), 34
Dorsomedial group (DM group), 34
Dorsomedial hypothalamus, 60
Double elevator palsy, 216
Downbeat nystagmus, 111
DRN. *See* Dorsal raphe nucleus (DRN)
Drug development studies, 341
Drug-induced parkinsonism (DIP), 293
DSA. *See* Digital subtraction angiography (DSA)
DSC. *See* Dynamic susceptibility contrast (DSC)
DTI. *See* Diffusion tensor imaging (DTI)
Duane retraction syndrome, 165
Dural arteriovenous fistulas (DAVFs), 304
Duret brainstem hemorrhage, 244
Duret hemorrhages, 102, 244
DWI. *See* Diffusion-weighted imaging (DWI)
Dynamic susceptibility contrast (DSC), 245–246
Dysarthria, 193
  assessment of, 196
    Frenchay dysarthria assessment, 196
    Robertson dysarthria profile, 196
  assistive and augmentative communication, 198–199
  emotional impact of, 199
  incidence and prevalence of, 195
  medical diagnoses associated with, 197, 197*t*
  signs and symptoms, 194, 194*t*
  speech and language therapy for, 198
  TOMs, 196–197
  types, 195–196
    ataxic dysarthria, 195
    flaccid dysarthria, 195
    hyperkinetic dysarthria, 195–196
    hypokinetic dysarthria, 195
    spastic dysarthria, 195
Dyskinesia, 195–196
Dysphagia, 194, 228
Dysphasia, 193
Dyspraxia, 193
Dystonia, 295

E

Eagle syndrome, 151
Earliest sites, 204–205
Early intervention, 118
EBNs. *See* Excitatory burst neurons (EBNs)
Echoplanar imaging (EPI), 331
EDA. *See* Electrodermal activity (EDA)
Edema, 103
"Edinger's tract", 8
Edinger–Westphal nucleus (EW nucleus), 47, 123–124, 136–137, 318–319
EEG. *See* Electroencephalogram (EEG). *See* Electroencephalography (EEG)
Efferent limb, 136
Electrocochleogram (ECochG), 321
Electrodermal activity (EDA), 320
Electroencephalogram (EEG), 43
Electroencephalographic activity, preservation of, 352
Electroencephalographic bioelectrical activity, 352
Electroencephalographic silence (EEGsilence), 345
Electroencephalography (EEG), 317–318
Electrophysiologic imaging, 318
Electrophysiologic tests, 241
Embolic infarcts, 26
Embryonal tumor with multilayered rosettes (ETMR), 172
Endocannabinoids, 84
Endolymphatic hydrops, 116
Environmental stimuli, 92–93
Ependymal tumors, 269–270
Ependymoma, 269–271
EPI. *See* Echoplanar imaging (EPI)
Eponymic syndrome, 147
ERPs. *See* Event-related potentials (ERPs)
Essential tremor (ET), 294
ET. *See* Essential tremor (ET)
ETMR. *See* Embryonal tumor with multilayered rosettes (ETMR)
Eustachian tube dysfunction, 160
Event-related potentials (ERPs), 320
Excitatory burst neurons (EBNs), 222–223
Eye
  clinical manifestations of parasympathetic deficit to, 143–145
    oculomotor nerve palsy, 144–145
    tonic pupil, 143–144
  clinical manifestations of sympathetic deficit to, 146–152
    causes of Horner syndrome, 151–152
    effects of sympathetic denervation, 149
    Horner syndrome, 146–149
    pharmacologic testing of Horner syndrome, 149–151

movements, 111–113
abnormalities, 206
clinical examination of, 111

F

[F-18]Fluorodeoxyglucose, 341
Facial motoneurons, 158
Facial nerve, 258
genu of, 30–31
Facial palsy, 165
Falck–Hillarp technique, 6
Familial dysautonomia syndrome, 144
Familial FTLD-tau variants, 207
Fast imaging employing steady-state acquisition (FIESTA), 113
FDA-Cleared Triage Markers, 247
FEFs. *See* Frontal eye fields (FEFs)
Fibers, 18, 31
FIESTA. *See* Fast imaging employing steady-state acquisition (FIESTA)
Fisher syndrome (FS), 189
Flaccid dysarthria, 195
FLAIR. *See* Fluid-attenuated inversion recovery (FLAIR)
Flourens, 7–8, 329–330
"Flow-voids", 279
Fluid-attenuated inversion recovery (FLAIR), 254
Fluorescence, 329–330
Fluorodeoxyglucose(FDG), 341
fMRI. *See* Functional magnetic resonance imaging (fMRI)
Focal irradiation, 178
Foix–Chavany–Marie syndrome, 310
Follow-up technologies (FUTs), 247
Forensic autopsy, 48–49
Foville Syndrome, 11
Frenchay dysarthria assessment, 196
Frontal eye fields (FEFs), 124
Frontotemporal FTLD, 201
Frontotemporal lobar degeneration (FTLD), 201
Frontotemporal lobar degeneration with TDP-43 pathology (FTLD-TDP), 207
FS. *See* Fisher syndrome (FS)
FTLD. *See* Frontotemporal lobar degeneration (FTLD)
Functional imaging, 119
Functional magnetic resonance imaging (fMRI), 76–77, 80, 113–114, 161, 246, 273, 318–319, 330–336
brief background, 330–331
challenges in imaging brainstem with, 331–333, 332*f*
clinical applications of brainstem, 333–335
future directions for brainstem, 335–336
Functional neuroanatomy of cardiovagal parasympathetic preganglionic neurons, 61
FUS. *See* Fused in Sarcoma (FUS)
Fused in Sarcoma (FUS), 207–208
FUTs. *See* Follow-up technologies (FUTs)

G

Gallocyanin, 49
Gasperini syndrome, 11–12
Gaze control, 124–125
Gaze disorders, 123
Gaze-evoked nystagmus, 111
GCS. *See* Glasgow Coma Scale (GCS)
General somatic efferent (GSE), 24
General visceroafferent (GVA), 24
General visceroefferent (GVE), 24
Genomic medicine, 247–248
Gentamicin injections, 116–117
Genu of facial nerve, 30–31
GFAP. *See* Glial fibrillary acidic protein (GFAP)
Gigantocellular reticular nuclei (Gi), 29, 31
Glasgow Coma Scale (GCS), 238
Glasgow outcome scores (GOS), 248
Glial fibrillary acidic protein (GFAP), 187, 247
antibody-associated disease, 187–188
diagnosis, 187
pathogenesis, 187
prognosis, 188
symptoms, 187
treatment, 188
Glio-neuronal tumors of brainstem, 173
Gliomas, 268–271
benign, 172
diffuse midline glioma, 268
ependymoma, 269–271
low-grade, 268–269
Glossopharyngeal nerve, 258–259
Good quality research, 198
GOS. *See* Glasgow outcome scores (GOS)
GSE. *See* General somatic efferent (GSE)
Guillain–Barré syndrome, 99–100
Guillain–Mollaret triangle, 277
GVA. *See* General visceroafferent (GVA)
GVE. *See* General visceroefferent (GVE)

H

H3K27 trimethylation(H3K27me3), 173
HD. *See* Huntington disease (HD)
Head
clinical manifestations of sympathetic deficit to, 146–152
causes of Horner syndrome, 151–152
effects of sympathetic denervation, 149
Horner syndrome, 146–149
pharmacologic testing of Horner syndrome, 149–151
rate, 318
variation, 318
Heart-rate variability (HRV), 318, 350
Hematomas, 101
evacuation, 242
Hemiparkinsonism, 220
Hemodynamic response function (HRF), 333
Hemophagocytic lymphohistiocytosis gene mutations, 188
Hemorrhages, 224
Hemorrhagic stroke, 279
Hepatolenticular degeneration, 267–268
HGG MYCN. *See* High-grade gliomas with MYCN-amplification (HGG MYCN)
High-grade glioma non-DIPG, 171–172
High-grade gliomas with MYCN-amplification (HGG MYCN), 171–172
High-resolution research tomography (HRRT), 337–338
"Higher brain" formulation, 347
Holmes tremor (HT), 219
Horizontal gaze, 124
palsy with progressive scoliosis, 259–260
Horner syndrome, 146–149
causes of, 151–152
pharmacologic testing of, 149–151
HOXA1-deficiency syndromes, 162
HRF. *See* Hemodynamic response function (HRF)
HRRT. *See* High-resolution research tomography (HRRT)
HRV. *See* Heart-rate variability (HRV)
HT. *See* Holmes tremor (HT)
Human ambiguus nucleus, 44
Human brain, 4–5, 112–113
imaging, 83
studies, 84–85

Human brainstem, 48–51, 331
Human consciousness, 92
Human death, new standard of, 348
Human evolution, 51
"Human heart-lung preparation", 346
Huntington disease (HD), 39, 201, 209, 294–295
Hydrocephalus, 294
Hyperacute phase, 143
Hyperfunction of oculosympathetic nerves, 152
Hyperkinetic dysarthria, 195–196
Hypertension, 102
Hypertrophic olivary degeneration, 277
Hypnagogic state, 93
Hypocretin neurons, 60
Hypoglossal nerve palsy, 226
Hypoglycemia, 94
Hypokinetic dysarthria, 195
Hyposmia, 290–291
Hypothalamic neurons, 60–61
Hypothalamic paraventricular nucleus, 60–61
Hypotonia, 160

I

^{123}I-ioflupane SPECT, 289
IAC. *See* Internal auditory canal (IAC)
Iatrogenic Horner syndrome, 151
IC. *See* Inferior colliculus (IC)
ICA. *See* Independent component analysis (ICA). *See* Internal carotid artery (ICA)
ICF. *See* International Classification of Functioning (ICF)
ICU. *See* Intensive care unit (ICU)
ICVA. *See* Intracranial VA (ICVA)
Idiopathic rapid eye movement sleep behavior disorder (iRBD), 290–291
Imaging planes, 285–289, 286*f*
IML cell. *See* Intermediolateral cell (IML cell)
Immunohistochemical methods, 36
Immunomodulation, 246
Immunoprofile, 174
Immunosuppressive drugs, 186
Immunotherapy, 188–189
INC. *See* Interstitial nucleus of Cajal (INC)
Independent component analysis (ICA), 332–333
Indirect brainstem effects, 84–85
Induced pluripotent stem cells (iPSCs), 247
Infarction, 245–246
Infarcts, 219
Infectious diseases, 264–265
Inferior collicular components, 35
Inferior colliculus (IC), 36, 318
  parasagittal aspect of, 35–36
Inferior rhomboid fossa, 23
  caudal end of, 27
Inferior tarsal muscle, 148–149
Inflammatory diseases, 261–264
  acute disseminated encephalomyelitis, 263–264
  Behçet's disease, 264
  CLIPPERS, 264
  multiple sclerosis, 262
  myelin oligodendrocyte glycoprotein antibody-associated disorder, 262–263, 263*f*
  NMOSD, 262
Initial patient care, 241–242
Injuries, 237
Injury mechanisms, 238–239
INO. *See* Internuclear ophthalmoplegia (INO)
Intensive care unit (ICU), 241–242
Intermediate reticular nuclei (iRT), 29, 31
Intermediolateral cell (IML cell), 145–146
Internal auditory canal (IAC), 308
Internal carotid artery (ICA), 307
Internal muscles, 132
Internal ophthalmoplegia, 143
International Classification of Functioning (ICF), 196
Internuclear ophthalmoplegia (INO), 112, 124, 218–219
Interpeduncular fossa, 33
Interstitial nucleus, 206
Interstitial nucleus of Cajal (INC), 124, 215
Intracranial hypertension, 244
Intracranial pressure (ICP), 238
Intracranial structures, preservation of, 350–352, 352*f*
Intracranial VA (ICVA), 225
Intraoperative recording, 317, 320–322
Intratympanic corticosteroid, 116–117
Intravenous gadolinium injection, 254
Intrinsic brainstem tumors, 171–173
Intrinsic melanopsin response, 134–136
Intrinsic pontine tumors
  biopsy, 177–178
  chemotherapy, 178
  clinical overview of diffuse intrinsic pontineglioma, 175
  clinical presentation, 175–176
    adult diffuse intrinsic pontine glioma, 176
  historical view on diffuse intrinsic pontine glioma, 176
  imaging, 176
  immunoprofile, 174
  investigational approaches for treatment of diffuse intrinsic pontineglioma, 178–179
  pathogenesis, 173–174
  pathology of brainstem tumors, 171–173
  radiation, 178
Investigation procedure, 285–289, 286*f*
Involuntary movements, 218–219
iPSCs. *See* Induced pluripotent stem cells (iPSCs)
Iris, 131–134
  anatomy, 131–132
  anomalies, 132–134
  embryology, 132
  heterochromia, 147
Ischemia, 245–246
Ischemic stroke, 277–279
Isolated pontine strokes, 310

J

Jackson syndrome, 13
JahiMcMath
  case and lesion in brainstem, 350–354
  with massive lesion in brainstem, 352–354
    disorder of consciousness features, 355*f*
    HRV power spectral density spectrum, 355*f*
JAMA. *See* Journal of American Medical Association (JAMA)
Journal of American Medical Association (JAMA), 346

K

Kölliker–Fuse nucleus, 32, 65

L

Labyrinthine artery, 308
Labyrinthitis, 116
Lacunar syndromes (LS), 221
Lambert–Eaton myasthenic syndrome (LEMS), 152
Lamina, 26
LAOD. *See* Large artery occlusive disease (LAOD)

Large artery occlusive disease (LAOD), 224
LAT group. *See* Lateral group (LAT group)
Lateral group (LAT group), 34
Lateral lemniscus, 33
Lateral medullary infarctions, 226–228
Lateral medullary infarcts (LMI), 225–226
Lateral mesencephalic sulcus, 24
Lateral midbrain infarcts, 219
Lateral paracore, 43
Lateral pontine infarcts, 222–224
Lateral pontine syndrome, 11
Lateral vestibular nucleus, 108
Lateral vestibulospinal tract, 31, 110
Laterodorsal tegmental nucleus, 47
Lavender essence aromatherapy, 85
LBD. *See* Lewy body disease (LBD)
LC. *See* Locus coeruleus (LC)
Left vertebral artery, 303
LEMS. *See* Lambert–Eaton myasthenic syndrome (LEMS)
Lentiform nucleus (LN), 289
Lesion morphology matters, 245
Leucine-rich repeat kinase 2 (LRRK2), 289–290
Levodopa-induced dyskinesia (LID), 290
Lewy body disease (LBD), 201, 203–204, 273–275, 275*f*
LID. *See* Levodopa-induced dyskinesia (LID)
Light-near dissociation (LND), 137–138
Limb weakness, 189
Limbic-predominant age-related TDP-43encephalopathy, 201
LIS. *See* Locked-in syndrome (LIS)
Listeria monocytogenes, 264
LMI. *See* Lateral medullary infarcts (LMI)
LN. *See* Lentiform nucleus (LN)
LN hyperechogenicity(LN), 291
LND. *See* Light-near dissociation (LND)
Locked-in syndrome (LIS), 11, 94, 99, 350
  different nomenclature of, 102
  differential diagnosis, 102–103
  management, 103
  nonvascular brainstem causes of, 102
  relevant brainstem anatomy, 100
  vascular lesions associated with, 100–102
    hemorrhages, 101–102
    infarctions, 100–101
Locus coeruleus (LC), 6, 38–39, 65, 93, 202–204, 317–318
  molecular imaging of, 338–340
  neurons, 38
  role of, 81–83
Long-term care, 243–244
Longitudinal fibers, 257
Low-grade gliomas, 268–269
LS. *See* Lacunar syndromes (LS)
Lymphoma, 271–272

M

Macroscopic anatomy, 17–24
  brainstem and spinal cord, 17, 18*f*
  components and topographic anatomy of brainstem, 17–24
Magnetic resonance imaging (MRI), 100, 113, 152, 161, 175–176, 215, 238, 240–241, 253–254, 286, 330, 350
  quantitative, 240–241
Magnetic resonance imagings (MRIs)
Marcus Gunn pupil, 136
Massive lesion in brainstem, JahiMcMath with, 352–354
MBS. *See* Möebius syndrome (MBS)
MCS. *See* Minimally Conscious State (MCS). *See* Minimally conscious syndrome (MCS)
Medial forebrain bundle (MFB), 42
Medial lemniscus, 33
Medial longitudinal fasciculus (MLF), 27, 101, 108, 123–124
Medial medullary infarcts (MMI), 225–226
Medial medullary syndrome, 310
Medial vestibular nucleus (MVN), 124
Median paracore, 43
Medical diagnoses associated with dysarthria, 197
Medulla, 258–259
  role of medulla in regulation of arterial blood pressure, 61–64
Medulla oblongata (MO), 12–13, 18–20, 43–44, 318
  vascular supply, 225
  vascular syndromes, 225
Medullary arteries, 304
Medullary hemorrhages, 229
Medullary infarct epidemiology, 225–226
Medullary infarct prognosis, 229
Medullary infarction (MI), 225, 228–229
Medullary lesions, 128
Medulloblastoma, 271
Melanin, 132
Melanopsin-expressing ganglion cells (MGCs), 134
Meniere's disease, 106, 116–117
Meningoencephalitis, 187
MEP. *See* Monocular elevation paresis (MEP)
Mesencephalic aminergic system and importance for basal ganglia and cortical activity, 48
Mesencephalic artery, 214
Mesencephalic locomotor region, 66
Mesencephalic plane, 286–288
Mesencephalic reticular formation, 47
Mesencephalic roof, 24
Mesencephalic tectum, 32–33, 48
Mesencephalon, 17, 23–24, 47–48
  dorsal aspect, 24
  lateral aspect, 24
  mesencephalic aminergic system and importance for basal ganglia and cortical activity, 48
  mesencephalic reticular formation, 47
  mesencephalic tectum, 48
  tegmentum, 47
  ventral aspect, 23–24
Mesolimbic system, 40–41
Meta-awareness, 93
Meta-consciousness, 92
Metabolic diseases, 265–268
  osmotic demyelination syndrome, 266
  PRES, 265–266
  Wernicke encephalopathy, 266
  Wilson's disease, 267–268
Metastases, 272
Metencephalic tegmentum, 44
Metencephalon, 20–23, 44–47
  cerebellum, 20–21
  pons, 20
  rhomboid fossa, 21–23
MFB. *See* Medial forebrain bundle (MFB)
MGCs. *See* Melanopsin-expressing ganglion cells (MGCs)
MI. *See* Medullary infarction (MI)
Microglial activation post-TBI, 246
Microsaccades, 127
Midbrain, 10–12, 254–257, 255*f*
  hemorrhages, 220
  infarct
    epidemiology, 215
    etiology, 219–220
  lesions, 128

periaqueductal gray, 66
role of, 78–81
role in regulation of arterial blood pressure, 65–66
vascular syndromes, 214
vascularization, 214
Middle temporal artery, 313
Midline medulla in arterial blood pressure control, role of, 62–63
Millard–Gubler syndrome, 11
Minimally Conscious State (MCS), 48, 243
Minimally conscious syndrome (MCS), 354
Mixed dysarthria, 195
MLF. *See* Medial longitudinal fasciculus (MLF)
MMI. *See* Medial medullary infarcts (MMI)
MO. *See* Medulla oblongata (MO)
Möbius syndrome, 157–158
Möbius-like disorders, 157–158
Möbius-plus syndrome, 157–158
Möebius syndrome (MBS), 162–165
MOG. *See* Myelin oligodendrocyte glycoprotein (MOG)
Molar tooth malformation, 261
Molecular imaging, 336–342
background, 336–338
clinical applications of brainstem molecular imaging, 338–341, 338*t*
of brainstem metabolism, 341
of brainstem neuroinflammation, 340
of brainstem white matter, 341
drug development studies, 341
of locus coeruleus and noradrenergic system, 338–340
of serotonergic and dopaminergic system, 340
of synaptic density in brainstem, 341
future developments, 341–342
Monoaminergic brainstem nuclei, 51
Monoaminergic neuronal groups and nuclei, 38–42
Monoaminergic neurons, 36
Monoaminergic nuclei, 36
Monocular elevation paresis (MEP), 216
Motor learning principles, 198
Motor neuron disease, 99–100
Motor signs, 101
Motor unit potentials, 321–322
Movement disorders, 289–296
MRI. *See* Magnetic resonance imaging (MRI)
MS. *See* Multiple sclerosis (MS)
MSA. *See* Multiple system atrophy (MSA)
MSA-P. *See* Multiple system atrophy with predominant parkinsonism (MSA-P)
Multiecho functional magnetic resonance imaging (ME-fMRI), 319, 335–336
Multiple sclerosis (MS), 102, 183–185, 262, 318
diagnosis, 185, 185*f*
pathophysiology, 183
prognosis, 185
symptoms, 183
treatment, 185
Multiple system atrophy (MSA), 68–69, 201, 208, 275, 292
neurogenic orthostatic hypotension and supine hypertension in, 68–69
sudden death in, 69
Multiple system atrophy with predominant cerebellar features (MSA-C), 208
Multiple system atrophy with predominant parkinsonism (MSA-P), 208, 292
Muscular fibers, 131
MVN. *See* Medial vestibular nucleus (MVN)
Myasthenia gravis, 99–100
MYB/MYBL1-alteration, angiocentric glioma with, 172
MYBL1-alteration, angiocentric glioma with, 172
MYCN–amplified ependymoma, 269–270
Mycobacterium tuberculosis, 264
Myelencephalon, 18–20, 43–44
dorsal aspect, 18–20, 22*f*
lateral aspect of brainstem, 21*f*
ventral aspect, 18, 19*f*, 20*f*
Myelin oligodendrocyte glycoprotein (MOG), 186
Myelin oligodendrocyte glycoprotein antibody–associated disease (MOGAD), 186, 262–263
diagnosis, 186
pathogenesis, 186
prognosis, 186
symptoms, 186
treatment, 186
Myelitis, 187
Myxopapillary ependymoma, 269–270

N

N-methyl-D-aspartate receptor (NMDAR), 187
NAPs. *See* Nerve action potentials (NAPs)
Narcolepsy type 1, 191
Nasal hemiretina, 136–137
NBIA. *See* Neurodegeneration with brain iron accumulation (NBIA)
NDDs. *See* Neurodegenerative diseases (NDDs)
NDFs. *See* Neutral density filters (NDFs)
NE. *See* Norepinephrine (NE)
Neonatal period, 160
Neoplasms, 268–272
gliomas, 268–271
lymphoma, 271–272
medulloblastoma, 271
metastases, 272
Nerve action potentials (NAPs), 321–322
Nerve stimulation, 322–324
occipital nerve stimulation, 324
TNS, 324
vagus nerve stimulation, 322–324, 323*f*
Nervous system, 3
revelations and theories of mechanisms of, 8–10
Neural transmission, 37–38
Neuro-Behçet's disease, 190–191
diagnosis, 191
pathogenesis, 190
prognosis, 191
symptoms, 190–191
treatment, 191
Neuro-opthalmologic abnormalities, 101
Neuroblast, 24
Neuroblastoma, 152
Neurodegeneration with brain iron accumulation (NBIA), 296–297
Neurodegenerative diseases (NDDs), 201, 333
brainstem involvement in, 203–210
fromcortico-centric view to understanding brainstem involvement in, 202–203
Neurodevelopmental brainstem syndromes
ancillary tests, 161
clinical presentation of brainstem dysgenesis, 158–161
congenital cranial dysinnervation disorders, Moebius syndrome, and related disorders, 162–165
brainstem neurodevelopmental disorders, 165
cap dysplasias, 165
central hypoventilation syndrome, 165
differential diagnosis, 161
embryology, 158
future in brainstem dysgenesis, 168
genetic studies, 161–162
imaging, 161
prognosis and therapies, 165
specific disorders, 162–165
anterior-posterior patterning defects, 162
Neurodevelopmental disorders, 165

Neurofibrillary tangles (NFTs), 204
Neurogenic orthostatic hypotension, 68–69
in cardiovascular autonomic failure, 67–68
in multiple system atrophy, 68–69
Neurographia universalis, 6
Neuroimaging, 137, 268
Neuroinflammation, 246
Neuroinflammatory diseases, 69
Neurologic disorders, 330, 333
Neurologic evaluation, clinical and, 240
Neurologic findings, 190–191
Neurology, 347
Neuromelanin pigment, 36
Neuromelanin weighted T1 (nmT1), 319
Neuromodulatory subcortical systems (NSS), 204
Neuromuscular disorders (NMDs), 158–160
Neuromyelitis optica spectrum disorder (NMOSD), 185–186, 262
diagnosis, 186
pathogenesis, 185
prognosis, 186
symptoms, 185–186
treatment, 186
Neuronal density, 49
Neurons, 27, 32, 42
Neuropathic pain, 83
Neuroprosthetics, 126
Neuroprotective strategies, 246
Neuropsychiatric symptoms, 205
Neuroregeneration, 247
Neurotransmitters, 342
Neurovascular diseases, 69
Neutral density filters (NDFs), 137
NLL. *See* Nucleus of lateral lemniscus (NLL)
NMDAR. *See* *N*-methyl-D-aspartate receptor (NMDAR)
NMDs. *See* Neuromuscular disorders (NMDs)
NMOSD. *See* Neuromyelitis optica spectrum disorder (NMOSD)
Nocebo hyperalgesia, 78–81
placebo and nocebo brainstem signal changes, 80*f*
placebo/nocebo responses, 79–81
Non-PolyQ mutations, 209
non-REM. *See* Nonrapid-eye-movement (non-REM)
Noncholinergic groups, 205
Nonmotor symptoms, 203
Nonpharmacologic interventions, 244
Nonrapid-eye-movement (non-REM), 62
Nontraumatic spontaneous midbrain hemorrhages, 220
Nonvascular brainstem causes of locked-in syndrome, 102
Noradrenergic cell groups and nuclei, 38
Noradrenergic neurons, 65, 329–330
Noradrenergic system, molecular imaging of, 338–340
Norepinephrine (NE), 204–205
NE-depleted transgenic mice, 338
Norepinephrine transporter (NET), 338
Nothnagel syndrome, 10–11
NSS. *See* Neuromodulatory subcortical systems (NSS)
NTS. *See* Nucleus of solitary tract (NTS)
Nucleus ambiguus of vagus, functional neuroanatomy of cardiovagal parasympathetic preganglionic neurons in, 61
Nucleus of lateral lemniscus (NLL), 320–321
Nucleus of solitary tract (NTS), 318
role in reflex control of arterial blood pressure, 63–64
Nystagmus, 111–113, 128
oculomotor control
brainstem as integrative center in, 124
detailed brainstem neuroanatomy in, 123–124
fixation system, 127
horizontal gaze, 124
medullary lesions and, 128
midbrain lesions and, 128
optokinetic system, 126–127
pontine lesions and, 128
saccades and gaze control, 124–125
smooth pursuit eye movements, 125–126
vergence system and brainstem contributions to binocular alignment, 127–128
vertical gaze, 124
vestibulo-ocular reflex, 126

O

Occipital nerve stimulation (ONS), 322, 324
Ocular tilt reaction (OTR), 223–224
Ocular trauma, 133
Ocular VEMP (oVEMP), 114
Oculo-motor nerve, 141
palsy, 144–145
Oculomotor complex, 33–34
Oculomotor disorders
fixation system, 127
horizontal gaze, 124
nystagmus, 128
oculomotor control
brainstem as integrative center in, 124
detailed brainstem neuroanatomy in, 123–124
optokinetic system, 126–127
saccades and gaze control, 124–125
smooth pursuit eye movements, 125–126
vergence system and brainstem contributions to binocular alignment, 127–128
vertical gaze, 124
vestibulo-ocular reflex, 126
Oculomotor information, 46
Oculomotor nucleus, 123–124, 206
Oculomotor system, 123
Oculosympathetic fibers, 146
Oculosympathetic nerves, hyperfunction of, 152
Olivary pretectal nucleus (OPN), 136
Olivocerebellar tract, 29
Omnipause neurons, 127
One-and-a-half syndrome (OAAHS), 216
ONS. *See* Occipital nerve stimulation (ONS)
Opalski syndrome, 228
OPN. *See* Olivary pretectal nucleus (OPN)
Optic nerve, 134, 185
Optokinetic eye movements, 126–127
Optokinetic nystagmus, 112
Optokinetic system, 126–127
Oral dorsal paramedian nucleus, 31
Oral steroids, 186
Orbitomeatal line, 285–286
Orexin neurons, 60
Orthostatic hypotension, 67–69
Oscillopsia, 117–118
Osmotic demyelination syndrome, 266
Osteoarthritis treatment, 82–83
Otolith organs, 105
OTR. *See* Ocular tilt reaction (OTR)
oVEMP. *See* Ocular VEMP (oVEMP)
Oxygen (O_2), 332–333

P

PAG. *See* Periaqueductal gray (PAG)
Pain, 75
control
brainstem analgesic circuits for pain relief, 83–85

conditioned pain modulation, 77*f*
future directions, 85–86
pain inhibits pain, 76–77
placebo analgesia and nocebo hyperalgesia, 78–81
stress-induced analgesia, 81–83
subnucleus reticularis dorsalis, 76*f*
Pallis, 347
Pancoast syndrome, 151–152
Parabrachial complex, 93
Parabrachial neurons, 80–81
Parabrachial nucleus, 65
role of, 78–81
Paramedian dorsal infarcts, 221
Paramedian midbrain infarctions, 215–219
clinical manifestations of midbrain infarctions, 217*t*
crossed brainstem syndromes associated with infarcts and hemorrhages, 218*t*
Paramedian pontine reticular formation (PPRF), 101, 124, 222
Parasomnias, 93
as model to understand brainstem contributions to consciousness, 93–94
Parasympathetic fibers, 141, 143
Parasympathetic innervations of pupil, 139–145
anatomy of parasympathetic pathway, 139–141
clinical manifestations of parasympathetic deficit to eye, 143–145
Parasympathetic nervous system, 318
Parasympathetic pathway, anatomy of, 139–141
Parasympathetic system, 132
Paraventricular nucleus (PVN), 145
Parietal eye fields, 124
Parieto-occipital artery, 313
Parinaud syndrome, 11, 137
Parkin (PRKN), 289–290
Parkinson disease (PD), 39, 48, 203, 289–291, 334–335, 339–340
Parkinson disease dementia (PDD), 201, 203
Parotid gland, 149
PART. *See* Primary Age-Related Tauopathy (PART)
Parvi-cellular reticular nuclei (PcRT), 29, 31
Patient-specific models and simulations, 248
Patients' quality of life, clinical importance and impact on, 237–238
PC. *See* Posterior commissure (PC)
PCA. *See* Posterior cerebral artery (PCA)
PChoA. *See* Posterior choroidal artery (PChoA)
PD. *See* Parkinson disease (PD)
PDD. *See* Parkinson disease dementia (PDD)
Peduncular hallucinosis (PH), 217
Peduncular segment, 311–312
Peduncular syndrome, 10
Pedunculo-quadrigeminal artery, 214
Pedunculopontine nucleus (PPN), 66, 93, 203
Pedunculopontine tegmental nucleus, 47
Penetrating trauma, 238
Periaqueductal gray (PAG), 32–33, 36, 76, 141
role of, 81–83
Peripheral nervous system (PNS), 195
lesions, 99–100
Peripheral vestibular disorders, 110
Peripheral vestibular structures, 107–108
Persistent hypoglossal artery (PHA), 307
Persistent postural-perceptual dizziness (PPPD), 117
Persistent proatlantal intersegmental artery, 307–308
Persistent trigeminal artery (PTA), 307
PET. *See* Positron emission tomography (PET)
PH. *See* Peduncular hallucinosis (PH)
PHA. *See* Persistent hypoglossal artery (PHA)
Phonation, 194
Physiologic anisocoria, 138–139
Physiologic imaging, 318
PICA. *See* Posterior inferior cerebellar artery (PICA)
Pick's disease (PiD), 205, 207
PiD. *See* Pick's disease (PiD)
Pilocytic astrocytoma, 172
Pineal gland, 288–289
Placebo analgesia, 78–81
placebo and nocebo brainstem signal changes, 80*f*
placebo/nocebo responses, 79–81
Placebo analgesic techniques, 85
Placebo hypoalgesia, 79
Planes, 289
Planimetric measurement, 287
PLL. *See* Posterior longitudinal ligament (PLL)
PMH. *See* Pure motor hemiparesis (PMH)
PNS. *See* Peripheral nervous system (PNS)
Polyglutamine expansions, 209
Pons, 20, 175, 193, 257–258
central parts of, 31
and midbrain role in regulation of arterial blood pressure, 65–66
Pontine arteries/perforators, 309–310
Pontine gliomas, 102
Pontine hemorrhages, 224–225
Pontine infarctions, 224
etiology and prognosis, 224
Pontine infarcts, epidemiology of, 221
Pontine lesions, 128
Pontine neurons, 31
Pontine tegmental cap dysplasia, 259
Pontine vascular syndromes, 220
Pontocerebellar hypoplasia, 260–261
Positional maneuvers, 113
Positional testing, 111–113
Positron emission tomography (PET), 80, 161, 245–246, 253, 289, 330, 336–337
Positrons, 337
Postemergence, 308
Posterior cerebral artery (PCA), 213, 310–313
Posterior choroidal artery (PChoA), 214
Posterior circulation, 213, 303
Posterior commissure (PC), 214
Posterior fossa ependymomas, 270
Posterior fossa group A (PFA), 269–270
Posterior fossa group B (PFB), 269–270
Posterior inferior cerebellar artery (PICA), 107–108, 225, 303–306
Posterior longitudinal ligament (PLL), 304
Posterior meningeal artery, 304
Posterior P2 segment (P2P), 311–312
Posterior reversible encephalopathy syndrome (PRES), 265–266
Posterior spinal arteries (PSA), 225, 304–305
Posterior temporal artery, 312–313
Posterolateral pontine artery, 309
Postganglionic sympathetic fibers, 146–147
Postganglionic sympathetic neurons, 146
Posttraumatic epilepsy (PTE), 246
Posturography, 114
PPPD. *See* Persistent postural-perceptual dizziness (PPPD)
PPRF. *See* Paramedian pontine reticular formation (PPRF)
Precision medicine, 247–248
Precommunicating segment, 214
Predominantly brainstem malformations, 259–260
Premotor nuclei, 34
PRES. *See* Posterior reversible encephalopathy syndrome (PRES)
Presympathetic neurons, 60–62
Primary Age-Related Tauopathy (PART), 204
Primary brainstem trauma, 280
Primary central nervous system lymphomas, 271–272
Primary injury mechanisms, 246
Primary medullary hemorrhages, 229

Primary supratentorial brain lesion, 348
Primary tauopathies feature tau dysfunction, 205
Progressive scoliosis, horizontal gaze palsy with, 259–260, 260*f*
Progressive supranuclear palsy (PSP), 205–206, 275–276, 291–292
PSP-corticobasal syndrome, 276
PSP-progressive nonfluent aphasia, 276
PSP-pure akinesia, 276
Progressive supranuclear palsy-parkinsonism (PSP-P), 276, 292
Progressive supranuclear palsy-Richardson's syndrome (PSP-RS), 292
Prosody, 194
Proton-to-neutron ratio, 337
PSA. *See* Posterior spinal arteries (PSA)
Pseudo-abducens palsy, 216–217
PSP. *See* Progressive supranuclear palsy (PSP)
PSP-P. *See* Progressive supranuclear palsy-parkinsonism (PSP-P)
PSP-RS. *See* Progressive supranuclear palsy-Richardson's syndrome (PSP-RS)
PSS. *See* Pure sensory stroke (PSS)
PTA. *See* Persistent trigeminal artery (PTA)
PTEN-induced putative kinase 1 (PINK1), 289–290
Pupil
afferent limb, 134–136
dilation, 319
efferent limb, 136
fibers, 144
light reflex, 134–139
light-near dissociation, 137–138
parasympathetic innervation of, 139–145
physiologic anisocoria, 138–139
relative afferent pupillary defect, 136–137
schema of pupillary light reflex, 134*f*
sympathetic innervation of, 145–152
Pupillary abnormality, 131
Pupillary disorders
iris, 131–134
parasympathetic innervation of pupil, 139–145
pupil light reflex, 134–139
pupillary dysfunction in autonomic neuropathies, 152
sympathetic innervation of pupil, 145–152
Pupillary dysfunction in autonomic neuropathies, 152
Pupillometry, 318
Pure midbrain infarcts, 219–220
Pure motor hemiparesis (PMH), 221
Pure sensory stroke (PSS), 220
Purkinje cells project, 30
PVN. *See* Paraventricular nucleus (PVN)
Pyramidal decussation, 26–27
Pyramidal fiber bundles, 18

Q

Quadrigeminal artery, 214
Quadriplegia, 101

R

Radiation, 178
Radio-induced glioma, 171–172
Radiologists, 176
RAPD. *See* Relative afferent pupillary defect (RAPD)
Raphe nuclei, 92
Raphe obscurus, 29
Raphe system, 48
Rapid behavior disorder (RBD), 203
Rapid-eye-movement (REM), 63, 203
Rapid-eye-movement sleep behavior disorder (RBD), 94
RAS. *See* Reticular activating system (RAS)
Raymond–Céstan syndrome, 11
RBD. *See* Rapid behavior disorder (RBD)
RDP. *See* Robertson dysarthria profile (RDP)
Rehabilitation, 243–244
Relative afferent pupillary defect (RAPD), 136–137
REM. *See* Rapid-eye-movement (REM)
Remote monitoring, 247
Renal enuresis, 191
3-repeat tau (3R tau), 205
4-repeat tau (4R tau), 205
Resonance, 194
Respiration, 194
Respiratory disorders, 239
Responsive unawakefulness syndrome (RUS), 350, 354
Restless legs syndrome (RLS), 295
Reticular activating system (RAS), 100
Reticular formation (Rf), 27, 29, 42–43
entrance of vestibulo-cochlear nerve and comparative anatomic aspects of, 27–29
subdivision of, 43
Retino-tectal tract, 134
Rett syndrome, 165
Revised Uniform Determination of Death Act (rUDDA), 347
Rf. *See* Reticular formation (Rf)
Rhombencephalitis, 264
Rhomboid fossa, 21–23
Riley–Day syndrome, 144
Risk ratio (RR), 116–117
RLS. *See* Restless legs syndrome (RLS)
Robertson dysarthria profile (RDP), 196
ROBO3 gene, 259–260
Romberg test, 114
Rose criteria, 341–342
Ross syndrome, 144
Rostral interstitial nucleus of medial longitudinal fasciculus (riMLF), 124, 206, 215
Rostral ventromedial medulla (RVM), 62–63, 78, 318
role of, 78–81
regions, 81–83
Rotational forces, 238–239
Rowland Payne syndrome, 151–152
RR. *See* Risk ratio (RR)
rUDDA. *See* Revised Uniform Determination of Death Act (rUDDA)
RUS. *See* Responsive unawakefulness syndrome (RUS)
RVM. *See* Rostral ventromedial medulla (RVM)

S

(S,S)[C-11]Methylreboxetine ([C-11]MRB), 338
Saccades control, 124–125
Saccadic eye movements, 112
SAD. *See* Small artery disease (SAD)
SAEPs. *See* Short-latency auditory evoked potentials (SAEPs)
Salivary α-amylase (sAA), 319
Saltzman variants, 307
SC. *See* Superior colliculus (SC)
SCA. *See* Superior cerebellar artery (SCA)
SCAs. *See* Spinocerebellar ataxias (SCAs)
SCP. *See* Superior cerebellar peduncles (SCP)
SCR. *See* Skin conductance response (SCR)
Secondary brain lesion, 348
Secondary brainstem hemorrhages, 102
Secondary injury mechanisms, 246
Secondary parkinsonism, 293–294
drug-induced parkinsonism, 293
hydrocephalus, 294
vascular parkinsonism, 294
Sectorial sphincter palsy, 143

See-saw nystagmus (SSN), 216
SEFs. *See* Supplementary eye fields (SEFs)
Selective serotonin reuptake inhibitors (SSRIs), 117
Semicircular canals, 105, 126
Sensory deficits, 227, 240
Sensory disturbance, 224
Serotonergic neurons, 329–330
Serotonergic system, molecular imaging of, 340
Serotonin, 42
Serotonin transporter (SERT), 340
Serotonin-norepinephrine reuptake inhibitors (SNRIs), 117
Serotoninergic neuronal groups, 42
Serotoninergic neurons, 62–63
Serum biomarkers in brainstem injury, 247
Short-latency auditory evoked potentials (SAEPs), 321
Single photon emission computed tomography (SPECT), 245–246, 289, 337
Skew deviation, 216
Skin conductance response (SCR), 320
Skin conduction, 318
Sleep
  consciousness and, 92–93
  disorders, 161, 206
  paralysis, 93
  sleep–wake cycle, 91
    disturbances, 205
  terrors, 93
  and wakefulness, 91–92
Sleepwalking, 93–94
SLTs. *See* Speech and Language Therapists (SLTs)
Small artery disease (SAD), 224
Smooth muscle cells, 132
Smooth pursuit eye movements, 111, 125–126
SN. *See* Substantia nigra (SN)
SN hyperechogenicity(SN), 289
SNRIs. *See* Serotonin-norepinephrine reuptake inhibitors (SNRIs)
SOF. *See* Superior orbital fissure (SOF)
Solitary tract in reflex control of arterial blood pressure, role of nucleus of, 63–64
Somnambulism, 93–94
Spastic dysarthria, 195
Spatial orientation, 106
Special somatic afferent (SSA), 24
Special visceroefferent (SVE), 24
SPECT. *See* Single photon emission computed tomography (SPECT)
Speech and Language Therapists (SLTs), 197
Speech and language therapy for dysarthria, 198
Speech disorders, 196
Sphincter pupillae muscle, 131
Spinal cord, 17, 24–26, 185
Spinal diamond, 305
Spinal ependymoma, 269–270
Spinal nerves' dorsal roots, 18–20
Spinal tracts, 31–32
Spinal trigeminal nucleus (SpV), 26–27, 75–76
Spino-parabrachial tract, 80–81
Spinocerebellar ataxias (SCAs), 209–210, 276–277
Spinothalamic tract, 33
Spontaneous dissection, 100
Spontaneous nystagmus, 111
"Spot sign", 279
SpV. *See* Spinal trigeminal nucleus (SpV)
SRD. *See* Subnucleus reticularis dorsalis (SRD)
SSA. *See* Special somatic afferent (SSA)
SSN. *See* See-saw nystagmus (SSN)
SSRIs. *See* Selective serotonin reuptake inhibitors (SSRIs)
Stem cell therapies, 247
Steroid pulses, 186, 191
Steroid therapy, 188
Stress-induced analgesia, 81–83
  brainstem pain modulation and analgesic clinical trials, 81–82
  brainstem pain modulation and chronic pain, 82–83
Stroma, 131
Subcoeruleus area, 93
Subcoeruleus nucleus, 93
Subcortical cerebellar nuclei, 30
Subcortical nuclei, 36
Subependymoma, 269–270
Subnucleus reticularis dorsalis (SRD), 76
  role of, 76–77
Substantia nigra (SN), 203, 286–287
Substantia nigra hyperechogenicity(SN), 287
Substantia nigra pars compacta dopaminergic neurons, 39–40
Substantianigra pars compacta (SNc), 203, 273
Sudomotor fibers, 146
Superficial pain, 78
Superior cerebellar artery (SCA), 214, 303, 310–311
  basilar artery above, 311–313
  basilar artery below, 306–311
Superior cerebellar peduncles (SCP), 214
Superior colliculus (SC), 35, 123–124, 206, 318
  parasagittal aspect of, 35–36
Superior orbital fissure (SOF), 141
Superior rhomboid fossa, 31–32, 44
Superior vermian artery, 310
Superior vestibular nucleus, 108
Superolateral pontine artery, 309
Superoxide dismutase 1 (SOD1), 207–208
Supine hypertension, 69
  in cardiovascular autonomic failure, 67–68
  in multiple system atrophy, 68–69
Supplementary eye fields (SEFs), 124
Supratentorial ependymoma, 269–270
Surgical decompression, 242
Surgical interventions, 242
Susceptibility-weighted imaging (SWI), 254
SVE. *See* Special visceroefferent (SVE)
"Swallow tail" sign, 255–256, 273–274
SWI. *See* Susceptibility-weighted imaging (SWI)
Sympathetic denervation, effects of, 149
Sympathetic fibers, 145–146, 149
Sympathetic ganglionic neurons, 59
Sympathetic innervations of pupil, 145–152
  anatomy of sympathetic pathway, 145–146
  clinical manifestations of sympathetic deficit to eye and head, 146–152
  hyperfunction of oculosympathetic nerves, 152
Sympathetic nervous system, 318
Sympathetic pathway, anatomy of, 145–146
Syphilitic tumors, 102

T

T2-weighted imaging, 254
TAI. *See* Traumatic axonal injury (TAI)
Tapia syndrome, 13
TAR DNA-binding protein 43 (TDP-43), 201
  proteinopathies, 207
Tarsal muscle supersensitivity, 150
Tauopathies, 205–207
taVNS. *See* Transcutaneous auricular VNS (taVNS)
TBI. *See* Traumatic brain injury (TBI)
TBSI. *See* Traumatic brainstem injury (TBSI)
TCBS. *See* Transcranial B-mode sonography (TCBS)
TDP-43. *See* TAR DNA-binding protein 43 (TDP-43)
Tectospinal tract, 29
Tectum, 257

Tegmental pedunculopontine nucleus, 33
Tegmentum, 20, 47
Telemedicine, 247
Temporal bone window, 286
Thalamic hematomas, 101
Therapeutic nihilism and early prognostication, 245
Therapy Outcome Measures (TOMs), 196–197
Third ventricle plane, 288–289
Thromboxane, 63
Tissue decay, 4
TNS. *See* Trigeminal nerve stimulation (TNS)
TOMs. *See* Therapy Outcome Measures (TOMs)
Tonic pupil, 143–144
Tonsillohemispheric trunk, 305
Topical cocaine, 150
Total bilateral ophthalmoplegia, 220
Total locked-in midbrain syndrome, 219
Toxic effects, 94
"Tracer principle", 337
Tracheotomy, 49
TRACK-TBI, 245
Transcranial B-mode sonography (TCBS), 285
  in disease, 289–297
  diseases/disorders, 296–297
    accumulation syndromes, 296–297
    limitations of, 297
  investigation procedure, imaging planes, and visible structures, 285–289, 286*f*
    mesencephalic plane, 286–288
    planes, 289
    third ventricle plane, 288–289, 288*f*
  method of, 285–289
    equipment and ultrasound settings, 285
  movement disorders, 289–296
    atypical parkinsonism, 291–293
    in deep brain stimulation, 295–296
    dystonia, 295
    essential tremor, 294
    Huntington disease, 294–295
    Parkinson disease, 289–291
    restless legs syndrome, 295
    secondary parkinsonism, 293–294
Transcranial color-coded duplex sonography, 285
Transcranial sonography, 285
Transcutaneous auricular VNS (taVNS), 323–324
Transcutaneous delivery of occipital nerve stimulation (tONS), 324
Translocator protein (TSPO), 339–340
Transneuronal degeneration, 49
Transverse fibers, 20, 49
Trauma, 280
  types of, 238
Traumatic axonal injury (TAI), 245
Traumatic brain injury (TBI), 47, 102, 237, 245–246
  structure and function of human brainstem after, 48–51
    drawn from systemic studies of, 49–51
Traumatic brainstem injury (TBSI), 237
  clinical importance and impact on patients' quality of life, 237–238
  complications, 244
    Duret brainstem hemorrhage, 244
  definition, 237
  diagnosis, 240–241
    angiography, 241
    cerebrospinal fluid analysis, 241
    clinical and neurologic evaluation, 240
    computed tomography, 240
    electrophysiologic tests, 241
    magnetic resonance imaging, 240–241
  future research and advances, 245–248
    advanced imaging techniques, 245–246
    biomarkers for diagnosis and prognosis, 246
    genomic and precision medicine, 247–248
    neuroinflammation and immunomodulation, 246
    neuroprotective strategies, 246
    neuroregeneration and stem cell therapies, 247
    patient-specific models and simulations, 248
    serum biomarkers in brainstem injury, 247
    telemedicine and remote monitoring, 247
  injury mechanisms, 238–239, 239*f*
  prognosis, 244–245
    therapeutic nihilism and early prognostication, 245
    TRACK-TBI, 245
  signs and symptoms, 239–240
    disorders of consciousness and cranial nerves, 239–240
  treatment, 241–244
    initial patient care, 241–242
    rehabilitation and long-term care, 243–244
    surgical interventions, 242
  types of trauma, 238
Traumatic dissection, 100
Traumatic unresponsive wakefulness syndrome, 243
Traumatism, 133
Trigeminal nerve stimulation (TNS), 324
Trigeminal tubercle, 18–20
Trigeminothalamic tract, 32
Trochlear nerve, 5
Trochlear nuclei, 124
Trochlear nucleus, caudal pole of, 32–33
Tuberculomas, 102

U

Ubiquitin C-terminal hydrolase-L1 (UCH-L1), 247
UDDA. *See* Uniform Determination of Death Act (UDDA)
Ultra-high-field fMRI, 85–86
Ultrahigh magnetic field strength, 331–332
Ultrahigh-field MRI systems, 331, 335–336
Ultrasound (US), 253–254
  equipment and ultrasound settings, 285
Uni-INO, 220
Uniform Determination of Death Act (UDDA), 346
Unilateral esotropia, 216–217
Unilateral upper motor neuron dysarthria, 195
Unilateral vestibular disorders, 110
Unresponsive wakefulness state (UWS), 354
Unresponsive wakefulness syndrome (UWS), 353
Urrets-Zavalia syndrome, 133–134
US. *See* Ultrasound (US)
Uveitis, 133
UWS. *See* Unresponsive wakefulness state (UWS). *See* Unresponsive wakefulness syndrome (UWS)

V

VA. *See* Vertebral arteries (VA)
Vaccination triggers, 187
Vagusnerve stimulation (VNS), 322–324
Vagusnerve stimulation via transcutaneous methods (tVNS), 323–324
Vascular diseases, 277–280
  arteriovenous malformation, 279–280
  cavernoma, 279
  hemorrhagic stroke, 279
  ischemic stroke, 277–279
Vascular imaging, 113
Vascular lesions associated with locked-in state, 100–102
Vascular medullary syndromes, 228

Vascular parkinsonism, 294
Vasomotor fibers, 146
VB system. *See* Vertebrobasilar system (VB system)
Vegetative State (VS), 94
Venous congestion, 280
Ventral tegmental area (VTA), 34, 40–41
Ventral tegmentum, 33–34
Ventrolateral medulla, 59–60
  role in control of arterial blood pressure, 61–62
Ventrolateral PAG (vlPAG), 78
Ventromedial medulla in arterial blood pressure control, role of, 62–63
Vermal cortex processes, 20–21
Vermis, 20–21
Vernet syndrome, 13
Vertebral arteries (VA), 213, 303–304
Vertebrobasilar perforators, 313
Vertebrobasilar system (VB system), 213, 224
Vertical gaze, 124
Vesalius, 5
Vestibular area, 23
Vestibular disorders, 119
Vestibular dysfunction, 126
  clinical assessment overview, 110–113
    clinical examination, 110
    gait and posture, 110
  clinical correlates of, 110–113
  nystagmus, eye movements, positional testing, 111–113
    clinical examination of eye movements, 111
    gaze-evoked nystagmus, 111
    optokinetic nystagmus, 112
    positional maneuvers, 113
    saccadic eye movements, 112
    smooth pursuit eye movements, 111
    spontaneous nystagmus, 111
    vestibulo-ocular reflex, 112
    vestibulo-ocular reflex suppression, 112–113
  rehabilitation strategies for, 118
Vestibular evoked myogenic potentials, 114
Vestibular function, objective measures of, 114
  posturography, 114
  quantitative analysis, 114
  vestibular evoked myogenic potentials, 114
  video head impulse test, 114
Vestibular information, 46
Vestibular interactions, 119
Vestibular migraine, 117
Vestibular neuritis, 106, 116
Vestibular nuclei, 31, 46, 106
Vestibular prosthetics, 118–119
Vestibular rehabilitation physiotherapy, 118
Vestibular system, 105–108
  anatomy of, 107–108
    brainstem anatomy, 107*f*
    central vestibular structures, 108
    peripheral vestibular structures, 107–108
  balance and posture, 106
  brainstem–vestibular integration, 110
  clinical correlates of vestibular dysfunction, 110–113
  diagnostic and assessment methods, 113–114
    objective measures of vestibular function, 114
    role of imaging in brainstem and vestibular system evaluation, 113–114
  emerging concepts and future directions, 118–119
    advances in vestibular prosthetics and their impact on central adaptation, 118–119
    insights from functional imaging studies of brainstem, 119
  gaze stabilization, 106
  integration of vestibular system within brainstem, 106–107
  Ménière's disease, 116–117
  oscillopsia, 117–118
  persistent postural-perceptual dizziness, 117
  rehabilitation strategies for vestibular dysfunction, 118
  spatial orientation, 106
  specific vestibular disorders and treatment options, 114–118
    BPPV, 114–116
  therapeutic implications, 114–118
  vestibular migraine, 117
  vestibular neuritis, 116
  vestibular physiology, 108–110
    integration of vestibular inputs with visual and proprioceptive information, 109–110
    vestibulo-ocular reflexes, 108–109
Vestibular therapy, 126
Vestibulo-cochlear nerves, 27–29, 107
Vestibulo-ocular reflex (VOR), 105–106, 108–109, 112, 124, 126, 215
  neuroanatomic pathways, 110
  and posture control, 110
Vestibulo-ocular reflex suppression (VORS), 112–113
Vestibulo-spinal reflexes (VSR), 110
Vestibulospinal tracts, 108
vHIT. *See* Video head impulse test (vHIT)
Video head impulse test (vHIT), 114
Vieussens center, 6
Villaret's syndrome, 13
Viral infection, 187
Visible structures, 285–289, 286*f*
vlPAG. *See* Ventrolateral PAG (vlPAG)
VNS. *See* Vagusnerve stimulation (VNS)
Voluntary saccades, 125
VOR. *See* Vestibulo-ocular reflex (VOR)
VORS. *See* Vestibulo-ocular reflex suppression (VORS)
VS. *See* Vegetative State (VS)
VSR. *See* Vestibulo-spinal reflexes (VSR)
VTA. *See* Ventral tegmental area (VTA)

W

Wakefulness, 91–92
Wall-eye bilateral internuclear ophthalmoplegia (WEBINO), 218–219, 223–224
Wallenberg syndrome, 12, 151–152, 244
Wallerian degeneration, 279
WBDP. *See* World Brain Death Project (WBDP)
Weber syndrome, 10
WEBINO. *See* Wall-eye bilateral internuclear ophthalmoplegia (WEBINO)
Wernicke encephalopathy, 266
Westphal–Pilcz syndrome, 8
WHO. *See* World Health Organization (WHO)
Whole brain
  formulation, 347
  standard BD/DNC criteria, 350
Wilson's disease, 267–268, 296
"Wing-shaped" appearance, 338
Withdrawal of life-sustaining treatment (WLST), 245
WLST. *See* Withdrawal of life-sustaining treatment (WLST)
World Brain Death Project (WBDP), 347
World Health Organization (WHO), 268

Y

YAP1 fusion–positiveependymoma, 269–270

Z

ZFTA fusion–positive ependymoma, 269–270

Blank Page for Index